ACS SYMPOSIUM SERIES **455**

Materials for Nonlinear Optics
Chemical Perspectives

Seth R. Marder, EDITOR
Jet Propulsion Laboratory, California Institute of Technology

John E. Sohn, EDITOR
AT&T Bell Laboratories

Galen D. Stucky, EDITOR
University of California, Santa Barbara

Developed from a symposium sponsored
by the Divisions of Organic Chemistry
and Inorganic Chemistry
at the 199th National Meeting
of the American Chemical Society,
Boston, Massachusetts, April 22–27, 1990

American Chemical Society, Washington, DC 1991

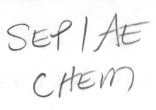

Library of Congress Cataloging-in-Publication Data

Materials for nonlinear optics: chemical perspectives / Seth R. Marder, editor, John E. Sohn, editor, Galen D. Stucky, editor.

p.   cm.—(ACS symposium series; 455)

"Developed from a symposium sponsored by the Divisions of Organic Chemistry and Inorganic Chemistry at the 199th National Meeting of the American Chemical Society, Boston, Massachusetts, April 22–27, 1990."

Includes bibliographical references and indexes.

ISBN 0–8412–1939–7

1. Materials—Optical properties—Congresses.   2. Stereochemistry—Congresses. 3. Nonlinear optics—Congresses.

I. Marder, Seth R. (Seth Richard) II. Sohn, John E., 1952–   . III. Stucky, Galen D., 1936–   . IV. American Chemical Society. Division of Organic Chemistry.   V. American Chemical Society. Division of Inorganic Chemistry.   VI. American Chemical Society. Meeting (199th: 1990: Boston, Mass.) VII. Series.

QD473.M29   1991
535.2—dc20   90–25768
                                         CIP

ACS Symposium Series

M. Joan Comstock, *Series Editor*

1991 ACS Books Advisory Board

Foreword

THE ACS SYMPOSIUM SERIES was founded in 1974 to provide a medium for publishing symposia quickly in book form. The format of the Series parallels that of the continuing ADVANCES IN CHEMISTRY SERIES except that, in order to save time, the papers are not typeset, but are reproduced as they are submitted by the authors in camera-ready form. Papers are reviewed under the supervision of the editors with the assistance of the Advisory Board and are selected to maintain the integrity of the symposia. Both reviews and reports of research are acceptable, because symposia may embrace both types of presentation. However, verbatim reproductions of previously published papers are not accepted.

We dedicate this book
to the memory of Donald Ulrich, a program manager
at the Air Force Office of Scientific Research
and a strong advocate and supporter
of the field of nonlinear optics.

Contents

COMPOSITE MATERIALS

MOLECULAR AND SUPRAMOLECULAR METAL-BASED SYSTEMS

INDEXES

Preface

MATERIALS RESEARCH AND DEVELOPMENT for nonlinear optical applications has rapidly progressed since the mid-1980s to the point where several systems are available commercially. A wide variety of materials—including inorganic and organic crystals, polymers, semiconductors, composites, and metal-based systems—possess nonlinear optical properties. No material system has proven to be the "silicon" of nonlinear optics, for each material has properties that are advantageous for certain applications but also properties that are disadvantageous for other applications. Thus considerable research is still needed to develop materials that can meet the critical requirements of devices used in information processing, optical frequency conversion, integrated optics, and telecommunications.

Our goal in organizing the symposium upon which this book is based was to expose the chemistry community to the critical role that chemistry can and must play in nonlinear optical research. In addition, we hoped to bring together those researchers who synthesize and characterize materials from the variety of systems mentioned above, with those who build devices. Previous symposia were typically confined to only a few of these materials, resulting in minimal interaction between those working with different material classes. If we were successful, those chemists, physicists, and engineers who attended the symposium now have a greater appreciation for the opportunities that lie ahead in understanding and developing nonlinear optical materials.

We put together this book to provide a permanent record of the talks presented at the symposium and to expose a wider audience to the chemistry in nonlinear optics. This monograph begins with a discussion of polarizability and hyperpolarizability from the view of a chemist. Having this background, we move into tutorial chapters dealing with the fundamental structures and properties of second- and third-order nonlinear optical materials, measurement and characterization of these systems, theoretical considerations, application of these systems to devices, and overviews of the current state of affairs in both organic and inorganic nonlinear optical materials. The remainder of the book is loosely organized into seven sections:

- progress toward understanding the structure–property relationships on the second-order microscopic susceptibility (β),

- preparation and characterization of poled polymers,

- organic and inorganic crystals,

- novel approaches to orientation of the molecular units,

- composite materials,

- molecular and supramolecular metal-based systems, and

- σ and π delocalized third-order nonlinear optical materials.

The breadth of participation in the symposium is due in great part to the support from a number of organizations. Their assistance allowed scientists from Asia, Europe, and North America to present and discuss their work. We thank the Divisions of Inorganic and Organic Chemistry of the American Chemical Society, the Air Force Office of Scientific Research, the Petroleum Research Fund, AT&T Bell Laboratories, E. I. du Pont de Nemours and Company, and Eastman Kodak Company for generous financial support.

In particular we acknowledge the Office of Naval Research and the Strategic Defense Initiative Organization/Innovative Science and Technology Office for financial support both for the preparation of this book and for the symposium. We are truly indebted to the numerous authors for their timely effort and to the referees for their critical evaluation of the manuscripts. Special thanks go to Tessa Kaganoff for her diligent coordination of both the symposium abstracts and the chapters contained in this volume. Also, the guidance and patience of Robin Giroux and her colleagues at the ACS Books Department have been essential in the publication of this book.

SETH R. MARDER
Jet Propulsion Laboratory
California Institute of Technology
Pasadena, CA 91109

JOHN E. SOHN
AT&T Bell Laboratories
Princeton, NJ 08540

GALEN D. STUCKY
University of California
Santa Barbara, CA 93106

November 6, 1990

INTRODUCTION TO NONLINEAR OPTICS

Chapter 1

Linear and Nonlinear Polarizability
A Primer

Galen D. Stucky[1], Seth R. Marder[2], and John E. Sohn[3]

[1]University of California, Santa Barbara, CA 93106
[2]Jet Propulsion Laboratory, California Institute of Technology, Pasadena, CA 91109
[3]AT&T Bell Laboratories, P.O. Box 900, Princeton, NJ 08540

In this introductory chapter the concepts of linear and nonlinear polarization are discussed. Both classical and quantum mechanical descriptions of polarizability based on potential surfaces and the "sum over states" formalism are outlined. In addition, it is shown how nonlinear polarization of electrons gives rise to a variety of useful nonlinear optical effects.

This chapter introduces the reader to linear and nonlinear optical polarization as a background for the tutorial and research articles that follow. We consider first how the passage of light changes the electron density distribution in a material (i.e., polarizes the material) in a linear manner, from both the classical and quantum mechanical perspectives. Next, we examine the consequences of this polarization upon the behavior of the light. Building on this foundation, we then describe, in an analogous manner, the interaction of light with nonlinear materials. Finally, we outline some materials issues relevant to nonlinear optical materials research and development.

Nonlinear optics is often opaque to chemists, in part because it tends to be presented as a series of intimidating equations that provides no intuitive grasp of what is happening. Therefore, we attempt in this primer to use graphical representations of processes, starting with the interaction of light with a molecule or atom. For the sake of clarity, the presentation is intended to be didactic and not mathematically rigorous. The seven tutorial chapters that follow this introduction as well as other works (1–5) provide the reader with detailed treatments of nonlinear optics.

Nonlinear Behavior and Nonlinear Optical Materials

The idea that a phenomenon must be described as nonlinear, at first, has inherent negative implications: we know what the phenomenon is not (linear), but what then is it? As a starting point, Feynman (6) has noted that to understand physical laws, one must begin by realizing they are all approximate. For example, the frictional drag on a ball bearing moving slowly through a jar of honey is linear to the velocity, i.e. $F = -kv$. However, if the ball is shot at high velocity the drag becomes nearly proportional to the square of the velocity, $F = -k'v^2$, a nonlinear phenomenon. Thus, as the speed of the

0097–6156/91/0455–0002$08.25/0

ball bearing increases, the form of the physical "law" describing the relationship between the drag and the velocity changes. Nonlinear dynamics forms the basis of the new discipline, chaos (7), a nearly universal phenomenon. A process is nonlinear when the response to an input (i.e., the output) changes the process itself. Nonlinear behavior is not unusual, and for most physical processes a linear formulation is generally just the lowest-order approximation to the actual process. This has been emphasized for the behavior of light by Chemla and Zyss (8): "The artificial distinction between linear and nonlinear optics is now obsolete. Optics is in essence nonlinear."

Nonlinear optics is concerned with how the electromagnetic field of a light wave interacts with the electromagnetic fields of matter and of other light waves. The interaction of light with a nonlinear optical material will cause the material's properties to change, and the next photon that arrives will "see" a different material. As light travels through a material, its electric field interacts with other electric fields within the material. These internal fields are a function of the time dependent electron density distribution in the material and the electric fields of the other light waves, if for example, two or more light sources are used. In a **nonlinear optical (NLO)** material, strong interactions can exist among the various fields. These interactions may change the frequency, phase, polarization, or path of incident light. The chemist's goal in the field is to develop materials that control this mediation so that one can modulate or combine photons (wave mixing). In addition, it is necessary to fine tune both the magnitude and response time of the optical processes. To effect this control, we must look more closely at how matter, and specifically the electronic charge density in matter, interacts with light.

Polarizability: A Microscopic View

What causes the electron density of an optical material to couple and polarize with the electromagnetic field of a light wave? To understand this process we need to consider more quantitatively what happens at the molecular level. How does light perturb or couple to the electrons in a molecule?

Light has an electric field, E, that interacts with the charges in a material producing a force (qE, where q is the charge). Figure 1 is a simple schematic that shows the instantaneous displacement (polarization) of the electron density of an atom by the electric field of the light wave. The displacement of the electron density away from the nucleus results in a charge separation, an <u>induced dipole</u> with moment μ (Figure 1b). For small fields the displacement of charge from the equilibrium position is proportional to the strength of the applied field.

$$\text{Polarization} = \mu = \alpha E \qquad (1)$$

Thus a plot of polarization as a function of the applied field is a straight line whose slope is the <u>linear polarizability</u>, α, of the optical medium (Figure 1c). If the field oscillates with some frequency, (i.e., electromagnetic radiation, light), then the induced polarization will have the same frequency if the response is instantaneous (Figure 1a). Polarization is a vector quantity with both direction and magnitude.

In this <u>classical</u> model of linear polarizability, the electrons are bound to the atoms by a <u>harmonic potential</u> (Figure 2), i.e., the restoring force for the electron is linearly proportional to its displacement from the nucleus:

$$F = -\kappa x \qquad (2)$$

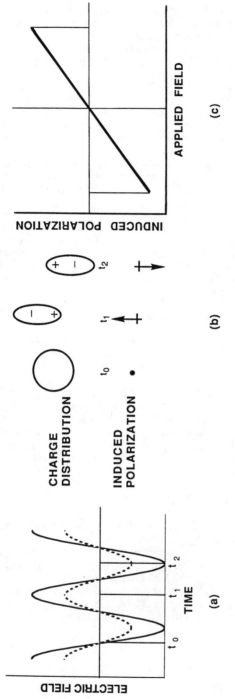

Figure 1. (a) Plots of the electric field of the applied light wave (solid) and the induced polarization wave (dotted) as a function of time for a linear material; (b) cartoon depicting the polarization of the material as a function of time; (c) plot of *induced* polarization vs. applied field.

The electrons see a potential energy surface:

$$V = 1/2 \, \kappa x^2 \tag{3}$$

Therefore, a symmetrical distribution of electron density exists around the atom with an equal ease of *charge* displacement in both the +x and −x directions.

The oscillating electric field of light affects all charges in the optical medium, not only the electron. For materials that contain electric dipoles, such as water molecules, the dipole themselves stretch or reorient in the applied field. In ionic materials, the ions move relative to one another (Figure 3). Dipolar and ionic motions involve nuclear reorientation.

The magnitude of the polarization depends on whether the charges can move fast enough to keep up with the oscillations of the electric field. Only electrons are efficiently polarized by optical frequency fields since they have small mass. Heavier, and thus more slowly moving, nuclei and molecules are efficiently displaced (polarized) at lower frequencies. If the charges fail to keep up with the field, a phase difference between the polarization and the electric field occurs, i.e., the charge displacement maximizes sometime <u>after</u> the electrons experience the maximum in the electric field of the light wave. The total polarizability, α, consists of an in-phase component, α', and an out-of-phase component, α'', that accounts for absorption ($\alpha = \alpha' + i\alpha''$). Figure 4 shows a plot of α' and α'' as a function of frequency. As a rough generalization, electron polarization, α_e, is the fastest, occurring on the femtosecond time scale (UV/visible); vibrational displacements, α_v, are slower, occurring in the picosecond regime (infrared or lower frequencies); and molecular dipole reorientation, α_d, is generally slower still, occurring on the nanosecond or slower time scale (microwave frequencies or lower). However, most third-order susceptibility measurements on organic based molecules show reorientational effects which are quite fast, sometimes sub nanosecond, which can dominate in this time regime. Heating effects are thus possible on the nanosecond time scale and <u>must</u> be taken into account since they can greatly modify the polarization response. The induced displacement of trapped charges at defect sites (space charges) within solids occurs at audio frequencies at the slowest end of the time scale. It is important, therefore, to write equation 1 as

$$\mu(\omega) = \alpha(\omega) \, E(\omega) \tag{4}$$

indicating that each of the variables is frequency dependent.

Figure 4 also illustrates that both α' and α'' change dramatically at certain frequencies. These are **resonant** frequencies, natural frequencies for transitions between quantum states. At these frequencies, transitions to higher energy rotational, vibrational, or electronic states lead to large charge displacements. In the search for large optical responses, resonant or near resonant optical frequencies have been used. Unfortunately, at or near-resonant frequencies, photons are no longer weakly perturbed as they travel through the optical medium; non-radiative decay from the high-energy or excited states to the ground state can result in sample heating and loss of photon efficiency.

The polarization for a medium may not be the same in all directions for a molecule or a collection of atoms in a thin film or crystal. For example, a field oriented along the long axis of 2,4-hexadiyne induces a polarization, and a field oriented perpendicular to the long axis induces a smaller polarization (Figure 5). Simply explained (ignoring conjugation effects), application of the electric field along a row of atoms (Figure 5a) induces a series of atomic dipoles in which the positive end of one induced dipole is attracted to the negative end of the neighboring dipole, thus reinforcing the polarization. Reinforcement of the dipoles does not occur with the field perpendicular to the row of atoms (Figure 5b).

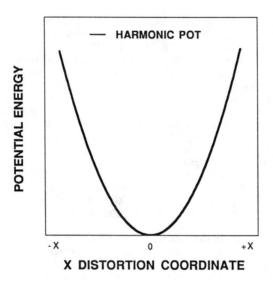

Figure 2. Plot of potential energy vs. distortion coordinate for a material with a harmonic potential.

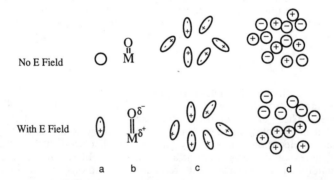

Figure 3. Different mechanisms for inducing polarization through (a) redistribution of electron density, (b) bond stretching, (c) alignment of dipoles, and (d) separation of ions.

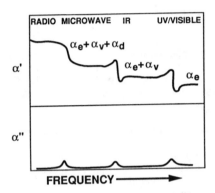

Figure 4. Plot of polarizability of a material as a function of the frequency of the applied field.Top:Real polarizability. Bottom: Imaginary polarizability.

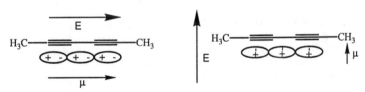

Figure 5. Cartoon illustrating the magnitude of induced polarization with the electric field applied along the long axis of a molecule (left) and induced polarization with the electric field applied perpendicular to the long axis of a molecule (right).

To understand the anisotropic nature of polarizability, consider a molecule as a balloon (Figure 6). Deforming it (applying a field) along the x axis changes the dimension (polarization) of the balloon not only in the x direction but also in the y and z directions. Therefore, three terms, each describing the polarizability along each of the three axes, describe the effect of a field along the x direction. Since the field can be applied in three orthogonal directions, a total of nine terms are required to fully define the polarizability. Taken together, these nine terms are called the polarizability tensor, α_{ij}. Mathematically, α_{ij} is written as a 3 x 3 matrix with elements α_{ij} (i,j = 1, 2, 3). Since both μ and E are vectors and α_{ij} is a tensor, equation 4 is more properly written in terms of the vector components, $\mu_i(\omega)$

$$\mu_i(\omega) = \sum_j \alpha_{ij}(\omega) E_j(\omega) \tag{5}$$

where $\alpha_{ij}(\omega)$ is the <u>linear polarizability</u> of the molecule or atom since it defines the linear variation of polarization (induced dipole moment) with the electric field. However, because of anisotropy in molecular electron density distribution, the direction of the induced dipole moment is not always the same as the electric field that generated the moment.

Polarizability: A Quantum Mechanical Point of View

So far we have described polarization in a classical manner, a useful approach from a phenomenological point of view, but quantitatively inadequate for quantum confined particles like the electrons in molecules (9, 10). We briefly outline the quantum mechanical approach for a small molecule. The polarization of electron density in a molecule by light can be viewed in terms of the electric field of the light wave modifying the ground-electronic state. The original ground- state of the molecule is no longer a quantum-state for the molecule when it is perturbed by the electric field of a light wave. We can write the new quantum-state as a linear combination of the ground- and excited-states of the undisturbed system. Generally, the unperturbed excited-states have different electron distributions than the ground-state, so that mixing the excited-states and the ground-states leads to net charge redistribution (polarization). Since the perturbation of the electron density by the electric field of light is caused by a time dependent field, E(t), we must start with the time-dependent Schrödinger equation to describe the time evolution of the polarization:

$$H\Psi = i\hbar(\partial/\partial t)\Psi \tag{6}$$

H is the Hamiltonian operator for the total energy, $\hbar$ = Planck's constant / 2π, t is the time, and Ψ is the wave function describing the electronic state. The electric field of the light adds another contribution to the Hamiltonian. Assuming that all the molecules are isolated polarization units, the perturbation part of the Hamilitonian is the electric dipole operator, $-\mu \cdot E$. Thus,

$$H = H^0 - \mu \cdot E \tag{7}$$

where

$$\mu = \sum er_i \tag{8}$$

with the sum over i, the number of electrons in the molecule; r_i is a vector to the ith electron.

The perturbation, $H'' = -\mu \cdot E$, represents the interaction of the light wave electric field with the molecule. The solutions to the wave equation 6 are given by

$$\Psi_m(r,t) = \sum C_{mn}(t) \, \Psi(o)_n(r,t) \qquad (9)$$

so that the $\Psi_m(r,t)$ eigenfunctions describe the new electronic states in terms of linear combinations of the original unperturbed $\Psi(o)_m(r,t)$ electronic states of the molecule. In this way the electric field serves to "mix" the unperturbed molecular states, $\Psi(o)_n(r,t)$. For example, the perturbed ground state is a combination of the unperturbed molecule's ground and excited-states and therefore has a different electron density distribution than the unperturbed ground-state. Furthermore, since the electric field is time dependent, the amount of excited-state character will vary with time, as will the electron density distribution. Thus an oscillating electric field induces a time dependent polarization in the molecule.

Suppose that the combination of original states includes the ground-state and an excited-state which has a large charge reorganization with a correspondingly large excited-state dipole. One then expects an increase in dipole moment from a light induced mixing of these states. Garito and coworkers (11) (Figure 7) have demonstrated that a large coupling exists between the ground-state and an excited-state in 2-methyl-4-nitroaniline (MNA). Application of an electric field to the molecule results in extensive charge displacement through the π orbital system and a change in dipole moment.

In summary, the polarization and distortion of electrons can be represented by a field induced mixing of states. This mixing introduces excited-state character but does not result in a long-lifetime population of any excited-states. The instantaneous formation of these polarized states has been referred to in terms of "virtual transitions". The polarization behavior can be written in terms of the ground- and excited-state dipole moments, e.g., <g|μ|g>, <e|μ|e>, and transition moments between the unperturbed molecular states, e.g., <g|μ|e> where g represents the initial ground-state wave function and e an excited state wave function for the unperturbed molecule. The mixing of ground- and excited-states is then described in perturbation theory by summing over the appropriate dipole and transition moments, i.e., a **"sum over states."**

The dipole moments describe the extent of charge separation, and the transition moments measure the mixing of the excited-states into the original ground-state. Extensive mixing of states with concomitant charge reorganization leads to a very soft potential well in the classical picture and, therefore, an increase in polarizability. This "sum over states" approach has been used extensively in the description of organic NLO properties (12, 13). Near resonance, the situation changes and real excited-states come into play.

Both the classical and quantum approaches ultimately lead to a model in which the polarizability is related to the ease with which the electrons can be displaced within a potential well. The quantum mechanical picture presents a more quantitative description of the potential well surface, but because of the number of electrons involved in nonlinear optical materials, theoreticians often use semi-empirical calculations with approximations so that quantitative agreement with experiment is not easily achieved.

Polarizability: A Macroscopic View

NLO characterization and device applications are based on bulk materials, not molecules. It is therefore necessary to look at what happens in the laboratory on the

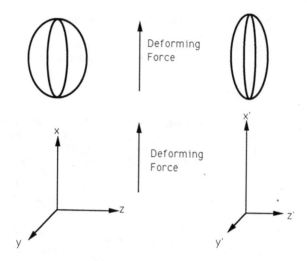

Figure 6. Top: Cartoon illustrating the effect of an arbitrary deforming force applied to a balloon. Bottom: Graphical representation of changes in the x, y and z dimensions when a force is applied along the x direction.

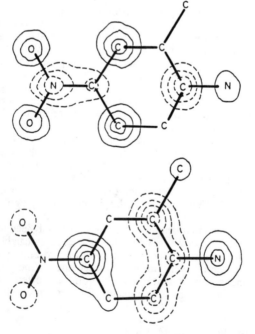

Figure 7. Contour diagram of the wave functions for the MNA ground state (bottom) and a principal excited state (top) showing charge correlations from Garito et. al.

(Reprinted with permission from ref. 11. Copyright 1984 Gordon & Breach.)

bulk, macroscopic level. In bulk materials, the linear polarization (cf. equation 5 for atoms or molecules) is given by

$$P_i(\omega) = \sum_j \chi_{ij}(\omega) \ E_j(\omega) \tag{10}$$

where $\chi_{ij}(\omega)$ is the linear susceptibility of an ensemble of molecules. To relate $\chi_{ij}(\omega)$ to the atomic or molecular polarization described above, it is usually assumed that the atoms or molecules making up the optical material are independently polarized by the light with no interatomic or intermolecular coupling (see below). Within this approximation $\chi_{ij}(\omega)$ is then given by the sum of all the individual polarizabilities, $\alpha_{ij}(\omega)$.

When the electronic charge in the optical material is displaced by the electric field (**E**) of the light and polarization takes place, the total electric field (the "displaced" field, **D**) within the material becomes:

$$\mathbf{D} = \mathbf{E} + 4\pi\mathbf{P} = (1 + 4\pi\chi)\mathbf{E} \tag{11}$$

Since $\mathbf{P} = \chi\mathbf{E}$ (Equation (10)), $4\pi\chi\mathbf{E}$ is the internal electric field created by the induced displacement (polarization) of charges. Usually, the induced polarization causes the spatial orientation of the internal electric field to differ from the applied electric field and, like $\alpha_{ij}(\omega)$, $\chi_{ij}(\omega)$ is a tensor quantity that reflects the anisotropy of the internal electric field.

The dielectric constant $\varepsilon(\omega)$ and the refractive index $n(\omega)$ are two common bulk parameters that characterize the susceptibility of a material. The dielectric constant in a given direction is defined as the ratio of the displaced internal field to the applied field ($\varepsilon = D/E$) in that direction. Therefore from equation 11,

$$\varepsilon_{ij}(\omega) = 1 + 4\pi\chi_{ij}(\omega) \tag{12}$$

The frequency dependence of the dielectric constant provides insight into the mechanism of charge polarization (see Figure 4). Until now, we were concerned with the effect of light on the medium (ε and χ). Since NLO addresses how the optical material changes the propagation characteristics of light, we must now ask what happens to the light as it passes through the medium.

Linear Polarization of Matter and Linear Optical Effects

As shown in Figure 1, the light wave moves electronic charge back and forth. This motion of charge in turn will re-emit radiation at the frequency of oscillation. For linear polarization, this radiation displays the same frequency as the incident light. However the polarization does change the propagation of the light wave.

We know from everyday experience that when light travels from one medium to another its path can change (Figure 8). For example, a straight stick entering the water at an angle appears to bend as it passes below the surface. This apparent bending is due to the fact that light takes the path of "least time", i.e. the fastest way to get from point A (the part of the stick that is under water) to point B (your eye). Since light travels faster in air than in water, the path of the light in water is shorter than that in air. The direction of the light paths in air and in water is determined by the ratio of the speed of light in air to that in water. The ratio of the speed of light in a vacuum, c, to the speed of light in a material, v, is called the index of refraction (n)

$$n = c/v \tag{13}$$

It is important to note that n is uniquely defined for every material or substance.

At optical frequencies in the absence of dispersion (absorption), the dielectric constant equals the square of the refractive index:

$$\varepsilon(\omega) \quad = n^2(\omega) \tag{14}$$

Consequently, we can relate the refractive index to the bulk linear (first-order) susceptibility:

$$n^2(\omega) = 1 + 4\pi\chi(\omega) \tag{15}$$

Since $\chi(\omega)$ is related to the individual atomic or molecular polarizabilities, this simple equation relates a property of light (its speed) to a property of the electron density distribution (the polarizability). Now we can see how the optical properties of a material depend on the electron density distribution, which is dictated by chemical structure. Therefore, as the chemist alters the structure, the optical properties change.

Remember that the electric field of light is perpendicular to its direction of propagation. A point source emitting light from the center of an isotropic crystal emanates light outward uniformly in all directions. The position of the wave front defines a sphere (wave surface) whose radius is increasing with time (Figure 9 top). If we chose another material with a larger susceptibility (more polarizable), its wave surface would expand more slowly because the susceptibility relates to the square of the refractive index (Figure 9 middle). At an arbitrary time (t), the wave surface shows a radius inversely proportional to the index of refraction. For non-cubic single crystals, such as a hypothetical crystal of 2,4-hexadiyne (Figure 9 bottom), the index of refraction, and hence, the polarizability varies with the direction that the light travels through the crystal. The hypothetical crystal of 2,4-hexadiyne is uniaxial, where the unique axis is referred to as the optic axis. Other noncubic crystals are characterized by two optical axes and are said to be optically biaxial. A material then, which has an index of refraction that depends on direction, is called birefringent, or doubly refracting. In the following discussion we limit the discussion of anisotropic crystals to those which are uniaxial.

When a beam of unpolarized light enters a birefringent crystal at normal incidence (but not along the optic axis), two light beams emerge (Figures 10 and 11). One ray, what we call the ordinary ray, passes straight through the crystal. The other ray, called the extraordinary ray diverges as it passes through the crystal and becomes displaced. The relative polarization of the two emerging beams is orthogonal.

This interesting result (first observed in calcite by the Vikings) is explained as follows. We can describe a beam of unpolarized light as the superposition of two orthogonally polarized rays (electric fields at 90 degrees to each other) traversing the same path. When both rays encounter the same index of refraction, as in an isotropic medium, or when the direction of propagation occurs along an optic axis (polarized in a direction perpendicular to the optic axis) of an anisotropic crystal, the beams remain collinear and in phase. For light polarized perpendicular to the optic axis, the material appears isotropic and the index of refraction is independent of the direction of propagation. We call this angularly independent index the ordinary index of refraction, n_o. If light is not polarized perpendicular to the optic axis, the index of refraction varies as a function of the direction of propagation. The angularly sensitive index becomes the extraordinary index of refraction, n_e. Generally, for light traveling through a birefringent material in an arbitrary direction, the two orthogonally polarized rays "see" different polarizabilities and thus different refractive indices. One ray

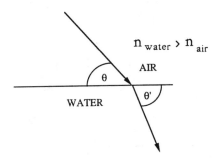

Figure 8. Apparent bending of a stick as it enters water at an arbitrary angle.

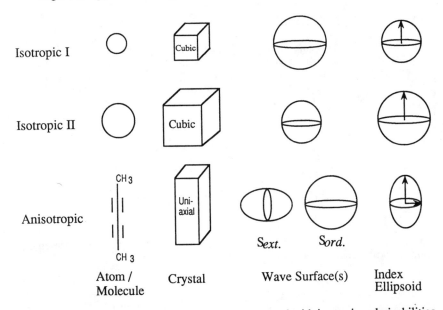

Figure 9. Top: An isotropic atom/molecule and crystal with isotropic polarizabilities will give rise to a spherical wave surface and index ellipsoid. Middle: Another isotropic atom/molecule and crystal with larger isotropic polarizabilities will give rise to a smaller spherical wave surface and a larger spherical index ellipsoid. Bottom: An anisotropic atom/molecule and crystal with anisotropic polarizabilities will give rise to an ordinary spherical wave surface and an ellipsoidal extraordinary wave surface. The index ellipsoid will have major and minor axes.

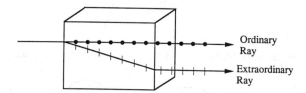

Figure 10. A light beam entering a birefringent crystal at normal incidence (but not along the optic axis) will be divided into an ordinary beam that passes straight through the crystal (line with dots, light polarized perpendicular to the plane of the page). The orthogonally polarized extraordinary beam path (line with dashes, light polarized in the plane of the page) diverges from the ordinary beam as it passes through the crystal.

Figure 11. (a) An image (b) appears to be doubled when viewed through a birefringent material.

becomes retarded relative to the other (introducing a phase shift) and may take a different path. A uniaxial material will therefore have two wave surfaces, a spherical ordinary wave surface and an ellipsoidal extraordinary wave surface. Note that the magnitude of the principle axes of the wave surface are inversely proportional to the refractive indices **normal** to the direction of propagation.

In optics, the optical indicatrix (Figures 9 and 12) is a useful construct that characterizes the birefringence of materials. The indicatrix is a surface that specifies the refractive indices of both the O and E rays traveling in any direction through the material. The indicatrix for a uniaxial material is defined by the equation

$$x^2/n_o^2 + y^2/n_o^2 + z^2/n_e^2 = 1 \qquad\qquad (16)$$

where n_e is the extraordinary index experienced by waves polarized parallel to the optic axis. The surface described by this equation is an ellipsoid with circular symmetry about the z axis and semiaxes equal to n_o for the circularly symmetric axes and n_e for the unique axis. For a given ray (OP) passing through the material, the intersection of the plane perpendicular to the path of the ray and the indicatrix defines the refractive index ellipse (Figure 12). One axis of the ellipse, which varies with θ (the angle between OP and the optic axis), represents the magnitude and polarization of the index of refraction for the extraordinary ray. The angularly insensitive axis is the index of refraction, n_o, for the ordinary ray. Note that the indicatrix is frequency dependent and that birefringence is critical for exploiting a process called second harmonic generation (see D. Williams's tutorial).

Nonlinear Polarizability: A Microscopic View

Until now we have assumed that the polarization of a molecule or material is a linear function of the applied electric field. In reality, the induced polarization generates an internal electric field that modifies the applied field and the subsequent polarization. This interrelationship is the origin of nonlinear polarization. In this section, we present a physical and chemical model for molecules and materials that describes the source of nonlinear behavior.

Figure 4 clearly illustrates that polarizability is a function of the frequency of the applied field. Changing the restoring force constant, κ (equation (2)) is another way to modify the linear polarizability. Another alternative is to add anharmonic terms to the potential to obtain a surface such as that shown in Figure 13. The restoring force on the electron is no longer linearly proportional to its displacement during the polarization by the light wave, it is now **nonlinear** (Figure 14). As a first approximation (in one dimension) the restoring force could be written as:

$$F = -kx - 1/2k'x^2 \qquad\qquad (17)$$

Now the magnitude of the polarization depends on the direction of displacement (Figure 14). For the covalent (e.g. titanyl or vanadyl) M=O bond, in general, one expects that the electron cloud would be more easily polarized towards the oxygen atom. This direction dependency means that the polarization coefficients must be described using tensor quantities.

Just as linear polarization leads to linear optical effects, such as refractive index and birefringence, nonlinear polarization leads to other and usually more subtle (nonlinear) effects. It is precisely these effects we hope to understand and exploit. In Figure 14, application of a symmetric field (i.e., the electric field associated with the light wave) to the anharmonic potential leads to an asymmetric polarization response. This polarization wave shows diminished maxima in one direction and accentuated

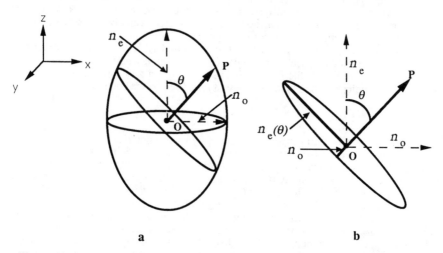

a   **b**

Figure 12. (a) Index ellipsoid defined by n_e and n_0 with a ray of light propagating in an arbitrary direction OP; (b) an ellipse that is formed by the intersection of the plane normal to OP and the index ellipsoid. The principal axes of this ellipse are the angularly dependent index $n_e (\theta)$ and the angularly independent index n_0.

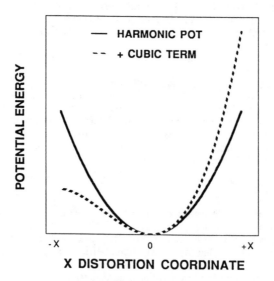

Figure 13. Plot of potential energy vs. distortion coordinate for a material with a harmonic potential and a material with an additional cubic anharmonic term.

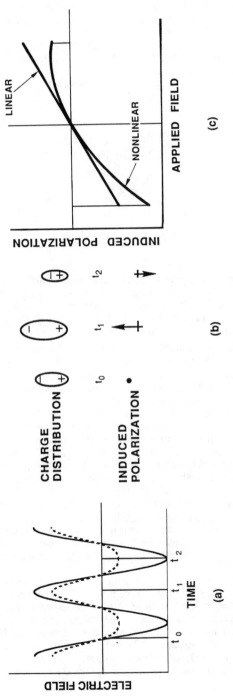

Figure 14. (a) Plots of the electric field of the applied light wave (solid) and the induced polarization wave (dotted), as a function of time, for a second-order nonlinear material; (b) cartoon depicting the polarization of the material as a function of time; (c) plots of induced polarization vs. applied field for both linear and second-order nonlinear materials.

maxima in the opposite direction. This asymmetric polarization can be decomposed into a DC polarization component and polarization components at the fundamental and second harmonic frequencies. This Fourier analysis of the resultant second-order NLO polarization is shown in Figure 15. Since only asymmetric polarization leads to second-order NLO effects, these effects can only be induced by molecules and materials lacking a center of symmetry.

The mathematical formulation of the nonlinear polarization is unknown but a common approximation is to expand the polarizability as a Taylor series:

$$\mu_i = \mu^0_i + (\partial\mu_i/\partial E_j)_{E_0} E_j + 1/2(\partial^2\mu_i/\partial E_j\partial E_k)_{E_0} E_jE_k +$$
$$1/6(\partial^3\mu_i/\partial E_j\partial E_k\partial E_l)_{E_0} E_jE_kE_l + ...(18)$$

or

$$\mu_i = \mu^0_i + \alpha_{ij}E_j + \beta_{ijk}/2 \ E_jE_k + \gamma_{ijkl}/6 \ E_jE_kE_l + ... \qquad (19)$$

where as indicated, μ^0_i is the static dipole in the absence of an electric field.

The series expansion breaks down as an approximation with increasing field strength, i.e. it is most accurate for relatively small fields and polarizations. Obviously, this description is not valid when the electric field strength (e.g. from a laser light beam) approaches the strength of the atomic fields that bind electric charges (10^8–10^9 V/cm). Fortunately, most nonlinear effects are observed at electric fields of 10^3–10^4 V/cm (laser intensities in the kilowatt to megawatt per cm^2 range) and the above expressions are generally applicable (1). The above expansion is not appropriate at or near a resonance frequency. The reader is therefore advised to use care in the application and interpretation of equation 18.

Physically, the higher-order (i.e., nonlinear) terms such as ß relate to the potential well anharmonicity. Miller has suggested that to a first approximation the second-order polarizability is directly proportional to the linear polarizability (first-order) times a parameter defining the anharmonic potential (14, 15). This relationship works best for inorganic materials. In organic molecules the relationship becomes complex because the linear polarizability and the anharmonicity are not necessarily independent variables (see tutorial by D. N. Beratan).

The terms beyond αE are not linear in **E**; they are referred to as the nonlinear polarization and give rise to nonlinear optical effects. Also note that at small fields the polarization will more nearly approximate a linear response; however, with increasing field strength, nonlinear effects become more important. Since $\alpha \gg \beta, \gamma$, there were few observations of NLO effects before the invention of the laser with its associated large electric fields.

Just as α is the linear polarizability, the higher order terms β and γ (equation 19) are the first and second hyperpolarizabilities, respectively. If the valence electrons are localized and can be assigned to specific bonds, the second-order coefficient, ß, is referred to as the bond (hyper) polarizability. If the valence electron distribution is delocalized, as in organic aromatic or acetylenic molecules, ß can be described in terms of molecular (hyper)polarizability. Equation 19 describes polarization at the atomic or molecular level where first-order (α), second-order (ß), etc., coefficients are defined in terms of atom, bond, or molecular polarizabilities. μ is then the net bond or molecular polarization.

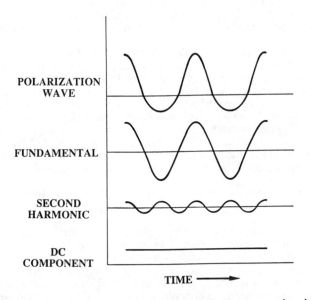

Figure 15. Fourier analysis of an asymmetric polarization wave showing that it is comprised of components at the fundamental frequency, second harmonic frequency, and zero frequency (DC).

Nonlinear Polarizability: a Macroscopic View

The observed bulk polarization is given by an expression analogous to equation 19:

$$\mathbf{P} = \mathbf{P_o} + \chi^{(1)}\cdot\mathbf{E} + \chi^{(2)}\cdot\mathbf{E}\cdot\mathbf{E} + \chi^{(3)}\cdot\mathbf{E}\cdot\mathbf{E}\cdot\mathbf{E} + \dots \tag{20}$$

where the $\chi^{(i)}$ susceptibility coefficients are tensors of order $i + 1$ (e.g., $\chi^{(2)}$ has tensor elements $\chi^{(2)}{}_{ijk}$). $\mathbf{P_0}$ is the built-in static dipole of the sample.

It is of particular interest from both NLO and chemical perspectives to understand how the polarizability changes in the evolution from an isolated atom to a molecule, a cluster of atoms or molecules, an extended array, and ultimately the bulk. Intuitively, one would expect that if the effective potential for the electron extends over several atomic sites, the polarizability and nonlinear optical coefficients might be larger. Indeed, the largest NLO coefficients have been found for semiconductors and unsaturated extended organic molecules both of which have highly delocalized electrons (2).

If the electronic coupling between local clusters of atoms is relatively weak, the bulk polarizabilities can be treated as a simple sum of the localized oscillators in each cluster. If the coupling is strong, a band structure approach should be used. A non-local polarization response must be considered in long conjugated chain compounds, semiconductors, or groups of small clusters that are spatially separated but electronically coupled via resonance tunnelling or similar phenomena. The intermediate situation, the interface between extended and molecular materials, is studied intensely since semiconductor clusters in this regime show unusual NLO behavior. At some point, as polarization dimensions increase, there is a transition from atomic or molecular linear and nonlinear polarizabilities to bulk susceptibilities.

Second-Order Nonlinear Polarization of Matter and Second-Order NLO Effects

Frequency Doubling and Sum-Frequency Generation. A static or oscillating electric field can induce polarization of the electronic distribution, but what has this to do with observable nonlinear optical properties? Earlier, we showed that the induced electronic charge displacement (polarization) by an oscillating electric field (e.g. light) can be viewed as a classical oscillating dipole that itself emits radiation at the frequency of oscillation. For linear first-order polarization, the radiation has the same frequency as the incident light. What is the frequency of this reemitted light for a nonlinear optical material? Recalling that the electric field of a plane light wave can be expressed as

$$\mathbf{E} = \mathbf{E}_0\cos(\omega t) \tag{21}$$

so that for an arbitrary point in space the polarization (equation 20) can be rewritten as:

$$P = P_0 + \chi^{(1)}E_0\cos(\omega t) + \chi^{(2)} E_0{}^2\cos^2(\omega t) + \chi^{(3)} E_0{}^3\cos^3(\omega t) + \dots \tag{22}$$

Since $\cos^2(\omega t)$ equals $1/2 + 1/2 \cos(2\omega t)$, the first three terms of equation 22 could be written:

$$P = (P_0 + 1/2\chi^{(2)} E_0{}^2) + \chi^{(1)}E_0\cos(\omega t) + 1/2 \chi^{(2)}E_0{}^2\cos(2\omega t) + \dots \tag{23}$$

Physically, equation 23 tells us the polarization consists of a second-order DC field contribution to the static polarization (first term), a frequency component w corresponding to the incident light frequency (second term) and a new <u>frequency doubling</u> component, 2ω (third term) (see Figure 15).

Thus, if an intense light beam passes through a second-order nonlinear optical molecule, light at twice the input frequency will be produced as well as a static electric field. The first process is called second harmonic generation (SHG) and the second is called optical rectification. Frequency mixing of this type is referred to as a three wave mixing process, since two photons with frequency ω have combined to generate a single photon with frequency 2ω. Since the nonlinear oscillating dipole re-emits at all its polarization frequencies, we will observe light at both ω and 2ω. We can extend this analysis could be extended to third- and higher-order terms. By analogy, third-order processes will involve four wave mixing.

As written, equation 22 depicts a simplified picture in which a single field, $E(\omega,t)$ acts on the material. The general picture of second-order NLO involves the interaction of two distinct waves with electric fields E_1 and E_2 with the electrons of the NLO material. Suppose for example that we use two laser beams with different frequencies. The second-order term of equation 23 with two interacting waves of amplitudes E_1 and E_2 at an arbitrary point in space becomes:

$$\chi^{(2)} \cdot E_1 \cos(\omega_1 t) E_2 \cos(\omega_2 t) \tag{24}$$

Trigonometry tells us that:

$$\chi^{(2)} E_1 \cos(\omega_1 t) E_2 \cos(\omega_2 t) =$$
$$1/2 \chi^{(2)} E_1 E_2 \cos[(\omega_1 + \omega_2)t] + 1/2 \chi^{(2)} E_1 E_2 \cos[(\omega_1 - \omega_2)t] \tag{25}$$

This equation shows that when two light beams of frequencies ω_1 and ω_2 interact with the atom(s) in the NLO material, polarization occurs at <u>sum</u> $(\omega_1 + \omega_2)$ and <u>difference</u> $(\omega_1 - \omega_2)$ frequencies. This electronic polarization will therefore, re-emit radiation at these frequencies, with contributions that depend on the relative magnitudes of the NLO coefficient, $\chi^{(2)}$. This combination of frequencies leads to sum frequency generation (SFG).

Phase-matching. To combine polarization waves efficiently, conditions must be met so that the fundamental <u>and</u> the second harmonic light waves reenforce each other. If this requirement is met, then the second harmonic intensity will build as the light propagates throughout the crystal. If this condition is not met, a periodic oscillation of second harmonic intensity occurs as the light travels through the crystal. Therefore, the refractive indices experienced by the interacting waves as they propagate through the medium must match, that is $\eta(\omega) = \eta(2\omega)$, to achieve efficient SHG.

As noted before, the polarizability of a material is frequency dependent so that the wave surfaces are frequency dependent. If one of the wave surfaces at the second harmonic frequency intersects one of the wave surfaces at the fundamental frequency, phase-matched SHG can occur. The ray passing through the origin of the ellipsoids and their point of intersection defines the direction of propagation for phased-matched SHG.

Two types of phase-matching exist for SHG: Type I, where the two fundamental photons are of the same polarization, and Type II, where they are orthogonally polarized. The phase- matching direction and one of the principal vibration directions of the crystal may be coincident in biaxial and uniaxial (birefringent) crystals. This situation, called noncritical phase-matching, is quite tolerant of the divergence of the incident beam from the most efficient phase-matching direction and hence is highly desirable. Alternatively, a propagation direction might be found in which the ordinary index at the fundamental frequency n_o (ω) is equal to the extraordinary index at the second harmonic frequency $n_e(2\omega)$.

Efficient SHG also requires a large projection of the β tensor along the phase-matching direction. Therefore molecules or atoms within the NLO material must be properly oriented to give the most efficient SHG response. It is generally believed that single crystals, or polymeric thin films doped with NLO chromophores which have been aligned by poling, are the most promising materials for phase-matched SHG. The phase-matching ability is a particularly critical property for the figure of merit for an NLO material (refer to D. J. Williams's tutorial and S. P. Velsko's article for further discussions of phase-matching). Fabrication of high optical quality single crystals or composites with proper phase-matching properties is often the bottleneck in the search for new second-order NLO materials.

In addition to sum-frequency generation (SFG), another important NLO property is optical parametric oscillation (OPO), the inverse of the SFG process. Pump photons decay into signal and idler photons such that $\omega_p = \omega_s + \omega_i$. The frequencies ω_s and ω_i are determined by the phase-matching condition, and output tuning is accomplished by altering the refractive indices experienced by ω_p, ω_s and ω_i. For oscillation to occur, ω_s and ω_i must change so that the pump and output beams phase-match. The phase-matching condition may be altered by changing the incident angle, temperature, or by applying an external potential (electrooptic tuning) as in the Pockels Effect.

Changing the Propagation Characteristics of Light: The Pockels Effect.
As noted above, refractive indices for different frequencies are usually not the same. Furthermore it is possible to change the amplitude, phase, or path of light at a given frequency by using a static DC electric field to polarize the material and modify the refractive indices.

Consider the special case $\omega_2 = 0$ (equation 24) in which a DC electric field is applied to the crystal. The optical frequency polarization (P_{opt}) arising from the second-order susceptibility is

$$\chi^{(2)}E_1E_2(\cos \omega_1 t) \tag{26}$$

where E_2 is the magnitude of the voltage applied to the nonlinear material. Remember that the refractive index is related to the linear susceptibility (equation 15), that is, given by the second term of equation 23

$$\chi^{(1)}E_1(\cos \omega_1 t) \tag{27}$$

so that the total optical frequency polarization becomes

$$P_{opt} = \chi^{(1)}E_1(\cos \omega_1 t) + \chi^{(2)}E_1E_2(\cos \omega_1 t) \tag{28}$$

$$P_{opt} = [\chi^{(1)} + \chi^{(2)}E_2] \ E_1(\cos \omega_1 t) \tag{29}$$

The applied voltage in effect changes the linear susceptibility and thus the refractive index of the material. This effect, known as the linear electrooptic (LEO) or Pockels effect, modulates light as a function of applied voltage. At the atomic level, the applied voltage is anisotropically distorting the electron density within the material. Thus, application of a voltage to the material causes the optical beam to "see" a different material with a different polarizability and a different anisotropy of the polarizability than in the absence of the voltage. Since the anisotropy changes upon application of an electric field, a beam of plane polarized light will (1) have its plane of polarization rotated by an amount related to the strength and orientation of the applied voltage and (2) travel at a different velocity and possibly in a different direction.

Quantitatively, the change in the refractive index as a function of the applied electric field is given by the general expression:

$$1/\underline{n}_{ij}^2 = 1/n_{ij}^2 + r_{ijk}E_k + s_{ijkl}E_kE_l + ... \tag{30}$$

where $\underline{n}_{ij}$ = the induced refractive indices, n_{ij} = the refractive index in the absence of the electric field, r_{ijk} = the linear or Pockels coefficients, and s_{ijkl} = the quadratic or Kerr coefficients. The optical indicatrix, therefore, changes as the electric field within the sample changes (Figure 16). Electrooptic coefficients are frequently defined in terms of r_{ijk} (5). The "r" coefficient is a tensor (just as is α). The first subscript refers to the resultant polarization of the material along a defined axis and following subscripts refer to the orientations of the applied electric fields. Since the Pockels effect involves two fields mixing to give rise to a third, r_{ijk} is a third rank tensor.

The LEO effect has many important technological applications. Light travelling through an electrooptic material can be modulated by refractive index changes induced by an applied electric field (Pockels effect). Devices exploiting this effect include optical switches, modulators, and wavelength filters. Modulators include the phase modulator that shifts the optical phase by altering an applied voltage, and the Mach-Zehnder intensity modulator. In the latter, the incoming field component is split between two waveguide "arms," where an applied voltage induces a relative phase shift between the optical paths, which results in either constructive or destructive interference upon recombination of the two beams (Figure 17).

Comments on NLO and Electrooptic Coefficients. Typically, the Pockels effect is observed at relatively low frequencies (up to gigahertz) so that slower nonlinear polarization mechanisms, such as vibrational polarizations, can effectively contribute to the "r" coefficients. The tensor used traditionally by theorists to characterize the second-order nonlinear optical response is χ_{ijk}. Experimentalists use the coefficient d_{ijk} to describe second-order NLO effects. Usually the two are simply related by equation 31 (16):

$$d_{ijk} = 1/2\chi_{ijk} \tag{31}$$

The "r" coefficient characterizes the low frequency electrooptic nonlinearity and the "d" coefficient the optical frequency nonlinearity. The conversion from "r" coefficients to "d" or "χ" coefficients must take into account the frequency dependence of the dielectric properties.

Thus, just as linear polarizabilities are frequency dependent, so are the nonlinear polarizabilities. Perhaps it is not surprising that most organic materials, with almost exclusively electronic nonlinear optical polarization, have similar efficiencies for SHG and the LEO effect. In contrast, inorganic materials, such as lithium niobate, in which there is a substantial vibrational component to the nonlinear polarization, are substantially more efficient for the LEO effect than for SHG.

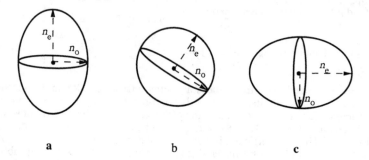

Figure 16. Index ellipsoid of an anisotropic material (a) in the absence of an applied field, (b) in a medium field, and (c) in a strong field.

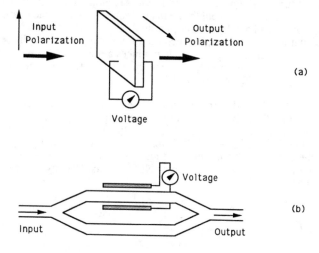

Figure 17. (a) Transverse electrooptic modulator that rotates polarization of an incoming light beam as a function of applied electric field and (b) a travelling Mach-Zehnder interferometer that introduces a phase shift to the light beam in one arm as a function of applied field.

Third-Order Nonlinear Polarization of Matter and Third-Order NLO Effects

Second-order optical nonlinearities result from introduction of a cubic term in the potential function for the electron, and third-order optical nonlinearities result from introduction of a quartic term (Figure 18). Two important points relate to the symmetry of this perturbation. First, while negative and positive β both give rise to the same potential and therefore the same physical effects (the only difference being the orientation of the coordinate system), a negative γ will lead to a different electron potential than will a positive γ. Second, the quartic perturbation has mirror symmetry with respect to a distortion coordinate; as a result, both centrosymmetric and noncentrosymmetric materials will exhibit third-order optical nonlinearities. If we reconsider equation 23 for the expansion of polarization of a molecule as a function of electric field and assume that the even-order terms are zero (i.e., that the molecule is centrosymmetric), we see that polarization at a given point in space is:

$$\mu = \mu_0 + \alpha E_0 \cos(\omega t) + \gamma/6 \, E_0^3 \cos^3(\omega t) + \ldots \tag{32}$$

If a single field, $E(\omega)$, is acting on the material, trigonometry reveals:

$$\gamma/6 \, E_0^3 \cos^3(\omega t) = \gamma/6 \, E_0^3 [3/4 \cos(\omega t) + 1/4 \cos(3\omega t)] \tag{33}$$

thus,

$$\mu = \mu_0 + \alpha E_0 \cos(\omega t) + (\gamma/8) \, E_0^3 \cos(\omega t) + (\gamma/24) \, E_0^3 \cos(3\omega t) \tag{34}$$

or:

$$\mu = \mu_0 + [\alpha + (\gamma/8) \, E_0^2] E_0 \cos(\omega t) + (\gamma/24) \, E_0^3 \cos(3\omega t) \tag{35}$$

The above equation states that the interaction of an intense beam with a third-order NLO material will create an electric field component at the third harmonic. In addition, there is an electric field component at the fundamental frequency, and we note that the $[\alpha + (\gamma/8) \, E_0^2]$ term of equation 35 is similar to the term leading to the linear electrooptic effect. Likewise, the application of an intense voltage will also induce a refractive index change in a third-order NLO material. These two effects are known as the optical and DC Kerr effects, respectively. The sign of gamma will determine if the third-order contribution to the refractive index is positive or negative in sign. Materials with positive gamma are called self focusing; those with negative gamma are known as self defocusing.

Degenerate four wave mixing is another interesting manifestation of third-order NLO materials. Two beams of light interacting within a material will create an interference pattern (Figure 19) that will lead to a spatially periodic variation in light intensity across the material. As previously noted, the induced change in refractive index of a third-order nonlinear optical material is proportional to the magnitude of the applied field. Thus if the beams are interacting with a third-order NLO material, the result will be an index of refraction grating. When a third beam is incident on this grating a new beam of light, called the phase conjugate beam, is diffracted from the grating. In short, three beams (two writing beams and one probe beam) create a fourth beam, i.e. four-wave mixing. A potential use of this phenomena is in phase conjugate optics, which takes advantage of a special feature of the diffracted beam: its path

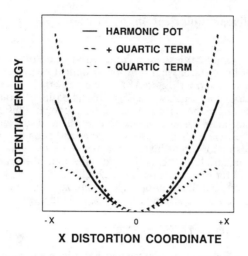

Figure 18. Plot of potential energy vs. distortion coordinate for a material with a harmonic potential and a material with positive and negative quartic anharmonic terms.

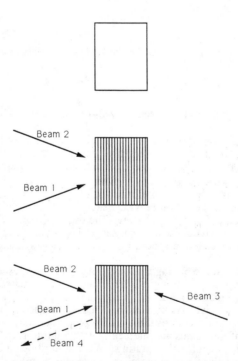

Figure 19. Top: A phase conjugate material in the absence of an applied field. Middle: Beams 1 and 2 create a refractive index grating. Bottom: Beam 3 interacts with the grating creating beam 4 that is the *phase conjugate* of beam 1.

exactly retraces the path of one of the writing beams. As a result, a pair of diverging beams "reflected" from a phase conjugate mirror will converge rather than diverge as from an ordinary mirror (Figure 20). This remarkable property means that distorted optical images can be reconstructed using phase conjugate optical systems (Figure 21).

Issues Relating to Nonlinear Optical Materials

A chemist attempting to design NLO materials is confronted with a variety of questions. Is the goal to map out structure/property relationships that will improve fundamental understanding of NLO phenomena or is it to fulfil requirements for a particular device application? These important goals have different challenges associated with them. We shall provide a brief introduction relating chemical structure to linear and nonlinear polarizability, then end with some comments relating to device issues.

Systematic studies of well-defined materials in which specific structural variations have been made, provide the basis for structure/property relationships. These variations may include the effect of charge, hybridization, delocalization length, defect sites, quantum confinement and anharmonicity (symmetric and asymmetric). However, since NLO effects have their origins in small perturbations of ground-state electron density distributions, correlations of NLO properties with only the ground state properties leads to an incomplete understanding of the phenomena. One must also consider the various *excited-state* electron density distributions and transitions.

It is necessary to understand how a structural change affects the variable to be studied. If possible the structural variation should have only a small effect on other variables. In addition, remember that results are not only sensitive to the material used in the study, but also to the method of measurement. It is often meaningless to compare measurements by different techniques or at different frequencies. The tutorials that follow, in particular the tutorial by J. Perry on characterization techniques, will provide the reader with some insight into these issues.

How does one engineer linear and nonlinear polarizability? If this were fully understood, this book would not exist. As chemists we bring some intuitive understanding of what factors affect polarizability. For example, the organic chemist knows that the electrons in polyacetylene are more polarizable than those in butane. Likewise, the inorganic chemist knows that semiconductors are more polarizable than insulators. These simple observations suggest that the extent of electron delocalization is related to the linear polarizability. In organic molecules the extent of delocalization is affected by the hybridization, degree of coupling, and number of orbitals (and electrons) in the electronic system of interest. We also see that molecules/materials with strong, low-energy, absorption bands tend to be highly polarizable. The linear polarizability derived from perturbation theory

$$\alpha \quad \sim \quad \sum_{\substack{\text{excited} \\ \text{states}}} \langle g|\mu|e\rangle^2/E_{ge} \tag{36}$$

is in accord with these observations. Therefore, it is not surprising that cyanine dyes and semiconductors with their large oscillator strengths and small HOMO–LUMO gaps are highly polarizable.

For second-order nonlinear polarization, the problem becomes more complex. As can be seen in Figure 13 the anharmonic polarization shows the largest deviation from the linear polarization with large distortion values. Therefore, if the material is not polarizable (i.e., if the electrons can only be perturbed a small distance from their equilibrium positions), then the anharmonicity will not be manifested. For large second-order nonlinearities we need a material that offers both a large linear

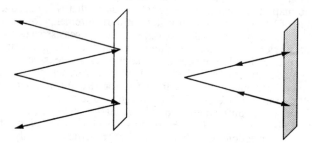

Figure 20. Left: A diverging set of beams reflected off of a normal mirror continue to diverge. Right: A diverging set of beams reflected off of a phase conjugate mirror exactly retrace their original path and are therefore recombined at their point of origin.

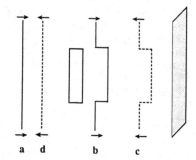

a   d   b   c

Figure 21. (a) A planar wave (b) passes through a distorting material that introduces an aberration and (c) the material interacts with a phase conjugate mirror creating the phase conjugate image. (d) When the phase conjugate wave passes through the distorting material on the reverse path the original aberration is cancelled producing an undistorted image.

polarizability and a large anharmonicity. In organic donor-acceptor molecules, it is easier to polarize the electrons toward the acceptor than towards the donor. Thus these systems have an asymmetric anharmonic term and not surprisingly organic donor acceptor molecules have some of the largest known values of β. It is not sufficient for a molecule to have a large β. In order to observe second-order NLO effects the bulk material must also be non- centrosymmetric. As noted before, the molecules must be oriented in a manner that optimizes the phase-matching of the fundamental and sum frequencies. Since ~75% of all non-chiral molecules crystallize in centrosymmetric space groups and thus have vanishing $\chi^{(2)}$, proper alignment of the chromophore in the bulk material is a major impediment to achieving the goal of engineering a material with a large $\chi^{(2)}$. The electronic driving force for crystallization in a centrosymmetric space group is the cancellation of destabilizing dipole-dipole interactions between adjacent molecules in the crystal lattice. Several approaches have been tried with varying degrees of success to overcome this obstacle (see tutorials by D. Williams and D. Eaton).

The structure/property relationships that govern third-order NLO polarization are not well understood. Like second-order effects, third-order effects seem to scale with the linear polarizability. As a result, most research to date has been on highly polarizable molecules and materials such as polyacetylene, polythiophene and various semiconductors. To optimize third- order NLO response, a quartic, anharmonic term must be introduced into the electronic potential of the material. However, an understanding of the relationship between chemical structure and quartic anharmonicity must also be developed. Tutorials by P. Prasad and D. Eaton discuss some of the issues relating to third-order NLO materials.

The synthesis of materials for device applications has very different requirements. Here, the most important questions are: What does the device do and what factors will affect its performance? The magnitude of the desired optical nonlinearity will always be one of many criteria that will ultimately dictate the material of choice. In many instances the magnitude of the nonlinearity will not be the most important parameter. Depending on the device applications, other considerations such as optical transparency, processability, one- and two-photon optical stability, thermal stability, orientational stability, and speed of nonlinear response will all be important. Our current understanding of NLO materials suggests that these variables are frequently interrelated and that there is often no ideal NLO material. The material of preference for a given application will typically be one that is the best compromise for a variety of variables. Tutorials by G. Stegeman and R. Zanoni, and by R. Lytel outline some of the NLO device applications and the related materials issues.

Many challenging materials issues addressable by chemists currently exist. Chemists bring to the field a chemical understanding of materials that physicists and engineers do not have. Chemists form the bridge between theoretical models for nonlinear polarizability and real NLO materials. As a result, they have an excellent opportunity to make a large, positive, impact on the field of nonlinear optics.

Acknowledgments

GDS wishes to thank the Office of Naval Research for financial support. This paper was in part written at the Jet Propulsion Laboratory, California Institute of Technology in conjunction with its Center for Space Microelectronics Technology. This center is supported by the Strategic Defense Initiative Organization, Innovative Science and Technology Office through an agreement with the National Aeronautics and Space Administration (NASA). The authors also thank Marsha Barr, David Beratan, Laura Davis, Eric Ginsburg, Christopher Gorman, Lynda Johnson, Joseph Perry, William

Schaefer, Bruce Tiemann, and David Williams for their input and critical reading of this manuscript. SRM in particular thanks Bruce Tiemann for many helpful discussions.

Literature Cited

1. *Nonlinear Optical Properties of Organic Molecules and Crystals,* Volumes 1 and 2; Chemla, D. S.; Zyss, J. Eds.; Academic Press: Orlando, 1987.

2. Williams, D. J.; *Angew. Chemie Int. Ed. Engl,* **1984,** *23,* 690.

3. *Nonlinear Optical Properties of Organic and Polymeric Materials;* Williams, D. J., Ed.; ACS Symposium Series 233; American Chemical Society: Washington, DC; 1983.

4. Williams, D.; Prasad, P.; *An Introduction to Nonlinear Optical Effects in Molecules and Polymers,* in press.

5. Kaminow, I. P.; *An Introduction to Electrooptic Devices;* Academic Press: New York, 1974.

6. Feynman, R. P.; Leighton, R. B.; Sands, M.; *The Feynman Lectures on Physics,* Volume 1; Addison Wesley: Reading, Massachusetts, 1963.

7. Steward, H. Bruce; Thompson, J. M.; *Nonlinear Dynamics and Chaos*; Wiley: Chichester, 1986.

8. Zyss, J.; Chemla, D. S.; In *Nonlinear Optical Properties of Organic Molecules and Crystals,* Volume 1; Chemla, D. S.; Zyss, J., Eds.; Academic Press: Orlando, 1987; pp 23–191.

9. Zyss, J.; Chemla, D. S.; In *Nonlinear Optical Properties of Organic Molecules and Crystals,* Volume 1; Liao, P. F.; Kelley, P., Eds.; Academic Press: New York, 1987; p 23.

10. Pugh, D.; Morley, J. O.; In *Nonlinear Optical Properties of Organic Molecules and Crystals,* Volume 1; Liao, P. F.; Kelley, P., Eds.; Academic Press: New York, 1987; p 193.

11. Garito, A. F.; Teng, C. C.; Wong, K. Y.; Kharmiri, Z.; *Mol. Cryst. Liq. Cryst.* **1984,** *106,* 219.

12. Ward, J. F. *J. Rev. Mod. Phys.* **1965,** *37,* 1.

13. Orr, J. B.; Ward, J. F.; *Mol. Phys.* **1971,** *20,* 513.

14. Miller, R. C. *Appl. Phys. Lett.* **1964,** *5,* 17.

15. Kurtz, S. K. In *Laser Handbook*; Arecchi, F. T.; Schulz-Dubois, E. O., Eds.; North Holland: 1972; p 923.

16. Zernike, F.; Midwinter, J. E.; *Applied Nonlinear Optics;* Wiley; p. 34.

RECEIVED October 8, 1990

Chapter 2

Second-Order Nonlinear Optical Processes in Molecules and Solids

David J. Williams

Corporate Research Laboratories, Eastman Kodak Company, Rochester, NY 14650-2110

In this paper, an overview of the origin of second-order nonlinear optical processes in molecular and thin film materials is presented. The tutorial begins with a discussion of the basic physical description of second-order nonlinear optical processes. Simple models are used to describe molecular responses and propagation characteristics of polarization and field components. A brief discussion of quantum mechanical approaches is followed by a discussion of the 2-level model and some structure property relationships are illustrated. The relationships between microscopic and macroscopic nonlinearities in crystals, polymers, and molecular assemblies are discussed. Finally, several of the more common experimental methods for determining nonlinear optical coefficients are reviewed.

Interest in the field of nonlinear optics has grown tremendously in recent years. This is due, at least partially, to the technological potential of certain nonlinear optical effects for photonic based technologies. In addition, the responses generated through nonlinear optical interactions in molecules and materials are intimately related to molecular electronic structure as well as atomic and molecular arrangement in condensed states of matter.

While much of the basic physics of nonlinear optics was developed in the 1960's and early 1970's, from both classical and quantum mechanical perspectives, progress in new materials designed to exhibit specific effects for various technologically important applications has been more recent and much remains to be done. While a variety of useful materials exist today for many of these applications, particularly inorganic crystals, many opportunities exist for new materials that can be fabricated and processed in thin film format for useful and potentially inexpensive devices. The potential for integration of devices on various substrates such as glass, Si, GaAs, etc., is a particularly attractive aspect of the organic polymeric approach.

This tutorial deals with nonlinear optical effects associated with the first nonlinear term in expression for the polarization expansion described in the next section. The first nonlinear term is the origin of several interesting and important effects including second-harmonic generation, the linear electrooptic or Pockels effect,

0097–6156/91/0455–0031$06.00/0
© 1991 American Chemical Society

various frequency mixing processes between coherent optical fields, and optical rectification where polarization of the medium at optical frequencies produces a DC response. Due to the scope of this chapter, the subject matter will be approached in a qualitative and conceptual manner, and the reader interested in learning the subject matter in greater depth will be referred to various sources of information from the literature on this subject (1).

The tutorial begins with a description of the basic concepts of nonlinear optics and presents illustrations from simple models to account for the origin of the effects. The microscopic or molecular origin of these effects is then discussed in more detail. Following this, the relationship between molecular responses and the effects observed in bulk materials are presented and finally some of the experimental methods used to characterize these effects are described.

Concepts of Nonlinear Optics

Lorenz Oscillator. Optical effects in matter result from the polarization of the electrons in a medium in response to the electromagnetic field associated with light propagating through the medium. A simple but illustrative model for these interactions is the Lorenz Oscillator described in a variety of texts (2). Here an electron is bonded to a nucleus by a spring with a natural frequency, ω_o, and equilibrium displacement, r (Figure 1). The electric component of the optical field felt by the electron is represented as a sinusoidally varying field. Considering only the linear response of this object, or equivalently the harmonic oscillator approximation, the equation of motion can be written as

$$eE - m\omega_o{}^2 r = m\frac{d^2 r}{dt^2} + 2\Gamma \frac{dr}{dt} \tag{1}$$

The terms on the left are the electric force and the restoring force that is linear in the displacement. A damping term with a proportionally constant Γ is added to the right hand side to account for dissipation of energy during the polarization response. The solution to this equation leads to an expression for the displacement, r, as

$$r = \frac{e}{m} \frac{1}{\omega_o{}^2 - 2i\Gamma\omega - \omega^2} E_o e^{i\omega t} + c.c. \tag{2}$$

This equation describes a sinusoidal response at frequency, ω, to the electric field component at ω. This is the basis for the linear optical response. To calculate the optical properties of the Lorenz oscillator the polarization of the medium is obtained as

$$P_L = -Ner = \chi^{(1)} E \tag{3}$$

where N is the density of polarizable units in the medium. The polarization is also often expressed in terms of a susceptibility $\chi^{(1)}$ whose value can be readily obtained by comparison with equation 2. In the linear regime the relationship between the susceptibility and two other quantities of importance the refractive index, n, and dielectric constant, ε, is given by

$$\varepsilon = n^2 = 1 + 4\pi\chi^{(1)} \tag{4}$$

and the optical properties of the Lorenz Oscillator as specified by the complex refractive index as

$$n^2 = 1 + \left(\frac{Ne^2}{m}\right)\frac{4\pi}{\omega_o{}^2 - 2i\Gamma\omega - \omega^2} + c.c. \tag{5}$$

The real and imaginary parts of the refractive index are plotted schematically as a function of frequency in Figure 2. For the case where $\Gamma = 0$ there is no damping and therefore no absorption, n is real and corresponds to the refractive index of the medium. The situation where Γ is not equal to zero corresponds to optical absorption. This model reasonably describes the linear optical properties, in the absence of vibronic coupling, for typical organic molecules.

Nonlinear optical effects can be introduced into this picture by postulating that the restoring force in equation 1 is no longer linear in the displacement and adding a term, say ar^2, to the left hand side of the equation, *(3)*. The differential equation can no longer be solved in a simple way but, if the correction term is assumed to be small relative to the linear term, a straightforward solution follows leading to a modification of equation 3.

$$P = P_L + P_{NL} \tag{6}$$

where

$$P_{NL} = -Ne\left(r_2(2\omega) + r_2(o)\right) \tag{7}$$

and r_2 is the correction to the linear displacement r. The correction term oscillates at 2ω rather than ω and induces a D.C. offset to the displacement. The linear and nonlinear contributions to the polarization of a molecule of general structure is shown

schematically in Figure 3 *(4)*. The oscillating polarization is clearly nonlinear relative to the driving field. This effect is due to the greater ease of displacement of electronic density in the direction from donor substituent, D, to acceptor, A, than vice-versa. The Fourier theorem states that a nonsinusoidal function can be expressed as a summation of sinusoidal responses at harmonics of the fundamental frequency with appropriate coefficients. A DC offset term is also associated with asymmetric functions. The polarization at the various frequencies, 2ω for example, act as the sources of a new electromagnetic fields at the appropriate frequency which can lead to substantial conversion of energy at the fundamental frequency to the new frequency. The D.C. offset is the source of the optical rectification effect mentioned above. Since the coefficients of the subsequent orders in frequency tend to be a few orders of magnitude smaller than the previous one, the largest effects will be at the first harmonic (sometimes referred to as second-harmonic generation) and zero frequency.

Constitutive Relations. A more general representation of the nonlinear polarization is that of a power series expansion in the electric field. For molecules this expansion is given by

$$\mu_i = p = \alpha \cdot E + \beta \vdots EE + \gamma \vdots EEE + ... \tag{8}$$

where p and E are the polarization and electric field vectors, respectively, the coefficients α, β, γ are tensors, and μ_i is the induced dipole moment of the molecule. This expression is valid in the dipolar approximation where the wavelength of the optical field is large compared to the dimensions of the polarizable unit. The tensors α, β, and γ relate the cartesian components of the electric field vectors to those of the

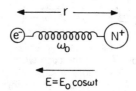

Figure 1. Representation of the Lorenz oscillator.

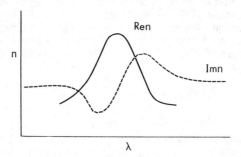

Figure 2. Illustration of the optical properties of the harmonic oscillator.

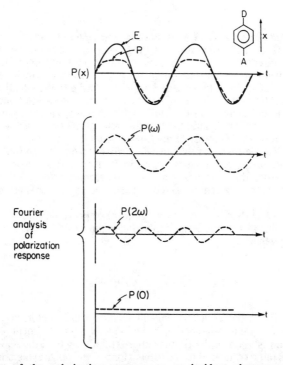

Figure 3. Plot of the polarization response to an incident electromagnetic field at frequency ω and the Fourier components of that response. (Reprinted with permission from ref. 4. Copyright 1975 John Wiley and Sons.)

polarization vectors. The array of tensor elements is therefore intimately related to the electronic structure of the molecule. For the hypothetical molecule illustrated in Figure 3 it might be anticipated that

$$\beta_{xxx}E_xE_x >> \beta_{xyy}E_yE_y \tag{9}$$

so that

$$p_x \sim \beta_{xxx}E_xE_x \tag{10}$$

since asymmetric charge displacement is the origin of the first nonlinear correction. Similarly β_{yyy} is expected to be zero since the molecule has a center of inversion symmetry along the y axis. An important point to note is that only molecules lacking a center of inversion can exhibit a nonzero value of β. The quantity γ, however, is not related to the asymmetry of the polarization response but its departure from nonlinearity at large values of the displacement occurring in either direction. All molecules and atoms regardless of symmetry exhibit nonzero values of γ tensor elements.

An expression similar to equation 8 can be written for the macroscopic polarization of a medium or ensemble of molecules as

$$P = \chi^{(1)} \cdot E + \chi^{(2)} \vdots EE + \chi^{(3)} \vdots EEE \tag{11}$$

For anisotropic media, the field and polarization terms must be treated as vectors and the coefficients as tensors for the reasons described previously. Even order terms such as $\chi^{(2)}$ are nonzero only in noncentrosymmetric media, whereas the $\chi^{(1)}$ and $\chi^{(3)}$ terms are nonzero in all media. The microscopic quantities α, β, γ and related to $\chi^{(1)}$, $\chi^{(2)}$, $\chi^{(3)}$ in a straightforward manner, as described in Section IV.

Propagation of Light Through the Medium. The relationship between the phases of the electric fields and polarization responses as they propagate through a nonlinear medium determines the amplitude of the generated fields. The coupled amplitude equation *(5)* formalism is often used to illustrate the coupling of the various fields to each other which is required if they are to exchange energy with one another. While this treatment is beyond the scope of this chapter, a schematic illustration of phase dependent coupling is shown in Figure 4. The top trace is the fundamental field propagating in the z direction with wave vector $(k_\omega = \frac{2\pi}{\lambda})$ The harmonic polarization response at 2ω propagates with wavevector 2ω and its phase is inextricably linked to the fundamental field. This is sometimes referred to as the "bound wave." The electric field generated by $P(2\omega)$ propagates with wavevector $k_{2\omega}$. Because of the dispersion in the refractive index as a function of frequency $n_{2\omega} \neq n_\omega$, or equivalently $k_{2\omega} \neq k_\omega$. As a result, the phase relationship between the "free wave" $E(2\omega)$ and $P(2\omega)$ varies along the z direction. The term "free wave" is used to signify that the electric field component at 2ω propagates with its own phase velocity which is determined by the refractive index at 2ω. The coupled amplitude equations predict the direction of power flow as a function of phase. As shown in the illustration, power flows into $E(2\omega)$ in the first coherence length and back into $E(\omega)$ during the second coherence length. In the example, the harmonic field amplitude is assumed to be a small fraction of the fundamental field so that the amplitude of the fundamental field does not vary significantly under these non-phase matched conditions. If phase matching can be achieved by some method, the field at $E(2\omega)$ can

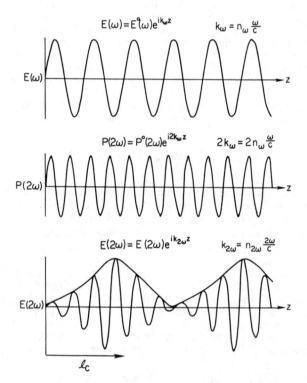

Figure 4. Illustration of the dependence of the amplitude of the harmonic field $E(2\omega)$ on its phase relationship with the polarization response $P(3\omega)$.

become quite large and $E(\omega)$ will be depleted. In a later section the circumstances under which the fundamental and harmonic fields propagate in phase will be illustrated.

Microscopic Description

A number of approaches have been described in the literature for calculating the microscopic nonlinear response and are reviewed in detail elsewhere *(1)*. In this section, these methods are briefly mentioned and a simplified version of time dependent perturbation theory is used to illustrate the intuitive aspects of the microscopic nonlinear polarization.

The approaches to this problem follow along two general lines. In the first approach, one computes derivatives of the dipole moment with respect to the applied field and relates them to the terms in the polarization expansion of equation 8. Inspection of equation 8 suggests that the second derivative of the dipole moment with respect to the field gives β. The choice of the exact form of the Hamiltonian, which incorporates the optical field and the atomic basis set, determines the accuracy of this procedure. In one popular version of this approach, the finite field method, the time dependence of the Hamiltonian is ignored for purposes of simplification and the effects of dispersion on β, therefore, cannot be accounted for.

A more widely used approach for organic molecules is based on second-order perturbation theory. Here the dipolar contribution to the field induced charge displacement is calculated by inclusion of the optical field as a perturbation to the Hamiltonian. Since the time dependence of the field is included here, dispersion effects can be accounted for. In this approach the effect of the external field is to mix excited state character into the ground state leading to charge displacement and polarization. The accuracy of this method depends on the parameterization of the Hamiltonian in the semi-empirical case, the extent to which contributions from various excited states are incorporated into the calculation, and the accuracy with which those excited states are described. This in turn depends on the nature of the basis set and the extent to which configuration interaction is employed. This method is generally referred to as the sum over states (SOS) method.

In a simplified version of this approach a molecule is considered to have a ground and single-charge transfer excited state which dominate the hyper-polarizability

This is referred to as the 2-level model *(6)* and was used in the past to understand trends in β with structural modifications of the molecule. With the computational methods and experimental accuracy available today this method is viewed as inadequate although it does serve to illustrate the essential features of molecular structure that control β. In the two level model, β_x (the component of the tensor along the charge transfer axis) is given by

$$\beta_x \approx \left(\frac{3\hbar^2}{2n}\right)\frac{\omega_{eg}f\delta\mu}{\left(\omega_{eg}^2 - \omega^2\right)\left(\omega_{eg}^2 - 4\omega^2\right)} \tag{12}$$

where ω_{eg} is the frequency of the molecular charge transfer transition, f is its oscillator strength, and $\delta\mu$ is the difference in dipole moment between the two states. Values of β calculated by the 2-level model and published experimental measurements of β are shown in Table I *(5)*. The calculated number in

Table I. Measured and Calculated Values of β Using Two-Level Model

Molecule	$\beta_x(10^{-30}$ esu) (2-level)	β_x (10^{-30}) (experimental)
I	19.6	16.2-345
II	10.9	10.2
III	4	6
IV	227 (69.1)	225
V	383	450

parenthesis was obtained by a SOS calculation and shows how the 2-level model over-estimates β in this case. In other cases under-estimation may occur. Nevertheless, the trends are clear and agree with the experimentally observed trends. Para-substitution provides for a strong resonance interaction and good charge separation. For ortho substitution, the resonance interaction is retained but the charge separation is clearly lower leading to the lower value of β. In meta substitution, charge transfer resonance is forbidden due to the symmetry relationship between the ground and excited state wave functions. An increase in molecular length results in a substantial increase in β_x. In many cases it is relatively straightforward to make qualitative predictions, based on physical organic principles of the effect of a change in the nature of a substituent, or some other aspect of molecular structure, on β.

Relationship of Macroscopic to Microscopic Nonlinearities

Crystals. In the discussion of equation 11, it was pointed out that the macroscopic nonlinear coefficients could be related to the microscopic ones in a relatively straightforward manner. For the hypothetical crystal shown in Figure 5, the relationship is (8)

$$\chi_{IJK}^{(2)} = 2/V \, f_I^{2\omega} f_J^{\omega} f_K^{\omega} \sum_{ijk} \left(\sum_{s=1}^{N_g} \cos\theta_{Ii}^{(s)} \cos\theta_{Jj}^{(s)} \cos\theta_{Kk}^{(s)} \right) \beta_{ijk} \qquad (13)$$

where V is the volume of the unit cell and N_g is the number of equivalent sites per unit cell (in this case 2), the $\cos\theta_{Ii}$ etc. are direction cosines between the molecular and crystal frames of reference and $f_I^{2\omega}$, f_J^ω, f_K^ω are local field factors that alter the value of the external fields at the molecular site due to screening effect. Zyss and Oudar (8) have tabulated the specific form of the symmetry allowed $\chi_{IJK}^{(2)}$ tensor elements for all of the known polar space groups. With a knowledge of the molecular hyperpolarizability, β, which can be obtained from EFISH measurements as described in the next section, and a knowledge of the crystal structure it is possible to calculate the macroscopic coefficients using estimated local field factors (9).

Poled Polymers. Polymeric materials offer many potential advantages (as well as some disadvantages) relative to crystalline materials for second order nonlinear optical applications. The main problem with polymer films prepared by solvent casting or other film forming techniques is that they tend to be isotropic or at best have axial symmetry due to strain fields developed in the preparation procedure. Both of these are centrosymmetric arrangements and the films do not exhibit second-order nonlinear optical properties. An approach to circumventing this problem discovered by Meredith et al. (10) is to apply an electric field to the polymer in a softened state and tends to align molecular dipoles associated with the chromophores in the direction of the field. The alignment forces are counteracted by thermal randomization associated with the kinetic energy of the system. A hypothetical situation is illustrated in Figure 6. When the field is applied and the orientation is quenched into the film by cooling or chemical crosslinking a polar axis is induced in the Z direction. The axis defines the symmetry of the medium as polar (uniaxial). Under these circumstances there are two unique directions; parallel and perpendicular to the polar axis. Using an oriented gas model with Boltzman thermal averages Meredith et al. (10) calculated the nonlinear coefficients applicable to poled polymers. These are

$$\chi_{11}^{(2)} = \chi_{zzz}^{(2)} = nf_z^{2\omega}\left(f_z^\omega\right)^2\left\langle\cos^3\theta\right\rangle\beta_z \tag{14}$$

and

$$\chi_\perp^{(2)} = \chi_{zxx}^{(2)} = \chi_{zyy}^{(2)} = Nf_z^{2\omega}\left(f_y^\omega\right)^2\left\langle(\cos\theta)(\sin^2\theta)\right\rangle\beta_z \tag{15}$$

where N is the concentration of the chromophore. In these expressions, θ is defined as the angle between the molecular dipole and the poling field direction. The brackets $\langle\rangle$ indicate thermal averages. In an isotropic sample $\langle\cos^3\theta\rangle$ is 0 and $\chi_{zzz}^{(2)}$ is likewise zero. If total alignment could be induced in the z direction, $\langle\cos^3\theta\rangle = >1$ and the maximum value of $\chi_{zzz}^{(2)}$ would be determined by N and β_z. Under these conditions, $\langle\sin^2\theta\rangle = 0$ and the perpendicular component, $\chi_\perp^{(2)}$, would approach 0. The problem of calculating these coefficients is one of evaluating the thermal average $\langle\cos^3\theta\rangle$. It should be apparent $\chi_\parallel^{(2)}$ and $\chi_\perp^{(2)}$ are not independent quantities and that a knowledge of one quantity implies a knowledge of the other. The Boltzman thermal average of $\cos^3\theta$ is given by

$$\left\langle\cos^3\theta\right\rangle = \frac{\int_0^\pi f(\theta)\cos^3\theta\sin\theta d\theta}{\int_0^\pi f(\theta)\sin\theta d\theta} \tag{16}$$

where $f(\theta)$ is the orientational distribution function

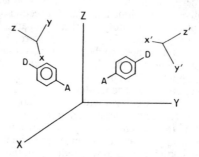

Figure 5. The relationship between molecular and crystal coordinate systems for a unit cell containing two molecules.

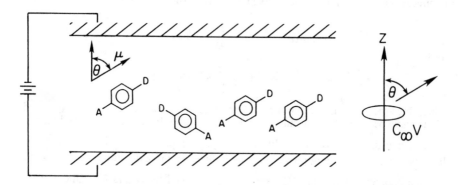

Figure 6. Schematic representation of molecular orientation in a polymer film with respect to the poling field in the z direction.

$$f(\theta) = e^{U(\theta)/kT} \tag{17}$$

and

$$U(\theta) = U'(\theta) - \mu \cdot E \tag{18}$$

$U(\theta)$ is the orientation dependent local potential energy and $U'(\theta)$ accounts for any anisotropy due to the local environment. For instance, if the sample were liquid crystalline, $U'(O)$ and $U'(\pi)$ would represent the potential well associated with nematic or smectic director. In addition to local potentials it is clear that $f(\theta)$ depends on the experimentally controllable quantities μ, E, and T. The solution to equation 16 is the series expansion referred to as the third-order Langevin function $L_3(p)$

$$\left\langle \cos^3 \theta \right\rangle = (1 + 6/p^2)L_i(p) - 2p + \ldots \tag{19}$$

and

$$L_i(p) = \left\langle \cos \theta \right\rangle = \frac{1}{3p} - \frac{1}{45p^3} + \frac{2}{945p^5} - \frac{2}{9450p^7} + \ldots \tag{20}$$

where $p = \mu E/kT$. A plot of $L_3(p)$ vs E is given in Figure 7. A linear relationship between $\chi^{(2)}$ and $L_3(p)$ is predicted and has been verified experimentally *(10,11)* for a number of polymeric systems.

Langmuir-Blodgett Films. In the Langmuir-Blodgett technique, monolayers of molecules containing a hydrophilic group at one end and a hydrophobic tail on the other are spread over the surface of water in an appropriate fixture. By controlling the surface pressure the molecules can be made to organize into highly oriented structures at the air-water interface. It is also possible to transfer the monolayers onto a substrate of appropriate polarity. Subsequent deposition cycles can be employed to fabricate multilayer films of macroscopic dimensions. With appropriate choices of head and tail units noncentrosymmetric films with polar cylindrical symmetry can be fabricated (Figure 8). Equations 14 and 15 apply to films fabricated by this method. Here however, the distribution of angles, θ, is expected to be much more tightly distributed around some central value, say θ_0. This being the case, it is possible to determine the value of θ_0 experimentally from the polarization and angular dependence of the second harmonic intensity generated by these films. Since the films can be fabricated layer by layer the concentration N is given by

$$N = nN_s \tag{21}$$

where n is the number of layers. A linear dependence of $\chi^{(2)}$ versus n is predicted if the film deposition process proceeds as anticipated and this has been observed *(12)*.

Experimental Methods

Electric Field Induced Second-Harmonic Generation. An essential aspect of the development of materials for second-order nonlinear optics is the determination of the β tensor components. The technique that has been developed to accomplish this is called electric field induced second harmonic generation (EFISH) *(13,14)*.

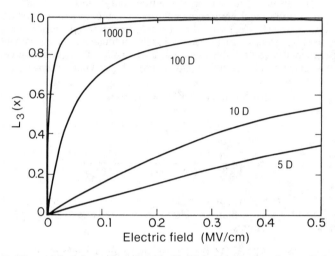

Figure 7. Plot of $L_s(p)$ versus electric field for various values of the dipole moment.

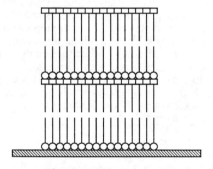

Figure 8. Schematic representation of a noncentrosymmetric Langmuir–Blodgett film. The nonlinear chromophore is incorporated into alternate layers, those represented by the squares, for example.

In the EFISH method, the molecule of interest is dissolved in an appropriate solvent and put into a cell of the type shown in Figure 9. Electrodes above and below the cell provide the means for a D.C. electric field, which orients the solute (and solvent) molecules through its interaction with the molecular dipoles. Similar to the poled polymer approach, the average molecular orientation is increased along the field direction and an oriented gas model used to extract β.

The EFISH experiment is formally a $\chi^{(3)}$ process since it involves the interaction of 4 fields as indicated in argument $\chi^{(3)}_{IJKL}$ ($-2\omega; 0, \omega, \omega$). The symbol Γ_{IJKL} is a shorthand notation for this process. The polarization at 2ω is given by

$$P_I^{2\omega} = \Gamma_{IJKL} E_J^\omega E_K^\omega E_L^o \tag{22}$$

An effective second-order nonlinear coefficient can be defined as

$$d_{IJK} = \Gamma_{IJKL} E_L^o \tag{23}$$

so that the generation of $P_I^{2\omega}$ resembles a second-order under nonlinear official effect. For a pure liquid, a microscopic hyperpolarizability γ can be defined by Γ

$$\Gamma = N\gamma^o f_o f_w^2 f_{2w} \tag{24}$$

where the f's are local field factors and

$$\Gamma = \Gamma_{zzzz} = 3\Gamma_{zzyy} \tag{25}$$

Designating the molecular axis parallel to the dipole moment as the z axis, γ^o can be written as

$$\gamma^o = \gamma + \frac{\mu_z \beta_z}{kT} \tag{26}$$

The first term on the right, γ, is the electronic contribution of γ^o to the polarization at 2ω and the second term the contribution from β_z. Note that β_z cannot be determined from this experiment without a knowledge of the dipole moment. In compounds exhibiting significant charge transfer resonance $\mu_z\beta_z \gg \gamma$ and the contribution of γ is often ignored.

The preceding discussion assumed a pure liquid was used for the measurement. Most molecules of interest, however, are not in the liquid state at room temperature. In this case it is common to dissolve the compound in an appropriate solvent and conduct the measurement. Contributions to the second harmonic signal are therefore obtained from both the solvent and solute. Since Γ and the local field factors that are related to ε and n, (the dielectric constant and refractive index respectively) are concentration dependent, the determination of β for mixtures is not straightforward. Singer and Garito *(15)* have developed methods for obtaining Γ_0, ε_0, and n_0, the values of the above quantities at infinite dilution, from which accurate values for β can be obtained in most cases.

Referring to Figure 9 again, it should be noted that the path length varies across the cell in a manner determined by the angle α. As indicated earlier, second harmonic generation is a phase dependent process and dispersion in the refractive index causes the phase velocities to be different at the fundamental and harmonic frequency. For a given pathlength, ℓ, the phase difference, $\Delta\phi$, is given by

$$\Delta\phi = \frac{\omega\ell}{c}\Delta n \tag{27}$$

where Δn is the difference in refractive index at the two frequencies. Typical coherence lengths for organic molecules are 10 to 20 μm. For a path length in the cell of ~1 mm the bound and free waves, referred to in Figure 4, will change phase by 2π many times over the path length. As the nonlinear interaction is terminated at the glass-solution interface the phase relationship at that boundary will determine the amount of second harmonic intensity that is obtained. By translating the cell across the laser beam, the phase coherence dependence of the process can be mapped out. In the experiment, a sinusoidally varying signal is observed and is illustrated schematically in Figure 10. The intensity of the harmonic signal, $I_{2\omega}$, is given as *(14)*

$$I_{2\omega} = I_{2\omega}^M 2\sin^2\frac{\omega\Delta n}{c}\bullet\ell \tag{28}$$

By measuring $I_{2\omega}^M$ relative to a reference $\mu_z\beta_z$ can be determined

$$\mu_z\beta_z = \frac{1}{f^o f^{\omega^2} f^{2\omega} N}\chi_{reference}^{(2)}\frac{I_{2\omega}^M}{I_{2\omega\ reference}^M} \tag{29}$$

where $\chi_{reference}^{(2)}$ is the effective value of $\chi^{(2)}$ for the reference.

Characterization of Crystals. Two methods used to measure $\chi^{(2)}$ in crystals are the Maker fringe technique and the wedge method *(16)*. The methods are related to one another. In the Maker fringe method a crystal is rotated about an axis perpendicular to the laser beam. An angular dependence of the phase mismatch occurs due to the differences in angles of refraction at the two wavelengths. As the crystal is rotated, fringes appear and $\chi^{(2)}$ tensor elements can be extracted if the signal can be compared with that from a crystal of known $\chi^{(2)}$.

If the unknown crystal can be shaped into a wedge similar to the EFISH cell geometry, a similar sinusoidally varying signal is observed from which $\chi^{(2)}$ tensor elements can be extracted provided a reference sample of known properties is available. Quartz crystals have been characterized extensively *(17)* and can be obtained in the form of a wedge. They are often used as a reference material for both EFISH measurements and other crystals.

Another technique for characterizing crystalline materials is the Kurtz powder technique *(18)*. In this method, a sample is ground into a powder, spread into a thin layer, and irradiated with a laser beam. The intensity of the beam is compared with that of a known reference material, (quartz powder is often used) and conclusions are drawn. Before discussing the nature of those conclusions, the concept of birefringent phase matching is discussed.

It was indicated numerous times that the SHG intensity is dependent on both the magnitude of $\chi^{(2)}$ tensor elements as well as the phase relationships between fundamental and harmonic fields in the crystal. Under certain circumstances, it is possible to achieve phase matched propagation of the fundamental and harmonic beams. Under these conditions, power is continually transferred from the fundamental to harmonic beam over a path length, which is only limited by the ability

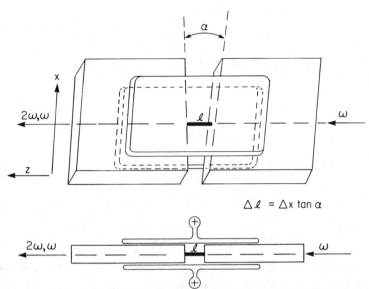

Figure 9. Diagram of the essential features of a cell used for EFISH experiments. (Reprinted with permission from Williams, D. J. *Angew. Chem. Int. Ed. Engl.* 1984, *23*, 690. Copyright VCH Publishers.)

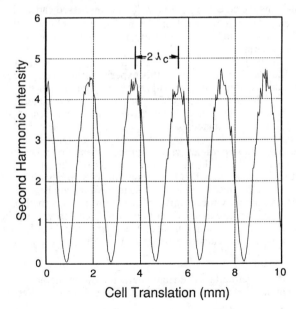

Figure 10. Experimental EFISH trace for nitrobenzene.

to maintain good overlap between the electric fields associated with the beams. Very efficient harmonic conversion can be obtained under these conditions. One method that will be discussed for doing this is birefringent phase matching. To understand this process refer to the refractive index ellipsoid, sometimes referred to as the optical indicatrix, in Figure 11. The intersections of the ellipse with the coordinates define the principal values of the refractive index tensor. For a positive uniaxial crystal the equation for the ellipse is

$$\frac{x^2 + y^2}{n_o^2} + \frac{z^2}{n_e^2} = 1 \tag{30}$$

where n_o and n_e are the ordinary and extraordinary values of the refractive index. The significance of the designations are as follows. For a light beam, S, propagating through the crystal at an arbitrary angle, θ, with respect to the optic axis only two propagating components are allowed. One is polarized orthogonal to S and the optic axis. The value of the refractive index governing its propagation is independent of θ and is designated as n_o. The other allowed polarization component is orthogonal to S and n_o. Inspection of the diagram indicates that its value is angular dependent and it is designated as $n_e(\theta)$. At some other wavelength, the shape of the ellipsoid will be different and a point might be found on the surface of the second ellipsoid where $n_e^{2\omega}(\theta) = n_o^{\omega}$. The angle θ at which this occurs is the phase matching angle. Since the electric vectors of the fundamental and harmonic beams are orthogonal for the polarization directions, the primary contribution to the process will have to come from the off diagonal tensor elements $\chi_{zxx}^{(2)}$ and $\chi_{zyy}^{(2)}$. As discussed earlier the magnitude of these coefficients depends in detail on the molecular packing in the unit cell. A number of other phase matching schemes are available but their discussion is outside the scope of this tutorial.

Returning to the discussion of the powder measurements, a schematic illustration of the possible outcomes of this experiment is given in Figure 12. In the figure, the second harmonic intensity is plotted as a function of the average particle size $<r>$ and the degree coherence length $<\ell_c>$. If the material is capable of birefringent phase matching the signal grows with particle size and eventually saturates. If it is not phase matchable the signal decays as the particle size increases substantially beyond the coherence length. In the phase matched case the behavior is easily rationalized. Those particles with the proper orientation relative to the beam for phase matched propagation will be the primary contributors to the signal as the particle size exceeds the coherence length. On the other hand, as the particles grow there will be less of them in the beam so that the increase in the intensity begins to saturate.

In practice, the only meaningful information that this method can provide is whether the material is phase matchable or not, which can provide guidance for single crystal growth strategies. A single measurement on a sample of known or unknown particle size provides no such information and comparison with a reference is ambiguous.

Electrooptic Measurements. The final characterization method to be discussed is the measurement of the electrooptic coefficient. The electrooptic effect is derived from the process indicated by $\chi_{IJK}^{(2)}(-\omega;o,\omega)$. Since this process is derived primarily from electronic polarization (as opposed to molecular reorientation) its value is expected to be very close to that for $\chi_{IJK}^{(2)}(-2\omega;\omega,\omega)$. This has in fact been observed by Morrell and Albrecht (19) for 2-methyl-4-nitroaniline crystals.

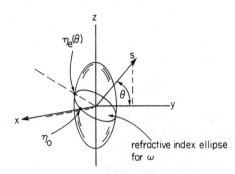

Figure 11. Schematic representation of the refractive index ellipsoid for a positive uniaxial material at frequency ω. (Reprinted with permission from Williams, D. J. *Angew. Chem. Int. Ed. Engl.* 1984, *23*, 690. Copyright VCH Publishers.)

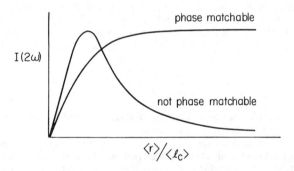

Figure 12. Dependence of the second harmonic intensity on the ration of the average particle size <r> to the average coherence length <l_c> after Kurtz (*16*). (Reprinted with permission from Williams, D. J. *Angew. Chem. Int. Ed. Engl.* 1984, *23*, 690. Copyright VCH Publishers.)

For the sake of illustration, the determination of the electrooptic coefficient for a uniaxial crystal is described below. Considering the nonlinear uniaxial medium of Figure 11, a D.C. electric field is applied in the z direction. The effect of the electric field is to modify the refractive index in the z direction by an amount proportional to the electric field. the modified ellipsoid is given as

$$\frac{x^2 + y^2}{n_o^2} + \left(r_{zzz} E_z + \frac{1}{n_e^2} \right) z^2 = 1 \tag{31}$$

where

$$\Delta \frac{1}{n_e^2} = r_{zzz} E_z \tag{32}$$

and r_{zzz} is a component of the electrooptic tensor. It is related to corresponding components of the $\chi^{(2)}$ tensor by

$$r_{zzz} = \frac{8\pi \chi_{zzz}^{(2)}}{n_e^3} \tag{33}$$

A way of measuring this tensor component is to split a coherent beam into two paths, directing one through the crystal and recombining them at a detector. The amplitude intensity pattern resulting from the interfering beams will vary according to their phase relationship, which in turn will depend on the applied field. Measuring the voltage required to shift from the minimum to maximum signal intensity is equivalent to measuring a π phase shift. Thus the coefficient r_{zzz} can be determined by

$$\pi = \frac{\omega \ell}{c} \frac{n_e^3 r_{zzz}}{2} \frac{V_\pi}{d} \tag{34}$$

where d is the sample thickness, V_π is the voltage required to provide retardation of π, and ℓ is the pathlength through the medium.

This method is quite flexible and can be extended to other tensor components by the appropriate choice of field and polarization directions. For example, rotation of the polarization to the direction perpendicular to the z axis allows determination of $r_{zyy} = r_{zxx}$.

A number of variations of this basic measurement exist for materials in waveguide and fiber formats. An understanding of the principles involved in the approach described above provides the basic framework for successfully understanding and utilizing these related approaches.

Literature Cited

1.   Prasad, P. N.; Williams, D. J., *Introduction to Nonlinear Optics in Molecular and Polymeric Materials,* John Wiley: New York, 1990.
2.   Hecht, E.; Zajac, A., *Optics*; Addison-Wesley, Reading, 1979; p 40.
3.   Zernike, F.; Midwinter, J. E., *Applied Nonlinear Optics;* John Wiley: New York, 1973; p 29.
4.   Yariv, A., *Quantum Electronics*; John Wiley: New York, 1975; p 419.
5.   Zernike, F.; Midwinter, J. E.; p 41.

6. Oudar, J. L., Chemla, D. S.; *J. Chem. Phys.* **1977**, *66*, 2664; Oudar, J. L.; Zyss, J., *Phys. Rev. A.* **1982**, *26*, 2016.
7. For the sources of data in Table I see references sited in Williams, D. J.; *Electronic and Photonic Applications of Polymers;* ACS Advances in Chemistry Series No. 218, M. Bowden and S. R. Turner, Eds.; American Chemical Society: Washington, 1988; p 307.
8. Zyss, J.; Oudar, J. L. *Phys. Rev. A* **1982**, *26*, 2025.
9. For a discussion of local field factors see Prasad, P. N. and Williams, D. J.; Chapter 4.
10. Meredith, G. R.; VanDusen, J. G.; Williams, D. J. *Macromolecules* **1982**, *15*, 1385.
11. Singer, K. D.; Kuzyk, M. G.; Sohn, J. E. *J. Opt. Soc. Am. B4,* 968 (1987).
12. Neal, D. B.; Petty, M. C.; Roberts, G. G.; Ahmad, M. M.; Feast, W. J.; Girling, I. R.; Cade, N. A.; Kolinsky, P. V., Peterson, I. R. *Electron. Lett.* **1986**, *22*, 460.
13. Levine, B. F.; Bethea, C. G. *J. Chem. Phys.* **1975**, *63*, 2666.
14. Oudar, J. L. *J. Chem. Phys.*, **1977**, *67*, 466.
15. Singer, K. D.; Garito, A. F., *J. Chem. Phys.*, **1981**, *75*, 3572.
16. Kurtz, S. K. *Quantum Electronics;* Editors, H. Robin and C. Tang, Academic Press, New York, 1975; Vol. 1, 209ff.
17. Jerphagnon, J.; Kurtz, S. K. *J. Appl. Phys.* **1970**, *41*, 1667.
18. Kurtz, S. K.; Perry, J. *J. Appl. Phys.* **1968**, *39*, 3798.
19. Morrell, J. A.; Albrecht, A. C. *Chem. Phys. Lett.* **1979**, *64*, 46.

RECEIVED August 28, 1990

Chapter 3

Third-Order Nonlinear Optical Effects in Molecular and Polymeric Materials

Paras N. Prasad

Photonics Research Laboratory, Department of Chemistry, State University of New York at Buffalo, Buffalo, NY 14214

This review is aimed at seeking the participation of the chemical community in the exciting new field of nonlinear optics. Chemists and chemical engineers with backgrounds ranging from synthesis to theory can make valuable contributions in this field as it offers challenges both for fundamental and applied research. The article focuses specifically on third order nonlinear optical processes in molecular and polymeric materials. Basic concepts are briefly reviewed along with a discussion of some structural requirements for third order effects. Some widely used measurement techniques are presented. The current status of third-order nonlinear optical material is reviewed along with a discussion of the relevant fundamental and technological issues. The article concludes with a discussion of the important areas in which chemists and chemical engineers can make important contributions (1).

Nonlinear optics is currently at the forefront of research because of its potential applications in the future technology of optical processing of information. For all optical processing which involves light control by light, third-order nonlinear optical processes provide the key operations of optical logic, optical switching and optical memory storage. Molecular and polymeric materials have emerged as an important class of nonlinear optical materials because of the tremendous flexibility they offer both at the molecular and bulk levels for structural modifications necessary to optimize the various functionalities needed for a specific device application (1-3). Since the nonlinear response of these molecular materials is primarily determined by their molecular structure, one can use molecular modeling and synthesis to design and custom tailor molecular structures with enhanced nonlinear responses simultaneously introducing other desirable functionalities. Polymeric structures have the additional advantage that one can incorporate structural

modifications not only in the main chain but also in the side chain
(3).

At the bulk level, molecular and polymeric materials also offer
flexibility of fabrication in various forms such as crystals, films,
fibers, as well as monolayer and multilayer Langmuir-Blodgett films.
In addition, one can make composite structures to introduce
multifunctionality at the bulk level.

The interest in this area is not only technological. This field
also offers a tremendous challenge for basic research. The focal
point of the nonlinear optics of molecular materials is a basic
understanding of the relationship between the molecular structure and
microscopic optical nonlinearity. Especially for third-order
nonlinear optical processes this understanding is in its infancy. We
have to significantly improve this understanding so that
theoretical/computational capabilities can be developed for
predicting structural requirements necessary for large
nonlinearities. The dynamics of various quantum states is another
important area since one, two, and multiphoton resonances
significantly influence the nonlinear optical response. The
relationship between microscopic nonlinearities and the corresponding
bulk effect is yet another area which also warrants detailed
investigation.

Chemists can play a vital role in making significant
contributions to the issues of both fundamental and technological
importance. Theoretical and synthetic chemists working together to
simultaneously develop reliable computational methods and synthetic
routes to systematically derivatized structures on which experimental
measurements are made can provide valuable input for developing a
microscopic understanding of optical nonlinearities. An experimental
study of dynamics of various resonances coupled with theoretical
analysis is another important contribution. Use of chemical
processing to make various forms of molecular assemblies for a given
class of compounds (crystalline, spin coated films, Langmuir-Blodgett
films) can yield useful insight into relationship between the
microscopic nonlinearities and the corresponding bulk nonlinearities.
Materials chemists can also make an important contribution by
developing processes using chemical synthesis routes by which high
optical quality films or fibers of a highly nonlinear material can be
fabricated.

This is a tutorial article written to stimulate the interest of
various chemists and chemical engineers in this exciting new field.
First, the basic concepts are reviewed. Then a survey of the current
status of third-order nonlinear materials is presented. This is
followed by a discussion of relevant issues and the valuable
contributions chemists can make to this field.

Basics of Nonlinear Optics

Optical response of a material is generally described in the
approximation of electric-dipole interaction with the radiation (4).
In this model, the oscillating electric field of radiation induces a
polarization in the medium. When a material is subject to a strong
optical pulse from a laser the electric field is intense and the

induced polarization shows a nonlinear behavior which can be expressed by the following power series expansion ($\underline{4}$).

$$P = \chi^{(1)} \cdot E + \chi^{(2)} : EE + \chi^{(3)} \ EEE + \ldots = \chi_{eff} \cdot E \qquad (1)$$

In Equation 1, $\chi^{(1)}$ is the linear susceptibility which is generally adequate to describe the optical response in the case of a weak optical field. The terms $\chi^{(2)}$ and $\chi^{(3)}$ are the second and third-order nonlinear optical susceptibilities which describe the nonlinear response of the medium. At optical frequencies ($\underline{4}$)

$$n^2(\omega) = \varepsilon(\omega) = 1 + 4\pi\chi(\omega) \qquad (2)$$

For a plane wave, we have the wavevector $k = \dfrac{n\omega}{c}$ and the phase velocity $v = \dfrac{c}{n}$. In a nonlinear medium, $\chi(\omega) \cong \chi_{eff}(\omega)$ of Equation 1 is dependent on E; therefore, n, k and v are all dependent on E.

The two important consequences of the third-order optical nonlinearities represented by $\chi^{(3)}$ are third-harmonic generation and intensity dependence of the refractive index. Third-harmonic generation (THG) describes the process in which an incident photon field of frequency (ω) generates, through nonlinear polarization in the medium, a coherent optical field at 3ω. Through $\chi^{(3)}$ interaction, the refractive index of the nonlinear medium is given as $n = n_o + n_2 I$ where n_2 describes intensity dependence of the refractive index and I is the instantaneous intensity of the laser pulse. There is no symmetry restriction on the third-order processes which can occur in all media including air.

Microscopic Optical Nonlinearities

At the microscopic level, the nonlinearity of a molecular structure is described by the electric dipole interaction of the radiation field with the molecules. The resulting induced dipole moment and the Stark energy are given as ($\underline{1,3}$)

$$\mu_{ind} = \alpha \cdot E + \beta : EE + \gamma \ EEE + \qquad (3)$$

$$\varepsilon_{Stark} = -E \cdot \mu_o -1/2 \ E \cdot \alpha \cdot E -1/3 \ E \cdot \beta : EE -1/4 \ E \cdot \gamma \ EEE - \qquad (4)$$

In the above equation α is the linear polarizability. The terms β and γ, called first and second hyperpolarizabilities, describe the nonlinear optical interactions and are microscopic analogues of $\chi^{(2)}$ and $\chi^{(3)}$.

In the weak coupling limit, as is the case for most molecular systems, each molecule can be treated as an independent source of nonlinear optical effects. Then the macroscopic susceptibilities $\chi^{(n)}$ are derived from the microscopic nonlinearities β and γ by simple orientationally-averaged site sums using appropriate local field correction factors which relate the applied field to the local field at the molecular site. Therefore ($\underline{1,3}$)

$$\chi^{(3)}(-\omega_4;\omega_1,\omega_2,\omega_3) = F(\omega_1)F(\omega_2)F(\omega_3)F(\omega_4) \sum_n \langle \gamma^n(\theta,\phi) \rangle \qquad (5)$$

The symbols used for $\chi^{(3)}$ indicate that three input waves $\omega_1, \omega_2, \omega_3$

generate an output of ω_4. For THG, $\omega_1=\omega_2=\omega_3=\omega$ and $\omega_4=\omega$. The terms $F(\omega_i)$ are the local field corrections for a wave of frequency ω_i. Generally, one utilizes the Lorentz approximation for the local field in which case (1,4)

$$F(\omega_i) = \frac{n_0^2(\omega_i)+2}{3} \tag{6}$$

In Equation 6, $n_0(\omega_i)$ is the intensity independent refractive index at frequency ω_i. The sum in Equation 5 is over all the sites (n); the bracket, $\langle \rangle$, represents an orientational averaging over angles θ and ϕ. Unlike for the second-order effect, this orientational average for the third-order coefficient is nonzero even for an isotropic medium because it is a fourth rank tensor. Therefore, the first step to enhance third order optical nonlinearities in organic bulk systems is to use molecular structures with large γ. For this reason, a sound theoretical understanding of microscopic nonlinearities is of paramount importance.

Structural Requirements for Third-Order Optical Nonlinearity

Electronic structural requirements for third-order nonlinear organic systems are different from that for second order materials. Although the understanding of structure-property relationships for third-order effects is highly limited, all microscopic theoretical models predict a large non-resonant third-order optical nonlinearity associated with delocalized π-electron systems (1-3). These molecular structures do not have to be asymmetric because γ is a fourth rank tensor. Conjugated polymers with alternate single and multiple bonds in their backbone structures provide a molecular frame for extensive conjugation and have emerged as the most widely studied group of $\chi^{(3)}$ organic materials. Examples of conjugated polymers are polydiacetylenes, poly-p-phenylenevinylene and polythiophenes.

The optical nonlinearity is strongly dependent on the extent of π-electron delocalization from one repeat unit to another in the polymer (or oligomer) structure. This effective delocalization is not always equally manifested but depends on the details of repeat unit electronic structure and order. For example, in a sequentially built structure, the π-delocalization effect on γ is found to be more effective for the thiophene oligomers than it is for the benzene oligomers (5).

The largest component of the γ-tensor is in the conjugation direction. Therefore, even though no particular bulk symmetry is required for nonzero $\chi^{(3)}$, a medium in which all conjugated polymeric chains align in the same direction should have a larger $\chi^{(3)}$ value along the chain direction relative to that in an amorphous or disordered form of the same polymer. Studies of $\chi^{(3)}$ in ordered or stretch-oriented polymers as discussed below confirm this prediction. Finally, the polymeric chains should pack as closely as possible in order to maximize the hyperpolarizability density and hence $\chi^{(3)}$.

Extensive π-conjugation is also often associated with enhanced conductivity in organic systems (6). Polyacetylene and polythiophene which in the doped state exhibit very high electrical conductivity also exhibit relative large third-order nonlinear optical effects in the undoped (nonconducting) state. However, it should be remembered that conductivity is a bulk property which is heavily influenced by intrachain as well as interchain carrier transports. In contrast, the origin of third-order nonlinearity in conjugated polymers is primarily microscopic, determined by the structure of the polymer chain. Therefore, a conjugated polymer may be a very good $\chi^{(3)}$ material but not necessarily a good conductor. Polydiacetylene is a good example; it exhibits a large non-resonant $\chi^{(3)}$ value but is a wide band gap semi-conductor.

Measurement Techniques for Third-Order Nonlinear Optical Effects

General Discussion. Experimental probes generally used to measure $\chi^{(3)}$ are based on the following effects: (i) Third Harmonic Generation; (ii) Electric Field Induced Second Harmonic Generation; (iii) Degenerate Four Wave Mixing, (iv) Optical Kerr Gate and (v) Self focusing. In addition, processes involving an intensity-dependent phase-shift due to intensity-dependent refractive index can also be used to measure $\chi^{(3)}$. The examples of these processes are found in nonlinear optical waveguides, Fabry-Perot etalons and surface plasmon optics. The intensity-dependent phase shift changes the resonance condition which defines the transmission characteristics of the wave through the waveguide, Fabry-Perot or the surface-plasmon coupling, as the intensity of the input (or pump beam) is changed.

Although one loosely uses a $\chi^{(3)}$ value for a material, in reality there are a number of relevant parameters which describe the third order optical nonlinearities. These parameters are:

(i) The $\chi^{(3)}$ tensor. $\chi^{(3)}$ is a fourth rank tensor which, even in isotropic media such as liquids, solutions, random solids or amorphous polymers, has three independent components $\chi^{(3)}_{1111}$, $\chi^{(3)}_{1212}$ and $\chi^{(3)}_{1122}$. They are defined by the relative polarizations of the four waves.

(ii) Response time of the nonlinearity. The response time of the nonlinearity relates to its mechanism. Therefore, its determination is of considerable value in establishing the mechanism of optical nonlinearity. In addition, the response time is a valuable parameter for device applications. The non-resonant electronic nonlinearity, which involves only virtual electronic states as intermediate levels for interaction, have the fastest response time, limited only by the laser pulse width. However, some resonant electronic nonlinearities can also have extremely fast response times when the excited state relaxation is ultra-fast. Therefore, one needs to use the best time resolution available, preferably subpicosecond, to study the time-response of the nonlinearity when investigating its mechanism.

(iii) Wavelength dispersion $\chi^{(3)}$. The $\chi^{(3)}$ value is dependent on the frequencies of the interacting waves. Therefore, strictly speaking one should specify the $\chi^{(3)}$ dispersion as $\chi^{(3)}(-\omega_4; \omega_1, \omega_2, \omega_3)$ while quoting a value. This feature also cautions one to be careful in comparing the $\chi^{(3)}$ values obtained by the various techniques. One measures $\chi^{(3)}$ (-3ω; ω, ω, ω) by third harmonic generation, and $\chi^{(3)}$ ($-\omega$;ω,$-\omega$,ω) by degenerate four wave mixing. The two values are not expected to be identical because of the dispersion effect. Still a qualitative correlation of the two values serves a useful purpose in identifying if one is measuring a non-resonant purely electronic nonlinearity.

(iv) <u>Sign of $\chi^{(3)}$</u>. The nonlinear susceptibility $\chi^{(3)}$ also has a sign which is an important fundamental property relating to the microscopic nature of optical nonlinearity.

(v) <u>Real or complex</u>. The $\chi^{(3)}$ value may not just be a real number. It can also be a complex number. This situation occurs when any frequency of the interacting waves approaches that of a one-photon, two-photon or three-photon electronic resonance (the latter only for third harmonic generation).

It is often difficult to get complete information on all the relevant parameters of third-order nonlinearity using one single technique. However, one can use a combination of techniques to probe the various aspects of the $\chi^{(3)}$ behavior. Here only two specific techniques to measure $\chi^{(3)}$ are discussed.

<u>Third Harmonic Generation</u>. For third harmonic generation (THG) one generally utilizes a Q-switched pulse Nd:Yag laser which provides nanosecond pulses at low repetition rates (10 to 30 Hz) (<u>7</u>). In making third harmonic generation measurements, the response time is not important; the pulse width of the laser, therefore, is not as crucial. However, if the longer pulses (higher photon flux) cause sample decomposition due to absorption, it may be advisable to use a CW-Q-switched and mode-locked Nd-Yag laser where the strongest pulses are selected through an electro-optic pulse selector. Usually, the organic systems have limited transparency towards the u.v. spectral range. The selection of wavelength should first be made so that the third harmonic signal does not fall in a u.v. region of high absorption. For this reason, either the fundamental output of the Nd:Yag laser is Raman-Stokes shifted in a H_2 gas cell to a longer wavelength in the near IR, or a dye is pumped and mixed with the green (or fundamental) from the Yag to generate the difference frequency. After proper selection of wavelength and polarization, the laser beam is split into two parts, one being used to generate the third harmonic in the sample and the other to generate third harmonic in a reference. For the THG technique, glass is generally taken as the reference. For the non phase-matched THG one uses the Maker fringe or wedge fringe method in which the path length of the sample (and the reference) is varied and the third harmonic signal is monitored as a function of the interaction length ℓ to obtain the fringes (<u>7</u>). From the fringes one determines the coherence length, l_c, for both the sample and the reference as the separation between two maximum corresponds to $2l_c$. The ratio of the third harmonic signals $I(3\omega)$ for the sample and the reference for the same input intensity and the same interaction length is given by (<u>7</u>).

$$\frac{I(3\omega)_{sample}}{I(3\omega)_{reference}} = (\frac{\chi^{(3)}_{sample}}{\chi^{(3)}_{reference}})^2 (\frac{l^{reference}_c}{l^{sample}_c})^2 \qquad (7)$$

From this expression one can determine $\chi^{(3)}$ of the sample by obtaining from the experiment, the third harmonic signals, $I(3\omega)_{sample}$ and $I(3\omega)_{reference}$, and coherence lengths, l^{sample}_c and $l^{reference}_c$. In the wedge fringe method which is simpler, a wedge shaped sample (a slab, or a cell) is used. The sample is simply translated to vary the pathlength which yields the wedge fringe.

The third-harmonic generation method has the advantage that it probes purely electronic nonlinearity. Therefore, orientational and thermal effects as well as other dynamic nonlinearities derived from excitations under resonance condition are eliminated (7). The THG method, however, does not provide any information on the time-response of optical nonlinearity. Another disadvantage of the method is that one has to consider resonances at ω, 2ω and 3ω as opposed to degenerate four wave mixing discussed below which utilizes the intensity dependence of refractive index and where only resonances at ω and 2ω manifest.

Degenerate Four Wave Mixing. Degenerate four wave mixing (DFWM) provides a convenient method of measuring $\chi^{(3)}$, which includes both electronic and dynamic resonant nonlinearities, and obtaining its response time (7). In a backward wave phase conjugate geometry for DFWM two waves I_1 and I_2 are counterpropagating and a third beam, I_3, is incident at a small angle; the signal, I_4, is the phase conjugate of I_3 as it is produced counterpropagating to I_3. In this arrangement the phase-matching requirement is automatically satisfied. Since all the input optical frequencies and the output optical frequency are of the same value, one measures $\chi^{(3)}$ ($-\omega$; ω, $-\omega$, ω). This $\chi^{(3)}$ value is an important parameter for the design of devices utilizing optical switching and bistability. Furthermore, as we have demonstrated from the measurements conducted in our laboratory, one can conveniently measure the anisotropy and time-response of $\chi^{(3)}$. Since dynamic nonlinearities such as thermal effects and excited state gratings produced by absorption of photons also contribute to the degenerate four wave mixing signal, the capability to go to time resolution of femtoseconds is helpful in separating the various contributions. An experimental arrangement which provides time-resolutions of ~350 femtoseconds and very high peak power consists of a CW mode-locked Nd-Yag laser, the pulses from which are compressed in a fiber-optic pulse compressor, and subsequently frequency doubled. The frequency doubled output is stabilized by a stabilizer unit, and then used to sync-pump a dye laser. The dye pulses are subsequently amplified in a PDA amplifier (from Spectra-Physics) which is pumped by a Quanta-Ray model DCR-2A pulse Nd-Yag laser. The resulting pulses are ~350 femtoseconds wide with a pulse energy of 0.3 mJ and at a repetition rate of 30 Hz. The effective $\chi^{(3)}$ values of a sample can be obtained by using CS_2 as the reference material.

The $\chi^{(3)}$ value then is obtained by using the following equation (7):

$$\frac{\chi^{(3)}_{sample}}{\chi^{(3)}_{CS_2}} = (\frac{n^o_{sample}}{n^o_{CS_2}})^2 \; (\frac{\ell_{CS_2}}{\ell_{sample}}) \cdot (\frac{I_{sample}}{I_{CS_2}})^{1/2} \; L \qquad (8)$$

In Equation 8, n^o_{sample} and $n^o_{CS_2}$ are the linear refractive indices of the sample and CS_2; ℓ_{CS_2} and ℓ_{sample} are the path lengths of the two media. I_{sample} and I_{CS_2} are the respective DFWM signals from the sample and CS_2. The term L is the correction factor for absorption and scattering losses in the sample. To obtain the time response of the nonlinearity, the backward beam is optically delayed with respect to the two forward beams.

Measurement of Microscopic Nonlinearities, γ

The measurement of $\chi^{(3)}$ of solutions can be used to determine the microscopic nonlinearities γ of a solute, provided γ of the solvent is known. This measurement also provides information on the sign of γ and (hence $\chi^{(3)}$) of the molecules if one knows the sign of γ for the solvent (5,7). Under favorable conditions one can also use solution measurements to determine if γ is a complex quantity. The method utilizes two basic assumptions: (i) the nonlinearities of the solute and the solvent molecules are additive, and (ii) Lorentz approximation can be used for the local field correction. Under these two assumptions one can write the $\chi^{(3)}$ of the solution to be

$$\chi^{(3)} = F^4[N_{solute}\langle\gamma\rangle_{solute} + N_{solvent}\langle\gamma\rangle_{solvent}] \qquad (9)$$

In Equation 9 the terms $\langle\gamma\rangle$ represent the orientationally averaged second hyperpolarizabilities defined as

$$\langle\gamma\rangle = 1/5(\gamma_{xxxx} + \gamma_{yyyy} + \gamma_{zzzz} + 2\gamma_{xxyy} + 2\gamma_{xxzz} + 2\gamma_{yyzz}) \qquad (10)$$

N is the number density in the units of number of molecules per cm^3. F is the Lorentz correction factor defined by Equation 6. If the solute is in dilute concentration, Equation 9 can be written as

$$\chi^{(3)} = F^4[N_{solute}\langle\gamma\rangle_{solute}] + \chi^{(3)}_{solvent} \qquad (11)$$

If the $\langle\gamma\rangle$ values for both the solute and the solvent have the same sign, Equation 11 predicts a linear dependence of $\chi^{(3)}$ with the concentration of the solution. By a least square fit of this concentration dependence, one can readily obtain $\langle\gamma\rangle$ of the solute molecule. If the signs of the nonlinearities are opposite but both are real quantities, a concentration dependence study would yield a behavior where the value of the resultant $\chi^{(3)}$ of the solution decreases and goes to zero at some concentration. In the case when $\langle\gamma\rangle$ of the solute is complex, Equation 11 yields the signal given by

$$I \propto |\chi^{(3)}|^2 = |F^4[N_{solute}\langle\gamma^R_{solute}] + \chi^{(3)}_{solvent}|^2$$
$$+ |F^4 N_{solute}\langle\gamma^{Im}_{solute}\rangle|^2 \qquad (12)$$

When the real part of $\langle\gamma\rangle$ for the solute has a sign opposite to that for the solvent, the resulting plot of the signal (or $\chi^{(3)}$) as a function of concentration does not show the $\chi^{(3)}$ value going through zero. To distinguish these situations, one must perform a concentration dependence study. Only then can one extract the value of $\langle\gamma\rangle$ for the solute. A one concentration measurement is likely to give an erroneous result.

Some Representative Measurements

Compared to a relatively large database existing for the second-order materials, that for the third-order molecular materials is rather limited. Since the third-order processes do not require any specific molecular or bulk symmetry, the scope of investigation therefore can be much wider to include many different types of structures and all the bulk phases. Measurements in solution or the liquid phase have been used to extract the microscopic nonlinear coefficient γ using the procedure discussed above. Here are quoted the γ values (orientationally averaged) of some representative molecular and polymeric structures measured in the solution phase. Table I lists these values, which are just randomly selected and do not necessarily represent the first measurement on these systems. Third harmonic generation (THG), degenerate four wave mixing (DFWM) and Kerr Gate methods have been used for the measurement of optical nonlinearity. In going from benzene to β-carotene (the second structure), the γ value increases by more than three orders of magnitude showing the importance of increase in the effective conjugation length. The third structure exhibits N-phenyl substitution in the benzimidazole type structures to introduce a two dimensional conjugation. The last structure, an organometallic polymer, has also been measured in the solution phase. Because of the resonance condition encountered, the γ value is complex.

The $\chi^{(3)}$ values for some representative polymers in thin film form are listed in Table II. From the polydiacetylene group, two specific examples chosen are PTS and Poly-4-BCMU. It should be point that the values listed for polythiophene, phthalocyanine and polyacetylene are resonant values. In resonant cases, the dynamic nonlinearities derived from the excited state population make the dominant contribution in the DFWM method (7). Consequently, the magnitude of effective $\chi^{(3)}$ and its response time can be dependent on the pulse width and the peak intensities. One has to investigate these features carefully before a meaningful conclusion can be drawn. If the excited state lifetime is longer than or comparable to the pulse width, a larger value of effective $\chi^{(3)}$ is measured using longer pulses. Polysilanes and polygermanes are inorganic polymers which have no π-electrons but show interesting photophysics attributable to delocalization of the σ-electrons (20,21). They have the advantage of having a wide optical transparency range but their $\chi^{(3)}$ values are smaller.

Table I

The microscopic nonlinearity γ for some molecular structures measured in the liquid phase

Structure	Wavelength of measurement	Method	γ (esu)
	1.907 μm (Fundamental)	THG	1.097×10^{-36} ref. 8
	1.89 μm	THG	4.8×10^{-33} ref. 9
	602 nm	DFWM	1.1×10^{-33} ref. 10
	602 nm	DFWM	$\sim 10^{-35}$ ref. 11
	$\lambda_{pump} = 1.064\ \mu$m $\lambda_{probe} = 0.532\ \mu$m	Kerrgate	$\gamma_{real} = 8.56 \times 10^{-34}$ $\gamma_{imag} = 3.57 \times 10^{-33}$ ref. 12

Table II

The $\chi^{(3)}$ values of some molecular and polymeric materials measured in the thin film form

Structure	Wavelength of measurement	Method	$\chi^{(3)}$ (esu) Film study
Poly-p-phenylenevinylene (PPV)	602 nm	DFWM	~ 4×10^{-10} (ref. 13)
Polydiacetylene			
R=$-CH_2-O-S(=O)_2-\langle O \rangle-CH_3$ PTS	$\lambda_{fundamental}$ =2.62 μ	THG	1.6×10^{-10} Parallel to polymer chain (ref. 14)
R=$-(CH_2-)_4-O-C(=O)-NH-CH_2COO(CH_2)_3CH_3$ Poly-4-BCMU	585 nm	DFWM	4.0×10^{-10} red form (ref. 15)
PBT	585 nm	DFWM	~ 10^{-11} (ref. 16)
Polythiophene	602 nm	DFWM	~ 10^{-9} (ref. 17)
Polyacetylene	1.06 μ	THG	4×10^{-10} (ref. 18)
	602 nm	DFWM	~ 10^{-9} (ref. 19)
Polysilane $R_1 = CH_3$ $R_2 = \langle O \rangle$	$\lambda_{fundamental}$ =1.064 μ	THG	1.5×10^{-12} (ref. 20)
Polygermane		THG	11.3×10^{-12} (ref. 21)

Several measurements have been reported for oriented polymers.
PTS polydiacetylene crystal was investigated by THG and the results
confirmed the largest value of $\chi^{(3)}$ along the polymer chain (14).
Uniaxially and biaxially oriented films of PPV as well as biaxially
oriented films of PBT have been investigated (13,16,22). Again the
largest component of $\chi^{(3)}$ in the uniaxial film is along the draw
direction, confirming the largest nonlinearity being along the
polymer chain which preferentially align along the draw direction. A
simple model which involves transformation of the fourth rank tensor
$\chi^{(3)}$ from a film based co-ordinate to the laboratory based coordinate
can explain the polar plot of $\chi^{(3)}$ obtained by rotating the film with
respect to the polarization vectors of the beam (13,16). For the
measurements reported here using DFWM, all the electric field vectors
were vertically polarized giving rise to measurement of $\chi^{(3)}_{1111}$ in the
laboratory frame.
The value of $\chi^{(3)}$ ranges from 10^{-9} esu to smaller values.

Relevant Issues and Opportunities for Chemists
===

Compared to materials for second-order nonlinear devices, the third-
order materials are even further away from being ready for device
applications. The relevant issues for third order materials are:

(i) Improvement in the currently achievable nonresonant $\chi^{(3)}$
values. Inorganic multiple quantum well semiconductors exhibit large
resonant nonlinearities (23). It is doubtful that molecular and
polymeric materials will offer any challenge to the inorganic
semiconductors for resonant nonlinearities. The attractive feature
of molecular materials such as conjugated polymers is the high
nonresonant $\chi^{(3)}$ they exhibit which naturally has the fastest
response time in femtoseconds. However, even the currently
achievable highest nonresonant value ($<10^{-9}$ esu) is not large enough
for any practical devices which would require only a small switching
energy. Therefore, there is a need for enhancing optical
nonlinearity. For achieving this goal, a better understanding of the
relation between molecular structure and microscopic nonlinearity, γ,
is required in order to identify structural units which enhance
optical nonlinearity. It will require improved theoretical modeling
of optical nonlinearity coupled with synthesis of sequentially built
and systematically derivatized structures and measurements of their
optical nonlinearities. Clearly, theoretical chemists, synthetic
chemists, physical chemists and analytical chemists all can play an
important role on this issue.
Our theoretical understanding of third-order optical
nonlinearity at the microscopic level is really in its infancy.
Currently no theoretical method exists which can be reliably used to
predict, with reasonable computational time, molecular and polymeric
structures with enhanced optical nonlinearities. The two important
approaches used are: the derivative method and the sum-over-states
method (7,24). The derivative method is based on the power expansion
of the dipole moment or energy given by Equations 3 and 4. The
third-order nonlinear coefficient γ is, therefore, simply given by
the fourth derivative of the energy or the third derivative of the
induced dipole moment with respect to the applied field. These

derivatives can be evaluated either by the finite field method
(numerically) or analytically. In the simplest form one uses only
time-independent Hamiltonian involving a static field. Consequently,
one computes only static nonlinearity at zero frequency which is not
the same as measured at an optical frequency. At optical
frequencies, the electronic transitions of a molecular structure may
provide pre resonance enhancements and large dispersion effects
(frequency dependence) in the Y value. The advantage of this method
is that only the ground state properties need to be computed. One
can, therefore, use an ab-initio approach for a reasonable size
molecule to compute optical nonlinearities. The calculations require
a careful choice of the basis functions such as inclusion of diffuse
and polarization functions which strongly influence the computed
nonlinearity (24). The ab-initio computations have a sound
theoretical foundation but they cannot realistically be conducted for
large molecules or heavy atoms. One has to rely on semi-empirical
methods which can be used for large molecules and polymers. However,
the parameters used in the various semi-empirical methods are
optimized for other properties and cannot be relied upon for
calculations of optical nonlinearity.

The sum-over-states method is based on the perturbation
expansion of the Stark energy term in which nonlinearities are
introduced as a result of mixing with excited states. For example,
the expression for $Y(-3\omega; \omega,\omega,\omega)$ which will be responsible for third
harmonic generation is given as (25)

$$Y_{ijkl}(-3\omega;\omega,\omega,\omega) = -\frac{e^4}{4\hbar^3} \sum_{m_1 m_2 m_3} \frac{\langle g|r_i|m_3\rangle\langle m_3|r_j|m_2\rangle\langle m_2|r_k|m_1\rangle\langle m_1|r_1|g\rangle}{(\omega_{m_1 g}-\omega)(\omega_{m_2 g}-2\omega)(\omega_{m_3 g}-3\omega)}$$

$$+ \ldots$$

In this calculation one computes the energies and various expectation
values of the dipole operator for various excited states. These
terms are then summed to compute Y. If one does an exact
calculation, in principle both the derivative and the sum-over-states
methods should yield the same result. However, such exact
calculations are not possible. The sum-over-states method requires
that not only the ground states but all excited state properties be
computed as well. For this reason one resorts to semi-empirical
calculations and often truncates the sum over all states to include
only a few excited states.

Electron correlation effects are expected to play an important
role in determining optical nonlinearities. Both the configuration
interaction and Moeller-Plesset perturbation correction approaches
have been used to incorporate electron-correlation effects (26,27).

Although both the ab-initio derivative method and the semi-
empirical sum-over-states approach have been used with some success
to predict qualitative trends, they are not sufficiently developed to
have predictive capabilities for structure-property relationship.
Clearly, there is a need to develop semi-empirical theoretical
methods which can reliably be used to predict, with cost-
effectiveness and with reasonable computational time, molecular and
polymeric structures with enhanced optical nonlinearity.

Another chemical approach to improve our microscopic understanding of optical nonlinearities is a study of nonlinear optical behavior of sequentially built and systematically derivatized structures. Most past work for third-order nonlinearities have focused on conjugated polymers. This ad hoc approach is not helpful in identifying functionalities necessary to enhance optical nonlinearities. A systematic study and correlation of γ values of systematically varied structure is an important approach for material development.

Zhao, et al have investigated the nonlinearities of the following series of oligomers (5,28,29):

Thiophene oligomers   Poly-p-phenyl oligomers   Pyridine oligomers

As a function on n, γ value increases much more rapidly for the thiophene oligomers than for the benzene and pyridine oligomers indicating that the π-electron delocalization from one ring to another is much more effective for thiophene units. In addition, the d-orbital of sulfur may be contributing to optical nonlinearity. The γ value for the thiophene oligomers follows a power law - n^4 which is close to what is predicted by an ab-initio calculation on polyenes (28). Zhao et al have also investigated systematically derivatized thiophenes and found that placing NO_2 groups at the end 2,2' positions of thiophene enhances the nonlinearity (5).

Recently, a joint effort of material Laboratory at Wright Research and Development Center and Photonics Research Laboratory at SUNY at Buffalo has resulted in a comprehensive study of structure-nonlinear optical properties of a large number of systematically varied aromatic heterocyclic compounds involving fused ring benzimidazole and benzthiazole structures (10,30).

This study has provided many useful insights some of which are as follows: (a) a sulfur ring in a conjugated structure is much more effective than a phenyl ring or other heteroaromatic ring such as furan or pyridine in increasing optical nonlinearity (b) an olefinic double bond provides a highly effective π-delocalization and consequent increase of the third-order nonlinearity, and (c) grafting of pendent aromatic groups through attachment to a nitrogen atom in a fused benzimidazole ring provides a means for producing two-dimensional π-conjugation leading to an enhancement of γ and also improved solubility.

The search of third-order materials should not just be limited to conjugated structures. But only with an improved microscopic understanding of optical nonlinearities, can the scope, in any useful way, be broadened to include other classes of molecular materials. Incorporation of polarizable heavy atoms may be a viable route to increase γ. A suitable example is iodoform (CHI_3) which has no π-electron but has a $\chi^{(3)}$ value (31) comparable to that of bithiophene

( ). Organometallic structures represent another vast
class of molecular materials which are largely unexplored.

(ii) <u>Improvement in Materials Processing through Chemistry</u>.
Another important issue concerning the conjugated polymeric structure
as third-order nonlinear materials is their processibility. The
conjugated linear polymeric structures tend to be insoluble and,
therefore, cannot readily be processed into device structures. The
lack of processibility may render a material totally useless for
practical application even if it may have a large $\chi^{(3)}$ value.
Synthetic chemists can play a vital role by designing chemical
approaches for processing of important nonlinear materials. Two
specific examples presented here are: (a) Soluble precursor route
and (b) Chemical derivatization for improving solubility.
In the soluble precursor route, a suitable precursor is
synthesized which can be cast into a device structure (i.e., film) by
using solution processing. Then it can be converted into the final
nonlinear structure upon subsequent treatment (such as heat
treatment). This approach has been used for poly-p-phenylenevinylene
(PPV) as shown below (<u>32</u>):

Precursor polymer   **PPV**

In the chemical derivatization approach one introduces a pendent long
alkyl or alkoxy group to increase solubility. Polythiophene itself

is insoluble but poly(3-dodecylthiophene),
is soluble in common organic solvents.

(iii) <u>Improving Optical Quality</u>. Optical quality of the
materials is of prime concern for integrated optics applications
which will involve waveguide configurations. Most conjugated
polymeric structures are optically lossy. There is a need for
chemical approaches which will provide a better control of structural
homogeneities so that the optical losses can be minimized. Another
approach is through the use of composite structures where both the
optical quality and $\chi^{(3)}$ can be optimized by a judicious choice of
the two components. The best optical quality medium is provided by
inorganic glasses such as silica. However, they by themselves have
very low $\chi^{(3)}$. A composite structure such as that of silica and a
conjugated polymer may be a suitable choice. Chemical processing of
oxide glass using the sol-gel chemistry provides a suitable approach
to make such composite structures. Using the sol-gel method, a
composite of silica glass and poly(p-phenylene vinylene) has been
prepared (<u>33</u>) in which the composition can be varied up to 50%. The

procedure involves molecular mixing of the silica sol-gel precursor and the polymer precursor in a solvent in which both are soluble. During the gelation, a film is cast. Subsequent heat treatment converts the precursor polymer to the conjugated poly(p-phenylene vinylene) polymeric structure. The optical quality of the film was found to be significantly improved and high enough to use them as optical waveguides at 1.06μ.

Conclusions

To conclude this article, it is hoped that the discussion of relevant issues and opportunities for chemists presented here will sufficiently stimulate the interest of the chemical community. Their active participation is vital for building our understanding of optical nonlinearities in molecular systems as well as for the development of useful nonlinear optical materials. It is the time now to search for new avenues other than conjugation effects to enhance third-order optical nonlinearities. Therefore, we should broaden the scope of molecular materials to incorporate inorganic and organometallic structures, especially those involving highly polarizable atoms.

Acknowledgments

The research conducted at SUNY at Buffalo was supported in part by the Air Force Office of Scientific Research, Directorate of Chemical and Atmospheric Sciences and Polymer Branch, Air Force Wright-Materials Laboratory, Wright Research and Development Center through contract number F49620-90-C-0021 and in part by the NSF Solid State Chemistry Program through Grant Number DMR-8715688. The author thanks his research group members as well as Drs. Donald R. Ulrich and Bruce Reinhardt for helpful discussions.

Literature Cited

1. Nonlinear Optical Properties of Organic Molecules and Crystals; Chemla, D. S.; Zyss, J., Eds.; Academic Press: Orlando, 1987; Vols. 1 and 2.
2. Nonlinear Optical Properties of Polymers; Heeger, A. J.; Orenstein, J.; Ulrich, D. R., Eds.; Materials Research Society Symposium Proceedings: Pittsburgh, 1987; Vol. 109.
3. Nonlinear Optical and Electroactive Polymers; Prasad, P. N.; Ulrich, D. R., Eds.; Plenum Press: New York, 1988.
4. Shen, Y. R. The Principles of Nonlinear Optics; Wiley & Sons: New York, 1984.
5. Zhao, M. T.; Samoc, M.; Singh, B. P.; Prasad, P. N. J. Phys. Chem. 1989, 93, 7916.
6. Handbook of Conducting Polymers; Skotheim, T., Ed.; Marcel-Dekker: New York, 1987; Vols. 1 and 2.
7. Prasad, P. N.; Williams, D. J. Introduction to Nonlinear Optical Effects in Molecules and Polymers; Wiley & Sons: New York, in press.
8. Kajzar, F.; Messier, J. Phys. Rev. 1985, A32, 2352.

9. Hermann, J. P.; Richard, D.; Ducuing, J. Appl. Phys. Lett. 1973, 23, 178.
10. Zhao, M. T.; Samoc, M.; Prasad, P. N.; Reinhardt, B. A.; Unroe, M. R.; Prazak, M.; Evers, R. C.; Kane, J. J.; Jariwala, C.; Sinsky, M. submitted to Chem. Materials.
11. Ghosal, S.; Samoc, M.; Prasad, P. N.; Tufariello, J. J. J. Phys. Chem. 1990, 94, 2847.
12. Guha, S.; Frazier, C. C.; Porter, P. L.; Kang, K.; Finberg, S. E. Opt. Lett. 1989, 14, 952.
13. Singh, B. P.; Prasad, P. N.; Karasz, F. E. Polymer 1988, 29, 1940.
14. Sauteret, C.; Hermann, J. P.; Frey, R.; Pradene, F.; Ducuing, J.; Baughmann, R. H.; Chance, R. R. Phys. Rev. Lett. 1976, 36, 956.
15. Rao, D. N.; Chopra, P.; Ghoshal, S. K.; Swiatkiewicz, J.; Prasad, P. N. J. Chem. Phys. 1986, 84, 7049.
16. Rao, D. N.; Swiatkiewicz, J.; Chopra, P.; Ghosal, S. K.; Prasad, P. N. Appl. Phys. Lett. 1986, 48, 1187.
17. Logsdon, P.; Pfleger, J.; Prasad, P. N. Synthetic Metals 1988, 26, 369.
18. Sinclair, M.; Moses, D.; Akagi, K.; Heeger, A. J. Materials Research Society Proceedings, 1988; Vol. 109, p 205.
19. Prasad, P. N.; Swiatkiewicz, J.; Pfleger, J. Mol. Cryst. Liq. Cryst. 1988, 160, 53.
20. Kajzar, F.; Messier, J.; Rosilio, C. J. Appl. Phys. 1986, 60, 3040.
21. Baumert, J. C.; Bjorklund, G. C.; Jundt, D. M.; Jurich, M. C.; Looser, H.; Miller, R. D.; Rabolt, J.; Sooriyakumaran, R.; Swalen, J. D.; Twieg, R. J. Appl. Phys. Lett. 1988, 53, 1147.
22. Swiatkiewicz, J.; Prasad, P. N.; Karasz, F. E., unpublished results.
23. Optical Nonlinearities and Instabilities in Semi-Conductors; Huag, H., Ed.; Academic Press: London, 1988.
24. Chopra, P.; Carlacci, L.; King, H. F.; Prasad, P. N. J. Phys. Chem. 1989, 93, 7120.
25. Ward, J. F. Rev. Mod. Phys. 1965, 37, 1.
26. Grossman, C.; Heflin, J. R.; Wong, K. Y.; Zamani-Khamari, O.; Garito, A. F. In Nonlinear Optical Effects in Organic Polymers; Messier, J.; Kajzar, F.; Prasad, P.; Ulrich, D. NATO ASI Series, Kluwer Academic Publishers: The Netherlands, 1989; Vol. 102, p 225.
27. Perrin, E.; Prasad, P. N.; Mougenot, P.; Dupuis, M. J. Chem. Phys. 1989, 91, 4728.
28. Zhao, M. T.; Singh, B. P.; Prasad, P. N. J. Chem. Phys. 1988, 89, 5535.
29. Zhao, M. T.; Perrin, E.; Prasad, P. N., unpublished results.
30. Prasad, P. N.; Reinhardt, B. A., submitted to Chem. Matls.
31. Samoc, A.; Samoc, M.; Prasad, P. N.; Willand, C.; Williams, D. J., submitted to J. Phys. Chem.
32. Gagnon, D. R.; Capistran, J. D.; Karasz, F. E.; Lenz, R. W.; Antoun, S. Polymer 1987, 28, 567.
33. Wung, C. J.; Pang, Y.; Prasad, P.N.; Karasz, F. E. Polymer, in press.

RECEIVED July 18, 1990

Chapter 4

Nonlinear Optical Properties of Molecules and Materials

Joseph W. Perry

Jet Propulsion Laboratory, California Institute of Technology, Pasadena, CA 91109

This paper is a tutorial overview of the techniques used to characterize the nonlinear optical properties of bulk materials and molecules. Methods that are commonly used for characterization of second- and third-order nonlinear optical properties are covered. Several techniques are described briefly and then followed by a more detailed discussion of the determination of molecular hyperpolarizabilities using third harmonic generation.

One purpose of this tutorial paper on optical characterization is to provide a brief introduction for chemists to the concepts and methods involved in studies of the nonlinear optical properties of molecules and materials. The intent is to familiarize chemists with the range of commonly used techniques and their physical basis. An attempt is made to provide some background on macroscopic nonlinear optics, relating to what is actually measured, and the connection to molecular nonlinear optical properties. This paper is not intended to be a detailed or comprehensive review. The reader is referred to introductory (1, 2) and advanced (3–6) texts on nonlinear optics for more detailed or complete coverage of the subject.

During the past fifteen years, there has been an increasing interest in the NLO properties of organic and polymeric materials (7, 8). This has led to an increased effort aimed at the synthesis of molecular based materials with improved properties for NLO applications(9). An important feature of the development of organic NLO materials is the attempt to control the primary NLO properties (the NLO susceptibilities or coefficients) and the secondary properties (solubility, processability, optical clarity, absorption, thermal stability, etc.) through molecular

0097–6156/91/0455–0067$06.50/0

structure. This approach requires a knowledge of the relationship between molecular structural features and the resultant microscopic and macroscopic properties of the material.

There has been a focus of attention on the electronic contribution to the NLO properties of conjugated π-electron systems since the realization that such systems can exhibit large ultrafast non-resonant NLO responses. Thus, a major effort has been underway to synthesize delocalized electronic structures of various geometries, both to act as the primary source of nonlinearity and to serve as an effective conduit or bridge between electronically active groups (donors and acceptors). Some recent research themes related to NLO properties of molecules and molecular materials are: 1) evaluation of the limits imposed by the trade-off between nonlinearity and transparency, 2) identification of the effect of charge transfer states and electronic asymmetry on nonlinearities and 3) exploration of the possible role of metal complexes and organometallic molecules as NLO active groups. Generally speaking, there is great interest in the identification of means for enhancing optical nonlinearities subject to certain constraints (that depend on the intended application).

While our theoretical understanding of the NLO properties of molecules is continually expanding, the development of empirical data bases of molecular structure-NLO property relationships is an important component of research in the field. Such data bases are important to the validation of theoretical and computational approaches to the prediction of NLO properties and are crucial to the evaluation of molecular engineering strategies seeking to identify the impact of tailored molecular structural variations on the NLO properties. These issues have led to a need for reliable and rapid determination of the NLO properties of bulk materials and molecules.

Commonly Used NLO Techniques

Out of the large range of possible nonlinear optical effects, chemists are likely to encounter only a limited number of measurement techniques. These include both second- and third-order NLO characterization methods. A brief listing of the different types of measurements, the nonlinear susceptibility involved and the related molecular nonlinear polarizabilities is given here.

Before proceeding, it is necessary to introduce the nonlinear polarizabilities and susceptibilities that we will be dealing with. The polarization of a molecule or a macroscopic material subject to an applied electric field is expanded as a power series in the applied field. The molecular polarization is given by:

$$\mathbf{p} = \mu + \alpha \cdot \mathbf{E} + \beta \cdot \mathbf{E} \cdot \mathbf{E} + \gamma \cdot \mathbf{E} \cdot \mathbf{E} \cdot \mathbf{E} + \ldots \tag{1}$$

where μ is the permanent dipole moment of the molecule, α is the linear polarizability, β and γ are the first and second hyperpolarizabilities and $\mathbf{E}$ is the applied electric field. $\mathbf{p}$, μ and $\mathbf{E}$ are vectors and α, β and γ are second, third and fourth rank tensors, respectively. The macroscopic polarization of a material is given by an analogous expansion:

$$P = P_0 + \chi^{(1)} \cdot E + \chi^{(2)} \cdot E \cdot E + \chi^{(3)} \cdot E \cdot E \cdot E + ... \qquad (2)$$

where P_0 is the spontaneous polarization of the material, $\chi^{(1)}$ is the linear susceptibility and $\chi^{(2)}$ and $\chi^{(3)}$ are the second and third-order susceptibilities, respectively.

Second-Order NLO Techniques. The two main second-order NLO techniques in use are linear electro-optic (LEO) modulation(*10*) and second harmonic generation (SHG)(*11*). In the LEO modulation measurement a light beam is passed through a material that is subjected to an electric field via attached electrodes. The electric field causes a change in the index of refraction, n, or birefringence (the difference between n in two directions in the material). This change in n leads to a change in phase or polarization of the light beam that can be converted into a change in intensity of the beam using interference (by interfering with a reference beam) or a polarizer, respectively. This measurement gives a value of the electro-optic coefficient, r, that is related to the second-order susceptibility, $\chi^{(2)}(-\omega;0,\omega)$, where $-\omega$ represents the frequency of the output light, 0 represents the D.C. electric field and ω represents the input light field. Being a second-order nonlinearity and a third (odd) rank tensor property, r, like $\chi^{(2)}$, can be non-zero only in materials that lack a center of symmetry. Such materials can be single crystals, poled polymers, or noncentrosymmetric Langmuir-Blodgett films, for example. The underlying molecular hyperpolarizability responsible for the LEO effect is β. In poled polymers, the noncentrosymmetric alignment of NLO active chromophores is achieved by using an electric field to align the dipoles while the polymer is held above its glass-rubber transition temperature. The alignment is then locked in by cooling the polymer with the field still applied. The LEO coefficient for a poled polymer is given by (*12*)

$$r \propto (|\mu \cdot \beta| E_p)/5 k T \qquad (3)$$

where μ is the dipole moment , β is the first hyperpolarizability, E_p is the electric field used for poling, k is Boltzmann's constant and T is the temperature during poling. $\mu \cdot \beta$ is the scalar product of the vector μ and the tensor β. From measurements of the LEO coefficient and μ, the vector part of β can be determined for molecules in a poled polymer sample.

The other common technique used for determination of second-order NLO properties is second harmonic generation. In an SHG measurement a laser beam at frequency ω illuminates a sample and coherent light at twice the frequency (2ω) is generated and detected. These measurements can be performed on a wide range of sample types including powders in addition to those mentioned above for the LEO measurements. SHG is therefore a very useful method for characterizing second-order materials. A technique involving SHG measurements on powders provides a rapid semi-quantitative characterization of second-order NLO response and can be used effectively to screen large numbers of compounds to find potentially interesting candidate materials that may then be evaluated in more detail.

The SHG intensity from a material can be expressed as

$$I(2\omega) \; \propto \; \left| d_{eff} \right|^2 \; I^2(\omega) \frac{\sin^2\left(\frac{\pi l}{2 l_c}\right)}{\left(\frac{\pi l}{2 l_c}\right)^2} \; l^2 \qquad (4)$$

where d_{eff} is the effective harmonic generation coefficient (that is proportional to $\chi^{(2)}(-2\omega;\omega,\omega)$), l_c is the SHG coherence length ($l_c = \lambda / [4 (n(\omega) - n(2\omega)]$) with λ the fundamental wavelength and n the refractive index at the indicated frequency), l is the interaction pathlength in the material, and I is the intensity at the appropriate frequency. If $n(\omega)$ approaches $n(2\omega)$ the coherence length tends towards infinity and the SHG would apparently increase dramatically. This situation is referred to as phasematching. In this case

$$I(2\omega) \; \propto \; \left| d_{eff} \right|^2 \; l_{eff}^2 \; I^2(\omega) \qquad (5)$$

where the effective length l_{eff} is generally smaller than the interaction length l due to "walk-off"(4). Phasematching can be achieved using the birefringence of uniaxial or biaxial crystals to compensate for the dispersion of the refractive index. Phasematched SHG in such crystals is of great technological significance and is widely used in laser systems to double the frequency of the laser output. As above for the LEO effect, the molecular hyperpolarizability responsible for SHG is β. Also as above, the nonlinear coefficient involved for a poled polymer sample, d_{eff}, is proportional to the $\mu\beta$ product, thus SHG measurements from such samples can also be used in the determination of β.

As mentioned above, the powder SHG method is a useful technique for the screening of second-order nonlinear materials. However, because of the sensitivity of the SHG coefficients of crystalline materials to the orientational aspects of the molecular packing and because the measurement is performed on an essentially random distribution of microcrystalline particles, the powder SHG method is not generally useful for obtaining information about molecular hyperpolarizabilities.

The original powder SHG method is due to S. Kurtz and T. T. Perry (13). There have been variations and improvements reported since that work(14). In the basic method, a fine powder is loaded into a holder such as a cuvette. The pathlength of the cell should be much larger than the particle size to ensure sampling of many particles. The particles may be immersed in an index matching fluid. Without index matching fluid, the SHG intensity is mainly backscattered into a broad range of angles. With index matching fluid, the intensity is concentrated into a narrow cone in the forward direction. The scattered second harmonic is isolated from the fundamental using filters or a monochromator and then detected using a photomultiplier tube. A relative measurement of the SHG intensity is made using a reference sample containing a powdered material with known second-order nonlinearity. Commonly, urea or quartz powders are used as reference materials.

There are several other considerations that are significant for powder SHG measurements. 1) Possible absorption of the second harmonic light or the fundamental light should be considered. Typically, a nanosecond pulsed Nd:YAG laser with a wavelength of 1064 nm is used for excitation. Absorption of the fundamental light is essentially negligible for most materials but absorption of the second harmonic light at a wavelength of 532 nm can be significant for highly colored materials. For these colored materials, fundamental light at longer wavelengths is needed so that the second harmonic falls below the absorption edge of the material. Usually, the output of the Nd:YAG laser is shifted to 1907 nm by passing the intense 1064 nm beam through a long (~1 m) cell filled with H_2 gas at moderately high pressures (~400 psi); this process involves stimulated Raman scattering from the H_2 vibrational mode. The second harmonic is then at a wavelength of 953.5 nm and is typically below the absorption edge for most materials encountered. 2) Spurious signals can be generated in the SHG measurement due to fluorescence (for example, fluorescence following two-photon absorption) or to broadband light generated on optical damage of the material. 3) Different SHG results can be obtained on the same compound recrystallized from different solvents. This results from crystallization of the material in various crystal structures. 4) Finally, and quite importantly, the SHG intensities are dependent on particle size. Different behavior of the SHG intensity with particle size is observed for phasematchable (PM) and non-phasematchable (NPM) materials. For NPM materials the SHG intensity grows linearly with particle size for small particles, goes through a maximum at some characteristic size (corresponding to an average coherence length), and then decreases as the inverse of the particle size for large particles. For PM materials the behavior is similar for small particles but as the particles become large the intensity flattens to a roughly constant value. The typical range of SHG coherence lengths is from a few to tens of μm and the usual range of particles sizes used for measurements is 50 - 150 μm. A study using graded particle sizes can be used to determine whether or not a material is PM.

Third-Order NLO Techniques. There is a wider range of third-order techniques commonly used to characterize materials, including: electric field induced second harmonic generation (EFISH) (*15, 16*), third harmonic generation (THG) (*17*) and degenerate four wave mixing (DFWM) (*18*). EFISH and DFWM will be discussed briefly then measurements using THG will be discussed in more detail as an example.

EFISH involves the generation of a second harmonic wave in a material that is subjected to an applied D.C. electric field. The process involves the third-order susceptibility $\chi^{(3)}(-2\omega;0,\omega,\omega)$. EFISH can be performed on gaseous, liquid or solid samples and is currently used extensively to determine the hyperpolarizability, β. For solutions, the EFISH process can be visualized as follows. Consider a solution of dipolar molecules subject to an applied D.C. field. The dipoles will partially align with the field leading to a net dipole moment and asymmetry in the solution. Now, a fundamental wave proceeds through the resulting asymmetric medium and generates second harmonic light. Even though the process involves SHG, EFISH is a third-order process and the molecular nonlinearity involved is γ^{EFISH} that is given by:

$$\gamma^{EFISH} = \gamma^e + \gamma^v + \frac{\mu\,\beta}{5\,k\,T} \qquad (6)$$

where γ^e and γ^v are the electronic and vibronic second hyperpolarizabilities, the sum of which is related to γ^{THG}, and here β is the vector component of $\underline{\beta}$ along the dipole moment. Thus, if the γ's and μ are determined from THG, as described below, and dielectric measurements, respectively, β can be determined from the EFISH measurement. If the molecules in solution are centrosymmetric then only the γ terms contribute and these are directly comparable to the γ involved in THG.

DFWM is a third-order process involving the nonlinear susceptibility $\chi^{(3)}(-\omega;\omega,-\omega,\omega)$. In the DFWM measurement, three beams at frequency ω intersect in the material. The nonlinear interaction produces a fourth beam at the same frequency and its intensity is proportional to the product of the input intensities and the absolute square of $\chi^{(3)}$. The molecular nonlinearity involved is the second hyperpolarizability, γ, but it should be noted that it is a different frequency dependent γ than those discussed above. The usual experimental configuration of the laser beams in DFWM measurements is the phase conjugation geometry. In the phase conjugate-geometry two of the beams are arranged to be counterpropagating and the third beam intersects the other two at some angle in the sample. The generated wave travels directly along the angled input beam but in the opposite direction as if the input beam were "reversed" in time. This apparent reversal also holds for the spatial shape of the beam which exactly retraces itself. Even if the beam were to pass through some distorting medium on the way to the sample, the generated beam would emerge undistorted after traveling a distance greater than that from the distortion to the sample. This behavior is a consequence of the fact that the generated wave is proportional to the complex conjugate of the third beam's electric field. One way to visualize four wave mixing is to realize that the input light beams modify the refractive index of the material in proportion to the intensity. Two beams intersecting at an angle in the medium interfere and form an intensity grating, therefore a grating of index modulation. This index modulation grating then diffracts the other beam in a direction that satisfies the Bragg condition for diffraction. Generally, in the phase conjugate geometry, two different gratings are formed (one due to beams one and three and the other due to beams two and three) that lead to diffraction (of beams one and two) in the direction opposite to beam three. The $\chi^{(3)}$ of a material can be determined from relative measurements of the intensity of the phase conjugate beam generated in a material compared to that generated in a material with a known $\chi^{(3)}$. Since many processes such as orientational motion, absorption of light followed by thermal index change, population gratings, and electrostriction can form an index grating in the material, in addition to the usual electronic nonlinearity, these measurements are usually performed with ultrashort (picosecond or subpicosecond) pulses. These pulses are short compared to the time scale of some of the complicating processes and measurements are performed as function of the time delay between the various input pulses to dynamically resolve the remaining contributions.

Characterization of Molecular Hyperpolarizabilities Using Third Harmonic Generation. Third harmonic generation (THG) is the generation of light at frequency 3ω by the nonlinear interaction of a material and a fundamental laser field at frequency ω. The process involves the third-order susceptibility $\chi^{(3)}(-3\omega;\omega,\omega,\omega)$ where -3ω represents an output photon at 3ω and the three ω's stand for the three input photons at ω. Since $\chi^{(3)}$ is a fourth (even) rank tensor property it can be nonzero for all material symmetry classes including isotropic media. This is easy to see since the components of $\chi^{(3)}$ transform like products of four spatial coordinates, e.g. x^4 or x^2y^2. There are 21 components that are even under an inversion operation and thus can be nonzero in an isotropic medium. Since some of the terms are interrelated there are only four independent terms for the isotropic case.

The third harmonic intensity from a transparent slab is

$$I(3\omega) \propto \left|\chi^{(3)}\right|^2 I^3(\omega) \frac{\sin^2\left(\frac{\pi l}{2 l_c}\right)}{\left(\frac{\pi l}{2 l_c}\right)^2} l^2 \tag{7}$$

where the coherence length $l_c = \lambda / (6\,\Delta n)$ is the distance over which the bound third harmonic wave and free third harmonic wave accumulate a phase mismatch of π; Δn is the difference in index of refraction at 3ω and ω. The bound wave is a nonlinear electric field in the material oscillating at 3ω that is being driven by the fundamental field. Even though its frequency is 3ω it propagates with the same velocity as the fundamental wave (i.e., it experiences the index of refraction at ω). The bound wave is the inhomogeneous part of the solution of the wave equation with a nonlinear source term. The free harmonic wave at 3ω propagates as usual with a velocity dependent on the index at 3ω. It corresponds to the homogeneous part of the solution to the wave equation. Its amplitude is determined by the boundary conditions at the interfaces of the material. It is proportional to the difference in bound wave amplitudes on the two sides of an interface. The expression above is appropriate to the usual case involved in THG studies of materials where Δn is nonzero, that is in the case of non-phasematched THG. In isotropic materials far from resonance the refractive index shows normal dispersion, i.e. $n(\omega)$ increases with increasing frequency, leading to $n(3\omega) > n(\omega)$. Phasematching ($n(3\omega) = n(\omega)$) in THG can be achieved in birefringent media or through anomalous dispersion associated with a resonance. Anomalous dispersion due to electronic resonances has been used to achieve phasematched THG in gases(4) and liquids(19). Phase-matched THG in gases is a useful technique for generating coherent vacuum-UV light. Measurements of the third harmonic intensity and l_c of a sample in comparison to that for a reference material with known $\chi^{(3)}$ and l_c can be used for a relative determination of $\chi^{(3)}$ for the sample. Methods for performing such a determination and some examples will be given below.

Generally speaking, non-phasematched THG measurements require a means for continuously varying the phase mismatch $\Delta\psi$ (i.e. the sample interaction length l since $\Delta\psi = 6\pi\Delta n\, l / \lambda$, where λ is the fundamental wavelength) in order to extract the maximum third harmonic intensity and l_c from the observed fringes. This can be accomplished easily by rotating a slab sample or by translating a wedged sample as illustrated in Figure 1. For the case of rotating a slab sample the fringes (often referred to as Maker fringes) become more closely spaced as the angle is increased because the sample length increases nonlinearly with the angle θ_i, i.e. $\Delta l = t / \cos\theta_i$, where t is the sample thickness and θ_i is the internal angle from the beam to the normal direction, $\theta_i = \sin^{-1}(n_0 \sin\theta / n_1)$, n_0 and n_1 are the indices of refraction outside and inside the slab, respectively, and θ is the external angle of incidence. Also, the intensities of the fringes decrease with increasing θ because of the increased reflection loss of the fundamental at larger angles of incidence. In contrast, for the wedge the sample interaction length increases linearly with the displacement, x, on translation, i.e. $\Delta l = 2 x \tan(\alpha / 2)$ where α is the wedge angle. The wedge fringes follow a simple periodic behavior as a function of x, allowing the amplitude and coherence length to be easily estimated. The wedge fringes will be described in more detail below.

THG has become an important technique for characterization of the second- and third-order nonlinearities of materials and molecules. This is largely due to the interest in determining the purely electronic nonlinearity of molecules without major complications due to orientational or other motional contributions to the observed signals. The

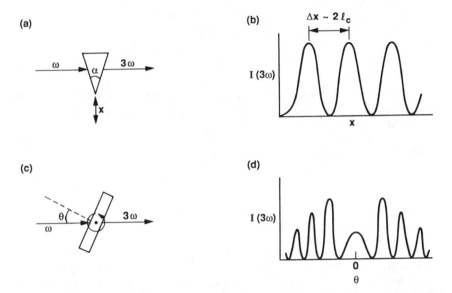

Figure 1. Schematic illustration of sample geometries and THG interference fringing patterns. a). Wedge sample geometry, x is the cell displacement direction. b). Wedge THG interference fringes as a function of x. c). Slab sample geometry, θ is the angle of incidence. d). THG interference fringes as a function of θ (Maker Fringes).

simplification results from the fact that the driving fields and the polarization giving rise to the third harmonic light oscillate at optical frequencies and these are much higher than the frequencies of the whole-molecule motional degrees of freedom. The THG technique gives a measure of the contributions of the high frequency motions of the system to the nonlinearity. The measured susceptibility is proportional to the sum of the electronic and intramolecular vibrational contributions to the hyperpolarizability:

$$\gamma^{THG} = \gamma^e + \gamma^v . \tag{8}$$

At optical frequencies γ^e is typically much larger than γ^v. For conjugated molecules the total electronic hyperpolarizability is expressed as a sum of π and σ electron contributions:

$$\gamma^e = \gamma^\pi + \gamma^\sigma . \tag{9}$$

For molecules containing several conjugated bonds γ^π becomes much larger than γ^σ. Of course, γ itself is a fourth rank tensor property (analogous to $\chi^{(3)}$) and can be specified in the molecular or laboratory reference frames. For an isotropic medium one measures an orientational average of the hyperpolarizability

$$<\gamma>_{XXXX} = 1/5(\gamma_{xxxx} + \gamma_{yyyy} + \gamma_{zzzz} + 2\gamma_{xxyy} + 2\gamma_{xxzz} + 2\gamma_{yyzz}) \tag{10}$$

where the lower case subscripts refer to the molecular frame components and the upper case to the lab frame. The XXXX indices represent the polarizations of the incident and output photons (in an ordering like the frequencies of $\chi^{(3)}$ above) and this term is appropriate to THG with a linearly polarized laser beam. In the remainder of this paper we will simply use γ to represent the average quantity defined above. Having introduced the basic phenomena of THG and its relevance to molecular electronic hyperpolarizability, we turn to a discussion of reliable, rapid methods for determination of γ for molecules in liquids and solutions using THG.

THG was first observed by Terhune, Maker and Savage in a pioneering study of third-order optical effects (20). The theory of THG in dielectric media was established by Bloembergen (3) and he showed how to derive an expression for the third harmonic field using the continuity of the electric and magnetic fields at the boundaries imposed by the interfaces. His group also extended THG studies to absorbing (and even opaque) media and they demonstrated nonlinear susceptibility determination using reflected third harmonic fields (21).

THG came into wider use in the mid- to late 1970s. In fact, THG was used to observe the large non-resonant $\chi^{(3)}$ of polydiacetylenes first reported in 1976 (22). However, in some of the early studies using THG the significance of environmental factors in the experiments were not fully appreciated. Meredith (23) and Kajzar and Messier (24) have discussed this problem in detail. Under the high field of a focussed laser beam the third harmonic amplitude generated in air before and after a sample cannot generally be ignored. While the nonlinearity of the air is smaller than that of condensed phase media, the coherence length for THG in air is substantially longer so the resulting third harmonic amplitude generated in the air can be comparable to that generated in a sample. As mentioned above, the free third harmonic field generated by propagation of the fundamental across a dielectric interface is in fact related to the difference in the bound wave amplitudes in the two media on either side of the interface. As an example of the significance of this effect, the observed third harmonic intensity from a thin (~1 mm) fused silica plate increases by a factor of four as the air around the sample is evacuated. The factor depends on the wavelength and the focal length. This complication can be circumvented by performing measurements with samples in an evacuated chamber. Alternatively, for studies of liquids and solutions, thick window cells can be designed such that the laser beam is tightly focussed in the cell but not on the external faces of the cell. Thus, the third harmonic signals generated in the air before and after the cell are rendered negligible. Such considerations have led to a variety of cell designs with an emphasis on simplification and reliability of the measurement. Kajzar has recently reviewed a variety of cell designs (25).

Meredith and coworkers have described several wedge cell designs (26, 27). These include a triple wedge (two wedged windows and a wedged liquid compartment) and a long liquid chamber cell with a wedged front window. While accurate results can be obtained with a variety of cell designs the pattern of the fringing resulting from the various interferences can be quite complicated and difficult to analyze. For example, the fringe pattern of the triple wedge cell involves a sum of four cosine terms. The long path length cells were a clever step towards reducing the complexity of the measurement. By using a relatively long path medium with a focussed laser, only the interfaces close to the beam waist (focus) make significant contributions to the third harmonic fields. The cell with the wedged front window and long liquid path length then has only two important interfaces, both involving the front window and one the window-liquid interface, thus the fringing was simplified. The coherence length of the liquid was determined in a separate cell with a smaller liquid path and two thick flat windows. Kajzar and Messier have combined the notion of using long paths to reduce the effective number of interfaces with use of a thin wedged liquid path (28). They have used a cell design with thick front and back wedged windows. The wedge of the windows defines a wedged compartment for the liquid that is thin relative to the depth of focus of the laser beam. A simple symmetric interference results from the two interfaces that bound the liquid. The cell is thick enough to allow accurate measurements in air. This is the cell design that we have employed and that will be described in more detail below. More recently, Meredith and coworkers have completely separated the measurement of the third harmonic amplitude and interferences using a cell with a single important interface (29). This was accomplished using a thick front window and a thick liquid compartment. Again, the coherence length is determined in a separate cell.

Kajzar and Messier have analyzed the THG from their cell described above. A brief overview of their analysis is given here. The cell is comprised of two thick wedge windows and a thin liquid wedge compartment. Since the windows are thick, they are considered to be infinite nonlinear media. Since the liquid chamber is thin, the laser field is treated as a plane wave in that region. The third harmonic field at the output of the cell is the resultant of the fields generated in the three media

$$E_R(3\omega) = E_{G1}(3\omega)\, t_{G1} + E_L(3\omega)\, t_L + E_{G2}(3\omega)\, t_{G2} \tag{11}$$

where G1 and G2 refer to the front and back windows and L the liquid compartment. The t's are the overall transmission factors described below. The harmonic field due to propagation of a focussed laser beam in an infinite medium has been discussed in the literature. The details will not be repeated here. However, it is important to note that the harmonic field generated before and after the focus differ in sign. Thus, in the absence of the liquid the fields destructively interfere leading to no observable third harmonic intensity. The presence of the liquid and its dispersion leads to a phase mismatch between the bound third harmonic waves and thus the harmonic fields generated in the windows interfere with the field generated in the liquid. The generated third harmonic field is proportional to the third order nonlinear polarization $P^{(3)}$ that is given by

$$P^{(3)} = \frac{1}{4}\, \chi^{(3)}(-3\omega,\omega,\omega,\omega)\, E^3(\omega). \tag{12}$$

The bound electric field amplitudes are given as

$$E(3\omega) \propto \frac{4\pi\, P^{(3)}}{\Delta\varepsilon} \tag{13}$$

Where $\Delta\varepsilon = \varepsilon(\omega) - \varepsilon(3\omega)$ is the dispersion of the dielectric constant , $1/\Delta\varepsilon = 6\, l_c /(\lambda\, [n(3\omega) + n(\omega)])$. Kajzar and Messier give for the fields from the windows

$$E_{G1}(3\omega) = 4\pi \left[\frac{\chi^{(3)}}{\Delta\varepsilon}\right]_G t_{G1}\, e^{i\varphi}\, E^3(\omega) \tag{14}$$

$$E_{G2}(3\omega) = -4\pi \left[\frac{\chi^{(3)}}{\Delta\varepsilon}\right]_G t_{G2}\, e^{i\Delta\psi}\, e^{i\varphi}\, E^3(\omega) \tag{15}$$

where the t's are transmission factors: $t_{G1} = t^{GL}(3\omega) \, t^{LG}(3\omega) \, t^{GO}(3\omega)$ and $t_{G2} = \{t^{GL}(\omega) \, t^{LG}(\omega)\}^3 \, t^{GO}(3\omega)$ and the superscripts on the t's indicate glass–liquid (GL), liquid–glass (LG) and glass–output (GO) interfaces, φ is the phase of the third harmonic field and $\Delta\psi$ is the phase mismatch accumulated between the bound and free third harmonic waves across the liquid path, $\Delta\psi = \pi / l_c \, (x \tan(\alpha / 2) + l_0)$, l_0 being the initial liquid pathlength. The third harmonic field from the liquid is given by

$$E_L(3\omega) = 4\pi \left[\frac{\chi^{(3)}}{\Delta\varepsilon}\right]_L t_L \, (e^{i\Delta\psi} - 1) \, e^{i\varphi} \, E^3(\omega) \tag{16}$$

where $t_L = \{t^{GL}(\omega)\}^3 \, t^{LG}(3\omega) \, t^{GO}(3\omega)$. One can write the resultant field:

$$E_R(3\omega) = c \left\{ 1 - e^{i\Delta\psi} + \rho \, (e^{i\Delta\psi} - 1) \right\} \tag{17}$$

where $\rho = \left[\dfrac{\chi^{(3)}}{\Delta\varepsilon}\right]_L \Big/ \left[\dfrac{\chi^{(3)}}{\Delta\varepsilon}\right]_G$, c is a constant containing transmission factors and a factor of $\left[\dfrac{\chi^{(3)}}{\Delta\varepsilon}\right]_G E^3(\omega)$, and all other transmission factors have been set to 1 for simplicity. It is easy to find the third harmonic intensity

$$I(3\omega) = |E_R(3\omega)|^2 = 4 \, |c|^2 \, (1-\rho)^2 \sin^2 (\frac{\Delta\psi}{2}) \tag{18}$$

that gives a simple sine squared fringing pattern as a function of cell displacement. The third harmonic signal can be calibrated with a reference liquid and the ratio of the intensity maxima is

$$\frac{I_L(3\omega)}{I_R(3\omega)} = \left[\frac{1 - \rho_L}{1 - \rho_R}\right]^2 \tag{19}$$

where the subscripts L and R refer to the liquid sample and the reference liquid. Noting that

$$\rho_{L(R)} = \left[\frac{\chi^{(3)}}{\Delta\epsilon}\right]_{L(R)} \Big/ \left[\frac{\chi^{(3)}}{\Delta\epsilon}\right]_G \approx \frac{(\chi^{(3)} l_c)_{L(R)}}{(\chi^{(3)} l_c)_G} \tag{20}$$

one can obtain a very simple approximate expression for $\chi^{(3)}_L$:

$$\chi^{(3)}_L = 1 / l_c L \left\{ (1 - R) \, (\chi^{(3)} l_c)_G + R \, (\chi^{(3)} l_c)_R \right\} \tag{21}$$

where R $= \sqrt{\dfrac{I_L(3\omega)}{I_R(3\omega)}}$.

It should be emphasized that these later expressions are valid for nonabsorbing media with a real susceptibility and refractive indices close to glass. A more general treatment of absorbing media and complex susceptibility is summarized below. The more complete expressions given by Kajzar and Messier, including the various transmission factors, should be used for accurate determinations.

Figure 2 shows some representative THG fringes using 1907 nm fundamental light for a solution of diphenylbutadiyne (DPB) in toluene as well as for the neat toluene reference (30). Also shown are the least squares fits of eq. 18 to the data, demonstrating excellent agreement with the observed fringes. The parameters of toluene used for calibration were $\chi^{(3)} = 10.0 \times 10^{-14}$ esu and $l_c = 18.7$ μm (for 1907 nm THG). From the least squares fit intensity maxima for the DPB solution and the reference, the coherence length for the DPB solution, $l_c = 13.8$ μm, and the expression above we obtain $\chi^{(3)} = 28 \times 10^{-14}$ esu for the 0.988 M DPB solution.

To treat an absorbing medium with a complex $\chi^{(3)}$ it is necessary to resort to eq. 17 including appropriate transmission factors accounting for absorption of the fundamental and/or third harmonic fields (31). In this case the third harmonic intensity is (again suppressing transmission coefficients, t)

$$I(3\omega) = |c|^2 \, e^{-\alpha(3\omega)l} \, |1 - a \, e^{i\Delta\psi} + \rho(a \, e^{i\Delta\psi} - 1)|^2 \qquad (22)$$

where a $= e^{(\alpha(\omega) + \alpha(3\omega))l/2}$ and $\alpha(\omega \text{ or } 3\omega) = 2.303 \, \varepsilon(\omega \text{ or } 3\omega) \, C$ with $\varepsilon(\omega$ or $3\omega)$ being the usual exctinction coefficient at ω or 3ω and C the concentration. From this expression it can be seen that there is an overall loss factor associated with absorption at the third harmonic frequency as well as a scaling of the interfering terms that depends on absorption at ω and 3ω. The effect of absorption on the THG fringing pattern is illustrated in Figure 3. The example shows a case where $\rho = 2$, there is no absorption at ω and the absorption at 3ω is varied. In the case of no absorption simple fringing with constant maxima (an amplitude of 4) and null intensity at the minima are observed, in accord with eq. 18. For weak absorption the amplitude of the fringes is damped and the minima no longer go to zero. For strong absorption the amplitude is heavily damped and approaches an asymptotic value that is independent of the liquid pathlength. In this regime the distance over which third harmonic light can escape (~ 1 / $\alpha(3\omega)$) from the liquid is less than l_c so no fringing is observed. Nonetheless, with knowledge of $\alpha(3\omega)$ the susceptibility can be estimated from the observed intensity, that is one fourth of that with no absorption and all other factors constant. This can be shown from eq. 22 by setting the phase factors equal to 1 (since the

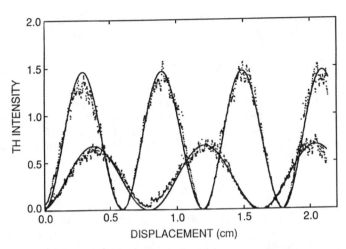

Figure 2. Wedge THG interference fringes for solution of DPB in toluene (larger signal) and for neat toluene using 1907 nm fundamental light. The wedge cell angle was 0.26⁰. (Reproduced with permission from ref 27. Copyright 1989 Royal Society of Chemistry)

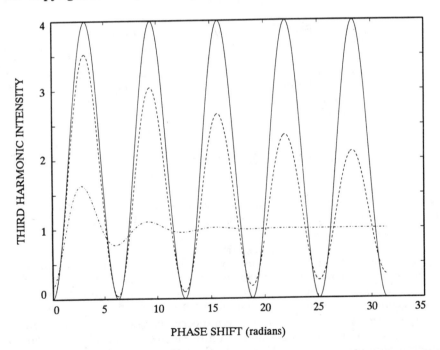

Figure 3. Effect of sample absorption at 3ω on wedge cell THG fringing pattern. Solid line: no absorption. Dashed line: weak absorption. Dot-Dash line: strong absorption.

effective length is $<<$ lc) and letting αl become large. This results in $I(3\omega)$ $= c^2 \mid 1 - \rho \mid^2$ rather than $4 c^2 \mid 1 - \rho \mid^2$ for the maximum intensity as in eq. 18. Now however, $(\chi^{(3)} l_c)_{(L)}$ in ρ should be replaced by $\left(\dfrac{2 \chi^{(3)}}{\alpha(3\omega)}\right)_L$.

To extract the molecular hyperpolarizability from the determined THG susceptibility for a given medium a suitable microscopic model for the susceptibility is needed. The common approach is to use an additive, noninteracting molecules model . The nonlinear susceptibility of a solution is given by a sum over contributions from the components. For a two component solution

$$\chi^{(3)} = N_1 L_1 \gamma_1 + N_2 L_2 \gamma_2 \tag{23}$$

where the index 1 refers to the solvent and 2 refers to the solute, N_i is the number density $N_i = N_A C_i / 1000$ with N_A being Avogadro's number and C_i the concentration of component i. L_i is a Lorentz local field factor that for THG is given by

$$L_i = \left[\frac{n_i^2 (\omega) + 2}{3} \right]^3 \left[\frac{n_i^2 (3\omega) + 2}{3} \right] \tag{24}$$

While distinct local field factors should, in principle, be used for each species in solution, L is usually taken as a uniform factor (defined in terms of the indices for the solution) for the whole solution. Limitations of microscopic models for nonlinear susceptibilities and the shortcomings of local field descriptions such as that above have been discussed by Meredith et al (26). Eq. 23 can be rewritten in terms of concentrations as

$$\chi^{(3)} = \frac{L N_A}{1000} \left[(C_1^0 - C_2 \frac{V_2^m}{V_1^m}) \gamma_1 + C_2 \gamma_2 \right] \tag{25}$$

where C_1^0 is the concentration of the pure solvent and V_1^m and V_2^m are the molar volumes of the solvent and solute. For a complex solute hyperpolarizability $\gamma_2 = \gamma_2' + i \gamma_2''$.

A complex γ^{THG} can result from one-, two-, or three-photon resonances. One-photon resonance occurs when the fundamental frequency ω is close to an allowed electronic transition. Two-photon resonance occurs when 2ω is close to a two-photon allowed electronic transition. For centrosymmetric molecules the two-photon selection rule couples states of like inversion symmetry, e.g. $g \leftrightarrow g$. For acentric molecules one-photon transitions can also be two-photon allowed. Three-photon resonance occurs when 3ω is close to the energy of an electronic transition; the same symmetry rules apply as for one-photon transitions.

The solute concentration dependence of the THG intensity is sensitive to whether γ for the solute is real or complex. This can be shown qualitatively as follows. The third harmonic intensity is $I(3\omega) \propto |\chi^{(3)}|^2$ so using eq. 25 with γ_2 being complex gives

$$I(3\omega) \propto C_1'^2 \gamma_1^2 + 2C_1' C_2 \gamma_1 \gamma_2' + C_2^2 (\gamma_2'^2 + \gamma_2''^2) \tag{26}$$

where $C_1' = C_1^0 - C_2 \left(\dfrac{V_2^m}{V_1^m} \right)$.

It can be seen that the sign of the term linear in C_2 depends on the sign of γ_2' (the real part of γ_2). Also, the relative value of the coefficients of the linear and quadratic terms in C_2 depends on the value of γ_2''. In other words the shape of the concentration dependence is sensitive to the sign of the real part and the magnitudes of the real and imaginary parts of γ_2. Kajzar (*31*) has derived a more complete expression for the third harmonic intensity using a complex γ_2 in eq. 25 and including this in eq. 22 for $I(3\omega)$. He applied this result to the analysis of THG data obtained with 1064 nm and 1907 nm fundamental light on a merocyanine dye in solution. He showed that γ was positive and real at 1907 nm, but at 1064 nm γ had a negative real part and was complex.

We have performed THG studies on a series of simple conjugated molecules in solution at 1064 nm and 1907 nm. The γ^{THG} results obtained are given in Table 1. For the oligoacetylenes with two and four triple bonds, the γ^{THG} values at the two different wavelengths are quite similar, indicative of non-resonant γ. However, for diphenylacetylene (DPA), DPB and stilbene the γ^{THG} values at 1064 nm are somewhat larger. While the third harmonic wavelength is somewhat longer than the absorption edge for these compounds there is some enhancement of γ that may reflect dispersion of the real part of γ or may reflect three-photon resonance on the wing of the $S_0 \to S_1$ transition and thus a complex contribution to γ. We have examined the concentration dependence of the THG intensity for DPA and DPB over a wide range. The experimental results are shown in Figure 4 along with the best fit theoretical curves calculated using eq. 22 along with eq. 25 for $\chi^{(3)}$. The best fit is obtained using positive, real γ for DPA and DPB. These results indicate that the enhanced γ^{THG} at 1064 nm is due mainly to the real part of γ. To illustrate the sensitivity of the measurement to the imaginary part of γ, several calculated concentration dependences are shown in Figure 5 with varying imaginary parts. The range of parameters used would correspond approximately to that for DPB in toluene. Under these conditions, the experiments would be sensitive to γ_2''/γ_2' of greater than about 0.3.

Table 1. Hyperpolarizabilities of some small conjugated molecules.
λ_f is the fundamental wavelength. Uncertainties in γ values are $\pm 15\%$

	γ(THG) 10^{-36} esu	
	$\lambda_f = 1064$ nm	$\lambda_f = 1907$ nm
$Me_3Si—C\equiv C—H$		5.4
$Me_3Si—C\equiv C—SiMe_3$	9	
$Et_3Si—C\equiv C—C\equiv C—SiMe_3$	15	16
$Et_3Si—(C\equiv C)_4-SiEt_3$	54	55
(phenyl—C≡C—phenyl)	27	18
(phenyl—C≡C—C≡C—phenyl)	68	38
(phenyl—CH=CH—phenyl)	55	29
(phenyl—CH=CH—CH=CH—phenyl)		68

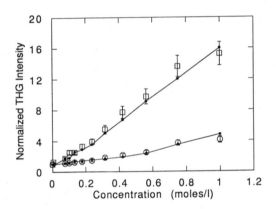

Figure 4. THG intensity as a function of concentration for DPB (squares) and DPA (circles) in toluene. Solid lines are the calculated dependences.

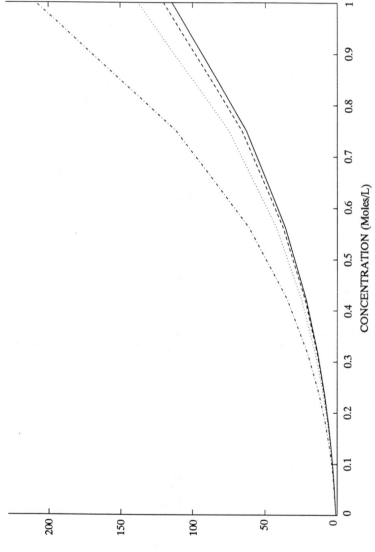

Figure 5. Effect of complex solute hyperpolarizability on the concentration dependence of the THG intensity. For all curves $\gamma_2'/\gamma_1 = 16$. Solid line: $\gamma_2''/\gamma_2' = 0$; dashed line: $\gamma_2''/\gamma_2' = 0.25$; dotted line: $\gamma_2''/\gamma_2' = 0.5$; dot-dashed line: $\gamma_2''/\gamma_2' = 1.0$. THG intensity is in arbitrary units.

Concluding Remarks

In this paper it has been attempted to provide an introductory overview of some of the various nonlinear optical characterization techniques that chemists are likely to encounter in studies of bulk materials and molecular structure-property relationships. It has also been attempted to provide a relatively more detailed coverage on one topic to provide some insight into the connection between the macroscopic quantities measured and the nonlinear polarization of molecules. It is hoped that chemists will find this tutorial useful in their efforts to conduct fruitful research on nonlinear optical materials.

Acknowledgments

The author thanks Dr. L. Khundkar, Dr. A. Stiegman and Mr. K. Perry for their contribution to the experimental work and Dr. F. Kajzar, Dr. S. Marder and Dr. D. Beratan for helpful discussions. This work was performed by the Jet Propulsion Laboratory, California Institute of Technology as part of its Center for Space Microelectronics Technology which is supported by the Strategic Defense Initiative Organization, Innovative Science and Technology Office through an agreement with the National Aeronautics and Space Administration.

Literature Cited

1) Yariv, A. *Introduction to Optical Electronics*; Holt, Rinehart and Winston: New York, NY, 1971.
2) Baldwin, G. C. *An Introduction to Nonlinear Optics*, Plenum, New York, NY, 1974.
3) Bloembergen, N. *Nonlinear Optics*, W. A. Benjamin, Reading, MA 1977.
4) Shen, Y. R. *The Principles of Nonlinear Optics*, John Wiley and Sons, New York, NY,1984.
5) Zernike, F. and Midwinter J. E. *Applied Nonlinear Optics*, John Wiley and Sons, New York, NY, 1973.
6) Yariv, A. and Yeh, P. *Optical Waves in Crystals*, John Wiley and Sons, New York, NY, 1984.
7) Williams, D. J.,, Ed.; Nonlinear Optical Properties of Organic and Polymeric Materials, ACS Symposium Series 233; American Chemical Society: Washington, D. C., 1983.
8) Chemla, D. S., Zyss, J., Eds.; Nonlinear Optical Properties of Organic Molecules and Crystals, Vols. 1 and 2; Academic Press, Orlando, FL, 1987.
9) Williams, D. J. *Angew. Chem. Int. Ed. Engl.* **1984**, 23, 690
10) Kaminow, I. P. *An Introduction to Electro-Optic Devices*, Academic Press, New York, NY, 1974

11) Kurtz, S. K. In *Quantum Electronics: A Treatise*, Rabin, H.; Tang, C. L., Eds.; Vol. 1A, Nonlinear Optics; Academic Press, New York, NY, 1975; 209.

12) Singer, K. D.; Kuzyk, M. G. Sohn, J. E., *J. Opt. Soc. Am. B* **1987**, 4, 968.

13) Kurtz, S. K.; Perry, T. T. *J. Appl. Phys.* **1968**, 39, 3798

14) Halbout, J. M.; Blit, S.; Tang C. L. *IEEE J. Quantum Elec.* **1981**, QE-17, 513.

15) Levine, B. F.; Bethea, C. G. *Appl. Phys. Lett.*, **1974**, 24, 445.

16) Singer, K. D.; Garito, A. F. *J. Chem Phys.*,**1981**, 75, 3572.

17) Maker, P. D.; Terhune, R. W. *Phys. Rev. A*, **1965**, 137,801.

18) Hellwarth, R. W. *Prog. Quant. Electon.*, **1977**, 5,1.

19) Bey, P. P.; Guiliani, J. F.; Rabin, H., *IEEE J. Quantum Elect.*, **1968**, QE-4, 932.

20) Terhune, R. W.; Maker, P. D.; Savage, C. M. *Phys Rev. Lett.* **1962**, 8, 404.

21) Burns, W. K.; Bloembergen, N. *Phys. Rev. B* **1971**, 4, 3437.

22) Sauteret, C.; Herman, J. P.; Frey, R.; Pradere. F.; Ducuing, J.; Baughman, R. H.; Chance, R. R. *Phys. Rev. Lett.* **1976**, 36, 956.

23) Meredith, G. R. *Phys. Rev. B* , **1981**, 24, 5522.

24) Kajzar, F.; Messier, J.; *Phys. Rev. A*, **1985**, 32, 2352.

25) Kajzar, F., In *Nonlinear Optical Effects in Organic Polymers*, Messier, J.; Kajzar, F.; Prasad, P.; Ulrich, D., Eds.; NATO ASI Series, Series E. Applied Sciences, Vol. 162; Kluwer Academic Publishers, Dordrecht, 1989, 225.

26) Meredith, G. R.; Buchalter, B.; Hanzlik, C. *J. Chem Phys.* **1983**, 78, 1533.

27) Meredith, G. R.; Buchalter, B.; Hanzlik, C. *J. Chem Phys.* **1983**, 78, 1543.

28) Kajzar, F.; Messier, J. *Rev. Sci. Instr.*, **1985,** 58, 2081.

29) Meredith, G. R.; Cheng, L.-T.; Hsuing, H.; Vanherzeele, H. A.; Zumsteg, F. C. In *Materials for Nonlinear and Electro-Optics*; Lyons, M. H., Ed.; IOP Publishing, New York, NY, 1989, 139.

30) Perry, J. W.; Stiegman, A. E.; Marder, S. R.; Coulter, D. R. In *Organic Materials for Nonlinear Optics*, Hann, R. A.; Bloor D., Eds.; The Royal Society of Chemistry, London, 1989,189.

31) Kajzar, F. In *Nonlinear Optics of Organics and Semiconductors*, Kobayashi, T., Ed.; Springer Proceedings in Physics, Vol. 36; Springer-Verlag, Berlin, 1989, 108.

RECEIVED September 28, 1990

Chapter 5

Electronic Hyperpolarizability and Chemical Structure

David N. Beratan

Jet Propulsion Laboratory, California Institute of Technology, Pasadena, CA 91109

The chemical structure dependence of electronic hyperpolarizability is discussed. Strategies for developing structure-function relationships for nonlinear optical chromophores are presented. Some of the important parameters in these relationships, including the relative ionization potential of reduced donor and acceptor and the chain length, are discussed. The correspondence between molecular orbital and classical anharmonic oscillator models for nonlinear polarizability is described.

Goals

The goal of this tutorial chapter is to give chemists a starting point to build an understanding of the chemical basis for electronic hyperpolarizability [1-4]. The discussion is intentionally oversimplified. Many important effects are suppressed to emphasize generic molecular structure - hyperpolarizability relationships. We will focus on single molecule properties. Complicating effects, described elsewhere in this volume, are essential for modeling bulk material properties, a process that just starts with an understanding of the molecular hyperpolarizability. The intent here is to present sufficient background to stimulate structure-function studies of this property.

The emphasis of the theoretical discussion is (1) derivation and interpretation of the sum on states perturbation theory for charge polarization; (2) development of physical models for the hyperpolarizability to assist molecular design (e.g., reduction of molecular orbital representations to the corresponding anharmonic oscillator description for hyperpolarizability).

What is electronic (hyper)polarizability?

Applied electric fields, whether static or oscillating, distort (polarize) the electron distribution and nuclear positions in molecules. Much of this volume describes effects that arise from the electronic polarization. Nuclear contributions to the overall polarization can be quite large, but occur on a slower time-scale than the electronic polarization. Electronic motion can be sufficiently rapid to follow the typical electric fields associated with incident UV to near IR radiation. This is the case if the field is sufficiently off resonance relative to electronic transitions and the nuclei are fixed (see ref 5 for contributions arising from nuclear motion). Relaxation between states need not be rapid, so

0097–6156/91/0455–0089$06.00/0

instantaneous response models are inadequate if the frequencies approach single or multi-photon electronic resonances with sufficient oscillator strength [6-8]. The term "nonlinear optics" suggests the physical response of interest: optical effects arising from nonlinear polarization of the molecule in an applied field. Since the electronic polarization is effectively instantaneous, the polarization at time t can be thought of as the response to a static field, $\mathcal{E} = \mathcal{E}_0 \cos(\omega t)$, that is nonlinear in $\mathcal{E}(t)$.

Polarization of springs and molecules

The calculation of linear and nonlinear polarizabilities for classical oscillators is a familiar textbook problem [9,10]. Analogies are, therefore, often drawn between the polarizability of molecules and the polarizability of harmonic and anharmonic classical springs. Harmonic springs display only linear polarizability. Nonlinear polarizabilities of classical springs are associated with the anharmonic spring constants [10]. One might not immediately expect molecular polarizabilities to be analogous to classical spring polarizabilities because of the anharmonic nature of screened coulombic interactions. Why then are these analogies so ubiquitous, and what chemical structures correspond to anharmonic springs?

The potential for a particle bound to a massless one-dimensional harmonic spring is

$$V(x) = \frac{1}{2} m \omega_0^2 (x - x_0)^2 \tag{1a}$$

For simplicity, the discussion is limited to polarization in one-dimension. V(x) is the energy cost of compressing or extending the spring beyond its equilibrium position x_0.

The force exerted by the spring on the particle at position x is

$$-\frac{d}{dx} V(x) \tag{1b}$$

A charge Z bound to the spring produces a polarization

$$\mathcal{P}_x = Z(x - x_0) \tag{1c}$$

The force due to an electric field $\mathcal{E}$ acting on the particle is $Z\mathcal{E}$. The polarization of the particle in an oscillating electric field is calculated by solving the classical equation of motion for x(t), neglecting damping effects (important near resonance)

$$m\ddot{x} = F \tag{2a}$$

$$m\ddot{x} = -\frac{d}{dx} V(x) + Z\mathcal{E}_0 \cos \omega t \tag{2b}$$

$$m\ddot{x} = -m\omega_0^2 x + Z\mathcal{E}_0 \cos \omega t \tag{2c}$$

$$x(t) = \frac{Z}{m(\omega_0^2 - \omega^2)} \mathcal{E}_0 \cos \omega t \tag{2d}$$

$$\mathcal{P}_x(t) = Z x(t) \tag{2e}$$

The polarization of the harmonic spring is linear in the applied electric field. Anharmonicities in $\mathcal{P}_x(t)$ are introduced by adding cubic or higher order terms in $(x - x_0)$ to V(x). In general,

$$\mathcal{P}_x = \sum_{n=0}^{N} a_n cos^n \omega t \tag{3}$$

The nonlinear polarizabilities in the classical spring problem arise from anharmonic contributions to the spring constant. Resolution of eq. 3 into harmonics of frequency $n\omega$ using trigonometric identities provides an understanding of how specific orders of anharmonicity in $V(x)$ lead to anharmonic polarizations at frequencies different from that of the applied field $\mathcal{E}(t)$. In the classical problem, the coefficients a_n are determined by the anharmonicity constants in $V(x)$ [10].

Classical anharmonic spring models with or without damping [9], and the corresponding quantum oscillator models seem well removed from the molecular problems of interest here. The quantum systems are frequently described in terms of coulombic or muffin tin potentials that are intrinsically anharmonic. We will demonstrate their correspondence after first discussing the quantum approach to the nonlinear polarizability problem. Since we are calculating the polarization of electrons in molecules in the presence of an external electric field, we will determine the polarized molecular wave functions expanded in the basis set of unperturbed molecular orbitals and, from them, the nonlinear polarizability. At the heart of this strategy is the assumption that perturbation theory is appropriate for treating these small effects (see below). This is appropriate if the polarized states differ in minor ways from the unpolarized states. The electric dipole operator defines the interaction between the electric field and the molecule. Because the polarization operator (eq 1c) is proportional to the dipole operator, there is a direct link between perturbation theory corrections (stark effects) and electronic polarizability [6,11,12].

Electric field dependent electronic polarization in a molecule yields a field dependent dipole moment:

$$\mu(\mathcal{E}) = \mu(\mathcal{E} = 0) + \alpha\mathcal{E} + \beta\mathcal{E}^2 + \gamma\mathcal{E}^3 + \cdots \qquad (4)$$

The molecular hyperpolarizabilities are β, γ, $\cdots$ and α is the molecular polarizability. Typical values of β are $\sim 10^{-30}$ esu (esu units mean that the dimensions are in CGS units and the charge is in electrostatic units, thus "β in esu" means β in units of $cm^3 esu^3/erg^2$) [1-4]. For an electric field typical of Q-switched laser light, $\sim 10^4$ statvolts/cm, the contribution to $\mu - \mu(0)$ from $\beta\mathcal{E}^2$ is 10^{-4} debye. These polarizations are infinitesimal on the scale of our usual chemical thinking. Yet, these small polarizations are responsible for the exotic effects described throughout this volume. The perturbation theory approach used to describe these properties is justified by the fact that so little charge actually moves.

Polarization, polarizability, perturbation theory, and stark effects

The square of the molecular wave function $|\Psi|^2$, defines the molecular charge distribution. The wave function of state i, $\Psi_i(\mathcal{E})$, can be calculated in the presence or absence of an electric field. Details of the zero field occupied and unoccupied states determine the size of the hyperpolarizability. Summing the expectation value of the electronic position over occupied states (1 to M) gives the polarization [11]

$$\mathcal{P}_x = -e \sum_{i=1}^{M} < \Psi_i(\mathcal{E})|x|\Psi_i(\mathcal{E}) > \qquad (5)$$

(We limit our discussion to polarization and polarizability in a single dimension for pedagogical reasons). The zero field wave functions substituted in the eq 5 yield the ground state dipole moment of the molecule.

When the effect of the field $\mathcal{E}$ on molecular eigenstates is sufficiently weak,

$$\mathcal{P}_x = \alpha\mathcal{E} + \beta\mathcal{E}^2 + \gamma\mathcal{E}^3 + \cdots \qquad (6a)$$

This simple polynomial expansion for $\mathcal{P}_x$ is appropriate because the Ψ_i's themselves can be written

$$\Psi_i(\mathcal{E}) = \Psi^{(0)} + \sum_{n>0} c_n \mathcal{E}^n \Psi^{(n)} \qquad (6b)$$

where c_1, c_2, $\cdots$ come from time-independent perturbation theory. Thus, the simple taylor's series in $\mathcal{E}$ form is carried through after the integration over the x-coordinate in eq 5. Since the wavelength of the radiation is long compared to the molecular size, the dipole approximation for the interaction between the field and the molecule is appropriate so

$$H = molecular\ hamiltonian, \qquad (7a)$$

$$H' = electric\ dipole\ operator = -e\mathcal{E}x \qquad (7b)$$

and the total electronic hamiltonian is $H + H'$.

As an example of the connection between perturbation theory wave function corrections and polarizability, we now calculate the linear polarizability, α_x. The states are corrected to first order in H'. Since the polarization operator (Zx) is field independent, polarization terms linear in the electric field arise from products of the unperturbed states and their first-order corrections from the dipole operator. The corrected states are [12]

$$\Psi_i \simeq \Psi_i^{(0)} + \sum_{j \neq i} \frac{< \Psi_i^{(0)} |H'| \Psi_j^{(0)} >}{E_i^{(0)} - E_j^{(0)}} \Psi_j^{(0)} = \Psi_i^{(0)} + \sum_{j \neq i} \frac{H_{ij}}{\omega_{ij}} \Psi_j^{(0)} \qquad (8a)$$

where $\omega_{ij} = E_i^{(0)} - E_j^{(0)}$.

$$\mathcal{P}_x \simeq -e \sum_{i=1}^{M} \left\{ \int \left[\Psi_i^{(0)} + \sum_{j \neq i} \frac{H_{ij}}{\omega_{ij}} \Psi_j^{(0)} \right]^* x \left[\Psi_i^{(0)} + \sum_{j \neq i} \frac{H_{ij}}{\omega_{ij}} \Psi_j^{(0)} \right] dx \right\} \qquad (8b)$$

for M occupied states, and

$$H_{ij} = -e\mathcal{E}X_{ij} \qquad (8c)$$

$$X_{ij} = < \Psi_i^{(0)} |x| \Psi_j^{(0)} > \qquad (8d)$$

Thus,

$$\mathcal{P}_x \simeq e \sum_{i=1}^{M} \left\{ -X_{ii} + 2e\mathcal{E} \sum_{j \neq i} \frac{X_{ij}^2}{\omega_{ij}} \right\} + \mathcal{O}(\mathcal{E}^2) \qquad (9a)$$

$$\alpha = 2e^2 \sum_{i} \sum_{j \neq i} \frac{X_{ij}^2}{\omega_{ij}} \qquad (9b)$$

The summartion in eq 9b is familiar from perturbation theory and can be identified with a second-order perturbation energy correction.

$$\alpha = 2\frac{\epsilon^{(2)}}{\mathcal{E}} \qquad (9c)$$

$\mathcal{O}(\mathcal{E}^2)$ means that the polarization expression (eq 8b) generates terms of order $\mathcal{E}^2$ as well. These contribute to β, along with terms that are generated from matrix elements arising from second order wave function corrections absent from eq 8a.

The linear polarizability α is proportional to the second-order stark energy, $\epsilon^{(2)}$, summed over occupied electronic states. If the occupied states are doubly occupied, a factor of two can be introduced and the summation understood to include each filled spatial orbital once. Higher-order polarizabilities are related to higher-order stark energies (β is proportional to a third-order energy and γ to a fourth-order energy) [6]. This is a significant observation and is explained by (1) the relationship between the wave function and energy corrections arising from perturbation theory and (2) the fact that the polarization operator and H' are both proportional to x.

Symbolically, we can write the perturbed wave functions as

$$\Psi_i \sim a_i(x) + b_i(x)\mathcal{E} + c_i(x)\mathcal{E}^2 + \cdots \tag{10a}$$

where a, b, c, $\cdots$ are functions of electronic position. The polarization of this state is

$$\mathcal{P} = -e \sum_{i=1}^{M} \int x[a_i^2 + 2a_i b_i \mathcal{E} + (b_i^2 + 2a_i c_i)\mathcal{E}^2 + (2b_i c_i + \cdots)\mathcal{E}^3 + (c_i^2 + \cdots)\mathcal{E}^4 + \cdots]dx \tag{10b}$$

We have added an ellipse in the $\mathcal{E}^3$ and $\mathcal{E}^4$ terms to emphasize that further corrections to terms of these orders arise from wave function corrections of higher-order than explicitly written in eq 10a.

To generate β more specifically, one writes the wave functions to second-order in the field

$$\Psi_i \simeq \Psi_i^{(0)} + \left[\sum_{j \neq i} \frac{< \Psi_i^{(0)}|H'|\Psi_j^{(0)} >}{E_i^{(0)} - E_j^{(0)}} \Psi_j^{(0)} \right] \tag{11a}$$

$$+ \left[\sum_{j \neq i} \sum_{k \neq i} \frac{< \Psi_i^{(0)}|H'|\Psi_j^{(0)} >< \Psi_j^{(0)}|H'|\Psi_k^{(0)} >}{(E_i^{(0)} - E_j^{(0)})(E_i^{(0)} - E_k^{(0)})} \Psi_k^{(0)} \right] \tag{11b}$$

and calculates $< \Psi_i|x|\Psi_i >$, grouping terms of order $\mathcal{E}^2$ as described in eqs 9-10. The result is

$$\beta = 3e^3 \sum_{i=1}^{M} \left[\sum_{n \neq i} \sum_{m \neq i} \frac{X_{in}X_{nm}X_{mi}}{\omega_{ni}\omega_{mi}} - X_{ii} \sum_{n \neq i} \frac{X_{in}^2}{\omega_{ni}^2} \right] \tag{12a}$$

In the low frequency limit this is equivalent to the time-dependent perturbation theory expression [1-4]:

$$\beta \propto \sum_{i=1}^{M} \sum_{n \neq i} \sum_{m \neq i} X_{in}X_{nm}X_{mi} [F_1(\hbar\omega, \omega_{in}, \omega_{im}) + F_2(\hbar\omega, 2\hbar\omega, \omega_{in}, \omega_{im})] \tag{12b}$$

where $\hbar\omega$ is the energy associated with an incident photon and $[F_1 + F_2]$ is an energy dispersion term. In the limit that $\omega \ll \omega_{in}, \omega_{im}$, this expression reduces to eq 12a.

In general, grouping the terms in

$$\mathcal{P}_x = -e \int x |\Psi^{(0)} + \Psi^{(1)} + \Psi^{(2)} + \cdots|^2 dx \tag{13}$$

of appropriate order in $\mathcal{E}$ and using the identity

$$< \Psi^{(s)}|H'|\Psi^{(r)} > = \sum_{q=0}^{s} \sum_{p=0}^{r} \epsilon^{(q+p+1)} < \Psi^{(s-q)}|\Psi^{(r-p)} > \tag{14}$$

to make the perturbation energy-polarizability connection gives the polarizability in terms of stark energies (where $\epsilon^{(j)}$ is the jth order energy correction due to the field). Expressions for the energy perturbation terms are well known. If the stark energy corrections are not small, this approach is limited. In such cases, perturbation theory may be useful, but the zero-order hamiltonian and perturbation may be different. Also, time-dependent treatments may be needed if the dependence of the polarizability on the light frequency is of interest [8].

Evaluation of molecular hyperpolarizabilities

Most numerical methods for calculating molecular hyperpolarizability use sum over states expressions in either a time-dependent (explicitly including field dependent dispersion terms) or time-independent perturbation theory framework [13,14]. Sum over states methods require an ability to determine the excited states of the system reliably. This can become computationally demanding, especially for high order hyperpolarizabilities [15]. An alternative strategy adds a finite electric field term to the hamiltonian and computes the hyperpolarizability from the derivatives of the field dependent molecular dipole moment. Finite-field calculations use the ground state wave function only and include the influence of the field in a self-consistent manner [16].

As with the solution of other many-body electronic structure problems, determination of the unperturbed eigenvalues is numerically challenging and involves compromises in the following areas: (1) approximations to the hamiltonian to simplify the problem (e.g., use of semi-empirical molecular orbital methods) (2) use of incomplete basis sets (3) neglect of highly excited states (4) neglect of screening effects due to other molecules in the condensed phase.

Methods that are known to calculate transition matrix elements reliably for the systems of interest (e.g., π-electron systems) have been used extensively [13,17]. Especially for β calculations, where relatively few electronic states often dominate the hyperpolarizability, numerical methods are reliable. However, γ calculations are more complicated because of the larger number of contributing terms and the possibility of subtle cancellations that can occur only when the full series is summed. General aspects of β and γ calculations are discussed in the next section.

Structure-function properties for β

A sum rule exists for electronic transitions

$$\sum_{ex} f_{g,ex} = N/2 \tag{15}$$

where $f_{g,ex}$ is the oscillator strength of each distinct one-electron transition, N is the number of electrons in the molecule, and each electronic state is doubly occupied [18-20]. Oscillator strength is defined in the usual way

$$f_{g,ex} = \frac{8\pi^2 mc}{3h} \nu_{g,ex} X_{g,ex}^2 \qquad (16)$$

where $\nu_{g,ex}$ is the energy of the transition in wavenumbers. For strong transitions ($\epsilon \sim 10^5 M^{-1} cm^{-1}$) $f_{g,ex} \sim 1$ and $X_{g,ex} \sim 2 \overset{\circ}{A}$ [18]. All $f_{g,ex}$'s are positive and $\sum f_{g,ex}$ has an upper bound defined by eq 15 (for example, the number of π-electrons in the β calculation of a typical organic chromophore). Molecules with large values of β often have intense charge transfer transitions with oscillator strengths ~ 1. We observe that: (1) eq 12 introduces numerators cubic in dipole matrix elements, with denominators quadratic in energy, (2) the transition matrix element between frontier orbitals in donor-acceptor π-electron systems is about as large as it can ever be in this family of compounds, and (3) the sum rule shows us that the frontier orbitals contain much of the total oscillator strength of the molecule. For these reasons, it is not surprising that eq 12 is often dominated by the frontier charge transfer states, i.e., those with the largest numerators and smallest denominators in the summation. Often, much of the quadratic (β) nonlinearity in large hyperpolarizable systems is dominated by contributions from the first few excited states and small corrections (of about a factor of two) occur on addition of the next 50 or so states, followed by apparent convergence of the calculations [21,22].

Eq 12a generates the two-state approximation when a single transition involving both a large oscillator strength and a significant dipole moment change exists. In this case, $i = g$ and $n = m = ex$ so the summations introduce a single dominant term. Such transitions, because of the dipole moment change, are often called charge transfer transitions. In this case eq 12a reduces to

$$\beta(2 - state\ approx.) = 6e^3 \left[\frac{X_{g,ex}\ X_{ex,ex}\ X_{ex,g}}{\omega_{ex,g}\omega_{ex,g}} - \frac{X_{g,g}\ X_{g,ex}^2}{\omega_{ex,g}^2} \right] \qquad (17a)$$

Regrouping terms the two-state approximation becomes

$$\beta(2 - state\ approx.) = 6e^3 \left[\frac{X_{g,ex}^2(X_{ex,ex} - X_{g,g})}{(E_g - E_{ex})^2} \right] \qquad (17b)$$

where $X_{ex,ex}$ and $X_{g,g}$ are the excited and ground state dipole moments. The molecular hyperpolarizability in the two-state limit is proportional to the oscillator strength of the charge transfer transition, the amount of charge moved, and the square of the transition wavelength (as the wavelength becomes long, the transparency of the material decreases and the nonresonant model may not be adequate). Ψ_g and Ψ_{ex} in the two-state model should be understood to include contributions from the donor, acceptor, and bridge; mixing between the sites is considerable.

A simple analysis of eq 17 shows the general aspects of the structure-function relations expected to control β. This demonstration is simplistic in its lack of explicit bridge orbital structure, but it demonstrates the compromises needed to optimize β. Consider two interacting orbitals, ϕ_D and ϕ_A coupled by the matrix element $t = <\phi_D|H|\phi_A>$, $<\phi_D|H|\phi_D> = \Delta$, and $<\phi_A|H|\phi_A> = -\Delta$. 2Δ is the relative ionization potential of the accessible donor and acceptor orbitals. Writing the donor and acceptor localized states as

$$\Psi_g = c_D^{(g)}\phi_D + c_A^{(g)}\phi_A \tag{18a}$$

$$\Psi_{ex} = c_D^{(ex)}\phi_D + c_A^{(ex)}\phi_A \tag{18b}$$

respectively, the Schrödinger equation in matrix form is

$$\begin{pmatrix} \Delta - E & t \\ t & -\Delta - E \end{pmatrix} \begin{pmatrix} c_D \\ c_A \end{pmatrix} = 0 \tag{18c}$$

The donor and acceptor orbitals are centered at $\pm a$, so $<\phi_D|x|\phi_D> = -a$, $<\phi_A|x|\phi_A> = +a$, and we make the usual assumption $<\phi_D|x|\phi_A> = 0$. Using these relationships, the constants in eq 18 can be calculated

$$\frac{c_A}{c_D}(\pm) = \pm\sqrt{1 + (\Delta/t)^2} - (\Delta/t) \tag{18d}$$

$$E_\pm/t = \pm\sqrt{1 + (\Delta/t)^2} \tag{18e}$$

where the $+$ and $-$ terms correspond to Ψ_g $(+)$ and Ψ_{ex} $(-)$. t is negative. Normalizing the states and calculating the three x matrix elements gives β as a function of the relative ionization potentials of the sites (in units of the coupling strength), $2\Delta/t$. Figure 1 shows β, $X_{g,ex}^2$, $X_{g,g} - X_{ex,ex}$, and $1/(E_g - E_{ex})^2$ as a function of $2\Delta/t$. Note that the transition matrix element and the energy terms peak for the totally symmetric system and decreases as the system becomes asymmetric. On the other hand, the dipole moment change peaks for asymmetric systems and vanishes for symmetric systems. The characteristic parameter, Δ/t, can be varied by changing relative ionization potentials of donor and acceptor. An important question to address is whether known systems are on the rising or falling side of the β plot. Calculations that include structural details of the bridge are needed, but this plot shows the interplay between charge localization and delocalization needed to maximize β.

The anharmonic oscillator - molecular orbital theory connection

We have shown the molecular orbital theory origin of structure - function relationships for electronic hyperpolarizability. Yet, much of the common language of nonlinear optics is phrased in terms of anharmonic oscillators. How are the molecular orbital and oscillator models reconciled with one another? The potential energy function of a spring maps the distortion energy as a function of its displacement. A connection can indeed be drawn between the molecular orbitals of a molecule and its corresponding "effective oscillator".

As an example, we return to the two orbital calculation of Figure 1 and eqs 17-18 for β. The strategy is calculate the states, Ψ_g and Ψ_{ex} as above, but subject to an additional constraint. This constraint is that the state have a fixed polarization. Technically, this is accomplished by introducing a LaGrangian multiplier (λ) into the Schrödinger equation. The eigenvalue problem is then solved in the usual way subject to this constraint. The polarization constraint is reflected in both the energies of the states and in the wave functions. Plotting the energy of this polarized state vs. its polarization defines the effective oscillator. The energy of the state is $< \Psi(\lambda)|H|\Psi(\lambda) >$. (The states are equivalent to those that would be found in a finite field calculation, but the energy is calculated using H in zero field.) This is a standard method of including external constaints in wave function calculations [23]. It is important to note that this energy function is summed over the occupied states and that this effective potential that defines the anharmonicity constants is <u>distinct</u> from the interaction potential in the hamiltonian.

For the two state system (as in eq 18c), the equation to solve with the multiplier λ in the two orbital system is [23]

$$\begin{pmatrix} \Delta - E - \lambda & t \\ t & -\Delta - E + \lambda \end{pmatrix} \begin{pmatrix} c_D \\ c_A \end{pmatrix} = 0 \qquad (19a)$$

Again, t is the coupling between the two orbitals that are located at ± 1 in our distance units, 2Δ is the relative ionization potential of the donor and acceptor orbitals and the c's are the orbital coefficients. Solutions of this equation for E are the variationally minimized energies subject to the polarization constraint. The λ dependent wave functions give the energy of the polarized state and its polarization $\mathcal{P}(\lambda)$. We will define a unitless polarization $\bar{\mathcal{P}}$ as $(-c_D^2 + c_A^2)$. The analytical result for the energy as a function of polarization is

$$\frac{E_\pm}{t} = \frac{1}{t} < \Psi_\pm|H|\Psi_\pm > = \frac{1}{t} < \Psi_\pm|H|\Psi_\pm > = \left(\frac{\Delta}{t}\right) \bar{\mathcal{P}} + \sqrt{1 - \bar{\mathcal{P}}^2} \qquad (19b)$$

H is the molecular hamiltonian in the absence of the field. This anharmonic energy profile is plotted in Figure 2 for three choices of $2\Delta/t$. A taylor series expansion of this equation around the equilibrium polarization, $\bar{\mathcal{P}}_0$, gives the effective cubic anharmonicity in the potential, where $\bar{\mathcal{P}}$ replaces the classical position x

$$V(\bar{\mathcal{P}}) \sim \cdots + t \left[\frac{-\bar{\mathcal{P}}_0/2}{(1 - \bar{\mathcal{P}}_0^2)^{5/2}}\right] (\bar{\mathcal{P}} - \bar{\mathcal{P}}_0)^3 + \cdots \qquad (19c)$$

Interestingly, this peaks for total polarization on the donor or acceptor ($\bar{\mathcal{P}} = \pm 1$) showing that the cubic anharmonicity is related to the ground state polarization in this two-orbital case. When the square of the ground state polarization is not too large, the anharmonicity is simply proportional to the polarization of the ground state so from eqs 9b and 17 in this simple example

$$\beta \propto anharmonicity \times linear\ polarizability \qquad (20)$$

Determination of the generality of this sort of expression and assment of whether analogous relations for γ are valid await further study. Related strategies for partitioning the hyperpolarizability have been discussed in the literature [24-29].

Structure-activity correlations for γ in pi-electron systems

While we have identified some of the design principles for β in a two-state model, design criteria for molecules with enhanced γ appear more difficult to state succinctly because single large matrix elements do not necessarily collapse the γ expression to

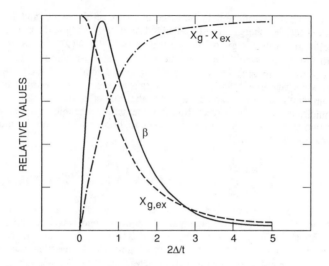

Figure 1. Contributions to β_{xxx} (solid) in a 2-orbital donor-acceptor molecule arise from a product of the terms $X_g - X_{ex}$ (dash) and $x_{g,ex}^2$ (dash-dot), and $1/(E_g - E_{ex})^2$ (dot).

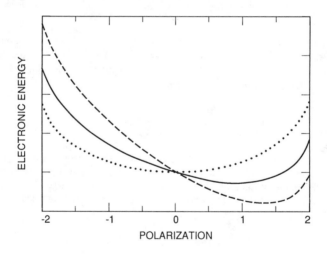

Figure 2. Effective anharmonic oscillator potential for a two-orbital donor-acceptor system with $\Delta/t = 2.0$, dashed line; $\Delta/t = 1.0$, solid line; $\Delta/t = 0.0$, dot line.

a small number of terms. For some model hamiltonians, the hyperpolarizability can be calculated unambiguously, and qualitative features of its structure understood. This section is devoted to such simple models. Refs. 24-29 also discuss the relevance of few state models for the calculation of the second hyperpolarizability.

Hydrogenic atoms (one electron bound by a nuclear charge Z) have γ proportional to the seventh power of the orbital radius [29]. Square well 1-D potentials with infinitely high walls and an appropriate number of filled states give γ proportional to the 5th power of the well width [29]. There is clearly a rapid increase expected in the second hyperpolarizability with system size for delocalized systems.

Square well models are not adequate to describe the length dependence of the spectroscopic properties in very long <u>bond alternated</u> systems since these molecules actually retain a nonzero HOMO-LUMO gap even for very long chains. This bond alternation leads to a hyperpolarizability (per repeat unit) that saturates as the chain length is extended beyond a critical length. Spatially varying 1-D potentials that exhibit this behavior (e.g., sine function potentials [29]) have been used to model long unsaturated hydrocarbons, as have tight-binding or modified Hückel hamiltonians [30-34]. The latter calculations are parameterized from spectroscopic data and predict the order of magnitude and sign of γ_{xxxx} correctly, as well as its dominance over the other tensor elements. These models predict a rapid rise of γ (per repeat unit) with increasing chain length followed by saturation as the chain becomes very long. Flytzanis predicted the value of the length normalized γ for long bond alternated polymers [34]. Finite chain length tight-binding calculations [30], shown in figure 3, demonstrate saturation to the Flytzanis value at long chain length following a rapid rise of γ/N for short chains. In polyenes and polyynes, the saturation is predicted to occur at roughly the chain length where the HOMO/LUMO

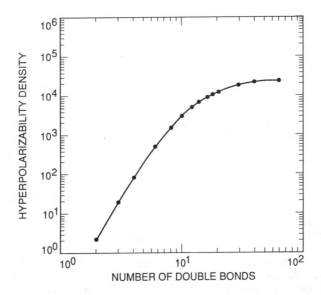

Figure 3. $\gamma_{xxxx}/N\gamma_0$ calculated from a Hückel-like bond alternated chain as a function of the number of sites (N). The ratio of the coupling between p-orbitals in single vs. multiple bonds determines the saturation of the $\gamma/N\gamma_0$ plot (here the ratio is 0.79 to model a polyene). γ_0 is the hyperpolarizability of an isolated double bond.

gap energy also has stabilized as a function of chain length $\sim 10 - 30$ double bonds). Numerically intensive methods are beginning to access the long chain length regime in which saturation occurs [35-38].

For chains with an inversion center, γ contains contributions from two terms of opposite sign [13]. Sign changes in γ can occur when there is a switching of dominance between these terms and strategies to force an unsaturated system into one limit or another have been discussed [31]. Noncentric materials are of increasing interest for γ studies [15].

Concluding remarks

The origin of molecular hyperpolarizability can be described in the off resonant limit using time-independent or time-dependent perturbation theory. We have emphasized the time-independent theory here for simplicity. Applications to the first hyperpolarizability, β, were described in some detail. The two-state limit was discussed and the structure dependence of each term in the two-state expression was analyzed in a two-orbital frame-work. Anharmonic oscillator descriptions of nonlinear phenomena are very common and form a useful starting point for theoretical studies of nonlinear optics. We showed how molecular orbitals can be reduced to their effective nonlinear oscillator potentials. Such analysis should allow determination of how "nonlinear" a particular charge cloud is compared to another. It may also allow separation of the molecular design problem into one of planning molecules with large anharmonicities together with large overall (linear) polarizabilities.

The development of design guidelines for molecules with large second hyperpolarizability, γ, is more difficult because of uncertainty in whether few or many state models are appropriate [24-28]. Some effects, such as saturation of γ with chain length, can be addressed with one-electron hamiltonians, but more reliable many-electron calculations (already available for β) are just beginning to access large γ materials [24,35-38]. Theoretical and experimental work in this area should hold some interesting surprises.

The theoretical problems associated with calculating nonlinear polarizabilities is closely linked to the field of charge transfer spectroscopy and reactivity as well as the field of multi-photon and excited state spectroscopy. It is likely that theoretical methods from these fields will contribute to a deeper understanding of nonlinear optical phenomena in organic, inorganic, and organometallic compounds.

Acknowledgments

I am grateful to the referees for their thoughtful comments on the manuscript. Thanks are due William Bialek (UC Berkeley) for suggesting the LaGrangian multiplier method to link molecular orbital and oscillator models. I also thank Joe Perry and Seth Marder for their enthusiastic collaboration on these problems over the last few years. This work was performed by the Jet Propulsion Laboratory, California Institute of Technology, as part of its Center for Space Microelectronics Technology which is sponsored by the Strategic Defense Initiative Organization Innovative Science and Technology office through an agreement with the National Aeronautics and Space Administration (NASA). It was also supported by the Department of Energy's Energy Conversion and Utilization Technology Program through an agreement with NASA.

References

1. Chemla, D.S; Zyss, J. *Nonlinear Optical Properties of Organic Molecules, vols. 1 and 2*, Academic Press: New York, 1987.
2. Williams, D.J., Ed. *Nonlinear Optical Properties of Organic and Polymeric Materials, ACS Symp. Ser No. 233*, American Chemical Society: Washington, DC, 1983.
3. Williams, D. *Angew. Chem. Int. Ed. Eng.* **1984**, 23, 690.
4. Williams, D. in *Electronic and Photonic Applications of Polymers*, Bowder, M.J.; Turner, S.R., Eds., ACS Symp. Ser. No. 218, American Chemical Society: Washington, DC, 1988.
5. Heeger, A.J. in Heeger, A.J., Orenstein, J., Ulrich, D.R., Eds., Materials Research Society Proceedings, Vol. 109, *Nonlinear Optical Properties of Polymers*, Materials Research Society: Pittsburgh, PA, 1988.
6. Flytzanis, C. in *Nonlinear behaviour of molecules, atoms, and ions in electric, magnetic, or electromagnetic fields*, Neel, L., Ed., Elsevier/North Holland, 1979.
7. Ward, J.F. *Rev. Mod. Phys.* **1965**, 37, 1.
8. Orr, B.J.; Ward, J.F. *Molec. Phys.* **1971**, 20, 513.
9. Marion, J.B. *Classical Dynamics, 2nd ed.*, Academic: New Yorek, 1970, Chapts. 3-5.
10. Hecht, E.; Zajac, A. *Optics, 2nd ed.*, Addison-Wesley: Reading, MA, 1987. Contains elementary discussion of electro-optical effects and nonlinear optics.
11. Mertzbacher, E. *Quantum Mechanics, 2nd ed.*; Wiley: New York, 1970.
12. Schiff, L.I. *Quantum Mechanics, 3rd Ed.*, McGraw Hill: New York, 1968.
13. Li, D.; Ratner, M.A.; Marks, T.J. *J. Am. Chem. Soc.* **1988**, 110, 1707.
14. McIntyre, E.F.; Hameka, H.F. *J. Chem. Phys* **1978**, 68, 3481.
15. Garito, A.F.; Heflin, J.R.; Wong, K.Y.; Zamani-Khamiri, O. *SPIE* **1988**, 971, 2.
16. Prasad, P.N. in *Nonlinear Optical Effects in Organic Polymers*, Messier, J.; Kajzar, F.; Prasad, P.; Ulrich, D., Kluwer: Boston, 1989.
17. Szabo, A.; Ostlund, N.S. *Modern Quantum Chemistry, 1st Ed., revised*, McGraw Hill: New York, 1989.
18. Suzuki, H. *Electronic Absorption Spectra and Geometry of Organic Molecules*, Sect. 4.1, Academic: New York, 1967.
19. Weissbluth, M. *Atoms and Molecules*, Academic: New York, 1978. Sect. 23.3
20. Green, B.I.; Orenstein, J.; Schmitt-Rink, S. *Science* **1990**, 247, 679.
21. Allen, S.; Morley, J.J; Pugh, D.; Docherty, V.J. *SPIE* **1986**, 20, 682.
22. Morley, J.O. and Pugh, D. in *Organic Materials for Nonlinear Optics*, Hann, R.A.; Bloor, D., Eds. Royal Soc. Chem. Special Pub No. 69, Royal Society of Chemistry: London, 1989.
23. Arfkin, G. *Mathematical Methods for Physicists, 3rd Ed.*, Academic: New York, 1985, Sect. 17.6.
24. Pierce, B.M. *J. Chem. Phys.* **1989**, 91, 791.
25. Rustagi, K.C.; Ducuing, J. *Opt. Commun.* **1974**, 10, 258.
26. Mehendale, S.C.; Rustagi, K.C. *Optics Comm.* **1979**, 28, 359.
27. Dirk, C.W.; Kuzyk, M.G., *SPIE* **1988**, 971, 11.
28. Meredith, G.R.; Stevenson, S.H. in *Nonlinear Optical Effects in Organic Polymers*, NATO ASI Series, Messier, J.; Kajzar, F.; Prasad, P.; Ulrich, D., Eds., Kluwer, Academic Publishers: Boston, 1988.
29. Ducuing, J. *International School of Physics, E. Fermi, LXIV - Nonlinear Spectroscopy*; North Holland: Amsterdam, 1977.

30. Beratan, D.N.; Onuchic, J.N.; Perry, J.W. *J. Phys. Chem* **1987**, 91, 2696.
31. Beratan, D.N. *J. Phys. Chem* **1989**, 93, 3915.
32. Beratan, D.N.; Lee, M.A.; Allender, D.W.; Risser, S. *SPIE* **1989**, 1080, 101.
33. Risser, S.; Klemm, S.; Lee, M.A.; Allender, D.W. *Mol. Cryst. Liq. Cryst.* **1987**, 150b, 631.
34. Cojan, C.; Agrawal, G.P.; Flytzanis, C. *Phys. Rev. B* **1977**, 15, 909.
35. Heflin, J.R.; Wong, K.Y.; Zamani-Khamiri, O.; Garito, A.F. *Phys. Rev. B* **1988**, 38, 1573.
36. de Melo, C.P.; Silbey, R. *J. Chem. Phys.* **88**, 2567 (1988).
37. Soos, Z.G.; Ramasesha, S. *J. Chem. Phys.* **90**, 1067 (1989).
38. Hurst, G.J.B.; Dupuis, M.; Clementi, E. *J. Chem. Phys.* **1988**, 89, 315.

RECEIVED August 14, 1990

Chapter 6

Electrooptic Polymer Waveguide Devices
Status and Applications

R. Lytel, G. F. Lipscomb, E. S. Binkley, J. T. Kenney, and A. J. Ticknor

Lockheed Research and Development Division, Palo Alto, CA 94304

Glassy nonlinear optical polymers containing molecules with second-order hyperpolarizabilities can be processed into channel waveguides. When poled, the channels become electro-optic, and can switch and modulate light. The materials exhibit primary optical properties such as large electro-optic coefficients and low dielectric constants that make them potentially attractive for fiber-optic applications, including CATV, LAN, and even connectors. Using lithographic and machining techniques familiar to the chip industry, it should even be possible to integrate large numbers of electro-optic switches into a board-level package or module, and thus provide the additional benefits of active switching and reconfiguration to passive hybrid optical interconnect modules. However, the primary properties of the current research materials are only a part of the complete picture. Secondary properties, such as thermal stability of the electro-optic effect, refractive indices, and other optical and electrical properties, require suitable development to permit the manufacture of reliable devices compatible with assembly for systems use. We highlight some of the primary properties of these extraordinary materials and attempt to provide some suggestions as to key developments required for device and systems applications.

The synthesis of glassy polymer films containing molecular units with large nonlinear polarizabilities has led to the rapid implementation of polymer integrated optics (1-4). These polymers may be guest-host systems (5), in which nonlinear optical molecular units are dissolved into a polymer "host", side chain polymers (6) with the active group covalently bound to the backbone of the polymer, or even crosslinked systems (7), where the nonlinear optical molecule is linked into the polymer chains. Films of nonlinear optical polymers, formed by either spinning, spraying, or dipping, are amorphous as produced, and can be processed to achieve a macroscopic alignment for the generation of second-order nonlinear optical effects by electric-field poling (8). Typically, the films are poled by forming an electrode-polymer-electrode sandwich, and applying an electric field to the heated sandwich, normal to the film surface. Films can

0097–6156/91/0455–0103$06.00/0
© 1991 American Chemical Society

also be corona-poled. Either way, the films are ideally suited for guided wave applications.

With the availability of certain classes of research formulations, many efforts are now underway to understand the design capabilities of devices based upon active polymers, and to attempt to develop suitable processes for fabricating prototype devices to evaluate both material and device performance. To date, many 'test structures' have been fabricated in E-O polymers, including Y-branch interferometers, directional couplers, novel evanescent switches, and travelling-wave modulators. With few exceptions, most of the structures represent a first attempt to make something that 'wiggles' on an oscilloscope. Recently, more advanced work in single-mode polymer channel waveguides has produced demonstration devices exhibiting some of the potential of the materials. In concert with this work, there has been a significant effort to understand the true requirements of polymer materials for integrated optical devices and packages. Much work remains, especially to produce stable, uniform, and reliable materials before practical, cost-effective products can be realized.

MATERIALS STATUS AND REQUIRED DEVELOPMENTS

Organic electro-optic (E-O) materials offer potentially significant advantages over conventional inorganic electro-optic crystals, such as $LiNbO_3$ and GaAs, in several key areas of integrated optics technology, as summarized in Table I, including materials parameters, processing technology and fabrication technology. The most striking advantage, and the reason for the intense interest in these materials, is due to the intrinsic difference in the electro-optic mechanism. Unlike inorganic ferroelectric crystals, where the electro-optic response is dominated by phonon contributions, the electro-optic effect in certain organic materials arises in the electronic structure of the individual molecules, yielding extremely large E-O coefficients with little dispersion from dc to optical frequencies (second harmonic generation) and low dielectric constants (9-11). Poled polymer organic materials have been demonstrated which exhibit electro-optic coefficients significantly larger than that of $LiNbO_3$ coupled with a dielectric constant nearly an order of magnitude smaller (12). The low dielectric constant is essential to the success of high bandwidth modulators due to the resulting lower velocity mismatch between the RF and optical waves, and could lead to an improvement of more than a factor of 10 in the bandwidth-length product over current $LiNbO_3$ devices. The microscopic molecular origin of the second order nonlinear susceptibility, $\chi^{(2)}$, and linear electro-optic coefficient, r, in organic NLO materials is now well understood theoretically and experimentally and the materials are ready for advanced development and formulation.

The materials research effort in recent years has centered on the inclusion of the nonlinear optical moiety in a guest/host or polymer system with appropriate linear optical, mechanical and processing properties and the artificial creation of the desired symmetry through electric field poling. These materials can then be simply and rapidly coated into high quality thin films, processed with standard photolithographic techniques and poled quickly and efficiently. Channel waveguides and integrated optic circuits can be defined by the poling process itself (12-14) or by a variety of other techniques (15-18), including photobleaching, excimer laser micromachining, and etching. These processes represent a considerable increase in fabrication flexibility and processing simplicity over current titanium indiffused $LiNbO_3$ waveguide technology, which requires processing at temperatures approaching 1000 °C after expensive and difficult crystal growth. The key potential advantages of these materials for device applications are summarized in Table II.

Table I. Comparison of Current LiNbO3 and Projected Organics Integrated Optics Technologies

PHYSICAL & DEVICE PROPERTIES	Ti:Lithium Niobate	Organic Polymers
Electro-optic coefficient (pm/V)	32	10-50*
Dielectric constant	28	4
Loss (dB/cm @ λ=1.3 mm)	0.1	0.2-0.5
Space-bandwidth product (GHz-cm)	10	120
Crystal Growth Temperature, °C	1000	NA
Waveguide Processing Temperature, °C	1000	150-200
Waveguide Processing Time	10 hr.	10 min.
Multiple Layers Possible	No	Yes
Fabrication & processing	difficult	simpler
Packaging	expensive	UNKNOWN
Maturity	30 years	12 years

* poling-field and wavelength-dependent

Table II. Potential Advantages of Polymer Electro-optic Materials

> • *Large, broadband electro-optic coefficients*
> • Current materials exhibit $r \leq 40$ pm/V @ $\lambda = 830$ nm
> • New materials with more active moieties may exhibit $r \sim 100$ pm/V
>
> • *Low DC and microwave dielectric constants*
> • $\varepsilon \sim 3\text{-}4$
> • high time-bandwidth products (~ 120 GHz-cm)
> • resonances located well above 20 GHz
>
> • *Ultrafast response times*
> • Materials have $\tau \sim$ fsec for electronic response
> • Waveguide switches have $\tau_{RC} \sim 50$ psec, and are capacitive
>
> • *Low optical loss*
> • $\alpha \leq 0.5$ dB/cm @ diode laser wavelengths, depending on material
> • Low loss channels producible by RIE, laser ablation, selective poling
>
> • *High levels of integration into optical packages*
> • Small single-mode waveguide dimensions ($\sim 3\text{-}5$ μm)
> • High waveguide packing density
>
> • *Potential for novel hybrid optoelectronic packages*
> • Multilevel polymer/metal interconnect packages
> • Hybrid optical multichip modules
> • Hybrid packaging and connection of sources, detectors, and electronics
> • Integrated Si substrates with detectors, amplifiers, and other circuitry

The application of electro-optic polymer materials to practical devices will require some significant advances in the development of the materials. The primary advance required is the achievement of suitable thermal stability of the poled state over temperature ranges determined by fabrication, assembly, and end use requirements. Table III illustrates some of the temperatures and conditions under which a poled polymer must remain poled to within a few percent of its initial value over the life of a device.

Table III. Temperature Ranges and Conditions for Stability of Poled Polymers, °C

MIL SPEC 883c LEVEL 2	TEMPERATURE RANGES
Use	-40 TO 125
Storage	200
COMMERCIAL	**TEMPERATURE RANGES**
Use	20-100
Storage	120
SHORT EXCURSIONS	**TEMPERATURE RANGES**
Fiber Attachment	≤ 250
Wire Bond	≤ 100
Hermetic Package	≤ 320
Die Attach	≤ 320
Subassembly	≤ 320

To leverage the key potential advantages listed in Table II, there are numerous other critical parameters that are not generally reported but must be carefully understood and controlled. These include stress-induced birefringence and its temperature dependence, temperature-dependent linear refractive indices, thermal coefficients of expansion, electro-migration effects, charge-trapping, conductivity, piezoelectricity, pyroelectricity, diffusion, and other parameters. Uncontrolled variations in all of these parameters affect the processing and the performance of a real device. Until materials with a sufficiently detailed data base are developed, research materials will produce research devices, viz., devices that 'switch or modulate', but do not exhibit reliable and reproducible performance.

Systems users will require cost/performance ratios that rival alternate approaches to solving their problems, and will not be favorably disposed to solving materials problems by the use of cooled packages with extra wires and power requirements, and reduced reliability. Moreover, such users, and component manufacturers and assemblers, will require that they be capable of using standard manufacturing and assembly equipment to build devices and assemble systems. The full assembly of a device, be it a fiber-pigtailed modulator or an integrated module, will require such processes as soldering for device or module packaging, fiber attachment, wire bonding, and other processing. The systems user will require that the finished device survive any final systems assembly (wave soldering, etc.) and that it be compatible with commercial and possibly military end use requirements, such as the thermal limits listed in Table III. This requires materials with good long-term thermal and environmental stability, large temperature operating ranges, and the existence of a significant reliability base, including failure-in-time (FIT) rates.

Many current research formulations probably cannot meet these requirements. For example, many of the current research E-O polymers, whether guest-host or side chain, are based on thermoplastic acrylate chemistry (i.e. PMMA) and, as a result, exhibit glass transition temperatures between 100 - 150 °C. This low T_g results in high polymer chain diffusion rates, of the order of 10^{-12} to 10^{-15}, and a calculable variation of at least 10% in the optical properties of the poled state over 5 years operation at ambient temperature. This rapid change is the natural consequence of the dynamic processes by which glassy polymers, operating close to their T_g's, undergo physical aging and relaxation to minimize free volume. When higher operating temperatures are considered (125 °C), the stability of the optical properties becomes even worse. The performance limitations of these materials are due to the fact that they have not yet been formulated for field use, and is not a result of any intrinsic scientific limitation. For example, recent work has shown that crosslinked systems may exhibit much better stability than guest-host or side-chain systems, and it is obvious that many higher temperature host materials could exhibit much better thermal stability, if properly formulated.

As an example, consider the sensitivity of the basic engineering performance parameters of an optical waveguide due to variations in poled optical properties that result from both thermal effects and processing variations over their expected temperature operating range of -40 to 125 °C. Specifically, consider the variation in half-wave voltage (V_π) for a polymer channel waveguide such as that shown in Figure 1. Assuming that the polymers adhere to the substrate (so that thermal expansion occurs in mainly the thickness), the fractional change in V_π is given by $\delta V_\pi/V_\pi = \delta\lambda/\lambda - 3\delta n_0/n_0 - \delta r/r + \delta\Delta/\Delta - \delta L/L$. The variables in the individual parameters that yield $\delta V_\pi/V_\pi$ are listed in Table IV for two different polymers, one with a high T_g (>400 °C) and the other with a low T_g (~150 °C), over a temperature range of -40 to 125 °C. The total variation in half-wave voltage over these conditions is then summarized below. As can be seen, the low temperature polymer, which is representative of most of the research materials currently under investigation, exhibits a substantial performance variation in this key poled optical property that is clearly dominated by material thermal effects. The large thermal coefficients of expansion (TCE) that characterize the thermoplastic acrylates make them impractical for any application, other than device research, that involves a wide range of temperature environments or operation close to their T_g's.

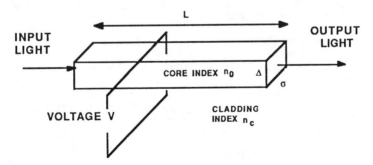

Figure 1. Parameters for a Polymer Channel Waveguide

Table IV. Parameters and Their Variation over Operating Ranges

VARIABLE	THERMAL		FABRICATION[1]
	LOW T_g (150 °C)[2]	High T_g (>400 °C)[3]	
$\delta\lambda/\lambda$	0	0	0
$\delta n_0/n_0$	1.5%	0.6%	2%
$\delta r/r$	10-50%	<1%	0.6%
$\delta\Delta/\Delta$	1.5%	0.6%	4%
$\delta L/L$	0.04%	0.04%	0%

TOTAL $\delta V_\pi/V_\pi$ FOR LOW T_g POLYMER: ~ 15 - 55%
TOTAL $\delta V_\pi/V_\pi$ FOR HIGH T_g POLYMER: ~ 5%
[1] 0.2 μm precision, substrate TCE = 3 ppm/°C
[2] TCE = 100 ppm/°C
[3] TCE = 30 ppm/°C

The simple analysis presented above confirms that new formulations are required to produce stable, reliable products for field use. Practical system requirements, as defined by Mil Spec conformity and the use of standard fabrication and assembly processes, definitely require that a electro-optic polymer system with better thermal properties than thermoplastic acrylates be developed. That this is true for optical interconnection boards and modules is not surprising because of their complexity. It is perhaps remarkable that it remains true for even simple devices, such as a packaged, pigtailed traveling-wave modulator. The ultimate success of electro-optic polymers will be their use in cost-effective products that are used by systems designers.

DEVICES AND APPLICATIONS

Device research with poled polymer materials is still in its infancy. For the most part, this is due to the requirement that the material must be poled to produce a nonzero electro-optic coefficient. However, it is also due to the evolution of the materials themselves, and the significant batch-batch variations that exist in formulations that are new today and obsolete tomorrow. The maturation process is rapid, and, consequently, device designs and processes must be continually updated for each improved generation of material. For example, it is possible to purchase large quantities of passive polymers (acrylates, polyimides, and others) with essentially the same uniformity and quality control as is available from the photoresist industry. With such materials, the precise measurement and control of the refractive index, stress and strain tensors, uniformity, and loss is possible, and, consequently, the design and fabrication of precise waveguide units is possible. With active polymers, the materials are evolving rapidly, and variations in all optical and electrical properties are expected and observed from one batch to another. Moreover, the requirement to pole a waveguide can add many mask steps to the process, and introduces the need to be concerned with the electrical properties of the core and cladding materials, and their compatibility with electrode materials. Consequently, essentially all reported device structures exhibit 'wiggles' on an oscilloscope and exhibit desired effects, but do not yet achieve the performance expected from these potentially important materials.

As the materials mature, it is expected that practical fabrication problems will be solved, and that eventually various grades of E-O polymers will be available, like photoresist is today, for a variety of different applications and needs. For this reason, polymers represent a unique, potentially powerful addition to the library of materials comprising optoelectronic components, and polymer devices provide new and novel approaches to optical interconnection, electronic packaging, and integrated optics.

Electro-optic polymer device research has proved many of the potential advantages of the materials. However, the device research is still maturing due to the rapid maturation of the materials. The fabrication techniques exist to produce very high-quality devices, including high-performance single-function components, such as modulators, switches, and active splitters. Good device performance demands both the use of high precision equipment and highly stable, consistent materials with predictable optical properties, such as refractive indices and E-O coefficients. Such materials will be formulated as research continues, and will probably emerge from the combination of highly active nonlinear optical moieties with electronic grade polymer hosts through proper formulation and synthesis.

Because of their switching properties, polymer E-O guides may find their most beneficial application to highly integrated packages, rather than single-function, pigtailed devices. These would include optical interconnection nodes and switches, and distribution networks. Since the application of these materials to such products represents a novel and powerful approach, we conclude this review with a summary of the use of E-O polymers in highly integrated packages for interconnection and switching.

Optical signals can carry information at very high bandwidths with much lower loss, power dissipation and crosstalk than electrical signals in metallic conductors (19-20). In order for these benefits to be realized, however, the improved transmission capability of optical signals must make up for the added difficulty of converting the information from electrical to optical form at the output pin of one chip and back to electrical form at the input pin of the receiving chip. This conversion process has been the strongest impediment to achieving optical interconnects because it requires an active function in the interconnection network. The most common solution suggested for this problem is to place an individual laser diode, either a discrete or monolithically integrated device, at each output line. In a typical multichip package or board with several hundred to several thousand output lines, the complexity of assembling discrete components becomes prohibitive. Monolithic integration of laser diodes onto high speed, high performance integrated circuits is a complex and difficult process. The chip real estate required for integrated laser diodes could probably be better used for logic gates.

Passive optical waveguides solve many of the problems confronting system designers attempting to interconnect electrical signals at speeds above 1 GHz. Optical waveguides reduce crosstalk, signal attenuation and the power necessary to drive the signals. There is a frequency (bandwidth) distance product where optical signal transmission offers advantages over electrical. However, the system design trade off analysis of optical and electrical comes up against the problem of the electrical to optical back to electrical conversion since the logic operations currently must be performed by electronic integrated circuits. An electronic system will have hundreds to thousands of signal lines interconnecting the integrated circuits. The conversion of each of these chip I/O's to optical lines is difficult and expensive with discrete sources and detectors. Additional factors hindering the use of passive optical interconnections are listed in Table V.

Table V. Limitations on Passive Optical Interconnects

Electrical to Optical Conversion Separate Laser for Each Line Laser Drive Circuitry Separate Detectors for Each Line Amplification for Detectors
Assembly Optical Alignment of Lasers Optical Alignment of Detectors
System Density Electrical and Optical in Same Board Multilevel Optical Interconnects
Material Thermally Stable Optical Materials Required Low Loss Waveguides
Reliability Discrete Laser: 50 - 100 FIT's Discrete Detector: 1 FIT's Many More Separate Components
Cost More Components More Assembly Steps More Complex Assembly

The alignment and assembly of individual sources increases the complexity of the interconnect system, reduces the reliability of the finished board, and significantly increases its cost. For high data rates the source lasers must be environmentally protected and temperature controlled, requiring individual packages or packaged laser arrays, and possible direct cooling. High speed laser packages are particularly real-estate inefficient, with the active device occupying less than 1% of the area of the package. This wasted space can lead to timing problems and data errors since the electrical signals from the control IC are thus forced to travel relatively long distances that may be several fractions of a wavelength (15 cm for 1 GHz). Multiplexing of multiple data lines is now used for optical interconnections between boards and may be appropriate as a way of reducing the number of lasers required for some applications. However, the electrical complexity of sending and receiving the signal at two or more times the data rate makes multiplexing impractical above a few hundred megahertz, and thus it may not be useful at high speeds where optical interconnects are the most appropriate.

The alignment of discrete detectors for each input is no less a difficult task than the source assembly problem. The extra traces associated with connecting discrete field effect transistors (FETs), PIN diodes, or avalanche photodiodes leads to degradation of signal, lower reliability, and greater cost. The integration of photo FETs onto the IC does provide a way to simplify the detector side of the problem, unfortunately at the expense of wavelength flexibility (850 nm). Integrated photo FETs are currently possible only in GaAs technology.

The optical assembly tolerances for single mode E-O polymer waveguides are much tighter than those found in electronics assembly, requiring device to waveguide alignment of ± 0.5 μm versus ± 10 to 25 μm for an electrical die assembly with pad pitch of 5 to 8 mils. The use of larger, multimode waveguides reduces the problem somewhat, requiring ± 5.0 μm accuracy, but this may still more difficult than electronic assembly. A modest size board with 10 IC's and 1,000 internal I/O's would require approximately 1,000 lasers and detectors. The assembly and alignment of all of these optical devices, even incorporating arrays and subassemblies, is a formidable undertaking. The total system complexity becomes greater because alignments must be made between two separate planar structures, with as many as six degrees of alignment required in assembly to very tight tolerances.

The direct modulation of laser diodes by each chip I/O requires additional electrical circuitry plus other elements to dissipate the power and noise associated with these high speed switching lines. The integration of lasers onto the IC solves many of these problems, and is thus a goal of the programs on monolithic integrated optics. This integration has been achieved in the laboratory, using the drive circuits on the chip to directly modulate the lasers. SRI has recently produced such a device that incorporates 100 lasers. However, complex IC's can have hundreds of outputs on the chip. Even on an IC, hundreds of direct drive lasers operating at multi GHz frequencies will produce more problems, such as crosstalk and power dissipation. Putting high power lasers on hot IC's that are close to their heat dissipation limit will certainly push the device over its thermal threshold, and thus force either a reduction in the number of on chip logic elements or an increase in chip size. The former is more likely since there is a practical limit to the size of a chip, dictated by fabrication and yield considerations.

The use of active polymer waveguides may simplify the signal drive and modulation problems in optical chip interconnections by providing external modulation of laser light and by using switched waveguides. With appropriate design, a single CW laser can service 100 I/O pins, reducing the number of lasers, and associated alignment processed during assembly, dramatically. This approach is comparable to an electrical hybrid and allows the integration on a module of various technologies of integrated optics and electronics without limiting optical interconnections to monolithic optical integration type devices. This would be accomplished by using the active nature of the switches to redistribute light within a package at high-speeds. It is worth noting, for example, that a typical E-O polymer waveguide switch would have an RC time constant of under 50 psec, and would provide very fast creation and distribution of optical signals within a package. This is particularly useful for active interconnection of high-speed GaAs IC's, but even more important for solving interconnection problems with Si CMOS at frequencies as low as 100 MHz. This latter fact, practically unstated in the literature, is the result of the much lower output buffer power dissipation when CMOS is driving an E-O capacitative switch, as opposed to an electrical interconnect line. The design of such packages is in its infancy at Lockheed, and will be reported at a later date.

ACKNOWLEDGMENT

The research, materials requirements, and applications reviewed in this article were developed by the Lockheed Photonic Switch and Interconnect Group.

LITERATURE CITED

1. Lytel, R.; Lipscomb, G.F.; and Thackara, J.I. In Nonlinear Optical Properties of Polymers; Heeger, A.J.; Orenstein, J.; and Ulrich, D.R., Eds.; Proc. Materials Research Society Vol. 109, 1988, p. 19.
2. Lytel, R.; and Lipscomb, G.F, In Nonlinear Optical and Electro-active Polymers; Prasad, P.N.; and Ulrich, D.R., Eds.; Plenum Press: New York, 1988; p. 415.
3. Lytel, R.; Lipscomb, G.F.; and Thackara, J.I. Proc.SPIE Vol. 824, 1987, p.152.
4. Lytel, R.; Lipscomb, G.F.; Elizondo, P.J.; Sullivan, B.J.; and Thackara, J.I. Proc. SPIE Vol. 682, 1986, p.125.
5. D.J. Williams In Nonlinear Optical Properties of Organic Molecules and Crystals, Vol. 1; Chemla, D.; and Zyss, J., Eds.; Academic Press: Florida, 1986; p. 405.
6. Demartino, R.N.; Choe, E.W.; Khanarian, G.; Haas, D.; Leslie, T.; Nelson, T.; Stamatoff, J.B.; Stuetz, D.; Teng, C.C.; and Yoon, H. In Nonlinear Optical and Electro-active Polymers; Prasad, P.N.; and Ulrich, D.R., Eds.; Plenum Press: New York, 1988; p. 169.
7. MHubbard, M.A.; Marks, T.J.; Yang, J.; andWong, G.K. Chemistry of Materials, 1989, 1, 167.
8. Singer, K.D.; Sohn, J.E.; and Kuzyk, M.G. In Nonlinear Optical and Electro-active Polymers; Prasad, P.N.; and Ulrich, D.R., Eds.; Plenum Press: New York, 1988; p. 189.
9. Nonlinear Optical Properties of Organic Molecules and Crystals, Vol. 1 and 2, Chemla, D.; and Zyss, J., Eds.; Academic Press: Florida, 1986.
10. Singer, K.D.; Garito, A.F. J. Chem. Phys., 1981, 75, 3572.
11. Lalama, S.J.; and Garito, A.F. Phys. Rev. A, 1979, 20, 1179.
12. Thackara, J.I.; Stiller, M.A.; Lipscomb, G.F.; Ticknor, A.J.; and Lytel, R. Appl. Phys. Lett. 1988, 52, 1031.
13. Thackara, J.I.; Stiller, M.A.; Lipscomb, G.F.; Ticknor, A.J.; and Lytel, R. Conference on Lasers and Electro-optics; CLEO Abstracts: Anaheim, CA, 1988; paper TuK4.
14. Ticknor, A.J.; Thackara, J.I.; Stiller, M.A.; Lipscomb, G.F.; and Lytel, R. Proc. Topical Meeting on Optical Computing '88, Toulon, France, 1989, p. 165.
15. Srinivasan, R. Science, 1986, 234, 559.
16. Beeson, K; Horn, K.A.; McFarland, M.; Nahata, C.W.; and Yardley, J., this volume.
17. McDonach, A. et. al Proc. SPIE Vol. 1177, 1989; p. 67.
18. G.H. Cross et. al Proc. SPIE Vol. 1177, 1989; p 92.
19. Hartman, D. H. Optical Engineering 1986, 25, 1086.
20. Feldman, M.R.; Esener, S.C.; Guest, C.C.; and Lee, S.H. Applied Optics, 1988, 27, 1742.

RECEIVED July 18, 1990

Chapter 7

Waveguiding and Waveguide Applications of Nonlinear Organic Materials

George I. Stegeman[1]

Optical Sciences Center, University of Arizona, Tucson, AZ 85721

Optical waveguides offer optimum conditions for nonlinear optical interactions involving, for example, nonlinear organic materials. In this tutorial we review the basic concepts of waveguiding, techniques for fabricating waveguides, and methods for exciting waveguide modes, concentrating on polymeric materials. In addition, we will discuss the operating principles of second harmonic generators and all-optical devices based on an intensity-dependent refractive index. The material requirements and figures of merit necessary for waveguide devices will be described.

The field of nonlinear optics has been active for more than 25 years. Frequency doublers for high-power lasers, usually used in research (but with a sizeable market, nonetheless) have represented the prime commercial application of nonlinear optics. In the last decade, however, data storage and duplicating applications have emerged for efficient doubling of GaAs lasers operating with 100-mW input powers. The pertinent nonlinearity is given by the third-rank tensor, $\chi((2))_{ijk}(2\omega;\omega,\omega)$. In response, there have been two developments in the area of nonlinear organics. New single-crystal organic materials have been developed. In addition, highly nonlinear molecules have been preferentially orientated (poled) in glassy polymer films, allowing the production of films with a second-order nonlinearity. Such poled polymers are ideal for electrooptic devices, which will be discussed in another chapter here. The most efficient application of both approaches is in waveguides, which will be discussed in this tutorial.

Recently, new device possibilities for materials with $\chi^{(3)}_{ijk\ell}$ nonlinearities have been projected for applications in optical computing, signal processing, and other areas. The key to these applications is that the local refractive index can be changed by the local optical intensity I, that is, $\Delta n = n_0 + n_2 I$. A well-developed field already exists that utilizes the electrooptic effect in integrated optics waveguides to perform switching. That is, the output channel can be changed by applying a voltage to a pair of electrodes. These devices can be made all-optical using media with $n_2 \neq 0$. The switching function is achieved by changing the intensity of the

[1]Current address: CREOL, University of Central Florida, 12424 Research Parkway, Orlando, FL 32826

incident field.   To date, a number of these switches have been demonstrated, including one using a nonlinear polymeric material.

The purpose of this tutorial is to introduce the material scientist to nonlinear optics in waveguides.   We begin by discussing the principles of waveguiding, waveguide fabrication techniques, and ways in which waveguide modes are excited. We then introduce nonlinear optics, make the argument that waveguide media are ideal geometries for efficient nonlinear interactions, and identify the key features of nonlinear optics in waveguides.   Finally, we summarize existing progress, and identify materials requirements.

Waveguiding

There are three basic types of waveguides, summarized schematically in Figure 1 (1). The simplest waveguide consists of a film whose thickness is comparable to the wavelength of light.   Beam confinement is achieved in one transverse dimension only, and the beam diffracts in the usual way in the plane of the film.   Essentially, a film of refractive index larger than the surrounding media (the cladding and substrate) is required.   If the surrounding media do not have the same index, the film must have a certain minimum thickness for waveguiding to occur.   Because it is the simplest, we will use this system to illustrate the basic concepts of waveguiding.

Planar waveguides can be fabricated by vacuum coating or spinning a film onto a substrate by material deposition, by transfer of a film onto a surface by dipping, or by the in- or out-diffusion of atomic or molecular species through the substrate surface (2-8).   Vacuum deposition includes electron or thermal evaporation (including MBE), RF sputtering, and MOCVD.   Current dipping techniques include LB monolayer deposition, and pulling substrates from molten liquids of plastics (4,5). The difference in chemical potentials can be used to make species near the surface of a substrate diffuse out of the substrate, and/or to make another species in solution adjacent to the surface diffuse into the substrate (6).   In the latter case, the index distribution is not step-wise, and usually decays with distance into the substrate.

Channel waveguides provide beam confinement in two transverse dimensions, so the light propagates totally in a diffraction-free manner.   That is, the beam cross-section remains the same for distances limited either by absorption or by scattering by waveguide inhomogeneities.   The characteristic $1/e$ attenuations vary from 0.1 to 10 cm^{-1}, depending on waveguide quality.   Channel waveguides are usually fabricated through thin-film techniques, but with a mask first deposited onto the substrate surface so that the film deposition or species exchange occurs only through the openings in the mask (2,3,7,8).   Alternatively, ion-milling or plasma etching can be used to produce ridges in thin films or substrates (2,3,7,8).   Another approach recently developed for channel waveguides in polydiacetylenes is to produce channel ion-exchanged regions in a glass, and to overcoat the waveguide with a poly-4BCMU film to obtain channel guiding in the polymer.   The appropriate design can result in up to 80% of the power guided in the polymer (9).   Most efficient guided wave devices will be made in channel waveguides.

There has been limited progress in the fabrication of fiber waveguides from nonlinear organic materials.   Although plastic fibers (highly multimode) have been made, to date such waveguides have not been made in single-mode form with interesting nonlinear dopants.   Some single-crystal fibers have been drawn with second-order active materials (10-12).   In general, short fibers of organic materials have been used, and their features are similar to those found in channel waveguides.   We will not, therefore, discuss the fiber case further.

The optical fields in the vicinity of a waveguide consist of guided modes and radiation fields that transport power away from the immediate vicinity of the waveguide. For optically isotropic media, the modes for a planar waveguide can be separated into a pure TE mode (E-field polarized along the y-axis, orthogonal to both the surface normal and propagation wavevector) and a pure TM mode (H-field polarized along the y-axis, and E-field components along the x- and z-axes). A finite number of discrete modes exist for a given film thickness, and the field distributions associated with the first few TE_m and TM_m waves, shown in Figure 2, exhibit oscillatory behavior in the film, decaying exponentially with distance into the cladding and substrate. Nonlinear interactions, therefore, can take place in any one of the three media. One of the unique features of thin-film guided waves (as compared to plane waves) is that the the guided wave wavevectors $\beta^{(m)}$, for each mode depend on film thickness. See, for example, solutions to the dispersion relations in Figure 3. The TM waves have similar characteristics, but the dispersion relations are different leading to different variation with film thickness.

The guided-wave field of a planar waveguide is written as (1)

$$E^{(m)}(r,t) = \frac{1}{2}E^{(m)}(x,y)a^{(m)}(z)e^{i(\omega t - \beta^{(m)}z)} + c.c. ,\qquad (1)$$

where "m" defines the mode number and β is the guided-wave wavevector. $a(z)$ is the amplitude coefficient, with the detailed cross-sectional dependence of the guided wave given by $E(x,y)$, normalized so that $|a(z)|^2$ is the guided-wave power in watts. This requires that

$$\frac{\beta^{(m)}}{2k_0 c\mu_0}\int_{-\infty}^{\infty}dx\int_{-\infty}^{\infty}dy\ E^{(m)}(x,y)\cdot E^{*(r)}(x,y) = \delta_{m,r} .\qquad (2)$$

Note that it is primarily this normalization which makes the subsequent formulae appear different from the well-known plane wave cases. For a planar waveguide, it is useful to assume that the guided-wave beam is very wide (D >> λ) and uniform along the y-axis, so that the field distribution is independent of y, and the integral over y just produces D, the beam width.

The term $\beta^{(m)}/k_0$ plays the role of a refractive index for propagation along the z-axis, and is called "the effective index," n_{eff}. Its value is obtained from an eigenvalue equation or dispersion relation by satisfying the boundary conditions across each interface. Typical variations in n_{eff} with normalized waveguide thickness are shown in Figure 3 for a thin film bounded by a substrate (s) and air (c $\equiv$ cladding). Near the minimum film thickness for a given mode, called cut-off, $n_{eff} \cong n_s$ and the field penetrates deeply into the substrate, resulting in low intensities for a given power. For thick films, $n_{eff} \rightarrow n_f$ and, although the field is localized within the film, the intensity again drops with increasing film thickness. There is, in fact, a film thickness which optimizes the intensity for a given power.

The situation is more complex for channel waveguides. Here D $\cong$ λ, and resonances across the y-dimension occur also. Typical fields are shown in Figure 2. Two sets of orthogonal normal modes remain in the sense of Equation 2. Both modes however, contain all three electric-field components [E_x, E_y, and E_z (usually small)]. The mode with E_y as the dominant field component is designated $TE_{m,n}$, and the mode for which E_x is dominant is called $TM_{m,n}$. Note that the modes are now designated by two integers because the field is confined and resonances occur in 2 dimensions. This is in contrast to the planar one-dimensionally confined modes described by a single integer. The dispersion relations are very complicated, and cannot be expressed in analytical form, requiring numerical techniques for evaluation. Here n_{eff} varies with two normalized thickness dimensions.

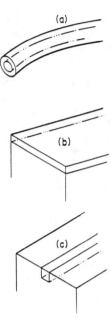

Figure 1. Three common types of waveguides: (a) fiber, (b) thin film, and (c) channel. In each case the guiding medium, fiber core, thin film, and channel region has a higher refractive index than the surrounding media.

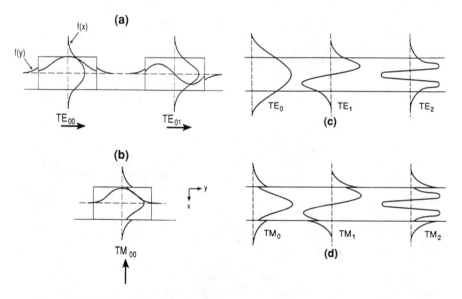

Figure 2. Typical field distributions for waveguides. For the channel case, the transverse distribution $f(x,y)$ is approximated by $f(x)f(y)$. The arrows indicate the dominant field component. (a) TE_{mn} channel modes, (b) TM_{mn} channel modes, (c) TE_m slab field distributions and, (d) TM_m slab waveguide fields.

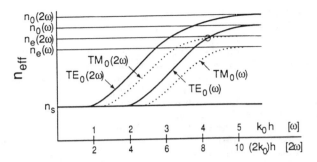

Figure 3.   The guided mode dispersion curves used to determine phase-matching possibilities (intersection of the fundamental and harmonic dispersion curves) for isotropic media.

Because $\beta^{(m,n)} > nk_0$, where n is the largest index of the surrounding media, waveguide modes cannot be excited simply by illuminating the waveguide surfaces. One of the three commonly used excitation methods shown in Figure 4 must be employed. Both prism ($n_p > \beta^{(m,n)}$; $\beta^{(m,n)} = n_p k_0 \sin\theta$) and grating coupling ($\beta^{(m,n)} = k_0 \sin\theta + 2\pi/\ell$) require wavevector conservation parallel to the surface for efficient coupling. In the prism case, the light is incident on the base of the prism at angles θ larger than the critical angle for the prism-air interface, and the mode is excited by the evanescent field in the air gap that penetrates the film. For channel waveguides (and for fibers), light is focused onto the end face of the channel, with the best efficiency obtained when the transverse spatial profile of the incident field matches that of the launched guided wave. Care must be taken to use single-mode waveguides, as all of the modes are excited to some degree in end-fire coupling.

It is difficult to judge the optical quality of a waveguide film by any simple technique, such as visual inspection. Waveguides typically are only a few wavelengths thick, and propagation of thousands of wavelengths down the film is required for useful waveguiding. It is necessary to excite waveguide modes and measure the lengths of their propagation "streaks."

Nonlinear Optics

Nonlinear optics entails the mixing of one (with itself in some cases) or more fields to produce a nonlinear polarization source term, which in turn can radiate a new electromagnetic wave (13,14). This term is usually written as

$$P^{NL}(r,t) = \frac{1}{2} P^{NL}(r,\omega_s) e^{i(\omega_s t - \beta_p \cdot r)} + c.c. \tag{3a}$$

Restricting ourselves to three input guided waves of frequency ω_a, ω_b and ω_c (of which two may be equal),

$$P^{NL}(r,\omega_s) = \epsilon_0 \chi^{(2)}(-\omega_s;\omega_a,\pm\omega_b):E^{(m,a)}(x,y)E^{(m',b)}(x,y) \, a^{(m,a)}(z)a^{(m',b)}(z) \tag{3b}$$

$$+ \epsilon_0 \chi^{(3)}(-\omega_s;\omega_a,\pm\omega_b,\pm\omega_c): E^{(m,a)}(x,y)E^{(m',b)}(x,y)E^{(m)}(x,y)$$

$$\times a^{(m,a)}(z)a^{(m',b)}(z)a^{(m'',c)}(z) ,$$

where $\chi^{(2)}$ and $\chi^{(3)}$ are the second- and third-order susceptibilities (material parameters), and a minus sign for a frequency corresponds to taking the complex conjugate of the appropriate field. Note that $\beta_p = \beta^{(m,a)} \pm \beta^{(m',b)} \pm \beta^{(m'',c)}$ is the wavevector associated with the nonlinear polarization source field, and is not necessarily equal to $\beta^{(n,s)}$, which is the value appropriate to a propagating field of that frequency (ω_s). The case $\beta_p = \beta^{(n,s)}$ corresponds to phase-matching, as will be discussed later. For the second-order processes, two possible input waves exist, with frequencies ω_a and ω_b which produce polarization and signal fields at $\omega_s = \omega_a \pm \omega_b$. For third-order processes, the nonlinear polarizations can occur at the frequencies $\omega_s = \omega_a \pm \omega_b \pm \omega_c$. Many of these interactions have been demonstrated in nonlinear guided-wave experiments (15-17). The processes and their nomenclature are listed in Table 1. Henceforth, we will discuss only second harmonic generation and intensity-dependent refractive index phenomena.

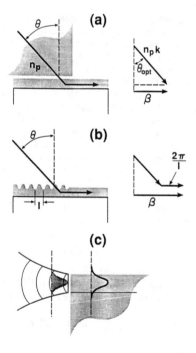

Figure 4.   Techniques for coupling an external radiation field into optical waveguides: (a) prism coupling, (b) grating coupling, and (c) end-fire coupling.

Table 1. Glossary of nonlinear interactions and their common nomenclature

Process	Input Beams	Frequencies
Second Harmonic Generation	1	$\omega + \omega \rightarrow 2\omega$
Sum and Difference Frequency Generation	2	$\omega_a \pm \omega_b \rightarrow \omega_c$
Optical Parametric Oscillator	1	$\omega_a \rightarrow \omega_b + \omega_c$
Third Harmonic Generation	1	$\omega + \omega + \omega \rightarrow 3\omega$
Intensity-Dependent Refractive Index	1	$\omega + \omega - \omega \rightarrow \omega$
Degenerate Four-Wave Mixing	3	$\omega + \omega - \omega \rightarrow \omega$
Coherent Anti-Stokes Raman Scattering	2	$\omega_a + \omega_a - \omega_b \rightarrow 2\omega_a - \omega_b$

The existence of the nonlinear polarization field does not ensure the generation of significant signal fields. With the exception of phenomena based on an intensity-dependent refractive index, the generation of the nonlinearly produced signal waves at frequency ω_s can be treated in the slowly varying amplitude approximation with well-known guided wave coupled mode theory (1). As already explicitly assumed in Equation 1, the amplitudes of the waves are allowed to vary slowly with propagation distance z, that is, $\frac{d^2}{dz^2}a^{(n',s)}(z) \ll \beta^{(n',s)} \frac{d}{dz}a^{(n',s)}(z)$, which leads to

$$\frac{d}{dz}a^{(n,s)}(z) = i\,\frac{\omega_s}{4} \int_{-\infty}^{\infty}dx \int_{-\infty}^{\infty}dy\; P^{NL}(r,\omega_s)\cdot E^{*(n,s)}(x,y)\; e^{-i(\beta_p - \beta^{(n,s)})z} , \qquad (4)$$

where we have generalized the result to include TM guided-wave field interactions. Once the nonlinear polarization field has been found for a particular process, Equation 4 allows the amplitude of the generated guided mode to be evaluated.

Second Harmonic Generation

Much can be learned about the advantages and characteristics of nonlinear guided wave interactions by treating the case of second harmonic generation. In the simplest case, a single input beam is assumed to produce a collinear SHG wave so that $\omega_a = \omega = k_0 c$ and $\omega_s = 2\omega$. In terms of the d_{ijk} tensor ($= \chi^{(2)}_{ijk}/2$), the nonlinear polarization field for the case of weak conversion to the second harmonic is (13,14):

$$P_i{}^{NL}(r, 2\omega) = \epsilon_0 d_{ijk}(2\omega;\omega,\omega)E_j{}^{(m,\omega)}(x,y)E_k{}^{(m,\omega)}(x,y)\,[a^{(m,\omega)}(z)]^2 . \qquad (5)$$

Substitution into Equation 4 and integration over the propagation coordinate gives (15-17)

$$|a^{(n,2\omega)}(L)|^2 = (k_0 L)^2 \left[\frac{c\epsilon_0}{2}\right]^2 |K|^2 \frac{\sin^2(\Delta\beta L)}{(\Delta\beta L)^2} |a^{(m,\omega)}(0)|^2 , \tag{6a}$$

$$K = \int_{\infty}^{\infty} dx \int_{-\infty}^{\infty} dy \; d_{ijk} \; E_j^{(m,\omega)}(x,y)E_k^{(m,\omega)}(x,y)E_i^{*(n,2\omega)}(x,y) , \tag{6b}$$

where $\Delta\beta = \frac{1}{2}(2\beta^{(m,\omega)} - \beta^{(n,2\omega)})$, and K is called the overlap integral. The $\sin^2\phi/\phi^2$ term describes the effect of phase mismatch, that is, $\Delta\beta L \neq 0$. Efficient conversion can be accomplished only when the phase mismatch $\Delta\beta L < \pi/4$. Optimum occurs conversion for $\Delta\beta = 0$, the phase-matched case.

From Equations 6, it is clear that there are a number of key factors governing efficient conversion. These factors are 1) phase-matching; 2) optimization of the overlap integral; 3) large nonlinear coefficients consistent with phase-matching; and 4) low waveguide losses (large L).

Phase matching requires a geometry for which $\beta^{(m,2\omega)} = \beta^{(n,\omega)}$. Figure 3 illustrated the problems associated with phase matching for a slab waveguide. Plotted is the effective index $n_{eff} = \beta/k_0$ for both fundamental and harmonic waves of both polarizations. The goal is to find an intersection between the dispersion curves for a fundamental and harmonic guided wave. In the absence of material birefringence, such crossings effectively occur only for $\beta^{(m,\omega)} = \beta^{(n,2\omega)}$, when n > m. That is, the mode numbers are different for the fundamental and harmonic waves, indicating that the respective field distributions have a different number of zeros inside the film. However, given thick enough films (and enough modes), phase-matching conditions can always be found, which is an advantage in using waveguides.

Tolerances on waveguide dimensions can also be understood from Figure 3. The propagation wavevector varies with the waveguide dimensions through the dispersion relations. The ideal situation, in terms of thickness tolerances and fluctuations, is for the relative angle between the crossing dispersion relations to be small, and for the curves to be parallel to the thickness axis. In terms of index uniformity, again a small crossing angle is advantageous, with the curves approximately parallel to the index axis at the phase-matching condition. The key point is the small crossing angle, which implies that little dispersion in the effective index with wavelength is desirable. Typically, tolerances on the order of 5 nm in waveguide dimensions are sufficient. Uniformity of refractive index and waveguide dimensions both pose problems in technology, and are not easily solved.

The overlap integral (K) can dramatically reduce the efficiency of a doubling phase-matching configuration, if the mode numbers of the interacting fields differ (15-17). This is clear from Figure 5 if both films constituting the waveguide are of the same material and are nonlinear. K is proportional to the integral over the two field distributions involved in the mixing interaction. For example, interference effects occur when m ≠ n, thereby reducing K. This property negates an apparent advantage in using guided waves, namely the availability of a range of βs at a given frequency for phase matching. Figure 5 also shows a solution to this problem. By making the guiding film a layered combination of linear and nonlinear films, the interference effects can be eliminated with a reduction (in Figure 5) in K of only a factor of 2 to 4 (18).

Material birefringence, or the natural birefringence arising from the dissimilar dispersion relations of the TE and TM modes can also be used to obtain phase matching with orthogonally polarized modal fields of the same mode number (15-17). An example is shown in Figure 6, where phase matching between the TE_0 and $\overline{TM_0}$ modes is illustrated. (Here material dispersion in the film only was assumed for simplicity.) Small birefringence, on the order of the material dispersion with wavelength, is desirable. The problem with this aproach is that it requires off-diagonal tensor elements which, unfortunately, tend to be smaller than the diagonal elements.

The advantages of using waveguides for SHG are numerous. Waveguides maintain high intensities (and therefore intense fields) for centimeter distances. As a result, the SHG conversion efficiency is proportional to L^2, rather than to L, as for focused plane-wave beams. The existence of discrete modes, each with its associated wavevector, allows greater flexibility in achieving phase matching. Although the use of modes with different mode numbers can lead to a large reduction in the overlap integral, this loss in efficiency can be reduced by using carefully engineered multilayer waveguides.

Although a large variety of nonlinear materials have been used in prototype harmonic doublers (15-17), only waveguide doublers in $LiNbO_3$ have been of usable quality. A list of previous waveguide doublers can be found in references 15-17. The key figure of merit is

$$\eta' = 100 \, \frac{|a^{(2\omega)}(L)|^2}{|a^{(\omega)}(L)|^2} \, W^{-1} \, . \tag{7}$$

This figure quantifies the intrinsic conversion efficiency of a material-waveguide combination, independent of the device length. Taniuchi and co-workers allowed the second harmonic to leave the waveguide region in the form of Cerenkov radiation (19). Here wavevector conservation parallel to the surface is achieved by the radiation field. When channel waveguides are used, the harmonic light appears in an arc, not all of which can be focused down to the diffraction limit. The best results correspond to $\eta \cong 40 \, W^{-1}cm^{-2}$. This finding implies that, for typical 40-mW semiconductor laser powers, 1.6 mW of blue light can be generated. Another approach is to use gratings with periodic reversals in $\chi^{(2)}$ to implement phase matching, called quasi-phasematching (20,21). The periodicity provides the wavevector component required for phase matching. The theoretical value for η is $> 300 \, W^{-1}cm^{-2}$, making this approach very promising. The flexibility associated with both the Cerenkov and quasi-phase-matching techniques allows any wavelength to be doubled, as long as both the fundamental and harmonic fall within the material's transparency band.

A number of nonlinear organic materials have been used in prototype waveguide doublers (15-17). In every case to date, the waveguides have exhibited too much loss and/or nonuniformity to achieve efficient doubling, especially in those cases where the organic material was vacuum deposited. The most promising result so far has been obtained in poled polymer films, the fabrication of which is discussed in detail elsewhere in this volume (22). In brief, a polymer host, charged with molecules exhibiting both large dipole moments and molecular hyperpolarizabilities, is heated above its glass transition temperature and a field is applied to orient the molecules via their dipole moment. The field is maintained through the cooling phase, locking in the molecular orientation. Phase matching is achieved by alternating the voltages applied to periodic electrodes--this results in an alternating $\chi^{(2)}$, with the period chosen specifically for phase matching (23).

The potential of organic materials for efficient doublers is clearly seen in Table 2, from a very simple figure of merit (15-17). The challenge is to produce

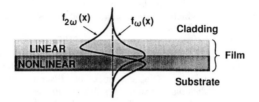

Figure 5.   Optimization of the cross-section for guided wave second harmonic generation.   Phase-matching occurs between two dissimilar modes and the overlap integral is optimized by causing the interference effects to occur in linear regions of the waveguide only.   For the case of a single nonlinear film, interference effects occur as the field product is integrated over the thickness dimension.

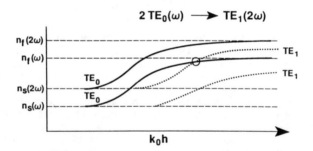

Figure 6.   The guided mode dispersion curves for a birefringent film and an optically isotropic substrate.   Both the fundamental and harmonic curves are shown. The TE mode utilizes the ordinary refractive index and TM primarily the extraordinary index.   Note the change in horizontal axis needed to plot both the fundamental and harmonic dispersion curves.   Phase-matching of the $TE_0(\omega)$ to the $TM_0(2\omega)$ is obtained at the intersection of the appropriate fundamental and harmonic curves.

single-crystal channels of appropriate orientation and uniformity, and poled films with larger net nonlinearities.   In fact, it is not clear whether or not larger molecular nonlinearities are required.   Material processing, instead, seems to be a more important issue.

Table 2.   Second harmonic generation figures of merit relative to LiNbO$_3$ (d$_{13}$ coefficient) for materials with potential for waveguide applications

Material	d$_{eff}$ (esu)	d^2/n^3
LiNbO$_3$	1.2×10^{-8}	1
LiNbO$_3$ (d$_{33}$)	8.5×10^{-8}	50
KTP	$\cong 10^{-8}$	1.5
MNA	7×10^{-8}	75
NPP	2×10^{-7}	600
(PS)O-NPP	2×10^{-8}	
DCV/PMMA	7×10^{-8}	85
HCC#1232	2.5×10^{-7}	

MNA - metanitroaniline
NPP - N-(4-nitrophenyl)-L-prolinol
(PS)O-NPP - chromophore functionalized polymer (29)
DCV/PMMA - dicyanovinyl azo dye in PMMA
HCC/^{1}232 - (CH$_2$)$_2$-amino-nitrostilbene

Third-Order Nonlinear Integrated Optics

One of the fastest growing areas in nonlinear integrated optics involves phenomena based on an intensity-dependent refractive index.   We restrict our discussion of third-order phenomena to this case (15-17,24).   The refractive index experienced by a wave of frequency ω can be changed by a second beam at a different frequency (cross phase modulation), by a wave of the same frequency (ω) but of orthogonal polarization (cross phase modulation), or by the original beam itself (self phase modulation).   Thus there are many ways of all-optically inducing a change in refractive index.   Restricting our discussion to self phase modulation and the pure cubic nonlinearity associated with an ideal Kerr-law medium, the total polarization can be written as

$$P_i(\mathbf{r},\omega) = \epsilon_0[\chi^{(1)}_{ii}(\omega;\omega) + \chi^{(3)}_{iiii}(\omega;\omega,-\omega,\omega)|E_j(\omega)|^2]\, E_i(\omega) \, , \tag{8}$$

where the quantity in the square brackets can be interpreted as an intensity-dependent dielectric constant, that is, n^2.   This leads to n = n$_0$ + n$_2$I, with

$$n_2 = 3\frac{\chi^{(3)}_{1111}(\omega;\omega,-\omega,\omega)}{n^2_0 c\epsilon_0} \, . \tag{9}$$

There are two limits in which such an optically induced refractive index change can be utilized in waveguide geometries (15-17,24).   The waveguide is defined in the first place by refractive index differences between the central guiding region and the bounding media, the largest being Δn_0.   For $\Delta n/\Delta n_0 \ll 1$, where Δn

is the optically induced change, the guided-wave field distribution corresponds to that of the normal mode, and only a small change in the propagation wavevector is induced. But for $\Delta n \geq \Delta n_0$, the field distributions depend on power, and it is necessary to solve the nonlinear wave equation with the boundary conditions, this time with a field dependent refractive index. We limit our discussion to the more common, small index change, case.

The power-dependent refractive index change is obtained by taking a suitable average of the nonlinearity over the intensity distribution associated with the guided wave. The propagation wavevector can be written as

$$\beta = \beta_0 + \Delta\beta_0 |a(z)|^2, \tag{10a}$$

$$\Delta\beta_0 = 3\omega\epsilon_0 \int_{-\infty}^{\infty} dx \int_{-\infty}^{\infty} dy \ \chi_{1111}^{(3)} |E_i(x,y)|^2 |E_i x,y)|^2 , \tag{10b}$$

$$\Delta\beta_0 \cong \frac{\omega}{c} \frac{n_2}{A_{eff}} , \tag{10c}$$

where A_{eff} is the effective cross-sectional area over which the nonlinear interaction occurs.

A number of all-optical integrated optics devices are summarized in Figure 7 (24). The output can be tuned by changing the input power. Although all of these devices require a power-dependent β, they can further be subdivided into two interaction geometries: 1) in which two guided wave modes interact with propagation distance, and for which the nonlinearity affects this interaction; and 2) in which a guided mode independently undergoes a nonlinear phase shift, which changes its interference condition with another optical field. The first category includes the nonlinear directional coupler (NLDC) and its variants, nonlinear distributed feedback gratings (NLDFB), and the nonlinear X- and Y-junctions. Belonging to the second category are the nonlinear Mach-Zehnder interferometer and the nonlinear distributed (prism or grating) coupler. It is noteworthy that the nonlinear coupled-mode devices exhibit the sharpest switching charactertistics.

The simplest device is a nonlinear Mach-Zehnder interferometer, as depicted in Figure 7d. Consider the two channels to be of equal length, with a nonlinearity in channel 1 only (difficult to implement). They are sufficiently well separated so that there is essentially zero field overlap. Thus the differential phase shift between the two channels of length L is $\Delta\phi^{NL} = \Delta\beta_0 L P_{in}/2$, where L is the length of the nonlinear region and P_{in} is the input power (half of which propagates in each channel). Whenever $\Delta\phi^{NL}$ changes by π, the output changes from a maximum to a minimum, producing the response sketched in Figure 7d.

The nonlinear directional coupler is potentially a useful device because it has four ports, two input and two output, and because the outputs can be manipulated with either one or two inputs. Optimally the two channels are identical, and the coupling occurs through field overlap between the two channels. As a result, when only one of the channels is excited with low powers at the input, the power oscillates between the two channels with a beat length L_b, just like what occurs in a pair of weakly coupled identical pendulii. As the input power is increased, a mismatch is induced in the wavevectors of the two channels, which decreases the rate of the power transfer with propagation distance. This leads to an increase in the effective beat length. There is a critical power associated with this device, for which an infinitely long, lossless NLDC acts as a 50:50 splitter, that is, $L_b \rightarrow \infty$. For higher input powers, the initial wavevector mismatch is too large to overcome, and the power effectively stays in the input channel. Therefore, if the device

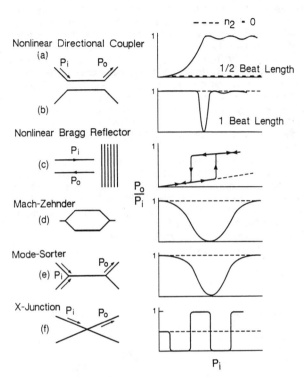

Figure 7. A number of all-optical guided wave devices and their responses to increasing power. (a) Half beat length directional coupler. (b) One beat length directional coupler. (c) Distributed feedback grating relector. (d) Nonlinear Mach-Zehnder interferometer. (e) Nonlinear mode mixer. (f) Nonlinear X-switch. For nonlinear media ($n_2 \neq 0$), the input power determines the output state.

length is terminated at $L_b/2$, at low powers the signal comes out of channel 2 and at high powers out of channel 1. This is essentially an all-optical switch. The net result is the response sketched in Figure 7a.

The operating characteristics of these devices are governed by a number of material parameters (24-26). The larger n_2 ($\propto \chi^{(3)}$) is, the smaller the switching power. The goal is switching with sub-milliwatt powers, which requires nonlinearities $> 10^{-15}$ m^2/W. The device lengths required are about 1 centimeter, implying that $\alpha < 0.5$ cm^{-1}, and preferably 0.1 cm^{-1}, to obtain good device throughput, and a nonlinearity recovery time < 1 ps. In many cases, the index change saturates (Δn_{sat}) with increasing power, which leads to the figure of merit $W = \Delta n_{sat}/\alpha\lambda$. W must be larger than 0.5 → 3.7, depending on the device (see Figure 7). In fact, it is the potentially large values for this figure of merit in nonlinear organics off resonance which has spurred interest in this material system for these switching applications. Finally, two-photon absorption can limit the effective length of a device through an absorption $\alpha = \gamma I$, where γ is the two photon-coefficient (25, 26). This yields a figure of merit $T = \gamma/\lambda n_2$, which must be less than unity for switching to occur.

At this time, many of the parameters listed above have not been measured for nonlinear organic materials. Measurement will be necessary for an assessment of the full potential of such materials. To date, only a nonlinear directional coupler has been implemented in a nonlinear polymer, (poly-4BCMU). For this device at 1060 nm, the nonlinear response was dominated by two-photon absorption, and some limited switching attributable to absorption changes was observed (27). More recent work at 1030 nm appeared to show switching attributable to refractive nonlinearities (28). Clearly there is much progress to be made.

Summary

We have presented a review of the salient features of nonlinear integrated optics. It appears that nonlinear organic materials can play an important role in second- and third-order guided-wave devices. This field requires a great deal of material characterization and processing, however, before significant advances are realized.

Literature Cited

1. Marcuse, D. Theory of Dielectric Optical Waveguides; Academic Press: New York, 1974.
2. Numerous articles in Integrated Optics; Tamir, T., Ed.; Topics in Applied Physics Vol. 7; Springer-Verlag: Berlin, 1975.
3. Lalama, S. J.; Sohn, J. E.; Singer, K. D. Proc. SPIE 1985, 578, 168.
4. Grunfield, F.; Pitt, C.W. Thin Solid Films 1983, 99, 249,
5. Carter, G. M.; Chen, Y. J.; Tripathy, S. K. Appl. Phys. Lett. 1983, 43, 891.
6. Goodwin, M. J.; Glenn, R.; Bennion, I. Electron. Lett. 1987, 22, 789.
7. Ulrich, R.; Weber, H.P. Appl. Opt. 1972, 11, 428.
8. Lipscomb, G. F.; Thackara, J.; Lytel, R.; Altman, J.; Elizondo, P.; Okazaki, E. Proc. SPIE 1986, 682, 125.
9. Schlotter, N. E.; Jackel, J. L.; Townsend, P. D.; Baker, G. L. Appl. Phys. Lett. 1990, 56, 13.
10. Nayar, B. K. In Nonlinear Optics: Materials and Devices; Flytzanis, C.; Oudar, J.L. Eds.; Springer-Verlag: Berlin, 1986, p 142.
11. White, K.I.; Nayer, B.K. Proc. SPIE 1987, 864, 162.
12. Kashyap, R. Proc. SPIE 1986, 682, 170.
13. Hopf, F.; Stegeman, G.I. Advanced Classical Electrodynamics Vol. II: Nonlinear Optics; John Wiley and Sons: 1986

14. Shen, Y.R. The Principles of Nonlinear Optics; John Wiley and Sons: New York, 1984.
15. Stegeman, G.I.; Seaton, C.T. Applied Physics Reviews (J. Appl. Physics) 1985, 58, R57–78.
16. Stegeman, G. I.; Zanoni, R.; Finlayson, N.; Wright, E. M.; Seaton, C. T. J. of Lightwave Technology 1988, 6, 953.
17. Stegeman, G.I. In Proceedings of Erice Summer School on Nonlinear Waves in Solid State Physics; Boardman, A,D.; Twardowski, T., Eds.; in press
18. Ito, H.; Inaba, H. Opt. Lett. 1978, 2, 139.
19. Taniuchi, T.; Yamamoto, K. Digest of CLEO'86; Optical Society of America: Washington, 1986; paper WR3.
20. Lim, E. J.; Fejer, M. M.; Byer, R. L. Electron. Lett. 1989, 25, 174.
21. Lim, E.J.; Fejer, M.M.; Byer, R.L.; Kozlovsky, W.J. Electron Lett. 1989, 25, 731.
22. Lytel, R.; Lipscomb, F. Electro-Optic Polymer Waveguide Devices: Status and Applications, this volume
23. Khanarian, G.; Norwood, R.; Landi, P. Proc. SPIE 1989, 1147, 129.
24. Stegeman, G. I.; Wright, E. M.; J. Optical and Quant. Electron. 1990, 22, 95–122.
25. Mizrahi, V.; DeLong, K. W.; Stegeman, G. I.; Saifi, M. A.; Andrejco, M. J. Opt. Lett. 1989, 14, 1140.
26. DeLong, K.; Rochfort, K.; Stegeman, G. I. Appl. Phys. Lett. 1989, 55, 1823.
27. Townsend, P. D.; Jackel, J. L.; Baker, G. L.; Shelbourne III, J. A.; Etemad, S. Appl. Phys. Lett. 1989, 55, 1829.
28. Townsend, P.D.; Baker, G.L.; Jackel, J.L.; Shelburne III, J.A.; Etemad, S. Proc. SPIE 1989, 1147, 256.
29. Ye, C.; Minami, N.; Marks, T. J.; Yang, J.; Wong, G. In Nonlinear Optical Effects in Organic Polymers; Messier, J.; Kajzar, F.; Prasad P.; Ulrich, D.; Eds.; NATO ASI Series; Kluwer Academic Publishers: Dordecht The Netherlands; Vol 162, p 173.

RECEIVED July 18, 1990

Chapter 8

Nonlinear Optical Materials
The Great and Near Great

David F. Eaton

Central Research and Development Department, E. I. du Pont de Nemours and Company, Wilmington, DE 19880-0328

This review summarizes the current state of materials development in nonlinear optics. The relevant properties and important materials constants of current commercial, and new, promising, inorganic and organic molecular and polymeric materials with potential in second and third order NLO applications are presented.

Historically, the earliest nonlinear optical (NLO) effect discovered was the electro-optic effect. The linear electro-optic (EO) coefficient r_{ijk} defines the Pockels effect, discovered in 1906, while the quadratic EO coefficient s_{ijkl} relates to the Kerr effect, discovered even earlier (1875). True, all-optical NLO effects were not discovered until the advent of the laser.

Second harmonic generation (SHG) was first observed in single crystal quartz by Franken and coworkers (1) in 1961. These early workers doubled the output of a ruby laser (694.3 nm) into the ultraviolet (347.15 nm) with a conversion efficiency of only about 10^{-4}% in their best experiments, but the ground had been broken.

Many other NLO effects were rapidly discovered. (2) The early discoveries often originated in two- or multiphoton spectroscopic studies being conducted at the time. Parametric amplification was observed in lithium niobate ($LiNbO_3$) by two wave mixing in temperature-tuned single crystals in 1965. (3) The first observation of SHG in an organic material was made in benzpyrene by Rentzepis and Pao. (4) Hexamethylenetetramine single crystal SHG was examined by Heilmeir in the same year, 1964. (5) Two other organic materials followed rapidly: hippuric acid (6) and benzil. (7) Benzil was the first material which proved relatively easy to grow into large single crystals.

0097–6156/91/0455–0128$08.25/0

At the end of the 1960's, the Kurtz and Perry powder SHG method was introduced. (8) For the first time, rapid, qualitative screening for second order NLO effects was possible. The stage was set for a rapid introduction of new materials, both inorganic and organic. The early history summarized here is based on a more extensive report presented by Zyss and Chemla (9) in their series "Nonlinear Optical Properties of Organic Molecules and Crystals."

Various reviews and monographs have been published during the last five years which report on various aspects of NLO theory, practice and materials properties. In addition to the Zyss and Chemla series cited above, the reader is referred to those of Williams, (10,11) Gedanken, (12) Basu, (13) Valley et al., (14) and Glass (15) for description of more materials related aspects of this field. In addition, several popular articles describe both materials and applications aspects: Pugh and Sherwood (16) treat aspects of organic crystal growth related to NLO; Chemla (17) discusses semiconductor quantum well devices; Feinberg (18) describes several photorefractive processes and materials; Hecht (19) describes ways in which NLO third order materials may be used in optical signal processing; and two articles discuss optical computing applications generally. (20,21) The conclusions of an Optical Society of America workshop to assess NLO materials held during April of 1986 were published. (22) Finally, the proceedings of two recent symposia are of interest. (23,24)

The current review will present selected results on materials which appear to this author to have promise in NLO applications. Most references will be recent. The review will follow a format in which materials for linear optics are first reviewed very briefly, followed by a discussion of second order, then third order materials.

Materials for Linear Optics Applications

All of our natural experience with optics occurs in the linear domain. In order to apply nonlinear optics in practice, light must first interact with the NLO material. In our laboratories, free space interconnections are usually employed for this purpose. That is, a laser beam is aimed at the material under examination. In any practical use of NLO, such simplistic solutions will not be possible, for reasons both of safety and rugged construction of the device. Light will need to be moved around in space within the device. In many second order devices, whether they are color-specific lasers, such as doubled diode or YAG lasers, or EO modulators such as spatial light modulators (SLM's) waveguide or fiber optic connections will be used. Aspects of these materials will not be reviewed.

In other integrated optics applications, beam steering can be accomplished with holographic optical elements (HOE's). Holograms of this type are produced by spatially modulating the optical index of refraction of a material. Tomlinson and Chandross have reviewed early efforts in this area.(25) Some materials systems are now commercial. Du Pont (26,27) has announced systems based on photopolymerization, in which the optical index is modulated through loss of double bonds during a radical intitiated crosslinking. Because these systems form volume phase holograms, high diffraction efficiencies are possible. Both transmission and reflection holographic modes can be employed. Polaroid Corporation has also developed a holographic photopolymer, based in this case on dye sensitized photopolymerization.(28,29) Several other holographic compositions based on photopolymerization have been reported.(30,31) Among inorganic materials, holograms can be formed in a variety of silver halide (AgX) photographic films, especially dichromated (selectively crosslinked) films. These system generally operated by modulating the absorption of the film rather than the refractive index, so their diffraction efficiency is lower. Similarly, a variety of photochromic materials (both organic and inorganic) can be used to form amplitude modulated HOE's.(25)

The function of the HOE is not simply to move light around in a two-dimensional plane, (i.e, to function as a waveguide). HOE's can also function as focussing devices (lenses) or frequency splitters (diffraction gratings) which demultiplex frequency mixed signals, or as other Fourier optical devices. Such functions will be critical to the eventual manufacture of efficient and small optical devices and signal processing elements. It is precisely because of the importance of this technology to the successful development of NLO devices that chemists should be aware of the nature of the materials involved.

Second Order Materials

Second order materials will be used in optical switching (modulation), frequency conversion (SHG, wave mixing) and electro-optic applications, especially EO modulators (SLM). The applications all rely on the manifestation of the molecular β of the material (the hyperpolarizability) and either the optical coefficient of nonlinearity (d_{ijk}) or the electro-optic coefficient (r_{ijk}), the orientationally averaged tensors that describe the expression of the second order hyperpolarizability, $\chi^{(2)}$ of the bulk material. Materials will be used as single crystals, thin polycrystalline films, or dispersed in polymers and oriented in strong electric fields to provide an acentric environment, in which $\chi^{(2)}$ is nonvanishing by

symmetry. Recently, the use of pure, organic glass-
forming materials has also been demonstrated.(*32*)   We next
review organic materials for second order NLO.

Organic Materials. While most SHG materials are ultimately
referenced to quartz,[1] urea is a common powder standard
used in the Kurtz-Perry testing protocol.   Urea(*33,34*) has
a molecular β of ~0.5x10^{-30} esu, determined at 1.06 μ by
EFISH (Electric Field Induced Second Harmonic Generation).
Powder SHG intensities are nearly 400x the value of
quartz.   The tetragonal crystal, space group P4_2$_1$m, is of
the point group 4_2m, where the underscore implies the
crystallographic overbar to symbolize lack of center of
symmetry.   The crystal packing is not especially favorable
for SHG, and the only term which contributes significantly
is β_{yxx}.(*35*)   The nonlinear coefficient of urea is
estimated to be **d$_{14}$** = 1.4 pm/V.(*36*)   Urea is a useful NLO
material because of its high optical transparency (to 200
nm), its high damage threshold (1-5 Gw/cm^2) in single shot
applications, and its high birefringence which makes phase
matching possible.(*36*)   It is however difficult to grow,
and it is mechanically soft and hygroscopic.   It was the
first organic in which optical parametric amplification
was demonstrated, and it can be phase matched to 238 nm,
and is a superior organic for SHG into the ultraviolet.

The prototypical organic material with high β (all
values of β quoted here are for 1.06 μ light unless noted)
is *p*-nitroaniline (**PNA**).   In electrostatic units (esu), β
for **PNA** is 34.5 x 10^{-30} esu.(*37*)   It has a dipole moment
of 6.8 D.   Extension of the chromophore by insertion of a
single double bond to produce 4-amino-4'-nitrostilbene
increases β dramatically: 248 x 10^{-30} esu.(*38*) However,
both crystalline materials are centrosymmetric, and so no
SHG is possible.   Most achiral molecules crystallize in
centrosymmetric space groups, but some do not.   Several
materials have been discovered during the last few years
which have combinations of hyperpolarizability, crystal
growth and habit, and linear spectral properties which
make them candiates for useful NLO materials.   As will be
obvious from the comparison, considerable compromise must
be made to yield a practical material.

Table I lists the properties of several organic
materials which have been studied as single crystals.   The
materials are listed alphabetically according to the
acronym applied to them in order to avoid the appearance
of prejudice with respect to any one material.   Listed are
the molecular β (if known), space group and point group of
the crystal, SHG powder intensity relative to urea, NLO
coefficients for SHG, its transparency cutoff, and the
figure of merit for SHG if known.   Other aspects are

Table I. Properties of Selected Organic NLO Crystals

Compound[a]	β[b]	SHG[c]	Space Group	Point Group	d Coefficient(s)[d]	Cutoff[e]	FOM
BMC		10-70	Pc	m	$d_{13} \sim 90$	420	–
COANP		50	Pca2$_1$	mm2	$d_{32} \sim 32$ $d_{31} \sim 15$ $d_{33} \sim 14$	470	100
DAN	36[g]	115	P2$_1$	2	$d_{eff} \sim 27$	~500	420
MAP	22	~5	P2$_1$	2	$d_{21} \sim 17$ $d_{22} \sim 18$ $d_{23} \sim 4$	~500	–
MBA-NP		25	P2$_1$	2	$d_{22} \sim 28$	430	–
MMONS	18 26[h]	750-1250	Aba2	mm2	$d_{33} \sim 140$ $d_{32} \sim 34$ $d_{24} \sim 82$	510	850
MNA	16 9.5[h]	22	Ia	m	$d_{11} \sim 67$	500	–
NPP	45[g]	50-150	P2$_1$	2	$d_{21} \sim 84$ $d_{22} \sim 29$	500	–
PNP		140	P2$_1$	2	$d_{21} \sim 48$ $d_{22} \sim 17$	470	–
POM	8.5	13	P2$_1$2$_1$2$_1$	222	$d_{25} \sim 10$	450	–

a. See Scheme I for structures
b. In esu x 10^{30} at 1.06μ unless noted.
c. Powder SHG vs urea.
d. In pm/V

e. Long wavelength cutoff in nm
f. Figure of merit, d_{eff}^2/n^3, in (pm/V)^2.
g. From Paley et al. (42).
h. At 1.9 μ.

discussed separately below.   Scheme I shows structures of
the materials listed in Table I.
       BMC (4-bromo-4'-methoxychalcone)   (*39*) is a
monoclinic material (lattice constants a =   15.898 Å, b =
7.158 Å, c = 5.983 Å; β = 97.19°) with a melting point (mp)
of 159.5° C.   It is yellow, with total transparency from
about 420 nm to near 2 microns, though the absorption is
anisotropic.   Powder SHG values depend on the solvent from
which the material is crystallized.   Type I phase matching
at 1.06 μ has been exhibited with d_{eff} ~ 90 pm/V. Crystals
are grown from solution by slow cooling over one month to
give material of about 4x10x10 mm^3.   The laser power
damage threshold is stated to be high without supporting
data.
       COANP (2-N-cyclooctylamino-5-nitropyridine) (*40*) is
a yellow,   orthorhombic crystal (a = 26.281 Å, b = 6.655
Å, c = 7.630 Å), mp = 70.9° C, transparent from about 470
nm to 1.6μ.   The Sellmeier equations (which describe the
frequency dependence of the refractive index) have been
solved for this material (at 1.064 μ, n_x = 1.638, n_y =
1.717, n_z = 1.610, and at 0.532 m, the values are 1.710,
1.872 and 1.697, respectively).   Type I SHG fails below
1.02 μ.   Noncritical phase matching SHG along the a-axes
of the crystal employs the full d_{32}, while along b it
utilizes d_{31}.   (The crystal a, b and c axes are
respectively the dielectric x, y and z axes.)   At 1.064 μ,
d_{eff} is ~24 pm/V.   Crystals are grown from either ethyl
acetate or acetonitrile by gradient mass transport methods
to give plates of ca. 10x5mm^2 several mm thick.
       DAN   is   4-(N,   N-dimethylamino-3-
acetamidonitrobenzene) (*41*) and it is a monoclinic crystal
(a = 4.786 Å, b = 13.053 Å, c = 8.732 Å, b = 94.43°).(*42*)
The crystal b axis is parallel to the optical y axis and z
is inclined about 40° from c in the xy plane.   Refractive
indices have been determined at 546 nm: n_x = 1.658, n_y =
1.712 and n_z = 1.775.   Type I phase matched SHG has been
demonstrated down to 954 nm input (477 nm out) where
absorption prevents further advance.   At 1.064 μ, d_{eff} is
27 pm/V.   Melt growth techniques are used to produce
crystals of size 10x5x2 mm^3.   Workers at British Telecom
have pioneered the growth of DAN in optical fiber cores
for use in integrated optics.(*43*)
       MAP (2,4-dinitrophenyl-(l)-alanine methyl ester)
(*44,45*)   is   a   comparatively   older   material   which
crystallizes on slow cooling from ~34% ethyl
acetate/hexane solutions over about three weeks as ~0.2
cm^3 monoclinic crystals (A = 6.829 Å, b = 11.121 Å, c =
8.116 Å, b = 95.59°), mp 80.9° C.   The yellow-orange
crystals are transparent from about 500 nm to 2.0 μ; at

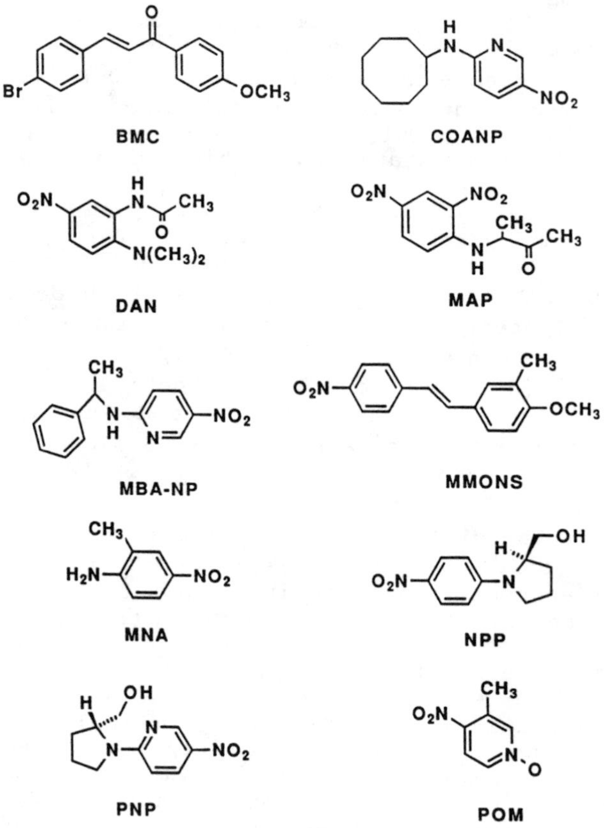

BMC

COANP

DAN

MAP

MBA-NP

MMONS

MNA

NPP

PNP

POM

Scheme 1

532 nm the absorption coefficient is 3.7 cm^{-1}. The Sellmeier equations are available for this material (at 1.064 μ, n_x = 1.5078, n_y = 1.5991, n_z = 1.8439, and at 0.532 μ, the values are 1.5568, 1.7100 and 2.0353, respectively). The optical z-axis is inclined ~36° from the crystallographic a-axis. Type I phase matching is possible at 1.06 μ (d_{eff} = 16 pm/V); type II phase matching is also possible (d_{eff} = 8.8 pm/V). At 1.06 μ, laser damage occurred at fluences of ~3 GW/cm^2 for 10 ns pulses.

The nitropyridine derivative **MBA-NP** [(-)-2-(α-methylbenzylamino)-5-nitropyridine] (*45,46*) is a chiral molecule which crystallizes in the monoclinic form (a = 5.392 Å, b = 6.354 Å, c = 17.924 Å, with β = 94.6°). It is very pale yellow, and transparent from 430 nm into the infrared at 1.8 μ. Damage thresholds near 1 GW/cm^2 are quoted at 1.06 μ where Type I phase matching occurs. Excellent crystals (up to 7x5x5 cm^3) of this material are reported to grow from methanol by slow cooling.

MMONS (3-methyl-4-methoxy-4'-nitrostilbene) (*47*) has the highest reported powder SHG value (1250 x urea) among simple stilbene derivatives, but its measured β value is relatively modest, 18-26 x 10^{-30} esu (dependent on wavelength of measurement). It was originally proposed that the high SHG activity was the result of extended interactions among stacked **MMONS** aggregates present in the crystal (*47*), but later results discount this possibility. (*48*) The powder SHG value observed depends strongly on the solvent from which it is crystallized. The orthorhombic orange crystal (a = 15.584 Å, b = 13.463 Å, c = 13.299 Å), mp 114° C, begins to absorb strongly at about 510 nm, and it has been fully characterized for SHG. (*48*) The crystal, in point group mm2, has its crystallographic axes coincident with the principle dielectric axes (x, y and z correspond to a, b and c). The Sellmeier equations are available (at 1.064 μ, n_x = 1.530, n_y = 1.642, n_z = 1.961, and at 0.532 μ, the values are 1.597, 1.756 and 2.312, respectively). The crystal is strongly birefringent and is Type II phase matchable across its transparency range, with d_{eff} ~ 68 pm/V at 1.06 μ. Conversion efficiencies (defined = d_{eff}^2/n^3) are ca. 120x KTP (potassium titanyl phosphate, *vide infra*). However its use is limited because of large "walkoff," which is the divergence of the fundamental and the doubled frequencies as radiation passes through the medium caused by the phase mismatch. In the case of MMONS it is a consequence of the large birefringence of the material. Crystals of over 1 cm^3 can be grown from chloroform/ethanol solutions by slow evaporation or by slow cooling. The FOM listed in Table I is the highest reported to date for a single crystal organic material. The electro-optic coefficients have also been determined (e.g., r_{33} = 40 pm/V). The extreme ease of crystal growth of **MMONS** may spur a search for other uses.

MNA (2-methyl-4-nitroaniline) (*49, 50, 51*) is another mature material which has been available since 1975 in Russia, and slightly later in the US. The material played a historical role in helping to establish the connection between molecular charge transfer and enhanced hyperpolarizability.(*44*) It is monoclinic (a = 11.57 Å, b = 11.65 Å, c = 8.22 Å), mp 131° C. The optical y axis is parallel to the crystallographic b axis, with -x inclined 68° from a and +x inclined 34.7° from c. At 1.06 μ, the yellow crystals have absorption coefficient about 1 cm^{-1}. The Sellmeier equations are available for **MNA**: at 1.064 μ, n_x = 1.763, n_y = 1.514, n_z = 1.359, and at 0.532 μ, the values are 2.291, 1.843 and 1.569, respectively. It is highly birefringent (n_x - n_y = 0.249 and 0.448 ; n_y - n_z = 0.155 and 0.274; n_z - n_x = 0.404 and 0.722 at 1.064 μ and 532 nm, respectively), and Type I phase matchable at 1.06 μ. The measured electro-optic coefficient is ~67 pm/V. Unfortunately the material is very difficult to grow in bulk form, though thin films have been grown from the melt.

NPP [N-(4-nitrophenyl)-(s)-prolinol] (*52*) is the levorotary isomer of this chiral nitroaniline analog. It is monoclinic (a = 5.275 Å, b = 7.178 Å, c = 14.893Å, β = 109.2°), mp 115.9° C. It is Type I phase matchable at 1.15 μ using d_{21}, and at 1.24 μ with d_{eff} ~ 42 pm/V. Crystals are grown from the melt by Bridgman methods to yield crystals of several cm^3. Femtosecond frequency conversion has been demonstrated in thin plates of **NPP**.

PNP [2-(N-(s)-prolinol)-5-nitropyridine] is another prolinol derivative, but in the nitropyridine family.(*45*) It is also monoclinic (a = 5.182 Å, b = 14.964 Å, c = 7.045 Å, β = 106.76°), mp 83° C. Its absorption is strongly anisotropic along various crystal axes: it is transparent from 0.49 to 1.21 μ when the electric field vector is aligned along either x or y, and from 0.47 to 2.3 μ with **E** along z. The Sellmeier equations are available (at 1.064 μ, n_x = 1.880, n_y = 1.732, n_z = 1.456, and at 0.532 μ, the values are 2.177, 1.853 and 1.471, respectively). Type I phase matched SHG was observed at 1.064 μ with d_{eff} ~ 16 pm/V. Crystals 0.5 cm^3 in size result by slow cooling methanol/water solutions.

POM (3-methyl-4-nitropyridine-N-oxide) (*53*) is the only commercially available organic crystal for SHG, other than urea. Crystals ca. 1 cm^3 in size are formed by solution growth from a variety of solvents. It has been very thoroughly characterized. It is orthorhombic (a = 21.359 Å, b = 6.111 Å, c = 5.132 Å), mp 136.2° C. It is transparent in the range 0.45 to 1.6 μ, and the crystallographic a,b and c axes correspond to the optical x, y and z axes, respectively. Sellmeier equations are available (at 1.064 μ, n_x = 1.663, n_y = 1.829, n_z = 1.625, and at 0.532 μ, the values are 1.750, 1.997 and 1.660, respectively). It is Type I phase matchable (d_{eff} = 31 pm/V^2), and Type II phase matchable (d_{eff} = 7.8 pm/V^2). Damage thresholds are reported to be ~1GW/cm^2 for nanosecond pulses of 1.064 μ light.

Many other organic and some organometallic materials have been examined by the powder SHG method and/or tested by EFISH techniques. They will not be discussed in detail here but are listed by structural classes for completeness. Several compendia of materials responses have been published. (*38,54,55*) Clearly, the largest single class of second order materials consists of donor-acceptor substituted aromatics. The class has been extended to stilbenes and diarylacetylenes and diacetylenes. (*56,57,58*) Biaryls have been studied. (*59*) Many other acceptor moeities have been described other than the now-classic nitro group: A variety of salts, especially pyridinium ions can function as acceptors. (*60,61*) Fulvenes, (*62*) dicyano- (*63*) or tricyanovinyl (*64*) groups, sulfones, (*65*) enone, (*66*) or imine (*67*) (Schiff base) fuctionallity can also be used. Donors other than the simple but effective amino or dimethyl amino group can also be employed. The virtue of the simple methoxy group is illustrated by the high SHG of **MMONS** (*47*) (*vide supra*). Various heterocyclic groups have been employed as donor groups: oxazole, (*68*) dithiafulvene, (*69*) and dihydropyrazole. (*59*)

In addition, various other classes of molecules have proved to be of special value. Amino acids are naturally occurring chiral species, some of which possess interest for NLO, especially arginine derivatives such as **LAP**, (*70*) and alanine derivatives and selected polypeptides containing nitrophenyl side groups. (*71*) Intermolecular interactions in halo-cyano substituted benzenes result in favorable crystal structures and make this class of aromatics attractive. (*72*) A growing number of organometallic compounds have been demonstrated to have good SHG powder intensities. (*73,74*) Metal-pyridine and -bipyridine complexes are effective for SHG. (*75*) Pd and Pt zero valent compounds were studied by Tam (*76*) and it was shown that the group (R$_3$P)$_3$M-, M = Pt or Pd, is a mimic for methoxy as a donor in D-A aromatics. The use of ferrocene as a donor in D-A aromatic analogs has also been

reported.(77)   The hyperpolarizability of metalloarene and
other   metal   complexes   has   been   demonstrated   by
incorporation of the complex as a guest into an inclusion
host   to   provide   polar   orientation   and   thus   SHG
capability.(78,79)

Polymeric and Thin Film Organic Materials. Second order NLO
applications that require crystalline materials limit the
scope of molecular types that can be employed to those
that   crystallize   in   acentric   space   groups.   Various
strategies have been reported to engineer molecules into
acentric   arrangements   so   that   bulk   second   order
susceptibility   can   be   usefully   expressed.   Chiral
auxiliaries   can   be   used   to   break   centers   of   symmetry.
Functional groups can be introduced into a basic nonlinear
optiphoric structure in order to encourage formation of
acentric crystals through intermolecular forces such as
the halo-cyano (72) or hydrogen bonding (48) interactions.
Several organic (74,78,79,80,81,82) or inorganic (83) guest-
host   inclusion   complexation   paradigms   can   be   used   to
orient a guest in a polar fashion in three-space.   But the
most widely practiced method of inducing polar alignment
is the poled polymer approach.(84,85,86)   Thin film samples
of the polymeric medium, containing a nonlinear additive
(either as a solute or as a chemically bound constituent
of the medium) which contributes hyperpolarizability, are
coated on an electrically conducting substrates.   The
medium is softened by heating above its glass transition
temperature, and an electric filed is applied.   The field
can be applied either by simply placing an applied voltage
across the medium using electrodes, or a corona charging
station,   similar   to   that   used   in   electrophotographic
copiers, can be used to apply charged particles to the
insulating surface to create the potential field.   The
hyperpolarizable guests align in the applied field, and
the   alignment   is   locked   in   by   lowing   the   medium
temperature below the glass transition point while the
medium is being poled.   To date, greatest success has been
attained using corona poling.   Additives which are
chemically bound to the polymer medium are found to
reorient slower than those which are simply dissolved in
the medium.
      Le Grange and coworkers (87) provide a general
discussion of the relation of three dimensional order and
the $\chi^{(2)}$ phenomenon.   The concepts developed are applied to
various ordered systems, including doped, poled polymer
glasses,   liquid   crystals,   liquid   crystal   side   chain
polymers and Langmuir-Blodgett films.   These authors also
discuss some of the limitations of the systems.   An
excellent review of the state of the art up until about
1987 is provided by a pair of monographs which present
papers given at an ACS Polymer Science: Materials and
Engineering   Symposium   on   Electronic   and   Photonic

Applications of Polymers.(*88*)   Sohn and coworkers have
reviewed the AT&T efforts recently.(*89*)

Several pure polymers have interesting $\chi^{(2)}$
properties. Poly(vinylidene flouride), PVF_2, and its
copolymers with poly(trifluoroethylene), PVF_3, have
piezoelectric properties, and are also capable of SHG,
which can be enhanced by poling the material in an
electric field.(*90*)   Similar effects are observed in
alternating copolymers of vinyl acetate and vinylidene
cyanide, co(PVCN-PVAc).(*91*)   The nonlinear polarization in
the latter copolymer is thought to reside in the CN side
groups of the vinyl acetate, while it is the highly polar
CF_2 group which contributes to polarization in the
fluorocarbon systems.

IBM workers (*92*) have prepared amorphous glasses
containing p-nitroaniline side groups on a poly(ethylene)
backbone and demonstrated SHG with d_{eff} = 31 pm/V after
corona poling above the glass transition temperature (125°
C) and subsequently cooling while still in the applied
field. After five days the activity had relaxed to 19
pm/V and was reported to remain stable at that figure.
The effect of chemically crosslinking a similar, but not
chemically identical, system containing **PNA** groups as the
nonlinear optiphore has also been reported.(*93*)   No
detectable decay of the initial SHG capability was noted
on standing even at temperatures above room temperature
(85° C). The crosslinking was provided by reacting 3,4-
diaminonitrobenzene with a diglycidylether of Bisphenol A
to give a presumably two-dimensionally crosslinked
network. A side chain poly(propylene) based polymer with
pendant p-nitrobiphenyloxy groups has been studied and
patented by Hoechst Celanese workers,(*94*) who have also
claimed systems comprised of poly(siloxane), poly(ester)
poly(amide) and other non-$(CH_2)_n$- backbones,(*95*) while
Kodak has patented a variety of systems containing
different donor-acceptor side chain compositions.(*64,96*)
Several main chain polymer compositions have been revealed
in the patent literature, especially those based on
poly(benzthiazole-phenylene) derivatives.(*97,98*)

Several unusual effects have been reported in
polymeric thin films. Wang (*99*) has shown that the head-
to-tail polycrystaline aggregate which forms when
polycarbonate films containing 2,6-diphenyl-4-(p'-
diethylaminophenyl)thiapyrilium tetrafulorobrate are
induced to crystallize is an extremely efficient source of
SHG. A thin film (5 μ) containing only 1.5 weight percent
of the dye can produce an SHG signal equivalent to 0.25 mm
of urea. Khanarian reports in a patent (*100*) that
inclusion of colloidal silver in a typical side chain
poled polymer composition enhances the SHG response of the
system. The enhancement mechanism may be similar to
surface enhanced resonance Raman effects.

A new potential area for exploitation of SHG is pure organic glass-forming materials. A preliminary report of the first example in this class has recently appeared from IBM laboratories.(*32*) Two initial examples were given. (S)-2-N-α-(methylbenzylamino)-5-nitropyridine (**MBA-BP**) and 2-N-(cyclooctylamino)-5-nitropyridine (**COANP**) both form glasses when cooled rapidly from above the melting temperature. If poling is conducted above the glass transition temperature, orientation is effected. Cooling freezes in polar order. Since no dilution effect occurs in pure glasses as it does in polymer host-guest glasses, this approach seems to offer promise of attaining high SHG levels.

At least one patent has issued dealing with Langmuir-Blodgett (L-B) films containing polarizable side groups. Uniaxial films of 7,7-bis(hexadecylamino)-8,8-dicyanoquinodimethane can be cast which exhibit SHG.(*101*) Orientation is assured by alternating successive transferred layers of the polar substrate with arachidic acid layers. A β value of 500x10^{-30} esu is claimed for the quinodimethane, but no information about SHG of the L-B films is provided.

Meredith has used the experimentally observed quadratic increase of SHG intensity with increasing layer thickness in an L-B film as a proof that the monolayer has the Z deposition structure.(*102*)

Inorganic Materials. Inorganic materials have been studied for NLO for a longer period than have organics, and they are consequently more advanced toward commercial application. In fact, most of the commercial materials available for frequency conversion are inorganic. Because they are more mature, inorganic second order materials will not be reviewed in depth. Rather, their merits will be compared and contrasted tabularly. Compared to organics, inorganics are generally more rugged, many allow anisotropic ion exchange which can be used to provide waveguide structures, and they have better purity in crystalline form than organics. For waveguide fabrication, it is easier to prepare thin film inorganics than films of crystalline organics. Organics, however, have lower dielectric constants and it is generally easier to prepare analogs of organic systems when an active class is found.

Table II lists the nonlinear coefficient, damage threshold and transparency cutoff for the most well known of the inorganic materials. Urea is included for comparison. As in Table I, materials are listed alphabetically. The data in Table II is taken from a tabulation by Bierlein (personal communication) and from one by Lin and Chen.(*103*)

These inorganics represent the current workhorse materials of NLO. Because they are more mature than

Table II. Properties of Selected Inorganic NLO Crystals

Compound	d Coefficient(s)[a]	Damage Threshold[b]	Cutoff[c]	FOM[d]
β-BaB$_2$O$_4$	$d_{11} = 4.1$	10000 MW/cm^2	198	26
Ba$_2$NaNb$_5$O$_{15}$	$d_{31} = 32$	1	370	
KH$_2$PO$_4$	$d_{36} = 1$	500	200	1
LiB$_3$O$_5$	$d_{32} = 3.1$	2000	165	~1
LiNbO$_3$	$d_{31} = 13$	20	400	
LiIO$_3$	$d_{31} = 10$	50	300	50
KTiOPO$_4$	$d_{31} = 15$	20000	350	215
Urea	$d_{14} = 3$	5000	210	6

a. Relative to KDP as 1. The d_{36} coefficient is quoted as 0.39 pm/V in Eimerl, D. Ferroel., 1987, 72, 95.
b. Damage thresholds are not absolute, since experimental conditions can be different.
c. Short wavelength absorption cutoff in nm.
d. The figure of merit, relative to KDP, is a function of the d coefficient, the acceptance angle for phase matchability ($\Delta\theta$), crystal length, L, wavelength of input, field strength and material index of refraction: FOM ~ $(d^2/n^3)(EL/\lambda^2)(\Delta\theta)^2$.

organic materials, they will be difficult to displace, technologically, as new devices are developed. For many of the materials listed, commercial samples are available and the materials are used in research or commercial medical or graphic arts laser systems. For some of the materials, waveguide fabrication techniques have been developed, and intergrated optic devices have been fabricated. However, the inherent diversity of organics, and their lower dielectric constants, which enhances their figures-of-merit for electrooptic applications compared to inorganics, provides the organic class of materials with some formidable attributes with which to challenge inorganics in the future.

Third Order Materials

Third order NLO effects include the Kerr effect, optical bistability, optical phase conjugation, photorefractivity, and third harmonic generation (THG). Practical applications of these effects are in optical switching, amplification, beam steering and clean-up, and image processing. The promise of "optical computing" was stated early in the 1980's by several authors.(104,105,106) Digital, serial uses of optics for computation may not be practical competition for cheaper, less power hungry electronic devices for many years, if ever, but optical image processing (107) has been a vigorous field for many years, and it has gained new impetus as new materials and processes are discovered and invented.

The main components of an "optical computer" are light sources, beam steering materials and light modulators. Semiconducting lasers are judged to be the ultimate light source, though most prototype devices are demonstrated at other wavelengths. Beam steering materials can be "fixed" (such as the holographic elements discussed above), or "reconfigurable," that is programmable. Reconfigurable holographic optical elements are known as photorefractives. They can be viewed as a class of phase conjugate materials, though actual physical mechanisms of action may not be identical. The time scale for the realization of the photorefractive effect is important: slow materials would be nearly permanent optical memories, while fast photorefractives with rapid decay would be transient, the optical equivalent of an electronic DRAM. Spatial light modulators (SLM's) spatially modify the phase, amplitude, polarization and/or intensity of light which passes through them. They have some properties similar to those of photorefractives, which can be used for modulation purposes. However, most SLM's today are electro-optic devices (liquid crystal light valves, optically addressed microchannel plate devices, electro-optic crystals, electrically actuated membranes or mirrors, etc.) which are relatively slow in operation (microseconds at the fastest). All-optical SLM's are desired but practically unknown.

This review will not discuss materials according to their application, since few materials have been demonstrated in practical devices, but will rather describe the third order properties generically. No symmetry requirements exist to express third order susceptibility, so that all kinds of materials are potential candidates: liquids, thin films, glasses and crystals. Eventually, practical considerations such as fabricability may become more important than the magnitude of the nonlinearity. Effects may be grossly classified as those which influence local index of refraction (so-called n_2 effects) and those which influence the absorption of a material (resonant or photochromic effects). Under action of either effect the presence of one light beam in a material influences the propagation of a second. For resonant effects, optical absorption changes so that the intensity of light is modulated, while for resonant or nonresonant n_2 effects, the phase of the transmitted signal varies. Third harmonic generation experiments performed far from resonance measure the inherent ability of a material to respond in a third order fashion.

Nonresonant Third Order Materials. The nonresonant Kerr effect is among the most rapid-response NLO effects known, capable of sampling "instantaneous" intensity of an optical pulse (femtosecond to picosecond response). However, high laser powers (MW to GW/cm^2) are required to induce the effect. It is now well established that the magnitude of the effect is proportional to the volume density of chemical bonds in the material and to their orientation with respect to the optical field. Thus, materials with the highest n_2 values are those with the largest number of conjugated, polarizable bond structures, e.g., poly(diacetylenes).(108) Glass discusses a variety of organic and inorganic materials and compares their n_2 coefficients.(15) A prototype, simple, organic n_2 material is CS_2, which has a quoted n_2 coefficient of 8 $x10^{-14}$ cm^2/W, which is ~100x more responsive that silica, but ~70x less active than the poly(diacetylene) PTS (1,4-bis-p-toluenesulphonyl1,3-dibutadiyne); the organic **MNA** (20x carbon disulfide), and the inorganic $LiNbO_3$ (~0.3x CS_2) are intermediate in response.(15) Among the best inorganics are semiconductors such as GaAs (12x) and InSb (600x CS_2), a reflection of the polarizability of conduction band electrons. The semiconductor responses, however, are limited to long wavelengths (>10m). It should be remarked that in this discussion, the n_2 of CS_2 is not a true electronic effect, but in fact an orientational effect. The other materials referred to above respond electronically.

A growing number of reports are appearing concerning $\chi^{(3)}$ of materials as determined by THG experiments. Among organics a variety of poly(diacetylenes), e. g., 4-BCMU (4-butoxycarbonylmethylurethane polydiacetylene) (15,109) have been studied as pure materials and as LB films, (110) Crystalline films of poly(4-BCMU) in the red form, were found to have higher $\chi^{(3)}$ than amorphous films, attributed to the orientational effect of crystallization. (109) The blue form also has been studied. (111) Maximum values of $\chi^{(3)}$ reported to date are ~1 x 10^{-10} esu at 1.3μ. Poly(phenylacetylene) THG at 1.06μ has recently been measured. (111) The 2- and 3-photon resonance enhanced value of $\chi^{(3)}$ determined is 7 x 10^{-12} esu.

Some simple organometallics have been studied. Poly(silanes) and germanes exhibit very interesting behavior, since they are photochromic and appear to possess excitonic, charge delocalized excited states involving the *sigma* electrons of the organometallic backbone. (112) THG measured susceptibilities of up to 1 x 10^{-11} esu (1.9μ) were reported by IBM workers. (113) Frazier and coworkers have measured molecular γ of some poly-yne platinum complexes using THG. (114) Metalated phthalocyanines, (115) such as chloro gallium- or fluoro aluminumphthalocyanines have $\chi^{(3)}$ = 0.2-0.5x 10^{-10} esu. An organic [poly(benzimidazole)]-inorganic (silver colloid suspension) composite is reported in a patent to exhibit enhanced THG over that without the metal inclusion. An enhancement of 1.3 at 1.2μ is quoted. (98) The nonresonant n_2 values for several metallocenes (hafnacene, ruthenocene and ferrocene) have been measured and reported (116) to be near that for nitrobenzene.

For inorganic compounds, nonresonant effects reported to date occur among either semiconductors (II-VI or III-V materials) or some of the colorless workhorses of second order materials, e.g. β-BaB_2O_4 (**BBO**). Bhar *et al*. demonstrated THG in the latter material as well as sum-frequency mixing as methods to produce UV wavelengths using **BBO**. (117,118) Semiconductors such as GaAs ($\chi^{(3)}$ ~4.8 x 10^{-11} esu) or Si (2.4 x 10^{-11} esu) were reported early. (119) Quantum well (supperlatice) structures of GaAs-GaAlAs are anticipated to have enhanced nonresonant $\chi^{(3)}$ values. (120)

Related to the quantum well structures, "quantum dot" materials, or size-quantized semiconductor particles, have also been recognized to have nonresonant $\chi^{(3)}$ properties that are attractive. A series of small (< 30 Å) capped (thiophenolate) CdS clusters has recently been shown to exhibit $\chi^{(3)}$ on THG from 1.9µ with values ranging from 0.05 to 3.3 x 10^{-10} esu, with larger clusters providing the highest susceptibility. (*121*) Particles of CdS with sizes 30-60 Å have been suspended in polymer glasses and examined by THG with exciting light from 1-1.5µ. (*122*) Larger cluster size glasses can be produced by conventional glass technology, and these doped glasses also exhibit THG. With an As_2S_3 glass, (*123*) a $\chi^{(3)}$ value of 2.2 x 10^{-12} esu was found. The earliest work was with CdS_xSe_{1-x} glasses. (*124,125*) Other semiconductors can also be doped into glasses, e.g. TiO_2. (*126*)

Resonant Third Order Materials. A variety of NLO effects are classified as resonant third order effects. Degenerate four-wave mixing (DFWM), optical limiting, optical bistablity, photorefractivity, optical phase conjugation, etc. all operate on the basis of an intensity dependent change in the refractive index of the absorbing material. Glass, (*15*) in a review of materials for optical information processing, compares many of the important materials discovered before 1983 by quoting the nonlinear absorptive n_2 coefficient. By this classification, GaAs quantum well structures are found to be the best among a variety of absorbing materials ($n_2 \sim$ 4 x 10^{-3} cm^2/W at 850 nm). Bulk GaAs is less active ($n_2 \sim$ 2 x 10^{-5} cm^2/W at 870 nm) (*15*) than the quantum well structures. Semiconductors of the II-VI family are also known to be good n_2 materials, e.g, the coefficient for absorption by CdS excitons (490 nm) is quoted by Glass as 1.6 x 10^{-6} cm^2/W. (*15*) Values for these materials are substantially higher than those for materials where the excited state involved in the resonance process is not highly delocalized. For example, sodium vapor has $n_2 \sim$ 5 x 10^{-11} cm^2/W at 590 nm. (*15*) Response times ("turn-on time) can be very rapid (picoseconds) since the origin of the effect is electronic. The absorptive n_2 effect persists for the lifetime of the excited state responsible for the nonlinear response. Thus, the "turn-off time" is related to the lifetime of the material; these values vary from ~20 ns (GaAs) to 2 ns (CdS exciton) for the materials described above.

A number of other materials have been discovered since 1983. Among inorganics, the best known are the optical glasses sold commercially by Schott and Corning as cutoff filters in the visible region. Originally reported by Jain and Lind, such CdS_xSe_{1-x} doped glasses (*127,128*) are extremely interesting composite materials. Their optical nonlinearities as determined by phase conjugation or power dependent optical bleaching are relatively large ($\chi^{(3)} \sim$ 10^{-8} esu) and they have fast response times (picoseconds). An absorption-normalized nonlinearity value, α_2/α_0, (that is, the nonlinear absorption coefficient) is a numerical datum which may have more value than an n_2 value or a $\chi^{(3)}$ value. Peyghambarian and coworkers report an a_2/a_0 value of -1×10^{-7} cm^2/W for the Corning 3-69 glass filter. (129) Because these materials are true glasses, they can be fabricated into shapes and polished in the same way as lenses and other optical elements are. However, the materials are susceptible to optical damage and the damage mechanism is intimately connected to the electronic mechanism which governs the optical nonlinearity. (130)

A new class of optical glasses which may circumvent these difficulties has been described by Wang, Herron, Mahler and coworkers. They have prepared size-quantized semiconductor particles, of sizes ultimately smaller than the commercial glasses, in a variety of polymeric and other porous host structures. Semiconductors of the II-VI class have been dispersed in poly(ethylene)-poly(acrylic acid) copolymer or Nafion® perfluorosulfonic acid polymer matrices and shown to have DFWM capabilities. (131) The nonlinearity of 50 Å CdS clusters composite at 505 nm was reported to be about one-half that of the commercial 3-69 glass at 510 nm. Later experiments determined α_2/α_0 to be -6×10^{-7} cm^2/W at 480 nm in the CdS exciton band, (132) a value ~6x larger than that for the 3-69 glass. A potential virtue of these composites over the pure glass composites is that there is considerable control over the chemical environment of the semiconductor cluster in the polymer system. Since surface electronic states dominate the linear and nonlinear optical properties of these materials, Wang and Herron suggest that manipulation of surface chemistry will be able to modify electronic properties in beneficial ways. (131)

Among organic materials, the premier material is undoubtedly poly(acetylene), **PA**. A commercial thermal precursor to **PA** is British Petroleum's "PFX" series of polymers. (133) These are based on the "Feast route" to **PA** (134) and represent a convenient way to prepare thin films of **PA** for integrated optics studies. The nonlinearity of stress-oriented **PA** has been documented to be as high as $\chi^{(3)} \sim 10^{-8}$ esu polarized along the polymer axis, arguably the highest nonresonant value reported to date, while its resonant nonlinearities rival those of inorganic semiconductors. Response times are in the femtosecond range. (135)

Poly(diacetylenes) also exhibit large $\chi^{(3)}$ values in resonant modes. Optical phase conjugation (DFWM) was demonstrated at 532 nm in solutions of a soluble diacetylene polymer, with $\chi^{(3)} \sim 5 \times 10^{-12}$ esu. (136) No saturation was observed up to 5GW/cm^2 in this study. However, the effect was judged to be a thermal grating effect, not an electronic response. This conclusion has been confirmed by time-resolved infrared and resonance Raman studies on a similar system. (*137*) Similar, thermal grating effects are reported in nematic liquid crystal compositions for use as phase conjugate mirrors. (*138*)

Another class of conducting polymer that has been examined for resonant nonlinear response are the poly(thiophene) polymers. DFWM has been reported for poly(alkylthiophene) and for several model oligomers. (*139,140*) $\chi^{(3)}$ values from DFWM are in the range 10^{-11}-10^{-10} esu at 1.06 μ. The values of $\chi^{(3)}$ increase quite rapidly with increasing degree of polymerization, but high molecular weight species, not yet obtained experimentally, would be required to produce useful nonlinearities. Jenekhe and coworkers reported a series of soluble precursor block copolymers consisting of separate segments of poly(thiophene) and its quinoid isomer poly(2,5-bismethylenethiophene). (*141*)   The reported $\chi^{(3)}$ values, measured by picosecond DFWM at 532 nm are in the range 10^{-8}-10^{-7} esu.

Since semiconductor materials appear to be general candidates for third order effects, it is not surprising that photoconductive charge transfer complexes such as poly(vinyl carbazole)-trinitrofluorenone (PVK-TNF) would exhibit modest DFWM. (*142*)   The observed nonlinearity at 602 nm, a wavelength absorbed by the CT complex, ranges from 0.2 - 2 x 10^{-11} esu, increasing as the molar fraction of TNF in the complex, PVK:TNF, increased. Charge transfer excitons are judged to be responsible for the resonant $\chi^{(3)}$.

Finally, solutions of organic dyes, especially dispersed in polymeric matrices, can function as resonant $\chi^{(3)}$ materials. Early examples are provided by Fayer. (*143*) A recently reported example is of rhodamine 6G in boric acid glasses. (*144*) Concentrated solutions of some dyes, e.g., fluorescein, are also known to be capable of optical bistability. (*145,146*) Dirk and Kuzyk report $\chi^{(3)}$ values for several azo dyes at 633 nm and for a squarylium dye at 799 nm dispersed in Poly(methylmethacrylate) glasses, and compare the results (obtained by electrooptic methods) to those for a model poly(thiophene) and a poly(diacetylene). (*147*)

Photorefractive Materials

The photorefractive effect is classified here as a special third order effect for several reasons. First, it is perhaps the least well understood, mechanistically. Second, it represents the area of greatest current opportunity for chemists among NLO effects today, since there are very few materials know to be photorefractive, and only one (possible) organic material among them. Understanding and improvement of photorefractive behavior is a prerequisite for successful implementation of many optical computing concepts, especially for all-optical systems.

Photorefractivity was first discovered (*148*) as an optical damage effect in LiNbO3. The effect is similar to optical phase conjugation (in fact it is sometimes called self-pumped phase conjugation). During the process, which occurs in crystalline materials, index of refraction patterns are developed in the internal structure of the subject material, so that input light is diffracted as it passes through the material. The index variations are developed by interference of the light as it is reflected internally throughout the crystal. An entertaining and enlightening review which presents the phenomenology of the photorefractive effect is given by Feinberg. (*18*)

The most useful of the known photorefractives are LiNbO3 and BaTiO3. Both are ferroelectric materials. Light absorption, presumably by impurities, creates electron/hole pairs within the material which migrate anisotropically in the internal field of the polar crystal, to be trapped eventually with the creation of new, internal space charge fields which alter the local index of refraction of the material via the Pockels effect. If this mechanism is correct (and it appears established for the materials known to date), then only polar, photoconductive materials will be effective photorefractives. However, if more effective materials are to be discovered, a new mechanism will probably have to be discovered in order to increase the speed, now limited by the mobility of carriers in the materials, and sensitivity of the process.

Potential applications of photorefractive materials are manifold. To date, demonstrated effects include real time holography, correlation filtering, and various "novelty' filter applications, one of which is the development of a microscope which distinguishes moving objects (such as living cells) from a stationary background.(*149*) The latter application employed BaTiO3 as the active material.

A general review of photorefractive materials was presented in 1988.(*150*) Also, two monographs in were published which detail theory, physical characterization and practice of the use of known photorefractives.(*151*) Three classes of inorganic materials dominate. Ferroelectric oxides, such as LiNbO3 and BaTiO3 mentioned above; compound semiconductors such as GaAs and InP, and the sillenite family of oxides, exemplified by Bi12SiO20 and Bi12TiO20. The semiconductors are sensitive only in the infrared, while the other materials operate in the visible. Many of the oxide materials are commercially available, though the available crystals have not been optimized for photorefractivity. The review by Hughes workers cited above should be consulted.(*150*)

Examples of organic photorefractives crystals, in the sense of the functional inorganic crystals cited above, are not actually known. However, Russian workers (*152*) report that a pyridinium ylide experiences reversible photorefraction which is attributed to local polarization caused by trapping of photoinduced charges at structural defects in the crystal. Russian workers have also reported reversible phase recording in liquid crystal layers by spatially modulating an adjacent organic photoconductor layer.(*153*) However, this hybrid device is not a true photorefractive but rather is a variation of a spatially addressed light modulator.

Conclusions

A variety of efficacious NLO materials now exist, and some are generally available. However, improvements are required in several classes of materials before applications will be feasible, and new materials opportunities still exist. There are also several areas where increased understanding of the fundamental structure-property relations among NLO materials is needed. Chemists will be important contributors if materials for NLO are to be successfully implemented.

1 Franken, P. A.; Hill, A. E.; Peters, C. W.; Weinrich, G. Phys. Rev. Letts., 1961, 7, 118.
2 Bloembergen, N., Nonlinear Optics; Benjamin: Reading, MA, 1965.
3 Giordmaine, J. A.; Miller, R. C. Phys. Rev. Letts., 1965, 14, 973.
4 Rentzepis, P. M.; Poa, Y. H. Appl. Phys. Lett., 1964, 5, 156.
5 Heilmeir, G. H.; Ockman, N.; Braunstein, N.; Kramer, D. A. Appl. Phys. Lett., 1964, 5, 229.
6 Orlov, R. Sov. Phys. Crytsallogr. , 1966, 11, 410 (English translation).
7 Gott, J. R. J. Phys. B., 1971, 4, 116.
8 Kurtz, S. K.; Perry, T. T. J. Appl. Phys., 1968, 39, 3798.
9 Zyss, J.; Chemla, D. S. Nonlinear Optical Properties of Organic Moleculaes and Crystals; Academic Press: New York, 1987, Vol. 1, Chapter 2.
10 Williams, D. J. Angew. Chem. Int. Ed. Eng., 1984, 23, 690.
11 Williams, D. J. in Nonlinear Optical Properties of Organic and Polymeric Materials, ACS Symposium Series No. 233; Am. Chem. Soc.: Washington, D. C., 1983.

12 Gedanken, A.; Robin, M. B.; Kuebler, N. A. J. Phys. Chem., 1982, 86, 4096.
13 Basu, S. Ind. Eng. Chem. Prod. Res., 1984, 23, 183.
14 Valley, G. C.; Klein, M. B.; Mullen, A. B.; Rytz, D.; Wechsler, B. Ann. Rev. Mater. Sci., 1988, 18, 165.
15 Glass, A. M., Science, 1984, 226, 657.
16 Pugh, D.; Sherwood, J. N. Chemistry in Britain, 1988 (October), 544.
17 Chemla, D. S. Physics Today, 1985 (May), 57.
18 Feinberg, J. Physics Today, 1986 (October), 46.
19 Hecht, J. High Technology, 1983 (October), 55.
20 Abraham, E.; Seaton, C. T.; Smith, S. D. Sci. Amer., 1983 (February), 85.
21 Abu-Mostafa, Y. S.; Psaltis, D. Sci. Amer., 1987 (March), 88.
22 Auston, D. H. et al., Appl. Opt., 1987, 26, 211.
23 Springer Proceedings in Physics, Vol 36: Nonlinear Optics of Organics and Semiconductors. Proceedings of the International Symposium in Tokyo, July 25-26, 1988, Kobayashi, T., ed.; Springer Verlag: Berlin, 1989.
24 Spec. Publ. - Royal Society Chem., 1989, 69 (Organic Materials for Nonlinear Optics).
25 Tomlinson, W. J.; Chandross, E. A. Adv. Photochem., 1980, 12, 201.
26 Monroe, B. M.; Smothers, W. K.; Krebs, R. R.; Mickish, D. J. SPSE Annual Conference and Symposium on Hybrid Imaging Systems, Rochester, N. Y., 1987, p. 131.
27 Smothers, W. K.; Monroe, B. M.; Weber, A. M.; Keys, D. E. SPIE OE/Laser Conference, "Practical Holaography IV," January, 1990, Los Angeles.
28 Ingall, R. T.; Fielding, H. L. Opt. Engg., 1985, 24, 808.
29 Ingall, R. T.; Troll, M. Opt. Engg., 1989, 28, 586.
30 Carre, C.; Lougnot, D. J.; Fouassier, J. P. Macromol., 1989, 22, 791.
31 Ingwal, R. T.; Troll, M. Proc. SPIE, 1988, 883, 94.
32 Eich, M.; Looser, H.; Yoon, D. Y.; Tweig, R. J.; Bjorklund, G. C.; Baumert, J. C. J. Opt. Soc. Am., B, 1989, 6, 1590.
33 Cassidy, C.; Halbout, J. M.; Donaldson, W.; Tang, C. L. Opt. Commun., 1977, 29, 243.
34 Ledoux, I.; Zyss, J. J. Chem. Phys., 1982, 73, 203.
35 Zyss, J.; Oudar, J. L. Phys. Rev. A, 1982, 26, 2028.
36 Halbout, J.-M.; Tang, C. L. in Zyss, J.; Chemla, D. S. Nonlinear Optical Properties of Organic Moleculaes and Crystals; Academic Press: New York, 1987, Vol. 1, Chapter II-6.
37 Oudar, J. L.; Chemla, D. S. J. Chem. Phys., 1977, 66, 2664.

38 Oudar, J. L.   J. Chem. Phys., 1978, 67, 446.
39 Zhang,G.J.; Kinoshita,T.; Sasaki,K.; Goto,Y.; Nakayama, A.   "Second Harmonic Generation in a New Organic Nonlinear Chalcone Derivative Crystal," CLEO, **1989**, THH1, Baltimore, MD.
40 Nicoud, J. F.; Tweig, R. J.; in Appendix I of Zyss, J.; Chemla, D. S.   Nonlinear Optical Properties of Organic Molecules and Crystals; Academic Press: New York, 1987, Vol. 2.
41 Tweig, R. J.; Jain, K. in Williams, D. J.   Nonlinear Optical Properties of Organic and Polymeric Materials, ACS Symposium Series No. 233; Am. Chem. Soc.: Washington, D. C., 1983, Ch 3.
42 Paley, M. S.; Harris, J. M.; Looser, H.; Baumert, J. C.; Bjorklund, G. C.; Jundt, D.; Tweig, R. J. J. Org. Chem., 1989, 54, 3374.
43 Holdcroft, G. E.; Dunn, P. L.; Rush, J. D. Mater. Res. Soc. Symp. Proc., 1989, 152, 57.
44 Oudar, J. L.; Hierle, R. J. Appl. Phys., 1977, 48, 2699.
45 Tweig, R. J.; Azema, A; Jain, K.; Cheng, Y. Y. Chem. Phys. Lett., 1982, 92, 208.
46 Tweig, R. J.; Jain, K.; Cheng, Y. Y.; Crowley, J. I.; Azema, A., Am. Chem. Soc., Div. Polym. Chem., 1982, 23, 147.
47 Tam, W.; Guerin, B.; Calabrese, J. C.; Stevenson, S. H. Chem. Phys. Letts., 1989, 154, 93.
48 Bierlein, J. D.; Cheng, L.-K.; Wang, Y.; Tam, W. Appl. Phys. Lett., 1990, 56, 423.
49 Koreneva, L. G.; Zolin, V. F.; Davydov, B. L., Molecular Crystals in NLO; Nauka: Moscow, 1975.
50 Levine, B. F.; Bethea, C. G.; Thurmond, C. D.; Lynch, R. T.; Bernstein, J. L. J. Appl. Phys., 1979, 50, 2523.
51 Lipscomb, G. F.; Garito, A. F.; Narang, R. S. Appl. Phys. Lett., 1981, 75, 1509.
52 Zyss, J.; Nicoud, J. F.; Coquillay, M. J. Chem. Phys., 1984, 81, 4160.
53 Zyss, J.; Chemla, D. S.; Nicoud, J. F. J. Chem. Phys., 1981, 74, 4800.
54 Okada S.; Nakanishi, H. Kagaku Kogyo, 1986, 37, 364.
55 Twieg, R. J. Organic Materials for SHG, Final Report, LLNL-2689405, March 1985. This report forms the basis for reference 41 above.
56 Fouquey, C.; Lehn, J.-M.; Malthete, J. J. Chem. Soc., Chem. Commun., 1987, 1424.
57 Wang, Y.; Tam, W.; Stevenson, S. H.; Clement, R. A.; Calabrese, J. C. Chem. Phys. Lett., 1988, 148, 136.
58 Tabei, H.; Kurihara, T.; Kaino, T. Appl. Phys. Lett., 1987, 50, 1855; J. Chem. Soc., Chem. Commun., 1987, 959.

59 Combellas, C.; Gautier, H.; Simon, J.; Thiebault, A.;
Tournilhac, F.; Barzoukas, M.; Josse, D.; Ledoux, I.;
Amatore, C.; Verpeaux, J.-N. J. Chem. Soc., Chem.
Commun., 1988, 203.

60 Meredith, G. R., in Williams, D. J. Nonlinear Optical
Properties of Organic and Polymeric Materials; ACS
Symposium Series No. 233; Am. Chem. Soc.: Washington, D.
C., 1983, p. 30.

61 Marder, S. R.; Perry, J. W.; Schaefer, W. P. Science,
1989, 245, 627.

62 Ikeda, H.; Kawabe, Y.; Sakai, T.; Kawasaki, K. Chem.
Phys. Lett., 1989, 157, 576.

63 Allen, S.; McLean, T. D.; Gordon, P. F.; Bothwell, B.
D.; Hursthouse, M. B.; Karaulov, S. A. J. Appl. Phys.,
1988, 64, 2583.

64 Katz, H. E..; Singer, K. D.; Sohn, J. E.; Dirk, C. W.;
King, L. A.; Gordon, H. M. J. Am. Chem. Soc., 1987, 109,
6561.

65 Ulman, A. et al., U. S. 4, 792, 208 ,1989.

66 Goto, Y.; Hayashi, A.; Nakayama, M.; Hirano, J.;
Watanabe, T.; Miyata, S. J. Photopolym. Sci. Technol.,
1988, 1, 330; also Europ. Pat. Appl. EP 262, 672 and
250, 099.

67 Palazzotto, M. C. U. S. 4, 733, 109, 1988.

68 Kaino, T.; Kurihara, T.; Matsumoto, S.; Tomaru, A. Jpn.
Kokai Tokkyo Koho JP 63 26, 638,1988.

69 Nogami, T.; Nakano, H,; Shirota, Y.; Umegaki, S.;
Shimizu, Y.; Uemiya, T.; Yasuda, N. Chem. Phys. Lett.,
1989, 155, 338.

70 Fuchs, B. A.; Syn, C. K.; Velsko, S. P. Appl. Opt.,
1989, 28, 4465.

71 Tokutake, S.; Imanishi, Y.; Sisido, M. Mol. Cryst. Liq.
Cryst., 1989, 170, 245.

72 Donald, D. S.; Cheng, L.-T.; Desiraju, G.; Meredith, G.
R.; Zumsteg, F. R. Abstract Q9.2, Fall Meeting of the
Materials Research Society, Boston, MA, Nov 27- Dec 2,
1989.

73 Frazier, C. C.; Harvey, M. A.; Cockerham, M. P.;
Chauchard, E. A.; Lee, C. H. J. Phys. Chem., 1986, 90,
5703.

74 Eaton, D. F.; Anderson, A. G.; Tam, W.; Wang, Y. J. Am.
Chem. Soc., 1987, 109, 1886.

75 Tam, W.; Calabrese, J. C. Chem. Phys. Lett., 1987, 133,
244.

76 Tam, W.; Calabrese, J. C. Chem. Phys. Lett., 1988, 144,
79.

77 Green, M. L. H.; Marder, S. R.; Thompson, M. E.; Bandy,
J. A.; Bloor, D.; Kolinsky, P. V.; Jones, R. J. Nature,
1987, 330, 360.

78 Anderson, A. G.; Calabrese, J. C.; Tam, W.; Williams, I. D. Chem. Phys. Lett., 1987, 134, 392.
79 Tam, W.; Eaton, D. F.; Calabrese, J. C.; Williams, I. D.; Wang, Y.; Anderson, A. G. Chem. Mats., 1989, 1, 128.
80 Tomaru, S.; Zembutsu, S.; Kawachi, M; Kobayashi, Y. J. Chem. Soc. Chem. Commun., 1984, 1207.
81 Wang, Y.; Eaton, D. F. Chem. Phys. Lett.,1985, 120, 441.
82 Weisbuch, I.; Lahav, M.; Leiserowitz, L.; Meredith, G. R.; Vanherzeele, H. Chem. Mats., 1989, 1, 114.
83 Cox, S. D.; Gier, T. E.; Bierlein, J. D.; Stucky, G. D. J. Am. Chem. Soc., 1989, 110, 2986.
84 Meredith, G. R.; Van Dusen, J. G.; Williams, D. J. in reference 11, p. 109.
85 Meredith, G. R.; Van Dusen, J. G.; Williams, D. J. , Macromol., 1982, 15, 1385.
86 Singer, K. D.; Sohn, S. E.; Lalama, S. J. Appl. Phys. Lett., 1986, 49, 248.
87 Le Grange, J. D.; Kuzyk, M. G.; Singer, K. D. Mol. Cryst. Liq. Cryst., 1987, 150b, 567.
88 a. Bowden, M. J.; Turner, S. J., Polymers for High Technology: Electronics and Photonics; ACS Symposium Series No. 346; Am. Chem. Soc.: Washington D. C., 1987.
   b. Bowden, M. S.; Turner, S. J., Electronic and Photonic Applications of Polymers; Adv. in Chemistry Series No. 218; Am. Chem. Soc.: Washington D. C., 1988.
89 Sohn, J. E.; Singer, K. D.; Kuzyk, M. G.; Holland, W. R.; Katz, H. E.; Dirk, C. W.; Schilling, M. L. Polym. Eng. Sci., 1989, 29, 1205.
90 Berge, B.; Wicker, A.; Lajerowicz J.; Legrand, J. F. Europhys. Lett., 1989, 9, 657.
91 Kishimoto, M.; Sato, M.; Gano, H. Springer Proc. Phys., 1988, 36, 196.
92 Eich, M.; Sen, A.; Looser, H.; Bjorklund, G. C.; Swalen, J. D.; Tweig, R.; Yoon, D. Y. J. Appl. Phys., 1989, 66, 2559.
93 Eich, M.; Reck, B.; Yoon, D. Y.; Willson, C. G.; Bjorklund, G. C. J. Appl. Phys., 1989, 66, 3241.
94 Leslie, T. M.; Yoon, H.-N.; DeMartino, R. N.; Stammatoff, J. B.; U. S. 4, 796, 976, 1989.
95 Leslie, T. M.; Yoon, H.-N.; DeMartino, R. N.; Stammatoff, Europ. Pat. Appl. EP 262, 672.
96 Robello, D. R.; Ulman, A., Willand, C. S.; U. S. 4, 796, 971, 1989.
97 Kuder, J. E., U. S. 4, 607, 095,1986.
98 Teng, C.-C.; Stammatoff, J. B.; Buckley, A.; Garito, A. F., U. S. 4, 775, 215,1988.
99 Wang, Y.; Chem. Phys. Lett., 1986, 126, 209.
100 Gillberg-LaForce, G. E.; Khanarian, G. U. S. 4, 828, 758, 1989.

101 Choe, E. W.; Khanarian, G.; Garito, A. F., U. S. 4,
    732, 783, 1988.
102 Popovitz-Biro, R.; Hill, K.; Landau, E. M.; Lahav, M.;
    Leiserowitz, L.; Sagiv, J.; Hsiung, H.; Meredith, G. R.
    J. Am. Chem. Soc., 1988, 110, 2672.
103 Lin, J. T.; Chen, C. Lasers and Optronics, 1987
    (November), 59.
104 Abraham, E.; Seaton, C. T.; Smith, S. D. Sci. Amer.,
    1983 (Feb.), 85.
105 Hecht, J. High Technol., 1983 (Oct.), 55.
106 Abu-Mostafa, Y. S.; Psaltis, D. Sci. Amer., 1987
    (March), 88.
107 Horner, J. L. Optical Signal Processing; Academic
    Press: N. Y., 1987.
108 Chemla, D. S. Rep. Prog. Phys., 1980, 43, 1191.
109 Hsu, C. C.; Kawabe, Y.; Ho, Z. Z., Peyghambarian, N.;
    Polky, J. N.; Krug, W.; Miao, E., submitted to Phys.
    Rev. Letts.
110 Ogawa, K.; Mino, N.; Tamura, H.; Sonada, N., Langmuir,
    1989, 5, 1415.
111 Neher, D.; Wolf, a.; Bubeck, C.; Wegner, G. Chem. Phys.
    Lett., 1989, 163, 116.
112 Miller, R. D.; Michl. J. Chem. Rev., 1989, 89, 1359.
113 Baumert, J.-C.; Bjorklund, G. C.; Jundt, D. H.; Jurich,
    M. C.; Looser, H.; Miller, R. D.; Rabolt, J.;
    Sooriyakumaran, R.; Swalen, J. D.; Tweig, R. J. Appl.
    Phys. Lett., 1988, 53, 1147.
114 Guha, S.; Frazier, C. C.; Porter, P. L.; Kamg, K.;
    Finberg, S. E. Opt. Lett., 1989, 14, 952.
115 Ho, Z. Z.; Ju, C. Y.; Hetherington, W. M., III J.
    Appl. Phys., 1987, 62, 716.
116 Winter, C. S.; Oliver, S. N.; Rush, J. D. in reference
    24, p. 232.
117 Bhar, G. C.; Chaterjee, U.; Das, S. J. Appl. Phys.,
    1989, 66, 5111.
118 Penzkofer, A.; Qiu, P.; Ossig, P., in reference 23 , p.
    312.
119 Wynne, J. Phys. Rev., 1969, 178, 1295.
120 Chang, Y.-C. J. Appl. Phys., 1985, 58, 499.
121 Cheng, L.-T.; Herron, W.; Wang, Y. J. Appl. Phys.,
    1989, 66, 3417.
122 Ohashi, Y.; Ito, H.; Hayashi, T.; Nitta, A.; Matsuda,
    H.; Okada, S.; Nakanishi, H.; Kato, M. in reference 23,
    p. 81.
123 Nasu, H.; Ibara, Y.; Kubodera, K. J. Non-Cryst. Sol.,
    1989, 110, 229.
124 Jain, R. K.; Lind, R. C. J. Opt. Soc. Am., 1983, 73,
    647.

125 Borrelli, N. F.; Hall, D. W.; Holland, H. J.; Smith, D. W. J. Appl. Phys., 1987, 61, 5399.
126 Friberg, S.; Smith, P. W. IEEE J. Quant. Electr., 1987, QE-23, 2089.
127 Jain, R. K.; Lind, R. C. J. Opt. Soc. Am., 1983, 73, 647.
128 Borrelli, N. F.; Hall, D. W.; Holland, H. J.; Smith, D. W. J. Appl. Phys., 1987, 61, 5399.
129 Olbright, O. R.; Peyghambarian, N.; Koch, S. W.; Banyai, L. Opt. Lett., 1987, 12, 413.
130 Van Wonterghem, B.; Saltiel, S. M.; Dutton, T. E.; Rentzepis, P. M. J. Appl. Phys., 1989, 66, 4935.
131 Wang, Y.; Mahler, W. Opt. Commun., 1987, 61, 233.
132 Wang, Y. ; Herron, N.; Mahler, W.; J. Opt. Soc. Am., B, 1989, 6, 809.
133 Gray, S. Photnics Spectrum, 1989 (Sept.), 125.
134 Edwards, J. H.; Feast, W. J.; and Bott, D. C. Polymer, 1984, 24, 395.
135 Heeger, A. J.; Moses, D.; Sinclair, M. Synth. Mets., 1986, 15.
136 Dennis, W. M.; Blau, W.; Bradley, D. J. Opt. Engg., 1986, 25, 538.
137 Wenzel, M.; Atkinson, G. H. J. Am. Chem. Soc., 1989, 111, 6123.
138 Khoo, I. C.; Normandin, R. Opt. Lett.,1984, 9, 285.
139 Zhao, M.-T.; Singh, B. P.; Prasad, P. N. J. Chem. Phys., 1988, 89, 5535.
140 Fukaya, N.Heinamaki, A.; Stubb, H. J. Mol. Electron., 1989, 5, 187.
141 Jenekhe, S. A.; Lo, S.; Flom, S. R. Appl. Phys. Lett., 1989, 54, 2524. I thank Prof. Jenekhe for preprints of this work.
142 Goshal, S. K.; Chopra, P.; Singh, B. P.; Swiatkiewicz, J.; Prasad, P. N. J. Chem. Phys., 1989, 90, 5070.
143 Gochanour, C. R.; Anderson, H. C.; Fayer, M. D. J. Chem. Phys., 1979, 70, 4254.
144 Kumar, G. R.; Singh, B. P.; Sharma, K. K. Opt. Commun., 1989, 73, 81.
145 Speiser, S.; Chisena, S. in reference 24, p. 211.
146 Zhu, Z. F.; Garmire, E. IEEE J. Quant. Electron., 1983, 19, 1495.
147 Dirk, C. W.; Kuzyk, M. G. Chem. Mater., 1990, 2, 4.
148 Ashkin, A.; Boyd, G. D.; Dziedzic, J. M.; Smith, R, G.; Ballman, A. A.; Nassau, K. Appl. Phys. Lett., 1966, 9, 72.
149 Cudney, R. S.; Pierce, R. M.; Feinberg, J. Science, 1988, 332, 424.
150 Valley, G. C.; Klein, M. B.; Mullen, R. A.; Rytz, D.; Wechsler, B. Ann. Rev. Mater. Sci., 1988, 18, 165.

151 Günter, P.; Huignard, H.-J. Photorefractive Materials
    and their Application, I and II; Topics in Applied
    Physics, 1988, 61 and 62, Springer-Verlag, Berlin.
152 Gailis, A.; Durandin, A. D.; Skudra, V.Latv. PSR Zinat.
    Akad. Vestis. Teh. Zinat. Ser., 1988, 119; Chem. Abs.,
    100:1463a.
153 Myl'nikov, V. S.; Grozonov, M. A.; Morozova, E. A.;
    Soms, L. N.; Vasilenko, N. A.; Kotov, B. V.;
    Pravednikov, A. N. Sov. Tsch. Phys. Lett., 1985, 11,
    16.

RECEIVED July 10, 1990

UNDERSTANDING STRUCTURE–PROPERTY RELATIONSHIPS ON THE SECOND-ORDER MICROSCOPIC SUSCEPTIBILITY

Chapter 9

Donor- and Acceptor-Substituted Organic and Organometallic Compounds

Second-Order Nonlinear Optical Properties

Wilson Tam[1], Lap-Tak Cheng[1], J. D. Bierlein[1], L. K. Cheng[1], Y. Wang[1], A. E. Feiring[1], G. R. Meredith[1], David F. Eaton[1], J. C. Calabrese[1], and G. L. J. A. Rikken[2]

[1]Central Research and Development Department, E. I. du Pont de Nemours and Company, Wilmington, DE 19880–0328
[2]Philips Research Laboratories, P.O. Box 80.000, 5600 JA Eindhoven, Netherlands

A systematic study has been undertaken to characterize the relationship between intrinsic molecular hyperpolarizabilities and structure-composition of donor- and acceptor-substituted organic and organometallic compounds. Donor-acceptor substituted benzenes, stilbenes and biphenyls have been examined by the electric field induced second harmonic generation (EFISH) technique to study donor–acceptor strength, transparency-nonlinearity trade-off and conjugation length. Suitable modification can lead to materials with relatively high short wavelength transparency yet high β. Several organometallic moieties serving as acceptors and donors have been evaluated. Molecular crystals of stilbenes have been shown to be unusually active in second harmonic generation. The linear and nonlinear optical properties of 3-methyl-4-methoxy-4'-nitro-stilbene (MMONS) single crystals will be presented.

We are developing a detailed understanding of structural factors that govern intrinsic molecular hyperpolarizabilities using an experimental approach that relies on solution-phase DC electric field induced second harmonic generation (EFISH) and third harmonic generation (THG) measurements. From this understanding, we expect to be able to synthesize materials which have optimized properties for various applications. One such application which will have significant technological implications in imaging and optical recording is the frequency doubling of

0097–6156/91/0455–0158$06.00/0
© 1991 American Chemical Society

short wavelength diode lasers. Efficient doubling of diode lasers requires a device which, considering the low powers of these lasers, utilizes large nonlinearity, finesse, and length in suitable proportions. Clearly, a material with very low absorption at the first and second harmonic wavelengths (near 800 and 400 nm) would be advantageous. One approach to fabricate a suitable device utilizes a medium obtained by doping or grafting nonlinear molecules into transparent polymeric matrices. Second order nonlinearity is achieved by the orientation directing influences of electric fields resulting from corona or electrode charging. Such an approach requires a large molecular dipole moment and hyperpolarizability product ($\mu\beta$). To identify suitable materials for the frequency doubling of these diode lasers, as well as to further our general understanding, we have studied the relationship between nonlinearity and transparency in donor-acceptor organics and organometallics.

Nonlinearity and Transparency Trade-Off

Solution-phase DC electric field induced second harmonic generation (EFISH) can be a rapid but approximate technique for investigating molecular hyperpolarizability; a vectorial projection of the hyperpolarizability tensor (referred to below as β) along the molecular dipole direction is determined. A lengthy set of physical and optical measurements are needed. The details of our experimental methodology and data analysis have been described (1,2). All measurements reported here, unless otherwise noted, have been performed with the longest conveniently available wavelength at 1907 nm to minimize dispersive enhancement due to low energy electronic excitations.

To determine the relationship between nonlinearity and transparency, we have determined the molecular hyperpolarizability of a series of donor-acceptor substituted benzenes, biphenyls, and stilbenes where the donor and acceptor are in the 1,4-, 4,4'-, and 4,4'- positions respectively. The choice of substitution pattern is chosen to insure that the charge-transfer (CT) axis is along the dipole direction. Interpretation of EFISH results may be hampered if the molecule lacks a clear CT axis along the dipole direction; other tensorial components inaccessible by EFISH may contribute to the bulk properties for such compounds. A log-log plot of β vs λ_{max}, the peak of the lowest intense feature in the optical absorption spectra, for the *p*-disubstituted benzenes and 4,4'-disubstituted stilbenes is shown in Figure 1. A similar correlation is seen with the biphenyl derivatives. Evidently, there is a strong correlation between transparency and β (1). Since the high β molecules are found to have generally high ground state dipole moments as a result of the strong donor and acceptor substituents, there is also a good correlation between $\mu\beta$ and λ_{max}. In addition, since substituent constants have long been used successfully to predict the position of the CT bands in color chemistry (3), a good correlation of β with Hammett's constants is expected; see Figure 2. Such a correlation (4), though not without exceptions, between $\sigma_p^{+/-}$ and β has also been observed.

The trade-off between transparency and nonlinearity is also seen in donor-acceptor organometallics as seen in Table I for a series of $W(CO)_5$(4-substituted pyridine) complexes.

Table I. Properties of $W(CO)_5$(4-X-pyridine)

X	solvent	λ_{max} (nm)	μ (10^{-18} esu)	β (10^{-30} esu)
NH$_2$	DMSO	290	8 ($\pm$1)	-2.1 ($\pm$0.3)
butyl	p-dioxane	328	7.3	-3.4
H	toluene	332	6.0	-4.4
phenyl	CHCl$_3$	330-340	6.0	-4.5
COMe	CHCl$_3$	420-440	4.5	-9.3
CHO	CHCl$_3$	420-440	4.6	-12

The low-energy metal to ligand charge-transfer (MLCT) excited state has been extensively studied and involves low-lying π^*-acceptor orbitals of the pyridine ligands (5). The negative sign of β indicates a reduction of the dipole moment upon electronic excitation. The use of 4-substituted pyridines with electron accepting groups leads to larger $|\beta|$ and lower dipole moments. Stronger accepting groups should lead to more back transfer of charge upon MLCT excitation, as well as to lower energy π^*-acceptor orbitals which reduce the energy gap for the CT transition (6). As in the organic case, increased conjugation results in higher nonlinearity; $W(CO)_5$(4-formylstyrylpyridine) has a $|\beta|$ of 20 x 10^{-30} esu compared to $W(CO)_5$(4-formylpyridine) with a $|\beta|$ of 12 x 10^{-30} esu. Similar nonlinearity and transparency trade-offs have been observed for other organometallics (Cheng, L.-T.; Tam, W.; Meredith, G.R.; Marder, S.R. Molecular Crystals and Liquid Crystals., in press.).

Modification of Stilbene Derivatives

The β for stilbene derivatives are much higher than their benzene analogs (1). To determine the effect of substitution on the nonlinearity, 4-methoxy-4'-nitrostilbene

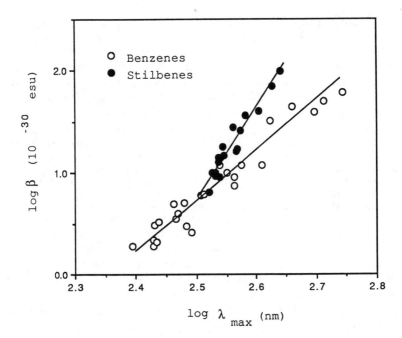

Figure 1. Log β vs log λmax of benzene and stilbene derivatives.

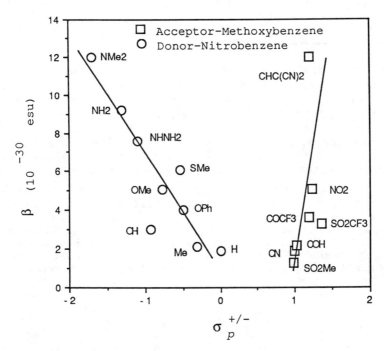

Figure 2. Correlation of β and Hammett's Constants.

(MONS) has been modified by heteroatom substitution, addition of other donors, and by addition of an acceptor in the ethylenic linkage. We have found that modification of MONS generally leads to a reduction of β. The only exception to this decreasing trend is provided by 2,4-dimethoxy-4'-nitrostilbene which shows a 20% increase in $\mu\beta$ product along with a red-shifted λ_{max} of 31 nm. However, the interpretation of the EFISH results for modified 4,4'-disubstituted stilbenes is less straight forward because the dipole direction may not be along the CT axis and geometric consideration must be taken into account.

We have found that crystals of substituted 4-nitrostilbenes have a high tendency to form non-centrosymmetric phases (7,8). Of the modified MONS we have examined, 3-methyl-4-methoxy-4'-nitrostilbene (MMONS) shows large powder second harmonic generation signals (1250 x urea at 1.064 μm) (9). Single crystals of MMONS have been grown and its crystal properties have been determined. MMONS is highly birefringent with n_z-n_x = 0.75 (0.52 μm), with large nonlinear optical coefficients (d_{33} = 184 pm/V and d_{24} = 71 pm/V (1.064 μm)) and it has large electrooptic coefficients (r_{33} = 39.9 pm/V (0.6328 μm)). It can be type II phase matched for efficient SHG of lasers emitting around 1 micron (10). However, MMONS is not suitable for frequency doubling of diode lasers because the absorption at 400 nm is too large.

Fluorinated Sulfone as an Acceptor

From the correlation of Hammett's constant and β, we might expect the fluorinated sulfone to be a more effective acceptor than the nitro group ($\sigma_p(NO_2)$=0.78 vs $\sigma_p(SO_2CF_3)$= 0.93) (11). A comparison between 4-N,N-dimethylamino-nitrobenzene,**6**, with 4-N,N-diethylaminophenylperfluoropropyl sulfone,**5**, as indicated in Table II suggests that the use of the fluorinated sulfone leads to comparable values of β (9.0 x 10^{-30} esu vs 12 x 10^{-30} esu in this case). More interestingly, the λ_{max} for the fluorinated sulfone is blue shifted from the nitro analogue (314 nm vs 376 nm) while the dipole moment is larger (7.3 vs 6.4 Debye). A more thorough study of the use of the fluorinated sulfone shows that materials with good nonlinearity and transparency can be obtained. Table II lists some of our data for the fluorinated sulfones along with comparison with nitro analogs.

Table II. Properties of Fluorinated Sulfones and Nitro
Analogs

Compound	A	D	λ_{max} (nm)	μ 10^{-18} esu	β 10^{-30} esu

1	$SO_2C_{10}F_{21}$	F	225	3.6	1.5
2	$SO_2C_{10}F_{21}$	Br	245	3.5	2.0
3	$SO_2C_3F_7$	OMe	290	5.4	3.3
4	NO_2	OMe	302	4.6	5.1
5	$SO_2C_{10}F_{21}$	NEt_2	314	7.3	9.0
6	NO_2	NMe_2	376	6.4	12

7	SO_2R_f*	OMe**	316	5.5	14
8	NO_2	OMe	352	4.6	17
9	SO_2R_f*	NMe_2	376	7.4	35
10	NO_2	NMe_2	440	6.5	50

**The OMe example has a methyl group adjacent to the OMe.
*$R_f = -(CF_2)_2C(CF_3)(OMe)(OCH_2CF_3)$

11	$SO_2C_3F_7$	OMe	305	6.0	9.1
12	$SO_2C_6F_{13}$	OMe	305	5.9	11
13	NO_2	OMe	332	4.5	9.2
14	$SO_2C_6F_{13}$	NMe_2	362	8.0	25
15	NO_2	NMe_2	390	5.5	50

| 16 | $SO_2C_6F_{13}$* | OMe | 347 | 7.8 | 14 |

*10:1 ratio of *trans* to *cis*

| 17 | NO_2 | OMe | 368 | 4.5 | 28 |

The benzene derivatives containing the fluorinated sulfone have been prepared either by nucleophilic substitution of the 4-fluorophenyl derivative (e.g. **1**) or by starting with the appropriately substituted sodium thiophenoxide and reacting with perfluoroalkyl iodide follow by oxidation with either MCPBA or chromium oxide (12-14). The biphenyl derivatives have been prepared by palladium catalyzed cross coupling chemistry of the 4-bromophenyl derivative (e.g. **2**) with substituted phenyl boronic acid (yields: 37-84%) (15,16). Compound **16** has been prepared by palladium catalyzed cross coupling of (4-bromophenyl)perfluorohexyl sulfone with vinyl anisole in 37 % yield (17). The vinyl sulfones, **7** and **9**, have been prepared by condensation of $CH_3SO_2R_f$ (18) with the appropriate aldehyde (yields: 70,and 73%) following a literature procedure (19). Yields were not optimized.

Before discussing the EFISH results on the fluorinated sulfones, one needs to determine the dipole direction relative to the CT-axis. The group moment of the SO_2CF_3 substituent has been determined to be 4.32 Debye with the angle between the direction of the dipole moment and the $C_{aryl}-S$ bond of 167° (20). A crystal structure shows that this geometry does not change substantially with longer perfluoroalkyl groups. From these determination, the total dipole (including the mesomeric and donor contributions) direction is less than 13° off from the CT-axis, which represents a negligible reduction in nonlinearity (less than 3%) when the hyperpolarizability tensor is projected along the dipole as determined from EFISH measurements. In contrast, the group dipole moment of the unfluorinated sulfone, SO_2CH_3, is 63° off from the CT-axis (21), which translates to nearly a 50% decrease in nonlinearity amenable to electric poling. Our measurements indeed showed significantly lower β values for the methylsulfone derivatives. Therefore, due to the polar nature of the trifluoromethyl group, the accepting strength of the sulfonyl group is enhanced and its dipole moment is better aligned for poled applications.

The use of the fluorinated sulfone in the phenyl compounds can lead to comparable nonlinearity with better transparency and higher dipole moments than the nitro analogs (Table II). However, the influence of the fluorinated sulfone group appears to be short range and decreases with conjugation length. The effect of conjugation length on nonlinearity is dramatic for the nitro group, in going from phenyl (5.1 x 10^{-30},**4**), vinylphenyl (17 x 10^{-30},**8**), biphenyl (9.2 x 10^{-30},**13**) to the stilbene derivative (28 x 10^{-30},**17**). For the fluorinated sulfone, the increase in β with conjugation length is more moderate: **3** (3.3 x 10^{-30}), **7** (14 x 10^{-30}), **12** (11 x 10^{-30}), **16** (14 x 10^{-30}). For the vinylphenyl and biphenyl derivatives, the fluorinated sulfones again give good nonlinearity for their transparency. The length of the fluorinated alkyl group appears to have a minor effect on β. Compounds **11** and **12** have comparable β (9.1 and 11 x 10^{-30} esu). In all cases, the fluorinated sulfone derivatives have larger dipole moments than their nitro analogs.

The use of dimethylamino as a donor has a larger effect on the nonlinearity of the nitro compounds than on the fluorinated sulfones as the conjugation length increases. When methoxy is replaced with the dialkylamino group in the phenyl derivatives, β increases by 6.9 x 10^{-30} esu (from 5.1 x 10^{-30} esu to 12 x 10^{-30} esu) for the nitro case while increasing by 5.7 x 10^{-30} esu in the fluorinated sulfone case. For the vinylphenyl derivatives, β increases by 33 x 10^{-30} esu (from 17 x 10^{-30} esu to 50 x 10^{-30} esu) for the nitro case while increasing by only 21 x 10^{-30} esu in the fluorinated sulfone case. A larger gap is seen for the biphenyl where differences in β are 40 x 10^{-30} esu for NO_2 and only 14 x 10^{-30} esu for SO_2R_f.

The observations of larger dipole moments, increased effectiveness with shorter conjugation length and the lower enhancement of the dimethylamino group as a donor suggest that the fluorinated sulfone group's contribution to nonlinearity is mostly inductive in nature. This conclusion is consistent with Hammett's constant studies which showed steady increase in inductive contributions as the accepting strengths of the sulfonyl side-groups increased. The modest resonance contributions of the sulfone allow for relatively high-lying excited states which are not strongly CT in nature. These factors lead to narrow absorption bands and relatively good transparency.

The nitro group is an effective acceptor partially because it has the proper hybridization to have effective overlap with the p orbitals of the benzene ring. Coupled with the resonance contribution of the dimethylamino group, N,N-dimethylamino-4-nitrobenzene has increased conjugation length and higher nonlinearity. Due to the modest resonance contribution of the fluorinated sulfone, the resonance contribution of the dialkylamino group to nonlinearity is not fully utilized. Therefore, a greater enhancement to nonlinearity is realized for the nitro acceptor over the fluorinated sulfone acceptor when the donor is changed from the methoxy to the dimethylamino group. However, the increase in nonlinearity from the resonance contribution is accompanied by a large red shift in the absorption band which makes the compound useless for frequency doubling of short wavelength diode lasers.

Enhancement of β by an inductive contribution can also be seen in fluorinated ketones. For example, 4-methoxybenzaldehye has a β of 2.2 x 10^{-30} esu and μ of 3.5 Debye while the trifluoromethyl ketone compound (21) as the acceptor has a β of 3.6 x 10^{-30} esu and μ of 4.0 Debye.   In this case, λ_{max} is red shifted (269 nm to 292 nm).

To examine the limits of using the inductive contribution to β, we have studied the sulfonylsulfimide (SSI) group, $S(R_f)=NSO_2CF_3$, as an acceptor. The SSI group has been determined to have the overall effect of two nitro groups with a σ_I which is 1.5 times larger than the σ_I constant of the SO_2CF_3 group (22). Reaction of p-F-$C6H4S(C_{10}F_{21})=NSO_2CF_3$ with pyrrolidine gives the nitrogen substituted product in 88% yield. This material has a β = 13 x 10^{-30} esu, μ = 9 Debye, and λ_{max} = 336 nm. Comparison with p-Et$_2$N-C$_6$H$_4$SO$_2$C$_{10}$F$_{21}$ (β = 9 x 10^{-30} esu, μ = 7.3 Debye, and λ_{max} = 314 nm) shows that β and dipole moment are larger for the SSI acceptor and the absorption edge is slightly red shifted. Due to the larger β and dipole moment, $\mu\beta$ increases by about 78% in going to the SSI acceptor. Although the EFISH results for the SSI compound is appropriate for poled polymer application, EFISH results can not conclude on the effectiveness of the inductive enhancement without further information on the group dipole direction. The largest component of β may not have been measured due to the geometry of the SSI group.

We have found that use of fluorinated sulfone and SSI acceptors leads to materials with better nonlinearity for their transparency compared with materials containing the conventional acceptors. Figure 3 summarizes our findings, and is a plot of $\mu\beta$ vs λ_{max} for donor-acceptor CT molecules of benzene, styrene, biphenyl, fluorene, and stilbene derivatives with fluorinted sulfone, SSI, and other common acceptors (SO$_2$Me, CN, COH, COMe, and NO$_2$) groups. All measurement results on molecules with λ_{max} below 376 nm have been included. The improved trade-off between nonlinearity and transparency of the fluorinated sulfonyl and SSI acceptors is evident.

Conclusions and Summary

The simplest approximation for describing β is a sum of two contributions, $\beta = \beta_{ADD} + \beta_{CT}$ where β_{ADD} is due to the substituents' inductive effects (23). This approximation, although not quantum mechanically correct, has allowed the physical organic chemist to design highly nonlinear molecules. Methods of increasing hyperpolarizabilities are modification of low-lying strong charge-transfer (CT) electronic transitions by changing donor and acceptor strengths and by increasing conjugation length. Both approaches are generally in conflict with the requirement

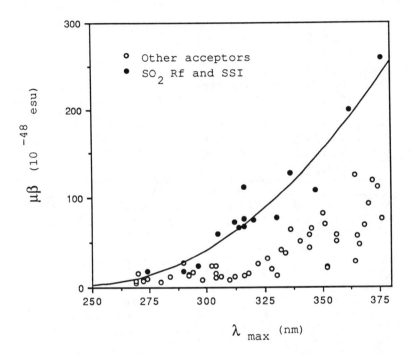

Figure 3. Transparency and nonlinearity trade-offs: $\mu\beta$ vs λ_{max}.

of transparency for the frequency doubling of near 800 nm light. Our approach for preparing materials with relatively high nonlinearity and transparency is to modify β_{ADD} by using acceptors with high inductive strength. The modest resonance contribution of the fluorinated sulfones results in better transparency, and the large inductive contribution results in relatively large β and μ.

Acknowledgments

We thank Todd W. Hunt, Edward R. Wonchoba, and Howard D. Jones for technical assistance.

Literature Cited

1. Cheng, L.-T.; Tam, W.; Meredith, G.R.; Rikken, G.L.J.A. Proc. of the Int. Soc. for Optical Eng.,1989, 1147, 61..
2. Meredith, G.R.; Cheng, L.-T.; Hsiung, H.; Vanherzeele, H.A.; Zumsteg, F.C. In Materials for Nonlinear and Electro-optics; Lyons,M.H., ed.; IOP Publishing, New York 1989, p 139.
3. Griffiths, J. Colour and Constitution of Organic Molecules; Academic Press, New York, 1976.
4. Katz, H.E.; Dirk, C.W.; Singer, K.D.; Sohn, J.E. Proc. of the Int. Soc. for Optical Eng., 1987, 824, 86.
5. Geoffroy, G.L.; Wrighton, M.S. Organometallic Photochemistry, Academic Press, New York, 1979.
6. Wrighton, M.S.; Abrahamson, H.B.; Morse, D.L. J. Amer. Chem. Soc 1976, 98, 4105.
7. Wang, Y.; Tam, W.; Stevenson, S.H.; Clement, R.A.; Calabrese, J. Chem. Physics Lett. 1988, 148, 136.
8. Tam, W.; Wang, Y.; Calabrese, J.C.; and Clement, R.A. Proc. of the Int. Soc. for Optical Eng.,1988, 971, 107.
9. Tam, W.; Guerin, B.; Calabrese, J.C.; Stevenson, S.H. Chem. Physics Lett. 1989, 154, 93.
10. Bierlein, J.D.; Cheng, L.K.; Wang, Y.; and Tam, W. Applied Physics Lett. 1989, **56**, 423.
11. Hansch, C.; Leo, A. Substituent Constants For Correlation Analysis in Chemistry and Biologhy; Wiley & Son, 1979.
12. Feiring, A.E. J. Fluorine Chem. 1984, 24, 191.
13. Popov, V.I.; Boiko, V.N.; Kondratenko, N.V.; Sampur, V.P.; Yagupolskii, L.M. J. Org. Chem. USSR (Engl. Trans.) 1977, 13, 1985.
14. Kondratenko, N.V.; Popov, V.I.; Kolomeitsev, A.A.; Saenko, E.P.; Prezhdo, V.V.; Lutskii, A.E.; Yagupolskii, L.M. J. Org. Chem. USSR (Engl. Trans.) 1980, 16, 1049.
15. Migaura, N.; Yanagi, T.; Suzuki, A. Synthetic Communications 1981, 11, 513.

16. Thompson, W.J.; Gaudino, J. <u>J. Org. Chem.</u> 1984, <u>49</u>, 5240.
17. Patel, B.A.; Ziegler, C.B.; Cortese, N.A.; Plevyak, J.E.; Zebovitz, T.C.; Terpko, M.; Heck, R.F. <u>J. Org. Chem.</u> 1977, <u>42</u>, 3903.
18. Krespan, C.G.; Smart, B.E. <u>J. Org. Chem.</u>, 1986, <u>51</u>, 320.
19. Sodoyer, R.; Abad, E.; Rouvier, E.; Cambon, A. <u>J. of Fluorine Chem.</u> 1983, <u>22</u>, 401.
20. Lutskii, A.E.; Yagupol'skii, L.M.; and Obukhova, E.M. <u>J. of General Chemistry of the USSR</u> 1964, <u>34</u>, 2663.
21. Creary, X. <u>J. Org. Chem.</u> 1987, <u>52</u>, 5026.
22. Kondratenko, N.V.; Popov, V.I.; Timofeeva, G.N.; Ignat'ev, N.V.; Yagupol'skii, L.M. <u>J. Org. Chem. USSR (Engl. Trans.)</u> 1985, <u>21</u> 2367.
23. Ulman, A. <u>J. Phys. Chem.</u> 1988, <u>92</u>, 2385.

RECEIVED July 10, 1990

Chapter 10

Use of a Sulfonyl Group in Materials for Nonlinear Optical Materials

A Bifunctional Electron Acceptor

Abraham Ulman, Craig S. Willand, Werner Köhler, Douglas R. Robello, David J. Williams, and Laura Handley

Corporate Research Laboratories, Eastman Kodak Company, Rochester, NY 14650-2109

In this study we describe semiempirical calculations, ground-state dipole moment measurements, and measurements of molecular hyperpolarizability coefficients (β) for new stilbene and azobenzene derivatives containing a methylsulfonyl group as the electron acceptor. We could show that theoretical calculations can be used to predict the *ratio* of molecular hyperpolarizabilities between similar compounds, and that these gas phase calculations underestimate β values, probably as a result of the valence basis set used in the calculations. Whereas the sulfone group has been demonstrated to give molecular hyperpolarizabilities less than that of similar nitro compounds, the difference becomes less as the degree of conjugation is increased. One would, therefore, expect that this difference will decrease further in more highly nonlinear systems. Even so, the increased visible spectrum transparency and the synthetic flexibility may make these compounds important for some applications.

The growing need for fast and efficient optical devices has made nonlinear optics (NLO) an area on the frontier of science today, and had generated great interest in second-order NLO materials. In such materials, an external stimulus produces a sudden, nonlinear modulation of light passing through the material. These materials generate three wave interactions, which are important physical phenomena that have many applications such as second-harmonic-generation (SHG), frequency-up and down conversion, parametric applications, and electrooptic modulation (*1*).

All of these applications require materials with high nonlinearity, high optical damage threshold, and high linear optical quality. Four major categories of nonlinear optical materials are receiving attention: electrooptic crystals (inorganic and organic)

0097–6156/91/0455–0170$06.00/0

(*2,3*), bulk semiconductors (*3*), polymeric organic materials (*4*), and molecular assemblies (*5*), each exhibiting a different mix of advantages and disadvantages.

A number of inorganic materials, e.g., KDP, KTP, $LiNbO_3$, $LiIO_3$, and borate crystals (*2*), are readily available on the market for parametric effects. However, these materials are very difficult to fabricate and are not easily integrated with semiconductor materials into monolithic circuits.

In recent years, much research has been directed to *organic* NLO materials because of several important advantages they possess (*6*). For example, the NLO response of many organic materials is extremely rapid (approaching *femtoseconds*) because the effects occur primarily through electronic polarization. In contrast, NLO effects in most liquid crystal materials operate via reorientation of whole molecules, and NLO effects in many inorganic materials operate through lattice distortions, which are comparatively slow processes. In addition, organic NLO materials offer simple processing techniques that are compatible with existing technologies for the fabrication of integrated optical or electrooptical devices.

In spite of the potential advantages, useful organic NLO materials have not yet been developed because the necessary molecular and macroscopic characteristics have only recently begun to be understood. However, because bulk NLO properties in organic materials arise directly from the constituent molecular nonlinearities, it is possible to decouple molecular and supramolecular contributions to the NLO properties. One can then semiquantitatively predict relative macroscopic nonlinearities based on theoretical analyses of the individual molecules (*7*). Reliable predictions of this kind are vital for the efficiency of a program aimed towards developing new organic materials with tailored NLO properties.

It is, by now, well known that molecules containing electron donor and electron acceptor groups separated by a large conjugated π-framework possess large values of the second-order molecular hyperpolarizability, β (*6*). However, while nitro and polycyanovinyl groups have been widely studied as acceptor groups in NLO, the sulfonyl group has not received much attention despite its strong acceptor properties. (For example, its σ_p and σ are +0.72 and +1.05, respectively; while for the nitro group these values are +0.79 and +1.24, respectively) (*8*). In addition, the sulfonyl groups is *bifunctional*, a feature that permits greater freedom in the design of compounds for specific applications and allows more flexibility for synthetically tailoring the physical properties.

This paper summarizes the theoretical analysis of some new molecules with methylsulfonyl group as the electron acceptor group, describes the syntheses of new stilbene and azobenzene systems, and presents the measurements of their optical spectra, ground-state dipole-moments, and molecular hyperpolarizability coefficients, β. We compare theoretical and experimental results and comment on the potential usefulness of these *chromophores* as components for NLO materials. The incorporation of sulfonyl-containing chromophores into polymers, and the NLO properties of the resulting materials, will be discussed in our forthcoming paper (*9*).

Detailed synthetic strategy for the new sulfonyl-containing chromophores will appear elsewhere (10).

Semiempirical Calculations

We have performed a series of semiempirical quantum-mechanical calculations of the molecular hyperpolarzabilities using two different schemes: the finite-field (FF), and the sum-over-state (SOS) methods. Under the FF method, the molecular ground state dipole moment μ_g is calculated in the presence of a static electric field E. The tensor components of the molecular polarizability α and hyperpolarizability β are subsequently calculated by taking the appropriate first and second (finite-difference) derivatives of the ground state dipole moment with respect to the static field and using

$$(\mu_g)_i = \alpha_{ij} E_j + \beta_{ijk} E_j E_k \tag{1}$$

In the SOS method, one uses an expression for the second-order hyperpolarizability β of the second-harmonic generation process derived from second-order perturbation theory (11):

$$\beta_{ijk} = \frac{1}{4\hbar^2} \sum_P \sum_{e,e'} \frac{< g |\tilde{\mu}_i| e > < e|\tilde{\mu}_j|e'> < e'|\tilde{\mu}_k|g >}{(\omega_{eg} - 2\omega)(\omega_{e'g} - \omega)} \tag{2}$$

where ijk are molecular Cartesian coordinates and ω is the incident frequency. $|g>$ denotes the molecular ground state; $|e>$, and $|e'>$ are excited states of the system having transition frequencies given by ω_{eg}, and $\omega_{e'g}$, respectively. The summation over P generates the six terms by permutation of the pairs $\{(-2\omega,i); (\omega,j); (\omega,k)\}$. The dipole difference operator $\tilde{\mu}$ is defined as

$$\tilde{\mu}_i = \mu_i - < g| \mu_i| g > \tag{3}$$

where μ_i is the dipole moment operator. The accuracy of these types of calculations are strongly dependent on the formalism employed in deriving the ground and excited state properties. Clearly the best correlation between theory and experiment should be obtained at the *ab initio* level. A number of such investigations have been published for small molecules at the coupled Hartree-Fock (12), and uncoupled Hartree Fock levels (13,14). However, similar analyses involving larger molecules (more than 20 heavy atoms) are not feasible due to the time and expense required to perform the calculations. In contrast, more approximate semiempirical algorithms such as INDO and CNDO can be implemented much more efficiently, and are therefore more suited to these types of calculations. The quality of the results from semiempirical procedures however depend on the atomic orbital (AO) basis set used as well as the parameterization of the AO interactions. We have chosen to use a valence basis set for the following reasons: (a) Valence basis set calculations have been shown to be fairly successful in predicting molecular structure and nonlinear optical properties (*12,15-19*). (*b*) There is insufficient experimental data to implement a properly parameterized calculation using an extended basis set, which includes diffuse polarization functions for our molecules.

For this work, the molecular structures derived using an AM1 semiempirical method (*20*) served as input for an INDO SCF procedure (*21*), which generated the unoccupied molecular orbitals. (Note: The summation in equation 2 over the vibrational subspace of each electronic state is approximated as unity and is valid for all off-resonance processes.) The dipole moment matrix and transition energies corresponding to these CI states were calculated and inserted directly into equation 2.

Formally, equation 2 requires knowledge of the complete state basis set (SBS). Realistically only a limited number of states can be used thereby requiring a truncation of the summation at some finite SBS size. Our results indicate that a SBS derived from the excited states mentioned above, is more than sufficient for our purpose; adequate convergence of the hyperpolarizability is typically obtained within the first few excited states.

The experimental methods utilized here, as is generally the case for electric-field poling, measure the Z-component of the vector component of the hyperpolarizability, β_z (equation 4),

$$\beta_z = \frac{1}{3} \sum_{i=xyz} (\beta_{zii} + \beta_{izi} + \beta_{iiz}) \qquad (4)$$

where z is defined to be parallel and in the direction of the ground state dipole μ_g.

The theoretically predicted values for μ_g and β_z of various nitro- and sulfonyl-containing compounds are listed in Table I.

Fundamentally both the SOS and FF methods should give similar results for the static second-order hyperpolarizability ($\omega = 0$). In fact, both methods predict similar trends in the hyperpolarizabilities for the molecules studied here. However, the FF method consistently predicts larger magnitudes for β_z than the SOS method. This is particularly true for the SOS calculations where doubly excited states have been included. Transitions requiring two electron promotions are not formally allowed and subsequently cannot directly contribute to equation 2. However, promotions of this type decrease the singly excited character of the excited state configurations yielding smaller transition moments and diluted molecular hyperpolarizabilities. Discussion of the relative accuracies of these methods will be continued in the section discussing EFISH results.

We started the analysis of the sulfonyl group as an acceptor with the calculation of 4-methylsulfonylaniline (I), which is an analogue of *p*-nitroaniline (II), and their methylated derivatives (III, and IV, respectively). We have found that while the ground state dipole moments are comparable for the nitro and sulfone derivatives (Table I), the β coefficients are different in magnitude, and depend on the method of calculation (Table I).

Table I. Theoretical Dipole Moments and Second-Order Hyperpolarizabilities for Selected Methylsulfonyl and Nitro Compounds ($\lambda = 1907$ nm)

Compound	β_z (10^{-30} esu)			
	$\mu_g(D)^a$	SOS-S^b	SOS-S/D^c	FF^d
I H_2N—⬡—SO_2CH_3	7.68	1.3 (1.1)	5.3 (0.2)	1.6
II H_2N—⬡—NO_2	7.33	11.3 (9.7)	4.2 (3.9)	10.3
III $(CH_3)_2N$—⬡—SO_2CH_3	8.68	2.4 (1.8)	1.6 (0.9)	3.1
IV $(CH_3)_2N$—⬡—NO_2	7.88	11.6 (13.0)	5.6 (5.4)	14.7
V H_2N—⬡—⬡—SO_2CH_3	9.47	5.5 (2.87)	3.2 (1.6)	9.1
VI $(CH_3)_2N$—⬡—CH=CH—⬡—SO_2CH_3	9.94	13.4 (7.54)	12.3 (6.44)	24.8
VII $(CH_3)_2N$—⬡—CH=CH—⬡—NO_2	8.62	42.5 (35.0)	15.7 (15.5)	62.1
VIII $(CH_3)_2N$—⬡—N=N—⬡—SO_2CH_3	9.70	18.0 (11.6)	13.0 (8.9)	27.8
IX $(CH_3)_2N$—⬡—N=N—⬡—NO_2	9.25	41.8 (35.1)	19.8 (19.8)	59.4

^{a}Dipole moments calculated using AM1 semiempirical method. (1D = 10^{-18} esu). ^{b}Sum-over-state method using only singly excited configurations (number in paranthese are β_{zzz}). ^{c}Sum-over-state method using both singly and doubly excited configurations (number in paranthese are β_{zzz}). ^{d}Finite-field method ($\omega = 0$).

Since the hyperpolarizability of a given molecule is a function of the donor and the acceptor properties, and nature of the conjugation path between them, we turned to the biphenyl system and analyzed the 4-amino-4'-methylsulfonylbiphenyl (V). The calculated ground state dipole moment of this molecule is smaller than expected for such an increase in the distance between the donor and the acceptor.

Thus, although the biphenyl compound has a large π-system located between the electron donor and acceptor, its calculated hyperpolarizability is substantially less than that of the stilbene analog (VI) and only slightly larger than that of substituted aniline (III, Table I). Steric interactions among the inner phenyl hydrogens cause the rings to be nonplanar and therefore reduce the electronic coupling between the sulfone and amino groups. On the other hand, analogous stilbenes are nearly planar giving rise to a longer effective conjugation length. Hence, biphenyl compounds are, in general, poorer candidates for nonlinear optics in so much as stilbene molecules are only slightly larger in volume, but display a much larger nonlinear response.

We have calculated μ_g and β_{zzz} for 4-dimethylamino-4'-methylsulfonylstilbene (VI), and compared these data with values for 4-dimethyl-amino-4'-nitrostilbene (DANS, VII). We have also included the azo-derivatives VIII and IX in the comparison (Table I). Calculated values of β_z for III, IV, VI, and VII as a function of basis set size are shown in Figure 1. It is evident that in all of these cases, there is a single excited state that provides the largest contribution to the hyperpolarizability.

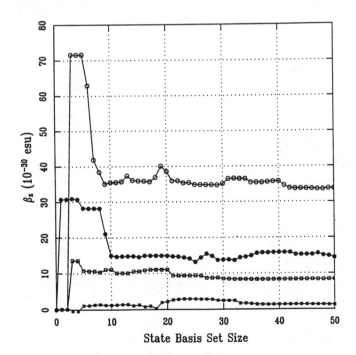

Figure 1. A comparison of β_z as a function of the basis set size for molecules VI (■), VII (□), VIII (•), IX (o) ($\lambda = 1907$ nm).

This contribution is the diagonal term ($e = e'$) involving the amino to acceptor charge transfer. The other significant contributions stem from off-diagonal coupling of higher lying excited states to the charge transfer state. One would expect therefore, as was noted above, that the details of the charge transfer interaction would strongly influence the nonlinear optical properties. Moreover, for both the phenyl and stilbene molecules the nitro group gives higher β_z values than the sulfonyl group, which is consistent based on the differing s values.

Our calculations predict only minor differences between the ground state dipole moments for molecules containing nitro electron acceptors versus those possessing methylsulfonyl. In contrast, the hyperpolarizabilities behave much differently, in that calculated β_z for the aminonitrostilbenes is about twice that of the aminosulfonylstilbenes and the nitroanilines are more than 5 times more nonlinear than the sulfonylanilines. The hyperpolarizabilities appear to be very sensitive to the details of the electron donors-acceptor interaction and hence accentuate the differences in the σ values for nitro and methylsulfonyl.

An interesting observation is that the differences between the nitro and sulfonyl electron accepting properties have less of an impact on the nonlinearity for the more highly conjugated stilbene compounds than they do in the case of the aniline systems. A similar effect also can also be seen in substituted polyenes. The calculated results for β_z of A-(CH=CH$_2$)$_n$-NH$_2$ (A = NO$_2$, SO$_2$CH$_3$) are shown in Figure 2 as a

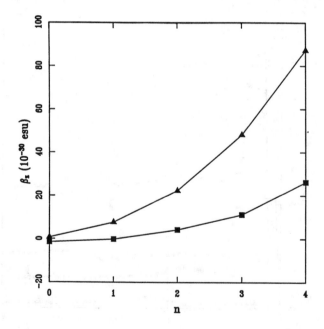

Figure 2. Calculated values (FF) of β_z for H$_2$N-(CH=CH)$_n$-NO$_2$ (▲), and H$_2$N-(CH=CH)$_n$-SO$_2$-CH$_3$ (■).

function of polyene chain length n. As expected, the hyperpolarizability increases as the number of double bonds increases. However, just as is the case for the substituted aminostilbenes and anilines mentioned above, the relative nonlinearity $\beta_{z(nitro)}/\beta_{z(sulfone)}$ shows a downward trend as the degree of conjugation increases having values of 5.3, 4.3, and 3.3 for $n = 2$, 3, and 4, respectively. Apparently there is a "saturation" phenomena associated with the impact of the electron acceptor properties on the hyperpolarizability; the nonlinearity of small molecules is strongly dependent on the electron acceptor strength whereas this dependency is weaker in larger, more conjugated systems. This has important consequences in the design of molecules for nonlinear optics. The choice of electron acceptors in the case of small molecules is largely restricted to those groups with large σ values; selection in highly conjugated systems can be made on the basis of other characteristics such as synthetic flexibility and optical absorption without drastically affecting the hyperpolarizability.

The calculations predict that azobenzene derivatives have nearly identical dipole moments and molecular hyperpolarizabilities as the stilbenes. Selection of compounds for use in specific applications can therefore, be based on linear optical properties (absorption) and photochemical stability requirements without sacrifice of nonlinear optical response.

The theoretical models discussed above indicate that the sulfonyl group, although slightly weaker in electron acceptor strength, is indeed a viable alternative to the nitro group. In particular, sulfonyl derivatives of stilbene and azobenzene display large molecular hyperpolarizabilities and can be used as bifunctional chromophores for the construction of materials with nonlinear optical properties.

The results presented above indicate that the sulfonyl group, although slightly weaker in electron acceptor strength, is indeed a good alternative to the nitro group, and that derivatives of stilbene and of azobenzene can be used as *bifunctional* chromophores for the construction of materials with nonlinear optical properties.

Optical Spectra

The optical spectra of the stilbene and azobenzene derivatives were studied in different solvents to establish the solvatochromic shifts of the new sulfonyl-containing chromophores, and to compare them to those found for the nitro analogues. Figure 3 shows the spectra in toluene and in DMF of 4-dibutylamino-4'-nitrostilbene, and of 4-dibutylamino-4'-methylsulfonylstilbene, and Figure 4 shows the spectra in the same solvents, but of 4-dibutylamino-4'-nitroazobenzene, and 4-dibutylamino-4'-methylsulfonylazobenzene. Table II summarizes the results.

The most important feature of the spectra is the large blue shift of the sulfones vs the nitro derivatives, i.e., 53 nm (3098 cm^{-1}) for the stilbene, and 30 nm (1400 cm^{-1}) for the azobenzene derivatives, in toluene). These large blue shifts suggest that sulfonyl compounds are more transparent in the visible region than their nitro

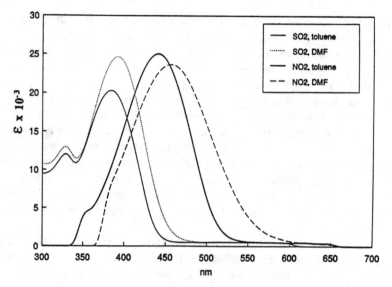

Figure 3. Optical spectra in toluene and in DMF of 4-dibutylamino-4'-nitrostilbene, and of 4-dibutylamino-4'-methylsulfonylstilbene.

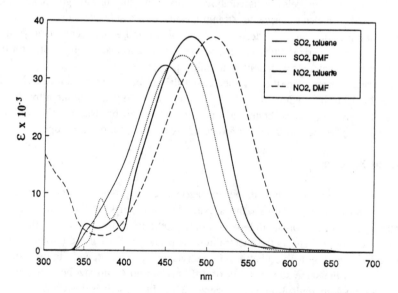

Figure 4. Optical spectra in toluene and in DMF of 4-dibutylamino-4'-nitroazobenzene, and of 4-dibutylamino-4'-methylsulfonylazobenzene.

Table II. Optical Spectra of Dibutylaminostilbene and Azobenzene Chromophores

Acceptor	Type	λ_{max} (nm)^a ε (x 10^{-4}) l mol^{-1} cm^{-1}					
		toluene	CHCl$_3$	CH$_2$Cl$_2$	aceton	DMF	DMSO
NO$_2$	N=N	478	498	500	492	504	513
		3.89	4.39	4.37	4.12	3.72	2.19
NO$_2$	CH=CH	441	452	453	444	455	463
		2.56	2.96	2.91	3.00	2.43	2.46
CH$_3$SO$_2$	N=N	448	461	464	455	469	475
		3.31	3.68	4.08	4.36	3.54	3.69
CH$_3$SO$_2$	CH=CH	388	391	390	389	390	395
		2.73	2.51	2.33	2.59	2.42	2.33

^{a}All reported λ_{max} and ε values are at 1 x 10^{-5} M concentrations.

analogues, a property that is crucial when considering the requirements for second harmonic generation and other parametric processes in the visible spectrum.

If we compare $\Delta\lambda$ values ($\Delta\lambda = \lambda_{DMSO} - \lambda_{Toluene}$) for the four dyes under study we get 35 nm (1428 cm^{-1}), and 27 nm (1269 cm^{-1}), for the nitro, and methylsulfonyl azobenzene derivatives, respectively, and 22 nm (1077 cm^{-1}), and 7 nm (457 cm^{-1}), for the nitro, and methylsulfonyl stilbene derivatives, respectively. Further examination reveals that the difference in $\Delta\lambda$ between the two stilbene derivatives is 15 nm (620 cm^{-1}), while that in the case of the azobenzene derivatives is 8 nm (159 cm^{-1}). Thus, two interesting conclusions can be drawn from this data: (*a*) the bathochromic shift is not only a function of the donor and acceptor groups, but also of the intermediate π-system between them; and (*b*) while the measured hyperpolarizability coefficients for the stilbene and azobenzene sulfonyl derivatives are very similar (see below), their solvatochromism behavior is different, and therefore solvatochromism is not an accurate prediction of β.

Measurements of Ground-State Dipole Moments

The measurement of ground-state dipole moments may help to establish the validity of the theoretical calculations. We measured two representative compounds, X, and XI (see below), where diallyl derivatives were used to increase solubility in nonpolar solvents (e.g., CCl$_4$).

The dipole moments were determined from the concentration dependence of the dielectric constant and the refractive index of the solutions in the low concentration limit (mol fraction *ca.* 0.001).

Osipov proposed a formula to calculate the dipole moment of a polar substance in a polar solvent and proved its applicability (*22,23*). In this case the molar orientation polarization of the solution (P_{sdu}^{or}) can be written as:

$$P_{sdu}^{or} = \frac{4}{3} \pi N \frac{\mu_S^2 (1 - x_D) + \mu_D^2 X_D}{3 kT} \tag{5}$$

$$= \frac{M_S (1 - x_D) + M_D x_D}{\rho} [\frac{(\varepsilon - 1)(\varepsilon + 2)}{8 \varepsilon} - \frac{(n^2 - 1)(n^2 + 2)}{8 n^2}]$$

where μ_S and μ_D are the ground state dipole moments of the solvent (subscript S), and the dye (subscript D), M_S and M_D are their molecular weights, and x_D is the mol fraction of the dye. The quantities ρ, ε, and n are the density, dielectric constant, and refractive index, respectively, of the solution. T is the temperature, k is Boltzmann's constant, and N is Avogadro's number. The molar orientation polarization of the dye, P_D^{or}, can be calculated from eq. 5:

$$\frac{4}{3} \pi N \frac{\mu_D^2}{3 kT} = P_D^{or} = \frac{P_{sdu}^{or} - P_S^{or}}{x_D} + P_S^{or} \tag{6}$$

$P_S^{or} = \frac{4}{3} \pi N \mu_S^2 / 3 kT$ is the orientation polarization of the solvent and given by equation 5 in the case $x_D = 0$. For low dye concentrations we can use a linear relationship of P_{sdu}^{or} in x_D and get:

$$P_D^{or} = \frac{dP_{solu}^{or}}{dx_D} \bigg|_{X_D = 0} + P_S^{or} \tag{7}$$

which together with equation 5 gives:

$$P_D^{or} = P_S^{or} [\frac{1}{f_S} (\frac{\varepsilon_S^2 + 2}{8 \varepsilon_S^2} \frac{\partial \varepsilon}{\partial x_D} - \frac{n_S^4 + 2}{8 n_S^4} \frac{\partial n^2}{\partial x_D}) - \frac{1}{\rho_S} \frac{\partial \rho}{\partial x_D} + \frac{M_D}{M_S}] \tag{8}$$

The subscript s refers to the values of the pure solvent and $f_S = P_S^{or} \rho_S / M_S$. $\partial \varepsilon / \partial x_D$ and $\partial n^2 / \partial x_D$ were determined by linear regression of ε and n^2 as a function of dye concentration for typically five different solutions of mol fraction less than 10^{-3}. In all cases, the linear approximation was fully justified within the experimental error. Figure 5 shows, as an example, ε as a function of concentration for X in chloroform. For molecules with large dipole moments, like the ones discussed in this paper, the contribution of $\partial \rho / \partial x_D$ can be neglected since even the extreme assumption of $\partial \rho / \partial x_D = 1$ g/cm^3 changes the value of μ_D by only 0.4%. Similarly, the term proportional to $\partial n^2 / \partial x_D$ contributes less than 4% to the

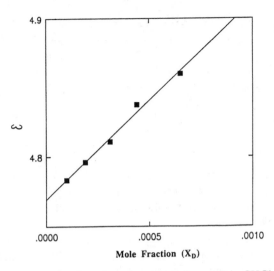

Figure 5. ε as a function of concentration of X in CHCl$_3$

determined dipole moment, and can thus also be neglected in a reasonable approximation, leaving the concentration dependence of the dielectric constant as the only experimental parameter. It is important to emphasize, however, that these approximations do not hold for molecules with low dipole moments, comparable to the ones of the solvent molecules.

The measured dipole moments for X and XI in different solvents are summarized in Table III. First, the experimental values of μ vary from solvent-to-solvent with a trend to higher values for more polar solvents. This may be partly due to the approximations mentioned above. It is also important to note that no attempt was made to account for the nonspherical shape of the dye molecule. We believe that this approximation is justified, since the local field factor used to calculate the hyperpolarizabilities in the EFISH experiment for the product $\mu\beta$ involves similar approximations. Thus, the effective dipole moment determined in these experiments,

in the respective solvent, and not the dipole moment *in vacuo*, is the quantity of interest, and should be used in the calculations of β from EFISH experiments.

From a chemical point of view, however, it is reasonable that the measured dipole moment increases with solvent polarity, simply because in these conjugated systems the intramolecular charge transfer from the electron donor to the electron acceptor is

Table III. Measured Dipole Moments (in Debye) for X and XI in Various Solvents

	μ_X	μ_{XI}	μ_{XI}/μ_X
Carbon tetrachloride	7.2	8.0	1.11
Toluene	6.9	7.3	1.06
m-Xylene	7.0	7.6	1.09
Chloroform	7.7	8.3	1.08

also enhanced with the increasing solvent polarity. An intramolecular electron transfer from the donor to the acceptor gives a quinonic charge-separated structure, which is more stable in more polar solvents, and hence contributes more to the overall dipole moment of the dye.

Second, the ratio μ_{XI}/μ_X is always roughly the same, about 1.1, both in the experiments and in the calculations for the sulfonyl-containing materials. This result, which has also been found for a number of other dyes not discussed in this report, is important because it indicates that theoretical calculations do predict the correct trend in dipole moments. Thus, a reliable evaluation of dipole moments of proposed dyes can be carried out before they are actually synthesized.

We have also measured the ground-state dipole moment of DANS (VII), in $CHCl_3$, and obtained the value 7.6 D. This, indeed, was an encouraging result since it further supported the credibility of our molecular design approach that was developed to speed up the search for new NLO materials.

β Measurements

Both theoretical analysis and dipole moment measurements indicated that sulfonyl-substituted compounds may have β coefficients similar in magnitude to their nitro analogues. Therefore, we have measured β for several sulfonyl- and nitro-substituted compounds using electric-field-induced second-harmonic generation method (EFISH) (11,25). In this experiment, one measures an effective third-order nonlinearity Γ_{EFISH} for a solution containing the compound of interest, given by

$$\Gamma_{EFISH} = f_0 f_\omega^2 f_{2\omega} \sum_{components} N[\frac{(\mu_g)_z \beta_z}{5kT} + \gamma_{el}] \qquad (9)$$

where N is the number density, kT is the thermal energy, and the summation is over all the components of the solution. γ_{el} is the effective third-order-hyperpolarizability for the pure electronic four-wave mixing process $\omega + \omega + 0 = 2\omega$. This quantity can be determined by examining the temperature behavior of Γ_{EFISH} or can be approximated from the results of four-wave mixing experiments. However, the magnitude of γ_{eff} is typically less than one tenth that of the β_z term in the case of second-order NLO materials and was therefore neglected. The local field factor f

relates the externally applied electric fields to that present at the molecular site. These local field factors can be approximated by (24):

$$f_0 = \frac{\varepsilon_0(\varepsilon_\infty + 2)}{\varepsilon_\infty + 2\varepsilon_0} \tag{10}$$

$$f_\omega = \frac{n_\omega^2 + 2}{3} \tag{11}$$

where ε_0 is the static dielectric constant, ε_∞ is the dielectric constant for frequencies faster than the dipolar relaxation, and n_ω is the index of refraction at frequency ω. We assumed that the dielectric constant and indices of refraction are identical to that of the pure solvent; the error introduced by this assumption is small compared to the other errors for the concentration ranges used here. The experiment was performed by measuring Γ_{EFISH} as a function of chromophore concentration (Figure 6), performing as least-squares analysis of the data, and extracting $(\mu_g)_z\beta_z$ by using equation 9, and assuming Γ_{EFISH} (chloroform) = 0.88 x 10^{-13} esu (25).

The results from EFISH experiments are listed in Table IV along with the FF predictions. We note that the our measured values for 4-dimethylamino-4'-nitrostilbene (DANS) and 4-dimethylamino-4'-nitroazobenzene are consistent with previously published work (26,27). The theoretical predictions discussed previously show an excellent correlation with the actual measured values for $(\mu_g)_z\beta_z$. As predicted the nonlinearities of nitro-containing compounds are larger

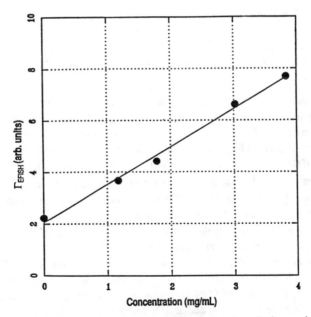

Figure 6. Γ_{EFISH} versus concentration for 4-dimethylamino-4'-nitroazobenzene in chloroform ($\lambda = 1907$ nm).

Table IV. Second-Order Hyperpolarizabilities for Sulfonyl and Nitro Compounds

Compound	$(\mu_g)_z\beta_z$ $(10^{-48}$ esu)	
	Experimental[a]	Calculated[b]
III $(CH_3)_2N$—⟨ ⟩—SO_2CH_3	26	26.9
IV $(CH_3)_2N$—⟨ ⟩—NO_2	143	116
X (allyl)$_2$N—⟨ ⟩—CH=CH—⟨ ⟩—SO_2CH_3	573	247
XII (allyl)$_2$N—⟨ ⟩—CH=CH—⟨ ⟩—NO_2	1150	535
XIII $(C_4H_9)_2N$—⟨ ⟩—N=N—⟨ ⟩—SO_2CH_3	512	270
XIV $(CH_3)_2N$—⟨ ⟩—N=N—⟨ ⟩—NO_2	1270	549

[a]All measurements are in $CHCl_3$. [b]Calculated from FF values of β_z in Table I.

than that of the corresponding sulfones with the relative differences between the two diminishing with increased conjugation. Also, the nonlinearities of the azobenzene derivatives are nearly identical to that of the stilbene compounds. While the overall trends in the theoretical predictions and the experimental data are similar, the magnitude of $(\mu_z)\beta_z$ are comparable for the anilines but differ by a factor of two for the stilbenes. This is most likely a result of the valence basis set used in our calculations. The introduction of diffuse polarization functions would afford the potential for larger charge-transfer interactions than is possible within the valence basis, and hence, larger hyperpolarizabilities for highly nonlinear molecules. Future work will be concerned with such calculations.

Theoretically, the SOS and FF methods should yield similar results for the static second-order hyperpolarizability ($\omega = 0$), since both are at similar levels of approximation. However, the SOS method requires information about many states of the system, while the FF method demands much more detailed information about one particular state (ground state). Evidently, the particular semiempirical algorithms implemented here are better suited towards the latter as the FF predictions are consistently closer to the actual experimental values than those of SOS method.

Conclusions

In this study we have described theoretical calculations, syntheses, optical spectra, ground-state dipole moment measurements, and measurements of molecular second-order hyperpolarizability coefficients (β) for new stilbene and azobenzene derivatives containing a methylsulfonyl group as the electron acceptor. We have shown that theoretical calculations can be used to predict the *ratio* of molecular hyperpolarizabilities between similar compounds, and that these gas phase calculations underestimate β, probably as a result of the valence basis set used in the calculations.

Whereas the sulfone group has been demonstrated to produce lower molecular hyperpolarizabilities than those of nitro groups, the difference becomes less as the degree of conjugation is increased. One would therefore expect that this difference will decrease further in more highly conjugated systems. The increased visible spectrum transparency and the synthetic flexibility may make these sulfonyl compounds important for some applications.

Literature Cited

1. Sohn, Y. R. The Principles of Nonlinear-Optics, Wiley: New York, 1985.
2. Twieg, R. J.; Dirk, C. W. J. Phys. Chem. 1986, 85, 3537
3. Henneberger, F. Phys. Status Solisi 1986, 137, 371.
4. Prasad, P. N.; Williams, D. J. Introduction to Nonlinear Optics in Molecules and Polymers, Wiley: New York, In press.
5. Williams, D. J.; Penner, T. L.; Schildkraut, J. S.; Tillman, N.; Ulman, A.; Willand, C. in Nonlinear Effects in Organic Polymers, Messier, J.; Kajzar, F.; Prasad, P.; Ulrich, D., Eds., NATO ASI Series No. 162, 1989, p 195.
6. Williams, D. J. Angew. Chem. Int. Ed. Engl. 1984, 23, 690.
7. Ulman, A. J. Phys. Chem. 1988 92, 2385.
8. Kosower, E. M. Physical Organic Chemistry, Wiley: New York, 1968.
9. Robello, D. R.; Schildkraut, J. S.; Dao, P.; Scozzafava, M.; Ulman, A., submitted for publication.
10. Ulman, A.; Willand, C.; Köhler, W.; Robello, D.; Williams, D. J.; Handley, L. J. Am. Chem. Soc. 1990, 112, 0000.
11. Levine, B. F.; Bethea, C. G. J. Chem. Phys. 1975, 63, 2666.
12. Chopra, P.; Carlacci, L.; King, H. F.; Prasad, P. N. J. Phys. Chem. 1989, 93, 7120.
13. Purvis, G. D.; Bartlett, P. J. Phys. Rev. 1981, A23, 1594.
14. Andre, J. M.; Barbier, C.; Bodart, V. P.; Delhalle, J. in Nonlinear Optical Properties of Organic Molecules and Crystals; Chemla, D. S.; Zyss, J., Eds., Academic Press: London, 1987, Vol. 2.
15. Hurst, G. J. B.; Dupuis, M.; Clementi, E. J. Chem. Phys. 1988, 89, 385.

16. Zyss, J. J. Chem. Phys. 1979, 70, 3333.

17. Lalama, S. J.; Garito, A. F. Phys. Rev. 1979, A20, 1179.

18. Waite, J.; Papadopoulos, M. G. J. Chem. Phys. 1985, 82, 1427.

19. Pugh, D.; Morley, J. in Nonlinear Optical Properties of Organic Molecules and Crystals; Chemla, D. S.; Zyss, J., Eds., Academic Press: London, 1987, Vol. 1.

20. See for example, Dewar, M. J. S.; Zoebisch, E. G.; Healy, E. F.; Stewart, J. J. P J. Am. Chem. Soc. 1985, 107, 3902, and references therein.

21. Ridley, J. E.; Zerner, M. C. Theor. Chim. Acta 1973, 32, 111; 1976, 42, 223; 1979, 53, 21, and references therein.

22. Minkin, V. I.; Osipov, O. A.; Zhdanov. Y. A. Dipole Moments in Organic Chemistry, Plenum Press: New York, 1970.

23. Osipov, O. A.; Panina, M. A. Zh. Fiz. Khim. 1958, 32, 2287.

24. Lorentz, H. A. The Theory of Electric Polarization, Elsevier: Amsterdam, 1952.

25. Qudar, J. L.; Chemla, D. S. J. Chem. Phys. 1977, 66, 2664.

26. Katz, H. E.; Singer, K. D.; Sohn, J. E.; Dirk, C. W.; King, L. A.; Gordon, H. M. J. Am. Chem. Soc. 1987, 109, 6561.

27. Singer, K. D.; Sohn, J. E.; King, L. A.; Gordon, H. M.; Katz, H. E.; Dirk, C. W. J. Opt. Soc. Am. B 1989, 6, 1339.

RECEIVED August 13, 1990

Chapter 11

Organic and Organometallic Compounds
Second-Order Molecular and Macroscopic Optical Nonlinearities

Seth R. Marder[1], Bruce G. Tiemann[1], Joseph W. Perry[1], Lap-Tak Cheng[2], Wilson Tam[2], William P. Schaefer[3], and Richard E. Marsh[3]

[1]Jet Propulsion Laboratory, California Institute of Technology, Pasadena, CA 91109
[2]Central Research and Development Department, E. I. du Pont de Nemours and Company, Wilmington, DE 19880–0356
[3]Division of Chemistry and Chemical Engineering, California Institute of Technology, Pasadena, CA 91125

Organic and organometallic stilbazolium cations can be crystallized with various counterions; some of the resulting salts exhibit large SHG powder efficiencies. Approximately linear stilbazolium cations have a greater tendency to crystallize in noncentrosymmetric space groups than do cations with substantial geometric asymmetry. The nonresonant quadratic molecular hyperpolarizabilities of several ferrocene and ruthenocene derivatives were studied by DC electric-field-induced second-harmonic generation (EFISH) experiments using fundamental radiation at 1.91 μm. Hyperpolarizabilities approaching that of 4-dimethylamino 4'-nitro-stilbene (DANS) were observed indicating that the ferrocene moiety can act as an effective donor. EFISH measurements indicate that indoaniline dyes with very polarizable π systems have large molecular hyperpolarizabilities (β), in one case approaching three times that of DANS.

Organic materials with second-order nonlinear optical (NLO) properties have been the subject of intense investigation owing to their potential use in a variety of technologies including telecommunications, optical information processing and storage.(1-3) Large second-order molecular hyperpolarizabilities (β) are associated with chromophores comprised of electron donors and acceptors linked by a conjugated π system.(4-6) The nonlinear chromophore must reside in a noncentrosymmetric environment if β is to lead to an observable bulk effect such as second harmonic generation (SHG) or the linear electrooptic effect (LEO). In this paper we focus on factors affecting each of the above design criteria for second-order NLO materials. We will first show that in many instances variation of the counterion in ionic structures leads to materials with large bulk second-order susceptibilities ($\chi^{(2)}$). In addition, we will show that metallocenes can function effectively as donors. Finally, we will explore the possibility of using highly polarizable π systems for enhancing β.

0097–6156/91/0455–0187$06.00/0

Results and Discussion

<u>Organic and Organometallic Salts with Large Macroscopic Second-Order Optical Nonlinearities.</u> Roughly 75% of all known, non-chiral, organic molecules crystallize in centrosymmetric space groups leading to materials with vanishing $\chi^{(2)}$.[7] Several strategies have been employed to overcome this major obstacle.[8-15] Meredith demonstrated that (4)-$(CH_3)_2NC_6H_4$-CH=CH-(4)-$C_5H_4N(CH_3)^+CH_3SO_4^-$ has an SHG efficiency roughly 220 times that of urea [16] which until recently was the largest powder SHG efficiency reported. He suggested that Coulombic interactions in salts could override the dipolar interactions which provide a strong driving force for centrosymmetric crystallization in neutral dipolar compounds.[16] The LEO coefficient of Meredith's molecule is 430 pmV^{-1}, roughly ten times that of lithium niobate.[17] We [18] and Okada [19] have recently extended this approach and demonstrated that various of 4-N-methylstilbazolium salts give large powder SHG efficiencies when combined with a suitable counterion (Table I).[18]

Table I. Powder SHG efficiencies for R-CH=CH-$C_5H_4NCH_3^+X^-$salts . The left value is for 1064 nm input and the right value is for 1907 nm input (Urea = 1)

R \\ X =	CF₃SO₃	BF₄	CH₃C₆H₄SO₃	Cl
4-$CH_3OC_6H_4$-	54 / 50	0 / 0	100 / 120	270 / 60
4-$CH_3OC_6H_4$-CH=CH-	0 / 0	2.2 / 1.0	50 / 28	4.3 / 48
4-$CH_3SC_6H_4$-	0 / 0	0 / 0	1 / -	0 / 0
2,4-$(CH_3O)_2C_6H_3$-	67 / 40	2.9 / 5.5	0.08 / 0	0.7 / 0.4
$C_{16}H_9$- (1-pyrenyl)	1.1 / 0.8	- / -	14 / 37	- / -
4-$(CH_2CH_2CH_2CH_2N)C_6H_4$-	0.06 / 0.5	0.05 / 5.2	0.03 / 0.2	0 / 1.1
4-BrC_6H_4-	0 / 0	0.02 / 0	5.0 / 1.7	100 / 22
4-$(CH_3)_2NC_6H_4$-	0 / 0	- / 75[14]	15 / 1000	0 / 0
4-$(CH_3)_2NC_6H_4$-CH=CH-	5 / 500	4.2 / 350	5 / 115	0 / 0

The data suggest that these ionic chromophores exhibit a higher tendency to crystallize noncentrosymmetrically than do dipolar covalent compounds. The efficacy of the approach is underscored by the observation that more than half of the compounds gave SHG efficiencies greater than urea. Also, of the nine chromophores discussed here, seven could be isolated with a counterion to give a compound with an SHG efficiency greater than 35 times urea. For example, the yellow compound $CH_3OC_6H_4$-CH=CH-$C_5H_4N(CH_3)^+Cl^-\cdot4H_2O$ was found to exhibit an efficiency roughly 270 times that of the urea reference standard; $(CH_3)_2NC_6H_4$-CH=CH-CH=CH-$C_5H_4N(CH_3)^+CF_3SO_3^-$ gave an efficiency 500 times the urea standard and $(CH_3)_2NC_6H_4$-CH=CH-$C_5H_4N(CH_3)^+CH_3C_6H_4SO_3^-$ gave a efficiency 1000 times the urea standard.

In an attempt to further explore the scope and limitations of the organic salt methodology, we have examined eleven 2-N-methyl stilbazolium compounds. (Marder, S. R.; Tiemann, B. G.; Perry, J. W.; Schaefer, W. P.; Marsh, R. E. Chem. Mater., In press) The 2-N-methyl stilbazolium compounds of the form R-CH=CH-(2)-$C_5H_4N(CH_3)^+X^-$, (where R=4-$CH_3OC_6H_4$, X=CF_3SO_3; R=4-$C_6H_4NCH_2CH_2CH_2CH_2$, X=CF_3SO_3; R=4-$C_6H_4N(CH_3)_2$, X= CF_3SO_3, BF_4; R=2,4-$C_6H_3(OCH_3)_2$, X=CF_3SO_3; R=2-$C_6H_4OCH_3$, X=CF_3SO_3;

R=$(C_5H_5)Fe(C_5H_4)$, X=CF_3SO_3, $CH_3C_6H_4SO_3$, I, Br) all gave negligible SHG efficiencies. The exception was (3)-$CH_3OC_6H_4$-CH=CH-(2)-$C_5H_4N(CH_3)^+$ $CF_3SO_3^-$, which has an efficiency of 25 times urea. The nonlinear optical properties of this compound are worthy of consideration since, contrary to simple resonance considerations for the design of NLO chromophores, the donor, the methoxy group, and the acceptor, the cationic alkylated nitrogen atom, are cross conjugated. This gives rise to enhanced transparency in the visible in comparison to the isomer 4'-methoxy-2-*N*-methyl stilbazolium triflate, in which the donor and the acceptor are conjugated. In methanol solution, 3'-methoxy-2-*N*-methyl stilbazolium triflate has a λ_{max} at 344 nm and a cutoff at 455 nm. In comparison, 4'-methoxy-2-*N*-methyl stilbazolium triflate has a charge transfer band at 368 nm. In the solid state the cutoff for 3'-methoxy-2-*N*-methyl stilbazolium triflate is at ~ 425 nm (for a ~100μm thick crystal). Although the molecular hyperpolarizability (β) of 3'-methoxy-2-*N*-methyl stilbazolium triflate is undoubtedly smaller than 4'-methoxy-2-*N*-methyl stilbazolium triflate, its SHG efficiency suggests that it is not necessary to have very strong donors and acceptors or for the donor and the acceptor to be strongly coupled in order to achieve significant macroscopic nonlinearities.(Cheng, L. T.; Tam, W.; Meredith, G. R.; Rikken, G; Marder, S. R. J. Am. Chem . Soc., submitted for publication.)

Crystal structures of several salts were determined in order to better understand how the chromophores align in the crystal lattice. Although it is difficult to generalize packing trends, in the nine crystal structures we have determined, a recurring structural motif is alternating parallel rows of cations and rows of anions. (Schaefer, W. P.; Marsh, R. E.; Marder, S. R., in preparation.) The compounds shown in Fig 1a-e follow this motif. In general, *neutral* dipolar molecules with geometrical asymmetry show a greater tendency to crystallize in noncentrosymmetric space groups than do more linear symmetric analogs. Thus, whereas crystals of 4-nitroaniline are centrosymmetric, 2-methyl-4-nitroaniline crystallizes in the noncentrosymmetric space group Cc. Similarly, although crystals of 4-methoxy-4'-nitrostilbene are most likely centrosymmetric (as surmised by no SHG activity), 3-methyl-4-methoxy-4'-nitrostilbene (20) and 2-methoxy-4'-nitro-stilbene (Grubbs, R. B.; Marder, S. R.; Perry, J. W.; Schaefer, W. P. Chem.Mater., Accepted for publication.) both crystallize in noncentrosymmetric space groups and give rise to large SHG efficiencies. The opposite trend is observed with the 2-*N*-methyl and the 4-N-methyl stilbazolium salts we have examined. Over half of the 4-N-methyl stilbazolium salts we have examined exhibit powder SHG efficiencies greater than urea, whereas only two of the 2-*N*-methyl stilbazolium salts we have studied had powder efficiencies substantially greater than urea. Thus, whereas molecular asymmetry may tend to favor crystallographic noncentrosymmetry in neutral molecules, it appears from our limited sampling that the opposite is true for ionic chromophores.

<u>Second-Order NLO Properties of Metallocenes.</u> Until recently,(21-26) the potential of organometallic compounds for quadratic nonlinear optics has been completely ignored. The observation that the ferrocene complex (Z)-{1-ferrocenyl-2-(4-nitrophenyl) ethylene} has an SHG efficiency 62 times urea demonstrates that organometallic compounds could exhibit large $\chi^{(2)}$.(27) Given this observation, we synthesized the new compound $(C_5H_5)Fe(C_5H_4)$-CH=CH-(4)-$C_5H_4N(CH_3)^+I^-$ and measured its SHG powder efficiency by a modification of the Kurtz powder technique.(27) Powder SHG efficiencies were determined using 1907 nm

fundamental radiation (SH at 953.5 nm) to avoid absorption of the SH by the dark colored salt. This salt has an SHG efficiency roughly 220 times urea, the largest efficiency known for an organometallic compound.(28) Furthermore, the magnitude of the powder SHG signal is sensitive to the nature of the counterion as shown in Table II. The crystal structure determination of the nitrate salt reveals the polar nature of the lattice (Fig. 1f).

The results obtained from the Kurtz powder test, although tantalizing, provide little insight into molecular structure-property relationships since they are almost entirely determined by crystallographic, linear optics (i.e. birefringence), and dispersive factors. In addition, since molecular structure modification is often accompanied by crystallographic changes, powder testing cannot be used to systematically probe molecular structure-property relations. Solution-phase DC electric-field-induced second-harmonic (EFISH) generation (29) is a more appropriate method for hyperpolarizability studies. It allows extraction of a vectorial projection of the hyperpolarizability tensor (β) along the molecular dipole (μ) direction. When experiments are carried out with radiation of sufficiently long wavelength, EFISH provides direct information on the intrinsic optical nonlinearity of a molecule.

For organic compounds, structure-property trends concerning donor-acceptor strengths and the effectiveness of different conjugated backbones have been topics of many studies.(30 and Cheng, L. T.; Tam, W.; Rikken, G. manuscript in preparation.) Our recent efforts have provided an extensive set of internally consistent results on many of the important molecular classes.(31) Organometallic compounds allow us to explore new variables. We can change the transition metal element, its oxidation state, the number of d electrons and examine the differences between diamagnetic and paramagnetic complexes and the effect of new bonding geometries and coordination patterns. Each of these factors creates new possibilities for the engineering of asymmetric polarizability.

The considerations outlined above coupled with the large observed powder efficiencies of several ferrocene complexes (23,26,28) motivated us to undertake a study of the molecular hyperpolarizabilities of metallocene complexes. Given the aromatic character of the cyclopentadienyl (Cp) ring and the propensity of the metal center to undergo redox chemistry, one may speculate on the potential for effective charge-transfer when a metallocene is conjugated to an electron acceptor. However, since the metal is centrally π-bonded to two Cp rings and the ring aromaticity also results in a formal divalence on the metal center, the donating ability of the metallocene is potentially complicated. At the least, it will be dependent on the oxidation potential of the metal center and additional substituents on both five-membered rings. To assess the effectiveness of using metallocene donors for nonlinear optics, we have characterized the hyperpolarizabilities of several ferrocene and ruthenocene derivatives and have examined various structural dependencies, summarized in Table III. Compounds III.1 and III.2 represent the cyclopentadienyl analogues of acceptor substituted benzenes. Compounds III.3 to III.7 carry structural resemblance to some nitrostilbenes whose nonlinear properties have been previously studied.(31) By comparing current results with those obtained for benzene and stilbene derivatives, the nonlinearities of the metallocene derivatives can be put into perspective. Several structural variations, including different metal centers, cis and trans isomers, extension of conjugation and symmetric electron donating substituents in the form of pentamethylcyclopentadienyl rings (Cp*) have been implemented.

Table II. Powder SHG efficiencies of (E)-$(C_5H_5)Fe(C_5H_4)$-CH=CH-(4)-$C_5H_4N(CH_3)^+X^-$ salts with 1907 nm input (Urea = 1).

X =	$B(C_6H_5)_4$	I	Br	Cl	CF_3SO_3	BF_4	PF_6	NO_3	$CH_3C_6H_4SO_3$
SHG eff.	13	220	170	0	0	50	0.05	120	13

Table III: Summary of linear and nonlinear optical data on metallocene derivatives

Compound	M	X	Y	solvent	$\mu \ 10^{-18}$(esu)	$\alpha \ 10^{-23}$(esu)	$\beta \ 10^{-30}$(esu)
III.1	Fe	H	$COCH_3$	p-Diox	3.0	2.6	0.3 ± 0.2
III.2	Ru	Me	NO_2	CH_2Cl_2	5.5	3.9	0.6 ± 0.2

Compound	M	X	n	λ_{CT}(nm)	$\mu \ 10^{-18}$(esu)	$\alpha \ 10^{-23}$(esu)	$\beta \ 10^{-30}$(esu)
III.3	Fe	H	1(*trans*)	356/496	3.6	4.4	34
III.4	Fe	H	1(*cis*)	325/480	3.4	4.1	14
III.5	Fe	Me	1	366/533	4.4	5.6	40
III.6	Ru	H	1	350/390	4.5	4.2	16
III.7	Ru	Me	1	370/424	5.3	5.3	24
III.8	Fe	H	2	382/500	4.1	5.3	66

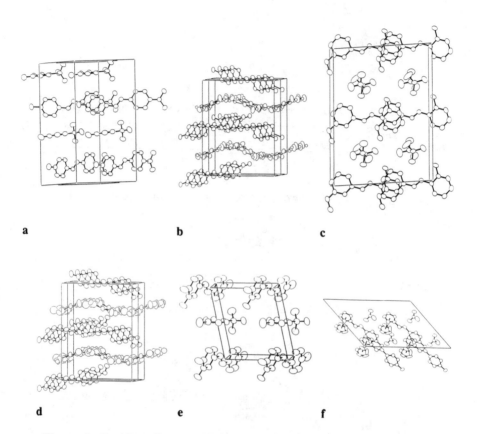

a   b   c

d   e   f

Figure 1. Packing diagrams for **a** (4)-(CH₃)₂NC₆H₄-CH=CH-(4)-C₅H₄N(CH₃)⁺p-CH₃C₆H₄SO₃⁻, **b** (4)-CH₃OC₆H₄-CH=CH-(4)-C₅H₄N(CH₃)⁺Cl⁻·4H₂O, **c** (4)-CH₃OC₆H₄-CH=CH-(4)-C₅H₄N(CH₃)⁺Cl⁻·4H₂O, **d** (4)-BrC₆H₄-CH=CH-(2)-C₅H₄N(CH₃)⁺CF₃SO₃⁻, **e** (3,4)-(OH)₂C₆H₃-CH=CH-(4)-C₅H₄N(CH₃)⁺p-CH₃C₆H₄SO₃⁻, and **f** (C₅H₅)Fe(C₅H₄)-CH=CH-(4)-C₅H₄N(CH₃)⁺NO₃⁻

The low energy spectra of simple metallocenes are dominated by two weak bands at 325 nm and 440 nm (the band at 425 nm is actually two unresolved bands) for ferrocene and 277 nm and 321 nm for ruthenocene.(32,33) The spectrum changes dramatically upon substitution of the Cp ring with conjugated and/ or acceptor groups. For example, in the spectrum of $(C_5H_5)Fe(C_5H_4)$-CH=CH-(4)-$C_5H_4N(CH_3)^+I^-$ there are two bands in the visible in acetonitrile, one at $\lambda_{max} = 380$ nm ($\varepsilon = 29,000$ $M^{-1}cm^{-1}$), the second at $\lambda_{max} = 550$ nm ($\varepsilon = 8,000$ $M^{-1}cm^{-1}$). Similar changes are observed for the ruthenocene derivatives, but are in general less pronounced. These changes are understood in terms of the changes of the molecular orbital (MO) picture upon substitution of the Cp with conjugated acceptors. The bonding in ferrocene is well understood.(32,33) Eight electrons reside in 4 strongly bonding orbitals which are largely π ring-orbital in character. Four electrons occupy two bonding orbitals which provide the key d -π interactions between the ring e_{1g} and the metal d_{xz} and d_{yz} orbitals. The remaining six electrons fill the largely nonbonding MOs which are essentially the d_{zz} (a_{1g},) and the d_{xy} and d_{x2-y2} (e_{2g}) of the metal center. Although there remains disagreement on their relative order, d_{zz} is generally accepted as the highest occupied molecular orbital (HOMO) of ferrocene. The lowest unoccupied molecular orbital (LUMO) is metal d_{xz} and d_{yz} (e_{1g}) and above this lie metal Cp antibonding orbitals derived from Cp π^* orbitals. The low energy bands in the electronic spectrum of ferrocene are assigned to two $^1A_{1g}$-$^2E_{1g}$ and a $^1A_{1g}$-$^1E_{2g}$ ligand field transitions.(32,33) Upon substitution of the Cp with conjugated acceptors, qualitatively one would expect that the low lying π^* ligand orbitals would shift to lower energy and there would be increased mixing of the ligand orbitals with the metal d orbitals. Extended Huckel molecular orbital calculations on III.4 are in agreement with this picture. (Green, J. C. Unpublished results) The HOMO is almost completely d_{zz} and nonbonding in character and the next lower energy occupied orbital has substantial metal as well as π ligand character. The LUMO is largely localized on the nitro group and the next highest unfilled orbital has coefficients distributed throughout the ligand π systems. We therefore tentatively assign the lowest energy transition in these systems as a metal (HOMO) to ligand (the orbital immediately above the LUMO) *charge transfer* band and the higher energy transition as being effectively a ligand π (immediately below the HOMO) to π^* (LUMO) transition *with some metal character*. Electron density is substantially redistributed in both transitions and therefore *both* transitions will likely contribute to β. Solvatochromic behavior has been observed for compounds III.3 and III.5. The lower lying bands are bathochromically shifted about 8-10 nm and the higher lying bands by somewhat less between *p* -dioxane and acetonitrile solutions, indicative of increased polarity in the Frank-Condon excited state. The π−π*CT transition is analogous to the CT transition in donor/acceptor substituted benzenes where electron densities move from a filled, bonding π-orbital of benzene perturbed by the donor, (here the iron atom) to an empty low-lying orbital of the substituent. The lowest energy MLCT is fundamentally different because an electron almost completely localized in one orbital is transferred upon excitation. The donated electron density involved in both charge transfer (CT) bands depends strongly on the metal center and it is not valid to consider the metal center as a counterion merely providing a full electron to form a 5-member aromatic Cp anion. We expect the higher energy band to be more sensitive to variations in the extended π system and the lower energy band to changes at the metal center. Using compound III.3 as a reference, pentamethyl substitution of one ring leads to 36 and 10 nm bathochromic shifts of the lower and higher energy bands respectively. Replacement of iron by ruthenium lowers the energy of the nonbonding d orbitals, thus increasing the meta'ls redox potential and lowering its donating strength. As expected the lower energy band is hypsochromically shifted by 106 nm but the higher energy band is only shifted by 6 nm. In contrast the higher energy band of Z, Z [1-ferrocenyl, 4-(4-nitrophenyl)-butadiene], III.8, is shifted bathochromically (26nm) and hyperchromically relative to III.3 and the lower energy band shifts slightly (4nm).

Compounds III.1 and III.2 show somewhat larger dipole moments but substantially lower β values compared to their benzene analogues. The dipole moment of the ruthenium compound III.2 (5.5 Debye) is particularly high given a value of only 4.0 Debye for nitrobenzene. Two independent factors contribute to the large dipole moment of this compound. First, the electron releasing Cp* enhances the donor strength of the metal center and therefore the donor strength of the acceptor-substituted Cp as well.(34) This effect is clearly seen among the compounds investigated (*e.g.* III.3 vs III.5 and III.6 vs III.7). The second factor is that the greater orbital extent of ruthenium 4d vs iron 3d orbitals could perhaps stabilize more charge-transfer in the ground state (III.6 vs III.3 and III.7 vs III.5) This latter rationale has been used to explain the increased stability of α ruthenocenyl cations relative to α ferrocenyl cations.(35)

The low β values for compounds III.1 and III.2 may be due to the poorly defined CT axes since the metal-ring bond is perpendicular to the ring substituent bond. Other derivatives, which have well defined charge transfer axes along the 4-nitrophenylvinyl group, show respectable nonlinearities in comparison with nitrostilbene (β = 9.1x10^{-30} esu), 4-4'-methoxynitrostilbene (β = 29x10^{-30} esu), and 4-4'-dimethylaminonitrostilbene (β = 75x10^{-30} esu).(30) Since these compounds have long wavelength absorption bands, the measured nonlinearity has a small dispersive enhancement. The *cis* compound III.4 is found to be less nonlinear than the *trans* compound III.3. It is expected that the *cis* geometry compound would exhibit a lower β for two reasons: the steric interactions between the ortho Cp and ortho benzene hydrogens preclude the two rings being coplanar (this was seen in the crystal structure of III.4 (23)) resulting in a diminution of coupling between the donor and the acceptor; the through-space distance between the donor and the acceptor is less in the *cis* compound and therefore the change in dipole moment per unit charge separated will be less. Permethylation of the opposite ring significantly increases both the dipole moment and the nonlinearity, resulting from the destabilization of the high-lying occupied orbitals as evidenced by the large spectral red shift and lowered oxidation potentials.(36) The ruthenium compounds are less nonlinear than their iron counterparts, which is consistent with the higher oxidation potential (36) of ruthenocene vs. ferrocene. In agreement with structural trends observed in stilbene derivatives, the effect of increased conjugation length is dramatic with compound III.8 exhibiting significantly higher β than III.3.

Our findings concerning quadratic hyperpolarizabilities can be summarized as follows: (1) The dipole projections of the β tensors of ferrocene and ruthenocene complexes are comparable to the methoxyphenyl system with similar acceptors. The donating strengths of the metallocenes are attributed to the low binding energy of metal electrons and their effectiveness is dependent on structural modifications which also influence their redox potentials. (2) Based on binding energies and redox potential alone the molecular hyperpolarizabilities might be expected to be larger than the observed values. *Poor coupling between the metal center and the substituent because of the π geometry most likely lowers the effectiveness of the metal center as a donor.* Future studies of organometallic systems for both second and third order hyperpolarizabilities should focus on improving the coupling between the metal center and the rest of the organic fragment.

Highly polarizable π systems for second-order NLO. The majority of systems which have been explored for second-order nonlinear optics fall into the categories of substituted: benzenes; biphenyls; stilbenes; tolanes; chalcones; and related structures. All of these compounds have have aromatic ground states. When an electric field polarizes the molecule and moves electron density from the donor to the acceptor, the π system develops some quinoidal character. Some of the stabilization associated with aromaticity is therefore lost upon charge transfer (Fig. 2a). The π system here does not act as a low resistance wire to transfer charge but rather serves to keep charge somewhat localized. Clearly if the donor and acceptor were connected by a polyene moiety, then to a first approximation there is little difference between the charge-separated and neutral forms (Fig. 2b) in so far as the π system is concerned. The two forms of the π system are degenerate and thus the π system does in fact act as a wire. The oft cited two state model predicts that β has a $1/\omega^2$ dependence (where ω is an energy denominator of the charge transfer transition).[4] It is well established from dye chemistry that chromophores with degenerate (or nearly degenerate) π systems will have very low energy absorptions. Another more classical way of looking at this is that an odd polyene system should be inherently more polarizable than an aromatic system of comparable length. The drawback of donor/acceptor polyene systems is their instability with respect to a variety of decomposition reactions including polymerization reactions, even at relatively short conjugation lengths. We therefore sought to examine a system which retained the degeneracy of the the π system and yet was amenable to handling in air for extended periods without significant decomposition. A structure such as that shown in Fig 2c was a good candidate for examination. As can be seen, one ring of the π system is aromatic, but the other ring is quinoidal. In the charge separated state, the nature of each ring is simply reversed such that the degeneracy of the π system is retained.

Dimethylaminoindoaniline (DIA), a commercially available dye, has the basic electronic features outlined above. We note that steric interactions between the ortho hydrogens of the two rings precludes both rings lying in the same plane. This undoubtedly results in a decrease in oscillator strength and polarizability. Nonetheless, this is a logical starting point to test our hypothesis. EFISH measurements of this compound in chloroform yielded a value for β of 190×10^{-30} esu. The low absorption edge leads to a dispersive contribution. Correcting for this using the two level model [4] gives the value, $\beta_0 = 95 \times 10^{-30}$ esu. In contrast, DANS gives $\beta = 75 \times 10^{-30}$ esu and $\beta_0 = 52 \times 10^{-30}$ esu. Thus even though DIA is two atoms shorter than DANS and is bent, its β is roughly a factor of three larger and β_0 is almost a factor of two larger. Another interesting comparison is to dimethylaminophenylpentadienal which is structurally similar to DIA but lacks a double bond which is critical for the degeneracy of the π system. β for this compound is 52×10^{-30} esu. These results strongly imply that the aromaticity gained by the quinone ring in the charge separated state is to a great extent responsible for the large nonlinearity of DIA. This can be viewed either as increasing the polarizability of the π system or increasing the acceptor strength of the carbonyl moiety.

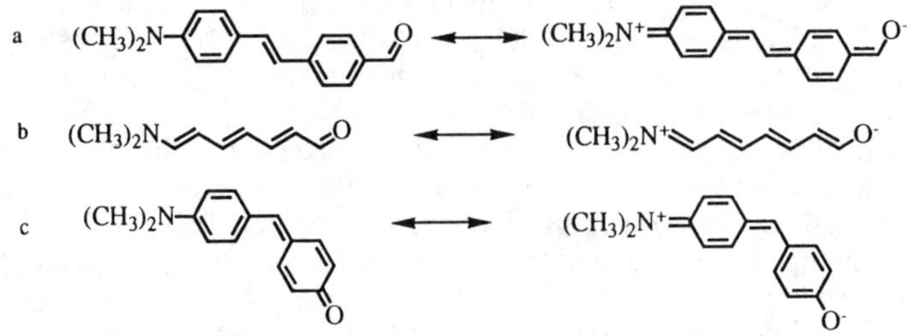

Figure 2. Neutral and charge separated resonance forms for various π systems.

DIA　　　　　　　IPB 1　　　　　　IPB 2

We hypothesized that structural changes which broke the degeneracy of the π system (but kept the overall length of the molecule constant) would in general lower the hyperpolarizability of the molecule. To test this hypothesis we examined indophenol blue (IPB). This dye is commercially sold as a mixture of isomers which can be readily separated by chromatography. The molecular structures are assigned on the basis UV-visible using steric arguments. The isomer with the aniline ring situated over the fused ring (IPB 1) would have severe steric interactions between the ortho hydrogen of the aniline ring and the hydrogen ortho to the ring juncture carbon leading to a large deviation from planarity; in contrast the other isomer has the aniline ring oriented away from the fused ring (IPB 2) and can therefore adopt a more planar configuration. We therefore assign the former structure, IPB 1, to the material which absorbs at higher energy (560nm) and the latter, IPB 2, to the material which absorbs at lower energy (610nm). We predicted that both IPB 1 and 2 would have lower β than DIA since one of the double bonds of the "quinone" is already involved in aromatic bonding. Thus the gain of aromaticity in the charge separated forms of IPB 1 and 2 would be expected to be less than in the case of DIA. The measured β values of 78×10^{-30} esu for IPB 1 and 90×10^{-30} esu for IPB 2 are consistent with the hypothesis. Further, the lower β value of IPB 1 as compared to compared to IPB 2 is indicative of diminished coupling between the donor and the acceptor, consistent with the assigned structures.

In conclusion we suggest a new methodology for enhancing β which does not rely of the use of "stronger" donors of acceptors in the normal sense of the word, but rather on the judicious tuning of the π system. We are currently exploring structure property relationships for this intriguing system.

Acknowledgments

The authors thank H. Jones, T. Hunt, Dr. L. Khundkar, and K. J. Perry for expert technical assistance. The research described in this paper was performed, in part, by the Jet Propulsion Laboratory, California Institute of Technology as part of its Center for Space Microelectronics Technology which is supported by the Strategic Defense Initiative Organization, Innovative Science and Technology Office through an agreement with the National Aeronautics and Space Administration (NASA). The diffractometer used in this study was purchased with a grant from the National Science Foundation #CHE-8219039.

Literature Cited

1. Williams, D. J. Angew. Chem. Int. Ed. Engl. 1984, 23, 690.
2. Nonlinear Optical Properties of Organic and Polymeric Materials; Williams, D. J., Ed.; ACS Symposium Series No. 233; American Chemical Society: Washington, DC, 1983.
3. Nonlinear Optical Properties of Organic Molecules and Crystals; Chemla, D. S.; Zyss, J., Eds.; Academic: Orlando, 1987; Vols. 1 and 2.
4. Oudar, J. L.; Chemla, D. S. J. Chem. Phys. 1977, 66, 2664.
5. Levine, B. F.; Bethea, C. G. J. Chem. Phys. 1977, 66, 1070.
6. Lalama, S. J.; Garito, A. F. Phys. Rev. A. 1979, 20, 1179.
7. Nicoud, J. F.; Twieg, R. W. In Nonlinear Optical Properties of Organic Molecules and Crystals; Chemla, D. S.; Zyss, J., Eds.; Academic: Orlando, 1987; Vol 1, p 242.
8. Zyss, J. S.; Nicoud, J. F.; Koquillay, M. J. Chem. Phys. 1984, 81, 4160.
9. Twieg, R. W.; Jain, K. In Nonlinear Optical Properties of Organic and Polymeric Materials; Williams, D. J., Ed.; ACS Symposium Series No. 233; American Chemical Society: Washington, DC, 1983, p 57.
10. Zyss, J. S.; Chemla, D. S.; Nicoud, J. F. J. Chem Phys. 1981, 74, 4800.
11. Tomaru, S.; Zembutsu, S.; Kawachi, M.; Kobayashi, M. J. Chem. Soc., Chem. Comm. 1984, 1207.
12. Tam, W.; Eaton, D. F.; Calabrese, J. C.; Williams, I. D.; Wang, Y.; Anderson, A. G. Chemistry of Materials, 1989, 1, 128.
13. Weissbuch, I.; Lahav, M.; Leiserowitz, L.; Meredith, G. R.; Vanherzeele, H. Chemistry of Materials, 1989, 1, 114.
14. Cox, S. D.; Gier, T. E.; Bierlein, J. D.; Stucky, G. D. J. Am. Chem. Soc. 1989, 110, 2986.
15. Singer, K. D.; Sohn, J. E.; Lalama, S. J. Appl. Phys. Lett. 1986, 49, 248.
16. Meredith, G. R. In Nonlinear Optical Properties of Organic and Polymeric Materials; Williams, D. J., Ed.; ACS Symposium Series No. 233; American Chemical Society: Washington, DC, 1983, p 30.
17. Yoshimura, T. J. Appl. Phys. 1987, 62, 2028.
18. Marder, S. R.; Perry, J. W.; Schaefer, W. P. Science, 1989, 245, 626.
19. Okada, S; Matsuda, H.; Nakanishi, H.; Kato, M.; Muramatsu, R. Japanese Patent 6 348 265, 1988, Chem. Abstr. 1988, 109, 219268w.
20. Tam, W.; Guerin, B.; Calabrese, J. C.; Stevenson, S. H. Chem. Phys. Lett. 1989, 154, 93.
21. Frazier, C. C.; Harvey, M. A.; Cockerham, M. P.; Hand, H. M.; Chauchard, E. A.; Lee, C. H. J. Phys. Chem. 1986, 90, 5703.
22. Eaton, D. F.; Anderson, A. G.; Tam, W.; Wang, Y. J. Am. Chem. Soc. 1987, 109, 1886.
23. Green, M. L. H.; Marder, S. R.; Thompson, M. E.; Bandy, J. A.; Bloor, D.; Kolinsky, P. V.; Jones, R. J. Nature 1987, 330, 360.
24. Calabrese, J. C.; Tam, W. Chem Phys. Lett. 1987, 133, 244.
25. Anderson, A. G.; Calabrese, J. C.; Tam, W.; Williams, I. D. Chem. Phys. Lett. 1987, 134, 392.
26. Bandy, J. A.; Bunting, H. E.; Green, M. L. H.; Marder, S. R.; Thompson, M. E.; Bloor, D.; Kolinsky, P. V.; Jones, R. J. In Organic Materials for Non-linear Optics; Hann, R. A.; Bloor, D., Eds.; Royal Society of Chemistry Special Publication No. 69; Royal Society of Chemistry: London, 1989.

27. Kurtz, S. K.; Perry, T. T. J. Appl. Phys. 1968, 39, 3798.
28. Marder, S. R.; Perry, J. W.; Schaefer, W. P.; Tiemann, B. G.; Groves, P. C.; Perry, K. J. Proc. SPIE. 1989, 1147, p 108.
29. Levine, B. F.; Bethea, C. G. Appl. Phys. Lett. 1974, 24, 445.
30. Cheng, L. ; Tam, W.; Meredith, G. R.; Rikken, G. L.; Meijer, E. W. Proc. SPIE. 1989, 1147, p 61.
31. Meredith, G. R.; Cheng, L. T.; Hsiung, H.; Vanherzeele, H. A.; Zumsteg, F. C. Materials for Nonlinear and Electro-optics; Lyons, M. H., Ed.; The Institute of Physics: New York, 1989; p 139.
32. See for example: Rosenblum, M. Chemistry of the Iron Group Metallocenes; Interscience, Wiley & Son: New York, 1965; Chapter 2.
33. Sohn, Y. S.; Hendrickson, D. N.; Gray, H. B. J. Am. Chem. Soc. 1971, 93, 3603.
34. Richmond, H. H.; Freiser, H. J. Am.Chem. Soc. 1954, 77, 2022.
35. Turbitt, T. D.; Watts, W. E. J. Chem. Soc., Perkin II 1971, 177.
36. Kuwana, T.; Bublitz, D. E.; Hoh, G. L. K. J. Am. Chem. Soc. 1960, 82, 5811.

RECEIVED August 14, 1990

Chapter 12

Chemistry of Anomalous-Dispersion Phase-Matched Second Harmonic Generation

P. A. Cahill[1] and K. D. Singer[2]

[1]Sandia National Laboratories, Albuquerque, NM 87185–5800
[2]AT&T Bell Laboratories, Princeton, NJ 08540

The anomalous dispersion associated with a strong
absorption in some carefully chosen asymmetric dyes
permits efficient phase-matched SHG at a given frequency
and concentration. One of these dyes was recently used
to demonstrate the validity of the two-state model for β,
and leads to a method of enhancing second harmonic
coefficients in poled polymer systems by 10^1 to 10^4. The
factor that primarily limits the utility of this process
is the residual absorbance in a nearly transparent window
on the high energy side of a charge transfer band. One
figure of merit for comparing dyes for this application
is the ratio between this minimum absorbance and ε_{max};
for many dyes this ratio is only 10^{-1} to 10^{-2}. Synthesis
of new dyes has led to $\varepsilon_{min}/\varepsilon_{max}$ ratios of 10^{-3} to 10^{-4}.

Organic materials for second order nonlinear optical (NLO)
applications were first investigated in the 1960's, and since that
time research has become divided along two lines: crystals for
second harmonic generation (SHG) and related applications, and thin
(aligned) films for integrated optics. The molecular basis of the
second order NLO coefficient β is well understood, and the challenges
associated with noncentrosymmetric alignment of molecules in crystals
and thin films have been addressed. Our work has focussed on a means
of doubling near-IR frequencies by using dyes which absorb between
the fundamental and second harmonic. This approach leads to a means
of efficient, collinear phase-matched second-harmonic generation
(SHG) through the anomalous dispersion associated with this
electronic transition, and results in an increase in the useful
magnitude of β, the microscopic second order hyperpolarizability.
The applications for this approach are in thin film devices for SHG
and electrooptic (EO) modulation.

Our recent report(1) on ADPM SHG (Anomalous-Dispersion Phase-
Matched Second-Harmonic Generation) addressed the physics of ADPM, a

0097–6156/91/0455–0200$06.00/0
© 1991 American Chemical Society

subject which was first discussed in 1962(2) and which has been
studied as a means of generating third (and higher odd order)
harmonic light many times. Phase-matched harmonic generation is
obtained by using the anomalous dispersion of an absorbing species to
cancel the normal dispersion of a host material, such that at a
particular concentration of dyes, the indices of refraction at the
fundamental and generated harmonic are equal. This method has been
the most successful in gases where absorption lines are narrow and
little residual absorption results;(3) it has been somewhat less
useful for third harmonic generation in solutions of organic dyes
because of problems associated with two photon absorption and
residual absorption at the third harmonic.(4) These problems are
absent or less severe with ADPM SHG.

Prior to our report, ADPM second order materials had not been
been investigated with the exception of one serendipitous discovery
of ADPM difference frequency generation in a noncentrosymmetric
semiconductor in the mid-IR.(5) However, experimental evidence and
theoretical arguments (vide infra) suggest that significant increases
(10^1 to 10^4) in the effective NLO coefficients in organic second
order materials are possible through this general approach. In
addition, this report includes the first work towards optimizing dyes
for these systems.

Consequences of the Origin of β in Organic Dyes

On the microscopic scale, overwhelming evidence suggest that, in the
absence of unusual delocalized excited states, for which there is
very little evidence in organic second order NLO materials, the
expression for β for almost all dyes is descibed by a simple two
state model via the following expression:(6)

$$\beta = \frac{[e^3 \mu_0^2 \Delta\mu]}{\hbar^2} \frac{3\omega_0^2}{(\omega_0^2 - \omega^2)(\omega_0^2 - 4\omega^2)}$$

where ω is the incident fundamental frequency
       ω_0 corresponds to the energy of the first excited state
       μ_0 is the transition moment, and
       $\Delta\mu$ is the difference between the dipole moments of the
             ground and excited states.
The conventional and very effective approach to increasing β has been
to use dyes that absorb at the longest feasible wavelength and to use
wavelengths as close as possible to the dye's absorption edge so that
the frequency factor (the second factor in the above expression) is
favorable. However, if the wavelength at which the device is
required to function is fixed, such as in the case of doubling a
diode laser to 400-450 nm, there is a very severe limit in the
conventional approach to the size of β in organic dyes because of the
inherent nature of chromophores that absorb to the blue of a desired
wavelength, i.e., β is limited by the magnitude of the first factor.
Furthermore, once this first factor in β is reasonably optimized, β
is dominated by the second (frequency) factor, which goes as $1/\omega_0^2$.

Our alternate approach for SHG is to design a dye which absorbs between ω and 2ω but which has a low absorption at both of these wavelengths.(7) Based on the expression for β, the following points are known: (1) the terms in the denominator of the frequency factor should change sign and if ω_0 is close to either ω or 2ω, the magnitude of the denominator should decrease, increasing β; (2) since both the transition moment and change in dipole moment terms also scale with increasing wavelength, β should further increase; (3) the ω_0^{-2} behavior of β would be expected to yield additional factor of approximately 4. Overall, a total increase in β of approximately an order of magnitude can be expected from an optimized dye which absorbs between ω and 2ω over the conventional approach of using a dye which absorbs only at wavelengths shorter than the second harmonic wavelength.

Consequences of the Origin of $\chi^{(2)}$ in Poled Polymers

Whereas the challenges of second order materials on a microscopic scale lie primarily in the nature of the electronic states of isolated molecules, the challenges on a macroscopic scale are associated with the noncentrosymmetric alignment of the NLO molecules. Phase mismatch for harmonic generation occurs due to the natural dispersion in all materials, but a phase-matched condition (long or infinite coherence length) is required for efficient transfer of energy from the fundamental to the harmonic. Phase-matching is often accomplished via the birefringence of crystalline materials which may be tuned by careful adjustment of the crystal relative to the beam. In poled polymer systems, which may be the most applicable to integrated optics applications, $\chi^{(2)}$ is proportional to the $\mu\beta$ product, and phase-matching in a waveguide might be accomplished by proper modal overlap.(8)

The use of a poled polymeric material incorporating a dye with an absorption between ω and 2ω (an "ADPM dye") leads to several advantages. Based on the discussion (above) on β, the macroscopic hyperpolarizability, $\chi^{(2)}$, can be expected to increase by at least an order of magnitude simply because it is proportional to the $\mu\beta$ product. In addition, if the dye is present in the proper concentration for ADPM SHG, this collinear process allows coupling to the largest component of the β tensor, which results in a further increase in $\chi^{(2)}$. Furthermore, because one can phase-match the diagonal components of $\chi^{(2)}$ by propagation along a principal dielectric axis, geometric problems such as beam walkoff, spatial dispersion, and beam overlap are reduced or eliminated; and geometric inefficiencies due to polarization changes typical of phase-matching in birefringent crystals are absent. In a waveguide configuration, ADPM SHG allows coupling from/to the lowest order modes and therefore maximizes overlap of the guided waves. In total, the magnitude of $\chi^{(2)}$ can reasonably be expected to increase by 10^1 to 10^2 over conventional materials, which, combined with the additional efficiencies gained from geometric considerations, leads to an increase of 10^2 to 10^4 or more in the efficiency of SHG by this technique. Such a large predicted increase in $\chi^{(2)}$ justifies a considerable effort in the synthesis of new dyes and polymers for this approach.

These general observations on the size and magnitude of β were confirmed by using Foron Brilliant Blue S-R (FBB) in an EFISH experiment.[1] In summary, β changes sign due to the frequency factor, while the frequency independent terms in β are the same within experimental error whether second harmonic light is generated above or below the first charge transfer absorption. The increase in the $\mu\beta$ product for ADPM dyes is apparent from Figure 1, in which the $\mu\beta$ product for several NLO dyes(9) is plotted against the shortest transparent wavelength (conventional NLO dyes) or the transparent "window" wavelength (for ADPM dyes). In this way, the relative $\chi^{(2)}$'s for a given desired (SHG) wavelength can be directly compared. A one order of magnitude increase in $\mu\beta$ is apparent even for these nonoptimal ADPM dyes. Even though the requirement for transparency for SHG is less severe than it is for third and higher order harmonic generation, the real limitation for SHG in FBB was still found to be absorption of the second harmonic. In FBB, the residual absorption is so great that useful amounts of SHG were not generated.

One challenge of ADPM SHG is therefore to design a system (dye(s) plus host) in which the absorption at the second harmonic is minimized. Two schemes have been devised to accomplish this goal. In the (less preferred) first scheme, two dyes are used -- one (generally symmetric) dye with an absorption between ω and 2ω for phase-matching and a conventional 2nd order NLO dye which absorbs to the blue of 2ω for SHG. Only the geometric factors leading to an increased $\chi^{(2)}$ are obtained from this scheme, but a symmetric dye with low residual absorption for phase-matching may be more easily prepared. The preferred scheme, which would potentially give rise to the greatest SHG efficiency, involves a dye, like FBB, which does both SHG and ADPM. Of course, fine tuning of the system via the addition of dyes which either add to or subtract from the dispersion may add considerable flexibility to both approaches.

Chromophore Design

The general constraints for the design of any dyes for ADPM SHG in poled polymer systems rapidly narrow the choice of chromophores. The dyes should be overall charge neutral (to facilitate poling) and highly soluble in polymer matrices. The first excited electronic state should be well separated from higher energy states for two reasons: (1) since the vibronic envelope associated with an electronic absorption often tails to the blue, a greater energy separation between excited states may give lower absorption, and (2) the next state's normal dispersion, if nearby, would subtract from the desired anomalous dispersion of the lower energy transition. Dyes that would be expected to show weak, low lying n-π^* transitions above the first excited state should therefore be avoided. Finally, the molar absorptivity of the dyes should be as large as possible (>50,000 1/mol-cm) in order to generate the largest possible anomalous dispersion (which is proportional to the area under the absorption curve) at a given concentration.

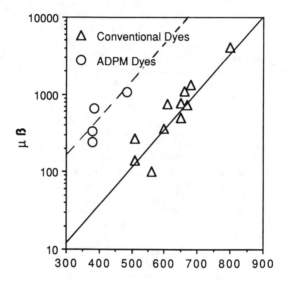

SHG Wavelength (nm)

Figure 1. Plot of $\mu\beta$ product vs. absorption edge wavelength (conventional NLO dyes($\underline{9}$)) or transparent "window" wavelength (ADPM dyes, ref. ($\underline{1}$) and this work). The lower line is a least squares fit to the data, the upper (dashed) line is a guide to the eye. An order of magnitude increase in $\mu\beta$ for a given wavelength has been observed. This is 10% of the predicted maximum increase possible for this technique.

The requirement for a strong charge transfer band well separated from other electronic absorptions immediately suggests the use of organometallic or coordination compounds with strong metal-ligand or ligand-metal charge transfer bands. An early attempt via this approach failed due to photoinstability of the compound. All subsequent work has concentrated on organic dyes.(10)

The next logical step toward chromophore design was to conduct a spectral survey of commercially available organic compounds in order to learn some general structure-property relationships for minimization of the residual absorbance. As an easily measured figure of merit, the ratio between the minimum and maximum molar absorptivities has been used. In many cases, this ratio (expressed in percent, or more conveniently, as the minimum molar absorptivity per 100,000 L/mol-cm of maximum absorbance) is 5-10% (5000-10,000 per 100,000). (The lower the number the better the dye.) An improved figure of merit would take into account the area under the absorption curve as well as the location of the transparent window relative to the peak in the absorption. This is tantamount to calculating the dispersion from the absorption spectrum, which was too complex for this type of survey.

Overall, the spectral survey provided more data on what kinds of dyes should not be used for ADPM, rather than what structural features lead to dyes with improved transparency. Among the symmetrical dyes, the carbo- and dicarbocyanines were among the best dyes surveyed, but these contain fundamentally cationic chromophores. Few of the popular laser dyes -- whether based on open chain merocyanine chromophores (like DCM), cyclic merocyanines (like the coumarins), or triarylmethine dyes (such as the rhodamines and related compounds) -- were sufficiently transparent above the first transition to yield insight into structure property relationships. However, among the merocyanine dyes, FBB S-R, a commercial fabric dye, showed the greatest figure of merit among the asymmetric dyes, at approximately 0.75% or 750 per 100,000 (the molar absorptivity of FBB S-R is approximately 62,000 L/mol-cm) residual absorbance. The wavelength of maximum absorption is moderately solvatochromatic, which is consistent with a moderate change in dipole moment in the first excited state. The wavelength of minimal absorbance is near 445 nm in CH_2Cl_2, and like most dyes that have been surveyed, λ_{min} is less solvatochromatic than λ_{max}.

Somewhat related to the (cationic) cyanines are the squarylium dyes which are overall charge neutral species derived from squaric acid. They are easily prepared, have high molar absorptivities (>100,000), but typically are unstable to hydrolysis in dipolar aprotic solvents. They are characterized by a sharp strong absorption which lies at wavelengths longer than 640 nm, with no other identifiable electronic transitions in the visible (Figure 2). The vibronic structure of these dyes may show only one shoulder corresponding to a reasonable value for a C-C or C-O stretch. We were able to systematically vary the structure of a series of squarylium dyes in order to test assumptions about the structure-property relationship for minimization of the residual absorption. An increase in the rigidity and symmetry of the structure of the dye was expected to lead to a gradual improvement in the figure of merit for the dyes in Figure 3 from diethylaminohydroxy- to

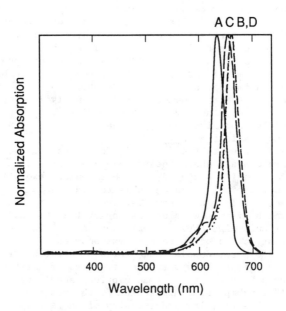

	λ_{max}	λ_{min}	ε_{min} per $10^5 \varepsilon_{max}$
A	643	480	100
B	670	508	124
C	663	502	85
D	670	421	601

Figure 2. Structure and spectral properties of squarylium dyes synthesized in this work.

A C B,D

Normalized Absorption

400   500   600   700

Wavelength (nm)

Figure 3. Normalized UV-Visible spectra of the dyes in Figure 2.

hydroxyjulolidino- to dihydroxyjulolidino- to julolidino-squarylium(11) dyes, but this was not observed.   The dye with the greatest symmetry and rigidity did yield the best figure of merit, but the Franck-Condon features (vibrational function overlaps in a vertical electronic transition model), that we believe give rise to the residual absorptions, cannot easily be predicted from the molecular structure.   The best that figure of merit observed with the squarylium chromophore was approximately 85 per 100,000, therefore alternate symmetric chromophores were investigated.   One symmetric chromophore with a figure of merit of approximately 10 per 100,000 was recently synthesized.

Furthermore, an assumption that dyes that absorb further to the red would give a better figure of merit than dyes that absorb to the blue was incorrect -- dyes that absorb in any part of the spectrum appear to be equally likely to be good for ADPM.   This appears to be due to counterbalancing effects: small chromophores, which may absorb near 450 nm, tend to have fewer available vibrational modes but at the same time have higher energy electronic states that are closely spaced, have about the same figure of merit as dyes that absorb near 700 nm, apparently because these chromophores generally contain more atoms and therefore necessarily have more vibronic states.

Variations on the structure of a basic chromophore were also pursued in asymmetric dyes, which are preferred for maximizing the NLO effects.   The starting chromophore was chosen was dimethyl-aminocinnamaldehyde (DACMA) and is shown in Figure 4 with several related dyes, including Foron Brilliant Blue S-R.   The spectra of FBB, and the closely related dyes I(12) and II are shown in Figure 5a.   Cyclization of the amine into a julolidine group results in a small bathochromic shift;   substitution of the ketone by a dicyanovinyl group leads to a much larger shift and moderately broadening of the vibronic envelope;   but neither change leads to a significant improvement in the figure of merit of these dyes.

Replacement of the sulfone by a ketone leads to the dyes III, IV, and V.(13)   The absorption spectra of III and IV (Figure 5b) are similar to FBB and its analogues, however the spectrum of V is anomalous.   The spectrum of V resembles that of the anion VI, and suggest that this dye has a zwitterionic ground state.   No improvement in the figure of merit of these dyes can reasonably be correlated with their structure.

The barbituric and thiobarbituric acid derivatives VII-X (Figure 6) are among the best (i.e. lowest residual absorption) dyes that we have synthesized.   Figures of merit for dyes VII-X are: 836 @ 349 nm, 350 @ 489 nm, 405 @ 374 nm, and 185 @ 403 nm.   These dyes have relatively high symmetry (similar to the indandione derivatives), but show much lower residual absorbances (Figure 7).   The transparent region in dye X is very close to the wavelength desired for doubling laser diode emissions.

Increasing the length of the chromophores does not directly improve the figure of merit for the series of barbituric acid derivatives shown in Figure 8.   The long flexible conjugated chain gradually increases the wavelength of maximum and minimum absorption, but is associated with a gradual increase in the residual absorbance in the near-UV.   This behavior is common to many chromophore series

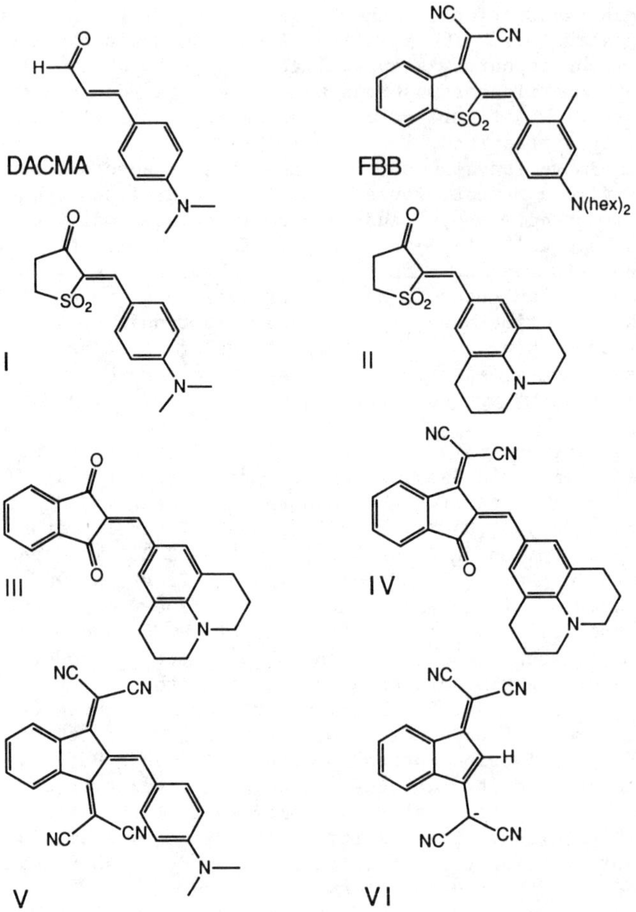

Figure 4.   Structure of some dyes related to dimethylamino-cinnamaldehyde whose spectra were studied.

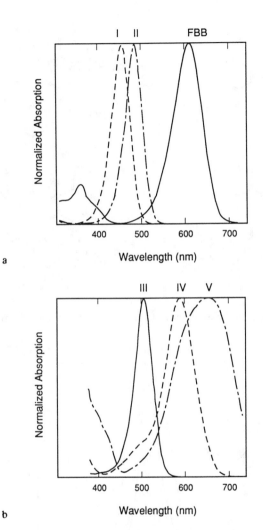

Figure 5. Normalized spectra of Foron Brilliant Blue S-R and analogs.
(a) Sulfones
(b) Ketones and derivatives

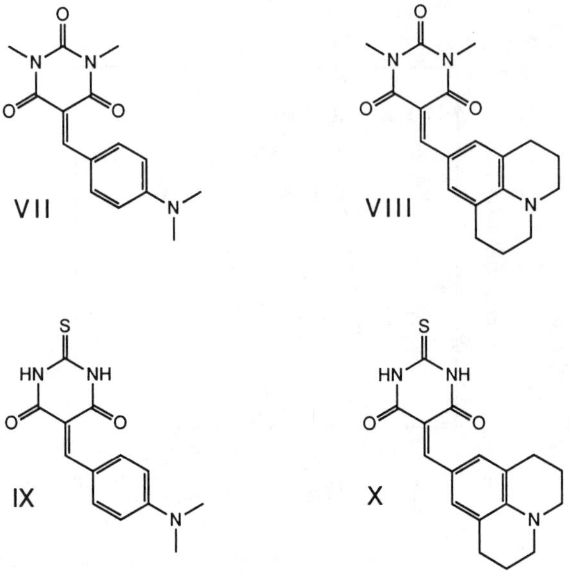

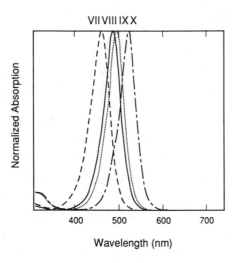

Figure 6.   Structure of barbituric acid dyes based on DACMA.

Figure 7.   Normalized spectra of barbituric acid dyes.

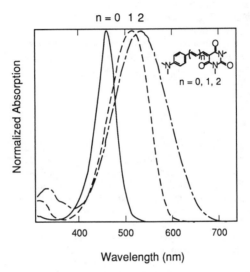

Figure 8. Normalized spectra of extended barbituric acid chromophores.

and it is not unusual to expect that the flexibility of the conjugated carbon linkage would give rise to a broader vibronic envelope.

Outlook

The ultimate limit of the transparency of a dye above its first electronic absorption is uncertain, but an order of magnitude improvement is not unreasonable in asymmetric chromophores, i.e. to levels of 10 per 100,000 of molar absorptivity. At this level, and assuming an ADPM dye concentration of 0.045 M (the same as observed for ADPM SHG with FBB by EFISH), a dye/host material would show a loss of 4.5 dB/cm. However, because of the tremendous increase in $\mu\beta$ for this material, it is likely to be 1 mm (or less) in thickness for a total loss of only 0.45 dB. This would be in the range of practical devices.

Conclusions

Based on the expression for β, a large increase in the useful NLO coefficient for a fixed wavelength is predicted in the case where the absorbance of the NLO dye lies between the fundamental and second harmonic. Residual absorption at the second harmonic is the limiting factor in the practical application of this technique, and has been addressed through the synthesis of new dyes. Improvement of 10x in reducing this absorbance has been achieved, and another factor of 5-10x is estimated to be required before practical devices can be fabricated. Franck-Condon effects (vibronic structure) appear to be responsible for this residual absorption because small, rigid chromophores are often correlated with the lowest amounts of absorption. Chromophores based loosely on dimethylaminocinnamaldehyde have been studied the most completely, and of these, barbituric acid derivatives have the highest figures of merit.

Further improvements in the magnitude of the $\mu\beta$ product which is proportional to $\chi^{(2)}$ are possible -- current ADPM dyes have only 10% of the theoretical maximum value. Considerable effort will be required to design and synthesize these advanced dyes, but is justified on the basis of theoretical SHG efficiency gains of 10^4.

Acknowledgments

A portion of this work was performed at Sandia National Laboratories and was supported by the U.S. Department of Energy under contract DE-AC04-76DP00789. The laboratory efforts of Don Strall (Sandia) and Lori King (AT&T) are gratefully acknowledged. Insightful discussions with Carl Dirk, Mark Kuzyk, and Howard Katz (AT&T) and Mike Sinclair (Sandia) have contributed to this project. FBB was a gift of the Sandoz Corporation.

Literature Cited

(1) P. A. Cahill, K. D. Singer, and L. A. King, Opt. Lett. 1989, 14, 1137.
(2) J. A. Armstrong, N. Bloembergen, J. Ducuing, and P. S. Pershan, Phys. Rev. 1962, 127, 1918.
(3) S. R. J. Brueck and H. Kildal, Opt. Lett. 1978, 2, 33.
(4) W. Leupacher, A. Penzkofer, B. Runde, and K. H. Dexhage, Appl. Phys. 1987, B44, 133.
(5) F. Zernike, Phys. Rev. Lett. 1969, 22, 931.
(6) J. L. Oudar and D. S. Chemla, J. Chem. Phys. 1977, 67, 446.
(7) Further enhancements might be obtained if the dye were to absorb even below ω_0, but at this time this appears to be very difficult.
(8) K. D. Singer and J. E. Sohn, in "Electroresponsive Polymers", T Skotheim, Ed., Marcel Dekker (1990).
(9) K. D. Singer, J. E. Sohn, L. A. King, H. M. Gordon, H. E. Katz, and C. W. Dirk, J. Opt. Soc. Am. B, 1989, 6, 1339 and references cited therein. See also reference (1).
(10) P. A. Cahill, Proc. Mat. Res. Soc., 1988, 109, 319.
(11) A. M. Morgan, P. M. Kazmaier, and R. A. Burt, U.S. Patent 4,507,480 (Mar. 26, 1985); G. Baranyi, R. A. Burt, C.-K. Hsiao, P. M. Kazmaier, K. M. Carmichael, and A. M. Horgan, U.S. Patent 4,471,041 (Sep. 11, 1984); M. S. H. Chang and P. G. Edelman, U.S. Patent 4,353,971 (Oct. 12 1982); K.-Y. Law, J. Phys. Chem. 1987, 91, 5184;
(12) K. G. Mason, M. A. Smith, E. S. Stern, and J. A. Elvidge, J. Chem. Soc. (C) 1967, 2171.
(13) These dyes are related to those first synthesized by K. A. Bello, L. Cheng, and J. Griffiths, J. Chem. Soc. Perkin Trans. II 1987, 815.

RECEIVED August 7, 1990

PREPARATION AND CHARACTERIZATION OF POLED POLYMERS

Chapter 13

Applications of Organic Second-Order Nonlinear Optical Materials

G. C. Bjorklund, S. Ducharme, W. Fleming, D. Jungbauer, W. E. Moerner, J. D. Swalen, Robert J. Twieg, C. G. Willson, and Do Y. Yoon

Almaden Research Center, IBM Research Division, San Jose, CA 95120-6099

The history of research on second order organic NLO materials is reviewed, with particular emphasis on crystals and poled polymers. Crystals are best for second harmonic generation applications, while polymers are best for electro-optic waveguide devices such as modulators and switches. Recent results on cw intracavity second harmonic generation using the organic nonlinear crystal DAN (4-(N,N-dimethylamino)-3-acetamidonitrobenzene) in an optically pumped Nd:YAG laser cavity are presented, demonstrating for the first time that laser grade optical quality can be achieved with organic NLO crystals. Progress toward high frequency electro-optic phase modulators using both organic crystals and poled polymers is discussed. A family of thermoset poled NLO polymers based on epoxy chemistry is reported. One of these materials has a second harmonic coefficient $d_{33} = 42$ pm/V that is stable for at least 14 days at 80C.

It is by now well recognized that organic nonlinear optical materials have the potential to supplant inorganic crystals as the materials of choice for frequency doubling, modulation, and switching (1). Key advantages of all types of organic NLO materials include the high intrinsic nonlinearities of individual organic molecules, the ability to use molecular engineering to tailor properties to specific applications, low dc dielectric constant, and low temperature processing.

0097–6156/91/0455–0216$06.00/0

Figure 1 schematically illustrates the history of research on 2nd order organic NLO materials.   Soon after the invention of the laser and the birth of the field on nonlinear optics, second harmonic generation and two photon absorption were observed in a variety of organic molecules.   Systematic studies of the relationship of molecular structure to molecular nonlinearities done during the 1970's brought out the importance of electron delocalization and charge transfer for high nonlinearity.   In the 1980's research began on ways to incorporate these highly nonlinear organic molecules into orientationally ordered bulk materials that would exhibit useful bulk NLO properties.   Two main approaches were initiated that are still being followed with great energy today: crystal growth and poled polymers.

Crystal growth is typically performed using a variety of techniques such as solution, melt, vapor phase or Bridgman to produce noncentrosymmetric crystals of pure NLO molecular chromophores. Unfortunately only a small fraction of the available NLO chromophores form suitable crystals with the necessary noncentrosymmetric orientational order.   However, in those cases where crystals with the proper symmetry can be grown, the high concentration of NLO chromophores can result in very large bulk nonlinearities.   For instance organic NLO crystals have figures of merit for second harmonic generation (SHG) that exceed the best inorganic materials by an order on magnitude.   In addition, these crystals also often have sufficient birefringence to allow angle tuned phase matching enhancement of the SHG conversion efficiency.

The poled polymer approach involves incorporating NLO chromophore molecules into a host polymer matrix and establishing orientational order by heating the polymer above its glass transition temperature, aligning the NLO chromophores using a strong dc electric field, and then cooling in the presence of the field.   The host NLO molecules can be simply doped into the host polymer, or in more advanced materials, chemically bonded to the polymer mainchain.   The great advantages of the poled polymer approach are the ability to use almost any NLO chromophore and the ability to cheaply and easily fabricate thin film optical waveguides on a variety of substrate materials, including electronics components.   One disadvantage of poled polymers is a gradual room temperature relaxation of the poling induced orientational ordering that occurs in many cases. (See Section IV)

Figure 2 shows the tradeoffs between crystal and polymer organic NLO materials for device applications.   Although either type of materials could in principle be used for both applications, crystals are best for second harmonic generation, and poled polymers are best for electro-optic waveguide devices such as modulators and switches.

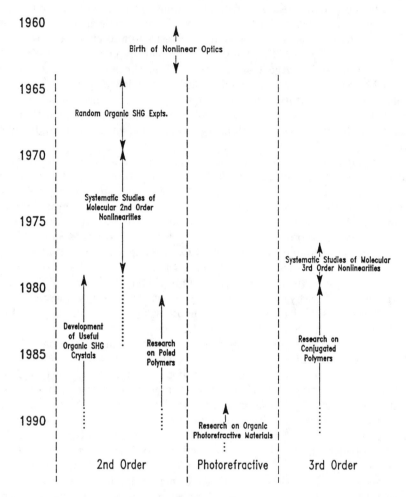

Figure 1. Schematic representation of the major themes in the history of research on organic nonlinear optical materials.

POLYMER		CRYSTAL
	General	
0	transparency	0
0	dielectric properties	0
	Fabrication	
+ + +	waveguide	−
− −	poling required	x
x	xtal growth required	− −
+	patterning by poling	x
+	dye molecule flexibility	− −
	Stability	
+ +	mechanical robustness	− −
0	optical (damage)	0
− −	nonlinear (orientational)	+ +
	EO	
+ +	level of integration	− −
+ +	$V_{1/2} \propto L/d$	− −
+ +	only d_{33} needed	−
	SHG	
− −	phase matching enhancement	+ +
+ +	waveguide enhancement	− −
− −	cavity enhancement	+ +
−	large off diagonal d_{ij}	+ +

Figure 2. Tradeoffs between polymer and crystal organic nonlinear optical materials. EO refers to applications for electro-optic waveguide devices such as modulators and switches. SHG refers to applications for frequency doubling of moderate and low power laser sources. A + indicates favored, - indicates disfavored, 0 indicates neither favored nor disfavored, and x indicates not relevant.

II. Intracavity Second Harmonic Generation Using an Organic Crystal

Intracavity second harmonic generation and frequency mixing using inorganic crystals has recently been demonstrated to be a practical means of obtaining milliwatts of cw blue or green laser light from infrared diode laser sources. A further enhancement in the achievable conversion efficiency could in principle be obtained using organic crystals with intrinsically higher figures of merit for SHG, provided that the optical quality is sufficient to allow cw intracavity laser operation.

We have recently conducted a set of intracavity second harmonic generation experiments using the organic nonlinear material DAN (4-(N,N-dimethylamino)-3-acetamidonitrobenzene) and an optically pumped cw Nd:YAG laser *(2)*. Figure 3 shows the experimental setup. Quasi-cw operation was achieved with crystal samples immersed in index matching fluid in an antireflection coated cuvette that was placed internal to the Nd:YAG laser cavity.

This technique permits rapid surveying of crystal samples obtained directly from solution growth without polishing or antireflection coating them. Up to 0.56mW peak power of 532nm light was generated for 2.3W of circulating intracavity 1064nm peak power using 0.5W of 810nm pump. Figure 4 shows the dependence of the SHG power on the fundamental power. In addition, we have achieved true cw operation using antireflection coated DAN crystals. These results represent the first cw intracavity application of an organic NLO material of any type and demonstrate that laser grade optical quality can be achieved with organic NLO crystals.

III. Electro-Optic Phase Modulators Using Organic Materials

In an attempt to demonstrate high frequency electro-optic phase modulation with organic NLO materials, we have tested several candidate crystals in a specially designed test fixture that incorporates a stripline electrode structure to produce a transverse traveling wave electrical field. The electrical response of the stripline structure was tested and found to be flat up to 3 GHZ. A single frequency laser beam was directed through the crystal and a high finesse scanning etalon was used to directly detect the optical power in the resulting FM sidebands. A schematic of the experiment is shown in Figure 5. Using a crystal of MNMA (2-methyl-4-nitro-N-methylaniline), a modulation index of M = .014 was achieved at a drive frequency of 400 MHz. This represents the first demonstration of high speed electro-optic phase modulation in an organic crystal.

Experiments are underway to fabricate a waveguide phase modulator using the thermoset poled NLO polymers described in Section IV. So far, metal bottom electrode / polymer buffer layer / NLO polymer layer /

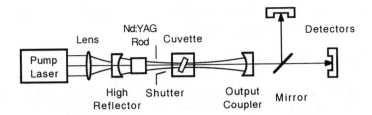

Figure 3. Experimental setup for intracavity SHG using an organic NLO crystal placed in the cuvette.

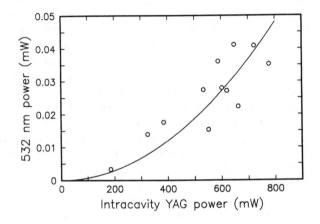

Figure 4. Peak SHG output vs peak intracavity power.

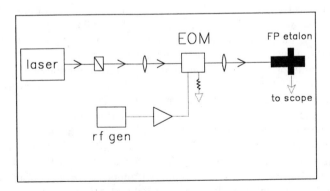

Figure 5. Experimental setup for electro-optic phase modulation.

polymer buffer layer / top metal electrode structures have been fabricated and waveguiding has been achieved.

IV. Thermoset Poled NLO Polymers

One of the problems that has plagued both guest-host and covalently functionalized linear mainchain types of poled NLO polymers has been relaxation of the poling induced orientational order and hence of the nonlinearity. The time scale of this relaxation in guest-host materials can be on the order of tens of hours at room temperature (3). Covalent attachment of the NLO chromophores to linear polymer mainchains improves the time scale to a few thousand hours at room temperature (4), but this is still far short of the ten year room temperature lifetime required for practical devices. In addition, compatibility with electronics requires long term stability at elevated temperatures in the 50 to 80C range as well as short term tolerance of processing temperatures that could exceed 200C.

Motivated by these concerns, we have pursued the development of thermoset poled NLO polymers where the NLO chromophores are covalently attached with multiple chemical bonds to a cross linked polymer matrix, as shown schematically in Figure 6. In our first experiments a tetrafunctional nonlinear chromophore, 4-nitro 1,2 phenylenediamine was reacted with an optically passive bifunctional epoxy monomer, diglycidylether of bisphenol-A to form a soluble prepolymer composed 21% by weight of NLO moieties that could be spin coated onto flat substrates (5). A precuring step at 100C was then necessary to increase the viscosity of the polymer to withstand the high poling electric fields without breakdown. The sample was then heated to 140C and subjected to > 1 MV/cm dc electric field from a corona discharge. After 16 hours the fully cured sample was cooled in the presence of the poling field and the SHG coefficient d_{33} measured using the Maker fringe technique with a 1.064 um fundamental wavelength. Figure 7 schematically illustrates these processing steps. A value of d_{33} = 14 pm/V was measured immediately after poling and found to be stable for at least 500 hours at room temperature and to exhibit no detectable decay after 30 minutes at 80C.

In subsequent experiments aimed at extending this approach to produce thermoset poled NLO polymers with higher nonlinearities (6), the epoxy monomer was also functionalized to contain an NLO moiety, as shown in Figure 8. The polymer thus formed by reacting bifunctional N,N-(diglycidyl)-4-nitroaniline and trifunctional N-(2-aminophenyl)-4-nitroaniline was composed 63% by weight of NLO moieties. However, the NLO moieties in this polymer are only singly attached to the crosslinked matrix as opposed to the double attachment of the previous polymer. After a processing sequence similar to that of Figure 6, except that the temperature was ramped up step by step for the final cure, a d_{33} = 50 pm/V was measured immediately after poling (for comparison, d_{33} = 30

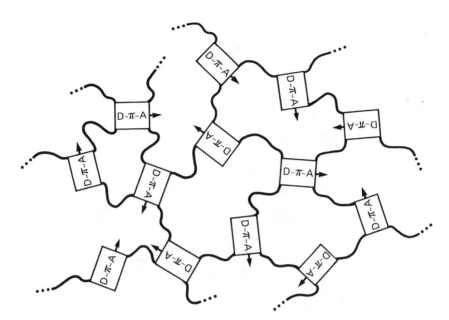

Figure 6. Schematic of a cross linked NLO polymer. The NLO chromophores are represented by the D-π-A boxes and the arrow represents the charge-transfer axis.

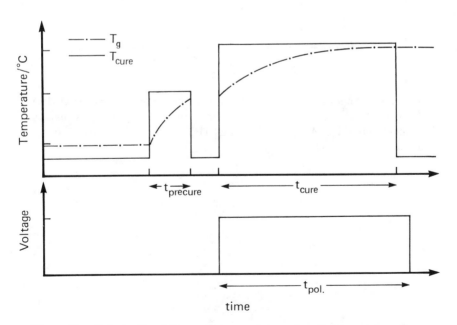

Figure 7. Schematic of thermal processing and poling procedure for a thermoset NLO polymer.

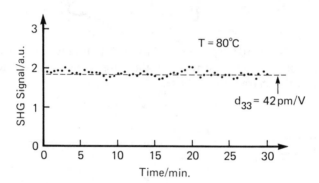

Figure 8. Schematics of the epoxy monomer N,N-(diglycidyl)-4-nitroaniline (a.) and of the amine monomer N-(2-aminophenyl)-4-nitroaniline (b.) Also shown is the SHG signal vs time at 80C for the already annealed sample.

pm/V for LiNbO$_3$). Upon heating to 80C, a small decay to 42 pm/V occurred in the first 15 minutes, but thereafter, as shown in Figure 7, no decay was observed for 30 minutes at 80C. Waveguide birefringence studies indicate that no further decay occurred after 14 days at 80C.

Literature Cited

1. Zyss, J.; Chemla, D. S., "Nonlinear Optical Properties of Organic Molecules and Crystals", *Academic Press*. 1987.
2. Ducharme, S.; Risk, W. P.; Moerner, W. E.; Lee, V. Y.; Tweig, R. J.; Bjorklund, G. C., accepted for publication in *Applied Physics Letters*.
3. Hampsch, H. L.; Yang, J.; Wong, G. K.; Torkelson, J. M. *Macromolecules*. 1988, *21*, 526.
4. Ye, C.; Minemi, N.; Marks, T. J.; Yang, J.; Wong, G. K. *Macromolecules*. 1988, *21*, 2899.
5. Eich, M.; Reck, B.; Yoon, D. Y.; Willson, C. G.; Bjorklund, G. C. *J. Appl. Phys.* 1989, *66*, 3241.
6. Jungbauer, D.; Reck, B.; Twieg, R. J.; Yoon, D. Y.; Willson, C. G.; Swalen, J. D., accepted for publication in *Applied Physics Letters*.

RECEIVED July 18, 1990

Chapter 14

Chromophore–Polymer Assemblies for Nonlinear Optical Materials

Routes to New Thin-Film Frequency-Doubling Materials

D.-R. Dai[1], M. A. Hubbard[1], D. Li[1], J. Park[1], M. A. Ratner[1], T. J. Marks[1], Jian Yang[2], and George K. Wong[2]

[1]Department of Chemistry and the Materials Research Center and
[2]Department of Physics and the Materials Research Center, Northwestern University, Evanston, IL 60208

The properties of polymer-based second harmonic generation materials are crucially dependent upon realizing high number densities of constituent chromophore moieties and upon achieving and preserving maximum microstructural acentricity. This article reviews recent progress toward these goals. Systems discussed include electric field poled, chromophore-functionalized polyphenylene ethers with second harmonic coefficients (d_{33}) as high as 65×10^{-9} esu, $T_g \approx 173°C$, and with superior temporal stability of the poling-induced chromophore orientation. Also presented are successful strategies for simultaneously poling and diepoxide cross-linking chromophore-functionalized poly(p-hydroxystyrene). The result is a significant improvement in the temporal stability of chromophore orientation. Two approaches to chromophore immobilization are then discussed which involve highly cross-linkable epoxy matrices. In the first, chromophore molecules are embedded in a matrix which can be simultaneously poled and thermally cured. In the second, a functionalized high-β chromophore is synthesized for use as an epoxy matrix component. Finally, a strategy is discussed in which robust, covalent, chromophore-containing self-assembled multilayers are built upon various surfaces. Very high chromophore layer second harmonic generation efficiencies are observed ($d_{33} = 300 \times 10^{-9}$ esu).

The great current interest in nonlinear optical (NLO) materials based upon π-electron chromophores stems from the demonstrated possibilities of large nonresonant susceptibilities, ultrafast response times, low dielectric constants, high optical damage thresholds, and the great intrinsic tailorability of the constituent structures (1-6). When such materials incorporate glassy polymeric architectures, the additional attractive characteristics of supermolecular organization, improved mechanical/dimensional stability, improved optical transparency, and processability into thin-film waveguide structures can be envisioned. Nevertheless, the progression from the above ideas to

0097–6156/91/0455–0226$07.00/0

efficient NLO materials has presented great challenges, and numerous obstacles remain to be surmounted.

For polymer-based second harmonic generation (SHG, χ^2) materials, the crucial synthetic problem is to maximize the number density of component high-β chromophore molecules while achieving and preserving maximum acentricity of the microstructure. One early approach to such materials was to "dope" NLO chromophores into glassy polymer matrices and then to align the dipolar chromophore molecules with a strong electric field (poling) (7,8). The performance of such materials is limited by the low chromophore number densities which can be achieved before phase separation occurs and the physical aging/structural relaxation characteristics of all glassy polymers (9-14), which lead to randomization of the poling-induced preferential chromophore orientation. Hence, the SHG efficiency of such "guest-host" materials is generally short-lived. In addition, we have observed that the chromophore constituents are not strongly bound in such matrices and that these materials readily undergo dielectric breakdown during poling. A second approach to the construction of efficient film-based SHG materials has been to incorporate NLO chromophores into Langmuir-Blodgett (LB) films (15-17). A priori, such an approach offers far greater net chromophore alignment than is possible in a poling field (where net alignment is statistically determined), temporal stability of the chromophore alignment, and controlled film thickness. While preliminary results with LB film-based NLO films have been encouraging (15-17), significant problems arise from the fragility of the films, the temporal instability of chromophore alignment, the problem of scattering microdomains, and the structural regularity of layer deposition that is possible (18-22).

With these results as a background, the goal of the present article is to briefly summarize recent research in this Laboratory aimed at the rational design, construction, and characterization of new types of polymer-based NLO substances. We discuss three classes of materials: i) chromophore-functionalized glassy polymers, ii) totally cross-linked matrices, and iii) chromophore-containing self-assembled organic superlattices. In each case, the goal has been to develop approaches to enhanced acentricity, chromophore number densities, and SHG temporal stability. In each case, chemical synthesis is also employed to test fundamental ideas about polymer structural dynamics, cross-linking processes, and monolayer/multilayer synthesis.

Poled Chromophore-Functionalized Polymers

A first step in ameliorating many of the deficiencies of the aforementioned guest-host materials has been to covalently bind NLO chromophores to selected polymer carriers (23-30). Initial work focussed upon functionalized polystyrene and poly(p-hydroxystyrene) systems (23-27). These materials provide greatly enhanced chromophore number densities, greater SHG temporal stability (tethering of chromophore molecules to massive polymer chains greatly restricts reorientational mobility), improved stability with respect to contact poling-induced dielectric breakdown (presumably a consequence of the restricted microstructural mobility), and enhanced chemical stability (chromophore molecules are more strongly bound within the matrix). It was found that contact poling fields as large as 1.8 MV/cm and d_{33} values as high as 19×10^{-9} esu (greater than the corresponding coefficient for $LiNbO_3$) could be realized. Considerably enhanced SHG temporal stabilities were also observed. Nevertheless, neither optimum chromophore number densities nor maximum chromophore immobilization could be achieved in these first-generation systems. In the following subsections, we describe the synthesis and properties of a functionalized polymer with greater than one chromophore moiety per polymer repeat unit and with a very high glass transition temperature (T_g - - one index of polymer chain mobility) (31). We next describe an approach to chromophore immobilization in which thermal cross-linking chemistry is effected in concert with electric field poling of a chromophore-functionalized polymer (32). We also compare contact to corona poling methodologies and results. The latter technique offers larger (but not precisely known) electric poling fields as well as far greater resistance to dielectric breakdown (33).

An NLO Chromophore-Functionalized Polyether. Poly(2,6-dimethyl-1,4-phenylene oxide) (PPO) is a high-strength, amorphous engineering thermoplastic with $T_g \approx 205\text{-}210°C$ and excellent film-forming characteristics (34). For the present work, PPO was prepared by oxidative coupling of 2,6-dimethylphenol and was purified as described in the literature (35) (Scheme I; $M_n = 27,000$). Bromination (36) in refluxing tetrachloroethane yields PPO-Br_x materials with functionalization levels on the order of 1.6-1.8 Br/repeat unit (predominantly benzylic bromination) as judged by elemental analysis and ^{1}H NMR. N-(4-nitrophenyl)-(S)-prolinoxy (NPPO-, Scheme I) was chosen as a model chromophore synthon since the optical properties have been extensively studied (37,38) and since it is readily amenable to NLO experiments at $\lambda = 1.064$ μm. Reaction of PPO-Br_x with NPPO⁻(from NPPOH + NaH) in dry N-methylpyrrolidone (NMP) yields the chromophore-functionalized material (PPO-NPP_x; Scheme I; 1.4 - 1.6 NPPO/repeat unit; $T_g \approx 173°C$) after precipitation with acetone, washing with H_2O, Soxhlet extraction with MeOH, and vacuum drying. Polymer films were cast in a class 100 laminar-flow clean hood onto ITO-coated conductive glass from triply-filtered NMP solutions. The solvent was then slowly evaporated at 80°C, and the films dried in vacuo at 150-170°C for 24 h. These PPO-NPP films have excellent transparency characteristics (vide infra; $\lambda_{max} = 405$ nm), adhere tenaciously to glass, and are insoluble in most organic solvents.

Contact poling of the PPO-NPP films was carried out at 160-170°C with 1.2 MV/cm fields using aluminum electrodes and techniques described elsewhere (23-27). After cooling the films to 30°C, the field was maintained for an additional 1.5 h. Corona poling was carried out at 180-190°C using a needle-to-film distance of 1.0 cm and a +4-+5 kV potential. After the film had cooled to room temperature, the field was maintained for an additional 1.5 h. Second harmonic data were measured at 1.064 μm in the p-polarized geometry using the instrumentation and calibration techniques described previously (23-27). Second harmonic coefficients (d_{33} values) were calculated from the angular dependence of $I^{2\omega}$ and the formalism of Jerphagnon and Kurtz for uniaxial materials, assuming additionally that $d_{31} = d_{24} = d_{15} = d_{33}/3$ (39). We have previously verified this latter assumption for other poled, chromophore-functionalized polymers (23).

In Table I are presented functionalization level and d_{33} data for representative PPO-NPP films. Assuming approximate additivity of PPO

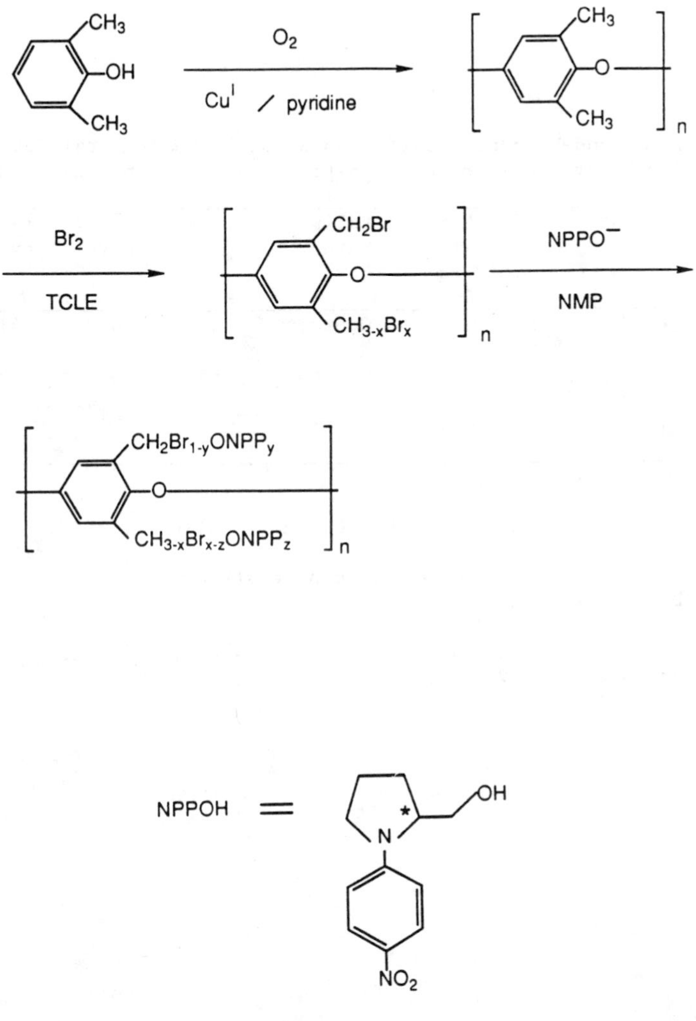

CH3
OH
CH3
O2
Cu^{I} / pyridine
CH3
O
CH3
n
Br2
TCLE
CH2Br
O
CH3-xBrx
n
NPPO^{-}
NMP
CH2Br1-yONPPy
O
CH3-xBrx-zONPPz
n
NPPOH
OH
N
*
NO2

SCHEME I

Table I. Second-Harmonic Coefficients (d_{33}) and Temporal Decay Data
for NPP-Functionalized Poly(2,6-Dimethyl-1,4-Phenylene Oxide)

Functionalization Level[a]	NPPO Number Density, $10^{20}/cm^3$	Poling Method	d_{33} $10^{-9}esu$[b]	τ_1 days[c]	τ_2 days[d]
1.4	~22	Contact	13	0.9	412
1.4	~22	Corona	65	0.3	39
1.6	~26	Corona	55		

[a]NPPO groups per PPO repeat unit.

[b]At $\lambda = 1.064$ μm; measured within 0.5 h of poling at 25°C.

[c]Short-term SHG decay constant from a least-squares fit to equation
   1. Data taken at 25°C.

[d]Long-term SHG decay constant from a least-squares fit to equation
   1. Data taken at 25°C.

and NPPOH densities, it can be noted that the present chromophore number densities (N) are substantial compared to typical guest-host NLO materials (N $\leq$ 2 x $10^{20}/cm^3$) and to most chromophore-functionalized polymers (ca. 8-15 x $10^{20}/cm^3$) (23-30). For comparative purposes, we note that the corresponding N value for crystalline NPPOH is 37 x $10^{20}/cm^3$ (37). In regard to SHG efficiency, the present d_{33} values are also rather large, with the corona-poled value of 65 x 10^{-9} esu comparing favorably with the highest values reported to date for any poled chromophore/polymer system (23-30). A slight but reproducible decline in d_{33} is also observed on increasing the chromophore functionalization level from 1.4 to 1.6. This may reflect unfavorable chromophore aggregation effects, although these are not obvious from uv-visible spectra. In viewing the present d_{33} results, we also note that $\mu_z\beta_{zzz}$ for NPPOH, 300 x 10^{-30} cm^5D/esu at λ = 1.064 μm (38), is relatively small. Even larger d_{33} values may be realizable with PPO and substituted chromophores having higher $\mu\beta$ values (e.g., $\mu_z\beta_{zzz}$ = 1090 x 10^{-30} cm^5D/esu for Disperse Red 1 at λ = 1.356 μm (40)).

The temporal characteristics of contact-poled and corona-poled PPO-NPP$_x$ NLO properties are shown in Figures 1A and 1B, respectively. As has been noted for other chromophore-functionalized NLO polymers (23-27), the present $d_{33}(t)$ data cannot be convincingly fit to a single exponential nor to the empirical Williams-Watts stretched exponential ($\exp[-(t/\tau)^\beta]$). This plausibly suggests that greater than one rate process is operative (e.g., different reorientation rates in matrix environments having access to free volume elements greater than or less than a certain threshold value (12)). More satisfactory fits of the present $d_{33}(t)$ data are found for a presently phenomenological biexponential expression (equation 1), and derived τ_1, τ_2 parameters are also given in Table I. The τ_2 for the contact-poled PPO-NPP$_x$ film appears to be the largest value reported to date and corresponds to a degradation in d_{33}, after the initial, rapid decay, of less than 10%

$$d_{33} = Ae^{-t/\tau_1} + Be^{-t/\tau_2} \qquad (1)$$

in 50 days (Figure 1A). Such decay processes are expected to be additionally impeded in systems with hydrogen bonding, cross-linking (vide infra), and more massive chromophores. We noted elsewhere that poled (PS)O-NPP film (PS = polystyrene) τ_1 and τ_2 values are significantly reduced as contact poling fields are increased (23), i.e., the system is driven further from thermodynamic equilibrium. The present $d_{33}(t)$ data for the corona-poled film support this trend (Figure 1B) in that both τ_1 and τ_2 are significantly diminished when higher poling fields are used. For the present materials, we also find that τ_1 is significantly increased by high temperature annealing, which removes solvent and other plasticizing impurities (23).

Preliminary waveguiding experiments were also performed on several PPO-NPP films to better define their optical properties. A planar waveguide configuration consisting of air/film/glass layers was employed (Figure 2). The waveguide modes were excited with a He-Ne laser (λ = 0.633 μm) using prism coupling techniques with SF6 glass prisms. The refractive index of the films was determined from the coupling angles of the various waveguide modes using previously described procedures (41). For a 1.4 μm thick, unpoled PPO-NPP film (1.4 NPP functionalization level), two TE modes are observed. The coupling angles for these modes (TE$_0$ and TE$_1$) are measured to be 29.6° and

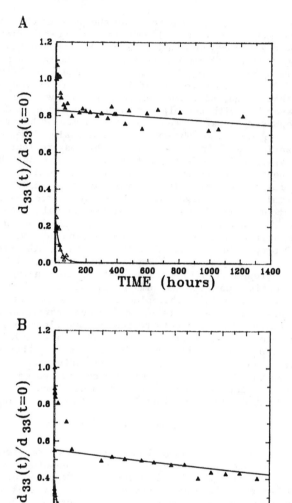

Figure 1.   (A) Time dependence of the second harmonic coefficient
of a PPO-NPP film (1.4 NPP moieties per polymer repeat unit)
contact-poled at 1.2 MV/cm.   Decay data taken at 25°C.   The data
points are shown as filled triangles.   The two curves describing
the biexponential fit to equation 1 are shown separately, with the
open triangles representing data points dominating the short-term
decay.   (B) Time dependence of the second harmonic coefficient of
a corona-poled PPO-NPP film (1.4 NPP moieties per repeat unit).
Decay data taken at 25°C.   The data points are shown as filled
triangles.   The two curves describing the biexponential fit to
equation 1 are shown separately, with the open triangles representing data points dominating the short-term decay.

25.1°, respectively, and from these angles, a refractive index of 1.584 ± 0.001 is calculated. The refractive index of a neat PPO film is similarly calculated to be 1.580 ± 0.001 at $\lambda = 0.633$ μm. Measurements of scattering loss were carried out with an output coupling prism. By measuring the output intensity as a function of the distance between the input and output coupling prisms, a loss coefficient of $\alpha < 1$ dB/cm is estimated for both the PPO-NPP and PPO films.

Thermal Cross-Linking of NLO Chromophore-Functionalized Poly(p-hydroxystyrene)

Poly(p-hydroxystyrene) ($M_W \approx 6,000$; $T_g \approx 155°C$) was functionalized with NPP (16% of phenol rings) as shown in Scheme II (27,32). The product was purified by repeated precipitation with benzene from THF solutions, followed by filtration through a short silica gel column. Purity was verified by elemental analysis, ^{1}H NMR, and FT-IR spectroscopy. In the aforementioned clean hood, 1-2 μm (PS)O-NPP films were cast onto ITO-coated conductive glass from multiply-filtered THF solutions also containing measured quantities of 1,2,7,8-diepoxyoctane (1) or 1,4-butanediol diglycidyl ether (2, Scheme II). Optimum thermal cross-linking conditions were established by parallel experiments in which the ring-opening process was monitored by FT-IR spectroscopy of films cast on KBr plates. Cross-linking is accompanied by disappearance of the epoxy ring mode in the infrared at 907-913 cm^{-1} and the simultaneous appearance of an ether C-O stretching transition at 1040-1048 cm^{-1} (42,43). The (PS)O-NPP/1,2 films were partially cured at 100°C for 1 h under a nitrogen atmosphere and then at 100°C/10^{-4} torr for 24 h. As judged by FT-IR spectroscopy, this procedure brings about partial cross-linking as well as removal of residual traces of solvent and other volatiles which deleteriously plasticize the polymer matrix (26,27). The annealed (PS)O-NPP films were next corona poled (+3.0 - +4.0 kV; 1.0 cm needle-to-film distance) at 180° C for 1 h. For optimum polymer/diepoxide stoichiometries (vide infra), such conditions induce high degrees of cross-linking, while as noted above, corona poling provides higher electric fields than contact poling techniques without ready dielectric breakdown. The poled (PS)O-NPP/1,2 films were cooled to room temperature and physically aged for 1 h prior to removal of the corona field. These films are impervious to organic solvents and are far more resistant to cracking than non-cross-linked films. Good transparency characteristics are illustrated by the successful fabrication of waveguides of the type already described for PPO-NPP.

Second harmonic data for the (PS)O-NPP/1 films were measured at $\lambda = 1.064$ μm using the techniques described above. No SHG was observed for unpoled specimens. SHG temporal decay data were fit by least-squares techniques to the aforementioned, phenomenological biexponential expression of equation 1. In Table II are set out d_{33}, τ_1, τ_2 data for (PS)O-NPP films as a function of cross-linking procedure. It can be seen that the present corona-poled d_{33} values are generally higher than previously achieved for contact-poled (PS)O-NPP samples at comparable or higher functionalization levels (27) and are also higher than values observed for cross-linked guest-host systems (vide infra). Equally important, Table II shows that SHG efficiency is not adversely affected by the cross-linking chemistry. In regard to the temporal stability of d_{33}, Figure 3 compares the

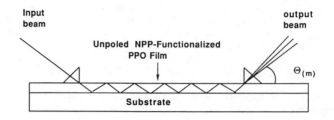

$$2kn_F \, d\cos\Theta_{INT} - 2\Phi_{AIR} - 2\Phi_{SUB} = 2m\Pi$$

$$\Theta_{(0)} = 29.62 \qquad n_F = 1.5844$$
$$\Theta_{(1)} = 25.07 \qquad n_F = 1.5847$$

Figure 2.   Schematic diagram of a PPO-NPP planar waveguide.

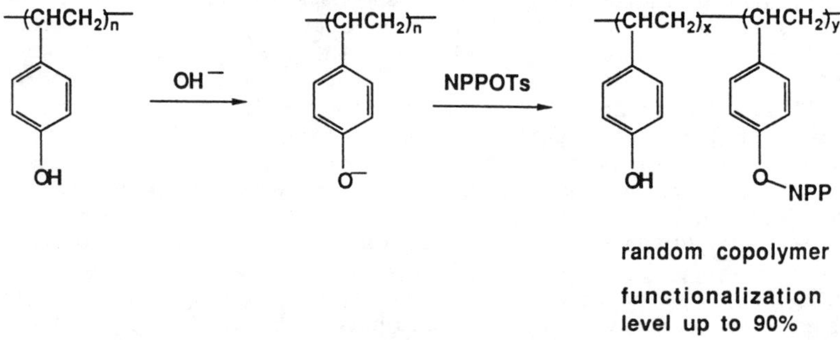

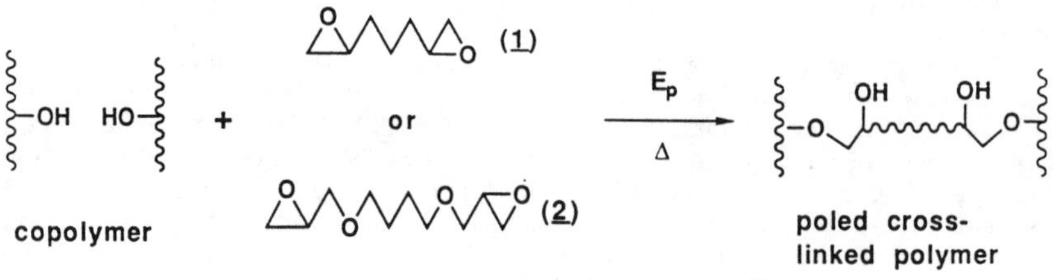

SCHEME II

Table II. Second-Harmonic Coefficients (d_{33}) and Temporal
Decay Parameters for Corona-Poled, NPP-Functionalized
Poly(p-hydroxystyrene) Films as a Function of
Thermal Cross-Linking [a]

Cross-Linking Agent	Stoichiometry Diepoxide/OH[b]	d_{33} 10^{-9} esu[c]	τ_1 days[d]	τ_2 days[e]
None	---	8.8	26	30
"	---	8.6[f]	36[f]	26[f]
1	0.50	7.0	79	100
2	0.25	3.8	18	74
2	0.50	5.5	20	53
2	0.75	2.1	11	51
2	1.00	1.4	9	46

[a]Films poled at 180°C unless otherwise indicated.

[b]Equivalents diepoxide cross-linking reagent per equivalent available phenol OH.

[c]At λ = 1.064 μm.

[d]Short-term SHG decay constant from a least-squares fit to equation 1. Data taken at 25°C.

[e]Long-term SHG decay constant from a least-squares fit to equation 1. Data taken at 25°C.

[f]Poled at 150°C.

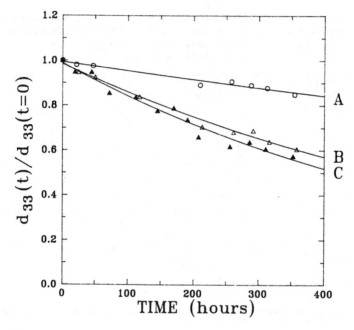

Figure 3. Time dependence of the second harmonic coefficient, d_{33}, for corona-poled (PS)O-NPP films. A. Simultaneously poled (180°C) and cross-linked with 0.50 equiv. 1,2,7,8-diepoxyoctane/phenol OH; B. Poled at 180°C; C. Poled at 150°C. The solid lines are least-squares fits to equation 1, yielding the decay parameters in Table II.

effects of simultaneously corona poling and cross-linking (PS)0-NPP with 0.5 equivalents 1 per available phenol OH group to two films poled in an essentially identical manner but without an epoxy cross-linking agent. The increase in SHG temporal stability is clearly evident and translates into 3.0-fold and 3.3-fold increases in τ_1 and τ_2, respectively. For the films that have not been cross-linked, we find that the present, higher poling temperatures (150-180°C) result in a decrease in the short-term decay component of $d_{33}(t)$ versus samples annealed at lower temperatures *(23,26,27)*.

The effect of the epoxy cross-linking/densification process on chromophore mobility should be a complex function of cross-linking temperature, stoichiometry, and diepoxide reagent. Figure 4 shows the effect on the $d_{33}(t)$ τ_2 parameter of increasing the stoichiometric ratio of cross-linking agent for constant poling methodology. It can be seen that τ_2 rises to a maximum at relatively low diepoxide/available phenol OH ratios, then declines at higher ratios. FT-IR spectroscopy reveals residual, unreacted epoxide groups (incomplete cross-linking) at the higher 2/OH ratios, and it is reasonable to suppose that dangling, unreacted epoxide sidechains would have a plasticizing effect on the chromophore/polymer matrix. At 2/OH ratios greater than ca. 0.5, the matrix becomes opaque after casting and curing, indicating phase separation. This results in a decrease in the measured d_{33} values. Differences in τ_1, τ_2 parameters for matrices cross-linked by 1 and 2 (Table II) can be tentatively related to differences in the chain flexibility of the diepoxide reagents.

SHG Temporal Stabilization by Embedding NLO Chromophores in Totally Cross-Linking Matrices

To determine whether NLO chromophores can be immobilized in highly cross-linked matrices, experiments were carried out in which the high-β molecules 4-(dimethylamino)-4'-nitrostilbene (DANS, 3) and Disperse Orange 1 (DO1, 4) were dispersed in an optical grade epoxy resin,

3   4

which could then be simultaneously poled and thermally cured *(45)*. In such experiments, dichloromethane or acetone solutions of the chromophores were mixed with the epoxy resin, and the solvent then stripped from the solution in vacuo. The chromophore-doped resin was next thoroughly mixed (using a vortex mixer with glass beads) with the appropriate quantity of amine cross-linker and the resulting fluid introduced between transparent ITO glass electrodes using capillary action (to exclude air bubbles). The spacing between the electrodes was maintained at 15-150 μm with Teflon or Mylar foil. Partial cross-linking of the matrix at 80°C prior to poling was necessary to avoid dielectric breakdown. Poling fields of 2×10^4-6×10^5 V/cm were next gradually applied and maintained for measured periods of time at 80-150°C. Films were cooled to room temperature prior to removal of the field.

Second harmonic coefficients of the poled films were measured at

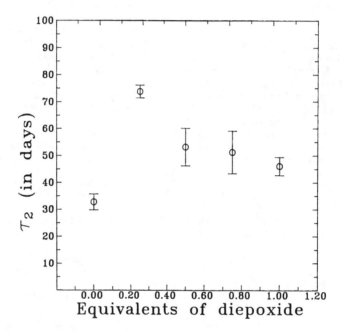

Figure 4.   Long-term decay parameters (τ_2, equation 1) for d_{33} of (PS)O-NPP films simultaneously corona poled and cross-linked with the indicated equivalents of 1,4-butanediol diglycidyl ether/equivalents available phenol OH groups.

λ = 1.064 μm using the instrumentation and data analysis procedures described above. Typical d_{33} values at zero time were found to be in the range 0.1 -1.0 x 10^{-9} esu. These magnitudes agree well with those expected for the chromophore number densities employed (N = 0.4-1.9 x 10^{19}/cm^3), assuming literature $\mu_z\beta_{zzz}$ values for the chromophore and the applicability of an isolated chromophore, molecular gas description of the field-induced chromophore orientation process (7,8).

The SHG temporal characteristics of the chromophore/epoxy matrices are found to be strongly dependent upon the thermal cross-linking conditions. In poling field/temperature cycling experiments carried out as a function of curing time, the decay rate of d_{33} following removal of the field gradually declines as progressive cross-linking (46-50) increases the degree of chromophore immobilization (45). After long curing times at 80°C, fitting of $d_{33}(t)$ to equation 1 yields τ_1 = 7, τ_2 = 72 days for the DANS-doped epoxy matrix and τ_1 = 8, τ_2 = 142 days for the DO1-doped epoxy matrix. The lower d_{33} decay rate of the latter matrix presumably reflects a slower reorientation rate for the larger DO1 chromophore. As shown in Figure 5, simultaneous curing and contact poling of a DO1-doped matrix at 150°C yields a more stable NLO material for which only minor decay in d_{33} can be detected over a period of many days.

While the above approach achieves substantial NLO chromophore immobilization, it also suffers from the low chromophore number densities which can be achieved. An attractive alternative would be to construct a cross-linkable matrix in which one active component was a high-β chromophore molecule. Simultaneous poling and thermal cross-linking would then lead to an acentric matrix with a very high NLO chromophore number density. One approach to such a chromophore molecule is illustrated in Scheme III (51). The synthesis of target molecule 5 has recently been achieved and poling/cross-linking experiments are presently in progress. Values of d_{33} as high as 70 x 10^{-9} esu at λ = 1.064 μm have been measured for corona-poled films.

NLO Chromophore-Containing Organic Superlattices

An alternative approach to poled polymers and LB films would be the construction of covalently linked, chromophore-containing multilayer structures. In principle, such materials offer greater net chromophore alignment and number densities than poled films and far greater structural control and stability than LB films. The general strategy we have employed for superlattice construction is shown in Scheme IV, where the basic siloxane technology follows from reactions developed by Sagiv (52-55). The exact chemistry employed is presented in Scheme V. Noteworthy features include the use of a stilbazole chromophore precursor in which the layer-building quaternization reaction simultaneously affords a high-β chromophore center (56-59) and readily monitored changes in the optical spectrum. In addition, soft, polymeric layers are introduced transverse to the stacking direction to enhance structural stability. The course of multilayer evolution on clean SiO$_2$ substrates is readily monitored by uv-visible spectroscopy (growth of the chromophore absorption); XPS spectroscopy (initial diminution of Si, O signals; growth and persistence of I, C, N signals); advancing contact angle measurements, which are in agreement with the expected properties of the surface functionalities (60-62) (step in Scheme V, θ_a(H$_2$O): clean glass substrate, 15°; a, 82°; b,

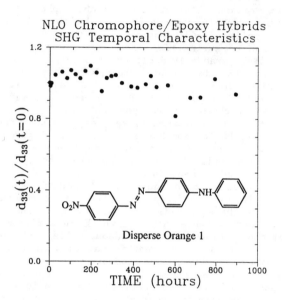

Figure 5. Time dependence at room temperature of the second harmonic coefficient, d_{33}, of epoxy films containing Disperse Orange 1 (4) after simultaneously poling and curing at 150°C.

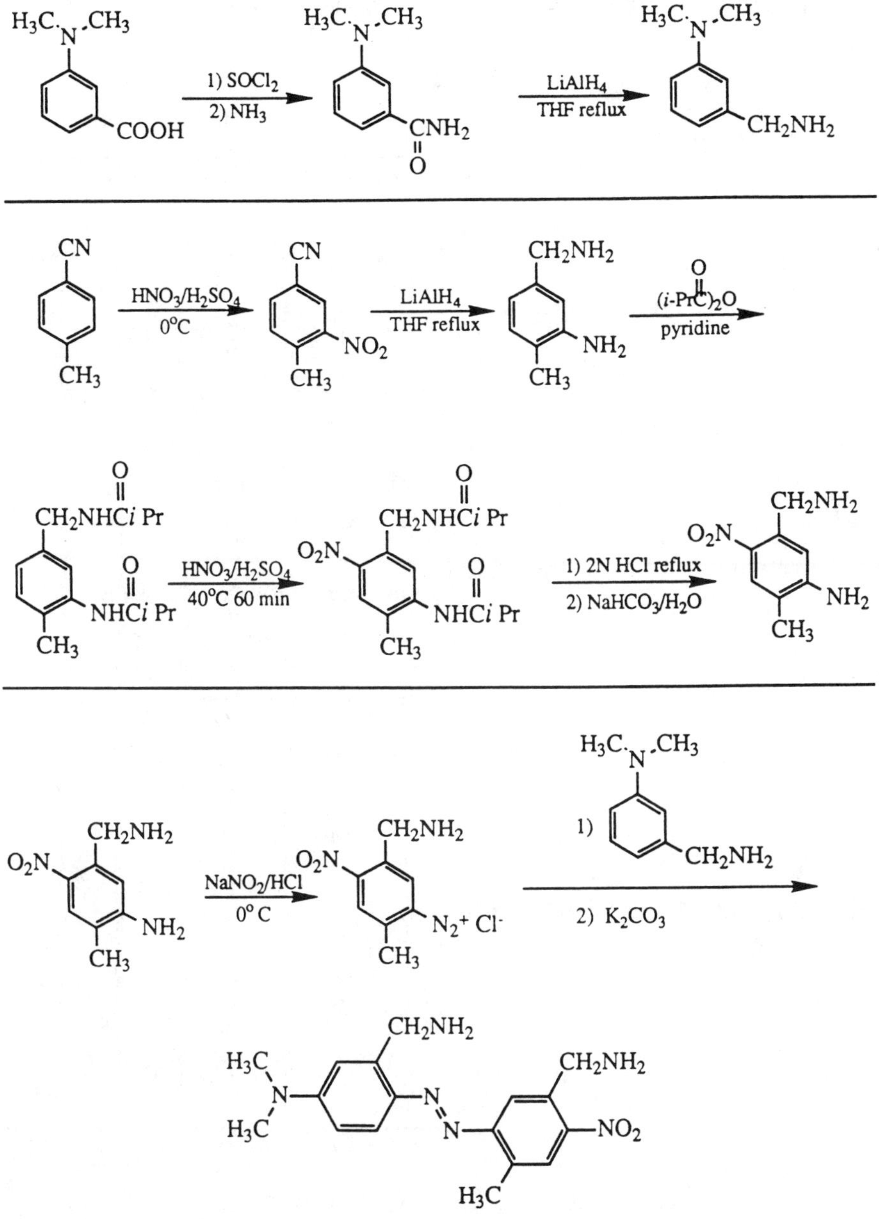
1) SOCl2
2) NH3
LiAlH4
THF reflux
HNO3/H2SO4
0°C
LiAlH4
THF reflux
(i-PrC)2O
pyridine
HNO3/H2SO4
40°C 60 min
1) 2N HCl reflux
2) NaHCO3/H2O
NaNO2/HCl
0° C
1)
2) K2CO3

SCHEME III

Strategy For Covalently Linked NLO Multilayers

a) **Defined Substrate**

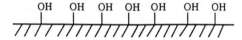

b) **Coupling Layer Formation**

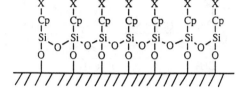

c) **Chromophore Layer Introduction**

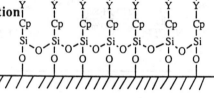

d) **Coupling Layer Introduction**

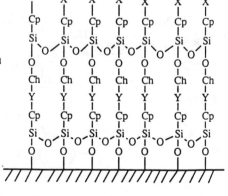

Cp = Coupling spacer   Ch = Chromophore

SCHEME IV

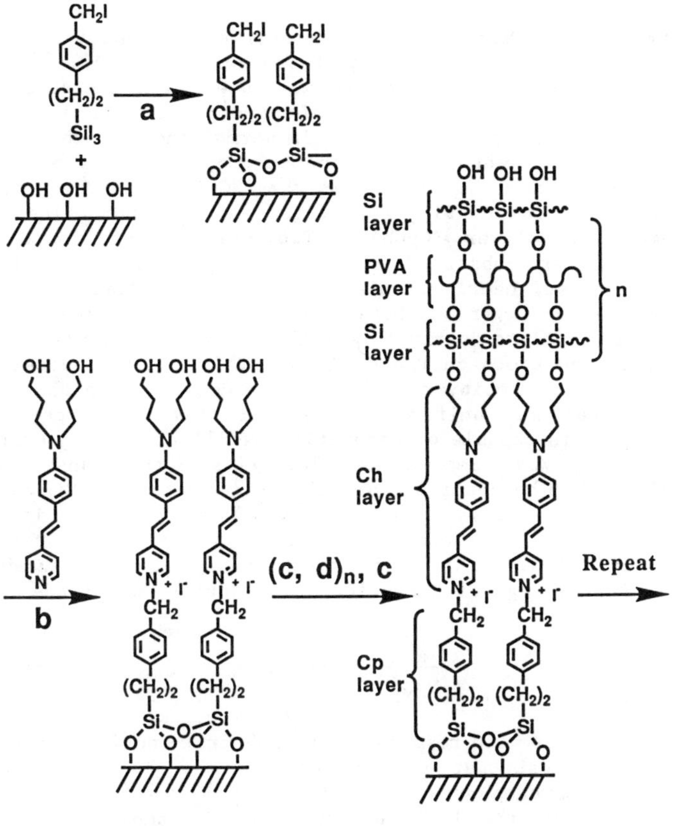

a. Benzene, 25°C   b. Reflux in n-PrOH

c. $Cl_3SiOSiCl_2OSiCl_3$ in THF   d. Polyvinylalcohol in DMSO

SCHEME V

55°; c, 17°; d, 17°; this pattern repeats in each cycle of layer construction); preliminary ellipsometry measurements, which are in accord with expected dimensions (approximate sublayer thicknesses in the notation of Scheme V: CpCh ≈ 22Å; Si ≈ 12Å; PVA ≈ 10Å); and NLO characteristics (vide infra). The multilayer structures adhere strongly to glass, are insoluble in common organic solvents as well as strong acids, and can only be effectively removed by diamond polishing.

Transmission SHG measurements were carried out in the p-polarized geometry using the 1.064 μm output of a Q-switched Nd-YAG laser. No in-plane anisotropy in the SHG signal was detected as the samples were rotated about the film normal. This indicates that these films possess uniaxial symmetry about the film normal and that the distribution of molecular orientations of the chromophores does not have an azimuthal dependence. Figure 6 shows the SHG intensity as a function of incident angle for an SiO_2 substrate coated on both sides with a self-assembled monolayer (CpCh). Two features of the data are noteworthy and are observed for all of the present multilayer structures. First, nearly complete destructive interference of SHG waves from the monolayer on the two sides of the glass substrate is evident. This complete destructive interference is observed over many randomly selected spots on these films. This result indicates that the quality of the monolayers on the two sides of the glass slide is nearly identical and uniform, suggesting that the present self-assembly method is capable of generating excellent quality monolayers in a reproducible way. Second, excellent fits of the transmission SHG data envelopes can be obtained for $\chi^{(2)}_{zzz}/\chi^{(2)}_{zyy} = 3-4$. For Figure 6, the fit is shown for $\chi^{(2)}_{zzz}/\chi^{(2)}_{zyy} = 3$. By assuming a one-dimensional chromophore (i.e., one characterized by a single, dominant β component) and minimal dispersion, the relationship in eq.(2) holds (63). Here, $\overline{\psi}$ is the average of the orientation angles, ψ, of the

$$\frac{\chi^{(2)}_{zzz}}{\chi^{(2)}_{zyy}} = 2 \cot^2\overline{\psi} \qquad (2)$$

chromophore dipoles with respect to the substrate surface normal. Our results thus suggest that $\overline{\psi}$ is in the range 35°-39° for the present self-assembled chromophoric superlattices.

By calibrating the 1.064 μm SHG from the superlattice samples against that from a quartz plate we obtain $d_{33} = 100 \times 10^{-9}$ esu for these multilayer structures and $d_{33} = 300 \times 10^{-9}$ esu for a ChCp monolayer of 22 Å estimated (from the ellipsometric data) thickness. These values are rather large compared to those observed in poled polymer films (vide supra). The possibility of obtaining such large nonlinearities in these self-assembled films is consistent with the higher chromophore number density and the high degree of noncentrosymmetric alignment of the chromophores. In regard to whether the above $\lambda = 1.064$ μm d_{33} results may include a very large resonant enhancement ($\lambda_{max} = 510$ nm for the chromophore), supplementary SHG measurements at $\lambda = 1.90$ μm yield a nonresonant d_{33} value only 40% smaller. Such large d_{33} values also suggest that the formation of aggregates with centric structures, which commonly occurs in LB films and lowers considerably the maximum possible value of d_{33} that can be achieved (18-22,64,65), is not important in these covalently connected, self-assembled films.

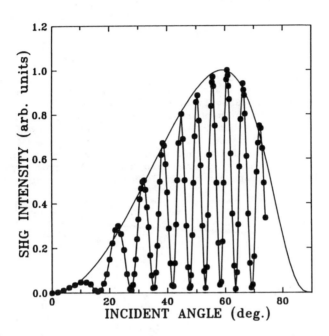

Figure 6. SHG intensity from a glass slide having self-assembled CpCh monolayers on both sides as a function of fundamental beam incident angle. The interference pattern is due to the phase difference between the SHG waves generated at either side of the substrate during propagation of the fundamental wave. The solid envelope is a theoretical curve generated for $\chi_{zzz}^{(2)}/\chi_{zyy}^{(2)} = 3$.

Since the present multilayers are extremely thin in comparison to the wavelength of light being employed and to the expected coherence length, the intensity of the SHG signal should scale quadratically with the number of chromophore layers (66). This is an NLO quality diagnostic commonly applied to LB films (15-22). As can be seen in Figure 7, the adherence of the present multilayer structures to quadratic behavior is good, indicating that it is possible to maintain the same degree of noncentrosymmetric chromophore ordering in the additions of successive layers.

Conclusions

These results illustrate the diversity of synthetic and processing approaches that can be taken in the synthesis of thin-film frequency doubling materials. Specifically, we have demonstrated that it is possible to assemble chromophore-functionalized polymers with greater than one chromophore substitutent per monomer subunit, with d_{33} values as high as 65 x 10^{-9} esu, with T_g values as high as 173°C, with improved temporal stability, and with good transparency characteristics at λ = 0.633 μm. We have also shown that known chromophore-functionalized polymers can be simultaneously poled and cross-linked

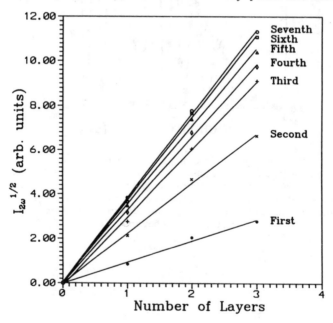

Figure 7. Plot of the square root of SHG intensity versus the number of chromophore layers in the multilayer superlattice. The straight lines are the linear least-squares fit to the experimental data. The number labels correspond to the interferogram maxima in Figure 6.

to significantly retard the rate of chromophore disorientation following electric field poling. In addition, NLO systems based upon highly cross-linkable epoxy matrices have been prepared and shown to be viable candidates for both improved SHG temporal stability and improved frequency doubling efficiency. Finally, we have shown that it is possible to sequentially construct robust, covalently linked, chromophore-containing organic superlatices with good structural regularity and high optical nonlinearity ($d_{33} \approx 300 \times 10^{-9}$ esu at $\lambda =$ 1.064 μm).

Acknowledgments

This research was supported by the NSF-MRL program through the Materials Research Center of Northwestern University (Grant DMR-8821571) and by the Air Force Office of Scientific Research (Contracts 86-0105 and 90-0071). We thank Mr. T. G. Zhang for helpful discussions and Drs. D. Lam and J. Parker of Argonne National Laboratory for assistance with the ellipsometry measurements.

Literature Cited

1. Messier, J.; Kajar, F.; Prasad, P.; Ulrich, D., Eds.; "Nonlinear Optical Effects in Organic Polymers," Kluwer Academic Publishers: Dordrecht, 1989.
2. Khanarian, G., Ed. "Nonlinear Optical Properties of Organic Materials," *SPIE Proc.* 1989, *971*.
3. Heeger, A. J.; Orenstein, J.; Ulrich, D. R., Eds. "Nonlinear Optical Properties of Polymers," *Mats. Res. Soc. Symp. Proc.*, 1988, *109*.
4. Chemla, D. S.; Zyss, J., Eds. "Nonlinear Optical Properties of Organic Molecules and Crystals," Vols. 1, 2; Academic Press: New York, NY, 1987.
5. Zyss, J. *J. Mol. Electronics* 1985, *1*, 25-56.
6. Williams, D. J. *Angew. Chem. Intl. Ed. Engl.* 1984, *23*, 690-703.
7. Singer, K. D.; Sohn, J. E.; Lalama, S. J. *Appl. Phys. Lett.* 1986, *49*, 248-250.
8. Singer, K. D.; Kuzyk, M. G.; Sohn, J. E. *J. Opt. Soc. Am.B* 1987, *4*, 968-975.
9. Hampsch, H. L.; Torkelson, J. M.; Bethke, S. J.; Grubb, S. G. *J. Appl. Phys.*, 1990, *67*, 1037-1041.
10. Hampsch, H. L.; Yang, J.; Wong, G. K.; Torkelson, J. M. *Macromolecules* 1988, *21*, 526-528.
11. Hampsch, H. L.; Yang, J.; Wong, G. K.; Torkelson, J. M. *Polymer Commun.* 1989, *30*, 40-43.
12. Yu, W-C., Sung, C. S. P.; Robertson, R. E. *Macromolecules* 1988, *21*, 355-364, and references therein.
13. Victor, J. G.; Torkelson, J. M. *Macromolecules* 1987, *20*, 2241-2250.
14. Struik, L. C. E. "Physical Aging in Amorphous Polymers and Other Materials"; Elsevier: Amsterdam, 1978.
15. Bosshard, Ch.; Küpfer, M.; Günter, P.; Pasquier, C.; Zahir, S.; Seifert, M. *Appl. Phys. Lett.* 1990, *56*, 1204-1206, and references therein.

16. Popovitz-Biro, R.; Hill, K.; Landau, E. M.; Lahav, M.; Leiserowitz, L.; Sagio, J.; Hsiung, H.; Meredith, G. R.; Vanherzeele, H. *J. Am. Chem. Soc.* **1988**, *110*, 2672-2674.

17. Cross, G. H.; Peterson, I. R.; Girling, I. R.; Cade, N. A.; Goodwin, M. J.; Carr, N.; Sethi, R. S.; Marsden, R.; Gray, G. W.; Lacey, D.; McRoberts, A. M.; Scrowston, R. M.; Toyne, K. J. *Thin Solid Films* **1988**, *156*, 39-52.

18. Allen, S.; McLean, T. D.; Gordon, P. F.; Bothwell, B. D.; Robin, P.; Ledoux, I. *SPIE* **1988**, *971*, 206-215.

19. Schildkraut, J. S.; Penner, T. L.; Willand, C. S.; Ulman, A. *Optics Lett.* **1988**, *13*, 134-136.

20. Lupo, D.; Prass, W.; Schunemann, U.; Laschewsky, A.; Ringsdorf, H.; Ledoux, I. *J. Opt. Soc. Am. B* **1988**, *5*, 300-308.

21. Ledoux, I.; Josse, D.; Vidakovic, P.; Zyss, J.; Hann, R. A.; Gordon, P. F.; Bothwell, B. D.; Gupta, S. K.; Allen, S.; Robin, P.; Chastaing, E.; Dubois, J. C. *Europhysics Lett.* **1987**, *3*, 803-809.

22. Hayden, L. M.; Kowel, S. T.; Srinivasan, M. P. *Optics Commun.* **1987**, *61*, 351-356.

23. Ye, C.; Minami, N.; Marks, T. J.; Yang, J.; Wong, G. K. in ref. 1, pp. 173-183.

24. Li, D.; Minami, N.; Ratner, M. A.; Ye, C.; Marks, T. J.; Yang, J.; Wong, G. K. *Synthetic Metals* **1989**, *28*, D585-D593.

25. Ye, C.; Marks, T. J.; Yang, Y.; Wong, G. K. *Macromolecules* **1987**, *20*, 2322-2324.

26. Ye, C.; Minami, N.; Marks, T. J.; Yang, J.; Wong, G. K. in ref. 3, pp. 263-269.

27. Ye, C.; Minami, N.; Marks, T. J.; Yang, J.; Wong, G. K. *Macromolecules* **1988**, *21*, 2901-2904.

28. Singer, K. D.; Kuzyk, M. G.; Holland, W. R.; Sohn, J. E.; Lalama, S. J.; Commizzoli, R. B.; Katz, H. E.; Schilling, M. L. *Appl. Phys. Lett.* **1988**, *53*, 1800-1802.

29. Eich, M.; Sen, A.; Looser, H.; Yoon, D. Y.; Bjorklund, G. C.; Twieg, R.; Swalen, J. D. in ref. 2, pp. 128-135.

30. Eich, M.; Sen, A.; Looser, H.; Bjorklund, G. C.; Swalen, J. D.; Twieg, R.; Yoon, D. Y. *J. Appl. Phys.* **1989**, *66*, 2559-2567.

31. Dai, D.-R.; Marks, T. J.; Yang, J.; Lundquist, P. M.; Wong, G. K. *Macromolecules* **1990**, *23*, 1894-1896.

32. Park, J.; Marks, T. J.; Yang, J.; Wong, G. K. *Chemistry of Materials* **1990**, *2*, 229-231.

33. Comizzoli, R. B. *J. Electrochem. Soc.* **1987**, *134*, 424-429.

34. Aycock, D.; Abolins, V.; White, D. M. in "Encyclopedia of Polymer Science and Technology," Wiley: New York, 1988, Vol. 13, pp. 1-30, and references therein.

35. White, D. M. *J. Org. Chem.* **1969**, *34*, 297-303.

36. Cabasso, I.; Jagur-Grodzinski, J.; Vofsi, D. *J. Appl. Polym. Sci.* **1974**, *18*, 196.

37. Zyss, J.; Nicoud, J. F.; Coquillay, M. *J. Chem. Phys.* **1984**, *81*, 4160-4167.

38. Barzoukas, M.; Josse, D.; Fremaux, P.; Zyss, J.; Nicoud, J. F.; Morely, J. *J. Opt. Soc. Am. B* **1987**, *4*, 977-986.

39. Jerphagnon, J; Kurtz, S. K. *J. Appl. Phys.* 1970, *41*, 1667-1681.
40. Singer, K. D.; Sohn, J. E.; King, L. A.; Gordon, H. M.; Katz, H. E.; Dirk, C. W. *J. Opt. Soc. Am. B* 1989, *6*, 1339-1351.
41. Tien, P. K. *Appl. Opt.* 1971, *10*, 2395-2413.
42. McAdams, L. V.; Gannon, J. A., in "Encyclopedia of Polymer Science and Engineering," Wiley: New York, 1986, Vol. 6, pp. 322-382, and references therein.
43. Mertzel, E.; Koenig, J. L. *Adv. Polym. Sci.* 1985, *75*, 74-112.
44. Hubbard, M. A.; Minami, N.; Ye, C.; Marks, T. J.; Yang, J.; Wong, G. K. in ref 1b, pp. 136-143.
45. Hubbard, M. A.; Marks, T. J.; Yang, J.; Wong, G. K. *Chemistry of Materials* 1989, *1*, 167-169.
46. Kloosterboer, J. G. *Advan. Polym. Sci.* 1988, *84*, 3-61.
47. Oleinik, E. G. *Advan. Polym. Sci.* 1986, *80*, 49-99.
48. Dusek, K. *Advan. Polym. Sci.* 1986, *78*, 1-59.
49. Rozenberg, B. A. *Advan. Polym. Sci.* 1986, *75*, 73-114.
50. Morgan, R. J. *Advan. Polym. Sci.* 1985, *72*, 1-43.
51. Marks, T. J., 1989 International Chemical Congress of Pacific Basin Societies, Honolulu, HI, Dec. 1989.
52. Netzer, L.; Iscovici, R.; Sagiv, J. *Thin Solid Films* 1983, *99*, 235-241.
53. Netzer, L.; Iscovici, R.; Sagiv, J. *Thin Solid Films* 1983, *100*, 67-76.
54. Pomerantz, M.; Segmuller, A.; Netzer, L.; Sagiv, J. *Thin Solid Films* 1985, *132*, 153-162.
55. Sagiv, J. *Israel J. Chem.* 1979, *18*, 339-345.
56. From perturbation theory and the PPP SCF MECI formalism (*57-59*) we estimate $\beta_{vec.} = 382 \times 10^{-30} cm^5 esu^{-1}$ at a frequency of 1.17 eV.
57. Li, D.; Marks, T. J.; Ratner, M. A. *Chem. Phys. Lett.* 1986, *131*, 370-375.
58. Li, D.; Ratner, M. A.; Marks, T. J. *J. Am. Chem. Soc.* 1988, *110*, 1707-1715.
59. Li, D.; Marks, T. J.; Ratner, M. A., to be published.
60. Chidsey, C. E. D.; Loiacono, D. N. *Langmuir* 1990, *6*, 682-691.
61. Wasserman, S. R.; Tao, Y.-T.; Whitesides, G. M. *Langmuir* 1989, *5*, 1074-1087.
62. Bain, C. D.; Troughton, E. B.; Tao, Y.-T.; Evall, J.; Whitesides, G. M. *J. Am. Chem. Soc.* 1989, *111*, 321-335.
63. Shen, Y. R. "The Principles of Nonlinear Optics," Wiley: New York, 1984, Chapt. 2.
64. Williams, D. J.; Penner, T. L.; Schildkraut, J. J.; Tillman, N.; Ulman, A. in ref. 1a, pp. 195-218.
65. Fang, S. B.; Zhang, C. H.; Lin, B. H.; Wong, G. K.; Ketterson, J. B.; Dutta, P., *Chemistry of Materials*, in press.
66. Bloembergen, N.; Pershan, P. S. *Phys. Rev.* 1962, *128*, 606-622.

RECEIVED July 18, 1990

Chapter 15

Novel Covalently Functionalized Amorphous χ^2 Nonlinear Optical Polymer

Synthesis and Characterization

Ayusman Sen[1], Manfred Eich, Robert J. Twieg, and Do Y. Yoon

**Almaden Research Center, IBM Research Division,
San Jose, CA 95120–6099**

The polymer, [-CH$_2$CH(CH$_2$NHC$_6$H$_4$NO$_2$-p)-]$_n$, PPNA, was synthesized by aromatic nucleophilic substitution reaction of poly(allylamine) hydrochloride) with p-fluoronitrobenzene. PPNA so prepared has mol. wt. ~100,000, T$_g$ = 125-140°C, and was stable in a N$_2$ atmosphere to 220°C. The ^{13}C NMR spectrum revealed the presence of π-interaction between those chromophoric groups that are in isotactic relationship to each other (~30-35% of total). The orientation of the chromophores in a PPNA sample (T$_g$ = 125°C) was achieved in a thin film by the corona poling technique at temperatures above T$_g$. The subsequent freezing process resulted in a polymeric film that exhibited an initial high second-order nonlinear coefficient, d$_{33}$=31 pm/V, as measured by Maker-fringe technique with 1.06 μm fundamental. However, a significant decay was observed until a stabilized value of 19 pm/V was obtained after 5d at room temperature.

Organic-based materials are attractive for nonlinear optics (NLO) applications, due to their very large nonlinear response over a broad frequency range, rapid response times, high laser damage thresholds, and the intrinsic tailorability of organic structures. (1-7) For second harmonic generation (SHG), a major synthetic challenge is to construct noncentrosymmetric molecular assemblies that have high structural integrity and suitable processability. Particularly attractive from this standpoint are amorphous polymers with covalently attached NLO-chromophores that can be aligned in an electric field. (1-4) The ideal polymeric material should: (a) be easily synthesized, (b) have a high concentration of NLO-chromophores, (c) have a glass transition temperature (T$_g$) well above ambient in order to stabilize the noncentrosymmetric alignment produced by electric field poling above T$_g$, (d) have good optical characteristics (e.g., low crystallinity, λ$_{max}$ at relatively short wave-lengths), (e) be stable to heat

NOTE: Ayusman Sen is the Paul J. Flory Sabbatical Awardee

[1]Current address: Department of Chemistry, Pennsylvania State University, University Park, PA 16802

0097–6156/91/0455–0250$06.00/0

and light, and (f) be easily processable into films, fibers, monoliths, etc. Herein, we report a new covalently functionalized amorphous NLO-polymer that meets many of these criteria. In particular, it is easily synthesized in one step from commercially available materials and, compared to the few polymers of this type that have been reported in the literature, (1-4, 8-11) it has the highest concentration of NLO-chromophores and among the highest SHG coefficients (initial d_{33} = 31 pm/V or 48 x 10^{-9} esu).

The polymer, $[-CH_2CH(CH_2NHC_6H_4NO_2\text{-p})-]_n$, PPNA, was synthesized by the reaction of poly(allylamine hydrochloride) (Aldrich Chemical, "low molecular weight") with a slight excess of p-fluoronitrobenzene (Aldrich Chemical) in the presence of an excess of base (anhyd. K_2CO_3), as shown in equation 1.

$$(-CH_2-CH-)_n \ + \ F \!-\!\!\langle O \rangle\!\!-\! NO_2 \ \xrightarrow[\substack{DMSO \\ 80°C}]{K_2CO_3} \ (-CH_2\text{-}CH\text{-})_n \qquad (1)$$

$$CH_2\text{-}NH_2 \cdot HCl \qquad\qquad CH_2\text{-}NH\!-\!\!\langle O \rangle\!\!-\! NO_2$$

(PPNA)

Dimethylsulfoxide (DMSO) was used as the solvent, and the reaction was allowed to proceed for 3d at 80°C. Following successive precipitations in water and methanol and a final base wash with methanolic pyridine, PPNA was obtained as a yellow solid. The two common organic solvents in which it has significant solubility are DMSO and N-methylpyrrolidinone (NMP). Elemental analysis (C,H,N) of a PPNA sample agreed with the formulation shown and is indicative of essentially complete derivatization of all of the amino groups present in the starting polymer. The absolute M_w obtained by light scattering in NMP was 94,000, and GPC analysis in dimethylacetamide/0.1 M LiBr gave M_n = 28,300, M_W = 114,000 (versus polymethylmethacrylate).

The ^{1}H-NMR spectra (DMSO-d_6) were too broad to be of any value. The ^{13}C-NMR spectra (DMSO-d_6 or NMP), however, had some interesting features (Figure 1). Each of the four inequivalent types of carbons of the aromatic ring appeared as two resonances in approximately 2:1 intensity ratio [e.g., ^{13}C-NMR $\{^1$H$\}$(DMSO-d_6)(60°C)(ppm): 154.2, 151.7 (2:1, -NH-$\underline{C}$); 141.5, 136.0 (1:2, O_2N-$\underline{C}$); 125.7, 120.9 (2:1, O_2N-C-$\underline{C}$); 125.1, 110.6 (1:2, -NH-C-$\underline{C}$)]. The more intense of the two absorptions in each case had a chemical shift that was similar to that observed for the corresponding carbon in the model compound, N-methyl-4-nitroaniline, [e.g., ^{13}C-NMR $\{^1$H$\}$(DMSO-d_6)(25°C)(ppm): 155.1 (MeNH-$\underline{C}$); 135.6 (O_2N-$\underline{C}$); 126.0 (O_2N-C-$\underline{C}$); 110.2 (MeNH-C-$\underline{C}$)]. We tentatively assign the second set of resonances to those chromophoric groups that are in isotactic relationship to each other. The ^{13}C-NMR spectrum of the nominally atactic precursor polymer, poly(allylamine hydrochloride) indicated the presence of ~30-35% isotactic segments. Models show that the chromophores in the isotactic segments of PPNA are sufficiently close to each other for π-interaction to occur. This, in turn, would be expected to result in significant shifts in the ^{13}C-NMR resonances of the aromatic carbons when compared to those observed in isolated chromophoric groups. Note that the ^{13}C-NMR spectral features were independent of concentration, temperature, and solvent.

Thermogravimetric analysis (heating rate: 10°C/min) of PPNA revealed that the polymer was stable to at least 220°C in an N_2 atmosphere. (Figure 2) Differential scanning calorimetric measurements (heating rate: 10°C/min) on several batches of PPNA indicated that T_g varied between 125-140°C. This variation may be a reflection of the amount of solvent (NMP) entrapped in the

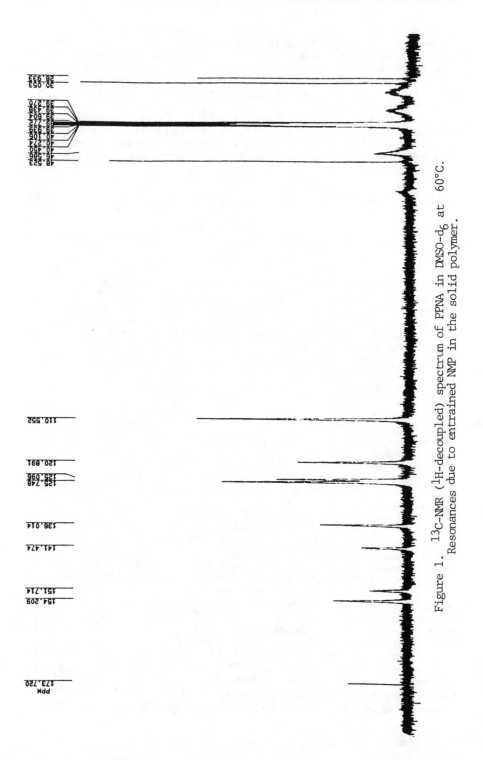

Figure 1. ^{13}C-NMR (^{1}H-decoupled) spectrum of PPNA in DMSO-d_6 at 60°C. Resonances due to entrained NMP in the solid polymer.

polymer (see Figure 1). Finally, wide angle X-ray diffraction measurements confirmed the completely amorphous nature of PPNA. (Figure 3) The combination of high thermal stability, high T_g, and noncrystallinity made PPNA a particularly attractive candidate for further SHG studies that were performed using a sample with T_g = 125°C. (12) The large difference between T_g and the decomposition temperature (~100°C) provided a convenient window for conducting the poling experiments.

In order to assess the orientational stability of the poled state, the temperature dependence of the dipole mobility of the side groups was examined through dielectric relaxation measurements. (13) No low temperature relaxation below T_g was observed in the frequency range studied (100 Hz-100 kHz). In addition, the dielectric constant was approximately equal to the square of the refractive index, indicating that below T_g only electronic and no significant orientational contributions to the dielectric displacement are present. Thus, it was expected that a given orientational state of the ensemble would be stable at temperatures significantly below T_g.

The orientation of the chromophores in thin films was achieved by corona poling (14) using a sharp tungsten needle. The advantage of this method lies in the fact that only the low conductivity polymer surface is charged, and impurities, defects, and pinholes cause only relatively small local currents. As a consequence, much higher breakdown field strengths are typically achieved (E>1MV/cm), compared to conventional poling with conductive electrodes. The SHG measurements were performed using a polarized Q-switched Nd-YAG laser beam (λ=1064nm). A single crystal quartz plate was used as a reference. The dynamics of the corona poling-induced noncentrosymmetry (15) was examined at temperatures >T_g by monitoring the SHG-signal while the electric field was alternately switched on and off (e.g., Figure 4). Total relaxation of the SHG-signal was observed when the field was off, and, in addition, the signal was reproducible through several switching cycles. This rules out electric field-induced structural changes as the possible origin of the SHG-response.

In order to determine the optimal poling conditions, the poling voltage was varied at a constant temperature (=140°C) and the SHG-response was monitored with the sample held at a fixed angle of 39° and with a distance of 50mm between needle tip and ground electrode. As shown in Figure 5, the second harmonic saturation intensity was found to level off at voltages greater than 15 kV. While competition between poling-induced orientation and Boltzmann distribution may lead to the observed leveling off, (16-18) it is also likely that charge saturation at the polymer surface leads to a limiting poling field resulting in the leveling off of the SHG-response.

The SHG coefficients (16, 17) were measured by Maker-fringe experiments after poling at 20 kV (Figure 6). Initial values of 31 and 5 pm/V were obtained for d_{33} and d_{31}, respectively. However, significant decay was observed in the first few hours following poling, and, after 5d, the values had stabilized at 19 and 4 pm/V, respectively. For comparison, the corresponding values for the well-known crystalline material, LiNbO$_3$, are 30 and 6 pm/V, respectively. The d_{33}/d_{31} ratio deviates significantly from the theoretical value of 3 expected for a poled isotropic system (19). One possible explanation is that the chromophores in the isotactic segments of the polymer behave like mesogens and pole together. This would cause the d_{33}/d_{31} ratio to increase towards infinity as predicted for Ising system (19).

The mechanism of the decay of nonlinear response in PPNA remains unclear; however, decay has also been reported for other functionalized polymers. (1-4, 8-11) It should be noted that the observed decay is significantly less than that seen for doped polymers, where nearly complete relaxation takes place. (16, 17) In

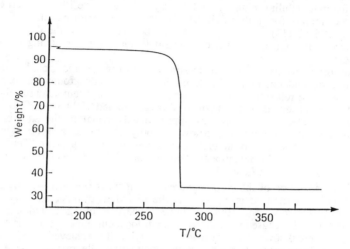

Figure 2. Thermogravimetric scan on a sample of PPNA under nitrogen atmosphere. Initial weight loss is due to entrained NMP in the solid polymer. (Reproduced with permission from Ref. 12. Copyright 1989 American Institute of Physics.)

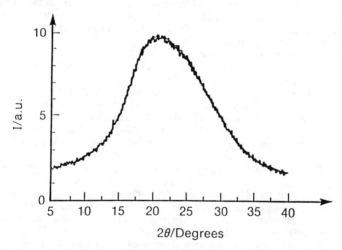

Figure 3. Wide angle X-ray goniometer scan of PPNA (λ=0.154nm). (Reproduced with permission from Ref. 12. Copyright 1989 American Institute of Physics.)

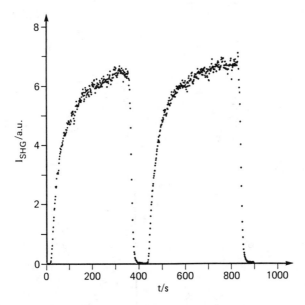

Figure 4. Time dependence of monitored second-harmonic generation in PPNA as the corona potential is switched on and off (T=148°C).

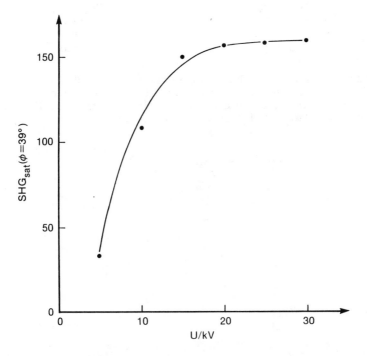

Figure 5. Second-harmonic saturation signal from PPNA as a function of the corona voltage. (Reproduced with permission from Ref. 12. Copyright 1989 American Institute of Physics.)

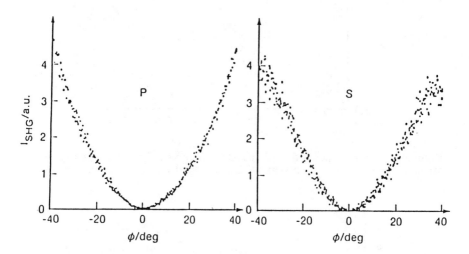

Figure 6. Plots of Maker-fringe data for a poled PPNA film of 1.2μm thickness at room temperature (P=p polarization, S=s polarization). (Reproduced with permission from Ref. 12. Copyright 1989 American Institute of Physics.)

conclusion, it is now clear that amorphous polymers with covalently attached NLO-chromophores are attractive candidates for the construction of SHG assemblies.

Acknowledgments. AS thanks IBM Almaden Research Center for a Paul J. Flory Sabbatical award. ME thanks IBM, Germany, for postdoctoral support. We also acknowledge E. Hadziioannou for DSC and TGA measurements, R. D. Johnson and W. W. Fleming for NMR measurements, and S. Kim for light scattering experiments.

Literature Cited

1. Messier, J.; Kajzar, F.; Prasad, P.; Ulrich, D., Eds. Nonlinear Optical Effects in Organic Polymers; NATO ASI Series, Series E, 1989; Vol. 162.
2. Hann, R. A.; Bloor, D., Eds. Organic Materials for Non-linear Optics; Royal Society of Chemistry, 1989; Special Publication No. 69.
3. Heeger, A. J.; Orenstein, J.; Ulrich, D. R., Eds. Nonlinear Optical Properties of Polymers; Materials Research Society Symposium Proceedings, 1988; Vol. 109.
4. Khanarian, G., Ed. Nonlinear Optical Properties of Organic Materials; SPIE, 1988; Vol. 971.
5. Prasad, P. N.; Ulrich, D. R., Eds. Nonlinear Optical and Electroactive Polymers; Plenum: New York, 1988.
6. Chemla, D. S.; Zyss, J., Eds. Nonlinear Optical Properties of Organic Molecules and Crystals; Academic: New York, 1987; Vols. 1,2.
7. Williams, D. In Electronic and Photonic Applications of Polymers, Bowden, M. J.; Turner, S. R., Eds.; American Chemical Society, Advances in Chemistry Series, 1988; Vol. 218, p. 297.
8. Singer, K. D.; Kuzyk, M. G.; Holland, W. R.; Sohn, J. E.; Lalama, S. J.; Comizzoli, R. B.; Katz, H. E.; Schilling, M. L. Appl. Phys. Lett. 1988, 53, 1800.
9. Hubbard, M. A.; Marks, T. J.; Yang, J.; Wong, G. K. Chem. Mater., 1989, 1, 167.
10. Ye, C.; Minami, N.; Marks, T. J.; Yang, J.; Wong, G. K. Macromolecules 1988, 21, 2901.
11. Ye, C.; Marks, T. J.; Yang, J.; Wong, G. K. Macromolecules 1987, 20, 2322.
12. For details concerning the poling experiments and the associated physical measurements, see: Eich, M.; Sen, A.; Looser, H.; Bjorklund, G.; Swalen, J. D.; Twieg, R.; Yoon, D. Y. J. Appl. Phys. 1989, 66, 2559.
13. McCrum, N. G.; Read, B. E.; Williams, G. Anelastic and Dielectric Effects in Polymeric Solids; Wiley: New York, 1967.
14. Sessler, G. M. Electrets; Springer-Verlag: Berlin, 1980; p. 30.
15. Boyd, G. T. Thin Solid Films 1987, 152, 295.
16. Singer, K. D.; Sohn, J. E.; Lalama, S. J. Appl Phys. Lett. 1988, 49, 248.
17. Singer, K. D.; Kuzyk, M. G.; Sohn, J. E. J. Opt. Soc. Am. B. 1987, 4, 968.
18. Williams, D. J.; Penner, T. L.; Schildkraut, J. J.; Tillman, N.; Ulman, A.; Willand, C. S. in reference 2, p. 195.
19. Williams, D. J. in reference 7, Vol. 1, p. 405.

RECEIVED July 18, 1990

Chapter 16

Second-Order Nonlinear Optical Polyphosphazenes

Alexa A. Dembek[1], Harry R. Allcock[1], Chulhee Kim[1], Robert L. S. Devine[2], William H. Steier[2], Yongqiang Shi[2], and Charles W. Spangler[3]

[1]Department of Chemistry, Pennsylvania State University, University Park, PA 16802
[2]Department of Electrical Engineering, University of Southern California, Los Angeles, CA 90089
[3]Department of Chemistry, Northern Illinois University, DeKalb, IL 60115

In this contribution we describe the synthesis and second-order nonlinear optical properties of a series of mixed-substituent poly(organophosphazenes) that possess covalently attached donor-acceptor substituted, conjugated moieties. The general structure of the polymers is $[NP(OCH_2CF_3)_x(OR)_y]_n$, where OR = $-O(CH_2CH_2O)_kC_6H_4-CH=CH-C_6H_4NO_2$, where k= 1-3, and $-OCH_2CH_2N(CH_2CH_3)C_6H_4-N=N-C_6H_4NO_2$, and x + y = 100%. The nonlinear optical properties of thin films of the polymers were investigated through measurement of second-harmonic generation, and exhibit second-harmonic coefficients, d_{33}, in the range 4.1-34 pm/V.

The development of polymeric nonlinear optical (NLO) materials is currently an area of intense investigation (1-4). Polymeric systems which show second-harmonic generation (SHG) have conjugated aromatic molecules with electron-donor and acceptor moieties in a noncentrosymmetric array. These nonlinear optical molecules can be "doped" into a glassy polymer matrix (5-7) or can be covalently attached to a polymer backbone (8-14). The noncentrosymmetric alignment of the nonlinear optical molecules in both approaches is achieved by heating the polymer to its glass transition temperature, at which point the chains have reorientational mobility, followed by application of a strong electric field. In this paper, we will discuss the synthesis and nonlinear optical properties of phosphazene macromolecules that possess covalently attached donor-acceptor substituted, conjugated moieties (15). The structures of the nonlinear optical side groups are illustrated in Figure 1. Polyphosphazenes offer a potential advantage in that the macroscopic properties of the polymer can be tailored by the incorporation of specific substituent groups (16-21).

0097–6156/91/0455–0258$06.00/0

Synthesis of Nonlinear Optical Side Groups

Our initial work involved the synthesis of side chains which have
the molecular characteristics required for a nonlinear optical
response. Compounds 1-3 were prepared by the use of
Horner-Emmons-Wadsworth Wittig methodology (22). Compound 4 was
commercially available (Aldrich) as the dye, Disperse Red 1. As
outlined in Scheme I, in the first step in the synthesis of 1-3,
4-hydroxybenzaldehyde was allowed to react with chloroethanol
derivatives in basic ethanol containing potassium iodide for 15 h
at reflux. The benzaldehyde product was then allowed to react with
diethyl(4-nitrobenzyl)phosphonate in the presence of potassium
tert-butoxide in ethylene glycol dimethyl ether for 15 h at room
temperature and 1 h at 85°C to yield the stilbene side groups.
Compounds 1-3 were purified by column chromatography and were
recrystallized from n-hexane/methylene chloride to yield yellow
solids.

Compounds 1-4 were characterized by conventional spectroscopic
techniques. For the stilbene compounds 1-3, the trans conformation
of the double bond was confirmed by ^1H NMR analysis. For example,
in the ^1H NMR spectrum of 1, the olefinic protons were detected as
a doublet of doublets with resonances at 7.23 and 7.01 ppm, with a
trans coupling constant of 16.3 Hz. In addition, the ^{13}C NMR
spectra of the stilbene compounds indicated the presence of a
single isomer that was consistent with the desired structures. The
UV/visible spectra in tetrahydrofuran solution showed a λ_{max} value
for 1-3 at 378 nm (ϵ 2.6 x 10^4) and for 4 at 490 nm (ϵ 3.1 x 10^4).

Synthesis of Nonlinear Optical Phosphazene Macromolecules

The overall synthetic pathway to mixed-substituent polyphosphazenes
5-9 is described in Scheme II, and the corresponding polymer structures and composition ratios are listed in Table I.
Poly(dichlorophosphazene) was prepared by the thermal ring-opening
polymerization of the cyclic trimer $(NPCl_2)_3$, as described in
earlier papers (16-18). The substitution reactions of
poly(dichlorophosphazene) were carried out in three steps. The
synthesis and purification of polymer 6 will be discussed as a
representative example. In the first step, sodium
trifluoroethoxide was added to poly(dichlorophosphazene) to replace
approximately 50% of the chlorine atoms. In the second step, a
stoichiometric deficiency of the sodium salt of 1 was allowed to
react with the partially substituted polymer. In the final step,
an excess of sodium trifluoroethoxide was added to replace the
remaining chlorine atoms in order to obtain a fully derivatized,
hydrolytically stable polymer. This three step synthetic procedure
was necessary because the direct addition of the sodium salt of 1
to poly(dichlorophosphazene) resulted in the formation of an
insoluble, incompletely substituted polymeric precipitate. Polymer
6 was isolated by precipitation from the concentrated
tetrahydrofuran reaction mixture into water and was purified by
dialysis against methanol/water (1:1 v/v) for 7 to 10 days.

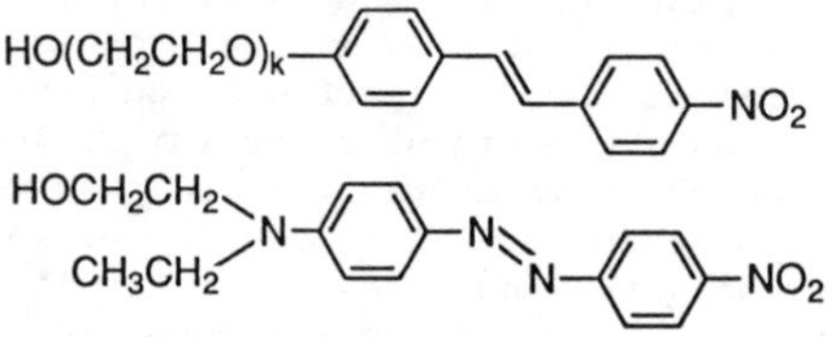

$HO(CH_2CH_2O)_k$—⟨⟩—CH=CH—⟨⟩—NO_2 **1 - 3**

$HOCH_2CH_2$\
CH_3CH_2 /N—⟨⟩—N=N—⟨⟩—NO_2 **4**

Figure 1. Structures of donor-acceptor substituted, conjugated side groups.

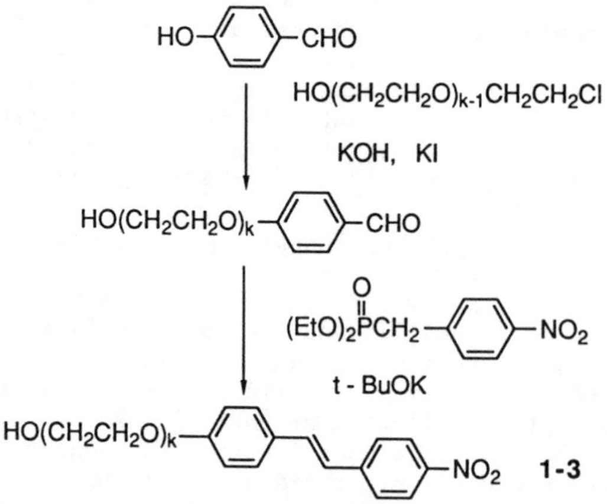

HO—⟨⟩—CHO

$HO(CH_2CH_2O)_{k-1}CH_2CH_2Cl$

KOH, KI

$HO(CH_2CH_2O)_k$—⟨⟩—CHO

$(EtO)_2\overset{O}{\overset{\|}{P}}CH_2$—⟨⟩—$NO_2$

t - BuOK

$HO(CH_2CH_2O)_k$—⟨⟩—CH=CH—⟨⟩—NO_2 **1-3**

1   k = 3
2   k = 2
3   k = 1

Scheme I

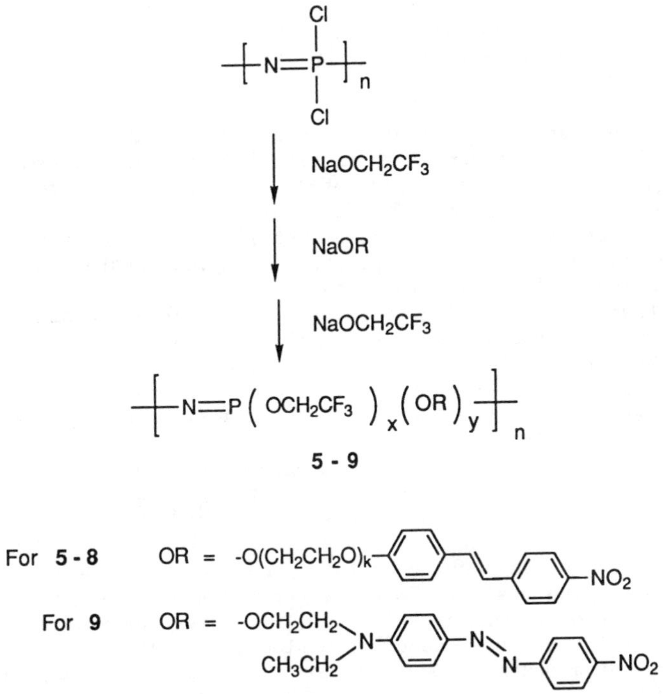

$$\left[-N{=}\underset{\underset{\text{Cl}}{|}}{\overset{\overset{\text{Cl}}{|}}{P}} \right]_n$$

NaOCH$_2$CF$_3$

NaOR

NaOCH$_2$CF$_3$

$$\left[-N{=}P \left(OCH_2CF_3 \right)_x \left(OR \right)_y \right]_n$$

5 - 9

For **5 - 8**   OR = -O(CH$_2$CH$_2$O)$_k$—⟨benzene⟩—CH=CH—⟨benzene⟩—NO$_2$

For **9**   OR = -OCH$_2$CH$_2$\
                    CH$_3$CH$_2$ ⟩N—⟨benzene⟩—N=N—⟨benzene⟩—NO$_2$

Scheme II

Table I. Polyphosphazene Structures and Composition Ratios

Compd[a]	Side Group Structure[b]	y[c], %
5	k = 3	26
6	k = 3	36
7	k = 2	39
8	k = 1	31
9	-	33

[a] See Scheme II for general polymer structure
[b] See Figure 1 for general side group structure
[c] x + y = 100%

The preparation of soluble, single-substituent polyphosphazenes that contained species 1-4 as side groups could not be accomplished because, as noted previously, the direct addition of the sodium salt of the chromophore to poly(dichlorophosphazene) resulted in the formation of a polymeric, incompletely substituted precipitate. The precipitate was insoluble in refluxing tetrahydrofuran as well as warm dioxane, N,N-dimethylformamide, dimethylsulfoxide, nitrobenzene and N-methylpyrrolidinone. This insolubility was attributed to both the extended rigid structure and the intrinsically high polarity of the donor-acceptor substituted, conjugated side chains. Both factors may induce extensive side group stacking and thus lead to the formation of insoluble polymers.

The preparation of soluble polymers containing species 1-4 was accomplished by the use of the polar trifluoroethoxy group as co-substituent. The partially substituted trifluoroethoxy polymer, prepared in the first step of the polymer synthesis (see Scheme II), provided a polar environment for the incorporation of the chromophoric side chains. However, the maximum loading of the polymers by the chromophores 1-4 was limited by the solubility of the polymeric products. Hence, the side group ratios for polymers 6-9 represent a maximum incorporation range of the chromophore side group by the use of this synthetic scheme.

The preparation of mixed-substituent polymers that contained co-substituents other than trifluoroethoxy groups was also explored. This part of the investigation was carried out in an attempt to tailor the macromolecular properties, for example, glass transition temperature, solubility behavior, morphology, and film-forming ability, in order to optimize the nonlinear optical behavior. However, the aryloxy substituents, including phenoxy, 4-methylphenoxy, and 3-ethylphenoxy, as well as the alkoxy substituent, methoxyethoxyethoxy, all yielded insoluble polymers, even with low incorporation ratios (10-15%) of the chromophore. These results suggest that the highly polar trifluoroethoxy group is a necessary co-substituent for the preparation of soluble polymers containing 1-4 as side chains.

Structural Characterization and Properties of Polyphosphazenes

Characterization of polymers **5-9** was achieved by ^{1}H and ^{31}P NMR spectroscopy, gel permeation chromatography, differential scanning calorimetry, UV/visible and infrared spectroscopy, and elemental microanalysis. All the polymers were soluble in common organic media, such as tetrahydrofuran, acetone, and methylethyl ketone.

A typical ^{31}P NMR spectrum consisted of a sharp, singlet resonance at -8 ppm, presumably a consequence of the similar environment at the trifluoroethoxy and ethoxy-ether substituted phosphorus atoms in the mixed-substituent system. In addition, the singlet resonance indicated a high degree of chlorine replacement. This was supported by the elemental microanalysis data.

The substituent ratios of the polymers were determined by ^{1}H NMR analysis by a comparison of the integration of the combined aromatic and vinyl resonances, which were generally between 8.4 and 6.8 ppm, with the trifluoroethoxy resonance at 4.5 ppm.

The molecular weights of polymers **5-9** were estimated by gel permeation chromatography to be in the range M_n = 9.4 x 10^4 to 3.2 x 10^5, M_w > 9.3 x 10^5, with M_w/M_n values in the region 4-7. UV/visible spectra in tetrahydrofuran showed the same trends as the corresponding side group compounds **1-4**, with λ_{max} values in the range 369-378 nm for **5-7** and 468 nm for **8**. Infrared spectroscopy of thin films cast on KBr for all of the polymers showed an intense P=N stretching vibration at 1250-1200 cm^{-1}. In addition, the absorbance for the NO_2 unit at ca. 1345 cm^{-1} was detected.

The glass transition temperature (T_g) of the mixed-substituent polyphosphazenes **5-9** varied with the loading of the chromophoric side chain and with the length of the connecting ethyleneoxy spacer group. Species with one ethyleneoxy unit comprising the spacer group generated the highest glass transition temperature. The T_g values were 19°C for **5**, 25°C for **6**, 25°C for **7**, 54°C for **8**, and 44°C for **9**. No evidence of T(1) or T_m transitions were detected for any of these polymer samples. Hence, the addition of the chromophoric substituent disrupts the microcrystallinity of the single substituent polymer [NP(OCH$_2$CF$_3$)$_2$]$_n$, which has a T_g at -66°C, a T(1) between 60 and 90°C, and a T_m at 240°C (*24*).

The colors of the polyphosphazenes corresponded to those of the chromophores employed. Thus, polymers **5-8**, which contained chromophores **1-3**, were yellow, while polymer **9**, which contained chromophore **4**, was red.

Evaluation of the Second-Order Nonlinear Optical Behavior

Films of polyphosphazenes **5-9** were spin cast onto indium-tin oxide coated glass from a concentrated solution in methylethyl ketone. The solution was first filtered to remove particulate impurities and the films were heated to 80-85°C to remove the solvent. The thicknesses and refractive indices of the polymers were obtained from ellipsometric measurements on calibration layers, which were spun on BK7 glass substrates. Measurements on each sample were performed at four different wavelengths (634.8 nm, 753.0 nm, 802.0 nm and 852 nm) in order to minimize the errors in the extrapolated

values at 532 and 1064 nm. The thickness of the layers examined
ranged from 70-250 nm, and were always much less than the coherence
lengths, as determined from the refractive index measurements.

The nonlinear optical properties of the films were subsequently investigated using second-harmonic generation. A Q-switched
Nd:YAG laser (λ = 1064 nm) with a pulse width of 8 ns and a pulse
energy of 10 mJ was used as the source of the fundamental, and a
reference sample of Y-cut quartz (d_{11} = 0.46 pm/V) was used for
calibration of the frequency-doubled signal.

Alignment of the nonlinear optical side groups in the films
was achieved by single-point corona poling, with the point source
held at +10 kV, at a distance of 1.5 cm from the surface. Increasing poling voltage led to an increase in the harmonic intensity,
i.e. maximum alignment was not achieved at this voltage. However,
higher voltages occasionally resulted in damage to the sample
surface. Hence, for comparison purposes, the voltage was limited
to 10 kV. Because of the low glass transition temperatures of
these polymers, the poling was carried out at room temperature,
concurrent with the second-harmonic generation measurements. This
arrangement led to the reproducibility of the measurement condition
for each layer. Following removal of the poling field, the
second-harmonic signal decayed to zero within a few minutes.

The values of the second-harmonic coefficient, d_{33}, for
samples 5-9 are listed in Table II. The values of d_{33} were obtained using the analysis of Jerphagnon and Kurtz (25), and were
calculated under the assumption that the degree of alignment of the
nonlinear optical chromophores can be described using the isotropic
model. Hence, we assumed $d_{33}=3d_{31}$ (4).

Table II. d_{33} Coefficients for Polyphosphazenes

Compd[a]	d_{33}, pm/V
5 (y=26%)	4.1
8 (y=31%)	4.7
6 (y=36%)	5.0
7 (y=39%)	5.0
9 (y=33%)	34

[a]See Scheme II and Table I for polymer structures and composition
ratios; Polymers 5-8 arranged in order of increasing value of y

In the series of polymers 5-8, which contain the nitrostilbene
side groups 1-3, the trend in the d_{33} value versus loading of the
chromophoric side group was well reproduced, with d_{33} values in the
range 4.1-5.0 pm/V. Note that the decrease in the spacer length
from three to one ethyleneoxy units appeared to have no effect on
the d_{33} value. For polymer 9, which contained the high β azo
chromophore 4, the d_{33} value was 34 pm/V, which was significantly
higher than for the stilbene substituted polymers that contained

equivalent side group incorporation ratios. This is partially a consequence of the greater resonant enhancement, given the longer wavelength of the azo chromophore absorption peak.

The d_{33} coefficient for $[NP(OCH_2CF_3)_{86\%}$ $(OCH_2CH_2OCH_2CH_2OCH_3)_{14\%}]_n$ (10) was examined in order to investigate the contributions of the phosphazene backbone and the trifluoroethoxy side groups to the second-harmonic signal. This mixed-substituent polyphosphazene was selected for study because of its amorphous morphology, as opposed to the single-substituent polymer $[NP(OCH_2CF_3)_2]_n$, which is microcrystalline (24). The d_{33} coefficient of 10 was less than that of quartz. Therefore, the contributions to d_{33} from the phosphazene skeleton and the trifluoroethoxy side groups were negligible in comparison to the contribution from the chromophoric side chain.

Conclusions and Future Prospects

The synthetic versatility offered by the phosphazene system has allowed the preparation of polymers that contain nonlinear optical units as pendant side chains. Our future research on nonlinear optical polyphosphazenes will focus on tailoring the macromolecular system to generate higher glass transition temperatures. This, and the stabilized alignment of the chromophoric side groups, should be attainable by the incorporation of a third co-substituent that contains a crosslinkable moiety. Thus, crosslinking of the polymeric matrix during the application of an electric field would be expected to stabilize the nonlinear optical character.

Acknowledgments

The work at The Pennsylvania State University was supported by the U.S. Office of Naval Research and the U.S. Air Force Office of Scientific Research. The work at the University of Southern California was supported by the U.S. Air Force Office of Scientific Research.

Literature Cited

1.   Chemla, D. S.; Zyss, J., Eds. Nonlinear Optical Properties of Organic Molecules and Crystals, Academic: New York, 1987; Vols. 1,2.
2.   Khanarian, G., Ed. Molecular and Polymeric Optoelectronic Materials: Fundamentals and Applications: SPIE: San Diego, 1986; Vol. 682.
3.   Williams, D. J., Ed. Nonlinear Optical Properties of Organic and Polymeric Materials; ACS Symposium Series 233; American Chemical Society: Washington, DC, 1983.
4.   Williams, D. J. Agnew. Chem., Int. Ed. Engl. 1984, 23, 690.
5.   Singer, K. D.; Sohn, J. E.; Lalama, S. J. Appl. Phys. Lett. 1986, 49, 248.

6.   Hill, J. R.; Dunn, P. L.; Davies, G. J.; Oliver, S. N.;
     Pantelis, P.; Rush, J. D. Electronics Lett. 1987, 23, 700.
7.   Hampsch, H. L.; Yang, J.; Wong, G. K.; Torkelson, J. M.
     Macromolecules 1988, 21, 526.
8.   Meredith, G. R.; VanDusen, J. G.; Williams, D. J.
     Macromolecules 1982, 15, 1385.
9.   Ye, C.; Marks, T. J.; Yang, J.; Wong, G. W. Macromolecules
     1987, 20, 2322.
10. Leslie, T. M.; DeMartino, R. N.; Choe, E.; Khanarian, G.;
     Haas, D.; Nelson, G.; Stamatoff, J. B.; Stuetz, D. E.; Teng,
     C.; Yoon, H. Mol. Cryst. Liq. Cryst. 1987, 153, 451.
11. Singer, K. D.; Kuzyk, M. G.; Holland, W. R.; Sohn, J. E.;
     Lalama, S. J.; Comizzoli, R. B.; Katz, H. E.; Schilling, M. L.
     Appl. Phys. Lett. 1988, 53, 1800.
12. Eich, M.; Sen, A.; Looser, H.; Bjorklund, G. C.; Swalen, J.
     D.; Tweig, R.; Yoon, D. Y. J. Appl. Phys. 1989, 66(6), 2559.
13. Hall, H. J., Jr.; Kuo, T.; Leslie, T. M. Macromolecules 1989,
     22, 3525.
14. Rubello, D. R. J. Polym. Sci.; Polym. Chem. 1990, 28, 1.
15. Dembek, A. A.; Kim, C.; Allcock, H. R.; Devine, R. L. S.;
     Steier, W. H.; Spangler, C. W. Chem. Mater. 1990, 2, 97.
16. Allcock, H. R.; Kugel, R. L. J. Am. Chem. Soc. 1965, 87, 4216.
17. Allcock, H. R.; Kugel, R. L.; Valen, K. J. Inorg. Chem. 1966,
     5, 1709.
18. Allcock, H. R.; Kugel, R. L. Inorg. Chem. 1966, 5, 1716.
19. Allcock, H. R. Chem. Eng. News 1985, 63, 22.
20. Allcock, H. R.; Allen, R. W.; Meister, J. J. Macromolecules
     1976, 9, 950.
21. Allen, R. W.; Allcock, H. R. Macromolecules 1976, 9, 956.
22. Wadsworth, W.; Emmons, W. J. J. Am. Chem. Soc. 1961, 83, 1733.
23. Allcock, H. R.; Kim, C. Macromolecules 1989, 22, 2596.
24. Ferrar, W. T.; Marshall, A. S.; Whitefield, J. Macromolecules
     1987, 20, 317.
25. Jerphangnon, J.; Kurtz, S. K. J. Appl. Phys. 1970, 41, 1667.

RECEIVED July 18, 1990

Chapter 17

Molecular Design for Enhanced Electric Field Orientation of Second-Order Nonlinear Optical Chromophores

H. E. Katz, M. L. Schilling, W. R. Holland, and T. Fang

AT&T Bell Laboratories, Princeton, NJ 08540

Three synthetic approaches to donor-acceptor-substituted conjugated molecules with enhanced orientability in electric fields, potentially applicable to the preparation of electro-optic polymers via electric field poling, are summarized. The three approaches are parallel attachment of chromophores to a common framework, embedding the chromophore in a zwitterion, and head-to-tail oligomerization of chromophores. The oligomerization method as well as the use of dyes as curing agents are briefly discussed in relation to the stability of electric field-induced polar order in polymer matrices.

Two of the most important nonlinear optical (NLO) processess, electro-optic switching and second harmonic generation, are second order effects. As such, they occur in materials consisting of noncentrosymmetrically arranged molecular subunits whose polarizability contains a second order dependence on electric fields. Excluding the special cases of noncentrosymmetric but nonpolar crystals, which would be nearly impossible to design from first principles, the rational fabrication of an optimal material would result from the simultaneous maximization of the molecular second order coefficients (first hyperpolarizabilities, β) and the polar order parameters of the assembly of subunits. (1)

The desire to increase β values above those of molecules used in the earliest materials has led to the exploration of organic compounds as the active components of second order NLO devices. Considerable effort has been expended in the synthesis and analysis of candidate molecules, which are largely donor-acceptor substituted conjugated π systems. (2) Examples of compound classes whose members display large nonresonant β include azo dyes, stilbenes, polyenes, merocyanines, stilbazolium salts, and quinoid

0097–6156/91/0455–0267$06.00/0

compounds. Structure-property relationships governing the effects of substituents, molecular length, and the nature of the molecular skeleton on β have been deduced for many of these types of compounds (3) and are fairly well established. >From these relationships, it might be possible to conjure molecules which might exhibit still larger β. However, there are diminishing returns in the increase in β as one increases the length of the π system or the donor-acceptor strength of the substituents beyond certain limits (4-6), and increases in low energy absorbances limit the utility of some chromophores exhibiting very high β, especially when the coefficient is enhanced primarily by resonance. In any event, the means of maximizing the second order hyperpolarizability are well laid out.

Methods for achieving the orientational order required for a second order NLO bulk material have also been intensively studied, but remain much more problematical. The most effective ways to impart this order consist of noncentrosymmetric crystallization, self-assembly at interfaces, and electric field poling of the active chromophores. Crystallization suffers from being difficult to model thermodynamically, although some intriguing results concerning rationally predicted polar crystallization have recently been reported. (7-9) Crystals are also troublesome to employ in waveguiding and integrated modes, although they are generally quite stable. The highest degrees of polar order are achieved in organized thin films, either deposited as Z-type Langmuir-Blodgett films (10,11) or chemisorbed from solution. (12,13) The main drawbacks to these systems are the tedium in building up enough mass to serve as a bulk device material and the fragility of the multilayer assemblies. More facile deposition techniques leading to more robust materials are currently being investigated. (14)

Electric field poling has been widely pursued as a means of orientation, generally in thin polymer films. (15-17) One advantage is that the process may be modeled thermodynamically. (1) A corresponding disadvantage is that the thermodynamic model (correctly) predicts maximum order parameters (excess projection of the principal molecular moment in one direction versus an isotropic ensemble) of only 10-20% for most chromophores considered. (18) Additional problems center on the stringent conditions necessary for producing films that retain optical quality and dielectric strength during poling, and that fully retain orientational order after poling. On the other hand, polymer thin films can be deposited and oriented in situ relative to other components with which they may be integrated, and are well suited to waveguide applications. (19) For these latter reasons, much of our research has been directed towards new compounds which may be useful in second order NLO materials prepared by electric field poling.

Very little effort has been devoted to the design and synthesis of compounds in which the susceptibility to electric field alignment has been enhanced without significantly perturbing the electronic states of the chromophoric moieties. Similarly, very few compounds have been prepared for the express purpose of improving orientational stability after poling through judicious functional group placement. The primary purpose of this presentation is to describe the syntheses of several new compounds which are aimed at these issues of electric field-induced orientational order, and to demonstrate the potential for traditional organic synthesis to further the advancement of this field.

Dipole Additivity and Increased Polar Order

The degree of orientation achievable by applying an electric field (E) to an ensemble of dipoles increases with the magnitude of the dipole moments (μ) involved. The increase is approximately linear when the product μE is substantially below kT. (18) Since most poling processes for NLO occur in the linear regime, increases in the effective μ should lead to improved ordering of the dipolar chromophores. We have examined three synthetic strategies for enforcing the superimposition or additivity of dipole moments coincident with the principal moments of established NLO-active chromophores. The strategies are 1) projecting two chromophores in parallel directions from a rigid molecular backbone, 2) surrounding a weakly dipolar chromophore with separated, full charges that define a much larger μ, and 3) linking dipoles head-to-tail so that the ordering force acts on a cumulative effective μ. All three strategies have been demonstrated by actual syntheses, and in some cases, physical measurements as well. However, extensive materials science would still be required in order to implement these schemes in actual bulk systems.

The 2,5-endo bonds of simple norbornanes are within 30° of being perfectly parallel, and the 1,4-trans bonds of substituted piperazines are even closer to being parallel, according to molecular models. Accordingly, we synthesized compound **1** as shown in Equation 1. (20) The reaction is highly stereoselective, with no exo substituents observed. An x-ray structure of the analogue without the nitro groups is shown in Figure 1. Unfortunately, the limited deviations from parallelism at each ring-ring bond and steric distortions of the norbornane skeleton force the aminophenyl residues outward to almost 90° angles. The dipole moment of **1** in dioxane is 8.9 D, compared to 6.8 D for N,N-dimethyl-p-nitroaniline. This reflects the vector addition of the two main moments of **1** at approximate right angles, which is mathematically identical to having two unattached chromophores with no enforced additivity at all. Even so, there could be some advantage to an arrangement like **1**. In a poled polymer, oriented **1** would have to sweep out a much larger volume in a disorientation process than would a monomeric chromophore, and thus might be more orientationally stable. Interaction of a polymeric or dipolar functional group with the basic sites in the cavity of **1** might further improve the magnitude or stability of orientation.

A more rigidly parallel pair of bonds for the projection of chromophores are the 1,8 positions of anthracene and anthraquinone. The respective 1,8-dichlorides undergo a limited substitution chemistry, which we extended as shown in Equation 2 to synthesize parallel-directed but weakly dipolar phthalimides. In principle, the use of donor-substituted phthalimide nucleophiles in the reaction of Equation 2 would give a fully additive pair of strong dipoles; however, this has not yet been accomplished.

(eq. 1)

1

(eq. 2)

The use of zwitterions in poling would be desirable because the dipole moment of a long zwitterion is many times that of most polarized π systems. Indeed, the high molecular figures of merit for some donor- acceptor substituted quinoid compounds is based in part on large dipole moments that result from ground state zwitterionic character. (K.D. Singer, unpublished) Concomitant with zwitterionic character, however, is intractability in all but the most polar media. Thus, it is only with difficulty that partially zwitterionic chromophores may be dispersed in moderately polar polymers. We sought to superimpose a full zwitterion dipole moment on an established NLO chromophore, cyanovinylaniline, such that the known NLO properties of the chromophore would be maintained while ordering consistent with a much larger dipole would be achieved.

The synthesis of such a molecule, 2, is illustrated in Scheme 1. Although the synthesis is multistep, most of the individual reactions are standard condensations or protecting group manipulations. The trifluoroacetyl protecting group was essential for the Vilsmeyer formylation to succeed, since more electron-rich amide groups reacted with the Vilsmeyer reagent. The two charged groups are rigidly separated, and define a dipole virtually parallel to the main moment of the chromophore. The 2-ethylhexyl substituent was selected for maximum lipophilicity and poor crystal packing without micelle formation. Despite this selection, the dipole-dipole intermolecular association apparently dominated the physical properties of pure, nonprotonated 2. It was not possible to dissolve 2 or a borane adduct of 2 in a sufficiently nonpolar solvent to measure a dipole moment, which is estimated from molecular models to be ca. 50 D (2 full opposite charges separated 1 nm). Compound 2 could not be dispersed in poly(methyl methacrylate) (PMMA), but films of 2 in poly(N-vinylpyrrolidone) were spun from aqueous N,N- dimethylformamide. Unfortunately, these films, while of good optical quality, were too conductive to pole, possibly because of partial protonation or low-level ionic contamination. Thus, the use of zwitterionic chromophores for second order NLO materials remains tantalizing; however, considerable material fine tuning will be required in order to disperse and pole such species in a continuous medium.

The dipole moment augmentation strategy that has come closest to realization in a material is the head-to-tail linkage of semirigid dipolar chromophoric subunits in an oligomer. This strategy has precedent in work published by Williams and Willand. (21,22) While the preceeding work focussed on high molecular weight polymers with limited registry among subunits, our thrust has been toward smaller assemblies in which the geometrical relationships among subunits are well defined. Examples of compounds we have studied in this light are dimers 3 and 4 and oligomers 5 and 6. (23,24) Once again, we have taken advantage of the piperazine ring to combine electron donating ability and conformational definition in as small a subunit as possible. The amide group provides electron withdrawing character and a linkage with little angular variability. A crystal structure of the non-

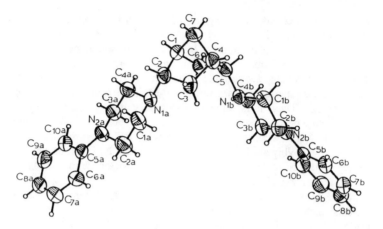

Figure 1. X-ray crystal structure of endo,endo-2,5-bis(4-phenylpiperazinyl)-norbornane.

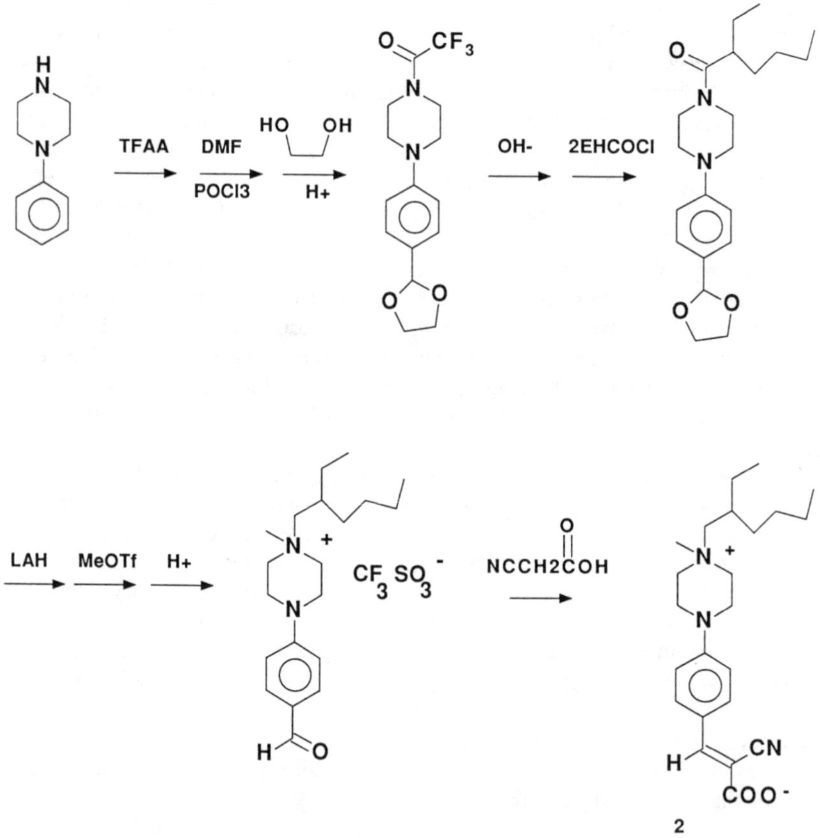

Scheme 1. Abbreviations: TFAA, trifluoroacetic anhydride; DMF, N,N-dimethylformamide; 2EHCOCl, 2-ethylhexanoyl chloride; LAH, lithium aluminum hydride; MeOTf, methyl trifluoromethanesulfonate.

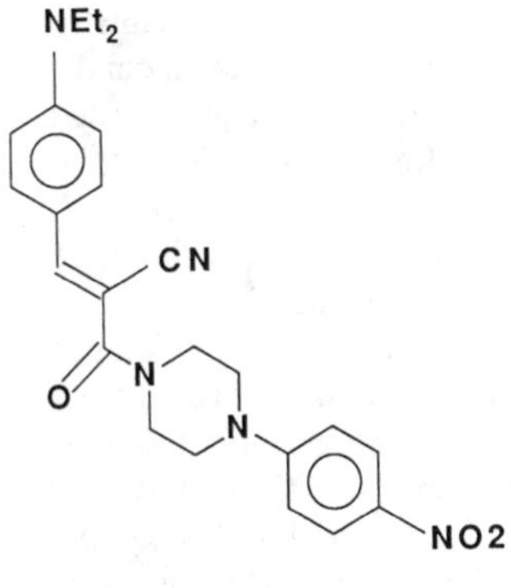
NEt$_2$
CN
O
N
N
NO2

3

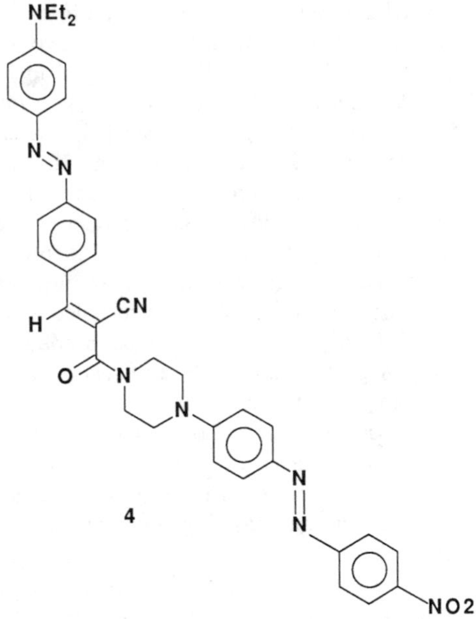
NEt$_2$
N
N
H
CN
O
N
N
N
N
NO2

4

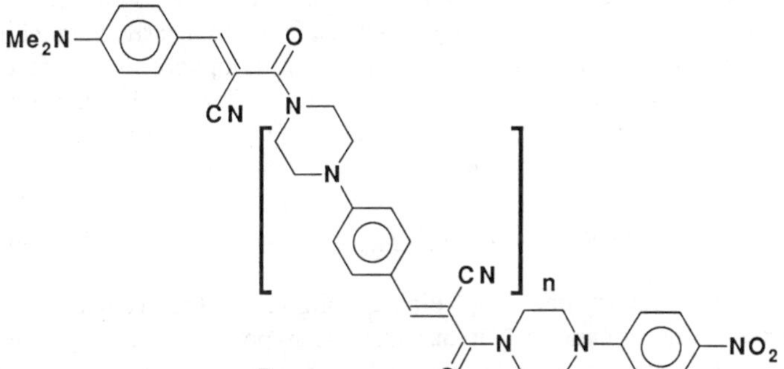
Me$_2$N
O
CN
N
N
CN
n
N
N
O
NO$_2$

5

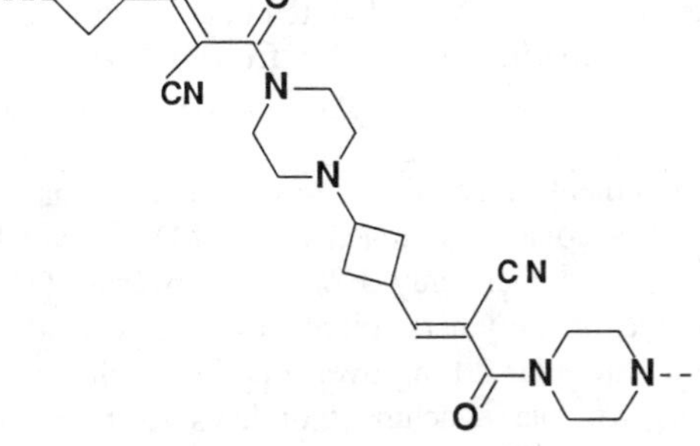
O
CN
N
N
CN
N
N
O

6,

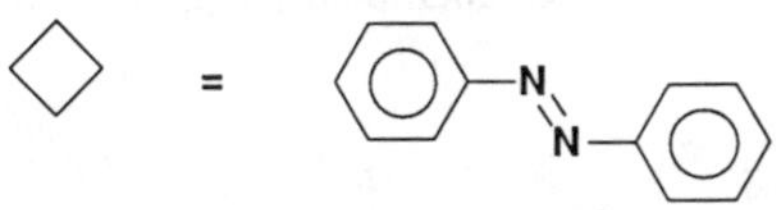
=
N
N

nitrated analogue of **3**, shown in Figure 2, clarified both the stereochemistry and relative conformations of the two chromophores in these compounds. Other related compounds were judged to be stereochemically analogous based on NMR spectra. The key steps in the syntheses were formation of the vinylic double bonds through Knoevenagel condensations. Dipole moments of **3**, **4**, and tetrameric **5** were 9.1, 13.0, and 16.6 D respectively, consistent with vector additivity of the chromophore moments at 110° angles as predicted by the crystal structure and molecular models. These data also indicate the preference of higher oligomers for extended conformations even though doubled-back conformations are available to these oligomers that are not accessible to dimers. The tetrameric material was also poled as a 10% mixture in PMMA at 120 °C in a field of 1 MV/cm; the electro-optic coefficient measured for this sample was 0.8 pm/V, and the birefringence was 0.006. These values are consistent with an extended conformation in that an ensemble of monomers poled at similar concentration should have had an electro-optic coefficient <0.6 pm/V and a birefringence of <0.003 based on their linear polarizabilities and values of $\beta\mu$, but are not precise enough to allow a rigorous assessment of the enhancement in ordering due to the dipole additivity in the oligomer. (10,11,21,22)

A major weakness in the piperazinamide oligomer approach is the angular connection between the subunits. Although the segment-to-segment angle is well defined and large enough to impart enhanced solution-phase dielectric susceptibility relative to separated monomers, a more nearly parallel linkage would be preferable. An example of a dimeric chromophore that incorporates a linkage significantly closer to 180° is **7**, whose synthesis is currently being attempted. Molecular models indicate that the angle between the bonds exocyclic to the tetrahydropyridine ring is more than 160°.

Molecular Length, Functional Group Placement, and Orientational Stability

Our motivation for synthesizing dipolar oligomers has primarily been to achieve enhanced orientation. It has also been postulated that longer active molecules will retain their poling-induced order for a longer time after poling than will very short chromophores. (25) In this respect, the head-to-tail oligomers might be advantageous even if enhanced initial ordering is not ultimately realized from them. The poled tetramer described above retained 60% of its electro-optic coefficient and birefringence over two months, as opposed to 20-40% for the simple azo dye Disperse Red 1 (**8**) dissolved in PMMA.

This modest improvement in stability is not as great as that observed in azo dyes sized similarly to **8** covalently attached to PMMA, which lose 10% of their alignment initially, and then only 10%/yr afterwards. (15) The covalent attachment of longer dyes to polymers might therefore lead to materials stable enough for testing in a particular setting over a period of years.

It has also been shown that attachment of dyes to a crosslinked polymer network greatly improves orientational stability, especially when the dye itself

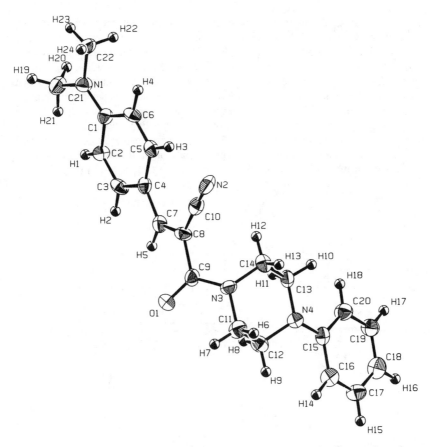

Figure 2. X-ray crystal structure of 4-dimethylamino-β-cyanocinnamic phenylpiperazinamide.

7

8

incorporates multiple points of attachment. (26) It has not yet been determined whether the location of these points of attachment along the length of the dye affects stability independently of the glass transition temperature of the polymer matrix. We therefore prepared compounds **9** and **10** to be employed as curing agents for epoxies. In **10**, the functional group on the nitrated ring, when bonded to the epoxy network, effectively lengthens the chromophoric unit in that a much longer segment must realign if the orientation of the dye is to be randomized. Furthermore, pivoting motions of the dye about the amino end should be inhibited for **10** relative to **9**. Preliminary results indicate that these carboxylic acid dyes do indeed cure epoxies, although slowly. Experiments to determine the molecular motions of the dye-cured epoxies are in progress.

Conclusion

This account has summarized several of our approaches to the preparation of electric-field-aligned chromophoric polymers for second order NLO applications. Molecular design has been employed wherever possible to arrive at structures that probe particular aspects of the polar orientation issue. The rich variety of accessible organic structures has enabled us to consider the orientation problem from a variety of points of view, and to indicate by example the manner in which multifunctional organic synthesis may play a role in the fabrication of oriented materials.

For the most part, the actual results presented here are reports of syntheses of compounds, accompanied by a limited amount of physical data. It would have been desirable to explore these compounds in much greater depth using materials science techniques. The addition of expertise and more intense experimental effort in materials engineering to the chemical studies conducted here and elsewhere would undoubtedly lead to more practical materials for second order NLO devices.

Acknowledgments

We are grateful for experiments and discussions with many collaborators, without whom this endeavor would not have been possible, and whose specific contributions are indicated in the references. We particularly thank K.D. Singer, M.G. Kuzyk, C.W. Dirk, A. Hale, M.L. Kaplan, and G.E. Johnson for helpful discussions related to this manuscript, and L.A. King and H.M. Gordon for assistance with compound characterization and film preparation.

Literature Cited

1. Singer, K.D.; Kuzyk, M.G.; Sohn, J.E. J. Opt. Soc. Am. B 1987, 4, 968-976.
2. Nicoud, J.F.; Twieg, R.J. in "Nonlinear Optical Properties of Organic Molecules and Crystals", Chemla, D.S.; Zyss, J., eds., Academic Press, New York, 1987, pp. 255-268.
3. Singer, K.D.; Sohn, J.E.; King, L.A.; Gordon, H.M.; Katz, H.E.; Dirk, C.W. J. Opt. Sci. Am. B 1989, 6, 1339-1350.
4. Barzoukas, M.; Blanchard-Desce, M.; Josse, D.; Lehn, J.-M.; Zyss, J. J. Chem. Phys. 1989, 133, 323-329.
5. Nicoud, J.F. Proc. SPIE 1988, 971, 68-75.
6. Li, DeQ.; Ratner, M.A.; Marks, T.J. J. Am. Chem. Soc. 1988, 110, 1707-1715.
7. Marder, S.R.; Perry, J.W.; Schaeffer, W.P. Science 1989, 245, 626-628.
8. Tam. W.; Wang, Y.; Calabrese. J.C.; Clement, R.A. Proc SPIE 1988, 971, 107-112.
9. Nicoud, J.F. Mol. Cryst. Liq. Cryst. 1988, 156, 257-268.
10. J. Am. Chem. Soc. 1988, 110, 2672-2674.
11. Blinov, L.M.; Dubinin, N.V.; Mikhnev, L.V.; Yudin, S.G. Thin Solid Films 1984, 120, 161-170.
12. Williams, D.J.; Penner, T.L.; Schildkraut, J.J.; Tillman, N.; Ulman, A.; Willand, C.S. In Nonlinear Optical Effects in Organic Polymers J. Messier et al, eds., Kluwer Academic Publishers: 1989, pp. 195-218.
13. Maoz, R.; Netzer, L.; Gun, J.; Sagiv, J. J. Chimie Phys. 1988, 85, 1059-1065.
14. Schilling, M.L.; Putvinski, T.; Katz, H.E.; Chidsey, C.E.D.; Mujsce, A.M.; Emerson, A.B. presented at the 199th National Meeting of the ACS, and submitted for publication.
15. Singer, K.D.; Kuzyk, M.G.; Holland. W.R.; Sohn, J.E.; Lalama, S.J.; Comizzoli, R.B.; Katz, H.E.; Schilling, M.L. Appl. Phys. Lett. 1988, 52, 1800.
16. Hampsch, H.L.; Yang, J.; Wong, G.K.; Torkelson, J.M. Macromolecules 1988, 21, 526-528.
17. Eich, M.; Sen, A.; Looser, H.; Bjorklund, G.C.; Swalen, J.D.; Twieg, R.; Yoon, D.Y. J. Appl. Phys. 1989, 66, 2559-2567.
18. Willand, C.S.; Williams, D.J. Ber. Bunsenges Phys. Chem. 1987, 91, 1304-1310.

19. Thackara, J.I.; Lipscomb, G.F.; Stiller, M.A.; Ticknor, A.J.; Lytel, R. Appl. Phys. Lett. 1988, 52, 1031-1033.
20. Katz, H. J. Chem. Soc. Chem. Commun. 1990, 126-127.
21. Green, G.P.; Weinschenk, J.I., III; Mulvaney, J.E.; Hall, H.K., Jr. Macromolecules 1987, 20, 722-726.
22. Green, G.P.; Hall, H.K., Jr.; Mulvaney, J.E.; Noonan, J.; Williams, D.J. Macromolecules 1987, 20, 716-721.
23. Katz, H.E.; Schilling, M.L. J. Am. Chem. Soc. 1989, 111, 7554-7557.
24. Schilling, M.L.; Katz, H.E. Chem. Mater. 1989, 1, 668-673.
25. Hampsch, H.L.; Yang, J.; Wong, G.K.; Torkelson, J.M. Polym. Commun. 1989, 30, 40-43.
26. Reck, B.; Eich, M.; Jungbauer, D.; Twieg, R.J.; Willson, C.G.; Yoon, D.Y.; Bjorklund, G.C. Proc. SPIE 1989, 1147, 74-83.

RECEIVED July 18, 1990

Chapter 18

Nonlinear Optical Chromophores in Photocrosslinked Matrices

Synthesis, Poling, and Second Harmonic Generation

Douglas R. Robello, Craig S. Willand, Michael Scozzafava, Abraham Ulman, and David J. Williams

Corporate Research Laboratories, Eastman Kodak Company, Rochester, NY 14650

Noncentrosymmetry required for second-order nonlinear optical effects has been achieved by photopolymerizing an acrylic matrix while simultaneously orienting NLO-active chromophores with a DC electric field. The chromophores were either dissolved in a NLO-inactive, polymerizable liquid binder or were themselves polymerizable. We synthesized several chromophores with substituents chosen to improve the solubility or to confer polymerizability on the molecules. After poling and polymerization, the induced orientation was probed using second harmonic generation (SHG). In all cases, the SHG signal was stable for short periods at room temperature, but dropped to nearly zero upon mild heating or prolonged storage. These results indicate that the films have relatively low glass transition temperatures, in agreement with studies on photocrosslinking of NLO-inactive polyacrylate monomers.

Recent advances in optical technology have created great interest in the construction of second-order nonlinear optical (NLO) devices for frequency conversion and electrooptic modulation. Although inorganic substances such as $LiNbO_3$ possess strong second-order NLO properties and have been well-studied, organic materials hold the promise of greater nonlinear susceptibility, higher laser damage threshold, increased operating speed, and improved processability (1-4).

In order to produce an organic material that would exhibit useful second-order NLO effects, two criteria must be met. First, the constituent molecules must possess a high molecular hyperpolarizability, β, which usually entails a large, unsymmetrically polarized, conjugated π-electron system. Second, the molecules must be arranged in a noncentrosymmetric orientation, which allows a nonzero bulk hyperpolarizability, $\chi^{(2)}$.

Molecules such as 4'-dimethylamino-4-nitrostilbene ("DANS", **1**) possess the necessary electronic structure, and have very large β

0097–6156/91/0455–0279$06.00/0

values. However, polar compounds like **1** have a propensity to crystallize in centrosymmetric structures with adjacent, opposing dipoles. The challenge lies in forcing such compounds into a noncentrosymmetric arrangement.

1

One of the most widely studied methods for achieving noncentrosymmetry in a bulk material is electric field poling. This technique has been applied to systems in which active small molecules were doped into polymeric hosts (5-10), to polymers with covalently-bound active chromophores (11-17), and to analogous thermally crosslinked systems (18-19). In a typical poling experiment, the material is heated to an elevated temperature in the presence of a strong DC electric field, then cooled to below the glass transition temperature (Tg) of the matrix with the field still applied. The Tg must be high enough to assure that alignment will be preserved for subsequent long term use under ambient conditions, so poling must be carried out at a high temperature as well. However, the achievable nonlinear susceptibility $\chi^{(2)}$ decreases with increasing temperature, as can be seen from the following equation (5, 6, 10):

$$\chi^{(2)} \propto L_3 \left[\frac{f_0 \, \mu_z \, E}{5 \, k \, T} \right]$$

in which L_3 is a third-order Langevin function, f_0 is a local field factor, μ_z is the molecular dipole moment, E is the applied electric field, k is the Boltzman constant, and T the temperature during poling. Also, heating might cause decomposition of sensitive components of the device. In addition, any mismatch in thermal expansion between the NLO film and its package can lead to strain, cracking, and delamination during cooling from high temperature.

We decided to explore the possibility of performing electric field poling of a NLO-active compound at room temperature and solidifying the material by in-situ photopolymerization. In essence, instead of cooling the sample below Tg, we proposed to raise the Tg of the material above the poling temperature. This paper reports on systems in which the active NLO species is present either as a dopant in a photocrosslinkable host matrix or as the matrix itself. We also describe the preparation and use of several new donor-acceptor stilbene compounds similar to **1**.

Synthesis of NLO-active dyes

Some representative examples of NLO-active guest compounds are shown in Figure 1. Some of these compounds, or their synthetic precursors, were

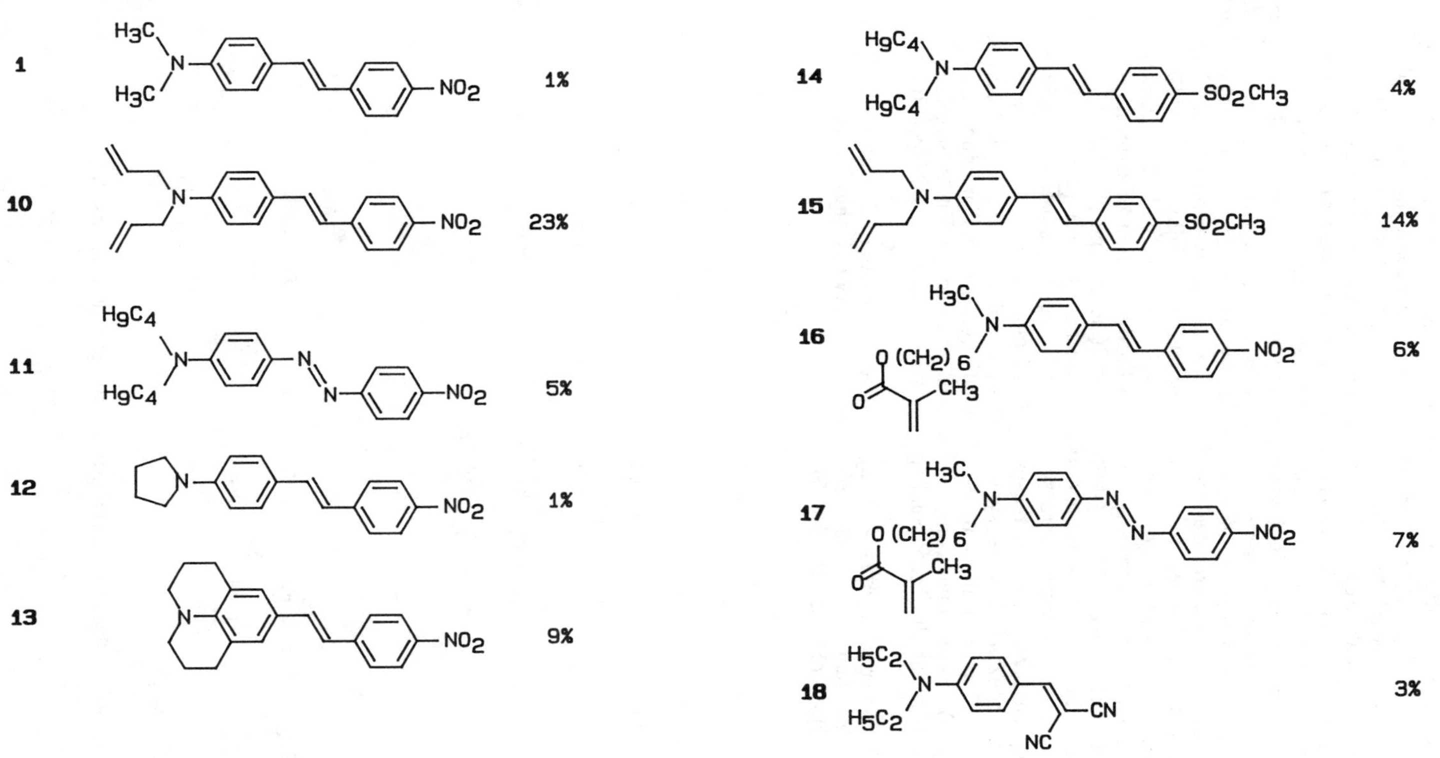

Figure 1.   Representative NLO-active guest compounds and maximum solubility (wt %) in host **7**.

obtained commercially.   Others were synthesized by routine methods as shown in the Figures 2-5.   In general, we synthesized the donor and acceptor portions of the chromophores separately, then joined the fragments in later steps.   This convergent approach enabled us to "mix and match" different donors and acceptors, in this way to produce many different compounds from a few common synthetic intermediates.

Stilbene compounds **4** were prepared according to Figure 2. *N,N*-Dialkyl anilines **2** were synthesized by direct alkylation of aniline. The aldehydes **3** were synthesized either by Vilsmeyer formylation of **2** or by reaction of 4-fluorobenzaldehyde with a secondary amine (20).   The stilbenes **4** were formed from the corresponding 4-dialkylamino-benzaldehydes either by the Horner-Emmons reaction with 4-nitrobenzyl-(diethyl)phosphonate (prepared by the Arbuzov reaction of α-bromo-4-nitrotoluene) or with 4-methylsulfonylbenzyl(diethyl)phosphonate (prepared in three steps from 4-methylthiobenzylalcohol) (21).   A few nitrostilbene compounds were synthesized by heating aldehyde **3** with 4-nitrophenylacetic acid in the presence of piperidine.

Azobenzenes **5** were synthesized by coupling of the dialkyl aniline derivative **2** with 4-nitro- or 4-methyl-sulfonylbenzene diazonium salt (Figure 3).   Vinylidine compound **6** was prepared by Knoevenagel condensation of the corresponding aldehydes with malononitrile (Figure 4).

Polymerizable dyes were synthesized by reacting chromophores bearing pendant primary alcohol groups with acryloyl chloride or methacryloyl chloride (Figure 5).

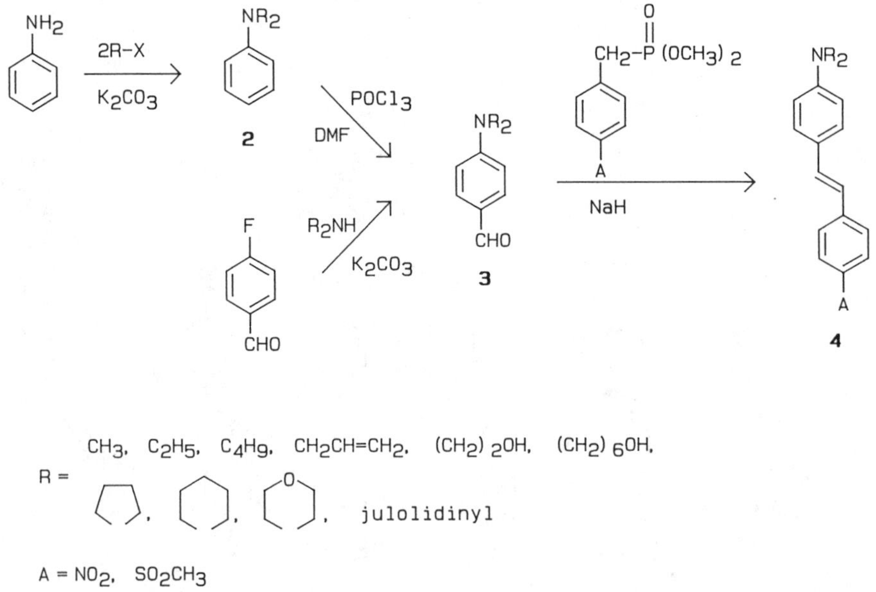

Figure 2.   Synthesis of stilbene derivatives **4**.

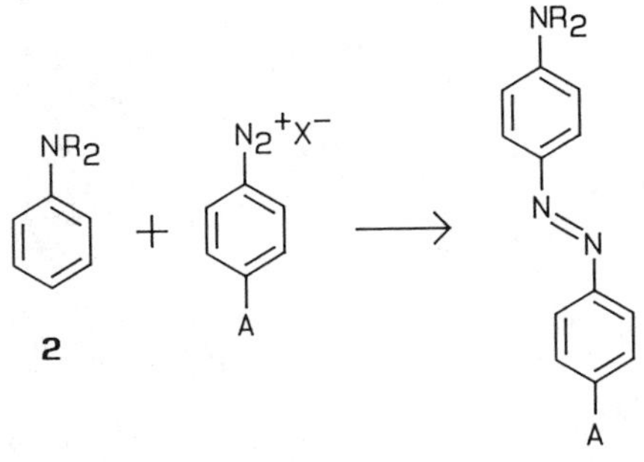

$R = CH_3, \quad C_4H_9, \quad (CH_2)_6OH$

$A = NO_2, \quad SO_2CH_3$

Figure 3. Synthesis of azobenzene derivatives **5**.

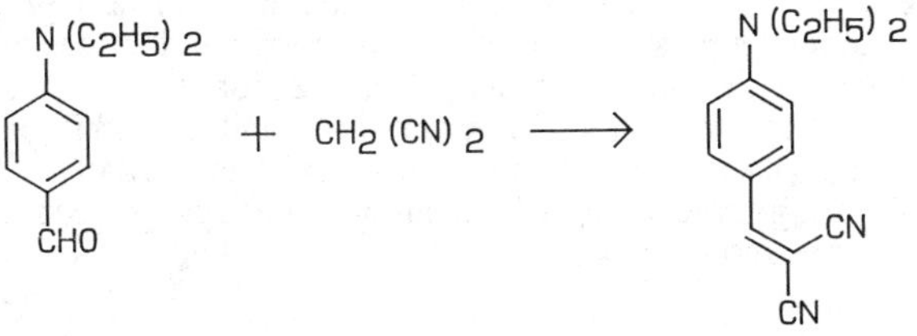

Figure 4. Synthesis of dicyanovinyl derivative **6**.

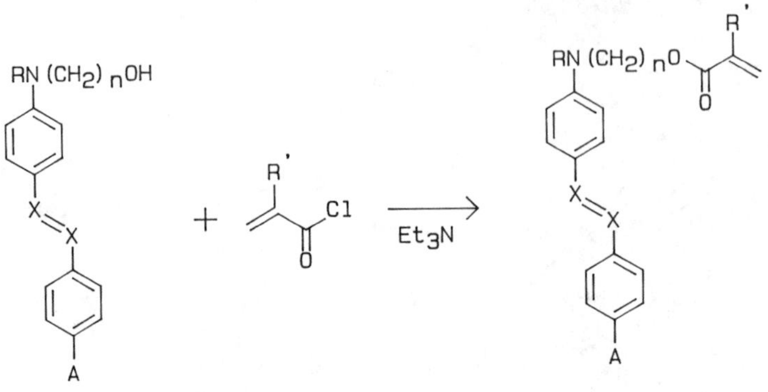

R = CH$_3$,   (CH$_2$)$_n$OH

n = 2, 6

R$^{\cdot}$ = H, CH$_3$

A = NO$_2$, SO$_2$CH$_3$

X = CH, N

Figure 5.   Synthesis of polymerizable dye compounds.

Electric Field Poling and Photocrosslinking of Host/Guest Systems

For host/guest systems, tri(acryloxyethyl)trimelletate (**7**), a liquid trifunctional monomer was employed.   This compound is easily and rapidly crosslinked to a tough solid by photopolymerization in the presence of a suitable sensitizer and activator, for example **8** and **9**.   This particular initiator system is useful because the sensitizer **8** is partially bleached out during irradiation, which allows the light to penetrate the sample as polymerization proceeds.

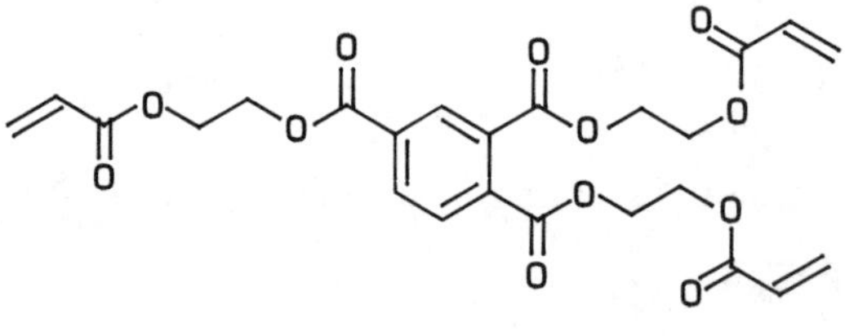

7

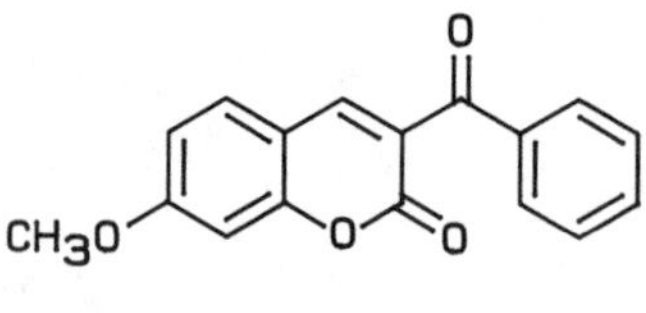

8

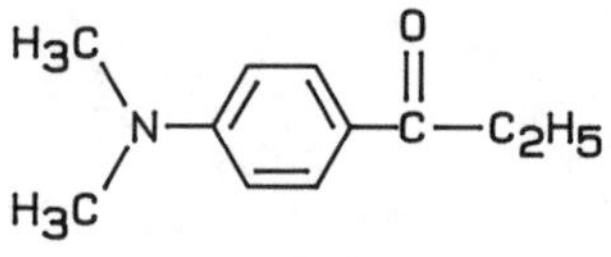

9

Unfortunately, **7** is a poor solvent for donor-acceptor compounds, and a maximum of 1 wt % concentration was achieved for **1**. Limited solubility is a common problem with host-guest systems; we found however, that the solubility of donor-acceptor stilbene and azobenzene compounds can be improved considerably by the attachment of bulky side-groups such as butyl or allyl to the chromophores. In this way, loading levels above 15 wt % were possible for certain compounds. On the other hand, simple cyclic side-groups such as pyrrolidino (**12**) led to poorly soluble materials (Figure 1).

Solutions to be used in the thin film formation were prepared as follows: The NLO chromophore was dissolved in dichloromethane with gentle heating and vigorous stirring or sonication. A second solution was prepared containing approximately 35 wt % triacrylate **7**, 1 wt % sensitizer **8**, and 2 wt % activator **9** in dichloromethane. The two solutions were mixed in varying proportions, the solvent was allowed to evaporate from a trial coating, and the resulting film was inspected for phase separation of the NLO chromophore.

For longitudinal electric field poling (where the electric field vector lies in the plane of the film), the solutions were spin-coated onto glass substrates on which chromium electrodes with 100-300 μm gaps had been previously deposited (Figure 6a). With the samples under a nitrogen atmosphere, a bias voltage was placed across the electrodes, inducing an acentric orientation of the dissolved NLO chromophores. This orientation was fixed by photopolymerization of the triacrylate host **7** by exposing the samples to filtered UV light from a 200 W high-pressure mercury lamp for several seconds.

In the case of transverse poling, the dichloromethane was allowed to evaporate, and the resulting viscous solution was injected via syringe into a cell comprised of two indium tin oxide (ITO) coated glass substrates separated by a spacer (Figure 6b). Poling and photocrosslinking of the samples were performed as described above. For thicker films (>10 μm), the cells were dismantled after crosslinking the mixture, producing free-standing, NLO-active films.

After the samples were sufficiently crosslinked, the electric field was removed and the successful alignment of the dye was tested using second harmonic generation (SHG). Light from a Quanta Ray Nd:YAG laser (1.06 μm, 10 Hz, ≤1 mJ/pulse) was focused onto the sample films. For longitudinally poled films, the light was polarized parallel to the poling field and normal incidence was used. The second harmonic signal was collected by a lens, focused onto a monochromator, and detected by a photomultiplier tube. Sample and poling uniformity was examined by monitoring the SHG while translating the sample parallel and perpendicular to the electrode gap. The transverse poled films were studied using p-polarized incident light and measuring SHG as a function of incident angle and position.

The second harmonic signal from a transverse poled and polymerized film containing 3 wt % of **18** is shown in Figure 7 along with that from a control sample containing no NLO-active compound. The SHG displays the expected rotational fringe pattern indicative of films thinner than the coherence length of the NLO process. The data clearly indicate that the SHG signal was solely a result of the NLO-active dopant and that successful alignment was achieved.

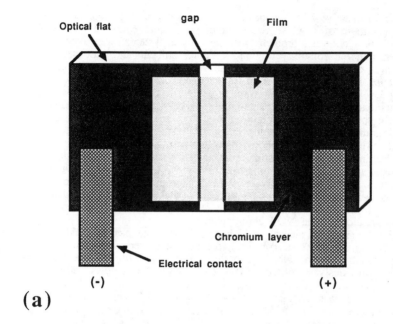

(a)

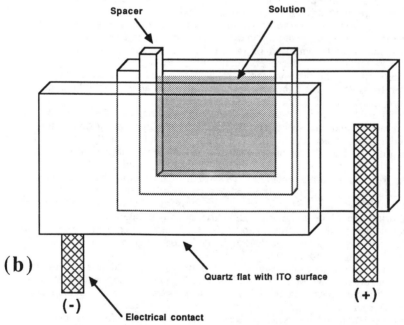

(b)

Figure 6. Cell design utilized in the (a) longitudinal and (b) transverse poling experiments.

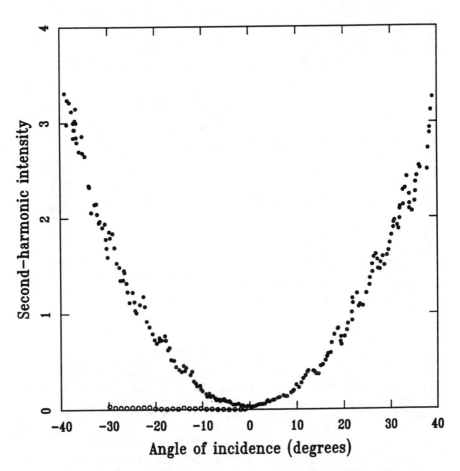

Figure 7. Second harmonic intensity as a function of angle of incidence for transversely poled and crosslinked films of triacrylate **7** (o) and 3 wt % **18** in triacrylate **7** (•).

Model Polymerization Reactions

A second disadvantage of host/guest systems is the possibility that the guest compound might leach out during subsequent processing steps. We therefore investigated the use of polymerizable NLO-active compounds, which might be covalently bound to the crosslinked matrix. To test the reactivity of these compounds, we conducted control copolymerizations with methyl methacrylate to form linear polymers whose molecular weight and composition could be easily determined. Results are summarized in Table I.

Table I. Copolymerization of polymerizable NLO dyes
with methyl methacrylate

Compound	m_{feed} [a]	$m_{polymer}$ [b]	Yield[c]
16	0.025	0.025	60%
16	0.073	0.084	65%
16	0.19	0.20	57%
17	0.026	0.034	77%
17	0.070	0.080	92%
18	0.19	0.20	92%[d]
10	0.134	0.01	33%[d]
15	0.122	0.009	45%[d]

a) Mole fraction of dye-monomer charge.
b) Mole fraction of dye-bearing repeat units in copolymer by ^{1}H NMR.
c) Yield based on total monomer charge.
d) Much dye-monomer was recovered unreacted.

For compounds bearing acrylate or methacrylate groups, the copolymer compositions were almost the same as the monomer feed compositions, and the molecular weights were nearly identical to that of poly(methyl methacrylate) (PMMA) synthesized as a control under the same reaction conditions. In addition, the dye-bearing repeat units were present uniformly in all molecular weights, as seen by comparing GPC molecular weight distribution curves determined by differential refractometry and by visible absorbance detection at the λ_{max} of the dyes (22). Although one should exercise caution in interpreting copolymer compositions at relatively high conversions, the above results taken together suggest that the dyes themselves have little effect on the polymerization, the pendant acrylate groups polymerize about as well as free methyl methacrylate, and the copolymers are probably random. Therefore, dyes bearing pendant acrylate groups can be covalently incorporated into the crosslinked matrix.

Notably, copolymers containing large percentages of stilbene dye-monomers were only partially soluble or had a significant high molecular weight fraction detected by GPC (22). However, copolymers containing lower amounts of stilbene dye monomers or polymerized in the presence of dodecanethiol as chain transfer agent were soluble in common solvents and exhibited less high molecular weight material by GPC. More significantly, copolymers that contained azobenzene dyes were completely soluble and showed no evidence of high molecular weight fractions.

We believe that the internal double bond in the stilbene-containing monomers participates in the vinyl polymerization to a minor but detectable extent, causing slight branching or crosslinking when this moiety is present in sufficient concentration. Such reactivity of a stilbene derivative with MMA appears to be unprecedented, although electron-withdrawing groups are known to activate carbon-carbon double bonds toward radical chain polymerization (23).

In contrast, diallylamino-substituted dyes copolymerized poorly with MMA, despite the reported polymerizability of other aliphatic (24-25) and aromatic (26) diallylamines. The concentration of dye-bearing repeat units in the polymer was far below the monomer feed concentrations. The possibility that these diallyl dye-monomers (or some unidentified impurity in them) acted as inhibitors of polymerization can be ruled out because the dye was found to be present uniformly in all molecular weight fractions and the molecular weights of the copolymers were again nearly identical to control samples of PMMA. Therefore, only a small amount of the diallyl amino substituted chromophores can be covalently incorporated into the crosslinked matrix because of the apparently unfavorable reactivity ratio. The balance of the chromophore remains simply dissolved in the mixture.

Electric Field Poling and Photocrosslinking of Polymerizable NLO Dyes

To increase the density of NLO-active chromophores in the crosslinked matrix, it was desirable to completely omit the inactive binder 7 and make the NLO dyes *themselves photocrosslinkable*. Therefore, we synthesized and studied chromophores substituted with two acrylate or methacrylate groups (27).

	A	n	R
19	NO_2	2	CH_3
20	NO_2	6	CH_3
21	NO_2	6	H
22	SO_2CH_3	6	CH_3

We found that when long (six methylene unit) spacers were used between the aromatic portion and the acrylic groups, the compounds were thick liquids at room temperature and could be spin-cast into high quality films. Poling and photopolymerization with sensitizer and activator, but *without* **7**, gave hard solids that exhibited strong second harmonic generation.

Although the nonlinear susceptibilities for these photopolymerizable dyes are much larger than those attainable with host/guest systems, the high concentration of strongly absorptive dye molecules prevents most of the light from reaching the sensitizer. This interference causes a drastic decrease in the rate of polymerization. A potential solution would be to employ a photoinitiating system which operates at long (>500 nm) wavelengths, where the dye molecules are transparent. Unfortunately, the common photoinitiators that operate in this region contain ionic components that contribute to high electrical currents during poling. This conductivity limits the attainable poling field and leads to sample decomposition during poling.

In order to circumvent problems associated with interfering absorption, we attempted to polymerize films composed of only the diacrylate dye (without photosensitizer and activator) using direct exposure to unfiltered UV light in a nitrogen atmosphere. For this single component system, we were able to obtain highly uniform, micron-thick films by vacuum deposition of the dimethacrylate dye directly onto the longitudinal poling cells (Figure 6a). Although the films fabricated in this manner exhibited excellent optical quality, the photocrosslinking was inefficient and difficult to reproduce; the polymerization was most likely initiated by trace impurities in the film. Nevertheless, we were able to produce a few hard films of crosslinked **22** in this way. The optical quality of the crosslinked films was tested by performing waveguide scattering loss measurements in which laser light (632 nm) was prism-coupled into the film. Propagation over several centimeters was observed, indicating fairly low loss levels and high optical quality.

The intensity of SHG for a poled (67 kV/cm) and crosslinked film of **22** in a longitudinal package is shown in Figure 8. The concave appearance of the signal across the gap is most likely caused by a field-induced modification of the film thickness, which can also be seen by optical microscopy. The thickness deviations appeared to be field-dependent and were absent in unpoled samples. The formation of this thickness gradient may be caused by flow as a consequence of hydrodynamic fluctuations stemming from ionic impurities.

The nonlinear susceptibility χ_{zzz} was determined to be 3.4×10^{-9} esu as measured relative to quartz crystal (d_{11}). This value is in good agreement with the calculated χ_{zzz} of 5.0×10^{-9} esu, which was derived assuming the thermodynamic model (5, 6, 10) and using data from electric field-induced second harmonic generation (21).

Stability of Alignment

The intensity of SHG from all of the poled and crosslinked films was unchanged for days at room temperature, but gradually declined to zero over many months. Apparently, orientation is lost at rather low temperature, indicating that the Tg of the material is quite low. This result is consistent with the temperature dependence of SHG, which we

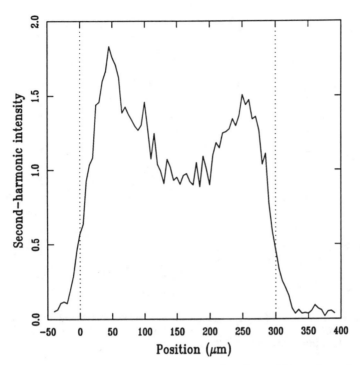

Figure 8. Second harmonic intensity as a function of inter-electrode position for a longitudinally poled, crosslinked film of **22**. The dotted lines indicate approximate positions of the electrodes.

have observed for films of **18** in **7** matrix, which showed a rapid decrease in the signal intensity above 60 °C (Figure 9).

This observation was corroborated by studies of the photocrosslinking of pure triacrylate **7** by infrared and Raman spectroscopy, and by dynamic mechanical analysis (28). Photopolymerization of **7** at 25 °C resulted in approximately 80% conversion of double bonds and a Tg of 70 °C. Apparently, the Tg of the material increased as crosslinking proceeded, but the reaction stopped at only partial conversion when vitrification occurred. The result was a Tg only 20-40 °C higher than the sample temperature. In addition, trapped free radicals were detected in crosslinked films of **7** by electron spin resonance (ESR) (28).

While it may be possible to carry out the photopolymerization at elevated temperature to achieve higher conversions and therefore higher glass transition temperatures, this heating negates the major advantage of performing simultaneous poling and photopolymerization.

Conclusions

We have successfully demonstrated the concept of simultaneous polymerization and poling of photocrosslinkable systems to produce high quality films with large second-order nonlinear optical susceptibilities and low scattering losses. The ability to process these materials under mild conditions may help avoid potential problems encountered in

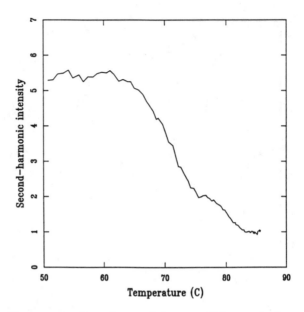

Figure 9.   Temperature dependence of the second harmonic intensity for a transversely poled, crosslinked film of 3 wt % **18** in triacrylate **7**. Heating rate was 1 °C/min.

standard thermoplastic processing caused by thermal decomposition while affording enhanced chromophore orientational parameters.   The compounds studied offer high NLO-chromophore film concentrations and are applicable to thin film vacuum technologies.   In addition, the potential for direct, photonegative patterning holds promise for the fabrication of integrated optical devices.

   The materials studied displayed relatively low glass transition temperatures and therefore poor orientational stability.   Future studies should be directed to solving this problem.   Additional research is required to improve the efficiency of photoinitiation for materials with strong absorptions in the visible region of the spectrum.   Other issues such as strain caused by shrinkage during crosslinking need to be addressed.

Acknowledgments

We express appreciation to John J. Fitzgerald and Samir Farid, both of Eastman Kodak Research Laboratories, for helpful discussions.

Literature Cited

1.   Williams, D. J. Angew. Chem. Intl. Ed. Eng. 1984, 23, 690.
2.   Williams, D. J., ed., Nonlinear Optical Properties of Organic and Polymeric Materials; ACS Symposium Series 233; American Chemical Society: Washington, DC, 1983.

3. Chemla, D. S.; Zyss, J. Nonlinear Optical Properties of Organic Molecules and Crystals; Academic Press: New York, 1987.
4. Williams, D. J. in Electronic and Photonic Applications of Polymers; Bowden, M. J.; Turner, S. R., Eds.; Advances in Chemistry 218; American Chemical Society: Washington, DC, 1989; p 297.
5. Meredith, G. R.; VanDusen, J. G.; Williams, D. J. Macromolecules 1984, 17, 2228.
6. Singer, K. D.; Kuzyk, M. G.; Sohn, J. E. J. Opt. Soc. Am. B 1987, 4, 968.
7. Singer, K. D.; Sohn, J. E.; Lalama, S. J. Appl. Phys. Lett. 1986, 49, 248.
8. Hampsch, H. L.; Yang, J.; Wong, G. K.; Torkelson, J. M. Macromolecules 1988, 12, 526.
9. Hampsch, H. L.; Yang, J.; Wong, G. K.; Torkelson, J. M. Polym. Commun. 1989, 30, 40.
10. Williams, D. J., in reference 3, p 405.
11. Singer, K. D.; Kuzyk, M. G.; Holland, W. R.; Sohn, J. E.; Lalama, S. J. Comizzoli, R. B.; Katz, H. E.; Schilling, M. L. Appl. Phys. Lett. 1988, 53, 1800.
12. Hill, J. R.; Pantelis, P.; Abbasi, F.; Hodge, P. J. Appl. Phys. 1988, 64, 2749.
13. Eich, M.; Sen, A.; Looser, H.; Bjorklund, G. C.; Swalen, J. D.; Twieg, R.; Yoon, D. Y. J. Appl. Phys. 1989, 66, 2559.
14. Ye, C.; Marks, T. J.; Yang, J.; Wong, G. K. Macromolecules 1987, 20, 2322
15. Dai, D.-R.; Marks, T. J.; Yang, J.; Lundquist, P. M.; Wong, G. K. Macromolecules 1990, 32, 1894.
16. DeMartino, R.; Haas, D.; Khanarian, G.; Leslie, T.; Man, H. T.; Riggs, J.; Sansone, M.; Stamatoff, J.; Teng, C. Yoon, H. Mat. Res. Soc. Symp. Proc. 1988, 109, 65.
17. DeMartino, R. N.; Yoon, H.-N. U.S. Patent 4 822 865, 1989; 4 865 430, 1989.
18. Hubbard, M. A.; Marks, T. J.; Yang, J.; Wong, G. K. Chem. Matls. 1989, 1, 167.
19. Eich, M.; Reck, B.; Yoon, D. Y.; Wilson, C. G.; Bjorklund, G. C. J. Appl. Phys. 1989, 66, 3241.
20. LeBarney, P.; Ravaux, G.; Dubois, J. C.; Parneix, J. P.; Njeumo, R.; Legrand, C.; Levelut, A. M. Proc. S.P.I.E. 1987, 682, 56.
21. Ulman, A.; Willand, C. S.; Köhler, W.; Robello, D. R.; Williams, D. J.; Handley, L. J. Am. Chem. Soc. 1990, in press.
22. Robello, D. R. J. Polym. Sci., Polym. Chem. Ed., 1990, 28, 1.
23. Odian, G. Principles of Polymerization; Wiley-Interscience: New York, 1981, Chapter 3.
24. Solomon, D. H. J. Macromol. Sci. Chem. 1975, 9, 97.
25. Butler, G. B. Pure Appl. Chem. 1970, 23, 255.
26. Mathias, L. J.; Vaidya, R. A.; Bloodworth, R. H. J. Polym. Sci., Polym. Lett. Ed. 1985, 23, 289.
27. Similar azo dyes bearing two acrylates have been previously reported: Brear, P.; Guthrie, J. T.; Abdel-Hay, F. I. Polym. Photochem. 1982, 2, 65.
28. Fitzgerald, J. J.; Landry, C. J. Appl. Polym. Sci. 1990, in press.

RECEIVED July 18, 1990

Chapter 19

Thermal Effects on Dopant Orientation in Poled, Doped Polymers

Use of Second Harmonic Generation

Hilary L. Hampsch[1], Jian Yang[2], George K. Wong[2], and John M. Torkelson[1,3]

[1]Department of Materials Science and Engineering, [2]Department of Physics and Astronomy, and [3]Department of Chemical Engineering, Northwestern University, Evanston, IL 60208

A wide variety of nonlinear optical (NLO) polymeric materials have been explored as possible candidates for device applications (2-3) including doped (4-11) and functionalized (12-14) polymers and polymer liquid crystals (15). To create useful NLO materials, not only is a large optical nonlinearity desired, but excellent thermal and temporal stability of dopant orientation are needed as well (2). In order to develop new NLO polymers that can meet the required specifications, a basic understanding of the polymer physics that governs the temporal stability of the optical signal, related to the dopant orientation in guest-host polymer systems must be obtained (4-6). Studies have shown that second harmonic generation (SHG), a second order NLO effect involving frequency doubling, is sensitive to the local microenvironment surrounding the optical dyes and any changes that occur in this environment in doped glassy films (4-9). Dopant orientation during and following electric field-induced poling can be studied continuously and in real time in order to examine the microenvironment surrounding the dopants in terms of the polymer relaxations and the applied corona field. In the results presented below, the SHG of 4-dimethylamino-4'-nitrostilbene (DANS) dispersed in polystyrene (PS) or poly(methyl methacrylate) (PMMA) matrices has been examined in corona poled films as a function of temperature in order to understand the effect of thermal conditions on the temporal stability of the dopant orientation.

The relaxation behavior of glassy polymers depends significantly on thermal history and temperature (16). As the temperature of the polymer matrix is increased, the mobility and local free volume surrounding the NLO dopant increases. This allows increased freedom for dopant rotation. Large scale mobility in the polymer matrix occurs at the glass transition temperature, T_g. However, at temperatures below T_g there is still limited molecular mobility described by secondary relaxations occurring down to and below a characteristic temperature T_β (16-18). These characteristic temperatures are a function of the polymer structure. T_g for the undoped PMMA and PS is about 100°C. Secondary relaxations in PS are probably due to wagging of the phenyl unit, with T_β measured at about 50°C via dilatometry (18). In PMMA, secondary relaxations are attributed to the rotation of the ester side group, and T_β has measured values between -40°C and 20°C depending on the experimental technique involved (18). Thus, it could be expected that PS has a lower degree of mobility than PMMA when examined at a given temperature below T_g. The dynamics of the dopant orientation during poling and disorientation following poling as observed via SHG should be sensitive to the local mobility, and thus should be sensitive to the polymer structure at a given temperature.

0097–6156/91/0455–0294$06.00/0
© 1991 American Chemical Society

Second Harmonic Generation

Second order NLO properties including SHG arise from the second order NLO susceptibility $\chi^{(2)}$ tensor in the relationship for the bulk polarization, **P**, such that (2-3)

$$\mathbf{P} = \chi^{(1)} \cdot \mathbf{E} + \chi^{(2)} \cdot \mathbf{EE} + \chi^{(2)} \cdot \mathbf{EEE} + ... \qquad (1)$$

The susceptibility tensors measure the macroscopic compliance of the electrons. Since the second order polarization is a second rank tensor, SHG is zero in a centrosymmetric or randomly oriented system. To make the material capable of SHG, the NLO dopants must be oriented noncentrosymmetricaly in the polymer matrix (2-3). When modeling the poled, doped films using a free gas approximation, the poled second order susceptibilities are given by (2,19)

$$\chi^{(2)}_{333} = NF\beta_{zzz}\langle\cos^3\theta\rangle \qquad (2a)$$

$$\chi^{(2)}_{311} = NF\beta_{zzz}\langle(\cos\theta)(\sin^2\theta)\rangle \qquad (2b)$$

where N is the number density of molecules, β is the second order hyperpolarizability (describing the microscopic second order NLO coefficient of the optical dopant), and $F = F(\omega)^2 F(2\omega)$ where the F's are local field factors (2-3,20). The subscript "3" refers to the direction of the poling field and "z" represents the axis of the molecule parallel to the molecular dipole moment (assuming $\beta = \beta_{zzz}$, implying one predominant component of the molecular tensor)(2-3). θ represents the angle between the z and 3 directions. Relationships between experimentally measured parameters and relative values for $\chi^{(2)}_{333}(-2\omega;\omega,\omega)$ (hereafter referred to as $\chi^{(2)}$) have been developed (6,11). The experimentally observed SHG intensity is given by (11,20)

$$I_{2\omega} = [l^2 I_\omega{}^2 \chi^{(2)2}/n_\omega{}^2 n_{2\omega}][\sin^2(\pi l/2l_c)/(\pi l/2l_c)^2] \qquad (3)$$

where $I_{2\omega}$ is the SHG intensity, l is the film thickness, I_ω is the fundamental intensity, n_ω and $n_{2\omega}$ are the index of refraction at frequency ω and 2ω, respectively, and l_c is the coherence length of the film. For a thin film, l/l_c approaches zero, and $\sin(\pi l/2l_c)$ approaches $\pi l/2l_c$. After simplification and assuming no dispersion ($n_{2\omega}=n_\omega=n$) (and using a quartz reference, Q), $\chi^{(2)}_{film}$ can thus be given by

$$\chi^{(2)}_{film} \sim (I_{film}/I_Q)^{1/2}(2/\pi)(n^3/l^2)^{1/2}{}_{film}(l_c{}^2/n^3)^{1/2}{}_Q\chi^{(2)}_Q \qquad (4)$$

where I is the SHG intensity, I_{film}/I_Q is the measured intensity ratio, and l_c is the coherence length of the quartz ($\sim 20\mu m$ (15)). $\chi^{(2)}$ values for the corona poled films are comparable to or larger than the values obtained from the quartz reference (6,9,15). The square root of the experimentally measured SHG intensity is thus directly related to $\chi^{(2)}_{film}$ and the NLO dopant orientation in the glassy matrix (6). More detailed and quantitative descriptions of these relationships are available (2-3,20).

Experimental

4-dimethylamino-4'-nitrostilbene (DANS) (Kodak), PS (M_n= 117,000, M_w= 430,000), and PMMA (M_n=46,400, M_w=93,900) (Scientific Polymer Products) were used as received and have been characterized (4-6). Films were spin coated onto soda lime glass slides from filtered chloroform (PMMA) or 1,2-dichloroethane (PS) solutions and carefully dried over a week long period, first under ambient conditions, then under vacuum at room temperature followed by temperatures above T_g in order to remove the maximum amount of solvent (since residual solvent acts as a plasticizer). The dopant concentration in the dried films (final thicknesses of 1-3μm) is 4 wt.%, and absorbance spectroscopy was used to verify that no aggregation of the dopants at these concentrations could be detected (5-6). It must be noted that the dopant concentration is too high for this to be a probe study of homopolymer behavior. However, the materials studied are similar to many practical plasticized systems (6). T_g for the doped material was measured via DSC (Perkin Elmer DSC-2, 10°C/min heating rate) as approximately 88±2°C for PS and 85±2°C for PMMA. Immediately prior to testing, the thermal history of all samples was erased by heating the film to 110°C for 3 hrs. Films were then placed in the beam path and allowed to equilibrate at the poling temperature for 1 hr prior to applying the field.

The experimental apparatus used to measure SHG intensities has been previously described (4-6). Light was generated by a q-switched Nd:YAG laser with a 1.064 μm fundamental (yielding the SHG signal wavelength of 532 nm). The beam was split so that a y-cut quartz reference and the sample were tested simultaneously.

Electric field-induced corona poling (21-23) is used to orient the NLO dopants (6-9,12) in the beam path. Poling involves applying a large electric field across the film normal to the surface so that the dopants align in the field direction. Poling can be performed at room temperature or at temperatures up to and above T_g, as long as the polymer has enough mobility and/or local free volume to allow the dopants freedom to rotate along the electric field vector (6-9). When the electric field is removed, the relaxation and mobility processes of the polymer release the dopant from its field-induced orientation. Dopants that are in a microenvironment with sufficient mobility are free to rotate out of their imposed orientation and thus their contribution to the SHG intensity diminishes (experimentally seen as a decay of the SHG signal with time) (4-9). The polymer mobility and free volume behavior and the effect of surface charge decay must be considered simultaneously when examining the overall temporal stability of corona poled films (6).

The corona discharge was generated by a tungsten needle biased with -3000V across a 0.6 cm air gap (relative humidity ~45% @ 25°C) normal to the polymer film. The corona current was limited to <5μA. Poling was performed in-situ with the sample oriented in the beam path. The reported $\chi^{(2)}$ is normalized to the sample thickness to allow comparisons to be made between a variety of films. Error limits due to noise in the SHG intensity measurements are less than twice the size of the symbols in the figures, with greater error at higher temperatures.

Surface voltage decay was measured with a Trek Model 341 high voltage electrostatic voltmeter (ESVM). The ESVM drives the potential of a probe body to the same potential as the film surface voltage. This device measures the voltage independent of distance within 5 mm of the film surface without causing current flow, which would modify the data. After poling, the corona needle was removed and the probe of the ESVM was placed 3 mm from the film surface.

Results and Discussion

The effect of temperature on the local microenvironment surrounding the NLO dopants can be examined using an in-situ SHG approach. Films are heated to an initial temperature of 110°C (about 25-30°C above T_g) in the beam path, and maintained at this temperature for 4 hours. This step erases previous thermal history, and ensures sample uniformity before the poling is performed (6). The films are then poled at 110°C and either maintained at 110°C, or cooled to 95°C or 60°C (over a 3 min period) with the field still applied. The field is applied for a total of 15 mins, and is then removed with the sample at its given final temperature.

The SHG intensity reached a maximum value for all of the films poled at 110°C within 1-2 mins after the field was applied. The dopants were free to orient in response to the applied field (thus increasing the SHG intensity) rapidly due to the high degree of mobility in the rubbery matrix. In all cases, the value of $\chi^{(2)}$ decreased during poling at 110°C. The film maintained at 110°C lost about 20% of its maximum signal and the film quenched from 110°C to 95°C lost about 15% of its maximum signal before the field was removed. $\chi^{(2)}$ decayed more rapidly during poling in PMMA films than in PS films poled under the same temperature treatments. This is consistent with previously observed results (6) and is due to a corona field effect. As discussed in the next section, increasing the sample (film+substrate) temperature increases charge penetration, which actually decreases the gradient across the thin film and thus decreases $\chi^{(2)}$ (5-6). The surface voltage magnitude, stability, and thermal dependence will affect the temporal stability of dopant orientation (5-9).

Figure 1 illustrates values of $\chi^{(2)}_{film}(t)/\chi^{(2)}_{film}(t=0)$ for PMMA and PS films doped with 4 wt.% DANS as a function of time after the applied field is removed. Once the field is removed, those dopants in regions of sufficient mobility and local free volume are free to rotate out of the poling imposed orientation. This causes the observed decrease in the value of $\chi^{(2)}_{film}(t)$. As the final temperature for the doped matrix decreases, the temporal stability of the dopant orientation following poling increases. This is due to the decreased mobility and local free volume observed at lower temperatures in the polymer matrix. The effect of temperature on SHG intensity has also been examined in contact poled PS functionalized polymers (24), with the SHG intensity decay much more rapid as the temperature increased.

Smaller magnitudes and greater temporal stabilities of $\chi^{(2)}_{film}/\chi^{(2)}_Q$ are observed in PS films when compared to PMMA films treated at the same temperatures. The maximum magnitude of $\chi^{(2)}_{film}/\chi^{(2)}_Q$ is 3 times greater in doped PMMA than in doped PS films poled at 110°C under the same field conditions. This is consistent with the results obtained when comparing the temporal stability of SHG intensity of doped PS and PMMA films prepared via high temperature (100°C) contact poling(4-5). Since PS is a more rigid matrix at room temperature ($T_{\beta,PS} > T_{\beta,PMMA}$) with less mobility than PMMA, the dopants have less freedom to rotate. In addition, there may be a small contribution to the observed differences in the magnitude of $\chi^{(2)}$ from a solvatochromic effect of the matrix on the dopant. However, it is probable that the majority of the changes in the magnitude of $\chi^{(2)}$ with the matrix are due to the differences in T_β. The secondary transitions are responsible for small scale motions in the glassy polymer host which are sufficient to allow the dopants to disorient with time. Since T_g for the two polymers is similar, this indicates that T_β should be considered an important parameter when examining the temporal stability of dopant orientation.

Attempts were made to fit the experimental $\chi^{(2)}$ decay data using simple and stretched exponentials. None of the decay data could be fit using a simple exponential, even at temperatures well above the glass transition. A variety of models, both phenomenological and molecular, have been examined in order to describe the relaxation in polymers near or slightly below the glass transition (16). The phenomenological models describe the kinetics of the relaxation process, and involve multiordering parameter equations with a variety of distribution functions. A distribution of retardation times generates the dependence of the relaxation time on structure (6,16). A phenomenological equation often used to fit polymer relaxation data obtained from bulk measurements such as dielectric relaxation is commonly known as the Williams-Watts stretched exponential and is of the form $y = \exp(-t/\tau)^\beta$. y represents a relaxing function, τ is a characteristic relaxation time, and β ($0 \leq \beta \leq 1$) is a parameter that reflects the cooperativity of the relaxations (6,16). It was found that over the time scale and temperature

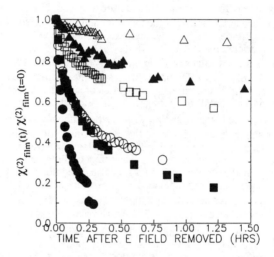

Figure 1. Effect of temperature on the temporal stability of $\chi^{(2)}{}_{film}$ (related to $[I_{film}]^{1/2}$ and the angle between the applied field and dopant optical axis) for corona poled PS and PMMA films doped with 4 wt.% DANS. $\chi^{(2)}{}_{film}(t)/\chi^{(2)}{}_{film}(t=0)$ versus time after the field is removed, where t=0 denotes the time that the applied field is removed. The initial temperature is 110°C, and the final temperatures after the field is removed are 110°C [PS (O),PMMA (●)], 95°C [PS (☐),PMMA (■)] or 60°C [PS (△), PMMA (▲)].

range of the experiments discussed here, the data were not fit accurately by this equation. A biexponential decay has been used to fit this type of data, and has been found to fit the data reasonably well except for the shortest time scales (<5 min) (6). However, no unique physical interpretation can be assigned to the fitting parameters. Some of the difficulty in fitting the data is due to the short time scale of the experiment and the additional contribution of the corona field persistence on the dopant orientation.

It is interesting to note that at temperatures well above T_g the dopant orientation is maintained for a long time period. For doped PS films poled and maintained at 110°C (~22°C above T_g), the magnitude of $\chi^{(2)}$ decreases to only about 75-80% one hour after the applied field is removed. If polymer relaxations were the only phenomena measured using this experiment, it might be expected that the dopants would disorient very rapidly in the rubbery matrix. This implies that a second effect, the surface charge persistence of the corona poling-induced field, is contributing to the experimentally measured stability of the dopant orientation. An attempt was made to short out the surface voltage following poling using a grounded copper brush. The magnitude of the surface voltage was unchanged, indicating that the charge may be trapped in surface sites.

A PS matrix is less polar than a PMMA matrix, and may interact differently with the applied electric field (6). For example, charge carrier mobility of undoped PMMA is about two orders of magnitude greater than that in undoped PS (23). This implies that the charge persistence should be greater in the PS matrix. This will affect the observed temporal stability of the SHG intensity. Figure 2 shows the surface voltage (SV) for the doped PS and PMMA films following corona poling at -3000V (0.6cm air gap). This experiment was performed at 25°C, with the films poled for 15 mins. In the thinner PS films (~ 1.5μm), higher initial SV magnitudes (~200V) are observed. The SV decays rapidly to about 150V in the first 10 mins, retains a value of ~110V 6 hrs after the poling field is removed and decays to zero after about 20 hrs. The PMMA films (~2μm) have an initial SV magnitude of about 75V, and slowly decay to a magnitude of about 25V over the following 6 hrs and to zero after about 15 hrs. When the same test is performed at 110°C similar trends are observed, with the initial magnitudes unchanged and more rapid decays. This different behavior relates to differences in the charge mobility in PS and PMMA, with the charge having greater difficulty penetrating into and moving through the PS film.

Charge penetration into the bulk depends on charge polarity, charge density, temperature, and surface characteristics (21). Corona discharge-induced charge penetration, which may occur at higher temperatures (21,23), is characterized by electrons or holes trapped at surface or near-surface sites. Traps are localized states within the polymer energy gap associated with impurities, disorder, or the electrical nature of the polymer (25). Trapped charges lead to a space charge layer at the film surface (25). Using charge transfer experiments, Bouattou and Lowell have found that polymers such as PS and PMMA in contact with metals indicate the presence of both permanent traps and "shallow" traps which retain charge for a short time. They also found that PMMA exhibited much less charge accumulation than PS (noting that the scatter in the PMMA data was greater) and a slightly more rapid decrease in surface charge over a several hour time scale. This was explained by assuming that only a fraction of the deposited charge is mobile, with the rest bound tightly in surface states. The time scale that the mobile charge in the PMMA moves away from the contact region is then shorter than that in the PS films, and was related to trap depth (25). This is consistent with the results observed in Figure 2. Future experiments will attempt to estimate the internal field during and following poling by measuring the voltage drop and current across the polymer film during contact poling. It would be expected that the internal field would be greater in the PMMA than in the PS and would be temperature dependent (22-23).

It is important to note the evidence is ambiguous as to the mechanism of charge transport within these materials. It has been noted "that where insulators are concerned the origin of the charge carriers is by no means clear" (23). Possible charge carriers can

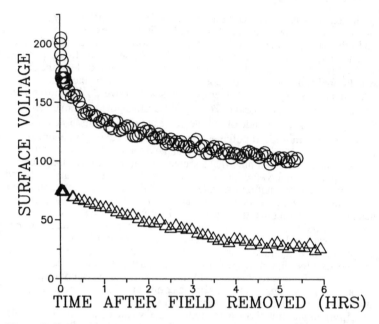

Figure 2. Surface voltage decay as a function of time for corona poled PS
(O) and PMMA (△) films doped with 4 wt.% DANS. Films poled at
-3000V (0.6 cm air gap) for 15 min at 25°C.

be extrinsic, such as ionic impurities, electrons, or holes injected from electrodes
(contact poling), or intrinsic to the polymer electronic structure (23). However, for the
purposes of this study, the exact nature of the charge carriers is less important than the
effect that they have on the temporal stability of the dopant orientation in the polymer
matrix. Thus, the examination of both surface voltage and second harmonic generation
should give at least general information on both polymer relaxations and electric field
effects in the corona poled, doped polymers.

It can thus be seen that both polymer relaxations and electric field effects
determine the temporal stability of dopant orientation in the corona poled, doped glassy
polymers as a function of temperature. Surface charge persistence is not a significant
issue in contact poled films or for studies performed over much longer time scales (6).
The two contributions discussed above for corona poled films must both be considered
when examining the overall temporal stability of the material as a function of temperature since both effects are temperature dependent (and have different temperature
dependencies) (6). For example, even though the initial magnitude of the applied field
is greater in the PS matrix than the PMMA matrix (approximately [200V]/[1.5μm+
1.2mm soda lime glass]= 0.0017 MV/cm for PS and [75V]/[2.5μm+ 1.2mm glass]=
0.0006 MV/cm for PMMA), fewer dopants can be oriented due to the greater hindrance
to rotation caused by the more rigid matrix. It is also important to note that the field
strengths used to contact pole NLO polymer films are up to three orders of magnitude
greater than the fields remaining due to the charge persistence (4-5,10-11). Thus, even
though the surface voltage is small in magnitude, it can significantly affect the dopant
orientation with time following removal of the applied field particularly at short times
following removal of the field. This reinforces the requirement that both electric field

and polymer relaxations be considered when examining the NLO properties of these materials with time and temperature over the time scales considered in this experiment. The effect of the surface charge on the measured dopant orientation becomes less significant at longer experimental time scales or when films are prepared by contact poling (4-6). PMMA and PS films doped with DANS and corona poled at room temperature show that over longer time scales the PS films show significantly greater temporal stability of the dopant orientation following removal of the applied field. For example, a PS+4% DANS film poled at 25°C for 50 mins lost approximately 15% of its initial $\chi^{(2)}$ after 20 hrs following removal of the applied field, whereas the PMMA+4% DANS film poled at 25°C for 40 mins lost about 65% of its initial $\chi^{(2)}$ after 12 hrs following removal of the applied field. Over these time scales the surface voltage decays to negligible values, and polymer relaxation effects dominate. In addition, the temporal stability of dopant orientation in contact poled PMMA and PS films doped with DANS has been examined. In contact poled films, surface voltage retention is negligible. These studies also indicate that the NLO dopants show improved temporal stability in PS as compared to PMMA (4-5). This is due to the decreased mobility and local free volume in the PS as compared to the PMMA. Thus, polymer relaxations can be studied using SHG.

Summary
        Polymer relaxations in doped amorphous polymers can be examined as a function of temperature using second harmonic generation. The effect of temperature above and below Tg on the dopant orientation yields information on the microenvironment surrounding the NLO chromophore. It has been shown that polymer relaxations and corona field-induced electric field effects can influence the magnitude and temporal stability of the dopant orientation. As the final temperature of the doped polymer system increases, the temporal stability of dopant orientation following poling decreases. At all temperatures, the temporal stability of the NLO dopant was greater in the PS than in the PMMA matrix. This is due to the less mobile structure of PS below T_g compared to PMMA. Surface voltage was found to be sensitive to polymer structure as well, with initial magnitudes greater and decreased temporal stability in PS matrices. Thus, the SHG and surface voltage experiments are sensitive methods for examining the microenvironment surrounding optical dopants in polymer matrices.

Acknowledgments
        The authors wish to acknowledge the financial support of the Materials Research Center at Northwestern University (Grant DMR88-20280); JMT also gratefully acknowledges the support of the National Science Foundation through its Presidential Young Investigator Program, and HLH gratefully acknowledges a Fellowship from the American Association of University Women. We appreciate helpful discussions with Dr. S. G. Grubb and Ms. S. J. Bethke of Amoco Technology Company and Dr. Gary Boyd of 3M. We wish to thank Trek, Incorporated for lending us a model 341 Electrostatic Voltmeter.

Literature Cited
1. a) Department of Materials Science and Engineering; b) Department of Physics and Astronomy; c)Department of Chemical Engineering.
2. Williams, D. J., Ed. *Nonlinear Optical Properties of Organic and Polymeric Materials*; ACS Symposium Series #233, American Chemical Society: Washington, DC, 1982.
3. Chemla, D. S., Zyss, J., Eds. *Nonlinear Optical Properties of Organic Molecules and Crystals Vol. 1 and 2*; Academic Press: New York, 1987.
4. Hampsch, H. L.; Yang, J.; Wong, G. K.; Torkelson, J. M. *Polym. Commun.* **1989,** *30*, 40.

5. Hampsch, H. L.; Yang, J.; Wong, G. K.; Torkelson, J. M. *Macromolecules* **1988,** *21*, 526.
6. Hampsch, H. L.; Yang, J.; Wong, G. K.; Torkelson, J. M. *Macromolecules*, in press.
7. Hampsch, H. L.; Wong, G. K.; Torkelson, J. M.; Bethke, S. J.; Grubb, S. G. *Proc. SPIE* **1989,** *1104*, 267.
8. Hampsch, H. L.; Torkelson, J. M.; Bethke, S. J.; Grubb, S. G. *J. Appl. Phys.* **1990,** *67*, 1037.
9. Bethke, S. J.; Grubb, S. G.; Hampsch, H. L.; Torkelson, J. M. *Proc. SPIE*, in press.
10. Singer, K. D.; Sohn, J. E.; Lalama, S. J. *Appl. Phys. Lett.* **1986,** *49*, 248.
11. Singer, K. D.; Kuzyk, M. G.; Sohn, J. E. *J. Opt. Soc. Am. B.* **1987,** *B4*, 698.
12. Singer, K. D.; Kuzyk, M. G.; Holland, W. R.; Sohn, J. E.; Lalama, S. J.; Comizzoli, R. B.; Katz, H. E.; Schilling, M. L. *Appl. Phys. Lett.* **1988,** *53*, 1800.
13. Ye, C.; Marks, T. J.; Yang, J.; Wong, G. K. *Macromolecules* **1987,** *20*, 2322.
14. Eich, M.; Sen, A.; Looser, H.; Yoon, D. Y.; Bjorklund, G. C.; Tweig, R.; Swalen, J. D. *Proc. SPIE* **1988,** *971*, 128.
15. Meredith, G. R.; VanDusen, J. G.; Williams, D. J. *Macromolecules* **1982,** *15*, 1385.
16. McKenna, G. B. In *Comprehensive Polymer Science Volume 2*; Booth, C.; Price, C., Eds.; Permagon Press: Oxford, 1989.
17. Struik, L. C. E. *Polym. Eng. Sci.* **1977,** *17*, 165.
18. Kolarik, J. In *Advances in Polymer Science Vol. 46*; Springer-Verlag: Berlin, 1982.
19. Willand, C. S.; Williams, D. J. *Ber. Bunsenges. Phys. Chem.* **1987,** *91*, 1304.
20. Shen, Y. R. *Principles of Nonlinear Optics*; John Wiley & Sons: New York, 1984.
21. Williams, E. M. *The Physics and Technology of Xerographic Processes*; Wiley Interscience: New York, 1984.
22. Sessler, G. M., Ed. *Electrets*; Springer-Verlag: Berlin, 1980.
23. Seanor, D. A. In *Electrical Properties of Polymers*; Seanor, D. A. Ed.; Academic Press: New York, 1982.
24. Hubbard, M. A.; Minami, N.; Ye, C.; Marks, T. J.; Yang, J.; Wong, G. K. *Proc. SPIE* **1989,** *971*, 110.
25. Bouattou, B.; Lowell, J. *J. Phys. D: Appl. Phys.* **1988,** *21*, 1787.

RECEIVED July 18, 1990

Chapter 20

Organic Polymers as Guided Wave Materials

Karl W. Beeson, Keith A. Horn, Michael McFarland, Ajay Nahata,
Chengjiu Wu, and James T. Yardley

Research and Technology, Allied-Signal, Inc., Morristown, NJ 07962

Organic polymers with chemically-engineered linear and
nonlinear optical properties offer great promise for the
integration of optical structures on silicon and gallium arsenide
semiconductors. Waveguide structures can be delineated in
organic polymers with a variety of processing techniques that are
compatible with the fabrication of microelectronic features. We
report that ultraviolet and visible radiation can be used to
photochemically delineate index of refraction profiles in solution
spin-coated organic films thereby generating a wide variety of
waveguide structures. The technique can use classical
photoresist type masks or laser writing techniques to produce the
desired features. The formation of single mode passive
waveguide structures is demonstrated using PMMA films
containing photochemically active nitrones. The methodology is
readily extended to poled polymer films with electro-optic
response, and is demonstrated in the preparation of a channel
waveguide electro-optic amplitude modulator.

The recent rapid development of powerful compact diode lasers operating at
wavelengths from the visible to the near infrared, the need for rapid information
processing and communications systems, and the microminiaturization of electronic
components have been key factors driving the field of integrated optics. In order
to implement many of the desired integrated optical devices, systems engineers
require new materials with large optical nonlinearities capable of being processed
into waveguide structures. To realize the promise of single-chip integrated
structures containing laser sources, passive and active spatially delineated
waveguides, detectors, and direct output for other applications, the processing

technologies for these new materials must be compatible with existing silicon and gallium arsenide electronic and optical devices. Organic polymer films offer interesting advantages for these purposes when compared to the existing waveguide technologies based on inorganic crystals such as lithium niobate and potassium titanyl phosphate. For example, organics have been shown to have optical nonlinearities (1-3) that can exceed those of the inorganic crystals. Since the large nonlinearities have their origin in electronic (4) rather than lattice motions, they have the potential for frequency response greater than that possible in the materials routinely used today. In addition, organics have low dielectric constants (5) and dielectric losses even at GHz frequencies (6). Another potentially key advantage of the organic polymers is that they can be chemically modified to vary their linear optical properties such as optical transmission (which is far greater than many of the semiconductors in the .7 to 2 micron wavelength region), refractive index, adhesion, mechanical properties, and thermal stability. In addition to these advantages, organic polymers can be processed at temperatures compatible with sensitive silicon and gallium arsenide electronic features using methodologies analogous to existing photoresist technologies.

Waveguide Fabrication

The properties of organic waveguides (7) which can be used to advantage in integrated optical applications include the confinement of light to micron dimensions, the diffractionless propagation of light in spatially delineated regions, and the generation of high intensities in small volume elements with long interaction lengths that can lead to novel nonlinear optical effects. Because of the potential offered by organic polymers in waveguide configurations, a variety of fabrication methods have been previously investigated for the creation of the necessary spatial refractive index profiles. Included in these methodologies are photolithography (8) followed by reactive ion or wet etching processes, gas or solution indiffusion (9,10), plasma polymerization (11), single crystal film formation (12,13), Langmuir Blodgett film preparation (14), electric field poling (15), and "photolocking" (16,17).

We have developed a new method for the spatial delineation of refractive index profiles which allows for the efficient and rapid generation of single-mode organic waveguides structures at room temperature using both coherent and noncoherent light sources. This "photodelineation" (18-21) method is based on the change in refractive index that results from the photochemical transformation of reactive chromophores mixed in polymeric matrices. The dispersion in refractive index of such organic materials is well described by the single oscillator Sellmeier equation (22) (Equation 1) where λ_o is the wavelength of the primary oscillator, λ is the measurement wavelength, A is proportional to the oscillator strength and B accounts for nondispersive contributions from all other oscillators.

$$n^2 - 1 = \frac{A}{\lambda_o^{-2} - \lambda^{-2}} + B \qquad (1)$$

If the material has a dominant long wavelength absorption with high oscillator strength, as is the case for many materials used for χ^2 processes, then the refractive index of the material is dominated by this single oscillator. If this absorption feature, or indeed any electronic absorption feature, is associated with a photoreactive state, irradiation into this band will result, after photochemical reaction, in a new absorption band either at longer or shorter wavelength than the original absorption maximum. This effects a corresponding increase or decrease in the refractive index for wavelengths longer than the resulting absorption wavelength ($\lambda > \lambda_o$).

Two photochemical technologies, lamp/mask patterning and laser direct writing, can be used to delineate waveguide structures in solution spin-coated organic polymer films. The first of these uses incoherent light sources and standard photoresist type masks with the desired positive or negative patterns. Both contact and projection methodologies can be used. The laser direct writing process uses a focussed coherent light source to generate the refractive index profiles. Analogous to photoresist systems, there are two methods of pattern formation for each of these methods. The photochemical generation of increased refractive index in regions to be used for waveguides is designated as a positive system, while systems in which the refractive index is reduced in regions adjacent to the waveguide are negative systems.

Nitrone/PMMA Waveguides: A Negative Pattern System

It is well established experimentally and theoretically that organic materials exhibiting large second-order nonlinear optical effects have a polarizable electronic structure (often π conjugated) with asymmetric charge distribution (aromatic charge-transfer states) and a noncentrosymmetric macroscopic orientation ([1]). In order for these same materials to be useful for the photochemical delineation of waveguides the materials must also have a photochemically reactive state characterized by a UV absorption band at a readily accessible wavelength, and a moderate to high quantum efficiency for reaction. It is also desirable that the reaction not generate secondary photoproducts. As a class of materials the nitrones (shown in Equation 2) exhibit many of the desired features. They are readily synthesized by the reaction of aromatic aldehydes with substituted phenylhydroxylamines, are thermally stable and undergo a photocyclization reaction to the corresponding oxaziridine with high quantum efficiency ([23,24]).

(2)

Since the cyclization results in the destruction of the conjugation between the two rings, the absorption maximum of the oxaziridine is blue shifted compared to that of the starting nitrone. The same photochemical conversion occurs efficiently in spin coated films consisting of (4-N,N-dimethylaminophenyl)-N-phenyl nitrone (DMAPN) in low molecular weight PMMA. Figure 1 shows the photochemically-induced spectral transformation of a 0.76 micron film consisting of DMAPN(23 wt%) in PMMA as a function of fluence (mJ/cm^2) at 361 nm. Table I shows the dispersion of the refractive index of the DMAPN/PMMA film before and after irradiation (1000 W xenon arc lamp, 361 nm interference filter, 1 h).

Table I. The dispersion of the index of refraction
of DMAPN(23 wt %)/PMMA films before and after photolysis
through a 361 nm interference filter

	Abs max (nm)	Refractive Index				Sellmeier		
		543nm	633nm	670nm	815nm	A(μm)^2	B	λ_o(nm)
DMAPN(23wt%)/PMMA (before irradiation)	380	1.5750	1.5573	1.5558	1.5524	0.504	1.294	404
DMAPN(23wt%)/PMMA (after irradiation)	274, 320	1.5394	1.5309	1.5310	1.5337	0.467	1.28	340

The intrinsic refractive index of DMAPN/PMMA films can be varied from that of PMMA (1.48) to greater than 1.57 by changing the weight percent of DMAPN. Figure 2 shows the measured refractive indices of DMAPN/PMMA films as a function of the weight percent of DMAPN both before and after irradiation through a 360 nm broad band filter. At 633 nm, a wavelength far from resonance, the observed changes in refractive index can be as large as 0.02.

Micron scale multi-mode waveguide structures were demonstrated in these films using both standard mask and laser writing techniques. Figure 3 shows optical micrographs (taken with crossed polarizers) of a multi-mode "Y splitter" and a "crossover" written using an argon ion laser writing apparatus. All UV lines of the argon laser were used and were focussed with a 10X microscope objective onto the sample which was translated under computer control. The writing process can be readily followed using a TV camera and monitor. The laser written lines are regions of decreased refractive index adjacent to the waveguide. Measured losses in these multimode structures were typically between 1 and 2 dB/cm at 815 nm.

The solution of Maxwell's equations for the propagation of optical radiation with the appropriate boundary conditions for an asymmetric step index channel waveguide with the structure shown in Figure 4a provides for a set of guided waveguide modes characterized by indexes j and k with corresponding propagation constants β_{jk}. An analysis of the mode structure for this geometry has been carried out according to the procedure of Marcatili (25). From this analysis the effective

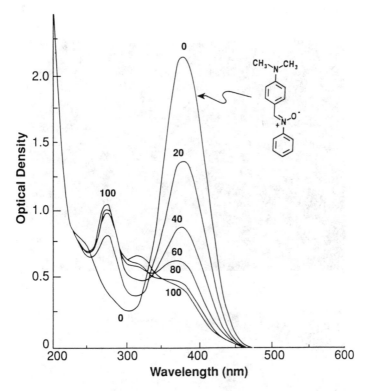

Figure 1. Spectral changes of a DMAPN(23wt%)/PMMA film as a function of fluence (mJ/cm^2).

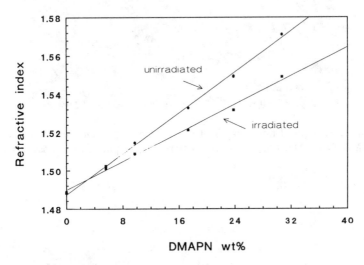

Figure 2. Refractive indexes (± 0.005, 633 nm) of DMAPN/PMMA films before (•) and after (■) irradiation vs. wt% DMAPN incorporated.

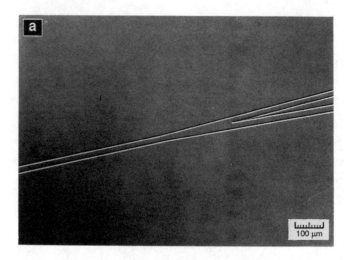

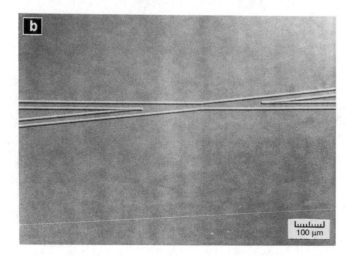

Figure 3. Optical micrographs of laser written waveguide structures in DMAPN/PMMA films on silicon wafers: a. "y splitter", b. "crossover".

Guided Modes in Polymer Waveguides

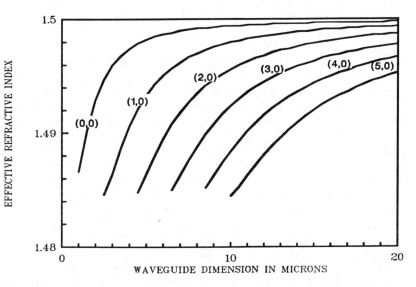

Figure 4. Mode structure of a 0.73 micron thick DMAPN/PMMA film calculated as a function of the waveguide width (a).

refractive index for propagation of the low-index-number "E_y-polarized" modes may be calculated for a 0.73 micron thick DMAPN/PMMA film on silicon dioxide as a function of the waveguide width, a (Figure 4b). This calculation suggests that for waveguides narrower than 2 microns only a single "E_y mode" should be supported. Experimentally, we have found that laser-written waveguides with an edge-to-edge width of about 6 microns support three "E_y" modes at 815 nm. As the edge-to-edge distance is reduced to 4 microns, only two modes are supported and for a width of 2 microns, only one mode is observed, in excellent agreement with the calculated mode structure shown in Figure 4b.

These nitrone/PMMA systems experimentally confirm that photochemical delineation of refractive index profiles in organic films can be used to create micron-sized single-mode, waveguide structures. The intrinsic refractive index and index change upon irradiation can be experimentally controlled and waveguide features can be designed with the ease, doses and resolution of standard photolithographic methods. The simplicity of this fabrication technique and its low processing temperature offer great promise for the use of organic polymers in integrated optic structures.

Photochemical Delineation of Waveguide Structures in χ^2 Materials

In the expansion of the macroscopic electric polarizability for a material in terms of the electric field, the first nonlinear susceptibility is a third rank tensor whose value is zero in materials containing a center of symmetry. Therefore a macroscopic noncentrosymmetric structure is a prerequisite for the generation of a second-order nonlinear optical effect in an organic polymer film. Electric field poling (26,27) has been demonstrated to be a convenient method for generation of the requisite polar order in amorphous polymer films for the preparation of electro-optic and second harmonic materials. If the technique of photodelineation is combined with poled polymers containing dye molecules with large second-order hyperpolarizabilities then active waveguide structures can be readily delineated on silicon and gallium arsenide substrates for integrated optics use. We have demonstrated the feasibility of this technology using copolymers of methylmethacrylate and methacrylate bound disperse red 1 dye (MA1), a material investigated by a number of groups (28,29) for use as a second-harmonic or electro-optic film. The standard synthetic route used to prepare these copolymers is shown in Scheme 1.

In the limit of the oriented gas model with a one-dimensional dipolar molecule and a two state model for the polarizability (30), the second order susceptibility $\chi_{33}^{(2)}$ of a polymer film poled with field E is given by Equation 4 where N/V is the number density of dye molecules, the f's are the appropriate local field factors, μ is the dipole moment, β is the molecular second order hyperpolarizability, and L_3 is the third-order Langevin function describing the electric field induced polar order at poling temperature $T_p \sim T_g$.

$$\chi_{33}^{(2)} (-\omega;\omega,0) - \frac{N}{V} (f^\omega)^2 f^{AC} \beta_{zzz}(-\omega;\omega,0) L_3(f^{DC} \frac{\mu E}{kT_p}) \qquad (4)$$

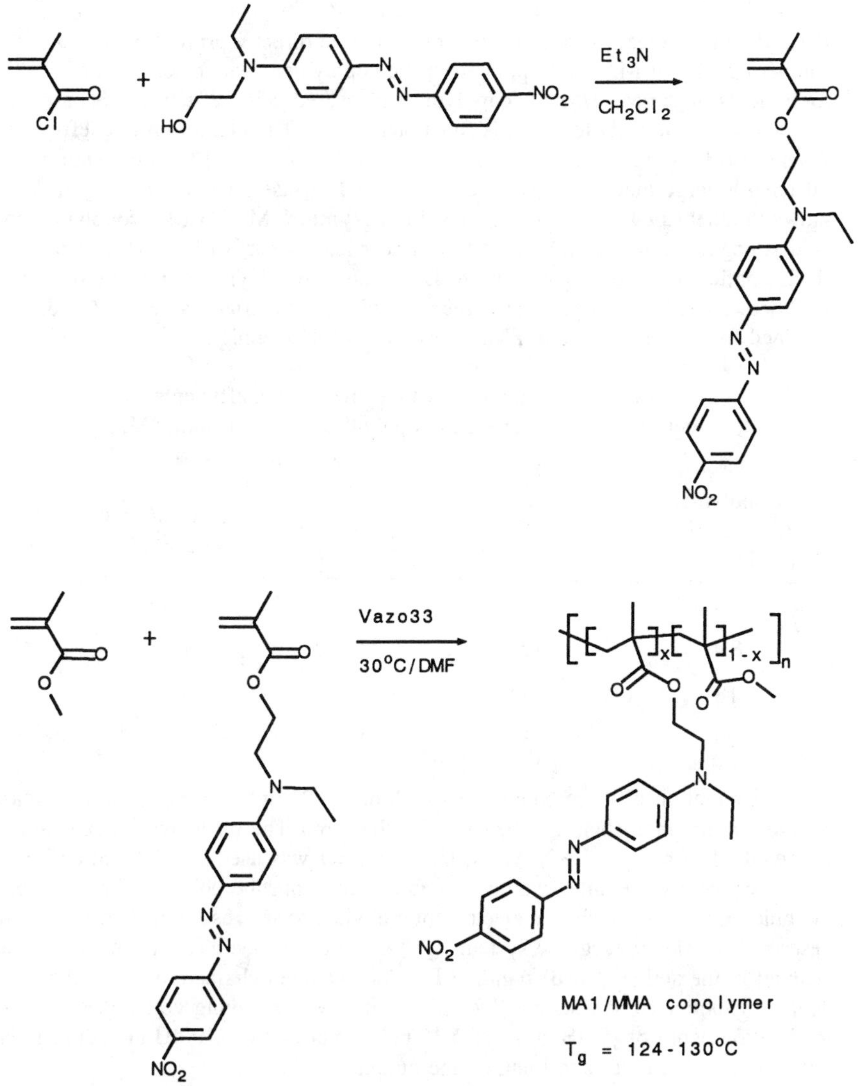

MA1/MMA copolymer

$T_g = 124-130^{\circ}C$

Scheme 1

The electro-optic coefficient in a poled polymer film can be related to the second-order susceptibility as shown in Equation 5.

$$r_{ijk} (-\omega;\omega,0) = -\frac{2}{n_i^2 n_j^2} \chi_{ijk}^{(2)} (-\omega;\omega,0) \tag{5}$$

Thus, the linear electro-optic coefficient should be directly proportional to the number density of MA1 incorporated in the copolymer. The electro-optic coefficients of the MA1/MMA copolymers measured using a standard Senarmont compensator (Table II) follow this functional form. The electro-optic coefficient should also follow a linear relationship with poling field until the field strength is sufficiently large that the higher order terms in L_3 cause a deviation. Figure 5 shows the measured r_{33} vs. poling field for a 20 mol% MA1/MMA copolymer for field strengths up to 2.5 MV/cm. A nonlinear least squares fit of the measured electro-optic coefficients according to Equations 4 and 5 yielded a value of $f^{DC}\mu$ of 16.0 debye. This is in excellent agreement with the reported value of 16.2 debye obtained for disperse red 1 in PMMA by the EFISH technique (3).

Table II. The measured electro-optic coefficients
and physical properties of copolymers of MA1 and MMA

Copolymer Composition (mol% MA1)	T_g (°C)	n at 815 nm	r_{33} (pm/V) at 815 nm (E = 0.5 MV/cm)
2.9	125	1.523	0.59
8.4	126	1.557	1.59
18	124	1.606	3.8
20	130	1.590	3.6 ± 0.3

Irradiation of these copolymers with deep UV light through a contact mask resulted in the photochemical bleaching of the film. The refractive index change observed when a 20 mol% MA1/MMA copolymer was used was 0.05 at 815 nm. Thus a channel waveguide was fabricated by bleaching the polymer surrrounding the guiding region. A simple electro-optic modulator was assembled as previously described for slab waveguide systems (31) using a photodelineated 3 micron wide channel in the active (poled) region. The demonstrated characteristics of the device include a 10 ns risetime (identical to that of the driving voltage), a modulation depth of -5 dB, a V_π of 5 V and a total loss of -10 dB including fiber butt coupling the input and imaging the output.

Conclusion

We have demonstrated that micron-scale passive waveguide structures including "Y splitters" and crossovers can be fabricated in organic films using a photodelineation technique. This methodology provides for the spatial delineation

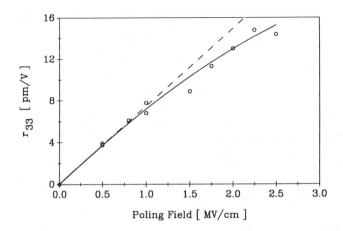

Figure 5. Electro-optic coefficient (r_{33}) vs. poling field for a 20 mol% MA1/MMA copolymer. --- Linear extrapolation; ——— Nonlinear least squares fit of Equations 4 and 5.

of structures with the dose, resolution and ease of standard photolithographic techniques. The same methodology has been extended to poled polymer films with the demonstration of the formation of a channel waveguide electro-optic modulator.

Literature Cited

1. Nonlinear Optical Properties of Organic Molecules and Crystals; Chemla, D. S.; Zyss, J., Eds.; Quantum Electronics - Principles and Applications; Academic: New York, 1987; Vol. 1 + 2.
2. Tam, W.; Guerin, B.; Calabrese, J. C.; Stevenson, S. H. Chem. Phys. Lett. 1989, 154, 93.
3. Singer, K. D.; Sohn, J. E.; King, L. A.; Gordon, H. M.; Katz, H. E.; Dirk, C. W. J. Opt. Soc. Am. B 1989, 6, 1339.
4. Garito, A. F.; Singer, K. D., Teng, C. C.; In Nonlinear Optical Properties of Organic and Polymeric Materials; Williams, D. J., Ed.; ACS Symposium Series No. 233; American Chemical Society: Washington, D.C., 1983; pp 1-26.
5. McCrum, N. G.; Read, B. E.; Williams, G. Anelastic and Dielectric Effects in Polymeric Solids; Wiley: New York, 1967.
6. Lytel, R.; Lipscomb, G. F.; Binkley, E. S.; Kenney, J. T.; Tickner, A. J. Proc. SPIE 1990, 1215-29.
7. Stegeman, G. I.; Seaton, C. T.; Zanoni, R. Thin Solid Films 1987, 152, 231.

8.   Baker, G. L.; Klausner, C. F.; Shelburne III, J. A.; Schlotter, N. E.; Jackel,
      J. L.; Townsend, P. D.; Etemad, S. Synth. Metals 1989, 28, 0639.
9.   Glenn, R.; Goodwin, M. J.; Trundle, C. J. Mol. Electr. 1987, 3, 59.
10.   Goodwin, M. J. Proc. SPIE 1987, 836, 265.
11.   Tien, P. K.; Smolinsky, G.; Martin, R. J. Appl. Opt. 1972, 11, 637.
12.   Tomaru, S.; Kawachi, M.; Kobayashi, M. Opt. Commun. 1984, 50, 154.
13.   Senmour, R. J.; Carter, G. M.; Chen, Y. J.; Elman, B. S.; Jagannath, C. J.;
      Rubner, M. F.; Sandman, D. J.; Thakur, M. K.; Tripathy, S. K.; In Proc.
      Symp. on Integrated Optical Engineering II, Sriram, S. Ed.; in Proc. Soc.
      Photo-Opt. Instrum. Eng. 1985, 578, 137.
14.   Carter, G. M.; Chen. Y. J.; Tripathy, S. K. Appl. Phys. Lett. 1983, 43, 891.
15.   Thakara, J.; Stiller, M.; Lipscomb, G. F.; Ticknor, A. J.; Lytel, R. Appl.
      Phys. Lett. 1988, 52, 1031.
16.   Tomlinson, W. J.; Weber, H. P.; Pryde, C. A.; Chandross, E. A. Appl.
      Phys. Lett. 1975, 26, 303.
17.   Schriever, R.; Franke, H.; Festl, H. G.; Kratzig, E. Polymer 1985, 26, 1423.
18.   Horn, K. A.; Beeson, K.; McFarland, M. J.; Nahata, A.; Wu, C.; Yardley, J.
      T. Abstracts of the 1989 International Chemical Congress of Pacific Basin
      Societies, Honolulu, Hawaii, Dec. 17-22, 1989. Macr. 0082.
19.   Rockford, K. B.; Zanoni, R.; Gong, Q.; Stegeman, G. I. Appl. Phys. Lett.
      1989, 55, 1161.
20.   Wells, P. J.; Bloor, D. In Organic Materials For Non-linear Optics; Hann,
      R. A.; Bloor, D., Eds.; Royal Society of Chemistry: London, 1989; pp398-
      403.
21.   McDonach, A; Copeland, M; Mohlmann, G. R.; Horsthuis, W. H. G.;
      Diemeer, M. B. J.; Suyten, F. M. M.; Trommel, E. S.; VanDaele, P.; Van
      Tomme, E.; Duchet, C. and Fabre, P. Proc. SPIE 1989, 1177, 67.
22.   Born, M.; Wolf, E. Principles of Optics; Pergamon: Oxford, 1980; p. 96.
23.   Splitter, J. S.; Calvin, M. J. Org. Chem. 1965, 30, 3427.
24.   Griffing, B. F.; West, R. U.S. Patents 4,677,049; 4,702,996.
25.   Marcatili, E. A. J. The Bell System Technical Journal 1969, 2071.
26.   Williams, D. G. Angew. Chem. Int. Ed. Engl. 1984, 23, 690.
27.   Singer, K. D.; Lalama, S. J.; Sohn, J. E. Proc. SPIE 1985, 578, 130.
28.   Singer, K. D. In Nonlinear Optical Properties of Materials: Opt. Soc. of
      Am. Technical Digest Series 1988, 9, 24.
29.   Brossoux, C.; Esselin, S.; LeBarny, P.; Pocholle, J. P.; Robin, P. In
      Nonlinear Optics of Organics and Semiconductors; Kobayashi, T., Ed.;
      Springer Proceedings in Physics; Springer-Verlag: New York, 1988, 36,
      126.
30.   Singer, K. D.; Kuzyk, M. G.; Sohn, J. E. J. Opt. Soc. Am. B. 1987, 4, 968.
31.   McFarland, M. J.; Wong, K. K.; Wu, C.; Nahata, A.; Horn, K. A.; Yardley,
      J. T. Proc. SPIE 1988, 993, 26.

RECEIVED July 18, 1990

ORGANIC AND INORGANIC CRYSTALS

Chapter 21

Functional Waveguides with Optically Nonlinear Organic Materials

K. Sasaki

Faculty of Science and Technology, Keio University, 3–14–1, Hiyoshi, Yokohama, 223 Japan

It is known many organic compounds containing pi electron systems exhibit remarkably large nonlinearities.

The intrinsic useful capabilities of those materials give many charming items to researchers.

Device-oriented research is an important field and the waveguides are essential structures for practial applications of those materials. Namely waveguide structures have several important virtues as follows;

(a)confined high energy density of the guided wave.

(b)clear selection of the guided wave-mode by waveguide thickness concerning with phase matched second harmonic generation (SHG).

(c)structural controllabilities of the coupling between the guided waves in two adjacent waveguides.

(d)high quality of the output light beam from the waveguide.

In this paper we report functional waveguides exhibiting SHG in tapered thickness 2-methyl-4-nitroaniline(MNA) thin film crystal waveguide, and all optical bistability,intensity dependent optical modulation in vacuum evaporated polydiacetylene(PDA) thin film waveguides.

Input couplings for all waveguide devices were performed by grating couplers.

Thin Film MNA Crystal Waveguides for Phase-matched SHG with Grating Couplers

As formerly reported[1][2] we prepared thin single crystal MNA in the narrow tapered gap of faced substrates as a hybrid waveguide structure together with previously rf sputtered Corning

0097–6156/91/0455–0316$06.00/0

7059 thin film on one fused quartz substrate for phase matched
SHG. The thin single crystal film was prepared by two step processes. First molten MNA at 131°C filled tapered gap of faced substrates by capillary effect. At this stage MNA was mosaic-like
micro-crystal.Then the faced substrates passed through a precisely thermo-controlled furnace for recrystalization. In this stage
the taper of faced substrate was parallel to equi-thermal line
inside furnace. By this method the largest SHG tensor element of
MNA crystal,d_{11} is oriented parallel to the tapered gradient.

The experiment of phase matched SHG had been carried out by
taking off one substrate and putting a coupling prism on the wave
guide with lateral translation of the system for thickness adjustment of the waveguide. The MNA film is unbearably weak for repe-
titive prism coupling and physically weak in ambient atmosphere.

In this study we adopted a grating coupler instead of the prism coupler. Faced substrates were kept so as to avoid bleaching
of MNA film after crystalization processing. In this case the phase matched SHG experiment was able to be done in a sealed hybrid
(MNA thin crystal film/rf sputtered Corning 7059 glass film) wave
guide inside the tapered gap of two faced substrates. The guide
length of the sample waveguide between two grating coupler is 5
mm as shown in Fig 1. The grating was prepared at one substrate
by holographic intereference with plasma dry etching.

In the experiment the phase matched SHG was realized again
translational shift of the tapered waveguide. There are many degenerate points for phase matching in the dispersion relation between waveguide thickness,d and propagation constants of both wave lengths($\beta_1/k_1, \beta_2/k_2$) as depicted in Fig 2. Degeneration of
the dispersion relation is connected to SHG conversion efficiency
in our waveguide as the following equation.

$$\eta = P_2/P_1 = (\omega d_{11}^2/16)\{S^{(n \cdot n \cdot m)}\}^2(P_1/w)l^2\{sin(l\Delta\beta)/(l\Delta\beta)\}^2$$
$$(1).$$

where,
k_1, β_1 are wave number and propagation constant of guided wave
for Nd;YAG laser(1064 nm) as a fundamental input respectively.
Also k_2, β_2 are wave number and propagation constant of guided
wave for the second harmonic frequency respectively.
P_1 and P_2 are the fundamental input power and the second harmonic
output.

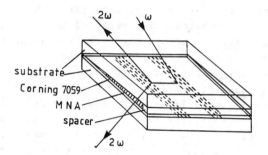

Figure 1.   Tapered slab-type MNA single crystal film waveguide
with grating couplers.

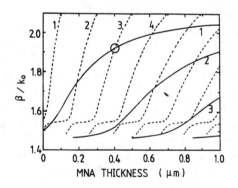

Figure 2.   Dispersion curves between propagation constant and
waveguide thickness(solid line;fundamental wave,dashed
line;doubled wave).

w is transerse beam width,l is guide length and $\Delta \beta = 2\beta_1 - \beta_2$.
$S^{(n,n,m)}$ is so called spatial coupling factor[3][4], meaning over
lapping of related fields,

$$S = \int e_y{}^{(n,1)} e_y{}^{(n,1)} e_y{}^{(m,2)} dx \qquad (2),$$

where $e_y{}^{(n,1)}$ is a cross-sectional TE field distribution of fundamental wave at nth mode and $e_y{}^{(m,2)}$ is the same kind of difini-
tion for SHG wave. x and y are Cartesian co-ordinate parameters.
The integration is carried out over the cross-sectional thicknesses existing both guided waves.
The phase matched SHG vitally corresponds to $\Delta \beta = 0$ in Eq.(1).
Furthermore various parameters affect the conversion efficiency.
The spatial coupling factor is the most important parameter in
the waveguide structure. By computational estimation the maximum
coupling factor can be given for m = 3 to n = 1 at thickness of
Corning 7059 film, 0.6 micron. The solid lines and the dashed
lines with integers to propagation modes in Fig 2 correspond to
the fundamental and the second harmonic waves respectively. The
circled point gives MNA thickness as about 0.4 micron.
    One of the remarked article in this study is use of grating
couplers. Coupling and decoupling gratings were directly etched
on a fused quartz substrate. Periodicity of a grating is 480nm
and its depth is 300 nm respectively as anexample. Experimentally
measured coupling and decoupling efficiencies are 20 % and 54 %
respectively.
    Measurement of SHG was carried out in a setup as shown in Fig
3 by a backward coupling at coupling angle 45°. The fundamental
wave is TE polarized, Nd;YAG pulsed laser at 1064 nm. Translational shift of the waveguide gave the maximum SHG peak by phase matching at corresponding waveguide thickness, 410 nm and waveguide
length,5 mm respectively as depicted in Fig 4. The total SHG output was 40W at the fundamental input of 10kW with conversion efficiency of 0.25 %. Inspite of phase matching and maximalization
of spatial coupling factor, conversion efficiency is lower for
expectation. Probably this comes from transverse de-confinement
of the guided waves.
    Finally we tried to prepare a channel type waveguide to persu-
ue the previously mentioned item (a) of the virtues. A conceptual
picture is sketched in Fig 5. The channel width is about 100 mic-

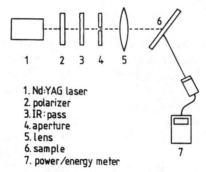

1. Nd:YAG laser
2. polarizer
3. IR:pass
4. aperture
5. lens
6. sample
7. power/energy meter

Figure 3.   Experimental setup for phase matched SHG in the waveguide.

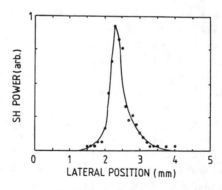

Figure 4.   Phase matched SHG at a certain waveguide thickness.

rons. In the device, the estimated input was 110 mW and the SHG output was 7.87 microW at the guide length 0.1mm. In this case conversion efficiency is about 7×10^{-2} %. If the guide length is 5 mm the estimated efficiency is 18 % at the same condition which is enough for operation by a semiconductor laser. Also narrower channels about 10 microns width can be easily realized for higher conversion efficiency in the near future.

Third Order Nonlinear Optical Effects in Vacuum Evaporated Poly-dyacetylene Film Waveguides

(1) All Optical Bistability of Vacuum Evaporated Polydiacetylene (PDA,12-8) Film Waveguide with a Grating Coupler

As previously reported[5] PDA Langmuir-Blodgett(LB) film waveguide showed all optical bistability with an incoupling prism for cw Nd;YAG laser 1064 nm. In the paper grainish PDA LB film performed as a hybrid waveguide in combination with rf-sputtered Corning 7059 glass film buffer layer on a fused quartz substrate.

In this report, vacuum evaporated PDA(12-8) film is used as an optically nonlinear layer with a grating coupler for nonlinear coupling for all optical bistability. Grating coupler on a substrate was prepared at the same periodicity and depth as the SHG devices. Vacuum evaporation of PDA on a substrate with previously rf-sputtered Corning 7059 buffer layer film were carried out at 5×10^{-5} torr with tungsten boat heater. Rapid evaporation can avoid thermal polymerization of the undesirable red phase PDA during the process. UV polymerization of the film for the useful blue phase PDA was carried out by Xe lamp 500 w for 20 min. at a distance of 40 cm. The absorption curve of the film is shwon in Fig 6. Refractive index of the PDA at Nd;YAG 1064 nm is 1.59.

Conceptual structure of the four-layered waveguide device is depicted in Fig 7. In figure dashed trapezoidal pattern was surrounded by grating. This is a method to realize appropriate waveguiding for various vacuum evaporated PDA film. Experimental setup for all optical bistability with a pulsed Nd;YAG laser 1064 nm is shown in Fig 8.

θ is the coupling angle at the grating coupler. The most suitable coupling angle at the low input power region is $\theta_0 = 34.9$ 5° for pulse width 200 ns. The input-output characteristic curve

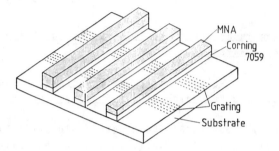

Figure 5.   Illustrative structure of the channel type waveguide
for high efficient phase matched SHG.

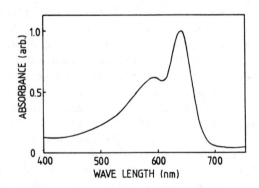

Figure 6.   Absorption curve of a vacuum evaporated PDA blue film.

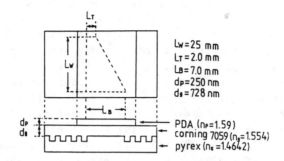

Figure 7. Conceptual structure of four-layered waveguide with grating coupler for all optical bistability.

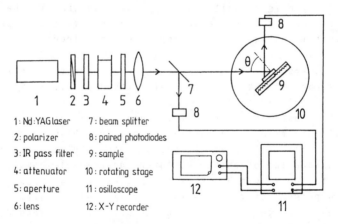

1: Nd:YAG laser   7: beam splitter
2: polarizer   8: paired photodiodes
3: IR pass filter   9: sample
4: attenuator   10: rotating stage
5: aperture   11: osilloscope
6: lens   12: X-Y recorder

Figure 8. Experimental setup for measurement of all optical bistability.

for θ_0 is shown in Fig 9 (a) without bistability. On the other
hand the input-output cu-rve at $\theta' = 34.77^0$ shows a clear bistability as in Fig 9 (b).
Moreover 20 ns pulse width laser was applied to the same device.
In this case, θ_0 is the same angle and all optical bistability
was observed at $\theta' = 34.85^0$ as shown in Fig 10. Input power levels was 400 W /pulse in Fig 9 (b) and 1.79 kW/pulse in Fig 10 respectively. In our previous report[5] cw-YAG laser 2 W was effective. The difference comes from clearly degrees of polymer orientations. In usual thin film waveguide with a coupling prism, a coupling angle of impinging light beam to the prism corresponds to
a propagation constant of a wave to be guided[6]. So the distributed coupling process is expressed as evanescently catching of in
coming light beam into the waveguide via an extremely thin air
gap with a finite beam size. This is a kind of resonace phenomenon. Insertion of nonlinear material makes the process more complex and interesting[7]. Power dependent effects of nonlinear materials in optical couplings developed for the case of a cw excit-
aion[8] and a pulsed excitation[9]. The reference [9] described
theoretically pulsed excitation of prism coupling. In the literature it is concluded that coupling efficiency varies pulse energy
and asymmetric output pulse is generated from input pulse in time
domain.

Now it seems that the above mentioned resonant state with nonlinearity for grating coupler creates the bistability with very -
sensitive angular dependence.

Fig 10 shows larger hysterisis loop than it of Fig 9 (a) for
the small change in the effective index. Probably thermal process
governs the bistability at the pulse width, 200 ns accompanying
thermal saturation of output at lower input level. On the other
hand for the pulse width, 20 ns thermal effect was remarkably reduced and pulse energy was kept in the waveguide system accompanying resonant state. In this case the pulse energy shuold be taken non-thermally. Appropriate changes in effective indices for
both pulse widths are not connected essentially to the bigger hysterisis loop.

In any way more precise development of this type of bistabilty
requires experiments with short pulses.

Recently R.Burzynski et al[10] reported ultra-high speed electronic bistabilities in polyamic acid waveguide with grating excitation. They classified electronic or thermal bistabilities usi-

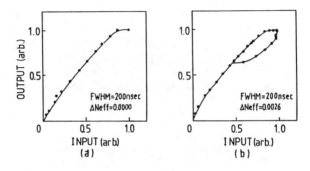

Figure 9. The output-input curve with all optical bistability at 200 ns pulse width Nd;YAG laser.

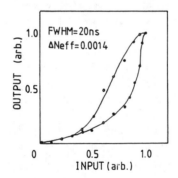

Figure 10. All optical bistability at 20 ns pulse width Nd;YAG laser.

ng three diferent wavelengths lasers with three different pulse widths. So our experiment also should be studied for the above mentioned item.

(2) Power-dependent Modulator and Limiter using Optically Nonlin-
     ear Directional Coupler

Evaporated PDA(12-8) film was used as a nonlinear optical med-
ium in a layered guided wave directional coupler. The directional coupling phenomenon happens in two adjacent waveguide by periodi-
cal energy transfer. The theory of linear directional coupler was exactly established [11]. It can be reduced to coupled mode equations;

$$da(z)/dz = i(\beta^a + K_{aa})a(z) + iK_{ab}b(z) \qquad (3),$$

$$db(z)/dz = i(\beta^b + K_{bb})b(z) + iK_{ba}a(z) \qquad (4),$$

where,
a(z) is the normalized amplitude of the guided wave in the waveg-
uide 1,b(z) is the normalized amplitude of the guided wave in the waveguide 2,z is propagation direction in Cartesian coordinate,
β^a,β^b are propagation constants of both guided waves,
K_{ab},K_{ba} are coupling constants between two waveguides,
K_{aa},K_{bb} are constants in respective waveguides.
    In this study we suppose nonlinear organic material shows opt-
ical Kerr effect as $n = n_0 + n_2|E|^2$ and $n_2 = X^{(3)}/(2n_0)$. Moreover for simplification, we suppose the waveguides allow single mode propagations and TE polarization. After appropriate handling we get the following nonlinear coupled mode equations[12];

$$-i[da(z)/dz] = Q_1a(z) + Q_2b(z) + [Q_3a^2(z) + 2Q_4b^2(z)]a(z) \qquad (5),$$

$$-i[db(z)/dz] = Q_1b(z) + Q_2a(z) + [Q_3b^2(z) + 2Q_4a^2(z)]b(z) \qquad (6),$$

where a(z),b(z) are again normalized amplitudes of electric field
Q_1,Q_2 are linear coupling coefficients, Q_3 and Q_4 are nonlinear coupling coefficients including optical Kerr effect. By solving (5) and (6) we can get transmittance of light energy from one

waveguide to another. The complex amplitudes, a(z) and b(z) can
be noted as follows;

$$a(z) = A(z) \exp\{i(\theta + Q_1 z)\} \qquad (7),$$

$$b(z) = B(z) \exp\{i(\theta' + Q_1 z)\} \qquad (8),$$

Those expressions make possible separate calculations for real
and emaginary parts.
Real part calculation give the expression for the power conservation as,

$$A^2(z) + B^2(z) = P_t \qquad (9).$$

As a initial condition, input power $A^2(0) = P_a(0)$ is low, then
completely exchange of the guided wave power between two waveguides can be realized as shown in Fig 11,the defined curve linear.
With increasing of input power, output power $P_b(1)$ from waveguide
B at a given coupling distance l is gradually suppressed according to increment of periodicity of energy delivery and at a certain critical input power P_c defined as giving equally energy separation for two waveguides. Moreover for $P_a(0) \gg P_c$, energy transfer is reduced(defined nonlinear in Fig 11). Energy ballance at
P_c is very sharp and self switching effect can be observed around
the value. This situation is shown in Fig 13 by theoretical curve
(solid curve). Reference [12] describes this kind of nonlinear
coupler in detail.
   The used waveguide dirctional coupler in our study is sketched
as in Fig 12. This device was basically prepared by facing of
two bistability waveguides in section (2).In this structure,coupl-
ing procees includes nonlinear effect as stated previously concerning optical bistability. Therefore intrinsic coupled power is
given by $I_1 + I_2 = P_a(0)$. The nonlinear interaction can be expressed
by a ratio $I_2/(I_1 + I_2)$. Experimental result is shown in Fig 13 together with theoretical curve by curve fitting with adjusting coupling length and third order succeptibility, where the coupling
length gives the phase difference of pi between the even and the
odd modes. Actually measured coupling distance was 2.9 mm and air
gap of facing was about 800 nm by optical interference. Also thickness of evaporated PDA film was 233 nm and rf sputtered Corning
glass film was 642 nm.
In theoretical calculation the author had two main ambiguities.
One was unmeasurable third order succeptibility of vacuum evaporated PDA film and another was effective coupling distance. Espec-

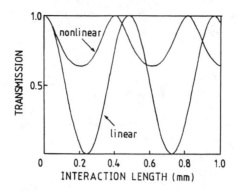

Figure 11. Power-dependent nonlinear coupling at a nonlinear directional coupler.

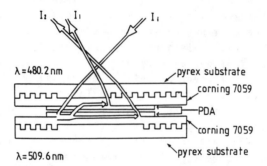

Figure 12. Schematical structure of the nonlinear directional coupler.

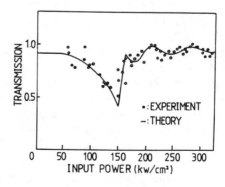

Figure 13. Switching behaviour of the nonlinear directional coupler with a theoretical curve.

ially vacuum evaporated PDA was grainish so as to reduce the effective coupling distance. Actually curve fitting gave that the
effective coupling distance was 0.55 mm. It looks that many origins give influences but the most effective factor for the difference between theory and experiment is the above situation at this
point.

Summary

In this paper some device-oriented studeis for optically 2nd
and 3rd order nonlinearities of organic materials are reported.
Especially strucural combinations of slab-type waveguides and
grating couplers are emphasized.

(A) Sealed type thin film MNA crystal waveguide for phase matched
SHG was realized with grating couplers. This type of structure is
advantageous because of separation from ambient atmosphere.
Some sample device is keeping the same crystal situation over half an year in atmosphere.
Moreover Channel type structure is effective for transversely
spreading of the coupled fundamental wave. This type of structure
can be applied sufficiently to semiconductor laser operation in
SHG. And this idea can be applied to other materials.

(B)All optical bistability device with grating coupler was realized at off-angle coupling for pulsed Nd;YAG laser. This bistability was very sensitive for the angle of incoming light beam.
Also thermal process governs the phenomenon for longer pulse operation. We used two pulse widths, 200 ns and 20 ns. It is said
the former corresponds to thermal process and the latter is mixture of electronic- and thermal-process[13]. In any way vacuum evaporated PDA is randomly oriented and more oriented polymer is
desired together with shorter pulse excitation to analize basic
process of this type of bistability.

(C)Slab-type waveguide nonlinear directional coupler was studied.
In this study the used directional coupler device was prepared by
facing of two bistability devices. Guided waves in two adjacent
waveguides are coupled via narrow air space. Vacuum evaporated
PDA was rather grainish. So effective coupling distance was very
small in comparison with actual measurement. The effective coupling distance was estimated by curve fitting to experimental values by adjustments of theoretical parameters (in this study, cou-

pling distance and third order succeptibility). Nonlinear dirctional coupler is useful for all optical switch and light intnsity
modulator. Organic materials having non-localized pi electron systems are expected because of their optically ultra-high speed
responses for the above devices.

Literature Cited

1. Itoh,H.;Hotta,K.;Takara.H.;Sasaki,K. Appl. Opt.1986, 25,1491.
2. Itoh,H.;Hotta,K.;Takara.H.;Sasaki,K. Opt. Commun.1986,59,299.
3. Zernike,F.;Midwinter,J. "Applied Nonlinear Optics" ; John-Wiley and Sons; New York, 1975.
4. Yariv,A. "Quantum Electronics"; John-Wiley and Sons; New York 1975.
5. Sasaki,K.;Fujii,K.;Tomioka,T.;Kinosita,T.J.Opt.Soc. Am. 1986, B5, 457.
6. Ulrich,R. J. Opt. Soc. Am. 1970, 60, 1337.
7. Stegeman,G.I.;Seaton,C.T. J. Appl. Phys. 1985, R57,58.
8. Liao,G.;Stegeman,G.I. Appl. Phys. Lett. 1984, 44(2),164.
9. Assanto,G.;Fortenberry,R.M.;Seaton,C.T.;Stegeman,G.T. J. Opt. Soc. Am. 1988, B5, 432.
10. Burzynski,R.;Singh,B.P.;Prasad,P.N.;Zanoni,R.;Stegeman,G.I. Appl. Phys. Lett. 1988, 53(21), 2011.
11. Chang,H.C. IEEE. J.Quantum Eledtron. 1987, 23(18), 1929.
12. Jensen,S.H. J.Quantum Electron. 1982, 18(10), 1580.
13. Private communications.

RECEIVED September 10, 1990

Chapter 22

Observing High Second Harmonic Generation and Control of Molecular Alignment in One Dimension

Cyclobutenediones as a Promising New Acceptor for Nonlinear Optical Materials

Lyong Sun Pu

Fundamental Technology Research Laboratory, Fuji Xerox Company, Hongo, Ebina-shi, Kanagawa, 2274 Japan

The promising new electron acceptor, cyclobutenedione, for nonlinear optical materials is proposed instead of conventional nitro (NO_2) group. The new acceptor prevents centrosymmetric crystal structures by an introduction of chirality and hydrogen bonding properties into acceptor itself. Asymmetric carbon and hydrogen bonds of amino acid derivatives may favor the formation of non-centrosymmetric crystal structures even in the case of molecules with large ground state dipole moments, which tend to form a centrosymmetric crystal structures. Several new second harmonic generation (SHG) active materials containing cyclobutenediones were synthesized. In particular, (-)4-(4'-dimethylaminophenyl)-3-(2'-hydroxypropylamino) cyclobutene-1,2-dione (DAD) shows high SHG by the powder method (SHG intensity; 64 times that of urea). An X-ray crystallographic analysis of DAD single crystals shows a triclinic system with the non-centrosymmetric space group P1 and that the direction of polarization of DAD in the molecular crystal is perfectly aligned in one dimension.

As great progress has been made in the field of electronics and semiconductors, optics will also evolve in the future to solve a variety of technological problems. Although optics and electronics have so far relied upon inorganic materials for fabrication of various components, organic materials have been focused on for future optical materials and devices for optical information processing, telecommunications and integrated optics (1). One of the important applications of organic materials for optical technology to be expected in the near future is frequency doubling. With a device based on the frequency doubled laser diode, recording capacity in optical memory will become four times larger than at present and printed patterns will be more fine and produce a higher resolution.

One of the biggest challenges that researcher must address to achieve these objectives at the present time is the control of molecular orientation in organic crystals and thin films. Molecular crystal structures will be generally determined by van der Waals' force, hydrogen bonding, dipole-dipole interaction and so on. For a frequency doubling, non-centrosymmetric molecular orientation is required. However, most organic molecules with large ground state dipole moments resulting from the introduction of Donor-Acceptor groups into π-conjugated systems tend to form centrosymmetric crystal structures due to electrostatic interaction between adjacent molecules.

Several methods have been known so far to prevent centrosymmetric crystal structures.

1. The most common method is the use of hydrogen bonds to make an alignment of molecular crystals for polar structures. As energy of hydrogen bond (3-6 Kcal/bond) is comparable with dipole-dipole interaction (5 Kcal/mol), it is effective to reduce the centrosymmetry of molecular crystal structure. Almost all known second harmonic generation(SHG) active organic crystals, such as urea (2), 2-methyl-4-nitro-N-methylaniline(MNMA) (3), etc have hydrogen bonds to make an alignment of molecules for polar structures.

2. Second method is the introduction of chilarity into molecules which ensure a non-centrosymmetric crystal structure. Several chiral materials such as methyl-(2,4-dinitrophenyl)-amino-2-propanoate (MAP) (4), N-(4-nitrophenyl)-(L)-prolinol (NPP) (5), etc are already known to be SHG active. Another interesting material of this type is trans-4'-hydroxy-N-methyl-4-stilbazolium paratoluenesulfonate which was recently presented by Nakanishi et al (6), although molecular salts of merocyanine dyes are already known to show large SHG intensity (7). An X-ray crystal structure determination of this materials shows space group P1 and one dimensional direction of molecular alignment. The origin of control for this molecular alignment is considered to be the chiral handle character of sulfonic anion to give the non-centrosymmetric space group.

3. Third method is to use recrystallization solvent effect (8). In certain materials, strong polar solvents have an effect to weaken dipole-dipole interaction between molecules, which causes centrosymmetric structure.

4. Another method is to use the molecules with relatively small ground state dipole moment (9). Weak dipole-dipole interaction can lead to non-centrosymmetric crystallization. However, it may decrease polarizability of molecules and result in low optical nonlinearities of molecular crystals.

5. The use of inclusion compound hosts is another interesting method for polar alignment of organic and organometallic compounds (10, 11). However, generally the fairly large size of molecular unit may be disadvantage to enhance optical nonlinearities of molecular crystal by reduction of number of molecules in unit volume. Most organic transition metal compounds have absorptions of d-d orbital transitions in visible region. This may be another concern we are to develop these materials for practical application to future optical devices.

In this paper, I propose a promising new electron acceptor of cyclobutenedione for nonlinear optical materials to prevent centrosymmetric crystal structures by the introduction of chirality and hydrogen bonding property into the acceptor itself. Compared with electron donative groups, electron acceptor is not yet well studied for nonlinear optical materials. The most commonly used electron acceptor is nitro (NO_2) group. Therefore, we evaluated the possibility of cyclobutenedione as a new electron acceptor for nonlinear optical materials. One of the most simple cyclobutenediones is squaric acid. Squaric acid is known to be soluble in water and show very strong acidity(12), as squarylium anion formed in water has a stable 2π delocalized electron system as shown below.

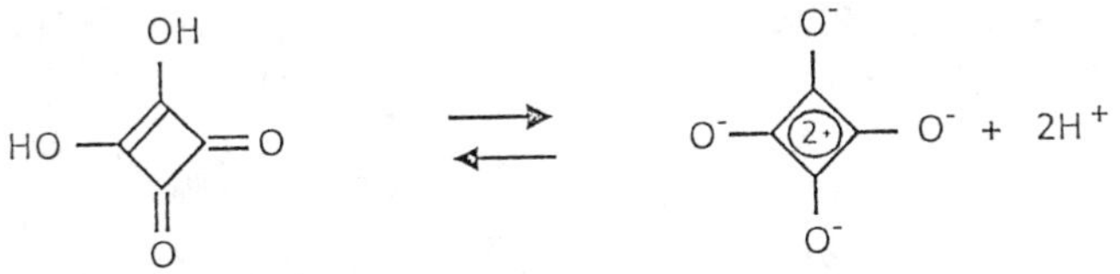

Squaric acid   Squarylium anion

Representative materials containing cyclobutenedione are squaraine dyes(13) which are known as functional dyes, such as photoconductive, photovoltaic materials and so on(14-17). They have strong intramolecular charge transfer bands in molecules and show strong absorptions at visible region, as cyclobutenolate in squaraine dye molecules plays a role of strong electron acceptor. Quite recently, large quadratic electrooptic nonlinearity for squaraine dye is reported by Dirk et al.(18).

Experimental

All materials containing cyclobutenedione listed in Table I were synthesized in my laboratory and identified by elementary analysis, mass spectrometry, melting point, infrared spectrum, and so on. Full details for the synthesis of cyclobutenediones will be reported separately. UV spectra were obtained with Hitachi spectrophotometer (U-3400).

Powder samples were sandwiched between two glass plates and set in optical sphere. They were irradiated with a pulsed Nd:YAG laser at 1.064µm (15nsec, 0.1mJ/pulse, 10Hz). SHG intensities of organic crystals were measured by detection of 532 nm generated from powder materials with a photomultiplier. This is known procedure as the powder technique developed by Kurtz and Perry (19). The intensity of SHG is always referred to that of urea in powder form (100-150µm). The measurement of molecular hyperpolarizability of cyclobutenediones by EFISH technique is now being in progress.

For the X-ray crystal structure determination, crystals were grown from methanol solution by slow evaporation at room temperature. Cell parameters and intensity data were derived from measurements on four-circle diffractometer ; Rigaku AFC5R. Molecular and crystal structures were determined by the direct

Table I. Spectroscopic Properties and Relative Powder SHG Intensities of Typical Cyclobutenediones (1-5) and Nitroanalogue (6)

	Material	$\lambda max(nm)$*1	ξ	$\lambda cut\ off(nm)$*2	SHG*4
1	CH3 —⟨ ⟩— (cyclobutenedione, OH, =O, O)	318.7	23072	426.7*3	2.5
2	(CH3)2N —⟨ ⟩— (cyclobutenedione, OH, =O, O)	379.2	23850	445.6	0.0
3	(CH3)2N —⟨ ⟩— (cyclobutenedione, NHCH2—CH(OH)—CH3, =O, O) [DAD]	396.5	38954	459.8	64.0
4	(CH3)2N —⟨ ⟩— (cyclobutenedione, piperidinyl-CO_2—t—C_4H_9, =O, O)	396.0	28780	467.8	8.0
5	(CH3)2N —⟨ ⟩— (cyclobutenedione, $C_2H_5CHCH_2OH$, NH, =O, O, C_2H_5) [DEAC]	396.5	18373	465.6	26.0
6	(CH3)2N —⟨ ⟩— NO_2	387.6	18557	476.0	0.0

*1. in C_2H_5OH solution.
*2. $\lambda cut\ off(nm)$ is obtained from the wavelength at transmittance,95% for 4×10^{-4} mol/l of each materials.
*3. It may be effected by impurities.
*4. Relative intensity to urea.

method by using the program of TEXSAN. Details will be published elsewhere.

Results and Discussions

Cyclobutenediones and N,N'-dimethylaminoparanitroaniline are listed in Table I together with their λmax, molecular absorption coefficients (ε) and SHG intensities in the relative scale to the urea. Visible spectra of cyclobutenedione 3 together with N,N'-dimethylaminoparanitroaniline 6 in CH_2Cl_2 are shown in Figure 1. It shows that λmax of cyclobutenedione 3 exists at the close position to the material 6 and shape of absorptions for cyclobutenedione 3 is rather sharper than nitro analogue 6, which reflects to the shorter cut off wavelength of the cyclobutenedione.

As shown in Table I , cyclobutenedione as an electron acceptor for nonlinear optical materials have several advantages. They are:
1. Accepting properties of cyclobutenedione seem to be as strong as nitro (NO_2) group , as these intramolecular charge transfer bands (λmax) are in almost same position as nitro analogue.
2. Molecular absorption coefficients are generally higher than the nitro analogue. It suggests the enhancement of oscillator strength and molecular hyperpolarizability of cyclobutenediones.
3. Various kinds of substituents such as amino acid derivatives can be easily introduced into cyclobutenedione resulting in the formation of variety of chiral nonlinear optical materials
4. Cyclobutenediones have OH and/or NH groups, which have hydrogen bonding properties as generally known.

We have synthesized several materials containing new acceptors instead of a conventional nitro (NO_2) group.

Material 1 shows fairly high SHG intensity in spite of less donative methyl substituent bonded to π-conjugated system corresponding to λmax in shorter wave length and smaller dipole moment. The fairly high SHG intensity of material 1 may be due to the ability of hydroxycyclobutenedione to form hydrogen bonded structure. Shapes of absorption spectra of 1 in C_2H_5OH and CH_2Cl_2 are different depending on solvents as shown in Figure 2. The protic solvent, C_2H_5OH seems to form hydrogen bond with 1 and changes the energetic structure of material.

However, in the case of the material 2, the more donating amino group, bonded to π-conjugated system, induces a larger ground state dipole moment than 1. Although 2 has a hydroxycyclobutenedione capable of hydrogen bonding, electrostatic interactions with adjacent molecules may favor a centrosymmetric crystal structure. Thus, SHG was not generated from material 2.

Even in such molecules with large ground state dipole moment, we observed the production of non-centrosymmetric crystal structures exhibiting SHG by introduction of asymmetric amino acid derivatives into the cyclobutenedione. (-)4-(4'-dimethylaminophenyl)-3-(2'-hydroxypropylamino) cyclobutene-1,2-dione (DAD) (3), (+)4-(4'-di-methylaminophenyl)-3-(2'-t-butoxycarbonylpyrrolidinyl) cyclobutene-1,2-dione(4) and (-)4-(4'-dimethylamino-2'-ethylphenyl)-3-(1'-hydroxybutyl-2'-amino) cyclobutene-1,2-dione (DEAC) (5) are the materials introduced chiral amino acid derivatives into the cyclobutenedione. They not only have hydrogen bonding but also have chiral centers in the molecules. These two factors prevent the

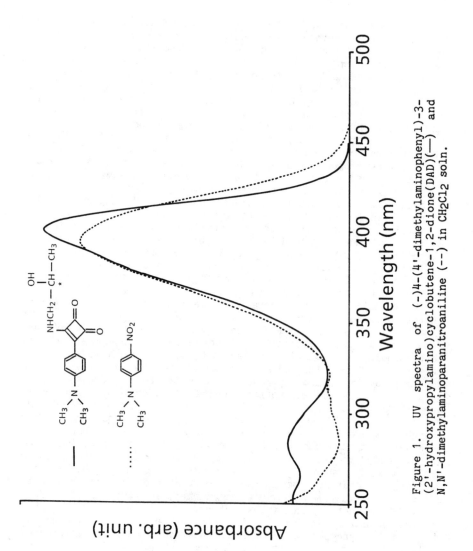

Figure 1. UV spectra of (-)4-(4'-dimethylaminophenyl)-3-(2'-hydroxypropylamino)cyclobutene-1,2-dione(DAD)(—) and N,N'-dimethylaminoparanitroaniline (--) in CH2Cl2 soln.

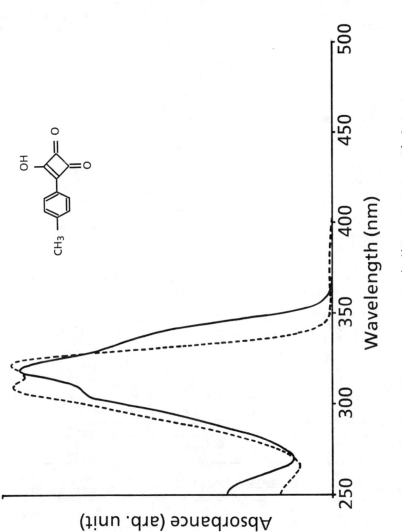

Figure 2. UV spectra of 4-(4'-methylphenyl)-3-hydroxy-
cyclobutene-1,2-dione. — in CH₂Cl₂, -- in C₂H₅OH soln.

formation of centrosymmetric molecular crystals, thereby leading to
strong SHG. Particularly, molecular crystals of DAD show high SHG
by the powder method (64 SHG intensity relative to urea).
An X-ray crystal structure determination was performed for DAD
molecular crystal. Table II shows crystallographic data of DAD.
It's selected bond lengths and angles are shown in Table III. The
sign of conformation angles is positive, if when looking from atom
2 to atom 3 a clock-wise motion of atom 1 would superimpose it on
atom 4.

Table II. Crystallographic Data of DAD

Formular	C(15)H(18)N(2)O(3)
FW	274.32
Crystal System	Triclinic
Space Group	P1
a (Å)	5.69
b (Å)	12.68
c (Å)	5.25
α (°)	93.8
β (°)	103.8
γ (°)	102.0
V (Å³)	357.3
R factor	0.041
Z value	1

Table III. Selected Bond Lengths and Angles for DAD Molecular
Crystal. The Atom Designations Refer to Figure 3

Intramolecular Bond Lengths (Å)

Atom 1	Atom 2	Length
C 1	N 1	1.366
C 1	C 2	1.409
C 2	C 3	1.374
C 3	C 4	1.402
C 4	C 7	1.435

Intermolecular Bond Lengths (Å)

Atom 1	Atom 2	Length
O 1	H 9	1.95
O 2	H 15	2.00

Torsion or Conformation Angles (deg.)

Atom 1	Atom 2	Atom 3	Atom 4	Angle
C 3	C 4	C 7	C 8	-3.3
C 5	C 4	C 7	C 10	-3.5

Molecular and crystal structure of DAD are shown in Figure 3 and
4. DAD crystallize in a triclinic system with the noncentrosymmetric space group P1. The direction of polarization of
DAD in the molecular crystal is perfectly aligned in one dimension
as shown in Figure 4. X-ray crystal analysis of DAD also shows the
existence of two hydrogen bonds between adjacent molecules, O 1--H

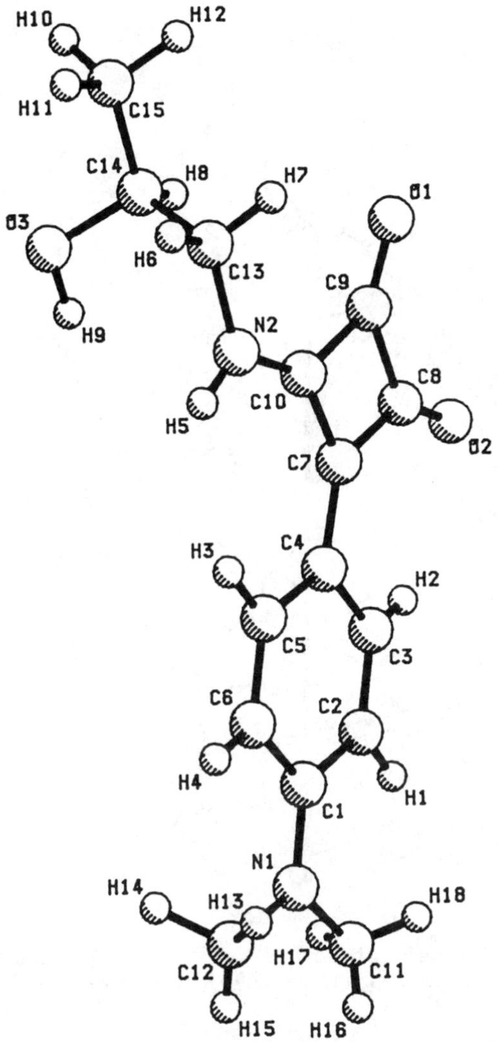

Figure 3. Molecular structure of (-)4-(4'-dimethylamino-phenyl)-3-(2'-hydroxypropylamino)cyclobutene-1,2-dione (DAD), as determined by X-ray crystallography.
C, carbon ; N, nitrogen ; O, oxygen ; and H, hydrogen.

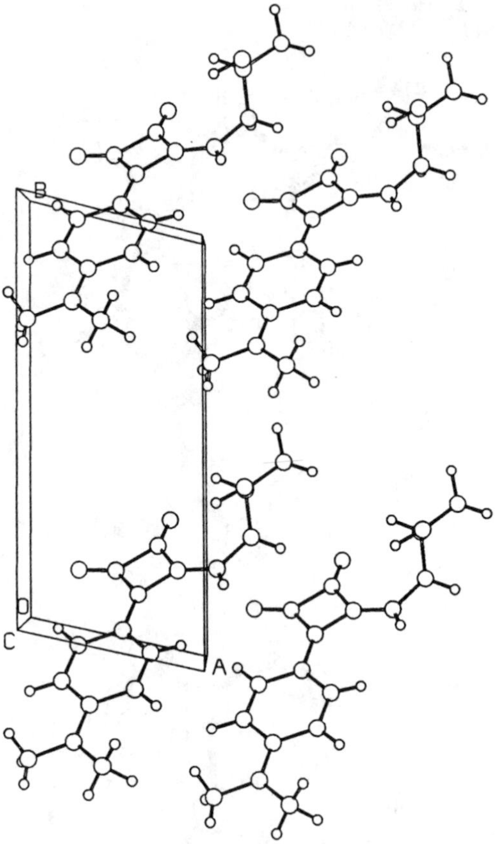

Figure 4.   Crystal   structure   of   (-)4-(4'-dimethylamino-
phenyl)-3-(2'-hydroxypropylamino)cyclobutene-1,2-dione (DAD).

9 and O 2--H 5 and the coplanarity of cyclobutenedione with benzene ring of dimethylanilino substituent. Table Ⅲ, Figure 2, 3 and 4 indicate that charge transfers in DAD molecule mainly ocurr from the dimethylanilino group to cyclobutenedione ring to form benzoquinoid structure, which is corresponds to λmax in UV spectrum, as discussed before and partly ocurr from the amino group bonded to the cyclobutenedione ring as described below.

It is evident that two hydrogen bonds between adjacent molecules and the chirality of molecule contribute to one dimensional molecular alignment in spite of the strong dipole-dipole interactions and that π-conjugated system of DAD molecule extend from amino groups to cyclobutenedione ring enhance second order nonlinearity of DAD molecular crystal.

From these results, As the particular tensor component of DAD molecular crystal is expected to be very high, more than 2-methyl-4-nitroaniline(MNA)(20), it would become one of the most suitable materials of highly efficient optical device for frequency doubler by using phase matching with optical wave guide. Electro-optical properties as well may be interesting.

Conclusion

Cyclobutenediones are shown to be excellent new electron acceptors for nonlinear optical materials as compared to the conventional nitro (NO_2) group as demonstrated by SHG measurements and X-ray crystallographic analysis of newly synthesized cyclobutenedione containing compounds. Cyclobutenediones with chiral amino acid derivatives are particularly effective for controlling the molecular alignments in the formation of molecular crystals even in the case of molecules with large ground state dipole moments. X-ray crystal analysis of DAD shows the compound to crystallize in triclinic system with the space group P1 and that the direction of donor-acceptor axis of DAD in the molecular crystal is perfectly aligned in one dimension.

Acknowledgments

We thank Mr. I. Ando for SHG measurement with the powder method and Dr. T. Hori and Dr. M. Furukawa in Rigaku Corp. for X-ray crystallography.

Literature Cited

1.   Williams, D. J., Ed. ; In Nonlinear Optical Properties of Organic and Polymeric Materials, ACS Symposium Series No.233; American Chemical Society: Washington, DC, 1983.
2.   Zyss, J. ; Berthier, G. J. Chem. Phys. 1982, 77, 3635.
3.   Sutter, K. ; Bosshard, C. ; Ehrensperger, M. ; Günter, P. ; Twieg, R. J. IEEE J. Quant. Electronics 1988, 24, 2362.
4.   Oudar, J. L. ; Hierle, R. J. Appl. Phys. 1977, 48, 2699.
5.   Zyss, J. ; Nicoud, J. F. ; Coquillay, M. J. Chem. Phys. 1984, 81, 4160.
6.   Nakanishi, H. ; Matsuda, H. ; Okada, S. ; Kato, M. Proc. of the MRS Internat. Mtg. on Advanced Materials, 1989, 1, p97.
7.   Meredith, G. R. In Nonlinear Optical Properties of Organic and Polymeric Materials; Williams, D. J., Ed.; ACS Symposium Series No.233; American Chemical Society: Washington, DC, 1983; pp 27-56.
8.   Tabei, H ; Kurihara, T ; Kaino, T. Appl. Phys. Lett. 1987, 50, 1855.
9.   Zyss, J. ; Chemla, D. S. ; Nicoud, J. F. J. Chem. Phys. 1981, 74, 4800.
10. Tomaru, S. ; Zembutsu, S. ; Kawachi, M. ; Kobayashi, M. J. Inclusion Phenom. 1984, 2, 885.
11. Eaton, D. F. ; Anderson, A. G. ; Tam, W. ; Wang, Y. J. Am. Chem. Soc. 1987, 109, 1886.
12. Park, J. D. ; Cohen, S. ; Lacher, J. R. J. Am. Chem. Soc. 1962, 84, 2919.
13. Sprenger, H. E. ; Ziegenbein, W. Angew. Chem. 1966, 78, 937.
14. Champ, R. B. ; Shattuck, M. D. U.S. Patent 3824099, 1974.
15. Kim, S. ; Tanaka, H. ; Pu, L. S. Japanese Patent L. O. 60-128453, 1985.
16. Merritt, V. Y. ; Hovel, H. J. Appl. Phys. Lett. 1976, 29, 414.
17. Furuki, M. ; Ageishi, K.; Kim, S. ; Ando, I. ; Pu, L. S. Thin Solid Films 1989, 180, 193.
18. Dirk, C. W. ; Kuzyk, M. G. Chem. Mater. 1990, 2, 4.
19. Kurtz, S. K. ; Perry, T. T. J. Appl. Phys. 1968, 39, 3978.
20. Levine, B. F. ; Bethea, C. G. ; Thumond, C. D. ; Lynch, R. T. ; Bernstein, J. L. J. Appl. Phys. 1979, 50, 2523.

RECEIVED July 10, 1990

Chapter 23

Strategy and Tactics in the Search for New Harmonic-Generating Crystals

Stephan P. Velsko

Lawrence Livermore National Laboratory, Livermore, CA 94550

Three basic questions must be answered to ensure success in the search for an optimized nonlinear crystal for a particular application: What are the most important optical properties which determine the crystal's figure of merit for the intended application? What is the best methodology for characterizing those optical properties so that materials of interest can be identified efficiently? Where in "materials space" can crystals with such properties be found with the highest probability? Answers to these questions will be discussed in the context of a program to find improved frequency conversion crystals for high power lasers.

It is generally recognized that practical high efficiency harmonic generation of very low power lasers (such as laser diodes) requires crystals with large nonlinear coefficients. This has spurred the search for such materials in many laboratories. It is less often appreciated that efficient conversion of very high power lasers is also materials limited. Even multimegawatt pulsed lasers are rarely frequency converted with more than about 60% efficiency in general practice, even though simple theoretical calculations might imply that much greater efficiency should be possible at those power levels.

Harmonic conversion efficiencies far less than unity are often, of necessity, accepted in laser design. However, in many cases this represents a severe blow to the overall ("wallplug") efficiency of a laser system, ultimately increasing the size and cost of a unit which must supply a certain desired amount of light at the harmonically generated wavelength. One area where the economic impact of frequency conversion efficiency is very clearly felt is in high power solid state laser systems used for inertial confinement fusion (ICF).[2] Here, it is currently believed that blue or near ultraviolet light is optimum for efficient compression of the fusion target, but the large aperture Nd:glass lasers used in these experiments produce near infrared light with a wavelength of 1.05 μm . To generate shorter wavelength light, nonlinear crystals are used to frequency double to 0.527 μm , and this doubled light is then mixed with the residual fundamental to produce light at 0.351 μm. This latter frequency mixing process is colloquially known as third harmonic generation (THG) or "tripling". KDP, whose optical properties and threshold powers have recently been reviewed,(2) is the material currently used for this purpose. On the NOVA fusion laser operated at Lawrence Livermore National Laboratory, for example, 70% conversion to the ultraviolet has been observed using KDP.(3,4) Thus, nearly 1/3 of the light energy generated by this 100KJ laser system remains in the form of longer wavelength photons less useful for target compression.

Less than unity conversion of high power lasers is an unavoidable consequence of laser beam divergence.(5,6) Eimerl has suggested recently on intuitive grounds that the performance limit for a given nonlinear crystal for a particular harmonic generating process is determined by a figure of merit called the "threshold power," which is a function of both the nonlinearity d_{eff} and angular sensitivity β of the crystal (5,6):

$$P_{th} \propto \left(\beta/d_{eff}\right)^2$$

The lower the threshold power, the easier it is to achieve high conversion efficiency of the noisy pulses and abberated beams which are typical of high-power lasers. Materials with threshold powers an order of magnitude lower than KDP could achieve conversion efficiencies closer to 90% for beam of similar quality to Nova. This has motivated us to undertake a broad and systematic search for such crystals.

The frequency conversion materials which are useful in Nd:glass based fusion lasers of recent design must satisfy several other criteria besides having a low threshold power. First, they must be transparent to wavelengths as short as 0.264 μm to preclude substantial two photon absorption of the intermediate 0.527 μm light, which can cause loss of THG efficiency and possibly photochemical damage to the crystals. (Strictly speaking, two photon absorption of the 0.351 μm light is also unwanted, but its effect on conversion efficiency is less deleterious.) Second, because the next generation of fusion lasers is designed to operate at (infrared) fluences as high as 40 J/cm^2, the frequency conversion crystals ideally should operate at these fluences without optical damage. Current laser designs also specify single segment apertures of 25 - 30 cm on a side.(4,7) While such large aperture frequency convertors could be constructed from many smaller sized crystals, this severely compromises the beam quality of the generated harmonic light, and could lead to unacceptable optical losses. Therefore, crystals for ICF lasers must be growable to sizes which allow 30 x 30 cm^2 plates to be fabricated. While there is no fundamental reason why high temperature flux or melt grown crystals could not be used for this application, substantial experience with crystal growth indicates that water soluble crystals are most likely to meet the growth and damage threshold requirements. In fact, KDP is the only frequency conversion material currently available in high quality pieces this size.

The transparency requirement quoted above immediately sets a limit on the maximum size of the nonlinearity we can expect to find in such materials. Data for known nonlinear crystals implies that it is unlikely to find crystals with phase-matchable d coefficients much greater than 2 pm/V if their UV edge is shorter than 250 nm. Although the size of the d coefficients is bounded by the transparency requirement, threshold powers substantially smaller than those of KDP can still be achieved by finding crystals which have much smaller angular sensitivities than KDP. This requires that either the crystals have low birefringences or that they have noncritical, or near noncritical phasematching orientations. This depends, in turn, on a fortuitous equality of birefringence and dispersion.

The probability of finding a new nonlinear crystal with such specific optical characteristics within a reasonable period of time depends on several factors. The first factor is the rate at which the relevant optical properties can be measured with sufficient accuracy to decide if the crystal is likely to be better than the ones already in hand. Second, there must be a reasonably high probability that the set of materials chosen for characterization contains crystals with the desired characteristics. Finally, assuming that crystals with the desired properties are only sparsely distributed among the set, the rate at which the crystals can be synthesized or otherwise obtained in characterizable form must be high, so that a large population of crystals can be scrutinized. The object of this paper is to review the strategy we have developed to optimize these factors in our search for improved ICF frequency convertors.

The next section briefly describes an efficient screening procedure which we have developed for identifying potential frequency convertors for fusion lasers. Section III discusses the chiral organic salt strategy which we have used to generate a large population of phasematchable crystals with many of the desired characteristics for ICF. Section IV presents a statistical model we used to estimate the probability of finding low threshold power harmonic generators in this class of materials. The specific phasematching properties of some new nonlinear crystals from this class are discussed in section V. Section VI contains some concluding remarks.

It is important to note that, with the exception of the extraordinarily large aperture size requirement, the criteria for a fusion laser frequency convertor are generally applicable to <u>any</u> high power laser. Moreover, many of the ideas presented here are applicable, mutatis mutandis, to the search for NLO crystals for <u>any</u> specific application, for example, diode laser doubling.

<u>Efficient Screening of Harmonic Generating Crystals</u>
The basic objective of our screening method is to identify those materials which are likely to be less angularly sensitive and more nonlinear than KDP for frequency doubling or tripling 1.05 μm light without the need to grow large, high quality crystals. In principle, it is possible to evaluate these properties by powder SHG experiments alone.(8-9) In practice, however, we have found that powder SHG tests for phasematching and for determining noncritical wavelengths are too often ambiguous to be useful in a large survey. Therefore, we have used powder SHG tests only as a means of identifying crystals with nonlinear coefficients as large or larger than those of KDP. A description of our powder apparatus and experimental method has been given elsewhere.(10-11) An important observation we have made is that, depending on particle morphology, a material with nonlinear coefficients the same size as those of KDP can easily exhibit a powder SHG signal as low as 1/2 or as high as 2x that of KDP itself. In this sense, the "uncertainty" of the powder measurement made on a new material is a factor of 2, and we accept for further characterization those materials which have a powder signal 1/2 of KDP or greater. While this sends some crystals with nonlinearities smaller than KDP to the next level of characterization, it ensures that we do not reject too many crystals which <u>do</u> have the desired nonlinearity.

The next step is to identify those crystals which also have low angular sensitivities for phasematched doubling or tripling. The most accurate method of determining such properties is to measure the wavelength dependence of the refractive indices and calculate the phasematching orientations and angular sensitivities using the exact phasematching equations which have been given by a number of authors.(12-13) Since the vast majority of crystals we study are biaxial, this would take too much time and effort to be useful for a rapid survey. Instead, we have developed an approximate method for determining phasematching properties which is rapid and complements the powder SHG screening step. The method relies on a determination of the refractive indices at a single visible wavelength or, more typically in practice, for white light.(11,14)

At a single wavelength, the refractive index ellipsoid of a biaxial crystal is specified by three principal refractive indices $n_\alpha < n_\beta < n_\gamma$ from which are derived the principal birefringence $\Delta = n_\gamma - n_\alpha$, and the optic angle V as defined in Figure 1. For phasematching processes which lie well within the transparency range of the crystal, there exists a simple approximate formula for the locus of phasematching orientations (θ, ϕ):

$$\text{Sin}^4\theta \, \text{Sin}^2(\phi + V) \, \text{Sin}^2(\phi - V) + \text{Cos}^2\theta \, \text{Sin}^2\theta \left(\text{Sin}^2(\phi + V) + \text{Sin}^2(\phi - V) \right)$$
$$+ \text{Cos}^4\theta = (\delta/\Delta)^2$$

which depends on the principle birefringence , and the optic angle V evaluated at the fundamental wavelength, and the difference between the average refractive indices at the fundamental and harmonic wavelengths, $\delta = n(2\omega) - n(\omega)$. We refer to the parameter δ as the "dispersion" of the crystal's refractive indices.

For nearly all the crystals we have studied, the principal birefringence and the optic angle vary only slightly with wavelength. Therefore, the white light indices give reasonably accurate values of these parameters over the entire transparency range of the crystals. The dispersion parameter δ can be estimated from from the magnitudes of the refractive indices themselves, using a crude empirical model.(11,14) From the estimated dispersion, the principal birefringence, and the optic axial angle we can generate representations of the phasematching loci for SHG and THG which usually lie within $\pm 10°$ of the true loci. This information is sufficient to indicate if angularly insensitive orientations exist. Examples of these approximate loci are shown in Figure 2, along with the exact (experimentally measured) loci for comparison.

If this approximate method predicts that the crystal has phasematching orientations which are angularly insensitive, then more complete and accurate determinations of the refractive indices as a function of wavelength are made using a unique microrefractometer developed in our laboratory.(15-16) Single crystallites as small as 50 μm are easily characterized using this technique, although currently it is limited to crystals which have refractive indices smaller than 1.7. The refractive indices determined this way are usually accurate to $\pm 2 \times 10^{-4}$, although larger errors are sometimes observed in the ultraviolet. For doubling 1.05 μm, the phasematching angles calculated using the refractive indices determined from microrefractometry are typically within 2° of those determined by direct phasematching measurements or by calculations using prism data. For sum frequency generation to produce 0.351 μm, the calculated phasematching angles are accurate to $\pm 5°$.

At this stage, we have screened out a set of materials which are very likely to be more nonlinear than KDP and are very certain to have phasematching orientations

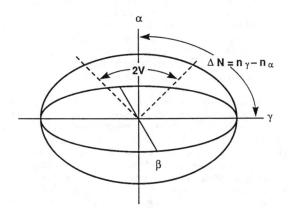

Figure 1. The biaxial indicatrix, whose principal axes correspond to the refractive indices $n\alpha \leq n\beta \leq n\gamma$. The optic axes are indicated by dashed lines and the optic angle is denoted 2V. N is the principal birefringence.

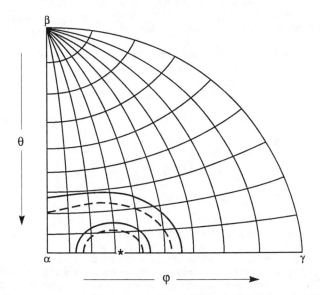

Figure 2. Comparison of measured phasematching loci (solid curves) with those predicted by the approximate formula (dashed curves) for type I and II doubling in L-arginine fluoride. The asterisk marks the optic axis position.

which are less angularly sensitive. However, for a crystal to have a low threshold power the large nonlinearity and small angular sensitivity must occur at the same phasematching orientation. It is more usual than not to find that crystal symmetry and optical orientation conspire to make the nonlinear coupling vanish in precisely the orientation which has the smallest angular sensitivity! Therefore, the next step in the screening procedure is to directly measure the phasematching properties and nonlinear coefficients in single crystals of the new materials.

Direct phasematching measurements (17) can be made on crystals as small as a few hundred microns, although we typically use crystals about 1 mm in cross section. This requires that the materials be recrystallized to the 1 mm range, if they are not already available in that size from the initial synthesis step. Nonetheless, this requires considerably less time and effort than growing the centimeter scale single crystals which would be required for traditional nonlinear coefficient measurements. As an example of the information generated by this technique, Figure 3 shows the results of direct phasematching measurements on a single crystal of dipotassium tartrate hemihydrate (DKT). The phasematching loci for type I doubling and tripling of 1.064 μm are plotted on a stereonet projection of a sphere, whose x, y and z axes correspond to the principal dielectric axes. The optic axes are designated by asterisks. Letters mark the positions of maximum harmonic generation intensity (maximum d_{eff}) while zeros mark the orientations where the effective nonlinear coefficient vanishes. The symmetry of the arrangement of the zeros and maxima is consistent with the monoclinic (2) symmetry of the crystal with the two-fold symmetry axis parallel to the high index (γ) axis. The values of the nonlinear coefficients and angular sensitivities at the marked positions are given in Table 1.

Table 1. Nonlinearity and angular sensitivity for type I doubling and tripling of 1.064 μm in dipotassium tartrate (DKT)

Position	θ	ϕ	d_{eff}(pm/V)	β(cm^{-1}/mrad)
2ω I/a	+ 23°	$\pm$ 34°	0.14	3.7
2ω I/b	- 10	$\pm$ 48	0.13	4.6
3ω I/a	+ 33	$\pm$ 52	0.16	6.5
3ω I/b	- 28	$\pm$ 60	0.12	6.3

In DKT, as in other monoclinic crystals we have studied (17-18), the largest nonlinear coupling occurs at orientations which do not lie in the principal planes of the dielectric ellipsoid. Therefore, measurements which do not explore the entire phasematching locus could significantly underestimate the nonlinearity of these crystals. The value of the direct phasematching technique is that a complete "global" picture of the phasematching properties of even the lowest symmetry crystals is obtained in a straightforward way.

The sequence of measurements described in this section are designed to increase the efficiency of the screening process by sending a crystal on to a more involved stage of characterization only if it is likely to have the desired properties on the basis of the simpler measurements. The accuracy of each of the techniques is limited, so that in practice there is considerable chance that useful crystals will be rejected - and non-useful crystals will be selected - for further characterization. Nonetheless, these methods greatly enhance our ability to identify useful crystals before any serious crystal growth efforts are necessary.

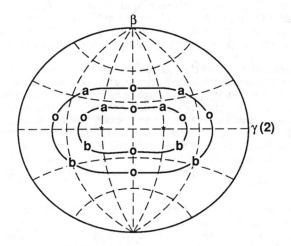

Figure 3. Phasematching loci for type I SHG and THG of 1.064 μm in dipotassium tartrate (DKT).

The Chiral Organic Salt Strategy

For a wide variety of ionic and molecular crystals, the nonlinearity and other optical characteristics can be attributed to the electronic properties of molecules or molecular ions which compose the crystal (19-20). The bulk nonlinearity is approximated as the sum of the contributions of each molecular or ionic unit, weighted by geometric factors depending on the orientational arrangement of the units within the unit cell. Nature provides a variety of small noncentrosymmetric molecular units with electronic excitations at wavelengths shorter than 300 nm from which NLO crystals for laser fusion can be "built". Examples include the planar borate, carbonate, nitrate, carboxylate and guanadinium ions, certain molecular ions with nonbonding electrons such as iodate and bromate, and certain fluoride coordination groups involving d^0 ions such as titanium (IV) and niobium (V). It has long been understood that a crystal composed of these (or any other) units will not exhibit an appreciable nonlinear response unless it is noncentrosymmetric. In addition, the crystals which show the largest nonlinearity will be those in which the units are mutually oriented to give the largest net projections of the molecular hyperpolarizability tensor onto the bulk second order tensor of the crystal (20). In this context, crystal structures whose macroscopic nonlinearities express the largest fraction of the microscopic hyperpolarizability can be called "optimized".

Once a particular class of chemical compositions has been chosen for study, there are several strategies for finding crystals with noncentrosymmetric structures which allow the nonlinearity of the units to be expressed as a bulk nonlinearity. This includes "passive" approaches such as utilizing the crystallographic literature to identify known acentric phases with the desired chemical composition and exploring substitutional analogues, and "active" approaches which attempt to increase the probability that noncentrosymmetric crystals will be formed in completely new materials. While a number of "active" approaches have been suggested (21-23), only the use of optically active enantiomers guarantees that the resulting crystal structure will be noncentrosymmetric.

We recognized (24-25) that many new NLO crystals which have properties favorable for ICF applications could be generated by a straightforward synthetic procedure, illustrated in Figure 4. These crystals consist of molecules which have at least one chiral center, and are optically active. The moieties attached to the chiral center are drawn from a large set of small charged or neutral fragments. At least one of these fragments should be a "harmonic generating unit", i.e. a noncentrosymmetric structure with delocalized bonding which will enhance the molecular hyperpolarizability. One of the fragments should be charged (this could be the harmonic generating unit itself, as in the case of the carboxylate anion.) This ionic chiral organic molecule can be co-crystallized with a variety of counterions, which might also be harmonic generating units themselves, to form a "chiral organic salt". It is also possible to have the chirality associated with a non-SHG active counterion, while the harmonic generating unit is non-chiral, e.g. formate ion co-crystallized with chiral amines. Often, such crystals will also incorporate one or more waters of hydration, depending on the conditions of crystallization. Finally, other chiral structures, such as those shown in Figure 5 could be utilized. It is clear that a wide variety of such crystals are possible, and that harmonic generating units which are much larger and more nonlinear than the ones we have listed could be used at the expense of a smaller transparency range, when the application warrants it.

The requirement that the molecule be charged, and the compounds be formed as salts is important for two reasons. First, at least for small molecules, it ensures water solubility. This is important for crystal growth, since it permits the techniques which allowed the scaling of KDP crystals to 30 x 30 x 60 cm^3 sizes to be applied without

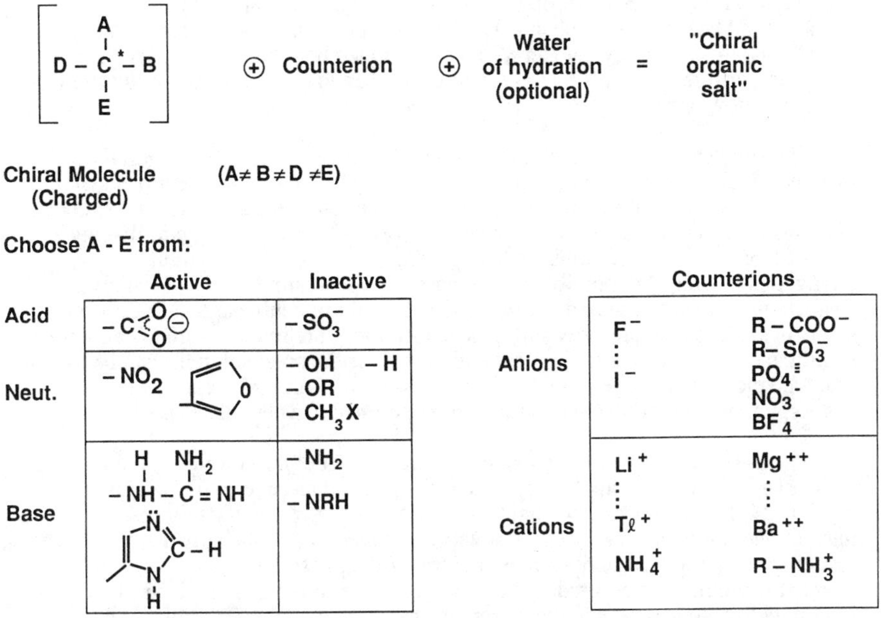

$$\begin{bmatrix} & A & \\ & | & \\ D - & C\overset{*}{} - B \\ & | & \\ & E & \end{bmatrix}$$ ⊕ Counterion ⊕ Water of hydration (optional) = "Chiral organic salt"

Chiral Molecule (Charged)　　(A≠ B ≠ D ≠E)

Choose A - E from:

	Active	Inactive		Counterions	
Acid	$-C\overset{O}{\underset{O}{\lessgtr}}\ominus$	$-SO_3^-$			
Neut.	$-NO_2$ 〔furan〕	$-OH$ $-H$ $-OR$ $-CH_3X$	Anions	F^- ⋮ I^-	$R-COO^-$ $R-SO_3^-$ $PO_4^{\equiv}$ NO_3^- BF_4^-
Base	H NH_2 $-NH-C=NH$ 〔imidazole〕	$-NH_2$ $-NRH$	Cations	Li^+ ⋮ Tl^+ NH_4^+	Mg^{++} ⋮ Ba^{++} $R-NH_3^+$

Figure 4. General scheme for the synthesis of chiral organic salts.

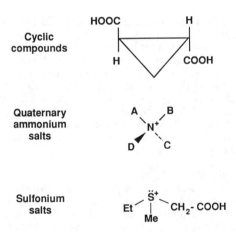

Figure 5. Other possible structures for making chiral salts.

any fundamental changes. Second, the resulting crystals are at least partly ionic in character and hence have mechanical and thermal properties far superior to those of crystals stabilized by only VanderWaals or hydrogen bonding interactions. Third, generating different crystals by ion substitution is an important aspect of the strategy, as we will now argue.

The structure of a typical chiral organic salt is determined by several forces, including ionic, VanderWaals, and hydrogen bonding. Moreover, the molecules usually have some conformational flexibility. As a result, the actual crystal structure of a given compound cannot be predicted a priori. Therefore, we regard making a series of salts as a way of empirically exploring the "space" of structural arrangements which are possible for the harmonic generating units in such crystals. This, in turn, allows us to explore the space of resulting nonlinear and linear optical properties. When certain isovalent ionic substitutions are mild enough to cause no significant structural change (e.g. bromide for chloride) solid solutions can be used to fine tune the optical properties, similar to the way that solid solutions of the KDP isomorphs can be used to cover a range of noncritical wavelengths (2).

The use of salt formation to expand the number of crystals which contain a single molecular type was first applied by Meredith (26), and more recently by Marder et. al. (22). In the latter work, ionic interactions are used to offset dipolar interactions among achiral molecules, which enhances the probability that the resulting crystal will be noncentrosymmetric. In our case, of course, noncentrosymmetry is ensured by the chirality of the molecules involved. It is important to note that, within the picture we have presented, neither the assurance of noncentrosymmetry, nor the enhanced hyperpolarizability of the chiral molecule guarantees that the nonlinearity of any particular chiral organic salt crystal will be large. These properties simply ensure that each crystal so formed has an equal opportunity to express the molecular hyperpolarizability in an optimized way.

Whereas Figure 4 is intended to represent schematically the synthesis of an arbitrary chiral organic salt, in fact there already exist many commercially available chiral organic molecules which fit the basic criteria. These are predominantly amino acids and alpha-hydroxy acids, and related derivatives. Nearly all of our studies so far have utilized these molecules, which we regard as models for the more general concept. (Not all of the crystals we studied were salts: some were zwitterionic or free base compounds where hydrogen bonding provides the dominant intermolecular forces.) It should be noted that, while the harmonic generating properties of a number of amino acid and hydroxy acid-containing crystals have been reported in the literature,(27) no systematic study of this general group of compounds has been previously attempted.

Statistical Model of the Search Process
Because the structure and optical properties of any particular compound cannot be predicted a priori, we have found it convenient to regard the search as a random process, sampling from a parent population with a fixed distribution of birefringences, nonlinearities, etc. To obtain an estimate of these distributions, we initially measured the powder SHG signals and refractive indices of more than 70 salts of commercially available amino and alpha-hydroxy acids and related compounds.(25) From the refractive indices, the principal birefringence and the optic angles of the crystals were computed. Approximate noncritical wavelengths were calculated for each crystal, using the empirical correlation between dispersion and refractive index discussed in the last section.

From this information, we have estimated($\underline{25}$) that between 0.5 and 1% of the chiral organic salts formed from amino acids and alpha-hydroxy acids have lower threshold powers than KDP for doubling or tripling 1.05 μm light. The probability P of finding such a crystal in a random sample of N crystals from the population of chiral organic salts is given by

$$P = 1 - (1 - p)^N$$

where p is the frequency of occurrence in the population. For p = 0.005, about 500 crystals must be examined to insure with 95% confidence that at least one low threshold crystal will be found. It is of some interest to compare this sample size with that which would be required if chirality were not utilized, assuming the same basic distributions of nonlinear and linear optical properties among the crystals of inorganic or achiral organic salts which were noncentrosymmetric. Figure 6 is a plot of N vs p for P = 0.95. Since only 30% of achiral organic crystals are noncentrosymmetric and about 20% of inorganic crystals, we would expect that the necessary sample size would increase to ~1000 and ~2500 crystals, respectively.

Strictly speaking, the empirical distributions given in reference 25 must be regarded as composites of subpopulations with varying types and densities of harmonic generating units because the molecules used to make the salts differed in size, type of unit, and number of units per molecule. Thus, for example, we would expect the distribution of powder intensities would be shifted towards larger values if we restricted the population to salts of chiral molecules having a higher ratio of harmonic generating units to total carbon atoms. Similarly, the distribution of noncritical wavelengths would be shifted towards longer wavelengths if we restricted the population to crystals containing heavier atoms, e.g. chloride, bromide or arsenate salts which raise the dispersion. To a large extent, chemical composition alone governs "average" optical properties such as the average refractive index and its dispersion. But chemical composition and molecular structure also determine the range of possible values of structure-dependent optical properties such as birefringence and nonlinearity in a set of crystals. Within this context, "molecular engineering" is regarded as a way of shifting the distribution to increase the probability of finding a crystal with particular properties.

Phasematching Properties of Chiral Organic Salts
    To date, we have screened more than two hundred chiral organic salts and related compounds. Table 2 lists crystals which have noncritical wavelengths for frequency doubling between 1.2 and 0.8 μm. The type of phasematching and the principal axis for noncritical phasematching are also listed. Because the crystals are biaxial, three noncritical wavelengths, corresponding to propagation down each of the three principal axes, are possible. For this reason several crystals appear more than once in this list. The value of the noncritical wavelength is accurate to about 0.02 μm. These compounds showed powder signals which were similar to, or larger than, KDP, but the value of the nonlinearity in the noncritical configuration is, in many cases, unknown. Symmetry and optic orientation will often cause the nonlinearity to vanish identically. It should also be noted that hydrogen vibrational overtones are likely to cause very high absorption coefficients at wavelengths longer than 1 μm. Therefore, the utility of undeuterated crystals at those wavelengths is questionable even if they have large nonlinearities.

We have investigated the phasematching properties of many, but not all, of the crystals given in Table 2 using the direct measurement technique. In general, we have found that, as expected, most of the crystals do not have threshold powers lower than KDP for generating 0.527 or 0.351 μm light. Most often this occurs

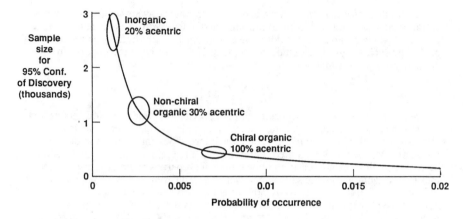

Figure 6. Sample sizes required to assure that a low threshold power crystal would be found among various crystal types.

Table 2. Crystals with non-critical wavelengths between 1.2 and 0.83 µm

NCPM λ (µm)	I[a] powder	Compound	Axis[b]/	Type
1.228	1.6	N-ACETYL HIS	n1	II
1.175	0.5	L-ARG OXALATE	n1	I
1.155	1.0	HIS CF3COOH	n1	II
1.149	0.6	L-ARG Cl H2O	n1	I
1.14	0.3	L-ARG ACETATE	n1	II
1.133	0.5	HIS CH3SO3H	n1	II
1.123	0.8	MET MALIEATE	n3	II
1.066	0.5	NaH TARTRATE	n3	I
1.03	1.0	Cd LACTATE	n1	II
1.03	1.0	Cd LACTATE	n2	II
1.024	0.7	N-ACETYL PRO · H2O	n1	I
1.012	1.0	HIS FLUOROBORATE	n1	II
1.003	1.2	N-ACETYL METHIONINE	n3	I
0.995	0.8	N-ACETYL TYROSINE	n1	II
0.959	0.8	MET NITRATE	n1	II
0.936	0.5	ALANINE CH3SO3H	n3	II
0.935	0.6	N-ACETYL ASN	n1	II
0.93	1.2	N-ACETYL METHIONINE	n2	II
0.926	1.4	L-ARG FLUORIDE	n1	I
0.923	0.3	L-ARG ACETATE	n2	II
0.923	0.6	N-ACETYL ASN	n3	I
0.91	1.0	N-ACETYL OH PRO	n3	II
0.906	0.8	NA L-VAL · NH3	n3	II
0.902	1.2	N-ACETYL METHIONINE	n1	I
0.885	1.9	Mg TARTRATE	n3	II
0.884	1.0	HIS ACETATE	n1	I
0.883	1.0	HIS CF3COOH	n1	I
0.868	1.0	L-ARG CF3COOH	n3	II
0.858	0.8	METHIONINE MALIEATE	n1	II
0.853	1.0	N-ACETYL OH PRO	n1	I
0.846	1.8	DIAMMONIUM TARTRATE	n3	II
0.837	1.0	ALANINE CH3SO3H	n2	II
0.831	1.9	MG TARTRATE	n2	II
0.831	0.5	HIS CH3SO3H	n1	I

[a] Powder SHG signal relative to KDP
[b] n1 = α, n2 = β, n3 = γ

because, although the crystal has an orientation with larger nonlinearity than KDP, the angular sensitivity is also substantially larger and results in a larger threshold power. Conversely, most of the crystals we have examined which do have orientations with substantially lower angular sensitivities have vanishing or nearly vanishing nonlinear coefficients in those orientations. This is very often a consequence of the crystal symmetry and the optic orientation.

L-arginine fluoride (LAF) is an example of a low threshold power doubler we have discovered. Figure 7 shows the phasematching loci for frequency doubling 1.064 μm determined by direct phasematching measurements. The type I loci intersect the α - γ plane at $\pm 12°$ from the α axis. At this orientation, the the angular sensitivity is slightly smaller than that for type I doubling in KDP (4.3 vs. 4.9 cm^{-1}/mrad) but the effective nonlinearity is almost 4 times larger (0.98 vs. 0.26 pm/V). As a result, the threshold power for type I doubling is 16 times smaller for LAF, making it an attractive substitute for KDP in the polarization insensitive type I/type II THG schemes recently proposed for solid state fusion drivers.(4)

In LAF the low angular sensitivity orientation on the type I doubling locus is also the point of maximum d_{eff}. By contrast, consider the phasematching loci of N-acetyl tyrosine (NAT), described in Figure 8 and Table 3. Here both the type I and type II doubling loci have the same topology as the type I locus in LAF. In this case, however, the type I locus lies nearly 30° away from the α axis. Because the birefringence of this crystal is so large (n_γ - n_α = 0.14) the resulting angular sensitivity at that orientation is 14 cm^{-1}/mrad - almost a factor of three larger than KDP! The type I nonlinearity is not significantly larger than that of KDP, and the resulting threshold power is much higher. While the type II locus comes much closer to the noncritical orientation ($\pm 15°$ from α), the nonlinear coupling is zero there because of symmetry.

L-arginine acetate (LAAc) is a low threshold power THG crystal which has emerged from our survey. The loci for type I SHG and type I THG of 1.064 μm

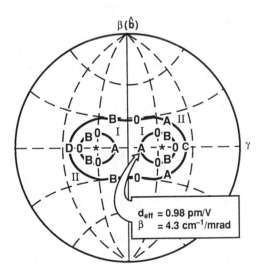

Figure 7. Phasematching loci for type I and II doubling of 1.064 μm in L-arginine fluoride.

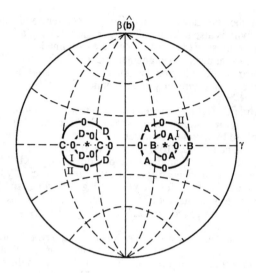

Figure 8. Phasematching loci for type I and II doubling of 1.064 μm in N-acetyl tyrosine.

light are shown in Figure 9. The type I THG locus intersects the α - β plane only a few degrees from the α axis. Because this is so close to the noncritical orientation, the angular sensitivity is very small. Unfortunately, the α axis coincides with the 2-fold symmetry axis in this monoclinic crystal, so the nonlinear coupling vanishes in this orientation. Nonetheless, two low threshold power orientations for type I tripling do exist in this crystal. The largest d_{eff} value (0.75 pm/V) is found in the β - γ plane, where the angular sensitivity is only 30% smaller than it is for type I THG in KDP (6.5 vs. 7.8 cm^{-1}/mrad). In addition, another (symmetry equivalent) pair of orientations exist with about 1/2 the angular sensitivity of KDP, and 20% higher nonlinearity (0.35 vs. 0.29 pm/V).

Table 3. Nonlinearity and angular sensitivity for type I and II doubling of 1.064 μm in N-Acetyl Tyrosine

Position	θ	ϕ	d_{eff}(pm/V)	β(cm^{-1}/mrad)
2ω I/A	± 6	43	0.18	15.0
2ω I/B	0	28	0.28	13.8
2ω I/C	0	- 28	0.41	13.8
2ω I/D	± 10	- 41	0.34	14.9
2ω II/A	± 14	20	0.22	4.5
2ω II/B	0	58	0.18	6.9
2ω II/C	0	- 58	0.12	6.9
2ω II/D	± 15	22	0.10	4.5

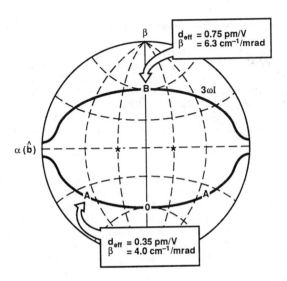

Figure 9. Phasematching loci for type I tripling of 1.064 μm in L-arginine acetate.

Concluding Remarks

Once a low threshold power crystal has been identified, the long and arduous task of growing larger crystals remains. Crystals with dimensions of the order of 1 cm are necessary to evaluate many of the other properties which are important for ICF or other high power laser applications. Among these are optical absorption, stimulated Raman scattering (SRS) and stimulated Brillouin scattering (SBS) thresholds, and optical damage thresholds. This latter property differs from the others because, even at the fluences conceived of for ICF lasers, it is apparently governed by extrinsic factors such as impurities and defects. Thus, it is more a function of the crystal growth process than a property of the crystal itself. On the other hand, the SRS and SBS are serious parasitic nonlinear processes which are primarily governed by the composition and structure of the material itself. Until more information about these properties is obtained, it is by no means certain that the class of chiral organic salts and non-chiral analogues will be useful alternatives to KDP for fusion lasers even if crystals with substantially lower threshold powers are found.

The program reviewed in this paper can be regarded as a paradigm for any directed search for a new nonlinear crystal. Chemistry plays a key role in defining the space of chemical compositions which produce materials which meet the basic requirement of optical transparency for the desired application. The magnitude of the nonlinearity is ultimately limited by this, so that the pertinent issue is finding materials with the largest nonlinearity consistent with the transparency requirement. Basic structural criteria such as chirality, possibly augmented by crystallographic intuition, can be used to enhance the chances of finding noncentrosymmetric crystals and thus reduce the number of crystals which have to be examined. Among the set of crystals with "optimum" nonlinear coefficients will be a subset with near noncritical phasematching for the desired wavelengths. Ultimately, the final choice of a crystal from this set will depend on crystal growth and other issues.

Given the large number of nonlinear crystals which have been characterized since the phenomenon of harmonic generation was first observed, the search for a new crystal can almost always be regarded as a search for an alternative to some currently available "best choice". (For example, a suitable "baseline" material for diode laser frequency doubling might be potassium niobate.) Because of this, the value of finding an "improved" material can be accurately gauged in a relative sense, and compared to the cost of both the search and the subsequent development of the material. To a large extent, the discovery and development of new crystals for harmonic generation now rests on such economic considerations.

Acknowledgments
    The work described in this paper represents the contributions of several people. Laura Davis is responsible for the linear optical property measurements and the development of the microrefractometer. Most of the nonlinear optical measurements on small single crystals have been made by Mark Webb, who is also responsible for several improvements in the apparatus and technique. Francis Wang synthesized the chiral organic salts and did the powder SHG measurements. David Eimerl was the source of much encouragement, advice, and support during the course of this work.

Literature cited
1. E. Storm, J. Fusion Energy, $\underline{7}$, 131-137, (1988).
2. D. Eimerl, Ferroelectrics $\underline{72}$, 95-139, (1987).
3. J. Hunt, Proc. SPIE Vol. $\underline{622}$, 10 - 17, (1986).
4. H. Powell, J. Campbell, J. Hunt, W. Lowdermilk, J. Murray, and R. Speck, in Inertial Confinement Fusion, (Proceedings of the Course and Workshop Held in Varrena, Italy, 1988), (A. Caruso and E. Sindoni, eds., Soc. It. di Fis., Bologna, 1989), p 197-216.
5. D. Eimerl, IEEE J. Quantum Elec. $\underline{QE-23}$, 1361 - 1371, (1987).
6. D. Eimerl, IEEE J. Quant. Elec., $\underline{QE-23}$, 575-592, (1987).
7. W.H. Lowdermilk, Lawrence Livermore National Laboratory, UCRL- JC - 103112; Laser and Particle Beams, in press.
8. S. Kurtz and T.T. Perry, J. App. Phys., $\underline{39}$, 3798-3813, (1968).
9. J. M. Halbout, S. Blit, and C.L. Tang, IEEE J. Quantum Electron. $\underline{QE-17}$, 513-517, (1981).
10. S. Velsko and D. Eimerl, Laser Program Annual Report, Lawrence Livermore National Laboratory, UCRL-50021-85, 7-69, (1985)
11 L. Davis, D. Eimerl, and S. Velsko, Lawrence Livermore National Laboratory, UCRL - 96109, (1987).
12. M. V. Hobden, J. Appl. Phys. $\underline{38}$, 4365-4372, (1967).
13. M. Kaschke, and C. Koch, Appl. Phys. $\underline{B49}$, 419-423, (1989).
14. S. P. Velsko, L.E. Davis, and F.T. Wang, in the Laser Program Annual Report, Lawrence Livermore National Laboratory, UCRL-50021-87, 5-33, (1987).
15. L.E. Davis, in the Laser Program Annual Report, Lawrence Livermore National Laboratory, UCRL-50021-87, 5-36, (1987).
16. L. Davis, Lawrence Livermore National Laboratory, UCRL-96102, (1987).
17. S. Velsko, Opt. Eng. $\underline{28}$, 76-84, (1989).
18. D. Eimerl, S. Velsko, L. Davis, F. Wang, G. Loiacono, and G. Kennedy, IEEE J. Quant. Electron. $\underline{QE-25}$, 179-193, (1989).
19. C. Chen and G. Liu, Ann. Rev. Mater. Sci. $\underline{16}$, 203-243, (1986).
20. J. Zyss and D. Chemla, in Nonlinear Optical Properties of Organic Molecules and Crystals, Vol 1, D. Chemla and J. Zyss, eds., Academic Press, Orlando, Fla. (1987), pp 23-191.

21. J. Zyss and D. Chemla, J. Chem. Phys. 74, 4800-4810, (1981).

22. S. Marder, J. Perry, and W. Schaefer, Science 245, 626-628, (1989).

23. M.C. Etter, Acc. Chem. Res. 23, 120-126, (1990).

24. S. Velsko, L. Davis, F. Wang, S. Monaco, and D. Eimerl, Proc. SPIE Vol. 824, 178-181, (1987).

25. S. Velsko, L. Davis, F. Wang, and D. Eimerl, Proc. SPIE Vol. 971, 113-117 (1988).

26. G. Meredith, in Nonlinear Optical Properties of Organic and Polymeric Materials, (ACS Symposium Series 233, American Chemical Society, Washington, D.C. 1983), pp. 27-56.

27. J.F. Nicoud and R.J. Twieg, in Nonlinear Optical Properties of Organic Molecules and Crystals Vol. 1, D. Chemla and J. Zyss, eds., (Academic Press, Orlando, 1987), pp 227-296.

RECEIVED September 14, 1990

Chapter 24

Development of New Nonlinear Optical Crystals in the Borate Series

Chuangtian Chen

Fujian Institute of Research on the Structure of Matter, Chinese Academy of Sciences, Fuzhou, Fujian, 350002 China

This review gives a brief presentation of the basic concepts and calculation methods of the "anionic group theory" for the NLO effect in borate crystals. On this basis, boron-oxygen groups of various known borate structure types have been classified and systematic calculations were carried out for microscopic second-order susceptibilities of the groups.

Through these calculations, a series of structural criteria serving as useful guidelines for finding and developing new NLO crystals in the borate series were found: (1) The planar six-membered ring $(B_3O_6)^{3-}$ and the planar trigonal $(BO_3)^{3-}$ group, each possessing a conjugated π-orbital system, are far more favourable for producing larger second-order susceptibilities' and anisotropy of linear susceptibilities than the non-planar tetrahedral $(BO_4)^{5-}$ group. (2) On the other hand, the ultraviolet absorption edges of non-planar groups, such as $(BO_4)^{5-}$, $(B_3O_7)^{5-}$ are shifted to shorter wavelengths than those of the $(B_3O_6)^{3-}$ and $(BO_3)^{3-}$ groups. (3) The SHG coefficients and birefringences of borate crystals can be adjusted to a certain extent by suitable arrangement of the 3- and 4-coordinated B atoms, e.g. $(BO_3)^{3-}$ and $(BO_4)^{5-}$, $(B_3O_6)^{3-}$ vs $(B_3O_7)^{5-}$ and $(B_3O_8)^{7-}$.

On the basis of these structural criteria, we have been successful in developing some excellent new NLO materials, including LiB_3O_5 (LBO).

The rapid development of laser science and technology which occurred after 1960 included the elucidation of the theoretical principles for designing nonlinear optical (NLO) devices. The major remaining problem that severly restricts progress in this field is the scarcity of appropriate NLO materials. As a result, the search for new NLO materials, particularly in the UV and FAR-IR regions, is still very active, even though intensive efforts in this field have been made for about 20 years[1].

Scientists searching for new NLO materials realise the importance of a thorough elucidation of the structure-property relationship between NLO effects and microstructure. Many attempts have already been made in this direction. Among them in particular we may cite the bond parameter methods, exemplified by the work of Bloembergen[2]; the anharmonic oscillator models of Kurtz and Robinson[3] and Garrett and Robinson[4]; the bond parameter methods of Jeggo and Boyd[5] and Bergman and Crane[6]; and the bond charge model of Levine[7] before the 1970s. Among these, the Levine [7, 8,] model is the most successful, and has been shown to be particularly useful in elucidating the structure-property relationship for NLO effects in A-B type semiconductor materials., the basic structure unit of which consists of SP^3-hybrid tetrahedrally coordinated atoms. However, this method has not been so successful for other types of NLO crystals in which the basic structural unit does not belong to the category of simple σ-type bonds. For example, if such a bond charge model should be extended to ferroelectric crystals consisting of oxygen octahedra with transition metal atoms as the centres, one must introduce new parameters that have some kinds of uncertainty [9, 10]. As a result, it would be difficult to use the model to understand the structure-property relation between NLO properties and microstructures of the crystals except the above A-B type semiconductor materials.

Since the 1970s, Several research groups have discovered that non-linear susceptibilities of crystals arise from basic structural units with delocalized valence electron orbitals belonging to more than two atoms, rather than with those localized around two atoms connected by a simple σ-type bond. Davydov et al [11] showed that non-linear susceptibilities of organic crystals arise from molecules as their basic structural units, and proposed that conjugated organic molecules with donor-acceptor radicals will exhibit large non-linear susceptibilities. This idea was further developed by Chemla et al. [12], Oudar and Chemla [13] and Oudar and Leperson [14], enabling them and others to discover a series of new organic NLO crystals exhibiting very large second-order susceptibilities, such as POM [15], NPP [16] ABP [17] as well as DAN [18]. Furthermore it helped to establish the scientific basis of a new approach in the field of organic NLO materials, known as 'molecular engineering'. During 1968 - 1970, DiDomenico & Wemple [19] found that the non-linear susceptibilities of perovskite and tungsten-bronze type materials are largely due to the distortion in BO_6 oxygen-octahedra. Thus, the latter is considered as the basic structural unit for the production of non-linear susceptibility in these crystals. But because they only used a parametric method, known as the polarization potential tensor β_{ij}, it is impossible to ascertain the relationship between the electronic structure of BO_6 oxygen-octahedra and their macroscopic second-order susceptibilities.

As early as 1967, during a very difficult period in China, we initiated an extensive study to develop a general quantum- chemical NLO-active group theory in order to make a systematic exploration of the structure-property relationship for NLO effects in some typical inorganic NLO crystals then known. This work has led to the establishment of the so-called "anionic group" theory [20,21] and an approximate method of calculation based on the second-order pertubation theory for NLO susceptibilities of crystals[22,23]. On the basis of this theoretical model, Chen's group succeeded in a systematic elucidation of the structure-property relationship for the NLO effect for almost all the principal types of inorganic NLO crystals, namely the perovskite and tungsten-bronze[24], phosphate [25], iodate[26], nitrite crystals [22] etc.

Since 1979 Chen's group has turned its attention to borates. They recognized that borate compounds have numerous structural types since borate atoms may have either three or four-fold coordination. This complex structural nature of borate compounds leads to a great variation in the selection of structural types favorable for the NLO effect, and the anionic group theory can be used to systematically elucidate which structural unit is most likely to exhibit large non-linearities[27]. This active theoretical analysis and systematic experimental work lead our group to discover BBO (barium metaborate, β-BaB_2O_4), which is a high-quality UV NLO borate crystal[28].

Following the discovery of BBO, much broader theoretical activities were conducted to extend structure-property relations from NLO phenomena to linear optical (LO) properties of the crystals as well[29]. Certain LO properties of crystals, such as transparency range and phase-matching range are important for sophisticated technical applications in optic-electronic fields. Extensive theoretical analyses made by our group in the past year involve calculations of the UV absorption edges and the birefringence of crystals. It was shown that using the Dv-SCM-Xα method the absorption edges of crystals in UV range may be evaluated in terms of the components that are the basic structural units of the crystals. This theoretical work enabled our group to appraise in a sophisticated way the UV properties of borate NLO crystals at a microstructure level. This led directly to the discovery of another new UV NLO crystal —— LiB₃O₅ (LBO) [30], which possesses some better NLO and LO properties than BBO. All these theoretical and experimental advances have encouraged us to try to set up a scientific basis for molecular engineering suitable for inorganic NLO materials as scientists have done for organic NLO materials.

In the following part we will give a brief description of the "anionic group theory" for the NLO effects in crystals, including the basic concepts and calculation methods adopted. In the next section we will discuss how to use this theoretical model to develop new UV NLO crystals in the borate series. Finally, the measurements and characteristic features of the NLO properties of these new borate crystals will be discussed.

I. The Anionic Group Theory and the Methods of Approximate Quantum-Chemical MO Theory Adopted for the Calculation of the NLO Susceptibilities of Crystals

Here we give only a brief description of this theory and the method of calculation used. For details the reader is referred to the literature [21,22,23].

In modern laser technology , second-order NLO effects such as SHG ($\chi_{ijk}^{(2\omega)}$) , sum or differenc frequency generation [SFG ($\chi_{ijk}^{(\omega_1+\omega_2)}$) or DFG ($\chi_{ijk}^{(\omega_1-\omega_2)}$)] and parametric oscillation and amplification are most commonly used. In this review, however, besides some linear optical (LO) properties, we confine ourselves to the discussion of only the SHG coefficients for most NLO crystals, since there is no significant difference between SHG and SFG, DFG, etc, if the dispersions of the second-order susceptibilities are not considered.

Physical properties related to the electron motion in crystals fall essentially into two categories. Some, such as the electrical properties of crystals, arise from long-range interactions in the lattice; here long-range forces from the electron--electron or the electron-core interactions play an important role. In these cases, the use of energy band theory is essential. On the other hand, in NLO effects the process of electronic excitation by the incident radiation does not make any important contribution. They essentially arise from the process of scattering, where the action of the incident photons on the electrons in the crystal serves only as a kind of perturbation.

In other words, the electrons confined to their ground state are only slightly disturbed by the incident photons. Hence the NLO effects should be classified into the second category where short-range forces play a decisive role. We therefore make the assumption that, in the NLO effects, the electron motion may be regarded as confined to small regions. In other words, any NLO susceptibility (or second-order susceptibility) in crystals is a localized effect arising from the action of incident photons on the electrons in certain orbitals of atomic clusters.

Therefore, what we need to do is to define at first the region of the localized motion of valence electrons in order to make reasonable estimates of the bulk second-order susceptibility of the crystal. For this purpose we have analyzed almost all principal types of NLO materials known, such as perovskite, tungsten-bronze type, iodate, phosphate and molybdate, nitrite and organic crystals containing substituted benzene as major NLO-active molecules. Much to our surprise, we found that in any type of the material with large NLO effects, the basic structure unit without exception is built up from anionic groups (or molecules) which are capable of producing large microscopic NLO effects, such as the $(MO_6)^{n-}$ coodination octahedron in perovskite and tungsten-bronze type materials; the $(IO_3)^-$ group in iodates; the $(PO_4)^{3-}$ and $(M_oO_4)^{2-}$ groups in phosphates and molybdates; the $(NO_2)^-$ group in nitrites; the substituted benzen molecules in most organic molecular crystals. On this basis we proposed a theoretical model called the "anionic group theory " for NLO susceptibilities, with the following two assumptions as basic premises: (i) The overall SHG coefficient of the crystal is the geometrical superposition of the microscopic second-order susceptibility tensors of the relevant ionic groups, and has nothing to do with the essentially spherical cations. The former can be expressed as

$$\chi_{ijk}^{(2\omega)} = \frac{N}{V} \sum_{p} \sum_{i'j'k'} \alpha_{ii'} \alpha_{jj'} \alpha_{kk'} \cdot \chi_{i'j'k'}^{(2\omega)}(p) \qquad (1)$$

Where V is the volume of a unit cell, N is the number of basic structural groups in this unit cell, $\alpha_{ii'}$, $\alpha_{jj'}$, $\alpha_{kk'}$ are the direction cosines between the macroscopic coordinates of the crystal and the microscopic coordinates of the pth group and $\chi_{i'j'k'}^{(2\omega)}(p)$ is the microscopic second-order susceptibility of this pth group. (ii) The microscopic second-order susceptibility of the basic anionic groups (or molecular structural units) can be calculated from the localized molecular orbitals of these groups (or molecules) by terms of the second-order perturbation theory of the SHG coefficient given by the ABDP theory of Armstrong and co- workers [31] and in Ref. [22].

The next step to reach to our aims is to determine the localized molecular orbitals of the anionic group. Of course, there are many methods available for the calculations of molecular orbitals in our theory, such as the various approximation methods and even the recently developed Dv-Xα method discussed in quantum chemistry. But, in view of the nature of the basic assumptions in our theory, the CNDO approximation seems to be suitable for calculations of SHG coefficients when the anionic groups consist of elements from the first, second and third families in the periodic table. EHMO type approximations are suitable for other elements, particularly if transition metal elements take part in the ionic groups or molecules. It is not necessary to use higher approximations.

Based upon the method of calculation adopted, a complete computer programme consisting of three main parts can easily be written for support of such calculations. The three parts are as follows: (a) the CNDO part or EHMO part with Madelung correction for calculation of the localized electron orbitals in the anionic group; (b) the transition matrix element calculation part; and (c) the second-order susceptibility part for the calculation of the microscopic susceptibility of the anionic group followed by the calculation of the macroscopic SHG coefficients of the crystal.

It is obvious that our 'anionic group theory' can be generalized into an 'NLO-active group theory', thus permitting a straightforward extention to the consideration of discrete uncharged groups (such as urea or substituted benzene) and even cationic groups as basic NLO-active structural units.

II. The Development of New Borates: from BBO to LBO

We now proceed to apply our anionic group theory to a systematic discussion of the NLO effects in borate crystals. The extension of our investigation into the NLO effects of borate crystals has great practical significance in two respects. On the one hand, most borate crystals are transparent far into the intermediate UV region and occasionally even farther because of the large difference in the electronegtivities of the boron and the oxygen atom on the B-O bond. This is one of the most interesting spectral regions in which laser material scientists are looking for new NLO applications. The intrinsic damage threshold of most borate crystals is very high on account of the wide band gaps and the difficulty of ion and electron transport in these compact lattices, even under very intense laser irradiation. On the other hand, most borate crystals can be grown from high temperature melts by top seeding methods, generally resulting in good yields of high optical quality crystals for making NLO devices.

According to the anionic group theory, the second-order susceptibilities in borate crystals should be mainly determined by boron-oxygen anionic groups and their alignment in space. The spherical cations contribute little to the NLO effects. Therefore, before considering the alignment, in order for a borate crystal to have large optical nonlinearities, at least the basic structural units or boron-oxygen groups in the crystals must be capable of exhibiting large microscopic second-order susceptibilities. From this point of view it is logical that, in order to identify and develop new UV NLO crystals among the borate compounds, a very important step is to carry out a systematic classification of the structures of the various kinds of boron-oxygen anionic groups found in borate crystals, and then to calculate the second-order susceptibilities for each of those groups. This step is essential to identify the structural units that are favorable for larger microscopic nonlinearities. Indeed, the structure classification and calculations have helped us to understand why d_{eff} of KB5 is so small and whether there are other boron-oxygen groups that may exhibit larger microscopic second-order susceptibilities.

From 1979 to 1984, the classification and calculations for various known boron-oxygen groups were performed in our research group[27]. Several major boron-oxygen groups, including trigonal group $(BO_3)^{3-}$, tetrahedral anionic group $(BO_4)^{5-}$, planar six-member-ring anionic group $(B_3O_6)^{3-}$, non-planar six-member-ring anionic group $(B_3O_7)^{5-}$, $(B_3O_8)^{7-}$, $(B_3O_9)^{9-}$, and siamese-twinned double six-member-ring anionic group $(B_5O_{10})^{5-}$ and $(B_4O_9)^{6-}$ are shown in Fig. 1-3. Some of the calculated results[27] for the nonlinearities of these anionic groups are listed in Table 1. Their relative orders of microscopic second-order susceptibilities are:

$$X(B_3O_6) \approx X(B_3O_7) > X(BO_3) > X(BO_4)$$

The calculated microscopic second-order susceptibilities listed in table 1 clearly show that

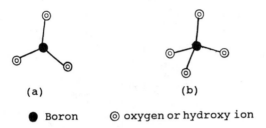

● Boron   ◎ oxygen or hydroxy ion

Fig. 1. The molecular configurations of (a) $(BO_3)^{3-}$, (b) $(BO_4)^{5-}$ groups. (Reprinted with permission from ref. 27. Copyright 1985.)

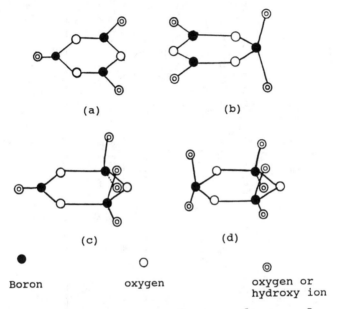

● Boron   ○ oxygen   ◎ oxygen or hydroxy ion

Fig. 2. The molecular configurations of (a) $(B_3O_6)^{3-}$, (b) $(B_3O_7)^{5-}$, (c) $(B_3O_8)^{7-}$, and (d) $(B_3O_9)^{9-}$ groups. (Reprinted with permission from ref. 27. Copyright 1985.)

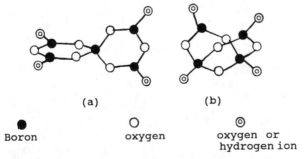

● Boron   ○ oxygen   ◎ oxygen or hydrogen ion

Fig. 3. The molecular configurations of the siamese-twinned double 6-ring $[B_5O_{10}]^{5-}$ (or $[B_5O_6(OH)_4]^-$) and $[B_4O_9]^{6-}$ (or $[B_4O_5(OH)_4]^{2-}$) groups. (Reprinted with permission from ref. 27. Copyright 1985.)

Table 1. The microscopic second-order susceptibilities of $(BO_3)^{3-}$, $(BO_4)^{5-}$, $(B_3O_6)^{3-}$, $(B_3O_7)^{5-}$, and $(B_5O_{10})^{5-}$ anionic groups

(units: 10^{-31} esu, $\lambda = 1.064$ μm)

$(BO_3)^{3-}$		$(BO_4)^{5-}$		$(B_3O_6)^{3-}$		$(B_3O_7)^{5-}$		$(B_5O_{10})^{5-}$	
111	0.641	123	-0.1578	111	2.9323	111	-2.9308	123	-1.335
122	-0.641	113	0.0335	122	-2.9323	122	0.8212	113	0.01732
		223	-0.0329·	133	0.0000	133	-0.6288	223	0.0458
								333	-0.0614

the planar $(B_3O_6)^{3-}$ anionic group with a six-ring conjugated π-orbital system is an ideal structural unit for large NLO effects, provided that the borate crystals would not be crystallographically centrosymmetric. This possibility is particularly attractive in view of the fact that all B-O bonds are capable of transmitting UV light due to the large difference in electronegativity between B and O atoms. These theoretical analyses and the other extensive efforts, including the synthesis of metaborate crystals with $(B_3O_6)^{3-}$ groups as basic structural units, some powder SHG tests, phase equilibrium studies, crystal growth, crystal structure determination, and a series of measurements of various physical properties, have led to the discovery and the establishment of BBO as a high quality NLO material[28].

While BBO crystals have been widely used as good UV NLO crystals for various NLO devices, three disadvantages of this crystal have been recognized in recent years:

(1) The absorption edge of BBO is only at 190 nm. Therefore, even though BBO has a large birefringence, and may be phase-matched down to 200nm, the phase-matching range is limited by the absorption edge.

(2) Its small angular acceptance (1mrad-cm, for SHG, at $\lambda = 1.064$ µm) limits its application in laser systems possessing large divergence and in cases where focussing is needed to increase the power density. Its sensitive angle tuning curve for optical parametric oscillation also limits the spectral stability achievable in that application.

(3) The small Z component of its SHG coefficients severely restricts the use of the BBO crystal at wavelengths under 200 nm and for 90^0 non-critical phase-matching.

If one uses the anionic group theory to analyze these deficiencies in view of the structure of BBO, it is not difficult to understand that the origin of all these disadvantages is the (B_3O_6) planar six-member-ring group itself[29,30].

First, there are two structural factors responsible for the small Z component of SHG coefficients of BBO. One is that $(B_3O_6)^{3-}$ groups do not have any Z component, as shown in the Table 1. The other is that the normal direction of the (B_3O_6) planes is parallel to the Z direction of the BBO lattice, as shown in Fig. 4. Therefore, in order to increase the Z component of SHG coefficients in borate crystals, two possibilities should be considered: (1) to tilt (B_3O_6) planar group relative to Z direction of the lattice; and (2) to select other structural boron-oxygen groups which have large Z component while the planar components are as large as $(B_3O_6)^{3-}$ group. Unfortunately, the orientation of a molecular group in the crystal lattice is not something we can control, and therefore, only the latter would be practical. From Table 1, the best candidate may be the $(B_3O_7)^{5-}$ group[30]. In this group, only one of the boron atoms in the $(B_3O_6)^{3-}$ planar group is changed from trigonal to tetrahedral coordination. As the result, while the χ_{111} and χ_{122} coefficients remain practically unchanged, the χ_{133} becomes numerically somewhat larger.

Secondly, both experimental and calculated results indicated that when cations are either alkali metals or alkaline earth metals the positions of the absorption edges of borate crystals are fully dependent on the anionic groups. In other words, cations contribute little to the band gap of crystals. Calculations which utilize the DV-SCM-Xα method[32], one of the best methods for calculating the electronic structure of clusters or anionic groups in a lattice, show that the absorption edge of the planar six-member-ring $(B_3O_6)^{3-}$ is in the 190-200 nm range. This is determined by the gap between dangling bond or π-conjugated orbitals and excited state anti-π-orbitals. However, the ultraviolet absorption edges of non-planar groups such as the $(B_3O_7)^{5-}$ group shift to shorter wavelengths: nearly 160nm, 30nm shorter than that of $(B_3O_6)^{3-}$ (see Fig. 5). This is because the tetrahedral coordination of boron atoms in non-planar groups destroys the π-conjugated electron system formed in the planar groups.

Finally, the anionic group of the crystals is also found to be useful in evaluating the anisotropy of linear susceptibility of the groups, and even of the crystals as a whole, since the essentially spherically symmetrical cations in the crystals shall only contribute isotropic values for the linear susceptibilities of the crystals, and the birefringence or anisotropy of the crystals mainly come from contribution of the anionic groups. In order to prove this point of view we have preliminary calculated some birefringence of crystals, like NaNO$_2$, BBO and LBO (see Table 2), using the localized molecular orbitals of the anionic groups, and first order perturbation theory of quantum mechanics. The calculated results show that although it is very difficult to reach accurately the absolute values for refractive indexes of the crystals, we have obtained quite close values of the birefringence in comparison with experimental data (also see Table 2). The reason is that using localized molecular orbitals to calculate microscopic linear susceptibilities of anionic groups or cations, you must

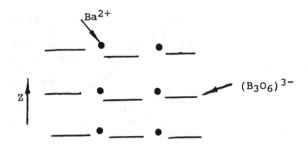

Fig. 4. Schematic drawing of BBO's Lattice net.

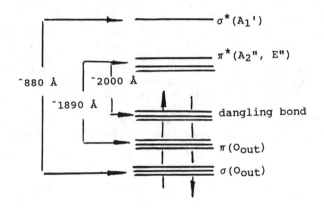

(a)

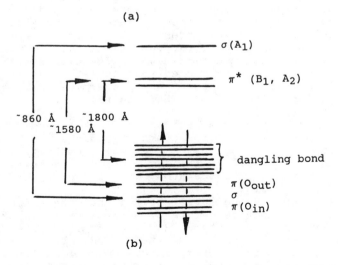

(b)

Fig. 5. Schematic picture of energy levels of (a) $(B_3O_6)^{3-}$, (b) $(B_3O_7)^{5-}$ groups.

face an infinite series, which unlike second-order susceptibility converges very slowly, but the situation is entirely different, when concerning the calculation for birefringence or anisotropy of linear susceptibility of the crystals, the latter is only determined by the frontier molecular orbitals of the anionic groups or molecules (for organic crystals) and the local optical frequency electric field acted on. Therefore, it would be possible to use localized molecular orbital methods to evaluate anisotropy of linear susceptibility for different kinds of crystals. For example, calculations for birefringence of some inorganic NLO crystals (see Table 2) predicted that the crystals constructed from $(B_3O_6)^{3-}$ or $(BO_3)^{3-}$ groups should generally possess a larger birefringence than crystals where basic structural units are $(BO_4)^{5-}$, $(B_3O_7)^{5-}$ group or other borate groups in which one or more boron atoms are tetrahedrally coordinated.

Based on this theoretical work, we predicted that the $(B_3O_7)^{5-}$ group is another ideal basic structural unit for UV NLO crystals which would improve upon the NLO and LO properties of BBO. It is this novel idea that motivated our research group to make extensive efforts which led to the discovery of a new UV NLO crystal - LiB_3O_5 (LBO)[30,34].

III. Measurements and calculations of the SHG coefficients for BBO, LBO and another borate crystals.

In the last section we briefly described how new NLO crystals in the borate were developed through a systematic classification and calculation of microscopic second-order susceptibilities and absorption edges for various kinds of B-O anionic groups. In this section, we will discuss how the anionic group theory is used to calculate and elucidate the NLO coefficients and properties for borate crystals, particularly for BBO and LBO.

1. BBO crystal.

Barium metaborate exists in both α and β phases with the transition temperature at 925±5°C· The α phase is centrosymmetrical and therefore exhibits no NLO response at all. The β phase belongs to a noncentrosymmetric space group and is very useful for UV NLO applications.

In 1969, Hubner[35] reported that the space group of the low-temperature form of barium borate is C2/C, a centrosymmetric structure. However, in 1979, by using a powder second-harmonic generation (SHG) test, my group found that the low-temperature form of barium borate possesses a large NLO coefficient, about six times

larger than that of d_{eff}(KDP). This result implied that
BBO had a noncentrosymmetric structure. The results of
work by various researchers [36,37,39] later on showed
that R3C is most likely the correct space group for
BBO, with cell dimensions a=1.2532 nm, c=1.2712 nm.

There are three non-vanishing SHG coefficients, d_{33},
d_{31} and d_{11}, for R3c. By means of the Maker fringe
technique and the phase-matching method, Chen et al.
[28] determined the values of these SHG coefficients
shown in table 3. The very large anisotropy of the SHG
effect is just as expected, with $d_{33}/d_{11} \approx 0.001$. Li
and Chen [23] made detailed calculations of the SHG
coefficients of BBO and their results are also given in
table 3. It is obvious that the agreement between the
experimental and the calculated values is satisfactory.
Moreover, it confirms our prediction that the planar
$(B_3O_6)^{3-}$ unit with the six-membered-ring conjugated
orbital system is mainly responsible for the large d_{11}
coefficient, whereas the very small d_{31} and d_{33}
coefficients arise mainly from the small deformation of
the π-orbital system due to the presence of an odd-
ordered crystal field along the 3-fold axis, arising
from the spontaneous polarization produced by the
arrangement of the Ba cations around the $(B_3O_6)^{3-}$
anionic groups.

2. LiCdBO3 and YAl3(BO3)4 crystals

We have shown in section II that the $(BO_3)^{3-}$
anionic group is also favourable for the production of
large second-order susceptibilities, although the NLO
effect will be expected to be smaller than that of
$(B_3O_6)^{3-}$. Powdered samples and tiny crystals of LiCdBO3
and YAl3(BO3)4, with the $(BO_3)^{3-}$ anionic group as their
structural unit, have been synthesized and grown in our
Institute. Powder SHG tests have been carried out on
these samples. In the light of our anionic group model,
with the help of the crystal structures determined by
Lutz [39] and Leonyuk and Filmonov [40], it is simple
to calculate their macroscopic SHG coefficients.
Although powder tests can only give rough relative
values for the SHG coefficients, the data [27] in table
4 show satisfactory agreement between theoretical and
experimental results, leading to the following sequence
of SHG coefficients (d_{eff}) for β-BaB2O4, LiCdBO3, and
YAl3(BO3)4: β-BaB2O4 > LiCdBO3 ≈ YAl3(BO3)4.

3. LBO (LiB3O5) crystal.

In section 3, it was shown that when one of the
three trigonally coordinated B atoms in the planar
$(B_3O_6)^{3-}$ anionic group is changed to tetrahedral
coordination, thus forming a B3O7 group, the z
components of the SHG coefficients, e.g. χ_{133} (which
plays an important role in the NLO effect of the UV
spectral range for 90^0 non-critical phase matching)

Table 2. Birefrigence and anionic group structure of $NaNO_2$, BBO and LBO

Crystal	$NaNO_2$ (NO_2)	BBO (B_3O_6)	LBO (B_3O_7)
Calculated	0.1843	0.1020	0.0538
Experimental	0.2422	0.1100	0.0457

Table 3. SHG coefficients of β-BaB_2O_4 crystal (units: 10^{-9} e.s.u. for SHG coefficient; λ= 1.079 μm for fundamental wavelength)

Anionic group	d_{ij}*	Calculated	Experimental	Relative value of $r^{(2\omega)}$ for the powder SHG effect test **
$(B_3O_6)^{3-}$	d_{11}	3.78	$\pm(4.60+0.30)$	$d_{eff} \approx 5\text{-}6\ d_{eff}$(KDP)
	d_{31}	-0.038	$>-(0.07+0.03)d_{11}$	
	d_{32}	-0.038	$= d_{31}$	
	d_{33}	-0.0038	0	

*). $d_{11}= \chi_{III}^{(2\omega)}$, etc.; data quoted from Standards on Piezoelectric Crystals, Proc. IRE, 1949, 37, 1378.

**). Standard sample: KDP powder.

will become larger, whereas χ_{111} or χ_{122} more or less retain the magnitudes found in $(B_3O_6)^{3-}$ (cf. table 1). This important result has been confirmed by our recent work on LBO[30].

LiB$_3$O$_5$ crystallizes in the space group P$_{na2_1}$ [41]. It is built up of a continuous network of endless (B$_3$O$_5$) spiral chains (running parallel to the z axis) formed from B$_3$O$_7$ anionic groups with each of the four exo-ring O atoms shared between the B$_3$O$_7$ groups in the same chain or neighbouring chains, and Li cations located in the interstices. There are five non-vanishing SHG coefficients, d_{33}, d_{31}, d_{32}, d_{15} and d_{24}, for the point group C$_{2v}$. Again with the help of the Maker fringe and phase-matching methods, we have been able to measure all these SHG coefficients, with the results listed in table 5. The macroscopic SHG coefficients of the LiB$_3$O$_5$ crystal have also been calculated on the basis of the anionic group model without any adjustable parameters by using the structural data reported by Konig and A.R.Hoppe[41]. The results are also shown in table 5. The agreement between the calculated and the experimental results is satisfactory. The discovery of this new UV NLO crystal LBO adds convincing support to our conclusion that the anionic group model is indeed a good working model for guiding the search for new NLO borate crystals and crystals of other structure types.

As pointed out by Chen et al. [30], the fact that one of the three trigonal B atoms in the $(B_3O_6)^{3-}$ anionic group has been changed to tetrahedral coordination to form the $(B_3O_7)^{5-}$ group is bound to weaken the conjugated π-orbital system to an appreciable extent and tends to shift the absorption edge to a shorter wavelength in the UV region, in fact down to 160 nm , ca. 30nm shorter than that of BBO, which is useful for applications as a UV NLO material.

4. KB5 (KB$_5$O$_8$ 4H$_2$O or K[B$_5$O$_6$(OH)$_4$]·2H$_2$O

The macroscopic SHG coefficients of the KB5 crystal have been calculated on the basis of our anionic group model [42], assuming that the [B$_5$O$_6$(OH)$_4$]$^-$ group is the primary active group responsible for the production of SHG effects (cf. figure 3 a). The calculated SHG coefficients of this KB5 crystal are shown in table 6. together with the experimental data. It has been pointed out that the largest component of the microscopic second-order susceptibility of B$_5$O$_{10}$ group is χ_{123} , which, unfortunately, does not contribute to the macroscopic SHG coefficients of KB5 since the point group of this crystal is C$_{2v}$, and thus the macroscopic SHG coefficient d_{14} ($=2\chi_{123}^{(2\omega)}$) vanishes identically. This accounts primarily for the fact that the macroscopic

Table 4. SHG coefficients of LiCd(BO$_3$) and YAl$_3$(BO$_3$)$_4$ crystals (λ=1.079um for the fundamental wavelength)

Anionic group	Crystal	Calculated SHG coefficient* (units:10^{-9} e.s.u.)		Relative value of I(2ω) from powder SHG effect test**
(BO$_3$)$^{3-}$	LiCd(BO$_3$)	d$_{11}$	2.7039 to 1.740	d$_{eff}$ 3.0 d$_{eff}$(KDP)***
		d$_{16}$	4.2144 to 2.715	
		d$_{12}$	- 2.7093 to -1.742	
		d$_{22}$	- 4.2144 to -2.715	
	YAl(BO$_3$)$_4$	d$_{11}$	3.4000 to 2.200	d$_{eff}$ 2.5 d$_{eff}$(KDP)
		d$_{12}$	- 3.4100 to -2.200	

*). The range of the calculated SHG coefficients is determined by the range of the strength of the odd-ordered crystalline field in the crystal.

**). Standard sample: KDP powder.

***). d$_{eff}$(KDP) = 0.66d$_{36}$(KDP) = 0.72x10^{-9} e.s.u. (d$_{36}$(KDP) is taken to be 1.1x10^{-9} e.s.u.)

Table 5. SHG coefficients of LiB$_3$O$_5$ crystal (units:10^{-9} e.s.u., λ=1.079 μm)*

	d$_{33}$	d$_{31}$	d$_{32}$	d$_{15}$	d$_{24}$
Calculated	0.61	-2.24	2.69	=d$_{31}$	=d$_{32}$
Experimental	±0.15(1+0.1)	∓2.75(1+0.12)	±2.97(1+0.01)	≈d$_{31}$	≈d$_{32}$

*). d$_{36}$(KDP)=1.1x10^{-9} e.s.u.

Table 6. The SHG coefficient of KB5 crystal (units: 10^{-9} esu for SHG coeff.; λ=1.064 μm for fundamental wavelength)

Anionic group	d_{ij}	Calculated	Experimental(1)	Experimental(2)*
	d_{31}	2.61	1.09	(-)2.53**
$[B_5O_6(OH)_4]^{-1}$	d_{32}	0.07	0.08	1.04
	d_{33}	3.26		2.88

* The second set of experimental measurements was made by the Institute of Crystalline Materials, Shangdong University (private communications).
** It is supposed to be uncertain.

SHG coefficients of KB5 amount only to one tenth that of KDP. In case there exists a crystal consisting of the same basic structural unit $[B_5O_6(OH)_4]^-$ but crystallizing in point group either C_2 or D_2, the component χ_{123} of the microscopic second-order susceptibility will make its contribution to the overall SHG effect of the crystal and exhibit as large an overall effect as half of that of KDP. This is left for further consideration.

Acknowledgments

    This work was supported by the Science Fund No. 1860823 of the national science fund committee. The author is very grateful to Wu Yicheng, Li Rukang and Lin Guomin for their assistance with the preparation of the manuscript.

References
[1] D.H. Auston, et al, Appl. Opt., 26 (1987), 211.
[2] N. Bloembergen, "Nonlinear Optics", pp. 5-8 (Benjamin, New York, 1965).
[3] S.K. Kurtz and F.N.H.Robinson, Appl. Phys. Lett, 10 (1967), 62.
[4] C.G.B.Garrett and F.N.H.Robinson, IEEE J. Quant. Elect., 2 (1966),328; C.G.B. Garrett, IEEE J. Quant. Elect., 4 (1968), 70
[5] C.R.Jeggo and G.D. Boyd, J. Appl. Phys., 41 (1974), 2471.
[6] J.G.Bergman and G.R.Crane, J. Chem. Phys., 60 (1974), 2470; B.C.Tofield, G.R.Crane and J.G. Bergman, Trans. Faraday Soc., 270 (1974), 1488.
[7] B.F. Levine, Phys. Rev. Lett., 22 (1969), 787; 25 (1970), 440.

[8] B.F.Levine, Phys. Rev.,   B7 (1973), 2600.
[9] Y. Fujii and T. Sakudo, ibid., B13 (1976), 1161.
[10] B.F.Levine,   ibid., B13 (1976), 5102.
[11] B.L.Davydov, L.D.DerKacheva, V.V.Duna,
     M.E.Zhabotinskii,
     V.F.Zolin, et al., Sov. Phys. JETP Lett., 12
     (1970), 16.
[12] D.S.Chemla, D.J.L.Oudar, and J.Jerphagnon,
                     Phys.Rev., B12 (1975), 4534.
[13] J.L.Oudar and D.S.Chemla, Opt. Commun, 13 (1975),
     164.
[14] J.L.Oudar and H.Leperson, ibid., 15 (1975),256.
[15] J.Zyss, D.S.Chemla, and J.F.Nicoud,
                     J.Chem. Phys., 74 (1981), 4800.
[16] J.Zyss, J.F.Nicoud and M.Coquillay
                     J.Chem. Phys. 81 (1984), 4160.
[17] C.C.Frazier, M.P.Cockererham, E.A.Chauchard and
     C.H.Lee
                     J. Opt. Soc. Am. B4 (1977), 1899.
[18] J.C.Baumert, R.J.Twieg, G.C.Bjorklund, J.A.Logan,
     and C.W.Dirk
                     Appl. Phys. Lett. 51 (1987), 1484.
[19] M.DiDomenico Jr., S.H.Wemple
                     J.Appl. Phys. 40 (1969), 720.
     S.H.Wemple, M.DiDomenico Jr.
                     J. Appl. Phys. 40 (1969), 735.
     S.H.Wemple, M.DiDomenico Jr., I. Camlibel,
                     Appl. Phys. Lett. 12 (1968), 209.
[20] C.T.Chen   Acta Phys. Sin., 25 (1976), 146.
                                   (in Chinese).
[21] C.T.Chen   Scientica Sin., 22 (1979), 756.
[22] C.T.Chen, Z.P.Liu and H.S.Shen
                     Acta Phys. Sin. 30 (1981), 715
                                   (in Chinese).
[23] R.K.Li, and C.T.Chen
                     Acta Phys. Sin., 34 (1985), 823
                                   (in Chinese).

[24] C.T.Chen   Acta Phys. Sin. 26 (1977), 486
                                   (in Chinese).
[25] C.T.Chen
     Commun. Fujian Inst. Struct. Matter. No. 2
                     (1979), 1. (in Chinese).
[26] C.T.Chen   Acta Phys. Sin. 26 (1977), 124
                                   (in Chinese).
[27] C.T.Chen, Y.C.Wu, and R.K.Li,
                     Chinese Phys.Lett. 2 (1985), 389
[28] C.T.Chen, B.C.Wu, A.Jiang and G.M.You
                     Scientia Sin. B18 (1985), 235.
[29] C.T.Chen   Laser Focus World Nov.(1989),129
[30] C.T.Chen, Y.C.Wu, A.Jiang, B.C.Wu, G.M.You
     and R.K.Li, S.J.Lin
                     J. Opt. Soc. Am. B6 (1989), 616.

[31] J.A.Armstrong, N.Blombergen, J.Ducuing and
P.S.Pershan
Phys. Rev. 127 (1962), 1918
[32] H.Sambe and R.H.Felton
J. Chem. Phys. 62 (3) (1975), 1122
B.Delley and D.E.Ellis
J.Chem. Phys. 76 (4) (1982), 1949
[33] J.Huang
"Calculations of birefringences of the crystals
using anionic group theory",
M.S.Treatise, Sep. 1987,
Fujian Institute of Research on the
Structure of Matter (in Chinese)
[34] S.J.Lin, Z.Y.Sun, B.C.Wu and C.T.Chen
J. Appl. Phys. 67 (1990) 634.
[35] K.H.Hubner, Neues Jahrb. Mineral, Monatsh,
(1969), 335.
[36] S.F.Lu, M.Y.Ho and J.L.Huang,
Acta Phys. Sci. 31 (1982), 948
[37] J.Liebertz and S.Stahr, Z.Kristallogr.,
165 (1983), 91.
[38] R.Frohlich, Z.Kristallogr.,
168 (1984), 109.
[39] F.Lutz,
Recent Dev. Condens. Matter Phys. Ist.,
3 (1983), 339
[40] N.I.Leonyuk and A.A.Flimonov
Krist. Tech., 9 (1974), 63.
[41] H.Konig and A.Hoppe,
Z. anorg. allg. chem. 439 (1978), 71.
[42] Y.C.Wu and C.T.Chen
Acta Phys. Sin. 35 (1986), 1 (in Chinese)

RECEIVED August 13, 1990

Chapter 25

Defect Chemistry of Nonlinear Optical Oxide Crystals

Patricia A. Morris

Central Research and Development Department, E. I. du Pont
de Nemours and Company, Wilmington, DE 19880–0306

The defect chemistry of a specific crystal is determined by both its structural characteristics and the growth, or processing, of the material. Structurally, the nonlinear optical oxides contain anionic oxide groups (i.e. TiO_6, NbO_6, PO_4, AsO_4, B_3O_6, B_3O_7) which are the basic structural units responsible for the second order nonlinear optical susceptibility. The relatively large contribution of covalent bonding in the anionic groups to the total lattice energy appears to allow the structures to accommodate nonstoichiometric defects on the other, more "ionic", cation sublattice or sublattices with a relatively small cost of energy. This enhances the incorporation of many isovalent and aliovalent impurities into the crystals. The nonlinear optical oxide crystals recently developed are grown by flux or solution techniques to prevent decomposition or to obtain a low temperature phase. The intrinsic nonstoichiometry and the impurity contents of the as-grown crystals are determined by the solutions and temperatures used for growth. Recent work on the defects present in KTP, KTA, BBO and LBO crystals shows that the intrinsic defect concentrations in these materials are relatively low, compared to the more traditional nonlinear optical oxides having the perovskite, perovskite-like and tungsten bronze type structures. As a result, their defect structures can be dominated by impurities present at relatively small concentrations. The defect chemistry of nonlinear optical oxide crystals can affect many of the materials' properties required for device applications and several examples are described.

The defect chemistry of nonlinear optical oxide crystals can affect many of the materials' properties required for device applications. Applications of these crystals, having high second order nonlinear optical susceptibilities ($\chi^{(2)}$), include frequency convertors for laser systems, electro-optic modulators and switches, and holographic and phase conjugate optics.(1-5) The materials' requirements for device applications include: 1) large $\chi^{(2)}$, 2) optical transparency in the wavelength range of interest, 3) low ionic and electrical conductivity for photorefractive, electro-optic and waveguide devices, 4) high optical damage threshold in frequency generation and Q-switching applications, 5) high sensitivity and fast response time of the photorefractive effect for

0097–6156/91/0455–0380$06.00/0

holographic and phase conjugate optics and 6) homogeneity with respect to the optical properties and conductivity. Both intrinsic (i.e. nonstoichiometry) and extrinsic (i.e. impurities) defects may be present in nonlinear optical oxide crystals which affect the materials' properties of interest. The defect chemistry of a specific crystal is determined by both its structural characteristics and the growth, or processing, of the material.

The purpose of this paper is to summarize the current understanding of the defect chemistry of nonlinear optical oxide crystals and specifically the relationship of the defects present to 1) the structure and growth, or processing, of the material and 2) the properties of interest for device applications. The defects in traditional nonlinear optical oxide crystals (i.e. $BaTiO_3$, $LiNbO_3$, $Sr_{1-x}Ba_xNb_2O_6$, $Ba_2NaNb_5O_{15}$, $K_3Li_2Nb_5O_{15}$) are reviewed. Our recent work on the defect chemistry of new nonlinear optical oxide crystals (i.e. $KTiOPO_4$, $KTiOAsO_4$, β-BaB_2O_4, LiB_3O_5) is then discussed.

Induced Polarization and Origin of the Second Order Nonlinear Susceptibility

A polarization is induced in a material when subjected to laser radiation or dc electric fields. The following expression (1-3),

$$P_i (\omega) = \mathcal{E}_0 \left[\chi_{ij}{}^{(1)} (\omega) E_j (\omega) + 2 \chi_{ijk}{}^{(2)} (\omega = \omega_1 + \omega_2) E_j (\omega_1) E_k (\omega_2) \right]$$

for the induced polarization (P_i) includes the first two terms in the series, where E is the electric field strength associated with the incident radiation or dc electric field, and ω is the frequency. The first term describes the linear optical effects: absorption, refraction, emission and reflection. The second term is responsible for the second order nonlinear polarization processes, such as second harmonic generation, parametric sum or difference mixing and the linear electro-optic or Pockels effect. Second harmonic generation and parametric generation are typically used to extend the frequency range of solid state lasers. The Pockels electro-optic effect is used in applications involving Q-switches for laser systems, optical modulators and switches, and the photorefractive effect for real-time holography and phase conjugation.

The basic structural units responsible for the second order nonlinear optical susceptibility in most oxide crystals are the acentric anionic groups. (4,6) The macroscopic $\chi^{(2)}$ is determined by the microscopic nonlinear susceptibility of the bonds in the acentric oxide group, the number and orientation of equivalent groups in a unit cell, and the number of unit cells per unit volume. The following are the acentric oxide groups contributing to $\chi^{(2)}$ in the nonlinear crystals discussed here: 1) MO_6, where typically M = Ti, Nb; $BaTiO_3$, $LiNbO_3$, $Sr_{1-x}Ba_xNb_2O_6$, $Ba_2NaNb_5O_{15}$, $K_3Li_2Nb_5O_{15}$, $KTiOPO_4$, $KTiOAsO_4$, 2) PO_4 & AsO_4; $KTiOPO_4$ and $KTiOAsO_4$, 3) B_3O_6; β-BaB_2O_4, 4) B_3O_7; LiB_3O_5.

Defect Chemistry : Structures, Growth and Properties

Review of Perovskite, Perovskite-like and Tungsten Bronze Type Crystals. Many of the traditional nonlinear optical oxide crystals containing MO_6 anionic groups have either the perovskite (e.g. $BaTiO_3$), perovskite-like (e.g. $LiNbO_3$), or tungsten-bronze (e.g. $Sr_{1-x}Ba_xNb_2O_6$, $Ba_2NaNb_5O_{15}$, $K_3Li_2Nb_5O_{15}$) type structures. The intrinsic defects present in typical as-grown crystals of these materials are shown in

Table I. An extensive amount of work has been done to investigate the defect structures of many of these materials and a full review of these results is not intended here. Only generalizations which are useful for understanding these materials and their defect structures are discussed.

Nonstoichiometry in nonlinear optical oxide crystals having these structures exists over the range of a few percent (e.g. $BaTiO_3$, $LiNbO_3$) or more (e.g. $Sr_{1-x}Ba_xNb_2O_6$). (7-17) The nonstoichiometry in these materials occurs at least in part due to the presence of the MO_6 anionic groups, which are also primarily responsible for the second order nonlinear optical susceptibility. As shown in Table I, as-grown crystals are typically nonstoichiometric with an excess of the M cation present. Nonstoichiometric crystals are grown using the Czochralski technique if congruently melting compositions are chosen (e.g. $Li_{(0.964)}NbO_3$, $Ba_2Na_{(0.72)}Nb_5O_{15}$). (10,18,15) When a congruently melting composition exists, it is typically M-rich, showing that the free energy curve is asymmetric and skewed to lower energies toward the nonstoichiometric M-rich phase. Many crystals require flux growth to obtain a low temperature phase (e.g. $BaTiO_3$ (19)) or prevent decomposition during melting and these crystals are also typically M-rich. The relatively large contribution of covalent bonding in the MO_6 anionic groups to the total lattice energy appears to allow the structures to accommodate the nonstoichiometric defects occurring on the other, more "ionic" cation sublattice or sublattices with a relatively small cost of energy.(10)

The crystal structures of these nonlinear optical oxides also have a relatively large degree of tolerance, especially with respect to the occupancy of the non-M, or non-anionic group, cation sublattices.(16,20) The tolerance of these crystal structures and the presence of nonstoichiometry in the crystals allows the formation of solid solutions with a variety of ions or ion pairs. (Table II). Impurity substitution in a crystal lattice depends on the size, charge misfit, bonding and coordination preference of the impurity.(24) Isovalent impurity substitution is relatively easy for many ions due to the tolerance typically found in these structures.(20,25) Aliovalent impurities are more readily incorporated in the lattice in the presence of charged nonstoichiometric defects, which can compensate the charge differences. The distribution coefficients of impurity ions, that can easily be accommodated in the lattices, are relatively high and these impurities are incorporated into the crystals during growth. For example, transition metals and rare earths can be incorporated during the growth of congruent $Li_{(0.964)}NbO_3$ crystals in amounts of up to several mole per cent and about 1 mole per cent, respectively.(22)

Both the intrinsic and extrinsic defects in these materials can affect the properties of interest for applications. An example of this is the observed decrease in the damage susceptibility or photorefractivity of $LiNbO_3$ with additions of H, Li or Mg to Li-deficient congruently grown crystals (Table I).(26) The additions produce a reduction in the octahedral vacancy concentrations in the crystals. Therefore, at least some photorefractive optical damage in $LiNbO_3$ is believed to be related to the concentration of octahedral vacancies present in crystals.

KTiOPO4 (KTP). KTP is a relatively new nonlinear optical material which has a unique combination of properties making it superior to many traditional materials for second harmonic generation and electro-optic applications. (27) The crystal structure is orthorhombic with the space group $Pna2_1$.(28,29) The framework is characterized by TiO_6 octahedra in chains oriented along the [011] and [0$\bar{1}$1] directions, linked together by PO_4 bridges. The K ions are located in two sites relative to the two-fold screw axis, along the z-axis. These positions along the z-axis provide "channels" for a high 1-dimensional K ion conductivity.(30,31) Hydrothermal and flux techniques are

Table I.   Intrinsic Defect Mechanisms in Nonlinear Optical Oxide Crystals

Crystal	Nonstoichiometry*	Defects**	References
$BaTiO_3$	$Ba_{(1-x)}TiO_{(3-x)} : x \cong 0.01$	V_{Ba}'', $V_O^{\cdot\cdot}$	(7-9)
$LiNbO_3$	$Li_{(1-x)}NbO_3 : x = 0.036$	$Nb_{Li}^{\cdot\cdot\cdot\cdot}$, $V_{Nb}^{/////}$	(10,11)
$Sr_{(1-x)}Ba_xNb_2O_6$	$Sr_{(1-x)}Ba_xNb_2O_6 : x = 0.39$		(12,13)
$Ba_2NaNb_5O_{15}$	$Ba_2Na_{0.72}Nb_5O_{15}$	$V_{Na}^{/}$	(14,15)
$K_3Li_2Nb_5O_{15}$	$K_{2.786}Li_{1.989}Nb_5O_{15}$	$V_K^{/}$, $V_{Li}^{/}$	(16,17)
$KTiOPO_4$	$K_{(1-x)}TiPO_{(5-x/2)} : x \cong 0.0005^+$	$V_K^{/}$, $V_O^{\cdot\cdot}$	(30,31)
$KTiOAsO_4^{++}$		$V_K^{/}$, $V_O^{\cdot\cdot}$, $\{As_{Ti}^{/}\}$	
β-$Ba_2BO_4^{++}$		$\{V_{Ba}''\}$, $\{V_O^{\cdot\cdot}\}$	
$LiB_3O_5^{++}$		$\{V_{Li}^{/}\}$, $\{V_O^{\cdot\cdot}\}$	

*   Representative of the range of intrinsic nonstoichiometry in as-grown crystals.
** The defects are presented as being fully ionized.
\+   Represents typical flux grown crystals.
\+\+ Defect structure in crystals presently grown are thought to be extrinsically controlled.
{ } Presence of this defect is suspected; insufficient data exists for confirmation.
Example: $V_K^{/}$ is a vacant potassium site with an effective negative one charge.

Example: $Nb_{Li}^{\cdot\cdot\cdot\cdot}$ is a niobium on a lithium site with an effective positive four charge.

presently used to grow KTP because the crystal decomposes upon melting.(32) Much work has been done to understand the defects in KTP crystals in the past two years and the results will be summarized here. For further discussion see references.(30,31,33)

The intrinsic defects present in KTP crystals are vacant potassium (V_K) and vacant oxygen (V_O) sites. This results in a very limited range of nonstoichiometry, relative to that observed in traditional nonlinear optical oxides (Table I). This mechanism of intrinsic defect formation in KTP has been confirmed by mass spectroscopic analysis of the gases evolved from typical KTP crystals. The intrinsic defect concentrations are dominant in typical KTP crystals grown by the flux technique and are very temperature dependent over the range of temperatures practical for flux growth. The calculated defect formation energy, using the bulk ionic conductivities of crystals grown by the flux technique over a range of temperatures (Figure 1) is approximately 5 eV per defect.

Protons are the dominant defects in hydrothermally grown KTP and are most likely the primary defect contributing to the formation of V_K sites in these crystals. Protons are present in KTP grown by both the flux and hydrothermal techniques and are considered to be present in the form of OH⁻. The relative amounts of OH⁻ found in KTP crystals as a function of the growth technique are high temperature hydrothermal > low temperature hydrothermal > flux, but all concentrations are within the same order of magnitude, estimated to be in the range of a hundred ppm.(33,34) Protons present as OH⁻ in KTP can be distributed in multiple sites on each of the eight inequivalent oxygen sites in the unit cell. The distribution of OH⁻ sites in KTP is a function of the technique and conditions used for growth (i.e. activity of H_2O and the K/P ratio, or effective pH, in the solution or flux, growth temperature, etc.). Protons in KTP can be charge compensated by the formation of $V_K^/$ or $Ti_{Ti}^/$ sites in the crystal. The compensating defect formed is believed to depend on the location of the oxygen site where the OH⁻ is present.

Several isovalent ions form solid solutions with KTP (Table II), showing that this structure is relatively tolerant, with respect to isovalent impurities, as are the traditional nonlinear optical oxide crystal structures. But due to the relatively limited range of nonstoichiometry in KTP, aliovalent impurities, such as divalent Ba, Sr and Ca introduced through ion exchange in nitrate melts, which substitute on the K site, are incorporated at concentrations less than one mole percent.(36) Typical impurity concentrations present in flux and hydrothermally grown KTP are shown in Table III.

Control of the ionic conductivity of KTP is important in both the processing of optical waveguides and to electro-optic waveguide device stability.(27) A moderate ionic conductivity is necessary to form waveguides in the material, but if excessive, the mode distribution in the waveguides can be altered during device processing. As discussed above, the intrinsic V_K defect concentration is primarily responsible for the potassium ion conductivity of typical flux grown KTP crystals. Protons present in specific sites in hydrothermal KTP are compensated by V_K sites, increasing the ionic conductivity above that expected due to the intrinsic V_K defect concentration formed at the growth temperatures involved. This is shown in Figure 1. The line drawn through the data for KTP crystals grown over different temperature ranges by the flux technique represents the ionic conductivity in KTP dominated by the intrinsic V_K defect concentrations. It is clear from the data for the high and low temperature hydrothermally grown KTP that some other defect mechanism is contributing to the ionic conductivity of these materials. The variations in ionic conductivities observed cannot be explained as due to conventional impurities. The concentrations of cation and anion impurities in flux and hydrothermal crystals of KTP, shown in Table III, are indistinguishable within the precision of techniques used. (The impurity concentrations in low temperature hydrothermally grown KTP are comparable to those grown using

Table II. Ions Forming Solid Solutions in Several Nonlinear Optical Oxide Crystals

Crystal	Solid Solution Forming Ions or Ion Pairs	References
$BaTiO_3$	Sr^{2+}, Pb^{2+}, $(Na^{1+}+Nb^{5+})$, $(La^{3+}+In^{3+})$, Ce^{3+}, Ca^{2+}, Si^{4+}, Zr^{4+}, Ge^{4+}, La^{3+}	(21)
$LiNbO_3$	Ta^{5+}, $(Mg^{2+}+Ti^{4+})$	(21)
	Mg^{2+}, Co^{2+}, Zn^{2+}, Cr^{3+}, Sc^{3+}, Sn^{4+}	(22)
$Ba_2NaNb_5O_{15}$	La^{3+}, Ti^{4+}, W^{6+}	(21)
	Li^{1+}, K^{1+}, Sr^{2+}, Ca^{2+}, Ta^{5+}	(23)
$KTiOPO_4$	Rb^{1+}, Tl^{1+}, NH_4^{1+}, As^{5+},	(32)
	Na^{1+}, Ag^{1+}, Cs^{1+}	(35)

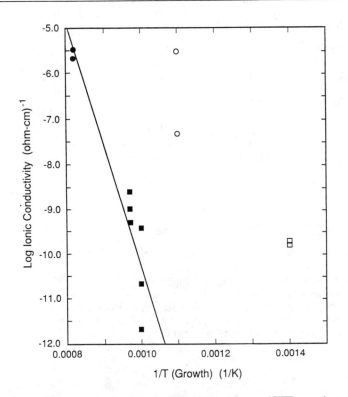

Figure 1. The room temperature bulk ionic conductivity, along z, of KTP crystals as a function of the reciprocal of their midpoint growth temperatures. ●, Philips flux; ■, DuPont flux; ○, Airtron high temperature hydrothermal; □, Airtron low temperature hydrothermal.

Table III. Typical Impurity Content of KTiOPO4 and KTiOAsO4 Crystals Grown by
the Flux and Hydrothermal Techniques
(Parts Per Million by Weight)

Element	Flux KTP	High Temperature Hydrothermal KTP	Flux KTA
B	0.05	0.1	4
F	0.5		2
Na	(8.3)	(5.1)	<80
Mg	(<1.0)	(<0.5)	<80
Al	3	2	35
Si	25	10	100
P	M	M	200
S	<3	<3	3
Cl	3	3	10
Ca	(4)	(2.3)	10
V	0.6	≤0.4	1
Cr	6	0.4	0.2
Mn	≤0.4	≤0.4	0.2
Fe	5	4	20
Co		≤1	0.4
Ni	<5	<5	1
Cu	3	1	0.6
As	2		M
Sr	≤1	≤1	
Y	0.2	0.2	
Zr	1	0.7	
Nb		≤4	6
Sn	≤0.2	0.5	
Sb	10	1	20
Ba		≤0.6	
Pt		2	
Total	82	48	414

Analyses done by spark source mass spectrography.
( ) Analysis of element done by atomic absorption.
"Blank" means that the element was below detection limits of 0.02 or 0.05 parts per
million (atomic) for KTP and KTA, respectively.
M means that the element is a major component.

the high temperature technique.) The V_K carrier concentrations calculated for high temperature hydrothermal KTP crystals are consistent with the estimated OH⁻ concentrations in these crystals and an increased ionic conductivity is correlated to the presence of specific OH⁻ sites. The measured activation energies of conduction for flux and high temperature hydrothermal KTP are 0.5 and 0.3 eV, respectively (Figure 2), indicating different mechanisms in the two types of KTP crystals. The 0.5 eV activation energy for flux grown KTP is believed to be composed of the energy of migration for potassium ions and a (V_O-V_K) defect dissociation energy. The 0.3 eV activation energy for high temperature hydrothermal KTP is believed to be composed of the energy of migration for potassium ions, an energy associated with proton motion between sites and the energy of dissociation of a V_K from the OH⁻ defect sites. Work is continuing to determine the specific mechanisms involved in the ionic conductivity in KTP grown by the flux and hydrothermal techniques. While it has been shown that the typical impurities present in KTP (Table III) are not responsible for the large variation in ionic conductivity of crystals grown by the flux and hydrothermal techniques, the influence on the ionic conductivity of divalent Ba impurities introduced by ion exchange can be dramatic. Ba at a concentration of 0.7 mole percent increases the ionic conductivity of KTP by several orders of magnitude. The observed increase is attributed to Ba^{2+} compensation in the lattice by $V_K{}^/$.(36)

Optical damage thresholds for KTP are reported to be 1 - 30 GW/cm^2.(27) Photochromic damage is observed as the creation of "gray tracks", which are dichroic and preferentially absorb both fundamental and second harmonic generated light with their electric vectors along the z-direction.(37) Damage with similar optical absorption properties can be created by electric fields or hydrogen annealing and is attributed to $Ti_{Ti}{}^/$, or Ti^{3+}, formation in the crystal. A correlation is observed between electric field induced damage and ionic conductivity indicating that the primary defects responsible for the ionic conductivity in KTP, described above, contribute to the damage process and are related to the concentrations of $Ti_{Ti}{}^/$, or Ti^{3+}, formed in the crystals. Further work is being done to determine the specific mechanisms of damage in flux and hydrothermally grown KTP crystals.

<u>KTiOAsO4 (KTA)</u>. Recent work on KTA shows a significant enhancement of the nonlinear optic and electro-optic coefficients versus those of KTP.(38) KTA is isostructural to KTP (39) and is also a 1-dimensional potassium ion conductor in the z-direction. KTA crystals have been grown using the flux technique.

The ionic conductivity of KTA crystals examined is typically approximately one order of magnitude less than KTP crystals grown by the flux technique over a similar temperature range. The activation energy of ionic conduction is the same as in flux grown KTP (0.5 eV), shown in Figure 2, indicating that the same mechanism is dominant in both crystals. Therefore, the V_K carrier concentration in KTA is approximately an order of magnitude lower than in KTP. The impurity concentrations in KTA crystals examined are shown in Table III. These crystals are less pure than typical KTP crystals and contain impurity ions (Al, Si) that, assuming the intrinsic V_K and V_O defect mechanism (Table I), could essentially lower the observed ionic conductivity by the amount observed. These impurities are present in the arsenate precursors used for growth. Further work is necessary to clarify the intrinsic defect structure of KTA. The intrinsic defects in KTA are presumed to be similar to KTP, but the range of intrinsic nonstoichiometry ($[V_K]$ and $[V_O]$) in KTA and the possibility of $As_{Ti}{}^/$ defects (Table I) should be investigated. The issues regarding the control of the ionic conductivity and damage in KTA are similar to those in KTP.

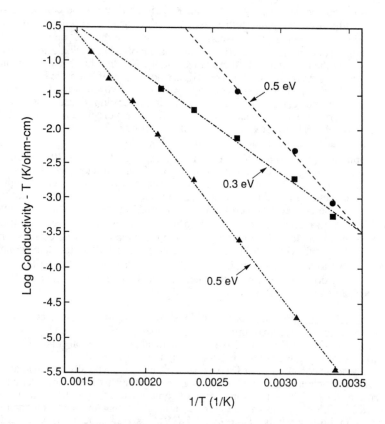

Figure 2.   Conductivity data on KTP and KTA, along z, as a function of temperature for determination of the activation energies of conduction. Key: ● KTP Philips flux; ■ KTP Airtron high temperature hydrothermal; ▲ KTA DuPont flux.

β-BaB$_2$O$_4$ (BBO). BBO is a crystal used for harmonic generation in the ultra-violet (UV) wavelength range.(4) The low temperature form, which exists below 920 °C, has a hexagonal crystal structure with the space group R3c.(40,41) It consists of nearly planar anionic B$_3$O$_6$ groups perpendicular to the polar axis. These groups are bonded ionically through the Ba ions.(41) Crystals of the low temperature phase are grown more readily using the flux technique from a solution containing BaB$_2$O$_4$: BaB$_2$O$_4$· Na$_2$O.(42) The defect structure of this material has not previously been studied. The dielectric constants have been measured and are reported to be 8.1-8.2 (c) and 6.5-6.7 (a) at low frequencies over the temperature range of -200 to 50 °C.(41,43) The loss tangent is reported to be < 0.001 for both orientations over this frequency and temperature range.

Recently, we have measured the dielectric properties of BBO from 10 Hz - 100 KHz over the temperature range of 22 - 500 °C. At room temperature the dielectric constants and loss measured in the a and c-directions were consistent with those previously reported. The b-direction was also measured and the loss was ~ 0.01 at 1 KHz. The c and b-directions were measured as a function of temperature. In both directions the loss increases with temperature and is significant at temperatures above 300 °C. The activation energies calculated are approximately 1 eV and 0.6 eV for the c and b-directions, respectively (Figure 3), using data at 1 KHz. An intrinsic defect structure consisting of V$_{Ba}$ and V$_O$ sites is suspected from comparison to the more fully studied, traditional nonlinear optical oxide crystals and KTP, described above. In these crystals the presence of the strongly covalent anionic groups in the structures would also presumably result in defect formation on the more "ionic" cation sublattice. Calculated carrier concentrations in BBO are on the order of 0.1 - 10 ppm (atomic). The impurity concentrations in the crystal examined are shown in Table IV. Considering the effect on conductivity, the impurities present are on the same level as the calculated carrier concentrations. Therefore, the loss mechanism is most likely extrinsically controlled in these crystals. The defect concentrations calculated are very low so only a very limited range of nonstoichiometry appears to exist in crystals of BBO. Samples of BBO annealed at 870 °C (approximate midpoint growth temperature) for 70 hours in air had comparable properties to the as-grown crystal indicating that the defect structure is in equilibrium at this temperature and is consistent with a mechanism of extrinsic control of the defect structure in the BBO crystals presently grown.

To use this material with high efficiency in the UV range the transmittance from 200 to 350 nm should be high. The UV intrinsic absorption edge at room temperature is 190 nm.(44) Investigation of the temperature dependence of the absorption edge of a typical BBO crystal from -200 to 27 °C shows that three absorption steps appear at about 10 °C and increase rapidly up to saturation with decreasing temperature.(44) The transmittance of a 1 cm thick crystal at 222 nm falls below 22% at 0 °C. This restricts the temperature range of use in the UV of BBO for practical applications. The low-temperature absorption steps are attributed to energy states induced by impurity centers in the crystal, but the specific impurities responsible were not reported. This indicates that the impurities present in BBO crystals presently grown are not only controlling the defect structure of the material, but the optical transparency as well.

LiB$_3$O$_5$ (LBO). The crystal structure of LBO has a continuous network of B$_3$O$_7$ anionic groups sharing one oxygen, which form chains along the z-axis. These chains are interlocked to form a three-dimensional orthorhombic structure with the space group Pna2$_1$.(45) The Li ions are located in the interstices. LBO has been developed to overcome the limitations of BBO crystals.(46) By having ōne of the B atoms in tetrahedral coordination, the crystal contains z-components of the nonlinear optical

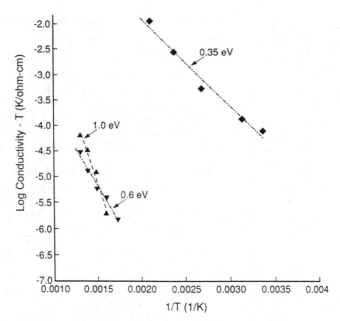

Figure 3.   Conductivity data on BBO and LBO as a function of temperature for determination of the activation energies of conduction. Key: ▲   BBO (c); ▼   BBO (b); ◆ LBO (z). Crystals grown by the flux technique at the Fujian Institute of Research on the Structure of Matter, China.

Table IV. Impurity Content in β-BaB$_2$O$_4$ and LiB$_3$O$_5$ Crystals
(Parts Per Million by Weight)

Element	BBO	LBO
F	2	4
Na	40	30
Mg	8	≤15
Al	30	55
Si	40	80
P	0.2	0.4
S	3	45
Cl	6	250
K	0.7	5
Ca	3	8
Cr	<0.9	<0.7
Fe	0.6	8
Ni	≤0.08	≤0.8
As		15
Ce	10	
Total	149	517

Analyses done by spark source mass spectrography.
"Blank" means that the element was below detection limits of 0.05 or 0.2 parts per million (atomic) for BBO and LBO, respectively.

susceptibility and the absorption edge moves to shorter wavelengths. LBO melts incongruently and is grown using a B_2O_3-rich flux, plus an additive to lower the viscosity of the solution.(47) The defect structure of this material has not been previously studied.

We have measured the dielectric properties of LBO crystals along the x, y and z axes from 10 Hz to 100 KHz over a temperature range of 22 - 500 °C to investigate the defect structure of this material. At room temperature and 1 KHz the dielectric constants and loss are: approximately 10 and 0.1 (x) and 7 and 0.05 (y), respectively. Along the z-direction at room temperature, the conductivity of various pieces of crystal ranged from 10^{-7} to 10^{-6} (ohm-cm)$^{-1}$. The y and z-orientations were measured as a function of temperature. The loss along y is significant above approximately 350 °C, but remains much lower than in the z-direction. The activation energy for conduction along z, using bulk conductivities, is 0.35 eV (Figure 3). An intrinsic defect mechanism consisting of V_{Li} and V_O is suspected, again based on comparison to other well studied nonlinear optical oxides. This crystal is grown from a B_2O_3-rich solution, so the crystalline compound grown would be on the Li-deficient side of the stoichiometric compound. The calculated carrier concentration, using a conducting mechanism involving Li ions moving through the interstices along the B_3O_7 chains in the structure, is several hundred parts per million (atomic). This is similar in magnitude to the concentration of V_{Li} sites expected from the impurity concentrations in the crystal (Table IV). Samples of LBO annealed at 805 °C (approximate midpoint growth temperature) for 70 hours in air had similar properties to as-grown crystals, indicating that the defect structure is in equilibrium at this temperature. The results thus far indicate that the defect structures of presently grown LBO crystals are dominated by the impurities present.

LBO has potential applications for second harmonic generation, especially involving a high pulse power or high average power.(48) The defects contributing to the conductivity in the crystals may affect several of the properties of interest for applications, discussed above. Further work is necessary to investigate the relationship of the defects to the properties of LBO.

<u>Summary</u>

The defect chemistry of a specific crystal is determined by both its structural characteristics and the growth, or processing, of the material. Structurally, the nonlinear optical oxides contain anionic oxide groups (i.e. TiO_6, NbO_6, PO_4, AsO_4, B_3O_6, B_3O_7) which are the basic structural units responsible for the second order nonlinear optical susceptibility. The relatively large contribution of covalent bonding in the anionic groups to the total lattice energy appears to allow the structures to accommodate nonstoichiometric defects on the other, more "ionic", cation sublattice or sublattices with a relatively small cost of energy. Nonstoichiometry ranges from a few hundred ppm (atomic) to a few per cent or more, depending on the specific structure. Isovalent impurities are typically incorporated readily into crystals on the non-anionic group cation sublattice due to the tolerance in the structures. The incorporation of aliovalent impurities is enhanced in crystals with nonstoichiometric defects which can compensate the differences in charges.

The nonlinear optical oxide crystals recently developed are grown by flux (and hydrothermal solution for KTP) techniques to prevent decomposition (KTP, KTA, LBO) or to obtain a low temperature phase (BBO). The intrinsic nonstoichiometry and the impurity contents of the as-grown crystals is determined by the solutions and temperatures used for growth. The intrinsic defect concentrations in these materials are relatively low, compared to the more traditional nonlinear optical oxides having the perovskite, perovskite-like and tungsten bronze type structures. As a result, their defect structures can be dominated by impurities present at relatively small concentrations. The defect chemistry of nonlinear optical oxide crystals can affect many of the materials' properties required for device applications. For further discussion of either the relationship between growth and the defects present or the effects of the defects on properties, please see reference (49).

Acknowledgments

I would like to acknowledge my collaborators in the work on KTP and KTA: J. D. Bierlein, M. Crawford, A. Ferretti, P. Gallagher, G. Gashurov, G. Loiacono and M. Roelofs. I would also like to thank J.D. Bierlein for samples of LBO and R.H. French for samples of BBO crystals grown at the Fujian Institute of Research on the Structure of Matter, China. The laboratory assistance of W.R. Greene, B. Jones and R. Harlow is greatly appreciated.

Literature Cited

1.   Shen, Y. R. The Principles of Nonlinear Optics; John Wiley & Sons, NY: 1984.
2.   Yariv, A.; Yeh, P. Optical Waves in Crystals; John Wiley & Sons: NY, 1984.
3.   Singh, A. In Handbook of Laser Science and Technology; Weber, M.J., Ed.; CRC Press: Baco Raton, FLA, 1986; Vol. 3.
4.   Chen, C.; Liu, G. Ann. Rev. Mater. Sci. 1986, 16, 203.
5.   Ballman, A.A.; Byer, R.L.; Eimerl, R.S.; Feigelson, R.S.; Feldman, B.J.; Goldberg, L.S.; Menyuk, N.; Tang, C.L. Appl. Optics 1987, 26, 224.
6.   DiDominco, M., Jr.; Wemple,S.H. J. Appl. Phys. 1969, 40, 720; 735.
7.   Rase, D.E.; Roy, R. J. Am. Cer. Soc. 1955, 38, 102.
8.   Eror, N.G.; Smyth, D.M. J. Solid State Chem. 1978, 24, 235.
9.   Baumard, J.F.; Abelard, P. Solid State Ionics 1984, 12, 47.
10.   Carruthers, J.R.; Peterson, G.E.; Grasso, M.; Bridenbaugh, P.M. J. Appl. Phys. 1971, 42, 1846.
11.   Abrahams, S.C.; Marsh, P. Acta Cryst. 1986, B42, 61.
12.   Carruthers, J.R.; Grasso, M. J. Electrochem. Soc. 1970, 117, 1426.
13.   Megumi, K.; Nagatsuma, N.; Kashiwada, K.; Furuhata, T. J. of Mater. Sci. 1976, 11, 1583.
14.   Carruthers, J.R.; Grasso, M. Mat. Res. Bull. 1969, 4, 413.
15.   Barraclough, K.G.; Harris, I.R.; Cockayne, B.; Plant, J.G.; Vere, A.W. J. of Mater. Sci. 1970, 5, 389.
16.   Lines, M.E.; Glass, A.M. Principles and Applications of Ferroelectrics and Related Materials; Clarendon Press: Oxford, 1977.
17.   Abrahams, S.C.; Jamieson, P.B.; Bernstein, J.L. J. Chem. Phys. 1971, 54, 2355.
18.   Neurgaonkar, R.R.; Kalisher, M.H.; Lim, R.C.; Staples, E.J.; Keester, K.L. Mat. Res. Bull. 1980, 15, 1235.
19.   Bellrus, V.; Kalnajs, J.; Linz, A.; Folweiler, R.C. Mat. Res. Bull. 1971, 6, 899.
20.   Goodenough, J.B.; Kafalas, J.A. J. Solid State Chem. 1973, 6, 493.
21.   Levin, E.M.; Robbins, C.R.; McMurdie, J.F. Phase Diagrams for Ceramists; Reser, M.K., Ed.; Am. Cer. Soc.: Columbus, OH, 1964.
22.   Rauber, A. In Current Topics in Mater. Sci.; Kaldis, E. Ed.; North Holland: Amsterdam, 1978; Vol. I.

23.   Van Uitert, L.G.; Rubin, J.J.; Bonner, W.A. J. Quantum Elec. 1968, 4, 622.
24.   Nassau, K. In Ferroelectricity; Weller, E.F., Ed.; Elsevier: Amsterdam, 1967.
25.   Bloss, F.D. Crystallography and Crystal Chemistry; Holt, Rinehart & Winston: NY, 1971.
26.   Birnie, D.P. SPIE Proc., 1988, Vol. 968, p. 81.
27.   Bierlein, J.D. SPIE Proc., 1988, Vol. 994, p. 160.
28.   Zumsted, F.C.; Bierlein, J.D.; Gier, T.E. J. Appl. Phys. 1976, 47, 4980.
29.   Torjman, I.; Masse, R.; Guitel, J.C. Z. Kristallogr. 1974, 139, 103.
30.   Morris, P.A.; Crawford, M.K.; Ferretti, A.; French, R.H.; Roelofs, M.G.; Bierlein, J.D.; Brown, J.B.; Loiacono, G.M.; Gashurov, G. Mat. Res. Soc. Symp. Proc., 1989, Vol. 152, p. 95.
31.   Morris, P.A.; Crawford, M.K. Roelofs, M.G.; Bierlein, J.D.; Gallagher, P.K.; Gashurov, G.; Loiacono, G.M. Mat. Res. Soc. Symp. Proc., 1989, Vol. 172., p. 283.
32.   Gier, T.E. U.S. Patent 4 231 838, 1980.
33.   Crawford, M.K.; Morris, P.A.; Roelofs, M.; Gashurov, G. Mat. Res. Soc. Symp. Proc., 1989, Vol. 172., p. 341.
34.   Ahmed, F.; Belt, R.F.; Gashurov, G. J. Appl. Phys. 1986, 60, 839.
35.   Stucky, G.D.; Phillips, M.L.; Gier, T.E. In Chemistry of Materials; Am. Chem. Soc.: 1989, p. 492.
36.   Roelofs, M.G.; Bierlein, J.D., J. Appl. Phys., submitted.
37.   Roelofs, M.G.; J. Appl. Phys. 1989, 65, 4976.
38.   Bierlein, J.D.; Vanherzeele, H.; Ballman, A.A. Appl. Phys. Lett. 1989, 54, 783.
39.   El Brahimi, M.; Durand, J. Rev. Chim. Miner. 1986, 23, 146.
40.   Lu, S.F.; Ho, M.Y.; Huang, J.L. Acta Physica Sinca 1982, 31, 948.
41.   Eimerl, D.; Davis, L.; Velsko, S.; Graham, E.K.; Zalkin, A. J. Appl. Phys. 1987, 62, 1968.
42.   Jiang, A.; Cheng, F.; Lin, Q; Cheng, Z.; Zheng, Y. J. Cryst. Growth 1986, 79, 963.
43.   Guo, R.; Bhalla, A.S. J. Appl. Phys. 1989, 66, 6186.
44.   Zhang, G.; Yang, Y.; Zhang, C. Appl. Phys. Lett. 1988, 53, 1019.
45.   Ihara, M.; Yuge, M.; Krogh-Moe, J. Yogyo Kyokai Shi 1980, 88, 179.
46.   Chen, C.; Wu, Y.; Jiang, A.; Wu, B.; You, G.; Li, R.; Lin, S. J. Opt. Soc. Am. B 1989, 6, 616.
47.   Jiang, A.; Chen, T.; Zheng, Y.; Zusheng, C.; Shengsheng, W. Guisuanyan Xuebao 1989, 17, 189.
48.   Lin, S.; Sun, Z.; Wu, B.; Chen, C. J. Appl. Phys. 1990, 67, 634.
49.   Morris, P.; J. Cryst. Growth 1990, to be published.

RECEIVED July 18, 1990

Chapter 26

Defect Properties and the Photorefractive Effect in Barium Titanate

Barry A. Wechsler, Daniel Rytz, Marvin B. Klein, and Robert N. Schwartz

Hughes Research Laboratories, Malibu, CA 90265

Barium titanate ($BaTiO_3$) is a photorefractive material
that exhibits high gain. However, significant
improvements in performance can be achieved through
proper control of defect properties. Our approach to
optimizing these properties involves growth of pure
and doped $BaTiO_3$ crystals. Annealing under controlled
atmospheres is used to control the valence state of
the dopants. The effects of growth conditions on the
defect and photorefractive properties are
characterized by optical absorption, electron
paramagnetic resonance, and two-beam coupling
measurements. Theoretical modeling of defect
equilibria aids in understanding the relation between
processing conditions and photorefractive behavior.
Substantial differences in behavior are observed for
various dopants and annealing conditions. Thus far,
cobalt-doped crystals are the most promising for
visible and near-infrared applications.

The photorefractive effect is a light-induced change in the index of
refraction of a crystal. Although referred to as "optical damage"
when the effect was first discovered ([1],[2]) it was soon realized that
refractive index gratings written and stored in such crystals could
be used for a wide range of optical applications. Photorefractive
crystals can be used to make simple phase conjugators with
applications in distortion correction, laser power combining, remote
sensing, and tracking systems. These materials may also play an
important role in various optical computing and signal processing
devices such as reconfigurable optical interconnects, associative
memories, and passive limiters for sensor protection.
    The most widely studied photorefractive materials can be divided
into three classes: ferroelectric oxides, including $LiNbO_3$, $KNbO_3$,
$BaTiO_3$, and various tungsten bronze-type crystals such as
$Sr_{1-x}Ba_xNb_2O_6$; oxides of the sillenite family, including $Bi_{12}SiO_{20}$,
$Bi_{12}GeO_{20}$, and $Bi_{12}TiO_{20}$; and semi-insulating compound

0097–6156/91/0455–0394$06.00/0

semiconductors, including GaAs, InP, and CdTe. The ferroelectric oxides generally display the largest refractive index changes due to their relatively large electro-optic coefficients. On the other hand, they have small charge carrier mobilities and large dielectric constants, which result in long response times and low sensitivities. In contrast with the sillenites and compound semiconductors, measured time constants for the photorefractive effect in the ferroelectric oxides are typically more than three orders of magnitude slower than the theoretical limit. Existing crystals display a great deal of variability among different samples and are photosensitive only over a limited wavelength range. For these reasons, efforts to optimize these materials through improved crystal growth and processing methods may prove worthwhile.

Because the photorefractive effect involves photocarrier ionization, transport, and recombination, this behavior is likely related to the presence of defects such as substitutional impurities and/or cation and anion vacancies. Our approach to optimization of $BaTiO_3$ involves both a theoretical understanding of the mechanisms involved in the grating formation process and an experimental effort on the growth, processing, and characterization of pure and doped crystals. In the following sections, we present some results of our recent studies aimed at probing the nature of the photorefractive centers in $BaTiO_3$ and altering the photorefractive properties through doping and heat treatment.

Physics of the Photorefractive Effect

The essential features of the photorefractive effect and the influence of defect centers on this behavior can be understood in terms of a band transport model ($\underline{3},\underline{4}$). Consider a hypothetical species with two possible charge states having energy levels within the band gap (Figure 1). Illumination with photons of sufficient energy causes carriers (holes and/or electrons) to be photoionized from filled sites. These carriers undergo transport by drift and diffusion and recombine at empty trapping centers. If the illumination is non-uniform, a space charge field is produced that modulates the index of refraction through the electro-optic effect.

A useful method of characterizing this behavior is two-beam coupling. Two interfering laser beams produce a periodic irradiance pattern in the crystal. A refractive index grating develops which is in general out of phase with respect to the irradiance, leading to energy exchange (gain) between the incident beams. The gain and response time of this grating can be measured as a function of the crossing angle of the beams (i.e., grating spacing or wave vector), intensity, wavelength, etc., and used to determine material properties related to the photorefractive effect. The gain coefficient, which is determined by measuring the ratio of transmitted intensities of a weak signal beam with and without a strong reference beam, is given by

$$\Gamma = (2\pi/\lambda)n^3 r_{eff}E_{sc} \ , \tag{1}$$

where λ is the wavelength, n is the background refractive index, r_{eff} is the effective electro-optic coefficient for the measuring geometry, and E_{sc} is the space charge field, given by:

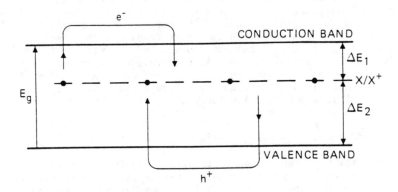

Figure 1.   Schematic energy-level model for a single species in two valence states.   (Reproduced with permission from Ref. 11.   Copyright 1988 Optical Society of America.)

$$E_{sc} = \frac{k_B T}{e} \frac{K}{1+(K/K_s)^2} \frac{(C-1)}{(C+1)} \quad . \tag{2}$$

Here, k_B is Boltzmann's constant, T is the absolute temperature, e is the electron charge, and K is the grating wave vector. K_s is the Debye screening wave vector, defined by

$$K_s = (e^2 N_e / \epsilon \epsilon_o k_B T)^{1/2} \quad , \tag{3}$$

where ϵ is the relative dielectric constant, ϵ_o is the permittivity of free space, and N_E is the effective empty trap density, given by

$$N_E = N_A N_D / (N_A + N_D) \quad . \tag{4}$$

N_D is the concentration of photoactive centers filled with electrons (species X), and N_A is the concentration of sites filled with holes (species X^+). The factor $\xi = (C-1)/(C+1)$ accounts for the competing photoconductivity contributions due to electrons and holes, and has a value of -1 when electron photoconductivity dominates and +1 when hole conductivity dominates. Intermediate values occur when mixed conductivity is present. C is defined by

$$C = \frac{s_h}{s_e} \frac{N_A}{N_D} \frac{(K^2+K_e^2)}{(K^2+K_h^2)} \quad , \tag{5}$$

where s_h and s_e are the photoionization cross sections for holes and electrons, respectively, and K_h and K_e are the inverse transport lengths for holes and electrons.

The response time, τ, is typically determined by measuring the grating build-up time or erasure rate. For large grating spacings, the response time is the inverse dielectric relaxation rate:

$$\frac{1}{\tau} = \gamma = \gamma_{die} \frac{1+(K/K_s)^2}{1+(K/K_e)^2} + \gamma_{dih} \frac{1+(K/K_s)^2}{1+(K/K_h)^2} \quad , \tag{6}$$

where

$$\gamma_{die} = \frac{s_e \mu_e}{\gamma_e} \frac{eI}{\epsilon \epsilon_o} \frac{N_D}{N_A} \quad \text{and} \quad \gamma_{dih} = \frac{s_h \mu_h}{\gamma_h} \frac{eI}{\epsilon \epsilon_o} \frac{N_A}{N_D} \quad . \tag{7}$$

γ_{die} and γ_{dih} are the dielectric relaxation rates for electrons and holes (neglecting the dark conductivity), μ_e and μ_h are the carrier mobilities, γ_e and γ_h are the recombination rates, and I is the incident intensity.

The gain and response time of a crystal are subject to variation through growth and processing techniques. In general terms, gain is enhanced by a large empty trap density, N_E, and large (positive or negative) values of the relative conductivity factor, ξ. The response time is made faster by higher absorption ($s_h N_A + s_e N_D$) and longer carrier lifetimes ($\tau_{Re} = 1/\gamma_e N_A$, $\tau_{Rh} = 1/\gamma_h N_D$). Within the context of the model described here, the effects of growth conditions on these properties can be understood. The refractive index, electro-optic coefficient, and dielectric constant are fundamental material properties that cannot generally be modified, although they may be sensitive to temperature and, in the case of mixed crystals, composition. The photoionization and recombination rates depend on the energy level structure of the photoactive species, and are therefore sensitive to the particular dopants, impurities, or other defects present. Although these can be altered by processing techniques, it is difficult to predict a priori the optimum species; therefore, various possible dopants must be tried. The carrier mobilities are fundamental properties of the material, although they may vary somewhat depending on the nature of the defects. Carrier transport can be altered by varying the temperature as well as by applying an electric field. Since hole and electron mobilities are in general unequal, it is also possible to modify the effective transport properties by selecting the dominant photocarrier through appropriate material preparation techniques. The properties most readily subject to control through growth and processing procedures are the concentrations of the filled and empty traps. The total concentration of photoactive species (N_D+N_A) can be altered by doping and/or other changes in growth conditions. Furthermore, the relative proportions of the two states (N_D/N_A) can be controlled by in-situ or post-growth oxidation-reduction treatments. It must be understood, however, that defect equilibria are complex and it is therefore usually not possible to vary any single parameter without also affecting other material properties.

Crystal Growth and Processing

$BaTiO_3$ is a perovskite-type crystal. At high temperature (between 1432 and 132°C), the structure is cubic and consists of corner-linked TiO_6 octahedra with the much larger Ba ions occupying the voids between these octahedra. At 132°C, a displacive phase transition occurs. The titanium ion moves slightly off the central position in the oxygen octahedron, giving rise to an electric polarization. This phase, which has tetragonal symmetry, is ferroelectric and ferroelastic: the direction of the spontaneous polarization can be reoriented by application of an electric field or a non-hydrostatic stress. Below room temperature, $BaTiO_3$ undergoes two additional phase transitions, from tetragonal to orthorhombic at about 10°C and from orthorhombic to rhombohedral at -90°C.

Between 1432°C and the melting point (1625°C), the stable phase of $BaTiO_3$ has hexagonal symmetry and differs significantly in structure from that of the cubic perovskite. The hexagonal-cubic transformation is reconstructive and would result in a serious degradation of the optical and mechanical integrity of single crystals grown from the pure melt and cooled through this phase transition. For this reason, $BaTiO_3$ crystals have been grown by a variety of flux methods. Of these, the top-seeded solution growth (TSSG) approach developed by Linz and co-workers (5) is the most successful for the production of large, high quality samples. This method uses an excess of TiO_2 as the flux, which has the advantage of not introducing any foreign ions as impurities. A melt composition near 65 mol% TiO_2 in the system $BaO-TiO_2$ is typically used, having a liquidus temperature near 1400°C. Growth is initiated on a seed lowered into the melt from above and is driven by cooling the melt at a rate of a few tenths of a degree per hour. The crystal is usually

rotated at about 60 rpm and withdrawn from the melt at a rate of 0.1-
0.2 mm/h during growth. The crystal is removed from the melt at
1335°C and then cooled to room temperature.

In order to optimize the photorefractive performance of $BaTiO_3$,
it is necessary to control the centers involved in the ionization and
trapping of photocarriers. This may be accomplished in part by
intentionally doping the crystal with various transition metal or
rare earth ions, which may exist in more than one valence state.
These may be added to the melt during growth in the form of oxide or
carbonate compounds. In our work, we have surveyed most of the
transition metals, using doping concentrations in the melt of 50 to
200 ppm of dopant ions per $BaTiO_3$ formula unit. Although even
undoped crystals with no more than about 10 ppm of unintentional
impurities present are still photorefractive, some studies (6,7) have
suggested that the photorefractive gain can be enhanced by the
addition of dopants (or by higher impurity concentrations). On the
other hand, there does not appear to be any simple correlation
between doping concentration and photorefractive gain; rather, the
behavior is more complex, being influenced by the inevitable presence
of other intrinsic and extrinsic defects.

Transition metal dopants and impurities are probably
incorporated substitutionally for Ti in $BaTiO_3$. Emission
spectrographic analyses indicate that the distribution coefficients
for Mn and Fe dopants are on the order of 1 to 2, i.e., the crystals
are slightly enriched relative to the melt. Cr and Ni may have
distribution coefficients slightly less than 1. For Co, the measured
concentrations in the crystals display considerable scatter; we
estimate that the distribution coefficient is on the order of 4. Fe
is the most prevalent transition metal impurity and is typically
present at a concentration of 10-15 ppm by weight. Si, Al, Mg, and
Cu are also typically present at 5-50 ppmw. Fe and Cr impurities
have also been observed by EPR spectroscopy, although Cr could not be
detected by emission spectroscopy, with a detection limit of 10 ppmw.

The two most important intrinsic defects in $BaTiO_3$ are oxygen
and barium vacancies. Both of these can be modified to some extent
by growth and processing conditions. Barium vacancies are introduced
as a result of the excess of TiO_2 in the TSSG melt. Phase
equilibrium experiments show that as much as 1 or 2 mol% excess TiO_2
may be incorporated in $BaTiO_3$ at temperature above 1500°C (8,9).
However, in the temperature range over which $BaTiO_3$ is normally
grown, i.e., below 1400°C, it is likely that the excess of TiO_2 is no
more than about 100 ppm (10). Nevertheless, this is similar to the
dopant concentrations, and therefore it would not be surprising if Ba
vacancies play an important role in the photorefractive behavior,
particularly in high purity undoped crystals.

Oxygen vacancies are the principal charge compensating defects
in $BaTiO_3$. Thus, their concentration is determined by the overall
balance of donor and acceptor type species. In most $BaTiO_3$ crystals,
acceptor-type impurities predominate, which raises the concentration
of oxygen vacancies. The oxygen vacancy concentration is also
dependent upon the oxygen pressure in the surrounding atmosphere
during growth and cooling. Because oxygen vacancies are highly
mobile at elevated temperatures, their concentration can be changed
by post-growth annealing. Each oxygen ion removed from the crystal
by annealing under low $P(O_2)$ conditions leaves behind two electrons.

These electrons are thermally ionized from the vacancy and may combine with an available acceptor, thus altering the charge state of the acceptor species. Experiments have shown that such a process can lead to a change in sign of the dominant photocarrier as well as modified gain and response time of the photorefractive effect.

Thermodynamic Point Defect Model

In order to obtain a better understanding of the effects of doping and oxidation-reduction processing on the defect populations and photorefractive properties of $BaTiO_3$, we have studied a thermodynamic point defect model (11). This model, originally developed for $BaTiO_3$ by Hagemann (12), can be used to calculate the concentrations of point defects as a function of temperature and oxygen pressure for various dopants. These populations determine the photorefractive trap densities (N_D and N_A) and can therefore be used to predict the beam coupling gain and response time. Although this model is probably too simplistic to give accurate quantitative results, it provides a useful qualitative picture of the way in which crystal growth and processing conditions can alter the behavior.

We assume that a single acceptor-type species is present that acts as both the source and trap for photocarriers. This species has at least two possible stable charge states with energy levels in the band gap. We consider the presence of eight defect species: oxygen vacancies in neutral, singly-, and doubly-ionized states; acceptor sites (e.g., transition metals substituting for Ti) also in neutral, singly-, and doubly-ionized forms; and free charge carriers (electrons and holes). The oxygen vacancy concentrations are governed by the exchange of oxygen between the crystal and its surrounding atmosphere, and are given by:

$$[V_0^x] = K_o' \, e^{-E_o'/k_B T} \, P_{0_2}^{-1/2} \tag{8}$$

$$[V_0^\cdot] = 2 \, [V_0^x] \, e^{(E^+ - E_F)/k_B T} \tag{9}$$

$$[V_0^{\cdot\cdot}] = [V_0^x] \, e^{(E^+ + E^{++} - 2E_F)/k_B T} \tag{10}$$

where the superscript "x" indicates a defect that is neutral, and the superscript dots indicate positive charges with respect to the ideal lattice. E_o' is the enthalpy of reduction, K_o' is a constant related to the entropy change of reduction, k_B is Boltzmann's constant and T is the absolute temperature. E^+ and E^{++} are the ionization energies corresponding to the removal of electrons from the neutral oxygen vacancy, and E_F is the Fermi level. The concentrations of the acceptor species are given by:

$$[A^x] = \frac{[A]}{1 + 1/2 \, e^{(E_F - E^-)/k_B T} + e^{(2E_F - E^- - E^{--})/k_B T}} \tag{11}$$

$$[A'] = 1/2 \, [A^x] \, e^{(E_F - E^-)/k_B T} \tag{12}$$

$$[A''] = [A^x] \, e^{(2E_F - E^- - E^{--})/k_B T} \tag{13}$$

where the superscript primes indicate negative charges relative to the ideal lattice and E^- and E^{--} are the energies corresponding to the ionization of holes from the neutral acceptor site. The free carrier concentrations are determined by thermal ionization across the band gap, and are given by:

$$np = N_C \, N_V \, e^{-E_g/k_B T} \tag{14}$$

$$n = N_C \, e^{-(E_C - E_F)/k_B T} \tag{15}$$

where N_C and N_V are the densities of states in the conduction and valence bands, respectively, E_C is the energy of the conduction band edge, and E_g is the band gap energy. One further equation is established by the condition that overall charge balance must be maintained:

$$[A'] + 2[A''] + n = [V_O^{\cdot}] + 2[V_O^{\cdot\cdot}] + p \; . \tag{16}$$

Here, all the negatively charged species appear on the left-hand side of the equation and the positively charged species are on the right.

These nine equations in nine unknowns (eight defect concentrations plus E_F) can be solved simultaneously for any temperature and oxygen pressure, given the values for the aforementioned energy parameters and densities of states. An example of such a calculation for $BaTiO_3$:Co is shown in Figure 2a. We have used the data of Ref. 12, showing an ionization level 1.45 eV above the top of the valence band, which involves the Co^{2+}/Co^{3+} states. A second ionization level at the valence band edge corresponds to the Co^{3+}/Co^{4+} ionization reaction. We assume that the total oxygen vacancy concentration is established and "frozen in" at the process temperature, but that the individual defect species are in thermal equilibrium at room temperature. In Figure 2b, the corresponding model predictions for the photorefractive gain and response time are shown. These were calculated using Equations 1-7, where the concentrations of Co^{2+} and Co^{3+} were used for the parameters N_D and N_A, respectively. Because the values of $s_{e,h}$, $\mu_{e,h}$, and $\gamma_{e,h}$ are not well known, it is necessary to make several assumptions to calculate the photorefractive properties (11). Therefore, these results should only be viewed as being qualitatively meaningful.

A number of interesting observations can be made from model calculations such as these. Referring to Figure 2a, it is seen that under oxidizing conditions, Co^{3+} is the predominant dopant species. Under reducing conditions, the cobalt valence state is lowered to 2+. Doubly ionized oxygen vacancies are present at half the Co^{3+} concentration in the oxidizing regime to provide charge compensation. At lower oxygen pressures, the oxygen vacancy concentration becomes equal to the Co^{2+} concentration. After all the cobalt has been converted to the divalent state, further reduction leads to an increasing population of electron-occupied oxygen vacancies. (Note that this calculation does not allow for the presence of Co^{1+}, although Co^{4+} is allowed at high oxygen pressures.)

The photorefractive gain, shown in Figure 2b, is predominantly determined by two factors: the trap density, N_E, and the relative conductivity factor, ζ. The latter factor accounts for the minimum

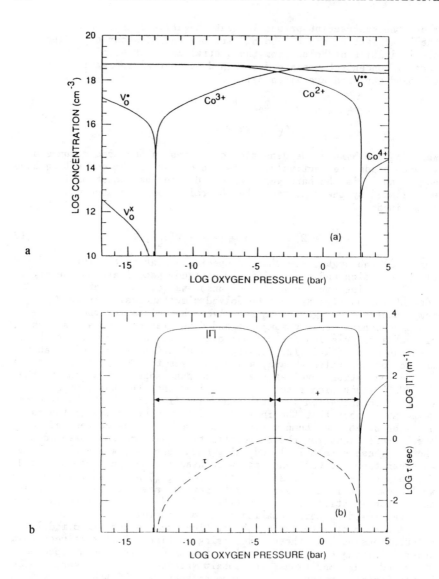

Figure 2. (a) Point defect concentrations as a function of oxygen partial pressure for BaTiO₃:Co processed at 800°C. (b) Calculated gain and response time for the model shown in (a).

in the gain and associated sign change near the middle of the $P(O_2)$
range. At this point, which actually corresponds to a maximum in the
trap density, the photoconductivity contributions due to electrons
and holes exactly cancel one another, resulting in no net
photorefractive effect. The gain also vanishes at higher and at
lower oxygen pressures due to the absence of traps (i.e., the
concentration of the minority species, which effectively determines
the trap density, becomes negligibly small). The response time is
slowest where the trap density reaches a maximum, because carriers
are trapped more quickly and hence their transport lengths are
shorter. With a smaller trap density, the response time is faster.
The model suggests that response times can vary by about 2 orders of
magnitude within the high gain regime.

Since the ionization energies of various transition metals in
$BaTiO_3$ have been determined (12), this model can be used to compare
the expected behavior for different dopants. For example, the model
indicates that Fe is present predominantly as Fe^{3+} throughout most of
the accessible region of $P(O_2)$ conditions. The minority species,
Fe^{4+} at high oxygen pressures and Fe^{2+} at low oxygen pressures, are
present at relatively low concentrations throughout most of the
available region of $P(O_2)$; thus, one would expect to find a smaller
trap density in Fe-doped $BaTiO_3$ than for the same concentration of
cobalt. Also, because the conversion between Co^{3+} and Co^{2+} occurs
under easily achievable conditions, the photorefractive behavior can
in principal be more easily controlled for this dopant. Calculations
for Mn and Cr also predict valence changes under readily accessible
$P(O_2)$ conditions. However, Ni is present only in the divalent state
except under high oxygen pressures and therefore would not be
expected to participate in the photorefractive effect. Model
calculations also suggest that the photorefractive behavior becomes
quite complex when two or more species can contribute to the effect.
For example, with two ionizable species present, each with two levels
in the band gap, seven sign changes are predicted as a function of
reduction. Even though one of these species may be present at low
concentrations (i.e., as an impurity in the background of a doped
crystal), it may nevertheless have a significant impact on the
expected behavior.

This model has a number of limitations, including the fact that
only point defects are considered. As discussed below, there is some
evidence that more complex defect pairs or clusters may occur in
$BaTiO_3$. In addition, the model requires that photocarriers are
associated with only a single partially filled level in thermal
equilibrium and cannot account for light-induced charge
redistribution among multiple levels. Several studies (13-15) have
suggested the possible importance of the latter effect, based on
observations of intensity-dependent absorption and the sublinear
intensity-dependence of the response time in many crystals.

Spectroscopic and Photorefractive Characterization

Optical Absorption. Figure 3 compares the optical absorption spectra
of undoped, Co-, Mn-, and Fe-doped $BaTiO_3$ crystals grown in air.
Transmission spectra were obtained on a Perkin-Elmer Lambda 9
spectrophotometer modified for use with polarized light and were
reduced to absorption coefficients by correction for Fresnel

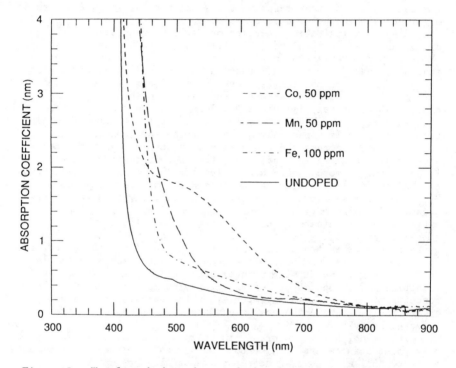

Figure 3. Wavelength dependence of the absorption coefficient (measured with light polarized perpendicular to the c-axis) of undoped and doped $BaTiO_3$ crystals. Concentrations refer to dopant atoms per $BaTiO_3$ formula unit in the melt.

reflection. In Co-doped $BaTiO_3$ a strong, broad absorption band is centered near 550 nm; the tail of this peak overlaps the intrinsic band edge near 400 nm. We assign this absorption to a charge transfer transition between the valence band (made up of oxygen 2p orbitals) and levels associated with cobalt. Theoretical calculations (16,17) suggest that neither Co^{2+} nor Co^{3+} in a complete octahedral environment has energy levels appropriate for a transition at this wavelength. However, a $Co^{3+}-V_O$ pair does have unoccupied mid-gap energy levels that could be responsible for this absorption. The peak absorption in Mn-doped $BaTiO_3$ appears to lie at shorter wavelengths than in $BaTiO_3$:Co. In $BaTiO_3$:Fe, a weak, broad absorption band is centered near 600 nm, and there is an apparent shift of the absorption edge from about 400 to 410 nm. Reduction at 800°C with a 99% CO_2/1% CO gas mixture induces a color change from reddish-brown to yellow in both $BaTiO_3$:Co and $BaTiO_3$:Cr.

Electron Paramagnetic Resonance. Electron paramagnetic resonance (EPR) spectroscopy is a powerful and extremely sensitive technique for characterizing the charge state and site symmetry of various defects in crystals. Unfortunately, it is difficult to perform EPR measurements on properly oriented, single domain samples of $BaTiO_3$ at low temperatures due to the phase transitions at 10 and -90°C. One way to avoid this difficulty is to use powder specimens. We have obtained spectra of $BaTiO_3$ powders doped with all the transition metals and annealed under a wide range of oxygen pressure conditions. We have also studied the spectra of doped $BaTiO_3$ crystals that have been ground into powder form. Although some information is inevitably lost as a result of the powder averaging, this technique provides a relatively rapid means of identifying the valence state of the dopants and monitoring the changes induced by oxidation-reduction processing.

Powder EPR spectra of $BaTiO_3$:Mn powders are shown in Figure 4. The spectrum in Figure 4a was taken on a powder that was synthesized in air at 1400°C and then cooled rapidly below 800°C. Three prominent features are observed: 1) a sextet at low magnetic field (g ≃ 4) with a hyperfine splitting constant |A| = 78.8 G that is attributed to Mn^{4+}; 2) a sextet centered at g = 2.0015 with hyperfine splitting |A| = 87.8 G that is attributed to Mn^{2+}; and 3) a strong resonance line at g ≃ 2 due to Fe^{3+} and Cr^{3+} impurities. Upon mild reduction (800°C, CO_2), the Mn^{4+} spectrum disappears and the Mn^{2+} spectrum grows substantially (Figure 4b). This result reveals the ease with which a valence change can be induced in this material by annealing in a mildly reducing atmosphere. Valence state changes in mildly reduced samples have also been observed in $BaTiO_3$:Co and $BaTiO_3$:Cr. However, the spectra on $BaTiO_3$:Fe show that Fe^{3+} remains predominant except under very reducing conditions (1200°C, 99% CO).

Photorefractive Measurements. Photorefractive gain and response time measurements have been carried out on a variety of samples using the two-beam coupling technique (18). Different samples of 50 and 100 ppm Co-doped crystals were processed at 800°C under oxygen and CO-CO_2 gas mixtures with various oxygen partial pressures. Samples of each dopant concentration were found to have gain vs. grating spacing curves that agreed with one another to within experimental uncertainty prior to the oxidation-reduction treatments. Theoretical

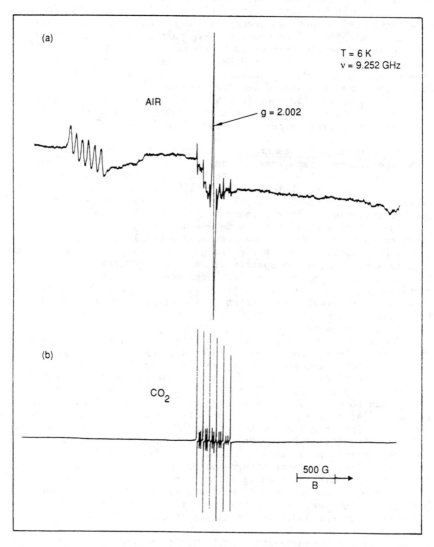

Figure 4. X-band EPR spectra of $BaTiO_3$:Mn powders. (a) Sample processed in air. (b) Sample processed in CO_2.

fits to the gain data using Equations 1-5 yield estimates for the trap density, relative conductivity factor, and transport lengths.

Gain data at 515 nm for as-grown samples of undoped, Mn-doped, and Co-doped $BaTiO_3$ are shown in Figure 5. As-grown Co-doped crystals have consistently high gains comparable to or better than those of any $BaTiO_3$ crystals previously reported. The high gains appear to be attributable both to a large trap density ($N_E = 6 \times 10^{16}$ cm^{-3} in both cobalt-doped crystals versus 9×10^{15} cm^{-3} in the undoped crystal and 2×10^{16} cm^{-3} in the Mn-doped crystal) and to relatively little electron-hole competition ($\xi = 0.74$ and 0.60 in 100 ppm and 50 ppm crystals, respectively). Both oxidized and reduced Co-doped crystals have lower gains than as-grown samples, due to smaller trap densities and increased electron-hole competition. The photoconductivity in as-grown crystals is hole-dominated. Co-doped crystals reduced in CO_2 are also hole-dominated, whereas crystals reduced in a 99% CO_2/1% CO atmosphere are electron-dominated. Preliminary measurements suggest that Cr doping also leads to enhanced gain relative to undoped crystals. However, Fe-doped crystals have gains similar to that of undoped $BaTiO_3$.

Response times of undoped and Co-doped crystals for an incident intensity of 1 W/cm^2 and a grating spacing of 0.7 μm are shown in Table 1. No systematic trend is apparent from these data, at least partially due to experimental uncertainties such as sample heating and variable reflection and absorption losses. Response times of as-grown and reduced Fe-, Mn-, and Cr-doped $BaTiO_3$ (18) range from 200 to 1200 ms under the same experimental conditions. Thus, it is clear that response times can vary considerably for crystals with different dopants and oxidation states. Additional experiments are needed to achieve better control over this key property.

Table 1. Response Times (ms) at 1 W/cm^2 ($\Lambda_g = 0.7$ μm)

Dopant	Atmosphere			
	O_2	Air	CO_2	1% CO
Undoped		420		
50 ppm Co		1400	800	55
100 ppm Co	170	730		610

Measurements of self-pumped phase conjugate reflectivity in $BaTiO_3$:Co in the wavelength range from 633 to 930 nm are indicative of a strong photorefractive response in the near infrared. Crystals doped with Co at concentrations above 25 ppm have high reflectivities when operated as self-pumped phase conjugate mirrors in the total internal reflection geometry, without the use of index matching fluids. This effect appears to be reproducible in Co-doped crystals, in contrast with undoped crystals, where a strong infrared response is observed in only a limited number of samples.

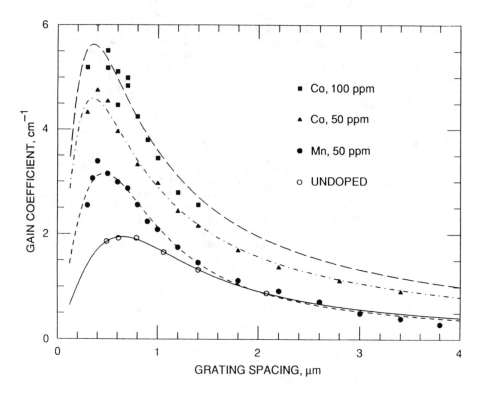

Figure 5. Photorefractive two-beam coupling gain coefficient of undoped and doped BaTiO$_3$ crystals. Curves are theoretical fits to the data. Experimental conditions: λ = 515 nm, I = 3-5 W/cm^2, beam ratio $\geq$ 200, s-polarization, grating wave vector parallel to c-axis. Concentrations refer to dopant atoms per BaTiO$_3$ formula unit in the melt.

Conclusions

Efforts to optimize the photorefractive performance of $BaTiO_3$ rely heavily on understanding and controlling the photosensitive defects. Theoretical modeling suggests that the behavior can most readily be influenced by doping crystals with aliovalent ions and through oxidation-reduction processing to control the valence state of the dopant. Undoped and transition metal doped crystals have been grown by the top-seeded solution growth method and characterizd by optical absorption, electron paramagnetic resonance, and photorefractive two-beam coupling. Of the crystals studied, cobalt doping gives the highest photorefractive gain in the visible wavelength region and appears very promising for applications in the near infrared as well. Co, Cr, and Mn in $BaTiO_3$ can have their valence state altered by oxidation-reduction processing and are therefore the most interesting dopants from the point of view of modifying the photorefractive response. Substantial progress has been made toward achieving reproducibly high-gain crystals. Based on our understanding of the importance of defect properties in this material, further improvements in response time are likely to be achieved.

Literature Cited

1. Ashkin, A.; Boyd, G.D.; Dziedzic, J.M.; Smith, R.G.; Ballman, A.A. Appl. Phys. Lett. 1966, 9, 72.
2. Chen, F.S. J. Appl. Phys. 1967, 38, 3418.
3. Strohkendl, F.M.; Jonathan, J.M.C.; Hellwarth, R.W. Opt. Lett. 1986, 11, 312.
4. Valley, G.C. J. Appl. Phys. 1986, 59, 3363.
5. Belruss, V.; Kalnajs, J.; Linz, A.; Folweiler, R.C. Mat. Res. Bull. 1971, 6, 899.
6. Godefroy, G.; Ormancey, G.; Jullien, P.; Lompre, P.; Ousi, W.; Semanou, Y. In Digest of Topical Meeting on Photorefractive Materials, Effects, and Devices; Optical Society of America: Washington, DC, 1988; p. 159.
7. Schunemann, P.G.; Pollak, T.M.; Yang, Y.; Teng, Y.-Y.; Wong, C. J. Opt. Soc. Amer. B 1988, 5, 1702.
8. Rase, D.E.; Roy, R. J. Am. Ceram. Soc. 1955, 38, 102.
9. Kirby, K.W. M.S. Thesis, U.C.L.A., Los Angeles, 1988.
10. Sharma, R.K.; Chan, N.-H.; Smyth, D.M. J. Am. Ceram. Soc. 1981, 64, 448.
11. Wechsler, B.A.; Klein, M.B. J. Opt. Soc. Amer. B 1988, 5, 1711.
12. Hagemann, H.-J. Ph.D. Thesis, Rheinisch-Westfalische Technische Hochschule, Aachen, Federal Republic of Germany, 1980.
13. Brost, G.A.; Motes, R.A.; Rotge, J.R. J. Opt. Soc. Amer. B 1988, 5, 1879.
14. Holtmann, L. Phys. Stat. Sol. (a) 1989, 113, K89.
15. Mahgerefteh, D.; Feinberg, J. Phys. Rev. Lett. 1990, 64, 2195.
16. Michel-Calendini, F.M.; Moretti, P.; Godefroy, G. Ferroelectrics Letters 1983, 44, 257.
17. Michel-Calendini, F.M.; Moretti, P. Phys. Rev. B 1983, 27, 763.
18. Rytz, D.; Wechsler, B.; Garrett, M.H.; Nelson, C.C. In Digest of Topical Meeting on Photorefractive Materials, Effects, and Devices II; Optical Society of America: Washington, DC, 1990; p. 2.

RECEIVED August 13, 1990

Chapter 27

What Is Materials Chemistry?

R. A. Laudise

AT&T Bell Laboratories, Murray Hill, NJ 07974

The increasingly chemical nature of forefront scientific opportunities and pressing technological needs has resulted in the development of a new interdisciplinary field, materials chemistry. Its principal characteristics are: a focus on solids with useful properties, an emphasis on synthesis, a long term goal of understanding structure-bonding-property relations, a frequent dependence on crystal growth for answering key questions and providing exploitable materials and a need to understand and optimize process physical chemistry. Selected subjects in electronic materials chemistry are used to illustrate these characteristics. Included are resist materials, III-V semiconductors, piezoelectric quartz hydrothermal single crystal synthesis, non-linear-optical potassium titanyl phosphate hydrothermal synthesis and optical fiber preparation. Finally, future challenges are discussed.

This paper is based in part on an invited talk given by the author as recipient of the ACS Materials Chemistry Prize, Boston, April 1990.

During the past two decades the science and engineering of electronic materials has become more and more intertwined with chemistry. Chemistry driven by the needs and opportunities of modern materials has lately come to be called *materials chemistry*. In this paper I would like to focus on electronic materials not only because of personal familiarity but also because they epitomize the interplay between chemistry and materials. From my experience, I would assert that the most important attributes of *electronic* materials chemistry are:

1. A focus on solids with interesting (and hopefully eventually useful) properties

0097–6156/91/0455–0410$07.00/0

2. An emphasis on synthesis

3. A long term goal of structure-bonding-property understanding and optimization

4. A frequent dependence on crystal growth and the control of perfection

5. A need to understand and optimize process physical chemistry

The most important determinant of problem choice in electronic materials chemistry is "are the properties or the process likely to fulfill a device or systems need". This consideration leads to a focus on synthesizing solids, understanding their properties in as fundamental a way as possible, growing crystals to enable understanding and permit devices and understanding the physical chemistry of the processes used so as to permit cost, yield and reliability improvement.

Novel synthesis often unlocks the door to novel materials, phenomena and applications. Materials processing in the laboratory and its transfer to the factory is the key to high quality, high yield, low cost materials. Indeed materials synthesis and processing were identified in the recent National Research Council "Materials Science and Engineering Study"*(1)* as key to U.S. industrial competitiveness in the 1990's.

In this review, in order to help define the field, I will present some of its recent achievements and describe some present exciting opportunities drawn principally from the work of my colleagues and myself at AT&T Bell Laboratories. These examples have been chosen to illustrate the attributes (1)-(5) which, I believe, distinguish materials chemistry. Work in Bell Labs is, of course, but a small fraction of materials chemical research world wide; my choices for examples have been based on personal knowledge. Neither value judgements nor historical balance are implied.

Resist Materials

Present electronic devices and systems are almost entirely based upon silicon integrated circuit chips. The chemistry involved in the preparation of chips becomes more involved and demanding as the scale of integration increases. Although not generally appreciated, a modern IC chip factory is essentially a chemical processing factory. The preparation of polysilicon, the growth of low defect single crystals, lithographic patterning, the deposition of epitaxial layers, etching, diffusion, and dielectric and conductor layer deposition are all essentially chemical processes. As the scale of integration increases, smaller feature sizes place even greater demands on the finesse of the chemistry needed.

One could choose any of a large variety of chemical achievements which have been essential to VLSI. Resists provide a representative example. To prepare very large scale integrated circuits new materials and processes are required, especially resist materials which are used to fabricate small scale features. Leading edge device fabrication today depends upon being able to etch patterns in Si with dimensions less than 1μm and features less than 0.5μm will be common place in the near future. This is accomplished by protecting areas which do not require removal with a polymer resist. The resist is patterned by altering its solubility by exposure to a defined uv, x-ray, electron or ion beam. When radiation absorbtion takes place, bonds with energy below the incident energy will be broken, formed or rearranged. If polymer chain scission takes place the molecular weight decreases with a concomitant increase in solubility (the material is a "positive" resist) and the resist region exposed to the beam is removed in subsequent processing. In negative resists solubility decreases and the exposed resist region is not removed during processing. Cross linking is one common mechanism. However, $\mu\upsilon$(250-400nm) energy is generally not sufficient to cause main chain cleavage so that uv positive resists frequently operate on a different principal. They consist of an aqueous alkali soluble polymer (e.g. a novalac resin) and a photo-active compound which inhibits the solubility. The inhibitor is ordinarily insoluble in aqueous base but with $\mu\upsilon$ exposure the inhibitor undergoes a rearrangement to a form which no longer inhibits dissolution causing the material to become insoluble. Following exposure, the imaged material is removed by emersion in a basic aqueous solution. The four major lithographic technologies and their chemistries are based largely upon the radiation used to expose the resist:

1. deep-uv photolithography

2. X-ray lithography

3. scanning electron beam lithography

4. scanning ion beam lithography

The reader should consult the literature (2) for discussions of the various lithographic technologies, materials advantages and limitations. Obviously the shorter the wavelength the smaller the ultimate resolution limit. Intensity of sources, ability to focus, sensitivity of results, scattering effects and costs are also important considerations so that it is by no means certain that the shortest wavelength is automatically the best. Even a cursory examination of resist materials suggests the synthetic and photochemical ingenuity which has been required for their development. Electron beam resist materials are particularly rich in their chemistry. They offer resolution limits which are less than 0.1µm. Some typical electron resist materials are listed in Table 1.(2) Resist design and the discovery of new resist materials have been key achievements in permitting increasing scales of integration. For further information see Thompson(2).

Improved processes which generally come about because of improved processing understanding are one of the main reasons why the costs of communication and computation have steadily decreased over the past four decades. Increased scale of integration coming about because of decreased feature size is the main driver for decreasing silicon IC costs. As feature sizes grow smaller, in addition to improved resists and exposure tools we require more discerning techniques for etching. Here the trend is to dry processing, especially plasma processing. The work of Dautremont-Smith et al.(3) epitomizes recent achievements in plasma chemistry understanding and their impact on real processes. Megabit memories are now articles of commerce and it is likely that with improvements in chemistry, present techniques and Si based materials can be made to produce gigabit devices well before the end of the century. Almost all the problems and obstacles to progress towards super VLSI are chemical in nature.

III-V Semiconductors

In contrast to Si, III-V semiconductors are direct band gap so that holes and electrons recombine to produce light. Consequently, they are the materials basis of semiconductor lasers. In addition the carrier velocity in many III-V's is higher than in Si resulting in higher speed devices. Solid solutions of, for example, $Ga_{1-x}In_xAs_yP_{1-y}$ can be band gap engineered to give lasers at frequencies corresponding to regions of high transparency in silica based optical communication fibers. At the same time these materials can be lattice parameter matched to convenient single crystal substrates thus enabling the preparation of high quality single crystal epitaxial films for devices. Typical crystal growth techniques include Czochralski melt - crystal growth for substrates and liquid phase epitaxy (LPE), vapor phase epitaxy (VPE), metal organic chemical vapor deposition (MOCVD) and molecular beam epitaxy (MBE) for thin film preparation. Quantum well devices require unprecedented control of the layer thickness and stoichiometry of, for example (Ga, Al)As. Such structures can now be made by molecular beam epitaxy (MBE) with essentially mono-layer control[4,5,6] so that stacking sequences like Fig 1 are near routine. Very high mobility devices can be made by, for example, Si doping a thin region of (GaAl)As next to a thin region of GaAs. A schematic of such a structure is shown in Fig. 2. Si is a donor in (GaAl)As, and because of the band structure and small dimensions the donor electrons contribute to the conduction of the adjacent GaAs in a region where there are no Si atoms to scatter them. The result is mobility at low temperatures orders of magnitude greater than in conventional III-V's.

Another class of III-V devices, SEED's (self electro-optic effect devices) show optical absorption which is strongly dependent on intensity. Comparatively small populations of electrons are effectively confined in separated potential wells made by MBE to control stoichiometry and layer thickness. The properties of such structures are significantly altered by the absorption of a comparatively small number of photons so properties change non-linearly with light intensity resulting in light activated absorbtion changes allowing one to make switches. Such switches are likely to be key elements in future optical communication and data processing systems.

Gas phase III-V preparation techniques require the volatilization of recalcitrant elements and are often accomplished by using gaseous arsine, phosphine, trimethyl gallium etc. Using these kinds of starting reagents crystal growth takes place by a chemical reaction. Consequently, the management of growth and the control of imperfections requires understanding of complex chemical equilibria in both the gas and solid phases. For example, to fully describe solid gallium arsenide one must consider equilibria involving not only Ga on Ga sites and As on As sites, but also antisite defects (Ga at As sites and As at Ga sites), interstitial Ga and As, and Ga and As vacancies. The equilibria involving dopants and accidental impurities must also be considered, as well as chemical reactions involving arsine, phosphine etc. Progress in III-V device and systems has been paced by our understanding of the chemical processes used in their preparation. Important challenges include achieving high yield, high quality large area device chips, eliminating poisonous materials like arsine and integrating Si and III-V devices. Clearly progress has

Table 1. Typical electron resist materials

Polymer	Type	Sensitivity at 10 kV ($\mu C\ cm^{-2}$)	Resolution (μm)
Polystyrene	negative	200	<0.5
Poly(glycidyl methacrylate-co-3-chlorostyrene) (GMC)	negative	1	<1.0
Poly(methyl methacrylate)	positive	60	0.2
Poly(butene-1-sulfone) (PBS)	positive	0.8	0.25
Cross-linked poly(methyl methacrylate)	positive	10	<0.5
Novolac/PMPS (NPR)	positive	2	<0.5

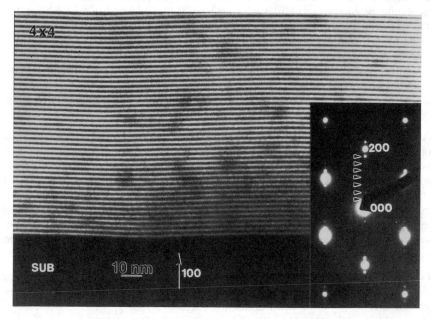

Figure 1: TEM of alternating layers of GaAs and (Ga,Al)As (Si doped) prepared by MBE (courtesy of A. Y. Cho).

been to a large degree due chemical achievements and just as clearly, limitations are very frequently due to chemical problems.

Quartz

I would like to describe hydrothermal quartz synthesis and commercialization as an example of synergistic interaction between synthesis and processing physical chemistry leading to better materials and more cost-effective devices. My discussions here and in the subsequent section on hydrothermal KTP will be at somewhat greater length because the role of chemists in hydrothermal synthesis and processing is perhaps not very well known and because of my familiarity with the subjects due to personal involvement.

Silica in the form of SiO_2, principally as quartz, is the most abundant compound of the earth's crust, but crystals more than a few mm in size occur in relatively few places. From the earliest days of radio the piezoelectric effect has been used to establish stable carrier frequencies and to multiplex and demultiplex signals. More recently quartz "tuning forks" have become essential for timing functions in electronic watches and in timing circuits for computers and telecommunications. Natural quartz was mined in Brazil (one of the few sources of large crystals) and used in electronic applications until World War II. Shortages in the U.S. caused by German submarine activity prompted efforts to synthesize quartz in the laboratory. Because the low temperature α-polymorph (stable below 573 ℃) is needed for applications, a low temperature crystallization process was required. The solution lay in *hydrothermal synthesis(7)* a novel low temperature preparative and crystal growth technique where ordinarily insoluble materials are dissolved (or where appropriate reacted) and crystallized in high pressure (10kpsi - 40 kpsi) water above its boiling point (300 ℃-600 ℃). Solubility is further improved by employing various complexing agents (e.g. OH^-, F^-). Similar agents are believed to aid in the formation of minerals in nature and hence such additives are frequently called "mineralizers". Foregoing lengthy historical review, typical quartz synthesis conditions based on work in Bell Labs are:(8,9)

mineralizer concentration-0.5-1.0M NaOH
pressure   -15.-25,000 psi
dissolving temperature   -425 ℃
growth temperature   -375 ℃
temperature difference
between dissolving and
growth zone (ΔT)   -50 ℃
% fill   -78-85%
growth rate in <0001>   -1.0-1.25mm/da

Growth (see schematic in Fig. 3) is typically conducted in cylindrical steel autoclaves (lab size 1" diameter x 1' length; factory 10" x 10' or more). Small (˜1") relatively inexpensive quartz chunks of feed stock (frequently called nutrient)

MODULATION DOPING

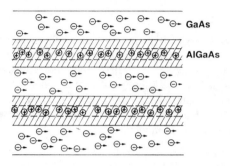

⊖⁻ **MOBILE ELECTRON**
⊕ **SILICON ATOM**

Figure 2:   Schematic of GaAs/(Al,Ga)As structures used to obtain high mobility (courtesy of A. Y. Cho).

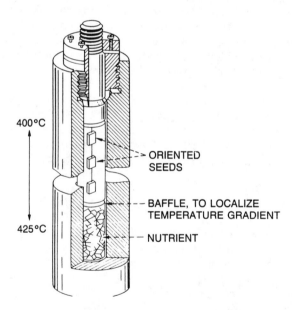

Figure 3:   Schematic of hydrothermal synthesis of quartz.

are placed in the bottom dissolving region of the autoclave. Suitably oriented seed plates are suspended in the growth region. The solution saturates in the bottom region and moves by convection to the top where because of the lower temperature it is supersaturated and growth occurs.

Careful studies of the physical chemistry of the growth process so as to understand the trade offs between growth rate, pressure, temperature and quality were essential in finding economically successful conditions. In order to understand the kinetics, solubility*(10)* and p-v-t*(11)* studies were necessary. The solubility in pure water was found to be too small for crystal growth (0.1 - 0.3 wgt %) but the solubility could be markedly increased by the addition of (OH)⁻ which acts as a mineralizer. We have studied mineralizers and their reactions for complexing various refractory oxides and sulfides.*(12-16)* A variety of complexers are known including (OH)⁻, Cl⁻, F⁻, Br⁻, I⁻ and acid media for the crystallization of Au and other noble metals. Frequently the ratio: (solubility/mineralizer concentration) is constant and independent of mineralizer concentration over wide ranges and sometimes it is a small rational number or fraction.

For instance complexing reactions such as

$$SiO_2 + 2OH^- \rightarrow SiO_3^{-2} + H_2O \tag{1}$$

go essentially to completion under particular conditions.*(10,12)* At these conditions, one observes that the ratio (solubility of SiO_2)/(OH⁻ concentration) = 1/2 which suggests that SiO_3^{-2} is a likely predominant species. Limitations in using solute/mineralizer ratios to determine species are discussed further in a series of papers by Laudise et. al.*(12-16)* Clearly the proper way to get species information is by direct measurement in solution. Recently Hosaka and Taki*(17)* have used Raman spectra to identify and quantify hydrothermal species. Their results agree with our earlier findings*(10)* which showed the preponderance of catenated species and reactions such as

$$2OH^- + 3SiO_2 \rightarrow Si_3O_7^{-2} + H_2O \tag{2}$$

in (OH)⁻. Systematic studies of the temperature dependence of quartz solubility in (OH)⁻ showed that log s is linear with $\frac{1}{T}$ (van't Hoff dependence).*(10)*

The dependence of solubility on pressure requires an understanding of the equation of state of hydrothermal solutions saturated with quartz. Systematic P-V-T studies have been conducted.*(11)* Fig. 4*(11)* summaries some of this data. As can be seen, behavior is qualitatively like that for pure H_2O but with pressures substantially reduced. Pressure in hydrothermal quartz synthesis is established by the initial fraction of the vessel volume (% fill) filled with (OH)⁻ solution at the beginning of the growth run.

It is worthwhile to outline the behavior of the fluid in a hydrothermal autoclave. Consider a case where the autoclave is filled to 80% full. As the autoclave is warmed the pressure is that indicated by the co-existence curve (g+*l*, A-B, in Fig. 4). The liquid level rises as the autoclave temperature increases until the autoclave

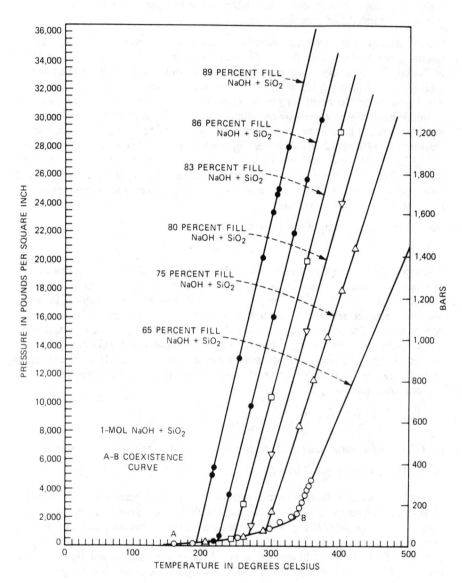

Figure 4: P-% fill-T relations for 1 M NaOH saturated with quartz

fills entirely with the liquid phase (for 80% fill at ≈270 ℃). At that point the pressure increases more abruptly (along the line marked 80% in Fig. 4) and there is only one fluid phase. At lower temperatures the phase is liquid. Above the critical temperature (for pure water 373 ℃) the fluid phase will, by definition, be a gas but a gas whose density, solvent power etc. are much like a liquid. The critical temperature and other properties of an NaOH solution saturated with quartz are not identical to those properties for water. However, for the relatively dilute solutions being considered here the qualitative behavior is similar. Further discussion can be found in references *(11)* and *(18)*. It is important to point out that for NaOH solutions saturated with quartz under hydrothermal conditions two properties are substantially different from water at ambient: the viscosity is markedly smaller and the density coefficient with temperature is markedly larger. and the density coefficient with temperature is markedly larger. Both of these differences result in vigorous fluid motion in the thermal gradients present under growth conditions and make diffusion problems less severe. Consequently hydrothermal reactions and crystal growth can take place at much faster rates than in solutions at ambient conditions.

The dependence of solubility on % fill in 0.5M (OH)⁻ solution is shown in Fig. 5.*(10)* The solubility in pure water is an order of magnitude smaller under similar conditions. In pure water, activity coefficient (actually fugacity calculated from appropriate compressibility) estimates enable one to get reasonably accurate values for the equilibrium constant. This treatment suggests that the solubizing reaction in pure water is:

$$SiO_2 + 2H_2O \rightarrow Si(OH)_4 \qquad (3)$$

Laudise et al.*(10)* have rationalized the % fill dependence of Fig. 5 using procedures based on their study of Eq. (3).

Systematic studies of growth rates on experimental parameters, including seed orientation, temperature, (OH)⁻, pressure (or % fill) and temperature difference between dissolving and growth zone have been conducted.*(19)* The growth was described by a simple rate equation:

$$R_{hkl} = k_{hkl} \alpha \Delta S \qquad (4)$$

Where R is the rate in a particular crystallographic direction, k is the velocity constant in that direction (with an Arrhenius temperature dependence), ΔS is the supersaturation and α is a dimensional conversion constant. ΔE^*, the activation energy typically varies between 10 and 20 kcal/mol depending on hkl.

Of particular importance is the role of chemistry in determining the perfection of the grown crystals. Defect chemistry and the equilibria which describe the incorporation of defects are essential to understanding and controlling the level of impurities in quartz. The most important impurity in quartz is H^+. H^+ easily fits interstitially in the large (1Å) channels which lie parallel to the c axis. It enters the lattice from the growth solution, as (OH)⁻, the $O^=$ being incorporated in the SiO_2 lattice. For device applications quartz must efficiently convert electrical to mechanical energy with minimal losses as heat. Inefficient piezoelectric conversion leads to heating and severe frequency stability problems. The figure of merit measuring conversion efficiency for a piezoelectric transducer is Q; 1/Q is a measure of loss. Dodd and Fraser[20] at Bell Labs showed that 1/Q was linearly related to (OH) as measured by the absorbtion at 2.86μm (an (OH) stretch frequency).

H^+ enters the lattice interstitially to charge compensate for Al^{+3} (an impurity unavoidably present in feed stock) which goes to a Si^{+4} site. The equilibrium reaction is:[21]

$$Al^{+3}_{solution} + OH^-_{solution} \xleftarrow{} Al^{+3}_{Si^{+4}\,site} \cdot OH_{interstitial}$$ (5)

The equilibrium constant K_o is defined as: S 3

$$K_o = \frac{[Al^{+3} \cdot OH]}{(Al^{+3})(OH)^-}$$ (6)

We early recognized[21] that K_o is a special case of the impurity distribution (or partition) constants whose behavior had been studied in great detail in order to understand impurity doping of semiconductors. Not too far from equilibrium (for example when growing a crystal at a moderate growth rate) one can treat impurity partition by solving the diffusion equation in front of the growing crystal interface. Fig. 6[21] shows the dependence of the effective partition coefficient K_{eff} for (OH)$^-$ impurity in quartz on growth rate. The effective partition coefficient is related to the equilibrium constant, K_o, for (OH)$^-$ incorporation by the Burton-Prim-Slichter Equation[22] which takes into account the diffusion of impurities toward the growing crystal:

$$K_{eff} = \frac{K_o}{K_o + (1 - K_o)e^{-Rt/D}}$$ (7)

Where R is the growth rate, t the thickness of the diffusion layer and D the diffusion constant. The understanding gleaned from studies such as those of Fig. 6 was used to improve the Q of synthetic quartz from initial Q's of 30,000 to present Q's in excess of 2×10^6. Thus systematic studies of the defect chemistry of quartz lead to procedures for defect control so as to provide device quality material equal and in some cases superior to natural quartz. Fig. 7 shows an early commercial quartz synthesis run in the AT&T factory facility in North Andover, Massachusetts.

Perfection control in quartz has lead to procedures for the preparation of dislocation free material[23]; controlled doping to prepare Fe doped synthetic citrine[24] and Fe doped X-irradiation induced color centers to make synthetic amethyst. Present activities center on procedures for preparing radiation hard quartz for space applications and material which etches uniformly for ease in preparing fine feature sized devices by lithographic techniques.[26] Domestic feedstock material has been shown to be equivalent to imported feed stock so that the U.S. is independent of foreign supplies.[27] However, increased capital investment in plant especially very large sized autoclaves has resulted in ever improving process efficiency on the part of Japanese quartz growers and has a substantial probability of eliminating the U.S. domestic quartz industry. Quartz processing competition epitomizes the difficulties of the U.S. electronic materials industry and has been discussed extensively elsewhere.[28]

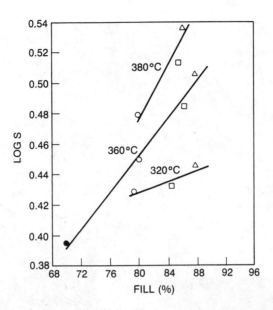

Figure 5: Quartz solubility dependence on % fill

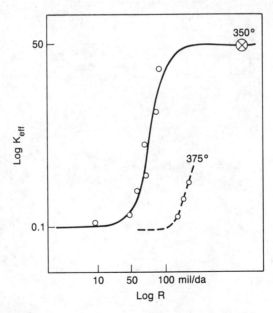

Figure 6: Dependence of the effective partition coefficient for (OH)⁻ impurity in quartz on growth rate

Figure 7: An early growth run in AT&T's Merrimac Valley, Mass., facility. (Photograph courtesy of AT&T Bell Laboratories.)

Potassium Titanyl Phosphate

An important application for lasers is their use in medical procedures. The high energy intensity, small spot size, cleanliness, self cauterization and easy steerability of laser light provide an ideal scalpel for many surgical procedures. A particularly useful laser for surgery is single crystal yttrium aluminum garnet doped with neodymium. Neodymium lasers efficiently emit high intensity near infrared light at 1.06μm. From the surgeon's viewpoint there is, however, a difficulty. The near infrared is not visible. The solution is to use a second crystal which efficiently converts the infrared to the visible - a harmonic generator. The requirement is for a material with high conversion efficiency and appropriate index of refraction characteristics to allow phase matching. Potassium titanyl phosphate (KTP) is ideal in this respect, converting the neodymium infrared to easily visible green light at .53 μm. The first report of KTP, $KTiOPO_4$ was the preparation of Ouvard(29) who melted TiO_2 and $K_4P_2O_7$ and K_3PO_4. KTP was prepared in polycrystalline form in 1971 by Masse and Grenier.(30) In both preparations the conditions were essentially anhydrous and only small crystallites were obtained. More detailed structural information was provided by Tordjman, Masse and Guitel.(31) Single crystals were first grown by Zumsteg, Bierlein and Gier(32) at the DuPont Experimental Station and they showed that KTP had large nonlinear optical coefficients. KTP's conversion efficiency is ~2x $LiNbO_3$ which is the most generally used 0.53μm harmonic generator and was first grown as a useful SHG material by Ballman(33) at AT&T Bell Labs. In addition the DuPont group discovered KTP to be phase matchable at 1.06μm, highly laser damage resistant and transparent over a wide range of wavelengths.$(32,34)$ The material decomposes at its melting point so that solution growth well above ambient is required. Thus, it is, so far, more expensive to prepare than $LiNbO_3$ which can be grown from the melt. For this reason up till now it has found uses mainly in medical and military applications. Lower cost preparative methods could have impact on its use and indeed more generally on the commercialization of non-linear optical materials (NLO's) and electrooptic devices and systems in general. Below we will discuss our work on improved preparative methods; but first we would like to use KTP to illustrate some of the bonding-structural-property aspects of materials chemistry.

Trying to answer the question "Why does KTP have such large NLO properties?" and "What sort of materials might be better?" is instructive in that it reveals our level of understanding of bonding-structure-property relations in the important area of optically useful oxides.

In the early days of electrooptic and nonlinear optic exploratory materials work in Bell Labs, A. A. Ballman and I(35) used the following rough hewn guide to search for materials:

1. For both electro-optic and NLO effects we would like the electric field (associated with an external applied voltage for the electro-optic effect and associated with the input light for NLO effects) to have a large effect on properties.

2. Therefore we look for materials whose polarizability is large and easy to change. Large electronic polarizability is associated with loosely bound electrons far from the nucleus, elements from the lower righthand side of the periodic table etc. Large ionic polarizability is associated with charge separation, a permanent dipole, piezoelectric materials with large coupling constants and ferroelectrics with large spontaneous polarization. Clues for good materials are a high index of refraction, high spontaneous polarization, and large dielectric constant.

3. Predictive power is poor, so confine searches to easily preparable materials. Use early screening techniques which work on powders like dielectric constant measurements, the Giebe-Scheibe circuit for piezoelectricity and the Kurtz[36,37] powder measurement test for SHG which was "invented" as a result of our pleas for help in finding materials.

The lack of rigor (and even complete accuracy and consistency) is apparent in such an approach, but it had the virtue of actually suggesting what to do and helping to guide Ballman in his discovery of LiNbO$_3$, LiTaO$_3$[33] and other practical optical materials.[38]

Today our understanding is more sophisticated. Chungstun Chin[39] in this volume reviews the recent history of the structure-bonding-property guidelines which have been developed for NLO's over the past twenty years and concludes with a discussion of his recent work on borates so we will forego a further extended discussion of recent structure-property-bonding work in NLO's. Suffice to say our _post facto_ explanations of NLO's are now vastly improved over what they were decades ago but in common with the theoretical situation in most of solid state chemistry our _a priori_ prediction of which new systems to look at for improved properties is still rudimentary.

Thus, we felt it particularly worthwhile to try to improve the preparation of KTP since we were willing to make the bet that drastically improved new materials were not just around the corner. Two KTP preparation methods are in general use, anhydrous flux growth as described by Loiacono and his co-workers[40] and by Ballman[41] and relatively high temperature hydrothermal growth as perfected by Belt and his colleagues.[42] We will not attempt a historical review of the development of these methods. Instead we will outline some features and preparative improvements[43,45] which we discovered as a result of a preliminary study of the process physical chemistry of hydrothermal KTP growth.

We begin with mineralizer screening by exposing weighed single crystals of KTP to various solvents in Pt capsules. If decomposition was severe as evidenced by the presence of non-KTP phases or if solubility were low, as evidenced by negligible weight loss, the solvent was ruled out. If solubility was a few percent then systematic phase equilibria and solubility measurements were made. It was our hope to find growth conditions for KTP below 450 °C where inexpensive, low carbon steel, large size autoclaves are available. The solubility in pure water and H$_3$PO$_4$ was too low to be interesting and KTP decomposed in (OH)$^-$. However, in potassium biphosphates and phosphates and their mixtures, solubilities above 1% were obtained at moderate temperatures with no apparent decomposition. Therefore, we investigated the phase stability relations in these solvents. Fig. 8[43] is a schematic of the water rich corner of the ternary phase diagram KPO_3-TiO_2-H_2O in 2M K_2HPO_4, the best mineralizer found. Everywhere along the line (A)-H_2O, the mole ratio KPO_3/TiO_2=1.00, the same as in KTP. The lines C-B and D-E correspond to phase boundaries. D-E is the boundary between the phase fields α+KTP and KTP. B-C is the boundary between the phase fields KTP and KTP + TiO_2 (anatase). It lies at stoichiometries richer in KPO_3 than (A)-H_2O and it moves to the left at lower temperatures. The line D-B schematically represents the solubility of KTP. It does not intersect (A)-H_2O, so KTP is incongruently soluble, i.e. at the mole ratio of KPO_3/TiO_2 in the liquid where KTP is the only stable solid phase KPO_3/TiO_2 in solution $\neq$ the ratio in KTP. We have measured the solubility of KTP in $1MK_2HPO_4$. If KPO_3 is absent, poor reproducibility indicative of slow decomposition was observed. If KPO_3 is present, as would be predicted by Fig. 8 no decomposition occurs, but because of the common ion effect, solubility is reduced as $(PO_3)^-$ increased as shown in Fig. 9.[45] However, conditions where solubility was high enough to permit growth could be established. KNO_3 is sometimes added to growth solutions to inhibit Ti reduction.

P-V-T measurements of 2M K_2HPO_4 and 2M K_2HPO_4 saturated with KTP were made. Fig. 10[45] shows the data for 2M K_2HPO_4. KTP saturated solutions give qualitatively similar data. It is important that growth be conducted in the absence of a gas-liquid interface, otherwise seeds in the gas phase will not grow and bubbling, boiling and bumping will contribute to poor quality deposition. In order to assure that growth was in the one-fluid phase region but still at as low a pressure as possible (to keep autoclave costs down) the P-V-T data were used to establish the temperature at which the autoclave fills with the one-fluid phase as a fraction of initial % fill (Fig. 11).[45] These data were essential in choosing growth conditions. Typical growth conditions are:[43,45]

Autoclave	Pt lined, 1 inch internal diameter x 6 inch inside length
Mineralizer	2M K_2HPO_4+0.5M KPO_3
% Fill	75%
Crystallization temperature	375 °C
Nutrient (dissolving) temperature	425 °C
Temperature differential (ΔT)	50 °C
Seed orientation (201), (010)	Rates (~0.07-0.14 mm/day), (010) rate generally higher, quality good

In the absence of KPO_3, TiO_2 inclusions close to the seed resulted in poor growth. These studies were used by Belt and Gasharov et. al.[44,45] to scale up the process at lower temperatures than ever used previously resulting in excellent crystals at practical rates. The defect chemistry of KTP has been studied by Pat Morris[46] and is reported in a paper in this symposium. An important characteristic of KTP made by low temperature hydrothermal growth is that it is superior with respect to

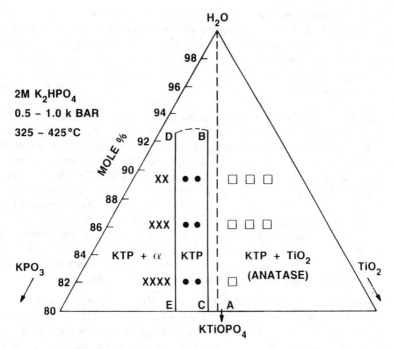

Figure 8: Part of the ternary phase diagram KPO$_3$-TiO$_2$-H$_2$O in 2MK$_2$HPO$_4$

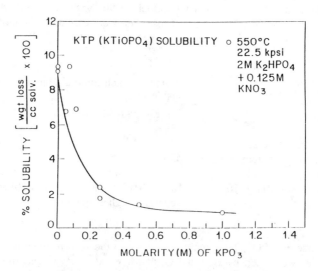

Figure 9: Solubility of KTP as a function of KPO$_3$ concentration

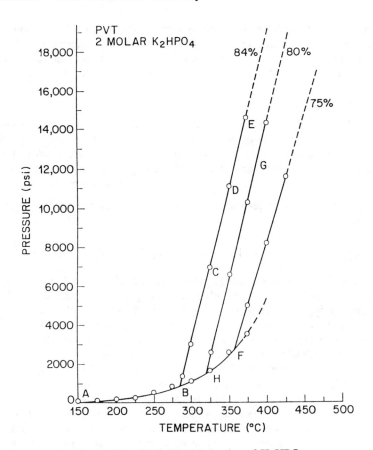

Figure 10: P-% fill-T behavior of K_2HPO_4

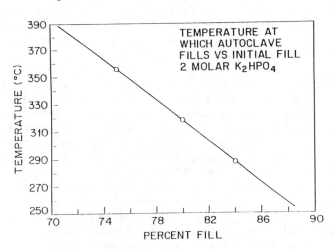

Figure 11: Temperature at which autoclave containing 2M K_2HPO_4 fills as a function of initial % fill

damage relative to material prepared at higher temperatures.[46] Although these studies are by no means complete they give us encouragement that process physical chemical studies can play a crucial role in the materials chemistry and commercialization of hydrothermal and other electronic materials.

Optical Fibers

Optical fibers are the materials basis of the optical communication systems which in the last five years have essentially replaced the backbone long distance communication network in the U.S. and are now in submarine systems that connect the U.S. to both Europe and Asia. Developing economical fiber preparation techniques has depended on process physical chemical understanding.

In the processes used to prepare communication optical fiber preforms the basic overall reaction is

$$nSiCl_4(g) + GeCl_4(g) + (n+1)O_2(g) \overset{\leftarrow}{\rightarrow} GeO_2(SiO_2)_n(\ell) + 2\,(n+1)\,Cl_2(g) \quad (8)$$

$GeO_2(SiO_2)_n(\ell)$ is supercooled germanium-doped silica liquid, i.e. the glass of appropriate composition to give the desired fiber properties. Reaction (8) is useful in an engineering sense in that it easily leads to the concept of process efficiency, but since it is a composite of a very large number of intermediate reactions, it does not aid appreciably in understanding whether a rate-limiting step is present which impedes the establishment of equilibrium. Other volatile compounds for example of P, B, Al, F, etc. are often introduced so as to dope the glass to produce the desired index of refraction, lack of dispersion etc. Since its discovery by MacChesney in 1974[47], the deposition process called Modified Chemical Vapor Deposition, MCVD, has achieved thorough study and exploration and is now the preparation method used for the production of many hundreds of thousands of kilometers of fiber. In this process the reactant gasses are introduced into a SiO_2 tube held in a glass lathe. A local region of the tube is heated with an external gas -O_2 torch initiating reaction (8). $GeO_2(SiO_2)_n(\ell)$ nucleates and grows as colloidal sized glassy particles (often called soot) in the gas phase. Deposition of the soot takes place down stream on cold regions of the tube. The hot region is caused to traverse the tube laying down a thin layer soot of composition fixed by the gas component concentrations, temperature etc. at which Eq. (8) takes place. As the hot zone moves down the tube it viscously sinters the soot to a thin layered glass deposit. Additional traverses are used to build up the layer of deposited material to the desired thickness. Composition changes in gas can be used to alter deposit composition and make a desired profile in composition and index of refraction so as to make useful light guiding structures. Following deposition the tube is heated and collapsed into a rod (usually called the preform). Fiber is drawn from the preform in a subsequent process. MCVD temperatures up to plasma temperatures have been employed. We[48] have been able to show that relatively straightforward thermodynamic modeling of the relevant equilibria taking into account the diversity of species expected for high temperatures can be used to describe the process. We used the method of element potentials since it easily deals with many species and cases (like melting) where a phase concentration goes to zero.

Figure 12[48] illustrates the wide range of species which occur as a function of temperature during fiber preform preparation. The input gas concentrations were fixed at a value typical of the real process. Temperatures where soot is formed, the soot is sintered and the tube is collapsed are indicated. Two particularly important results can be observed. The quantities of $SiCl_4(g)$ at equilibrium are negligible everywhere over the temperature range where Eq. (8) is used. Thus although Eq. (8) goes to essential completion Eq. 8 clearly does not describe the concentration of important species at equilibrium. One need only note that gaseous Cl, Cl_2, GeO, ClO, O and $GeCl$ are at substantial concentrations at for example, 2000 °K.

Thermodynamic calculations of this sort together with experimental measurements of glass and gas compositions have been used to establish that the Ge content of the soot is initially fixed at the temperature for soot formation and is essentially an equilibrium property. The Ge-Si ratio is initially established when $(Ge,Si)O_2$ solid nuclei form and grow in the gas phase to colloidal aerosol particles. The final Ge-Si ratio in the glass is established by the same equilibria operating at the sintering temperature. Understanding of thermodynamics gives a solid basis for understanding kinetics, which is essential to rate control and cost reduction. The kinetics of deposition have been shown to be limited by *thermophoretic* movement of the colloidal $(Si, Ge)O_2$ particles from the hot gas to the cold wall of the deposition tube.[49][50] Thermophoresis is the movement of a colloidal particle in hot gas down a temperature gradient. It is driven by the extra momentum exchange on the hotter side of a particle in a temperature gradient relative to the momentum exchange from collisions on its cooler side. Studies of the thermodynamics and kinetics established the physical chemical basis of MCVD, improved the rate and quality of preforms and were important to developing a commercially economic process. Space prevents discussion of the achievements and chemical aspects of optical loss reduction and strength enhancement. Today it is not unfair to say that optical communications rests to a very large degree on materials chemical discoveries, understanding and achievements. Future progress such as an all optical fiber amplifier will be paced by chemical understanding and control.

Future Challenges

This brief review has attempted to give a personal sampling of some modern electronic materials chemical achievements. Modern electronics is perhaps more than any other technology the driving engine for productivity and standard of living improvements in our society. The increasing scale of integrated circuit integration has been brought about largely by our chemical manipulation of Si and is at the root of the decreasing cost of computing and communication. The result is the information society, the second industrial revolution.

What does the future need from materials chemists? I would assert that any agenda must include the following:

1. Make electronic materials processing a science and improve U.S. competitiveness. Processing is clearly mainly chemical and poor processing

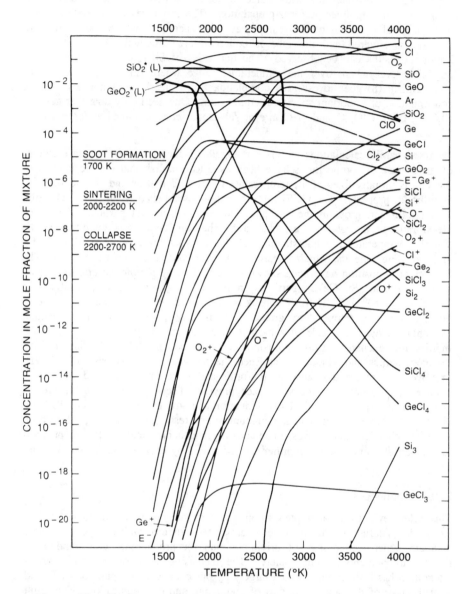

Figure 12:   Species concentrations as a function of temperature in the oxidation of $SiCl_4$ and $GeCl_4$.

is at the root of most U.S. electronics competition problems. The role of chemistry is self evident. The need is crying.

2. Find a new generation of materials and devices beyond VLSI. Si will run out somewhere in the sub-micron region. Feature size will eventually be too small to allow processing by present methods. New materials and new processes will be required or else electronics will go stagnant - the driving force for living standard improvement will die. Possible solutions include: MBE based structures, all in vacuum continuous processing, new materials, and electronic logic and memory based on new principals such as the "molecular memory". All will depend upon chemical progress and understanding and almost certainly will require new materials and processes.

3. True Molecular Engineering - At last understanding bonding-structure-properties so we can *a priori* choose the right atoms, compounds and structures to possess the engineering properties we need. Molecular engineering will perhaps always be a not fully accessible holy grail but much improvement of our predictive capability has been made and with commitment continued progress can be certain.

4. Exploit basic science findings on real materials. Examples include:

    — Nanostructures and clusters

    — MBE for inorganics

    — A new generation of optical materials for optical regeneration and eventually optical information processing.

5. Probably our greatest challenge is to insure a new generation of the best and brightest in science, engineering, materials and chemistry. It should be the purpose of all of us in science and technology to convey the excitement, relevance and contributions of our activities. I hope this review has in a small way helped this purpose.

Acknowledgments

I would like to thank my colleagues and collaborators especially E. D. Kolb, R. L. Barns, R. Belt and G. Gasharov, A. A. Ballman, K. B. McAfee, K. Nassau and W. A. Sunder for their invaluable contributions to quartz and KTP. I would like to thank J. B. MacChesney, A. Y. Cho, L. F. Thompson and R. Gottscho for patiently explaining their work and helping me to extract materials chemical examples from their achievements.

LITERATURE CITED

[1] "Materials Science and Engineering for the 1990's", National Research Council, National Academy Press, Washington, 1989.
[2] Thompson, L. F., in "Encyclopedia of Materials Science and Engineering", Michael Bever, Ed., MIT Press, 1986, p. 4207, for an introduction.

[3] Dautremont-Smith, W. C., Gottscho, R. A. and Schwartz, R. J. in "Semiconductor Materials and Process Technology Handbook", McGuire, Gary E., Ed Noyes Publications, Park Ridge, N.J. 1988 p. 191. See also other papers in this volume for discussions of other semiconductor synthesis and processing subjects.

[4] Cho, A. Y., Thin Solid Films, 100, (1983) 291.

[5] Cho, A. Y., J. Cryst. Growth, 95 (1989) 1.

[6] Capasso, F., Sen, S., Beltram, F., Lunardi, L. M., Vengurlekar, A. S., Smith, P. R., Shah, N. J., Malik, R. J., Cho, A. Y., IEEE Trans. on Elec. Dev., 36 (1989) 2065.

[7] Laudise, R. A., Hartman, P., ed., North-Holland Pub. Co., p. 162-197 (1973).

[8] Laudise, R. A., Published in Proceedings of ISSCG6, "Advanced Crystal Growth", Ed. by Dryburgh, P. M., Cockayne, B., and Barraclough, K. G., Prentice Hall, New York, 1987, p. 268.

[9] Laudise, R. A., Sullivan, R. A., Chemical Engineering Progress, Vol. 55, No. 5, pages 55-59 (1959).

[10] Laudise, R. A., Ballman, A. A., J. Phys. Chem., Vol. 65, 1396-1400 (1961).

[11] Kolb, E. D., Key, P. L., Laudise, R. A. and Simpson, E. E., Bell System Technical Journal, Vol. 62(3), 639-55, March 1983.

[12] Laudise, R. A., Kolb, E. D., Endeavour, Vol. XXVIII, No. 105, 114-117 (Sept. 1969).

[13] Laudise, R. A., Kolb, E. D., American Mineralogist Vol. 48, 642-648 (1963).

[14] Barns, R. L., Laudise, R. A., Shields, R. M., J. Phys. Chem. Vol. 67, 835-839 (1963).

[15] Kolb, E. D., Barns, R. L., Laudise, R. A. and Grenier, J. C., J. of Crystal Growth, Vol. 50, No. 2, 404-418, (1980).

[16] Kolb, E. D., Grenier, J. C. and Laudise, R. A., J. Crystal Growth, Vol. 51, 178-182 (1981).

[17] Osaka, M. and Taki, S., J. Cryst. Growth, 100 (1990) 343.

[18] Laudise, R. A., Chemical and Engineering News, Vol. 65, (39) 30-43 (1987).

[19] Laudise, R. A., J. Am. Chem. Soc., Vol. 81, 562-566 (1959).

[20] Dodd, D. M. and Fraser, D. B., J. Phys. Chem. Solids, 26, 673 (1965).

[21] Laudise, R. A., Kolb, E., Lias, N. C. and Grudenski, E., Crystal Growth, Vol. XI Soviet Academy of Sciences, Moscow, 352 (1975).

[22] Burton, J. A., Prim, R. C. and Slichter, W. P., J. Chem. Phys., 21 (1953)1987.

[23] Barns, R. L., Freeland, P. E., Kolb, E., Laudise, R. A. and Patel, J. R., J. of Crystal Growth, 43, 676-686 (1978).

[24] Ballman, A. A., Am. Min. 46 (1961) 439.

[25] Laudise, R. A., Barns, R. L., Stevens, D. S., Simpson, E. E. and Brown, H., Proceedings of the 42nd Annual Symposium on Frequency Control, 116-126 (1988).

[26] Laudise, R. A. and Barns, R. L., IEEE Transactions on Ultrasonics, Ferroelectrics and Frequency Control, Vol. 35, Number 3, (1988).

[27] Kolb, E., Nassau, K., Laudise, R. A., Simpson, E. E. and Kroupa, K., J. of Crystal Growth 36, 93-100 (1976).

[28] Laudise, R. A., National Materials Advisory Board (NMAB) #424, National Academy Press, Washington, D.C. (1985).

[29] Ouvard, L., Compt. Rend. (Paris) 121 (1890) 117, as reported in The Gmelin Handbook, Vol. 41, Ti, 8th ed. (1951).

[30] Masse, R. and Grenier, J. C., Bull. Soc. Franc. Mineral. Crist., 94 (1971) 437.

[31] Tordjman, I., Masse, R. and Guitel, J. C., Z. Krist. 139 (1974) 103.

[32] Zumsteg, F. C., Bierlein, J. D. and Gier, T. E., J. Appl. Phys. 47 (1976) 4980.

[33] Ballman, A. A., J. Am. Ceram. Soc. 48 (1965) 112.

[34] Bierlein, J. D., assigned to E.I. Du Pont de Nemours and Company, US Patent 3, 949,, 323; April 6, 1976. See also Gier, T. E. assigned to E.I. Du Pont de Nemours and Company, US Patent 4, 231, 838; November 4, 1980.

[35] Laudise, R. A., Bell Laboratories Record, 3-7 (Jan. 1968).

[36] Kurtz, S. K. and Perry, T. T., "A Powder Technique for the Evaluation of Nonlinear Optical Materials, J. Appl. Phys., 39, 3798-3813 (July 1968).

[37] Kurtz, S. K. and Dougherty, J. P., "Methods for the Detection of Noncentrosymmetry in Solids", Chapter 38 in Systematic Materials Analysis, Vol. IV. Academic Press, Inc. (1978), pp. 269-342.

[38] Laudise, R. A., Reprinted from Crystal Growth, Supplement to J. Phys. Chem. Solids, Pergamon Press, 3-16 (1967).

[39] Chen, Chungstun, Presented at Boston ACS Meeting, 1990, Published in this volume, p 00.

[40] Jacco, J. C., Loiacono, G. M., Jaso, M., Mizell, G. and Greenberg, B., J. Crystal Growth, 70, 484 (1984)

[41] Ballman, A. A., Brown, H., Olson, D. H. and Rice, C. E., J. Crystal Growth, 75, 390 (1986).

[42] Liu, Y. S., Drafall, L., Dentz, D. and Belt, R., Proc. Tech. Conf. on Lasers and Electrooptics, Washington, DC, June 1981, p. 26. See also U.S. Patents 3949323 and 4305778 (early DuPont patents).

[43] Laudise, R. A., Cava, R. J. and Caporaso, A. J., J. Crystal Growth, Vol. 74, 275-280 (1986).

[44] Belt, R. F., Gashurov, G. and Laudise, R. A., Proceedings of SPIE's 32nd International Technical Symposium, Vol. 968, 100-107, (1988).

[45] Laudise, R. A., Sunder, W. A., Belt, R. F. and Gashurov, G., submitted to Journal of Crystal Growth.

[46] Morris, Patricia, Presented at Boston ACS Meeting, 1990, Published in this Volume, p. 00.

[47] MacChesney, J. B., O'Connor, P. B. and Presby, H. M., Proc. IEEE, 62, (1974) 1280.

[48] McAfee, K. B., Laudise, R. A. and Hozack, R. S., J. Am. Ceram. Soc., 67, (1984) 6.

[49] Simpkins, P. G., Kosinski, S. G. and MacChesney, J. B., J. Appl. Phys., 50, (1979) 5676-81.

[50] Walker, K. L., Geyling, F. T. and Nagel, S. R., J. Am. Ceram. Soc. 63 (1980) 96-102.

RECEIVED August 28, 1990

NOVEL APPROACHES TO ORIENTATION OF MOLECULAR UNITS

Chapter 28

From Molecular to Supramolecular Nonlinear Optical Properties

J.-M. Lehn

**Collège de France, 11, Place Marcelin Berthelot, 75005 Paris
and Université Louis Pasteur, 4, Rue Blaise Pascal,
67000 Strasbourg, France**

The arrangement of photoactive molecular species into organized polymolecular architectures may lead to novel optical properties at the supramolecular level. Push-pull conjugated molecules have been synthesized, that contain diacetylenic and polyolefinic units. They display interesting absorption and emission properties as well as pronounced nonlinear optical (NLO) effects. Derivatives incorporating photosensitive and redox-active groups have been obtained. The observation of second harmonic generation from non-dipolar non-centrosymmetric aromatic charge transfer molecules may lead to dipole independent NLO materials. Supramolecular engineering has been performed by incorporation of push-pull molecular components into organized systems such as liquid crystals and Langmuir-Blodgett films. Organized materials may be generated by molecular recognition induced self-assembling of liquid crystalline phases or polymers and of ordered solid state structures. The molecular components involved may be designed so that the formation of the organized phases depends on interaction between complementary molecular subunits, thus leading to molecular recognition induced nonlinear optical properties at the supramolecular level.

Nonlinear optical (NLO) properties are usually considered to depend on the intrinsic features of the molecule and on the arrangement of a material. An intermediate level of complexity should also be taken into account, that of the formation of well-defined supermolecules, resulting from the association of two or more complementary components held together by a specific array of intermolecular interactions (1). Such intermolecular bonding may yield more or less pronounced NLO effects in a variety of supramolecular species (2). Thus, three levels of nonlinear optical properties may be distinguished: the molecule, the supermolecule and the material. The molecular and supramolecular levels involve respectively - intramolecular effects and structures, -

0097–6156/91/0455–0436$06.00/0

intermolecular interactions and architectures; both aspects intervene in the material together with collective effects.

Supramolecular chemistry (1), the chemistry of the intermolecular bond, involves both organization and function, depending in particular on molecular recognition events. It extends towards the design of molecular devices that operate on photons, electrons or ions and could form through self-assembling of their components (1,3). We have for instance studied photonic devices performing light conversion and photoinduced charge separation (1,4), electronic devices acting as molecular wires (1,5) and ionic devices for ion translocation through membranes by means of mobile carriers (1,6) or channel type structures (1,3,7). It has been pointed out earlier that supramolecular species of various types (organometallic and coordination compounds, supermolecules involving hydrogen bonding, charge transfer interactions, etc) may present specific NLO features (2). I herewith briefly describe the results obtained on the NLO properties of species related to our work on molecular devices and point out the role of processes directing supramolecular arrangements, in particular by recognition induced self-organization. More detailed descriptions and references may be found in the papers cited.

The *design* of molecules, supermolecules and materials presenting NLO activity involves *molecular and supramolecular engineering*. At the molecular level, a *high polarisability*, described by large quadratic β and cubic γ hyperpolarisability coefficients, is sought for. At the supramolecular level, it is necessary to achieve a *high degree of organization*, such as found in molecular layers, films, liquid crystals, solid state, which may be induced by molecular recognition and inclusion complex formation. Both features are required for materials to display pronounced second order NLO effects; the structure must also be non-centrosymmetric, due either to the molecular components or to their arrangement in condensed phase. On the other hand, centrosymmetric species present third order but no second order NLO properties. In addition, bulk characteristics such as stability, preparation and processing, mechanical features will determine the practical usefulness of a given material (for general presentations see for instance ref. 8-10 as well as the book cited in ref. 2).

Push-Pull Polyconjugated Molecules

Modification and functionalisation of natural polyenes, the carotenoids, is an efficient way for the molecular engineering of polyenic chains. Terminal bis-pyridinium carotenoids, termed *caroviologens*, represent an approach to electron conducting molecular wires (5). Fitting polyconjugated chains with an electron donor group on one end and an electron acceptor on the other end yields push-pull systems of type **1** that may be considered as polarized, unidirectional (oriented) molecular wires and also possess marked NLO properties.

Indeed such compounds containing long conjugated chains, are expected to be highly polarizable, to possess high lying π orbitals and low lying π* orbitals, to display solvatochromism and charge separation excited states.

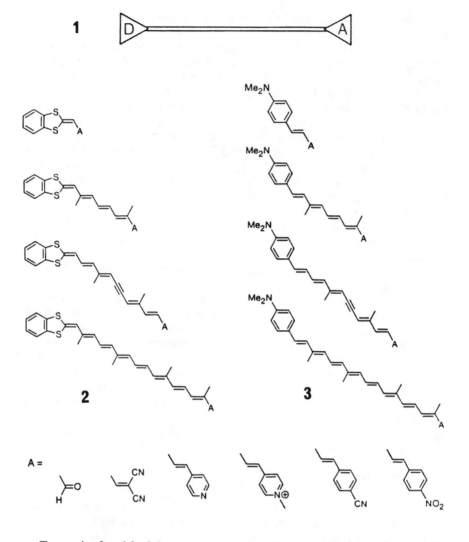

Two series **2** and **3** of donor-acceptor polyconjugated molecules bearing either a benzodithia (**2**) or a dimethylaminophenyl (**3**) group as donor and a variety of acceptor groups **A**, have been synthesized and their photophysical properties have been studied (11).

The electronic absorption, fluorescence and excitation spectra of these compounds indicate the presence of an internal charge transfer (ICT) excited state giving rise to a fluorescence band that displays strong solvatochromism. Both the emission wavelengths and the Stokes shifts increase with solvent polarity, in agreement with a large increase in dipole moment in the excited state. As the chain length increases the absorption undergoes both bathochromic and hyperchromic changes. The emission also shows a bathochromic shift. These effects level off with increasing length of the conjugation path, insertion of the triple bond leading to weaker conjugation. They indicate an increase in delocalisation with chain length and point to a long distance internal charge transfer in the excited state. Thus, the molecules of series **2** and **3** behave as highly polarizable molecular wires.

A detailed investigation of the NLO properties of molecules of series **2** and **3** has been performed in the powder state (12) as well as in solution by the electric-field-induced-second-harmonic (EFISH) generation method (13-15). It has allowed the analysis of the influence of the molecular parameters on the NLO features (12-15).

There is a sharp increase in quadratic hyperpolarisability β with the number of double bonds n. The variation of $\mu\beta(0)$ as a function of n is not exponential, the dependence being about $\mu\beta(0) \sim n^{2.4}$ for the molecules of series **2** (A=CHO). Similar trends hold for series **3** (A=CHO). These results are in agreement with earlier experimental data (giving a nearly quadratic dependence) (16) and with calculations yielding a dependence in $n^{2.6}$ for a series of type **2** (A=CHO) (17) and in $n^{2.5}$ or $n^{2.7}$ for compounds of type **3** (A=NO$_2$ or CHO) (18). For the longest compounds ever measured (eight conjugated double bonds), the static $\mu\beta(0)$ values are exceptionnally large, about 50 times that of 4-nitroaniline, and the effect does still not level off. The introduction of a triple bond causes a hypsochromic shift and a drop in $\mu\beta(0)$ with respect to the expected value, indicating a reduction in the electronic transmission ability.

The following sequences of donor and acceptor strengths have been obtained:

The difference in donating (or accepting) power between donor (or acceptor) groups falls off as the chain length increases. The efficiency of the donor → acceptor interaction, for different donor-acceptor pairs, appears to level out with the lengthening of the conjugation path (saturation effect). Preliminary data from measurements of symmetrical, polyene α,ω-dialdehydes, have yielded large third order hyperpolarisabilities $\gamma(0)$, that increase also markedly with chain length (Puccetti, G.; Ledoux, I.; Zyss, J., unpublished data).

Switchable Molecular Wires and NLO effects

The insertion of a photosensitive group or of a redox active unit into the push-pull system 1 yields switchable molecular wires and push-pull molecules that contain a photo-switch or a redox switch S, as represented in 4. Compounds of such type containing for instance electroactive ferrocene groups and photosensitive metal complexes, have been synthesized. Some of them are shown in series 5 (Marczinke, B. ; Przibilla, K.J.; Lehn, J.-M., unpublished data).

The electronic and optical properties of such compounds may be expected to respond to external electrical or light stimuli. The NLO properties of a ferrocene containing molecule have been reported (19).

Second-Harmonic Generation from Non-dipolar Non-centrosymmetric Aromatic Charge-Transfer Molecules

A study of second-harmonic generation (SHG) from powders of 1,3,5-triamino-2,4,6-trinitro benzene type compounds showed that the parent molecule 6 had an activity of about 3 x urea, whereas N-substituted derivatives were inactive (Ledoux, I.; Zyss, J.; Lehn, J.-M.; Siegel, J.; Brienne, M.-J., unpublished data).

Compound 6 may be considered as a generalized 4-nitroaniline of D_{3h} symmetry. It has a non-dipolar ground state and is non-centrosymmetric, whereas 4-nitroaniline is dipolar. The observation of SHG for **6** is compatible with a crystal structure departing from a hexagonal, centrosymmetric lattice. The absence of a dipole moment due to the symmetric structure of **6** precludes any molecular nonlinear contribution of vectorial nature. The origin of the SHG will be addressed elsewhere..

The observation of SHG from a molecule deprived of vectorial features, opens new perspectives in molecular engineering towards quadratic NLO properties. Non-dipolar non-centrosymmetric molecular moieties could serve as building block for novel types of NLO materials in which the organization is not influenced by dipole-dipole interactions.

Supramolecular Engineering of Organized NLO Materials

The realization of efficient SHG materials involves performing supramolecular engineering on the compounds presenting pronounced NLO properties obtained by molecular engineering. This may be achieved by introducing the molecules into organized phases such as molecular films, liquid crystals or solid state structures, by suitable derivatization or mixing with host substances.

Well organized *Langmuir-Blodgett (LB) films* have been obtained from mixtures of a push-pull carotenoid and ω-tricosenoic acid as shown in **7**. These mixed films exhibit a very good cohesion, with an area of about 25 Å^2 per carotenoid molecule. They can easily be transferred onto solid substrates. Examination by UV-visible linear dichroism measurements confirms that the carotenoid chains are oriented perpendicularly to the surface of the substrate in card-packed aggregates, in which the polyenic chains interact via excitonic coupling, as indicated by the large hypsochromic shift of the π-π* transition (20).

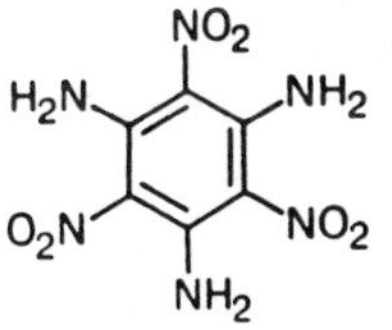

6

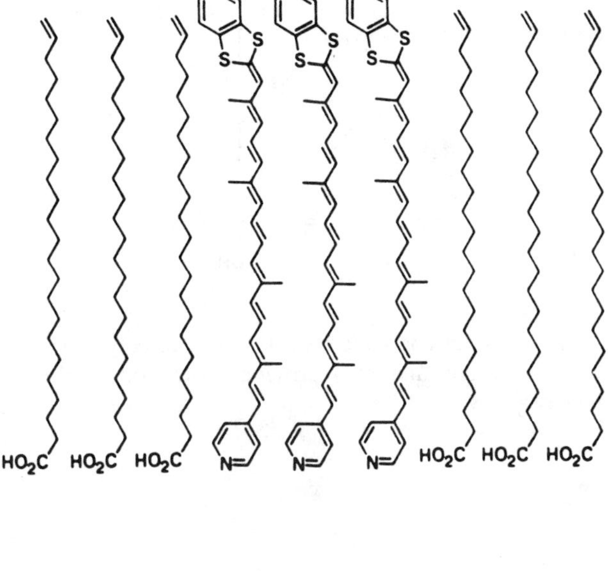

7

The same push-pull carotenoid can also be introduced into *LB multilayers* built from 1/1 mixtures with an *amphiphilic cyclodextrin*. The polyenic chains are again perpendicular to the substrate and some carotenoid molecules do not aggregate, but are isolated by the host substance (21).

Liquid crystals may be obtained from suitably modified molecules presenting NLO properties. Thus, various push-pull stilbene **8** and diacetylene **9** derivatives bearing long chains R have been shown to display nematic and smectic mesophases (22).

8

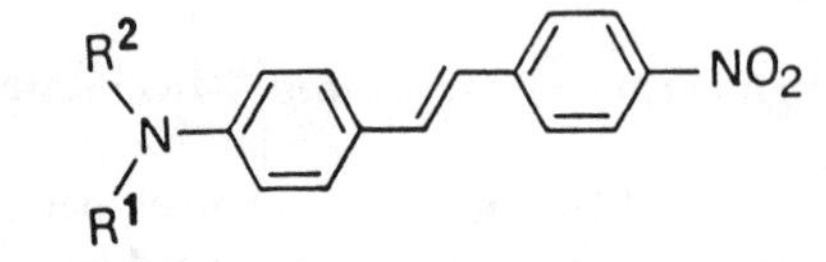

R^2
R^1
—NO_2

9 RNH—$C\equiv C$–$C\equiv C$—NO_2

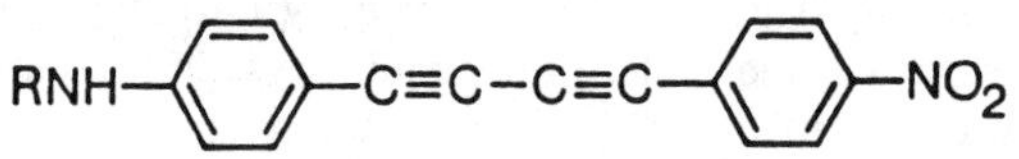

Preliminary measurements on some compounds of type **8** and **9** gave SHG effects similar to those of urea and quartz respectively.

Mesophases of columnar nature are formed by derivatives of the symmetrical triamino-trinitro benzene unit **6** of the type **10** (R = $C_{12} H_{25}$ for instance) Brienne, M.-J.; Lehn, J.-M., unpublished data).

10

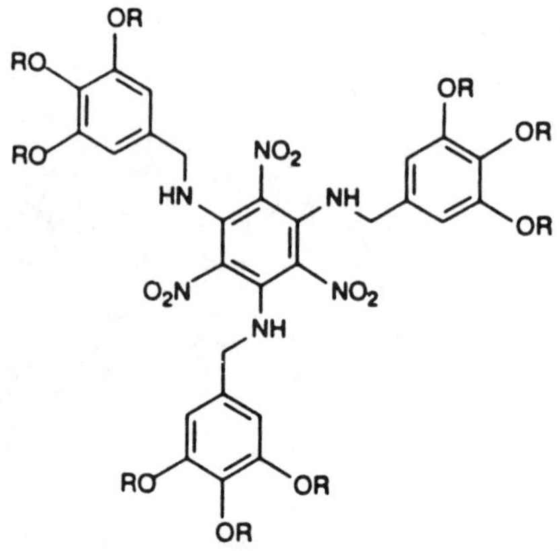

OR
RO
OR
RO
OR
HN NO_2 NH   OR
O_2N NO_2
NH
RO   OR
OR

Such LB films or liquid crystals as well as polymeric structures may yield efficient NLO materials and also provide protection of the SHG active species and processability.

Molecular Recognition Induced Self-organization and NLO Effects

Molecular recognition processes rest on selective intermolecular interactions between complementary components. They may affect the properties of the system at the molecular, the supramolecular and the material levels by respectively 1) perturbing the electronic and optical properties of the components; 2) generating supramolecular species; 3) inducing organization in condensed phase. All three effects are of importance with respect to the NLO properties of the material and its constituents.

The binding of two complementary components, respectively of donor D and acceptor A type, has two consequences: 1) it yields a push-pull supramolecular species, in which, 2) the interaction may be expected to modify the initial D, A features of the isolated units. Such a process is represented in scheme **11**. Both factors should influence the SHG activity, leading to recognition dependent NLO properties or, conversely, to the expression of the recognition event through NLO features. Scheme **12** illustrates the case of hydrogen bonding association; other interactions (electrostatic, D-A, ...) may also be envisaged.

Molecular recognition directed self-assembling of organized phases has been described recently in the formation 1) of mesophases by association of complementary molecular component, as in **13** (_23_); 2) of supramolecular liquid crystalline polymers of type **14** (_24_) and 3) of ordered solid state structures, such as that represented by **15** (_25_). In all these cases, the incorporation of NLO active groups may be expected to produce materials whose SHG properties would depend on molecular recognition induced self-organization.

Conclusion

The results presented above illustrate how combining the design of NLO active molecules with the manipulation of selective intermolecular interactions may produce novel NLO materials. Bringing together two basic features of supramolecular chemistry -molecular recognition and self-organization- with the optical properties of the components, opens ways towards the design of *supramolecular photonic devices*.

13

14

15

Literature cited

1. Lehn, J.-M. *Angew. Chem. Int. Ed. Engl.* 1988, *27*, 89.
2. Lehn, J.-M. In *Nonlinear Optical Properties of Organic Molecules and Crystals*; Chemla, D.S.; Zyss, J., Eds.; Academic: New-York, 1987; Vol. 1, 215.
3. Lehn, J.-M. *Angew. Chem. Int. Ed. Engl.* 1990, *29*, in press.
4. Lehn, J.-M. In *Supramolecular Photochemistry;* Balzani, V., Ed.; Reidel: Dordrecht, 1987; p. 29.
5. Arrhenius, T.S.; Blanchard-Desce, M.; Dvolaitzky, M.; Lehn, J.-M.; Malthête, J. *Proc. Natl. Acad. Sci* USA 1986, *83*, 5355.
6. Lehn, J.-M. In *Physical Chemistry of Transmembrane Ion Motions*; Spach, G., Ed.; Elsevier: Amsterdam, 1983; p. 181.
7. Jullien, L.; Lehn, J.-M. *Tetrahedron Lett.* 1988, 3803.
8. *Nonlinear Optical Properties of Organic and Polymeric Materials*; Williams, D.J., Ed.; A.C.S. Symp. Ser. 233, Washington, 1983.
9. Williams, D.J. *Angew. Chem. Int. Ed. Engl.* 1984, *23*, 690.
10. Garito, A.F.; Teng, C.C.; Wong, K.Y.; Zammani'Khamiri, O. *Mol. Cryst. Liq. Cryst.* 1984, *106*, 219.

11. Slama-Schwok, A.; Blanchard-Desce, M.; Lehn, J.-M. *J. Phys. Chem.* 1990, *94*, 3894.
12. Blanchard-Desce, M.; Ledoux, I.; Lehn, J.-M.; Malthête, J.; Zyss, J. *J. Chem. Soc., Chem. Commun.* 1988, 737.
13. Blanchard-Desce, M.; Ledoux, I.; Lehn, J.-M.; Zyss, J. In *Organic Materials for Nonlinear Optics*, Hann, R.A.; Bloor, D., Eds; Royal Society of Chemistry: London, 1989; special publication N° 69, p. 170.
14. Barzoukas, M.; Blanchard-Desce, M.; Josse, D.; Lehn, J.-M.; Zyss, J. *Inst. Phys. Conf. Ser. N° 103: Section 2.6*, 1989, 239.
15. Barzoukas, M., Blanchard-Desce, M.; Josse, D.; Lehn, J.-M.; Zyss, J. *Chem. Phys.* 1989, *133*, 323.
16. Dulcic, A.; Flytzanis, C.; Tang, C.L.; Pépin, D.; Fétizon, M.; Hoppilliard, Y. *J. Chem. Phys.* 1981, *74*, 1559.
17. Toussaint, J.M.; Meyers, F.; Brédas, J.L. In *Conjugated Polymeric Materials. Opportunities in Elecronics, Optoelectronics and Molecular Electronics*, Brédas, J.L.; Chance, R.R., Eds; NATO-ARW Series E; Kluwer: Dordrecht, 1990; Vol. 182, p. 207.
18. Morley, J.O.; Docherty, V. J.; Pugh, D. *J. Chem. Soc., Perkin Trans II*, 1987, 1351.
19. Green, M.L.H.; Marder, S.R.; Thompson, M.E.; Bandy, J.A.; Bloor, D.; Kolinsky, P.V.; Jones, R.J. *Nature* 1987, *330*, 360.
20. Palacin, S.; Blanchard-Desce, M.; Lehn, J.-M.; Barraud, A. *Thin Solid Films* 1989, *178*, 387.
21. Palacin, S. *Thin Solid Films* 1989, *178*, 327.I
22. Fouquey, C.; Lehn, J.-M.; Malthête, J. *J. Chem. Soc., Chem Commun.* 1987, 1424.
23. Brienne, M.-J.; Gabard, J.; Lehn, J.M.; Stibor, I. *J. Chem. Soc., Chem. Commun.* 1989, 1868.
24. Fouquey, C.; Lehn, J.-M.; Levelut, A.-M. *Adv. Mater.* 1990, *2*, 254.
25. Lehn, J.-M.; Mascal, M.; De Cian, A.; Fischer, J. *J. Chem. Soc., Chem. Commun.* 1990, 479.

RECEIVED August 2, 1990

Chapter 29

Control of Symmetry and Asymmetry in Hydrogen-Bonded Nitroaniline Materials

M. C. Etter, K. S. Huang, G. M. Frankenbach, and D. A. Adsmond

Department of Chemistry, University of Minnesota,
Minneapolis, MN 55455

The use of intermolecular hydrogen-bond formation between nitro groups and amino groups in nitroaniline crystal and cocrystal structures is explored. A set of hydrogen-bond rules is derived and used in the design of novel acentric molecular aggregates. A study of the frequency of occurrence of acentric space groups among nitroaniline structures indicates that the presence of acentric hydrogen-bond aggregates in a crystal may bias the final three-dimensional crystal structure to also be acentric.

Controlling the symmetry properties of a molecule means finding ways to assemble *covalent bonds* so the resulting set of *atoms* has the desired symmetry. Controlling the symmetry properties of aggregates means assembling *noncovalent bonds* so the resulting set of *molecules* has the desired symmetry. Organic chemists have focused generations of effort on the former problem, while neglecting the latter. This neglect has been acutely felt in recent years as the demand for bulk organic materials with specific electrical and optical properties has increased. Solving the general problem of correlating bulk properties with molecular packing patterns demands that we learn how to take advantage of molecular self-assembly mechanisms. Designing materials with second order nonlinear optical effects demands that we also learn how to use these self-assembly processes to bias bulk materials to be acentric.

There have been several approaches to this problem, including the use of chiral compounds (*1*), oriented host matrices (*2*), and the use of charged species (*3*). Our approach to making acentric materials that are useful for second harmonic generation is to choose molecules with large molecular hyperpolarizabilities that have functional groups capable of hydrogen bonding to one another to form acentric hydrogen-bonded networks. These networks constitute subsets of the final crystal structure. Crystal structures of a large series of related compounds, such as nitroanilines, tell us about preferred hydrogen-bond patterns and about symmetry consequences of these patterns. Cocrystal formation provides an experimental means for testing selectivity during self-assembly of heterogeneous systems and opens up a new array of useful solid-state materials. Our task is to first learn whether and how hydrogen bonds can be used as a tool for carrying out intermolecular "syntheses", and then to learn how to induce asymmetry during these processes.

0097–6156/91/0455–0446$06.00/0
© 1991 American Chemical Society

Nitroaniline Self-Assembly Properties

Nitroanilines are polarizable molecules with large hyperpolarizabilities (β) due to internal charge transfer between the electron-withdrawing nitro group and the electron-donating amino group (*4*). These intramolecular properties make nitroanilines good choices for nonlinear optical materials. But what about the intermolecular properties of nitroaniline molecules? More importantly, do nitroanilines aggregate, and if they do, what features of these aggregate patterns are useful?

A study of all known small molecule crystal structures of nitroanilines was carried out (*5*), and their packing patterns were analyzed for the presence of recurring and recognizable hydrogen-bond motifs (*6*). Despite considerable scatter in the lengths of the contacts between amino protons and nitro oxygens, several patterns were so common that we now use them as design tools for preparing new nitroaniline patterns. In the 30 structures examined the amino protons were always positioned near at least one nitro oxygen, usually on the *syn* side, as found for carboxylate hydrogen bonds (*7*). The nitro group traps amino protons in a three-center interaction which has quite variable geometry, shown as the cross-hatched region below. The potential well for these hydrogen bonds is shallow so the final position of the amino proton is strongly influenced by packing forces. These other forces are not usually sufficient to dislodge amino protons from the vicinity of a nitro group even when the NO2⋯HN distances are several tenths of an angstrom longer than the expected van der Waal's sums (shown as solid lines in the sketch below).

Having established that nitro groups and amino groups do associate with one another through hydrogen bonding, we now address the symmetry consequences of such interactions. Two molecules containing one nitro group and one amino proton each should associate with one another to form an acentric dimer. Additional molecules that hydrogen bond to this dimer use their nitro groups and amino groups in a similiar way, leading to an acentric chain. This polar chain motif is by far the most common pattern in nitroaniline crystal structures, even when the resulting crystal structure is centric.

An Acentric chain

Addition of a second proton donor, as in p-nitroaniline (PNA), serves to link acentric chains together (*8-10*). These chains form an acentric hydrogen-bonded array, as shown. PNA is a centric crystal because the hydrogen-bonded sheets pack with an inversion center between them. There are no inversion centers within the layer.

An Acentric Network

From analyses of these nitroaniline crystal structures we have derived a set of hydrogen-bond rules shown in Table I (*11*). These rules are based on connectivity patterns as well as on observations about symmetry relations and stereoelectronic effects in nitroaniline crystal structures. These rules do not address the question of how aggregates align themselves into their final crystal packing patterns. They deal only with the hydrogen-bonded subsets of a crystal. The rules are intended to be used as a guideline for predicting the structures of hydrogen-bonded sets of nitroaniline molecules within crystal structures, in solution, or on surfaces. The rules are empirical and are intended to be modified as larger data bases become available.

Table I. Hydrogen-bond Rules for Organic Compounds

General Rules
1. All good proton donors and acceptors are used in hydrogen bonding.
2. Six-membered ring intramolecular hydrogen bonds form in preference to intermolecular hydrogen bonds.
3. The best proton donors and acceptors remaining after intramolecular hydrogen-bond formation form intermolecular hydrogen bonds to one another.

Additional Rules for Nitroanilines
4. Amino protons will hydrogen bond to nitro groups.
5. One or more intermolecular amino-nitro hydrogen bonds will form.
6. The aggregate patterns formed from intermolecular hydrogen bonds between substituents in *meta* and *para* positions will be acentric.

7. The amino-nitro interaction is usually a three-center hydrogen bond.

8. Ortho substituted primary nitroanilines usually form two-center intermolecular hydrogen bonds, rather than three center.

It may not be a coincidence that nitroanilines have been a useful class of compounds for second harmonic generation (*12-13*). They naturally associate into polar arrays, which are manifested in their crystal structures. The important question is whether or not these polar arrays are likely to bias the resulting crystal structures to be acentric. Although we do not have large enough data sets to make a statistically convincing case, from a set of 32 primary and secondary nitroanilines and analogs, about 40% of their crystal structures are acentric. These compounds are all capable of forming acentric hydrogen-bonded chains and/or layers. On the other hand less than 20% of the 36 known tertiary nitroaniline structures are acentric (compared to about 25% for all known organic crystals, including enantiomers).

The other important role that hydrogen bonds play in nitroaniline crystal structures is related to charge redistributions, and hence changes in β, occurring during hydrogen-bond formation. Dannenberg has recently published a detailed study of nitroanilines showing preferred geometries of hydrogen-bonded dimers (*14*). He has also found that nitro groups and aniline hydrogens are stabilized by association, and that charge redistributions during association increase the ground state polarization of the individual molecules. Work is in progress to evaluate how aggregation affects the values of β.

Nitroaniline Cocrystallization Properties

The nitroaniline crystal structures discussed above were homomolecular. For most of these structures only one polymorph is known. If the bulk structure of a particular molecule, say PNA, is not a useful one then few options are known for modifying its packing pattern. For example, we tried to force PNA into a new polymorph by recrystallizing it from many different solvents. We obtained an interesting array of different crystal morphologies, none of which were polymorphic. The crystal morphologies that were obtained are shown in Figure 1.

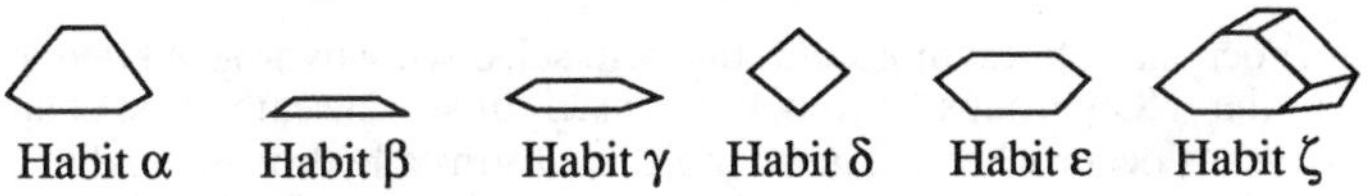

Habit α   Habit β   Habit γ   Habit δ   Habit ε   Habit ζ

Solvent	Habits Found
acetonitrile	ε
acetone	α,β
benzene	α
butanone	α,β
chloroform	α,β
cyclohexane	none
diethylether	δ
N,N-dimethyl formamide	γ
ethanol	α,β
ethylacetate	α
methanol	α,β,γ
methylene chloride	ζ
nitrobenzene	none
tetrahydrofuran	α,ε
water	α

Fig. 1 Crystals of p-nitroaniline obtained from different solvents by evaporation, showing multiple morphologies, but no polymorphism.

Having learned about the self-assembly properties of the entire class of nitroanilines, however, we have other tools besides polymorphism to use for altering the solid-state structure of PNA. A particularly useful tool is cocrystallization, whereby PNA can be forced into a limitless number of new structures. To predict what kinds of compounds will cocrystallize with PNA, its competitive hydrogen-bond properties were evaluated. The molecule has two proton donors (-NH2) and a bivalent proton acceptor site (-NO2). The first -NH proton has a solution pKa value of 18.35 (15), indicating that it is a weak proton donor. Solution pKa values do not directly measure hydrogen-bonding ability; rather, they measure proton-transfer ability in aqueous solution. Nevertheless, they are indicators of hydrogen-bonding abilities, particularly when used for comparison within a class of similiar kinds of structures (16-17). A range of hydrogen-bond strengths is available to both the -NH2 and -NO2 groups depending on other substituents in the molecule. The -NO2 group was shown to be a good proton acceptor, particularly when it was *para* to an electron-donating group (e.g., as in PNA). It can accept one proton in a *syn* position, and secondarily can accept another proton *anti* if there are extra protons available.

syn   H *syn* and *anti*

To design cocrystals, guest molecules that will selectively hydrogen bond to the -NH groups, to the -NO2 groups, or to both are used. In any case, the best chances for obtaining cocrystals occur when the guest molecule forms a hydrogen bond to PNA that is stronger than any of the hydrogen bonds present in either PNA or in the guest molecules by themselves (18-20). In addition, if the cocrystallizations are being performed in solution, the two reagents should be nearly equal in solubility in order to promote cocrystal formation. An alternative method for preparing these cocrystals involves simply grinding the two compounds together. Details of this method have been presented elsewhere (15-16). Nitroaniline cocrystals and their hydrogen-bond patterns are listed in Table II.

Nitroaniline Analogs

Cocrystals that have -NO2 groups on one component and -NH2 groups on the other can form "extended" analogs of nitroanilines which aggregate according to the nitroaniline hydrogen-bond rules. The host/guest components will first associate with one another by a strong hydrogen bond that does not compete with -NO2···H2N hydrogen bonds. Secondly, the remaining -NO2 and -NH2 will bond to one another as above. For this purpose, a very strong donor is placed on one of the components of the cocrystal pair and a very strong acceptor on the other. Since the driving force for cocrystal formation is to establish the strongest possible hydrogen bond, the host-guest pair should first dimerize to form this strong bond leaving the -NO2 and -NH2 groups to subsequently associate and link the dimer into an extended acentric structure, as shown.

Table II. Nitroaniline Cocrystals[a]

#	O₂N—⬡—NH₂	A—▢	S.G.	H-Bond Pattern
1	p-nitroaniline	tripiperidinophosphine oxide	P 1̄	
2	p-nitroaniline	triphenylphosphine oxide	P2₁/n	
3	p-nitroaniline	triphenylarsine oxide	P2₁/c	
4	m-nitroaniline	triphenylphosphine oxide	P2₁/n	
5	o-nitroaniline	triphenylphosphine oxide	P2₁/c	

#	O₂N—⬡—NH₂	A—▢	S.G.	H-Bond Pattern
6	p-nitroaniline	4-pyrrolidinopyridine	P 1̄	
7	p-nitroaniline	1,2-bis(4-pyridyl)ethylene	P 1̄	
8	m-nitroaniline	4-dimethylaminopyridine	P2₁/c	

#	O₂N—⬡—NH₂	A—▢—D	S.G.	H-Bond Pattern
9	p-nitroaniline	1,3-bis(m-nitrophenyl)urea	C2/c	⋯⋯ O₂N—⬡—NH₂ ⋯⋯ A—▢—D ⋯⋯

[a]Etter, M. C.; Huang, K. S. unpublished data.

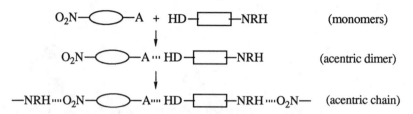

We have formed many donor/acceptor pairs which preferentially hydrogen bond to one another in the presence of nitro or aniline groups, and which form polar hydrogen-bonded chains. These pairs are shown in Table III.

Table III. Cocrystals of Nitroaniline Analogs

$$\text{... } O_2N-\bigcirc-A \text{ """ } HD-\bigcirc-NRH \text{ """}$$

#	$O_2N-\bigcirc-A$	$HD-\bigcirc-NRH$	S.G.	Ref.
10	4-nitropyridine N-Oxide	4-aminophenol	Cc	21
11	4-nitropyridine N-Oxide	4-aminobenzoic acid	$P2_12_12_1$	a
12	4-aminobenzoic acid	3,5-dinitrobenzoic acid	Fdd2	19
13	4-aminobenzoic acid	4-chloro-3,5-dinitrobenzoic acid	-	b
14	4-aminobenzoic acid	3,5-dinitrosalicylic acid	-	b
15	4-aminobenzoic acid	4-methyl-3,5-dinitro benzoic acid	-	b
16	4-aminobenzamide	3,5-dinitrobenzoic acid	$P2_1/c$	b
17	4-aminobenzamide	3,5-dinitrobenzoic acid	$P2_1/n$	b
18	3,5-dinitrobenzamide	4-aminobenzoic acid	-	b
19	4-aminobenzamide	3,5-dinitrobenzamide	-	b

[a]Lechat, J. R. personal communication.
[b]Etter, M. C.; Frankenbach, G. M. unpublished data.

An illustrative example is the cocrystal of 4-aminobenzoic acid and 3,5-dinitrobenzoic acid (*18*). These molecules preferentially form heterodimers through cyclic hydrogen-bonded carboxylic acid pairs. This dimer has an amino group at one end and two nitro groups at the other. The dimers self assemble by $-NO_2\cdots HN-$ bonding to give the acentric two dimentional array shown below. For many structures we have formed that it is not necessary to control the orientation of the third dimension by specific hydrogen bonds, since the acentric one and two-dimensional arrays in the structure frequently bias the final structure to be acentric anyway. The crystal structure of the acid heterodimer shown here has space group Fdd2.

Cocrystallization and the Formation of Large High Quality Crystals

In the course of preparing cocrystals of nitroaniline dimers and of related compounds from solution we observed that large high-quality cocrystals often form even when one or both of the component molecules give poor crystals by themselves. This observation was first made for cocrystals of triphenylphosphine oxide with small proton-donating molecules like acids, amides, phenols, and anilines (22).

best donor

best acceptor

In those cases triphenylphosphine oxide imparted its own favorable crystal growth properties to the other component. While preparing cocrystals of nitroaniline dimers we have found several additional examples of improved crystal growth and quality (as determined by low power microscopic investigation of crystal clarity, edge development, fracture planes, and defect structures). An example of the improved size and quality of a cocrystal of 3,5-dinitrobenzoic acid and 4-aminobenzoic acid juxtaposed with crystals of the starting materials is shown in Figure 2. Note that the polar nature of the cocrystal is also evident in its well-developed noncentric morphology.

Figure 2. Single crystals of p-aminobenzoic acid (on the left), 3,5-dinitrobenzoic acid (on the right), and their 1:1 cocrystal (center). The cocrystal is acentric and grows as large chunky crystals from methanol.

Biomimetic Design of Acentric Materials

Using polar chains and polar arrays to bias the formation of acentric bulk materials is a promising and potentially useful approach to designing acentric solids, but is somewhat unsatisfying because the nature of the bias is not well understood and is thus not easy to control. In searching for a more definitive and logical mechanism for preparing acentric bulk materials, we have borrowed one of nature's tricks.

The nucleotide base pairs of DNA specifically direct the two DNA helices in a predictable orientation. The predictability arises because the base pairs are complementary in only those arrangements that pair donors with acceptors. Unlike carboxylic acids which are free to rotate after dimer formation, the base pairs are constrained by their heterocyclic backbones to retain their mutual orientations, with both X-substituents pointing in the same direction, as shown.

Acid Dimers Have Non-constrained Orientations

Base Pairs Have Constrained Orientations

By preparing mixed dimer pairs, analogous to DNA base pairs, it is possible in principle to control the orientation of neighboring molecules, and of neighboring arrays. A biomimetic nitroaniline analog could conceivably be prepared by cocrystallizing derivatives of 2-aminopyrimidine.

If the best donors and acceptors couple, and the common aminopyrimidine bidentate hydrogen bonds form, then the following acentric structure is conceivable.

We have been unsuccessful in preparing this 1:1 cocrystal of 2,5-diaminopyrimidine and 2-amino-5-nitropyrimidine from solution but we have prepared a 1:1 cocrystal of the two components by solid-state grinding and heating. Neither the crystal structure nor the hydrogen-bond pattern is known at this point, but the X-ray powder pattern of the cocrystal product is distinctly different from those of the starting materials.

Conclusion

The use of intermolecular hydrogen bonds to control the acentricity of self-assembled sets of organic molecules has been demonstrated. These arrays, which may be homomeric or heteromeric, are subsets of the final crystal structure. The acentricity of the subsets is shown to bias somewhat the final three dimensinal crystal structures to be acentric also. Nitroaniline cocrystals and solid-state hydrogen-bonded analogs of nitroanilines were prepared. New materials based on other donor-acceptor pairs could also be prepared using the same design concepts. Although nonlinear optical materials have further materials requirements besides acentricity (such as high polarizability and phase matching), the ability to control acentricity has been one of the limiting steps in progress towards development of organic nonlinear optical materials.

Acknowledgments

We thank John MacDonald, a graduate student at the University of Minnesota (UM), for his help in understanding PNA crystal structures, and Crystal Hanscome, an undergraduate student at UM for her study of PNA morphology. We are also grateful to the Office of Naval Research (Grant No. N00014-89-K-1301) for their support of this work.

Literature Cited

1. Oudar, J. L.; Hierle, R. *J. Appl. Phys.* **1977**, *48*, 2699-2704.
2. Stucky, G. D.; Philips, M. L. F.; Gier, T. E. *Chem. Mater.* **1989**, *1*, 492-509.
3. Marder, S. R.; Perry, J. W.; Schaefer, W. P. *Science,* **1989**, *245*, 626-628.
4. Williams, D. J. *Angew. Chem. Int. Ed. Engl.* **1984**, *23*, 690-703.
5. Allen, F. H.; Bellard, S.; Brice, M. D.; Cartwright, B. A.; Doubleday, A.; Higgs, H.; Hummelink, T.; Hummelink-Peters, B. G.; Kennard, O.; Motherwell, W. D. S.; Rodgers, J. R.; Watson, D. G. *Acta Crystallogr.* **1979**, *B35*, 2331-2339.
6. Panunto, T. W.; Urbańczyk-Lipkowska, Z.; Johnson, R.; Etter, M. C. *J. Am. Chem. Soc.* **1987**, *109*, 7786-7797.
7. Gandour, D. R. *Bioorganic Chemistry* **1981**, *10*, 169-176.

8. Colapietro, M.; Dirienzo, F.; Domenicano, A.; Portalone, G.; Riva di Sanseverino, L. *Eur. Cryst. Meeting* **1977**, 517.
9. Colapietro, M.; Domenicano, A.; Marciante, C.; Portalone, G. *Acta Crystallogr.* **1981**, *Sec. A, 37*, C199.
10. Colapietro, M.; Domenicano, A.; Marciante, C.; Portalone, G. *Z. Naturforsch.* **1982**, *Teil B, 37*, 1309.
11. Etter, M. C. *Acc. Chem. Rev.* **1990**, *23*, 120-126.
12. Zyss, J.; Nicoud, J. F.; Coquillary, M. *J. Chem. Phys.* **1984**, *81*, 4160-4167.
13. Zyss, J.; Berthier, G. *J. Chem. Phys.* **1982**, *77*, 3635-3653.
14. Vinson, L. K.; Dannenberg, J. K. *J. Chem. Soc.* **1989**, *111*, 2777-2781.
15. Cox, R. A.; Stewart, R. *J. Am. Chem. Soc.* **1976**, *98*, 488-494.
16. Kamlet, M. J.; Abboud, J. M.; Abraham, M. H.; Taft, R. W. *J. Org. Chem.* **1983**, *48*, 2877-2887.
17. Abraham, M. H.; Duce, P. P.; Prior, D. V.; Barratt, D. G.; Morris, J. J.; Taylor, P. J. *J. Chem. Soc. Perkin Trans. 2* **1989**, 1355-1375.
18. Etter, M. C.; Adsmond, D. A. *J. Am. Chem. Soc. Comm.* **1990**, 589-591.
19. Etter, M. C.; Frankenbach, G. M. *Chem. Mat.* **1989**, *1*, 10.
20. Etter, M. C.; Panunto, T. W. *J. Am. Chem. Soc.* **1988**, *110*, 5896-5897.
21. Lechat, J. R.; de A. Santos, R. H.; Bueno, W. A. *Acta Cryst.* **1981**, *B37*, 1468.
22. Etter, M. C.; Baures, P. W. *J. Am. Chem. Soc.* **1988**, *110*, 639-640.

RECEIVED July 10, 1990

Chapter 30

Molecular Orbital Modeling of Monomeric Aggregates in Materials with Potentially Nonlinear Optical Properties

J. J. Dannenberg

Department of Chemistry, Hunter College and The Graduate School, City University of New York, New York, NY 10021

There has been extensive theoretical work aimed at understanding and predicting molecular hyperpolarizabilities. Much less theoretical attention has been focused upon understanding the hyperpolarizabilities of crystals or other materials, where the manner in which the molecules aggregate becomes extremely important. The AM1 molecular orbital method is used to model the interactions between molecular units in aggregates starting with dimers and going to trimers, tetramers, etc. Particular attention is devoted to hydrogen-bonding interactions between individual molecular units, as in the nitroanilines. Molecules that have the potential for co-crystallization, such as variously substituted benzoic acids, are considered in various combinations in an attempt to predict which pairs will co-crystallize.

In this paper, we make use of molecular modelling techniques, particularly the AM1 semiempirical molecular orbital method, to study the intermolecular interactions that are important for determining the manner in which crystal formation takes place. We are particularly interested in compounds that can potentially exhibit nonlinear optical properties. The calculational techniques are directed towards providing insight into the manner in which the desired nonlinear optical properties can be optimized in the macromolecular crystal state.(1)

Nonlinear optical properties depend upon the molecular environment as well as the individual molecular properties. In particular, at the molecular level, second harmonic generation depends upon the magnitude of β (the quadratic hyperpolarizability), which is the coefficient

0097–6156/91/0455–0457$06.00/0

of the second order term in the nonlinear optical expansion (see equation 1). For a crystalline material, $\chi^{(2)}$
has the same sense as β on the macromolecular level in
the analogous macromolecular equation (2) (equation 2).
It has been shown that $\chi^{(2)}$ will vanish exactly when the
environment is centrosymmetric. When $\chi^{(2)}$ is non-vanishing, crystals can have up to 38% efficiency in second
harmonic generation, depending upon the orientation of
the average molecular polarizabilities with respect to
the crystalline optical axes.(3)

$$p_i = \alpha_{ij}E_j + \beta_{ijk}E_jE_k + \gamma_{ijkl}E_jE_kE_l + \ldots \tag{1}$$

$$P = P_0 + \chi^{(1)}_{IJ}E_j + \chi^{(2)}_{IJk}E_jE_k + \chi^{(3)}_{IJkl}E_jE_kE_l + \ldots \tag{2}$$

While molecular orbital calculations have been used
to calculate the quadratic hyperporaizability, β, (4-5)
we believe this to be the first attempt to approach both
the molecular and macromolecular aspects of the design
of these materials in a consistent manner using molecular orbital theory.
    The two studies described here involve A: the inter-
molecular interactions in the crystal structures of meta-
and para-nitroaniline; and the predictable formation of
stoichimetic cocrystals which may potentially be used to
create new materials of interest.
    The comparison of the two isomeric nitroanilines is
that of two similar molecules, one (para) crystallizes
in a centrosymmetric, the other (meta) in a non-centrosymmetric manner. (6) Thus, only the meta exhibits nonlinear optical properties. Dimers of these (and several
other) nitroanilines were calculated using AM1, and
found to be in good agreement with the experimentally determined structures. (7) Here we focus on how these dimers can interact with additional molecules.
    Etter has shown that cocrystallization of mixed di-
mers of differently substituted benzoic acids can easily
be achieved. We present here calculations that predict
the relative stabilities of several possible mixed benzoic acid dimers that are variously substituted.

Methodology

The AM1 (8) approximation to molecular orbital theory
has been used for these studies. This method overcomes
the problems that previous semiempirical methods (notably, MNDO) (9) have in describing hydrogen-bonds. It has
been used with success in several hydrogen-bonding studies. (10-12) Ab initio studies of H-bonding systems are
very sensitive to basis set and correction for electron-

correlation, as exemplified in studies of the water dimer. (5) Calculations of sufficient accuracy on molecular complexes of the size to be considered here are not practicable using such costly methods.

All geometrical parameters for each of the monomers were optimized. For the dimers of the nitroanilines, all of the geometrical parameters for the second monomer unit were set equal to those of the first except for the parameters (bond lengths, angles and dihedral angles) of the amino-hydrogens and the nitro-oxygens directly involved in the intermolecular hydrogen bonds. In order to better approximate the crystal environment, the aromatic rings of the two monomer units were constrained to be coplanar.

Three general dimer types were considered for both nitroanilines: A, the optimal dimer with two distinct H-bonds, each between one amino-hydrogen and one nitro-oxygen; B, a relaxed geometry, with at least one bifurcated H-bond that is the local minimum closest to the crystal structure, and, C, the structure obtained by fixing the H-bonds at their experimental (crystal structure) distances and optimizing the rest of the dimer within the same constraints as A and B. Structure C is closest to the experimental structure.

Crystal structures of the nitroanilines

The geometries of the three different nitroaniline interactions for both species studied are presented in Figures 1-2. For p-nitroaniline, I, the optimal interaction, IA, has two distinct H-bonds, one between each amino hydrogen and a corresponding oxygen on the nitro group of the other monomer. The hydrogen H-bonding distances are 2.25 Å each. The relaxed structure, IB, has bifurcated interactions between one of the amino hydrogens and both of the nitro oxygens. Hydrogen bond distances are 2.29 and 2.31 Å. Additionally, one of the nitro oxygens is 2.47 Å from an ortho hydrogen of the ring. The crystal structure, IC, resembles structure B. The bifurcated bond is now unsymmetrical with distances being 2.34 and 3.22 Å while the ortho bond distance shortens to 2.03 Å (see figure 1).

The m-Nitroaniline, II, dimer differs from the para-isomer. Although the optimal interaction, IIA, again contains two distinct H-bonds, and the relaxed structure, IIB, has a bifurcated structure involving a H-bond to an ortho hydrogen, the crystal structure has no interaction between a nitro oxygen and an ortho hydrogen. Instead, one amino hydrogen H-bonds with both oxygens on the other monomer, with distances of 2.30 and 2.55 Å and one of the oxygens H-bonds with both hydrogens of the amino group (see figure 3).

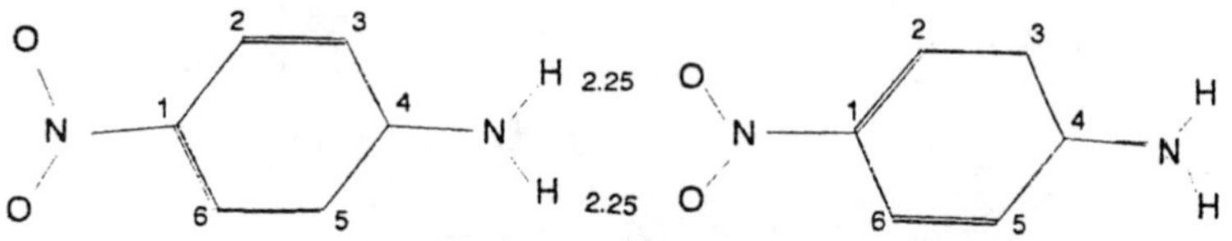

Figure 1A. Structures for p-nitroaniline dimers. Optimal dimer.

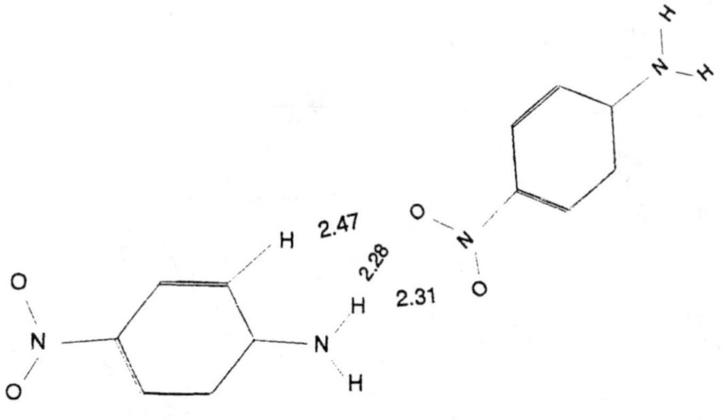

Figure 1B. Structures for p-nitroaniline dimers. Relaxed dimer with at least one H-bond that is the local minimum closest to the crystal structure.

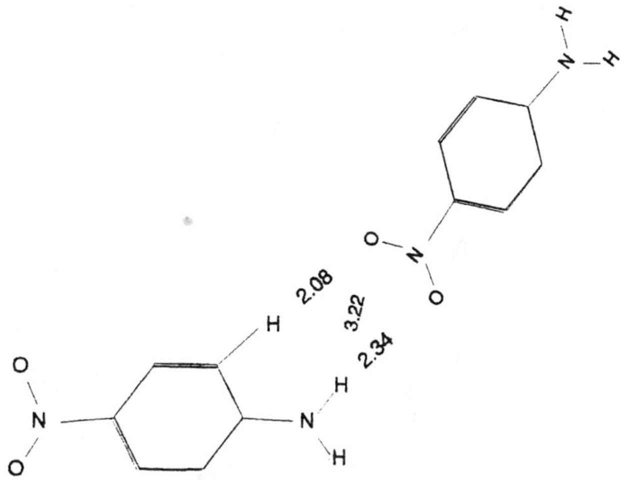

Figure 1C. Structures for p-nitroaniline dimers. Structure obtained by fixing the H-bonds at experimental distances and optimizing the rest. This is not a minimum (due to the constraint).

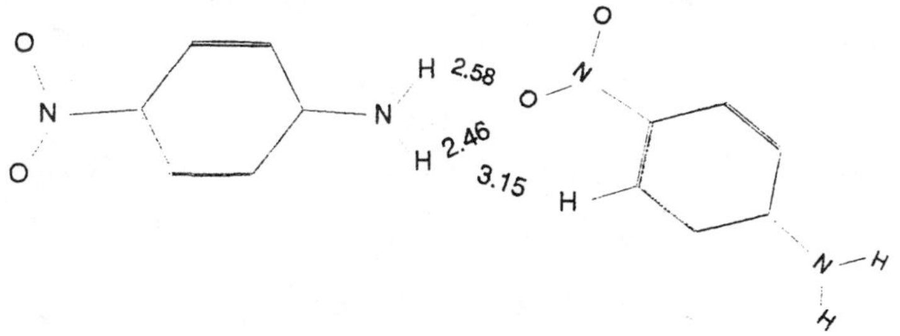

Figure 1D. Structures for p-nitroaniline dimers. See text for details.

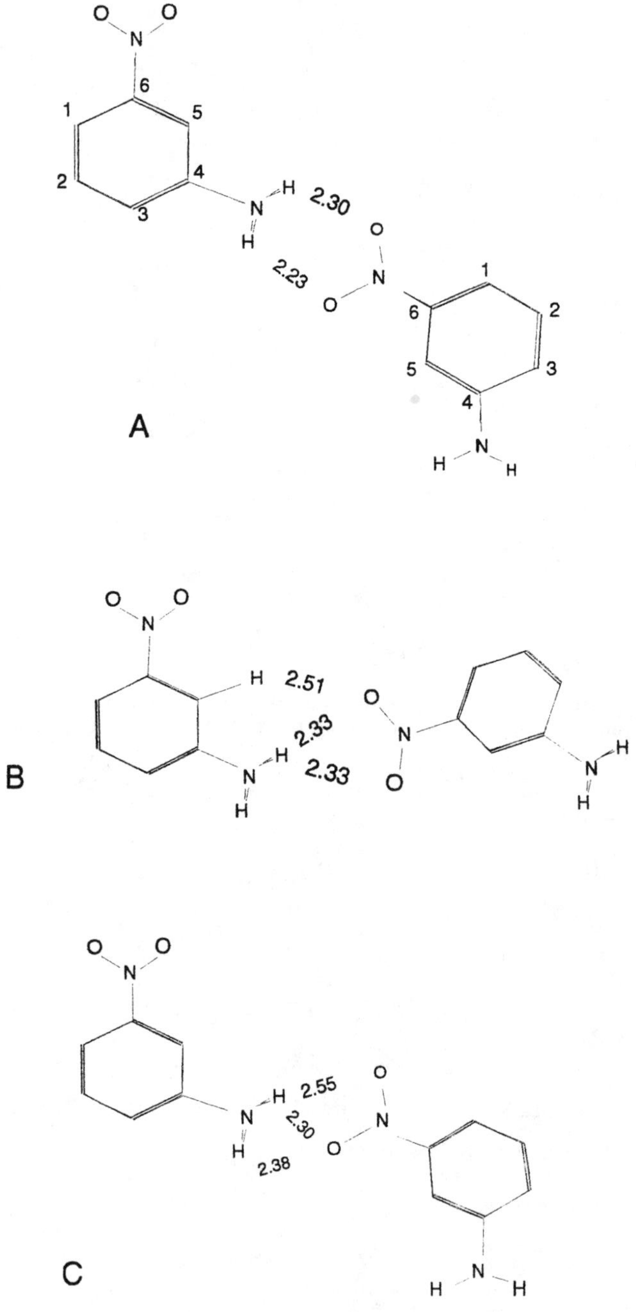

Figure 2. Structures for m-nitroaniline dimers, IIIA-C (structures are labeled as in figure 1).

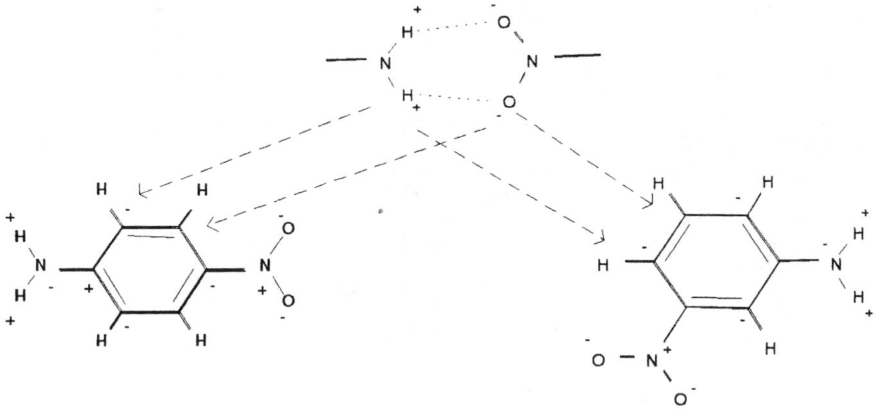

Figure 3. The arrows indicate how the local dipoles of the H-bonds might interact with the local dipoles of the p- and m-nitroanilines to give 'head to tail' and 'head to head' interactions. The '+' and '-' signs indicate centers of charge. Unmarked carbon atoms are neutral.

Discussion

The heats of formation and hydrogen bonding energies for each of the interactions are presented in Table I. The optimal dimer interactions have bonding energies of -6.9 and -5.0 kcal./mole for the para and meta isomers respectively. These values are of similar magnitude to hydrogen bond energies for the water dimer.

Table I. Heats of Formation and Interaction
Energies (kcal /mole)

	Monomer	Dimer			Interaction Energy[a]		
		A	B	C	A	B	C
I	21.6	36.3	37.5	39.4	-6.9	-5.7	-3.7
II	24.1	43.1	44.1	45.6	-5.0	-5.7	-2.5

[a]Defined as the appropriate dimer energy
minus twice the corresponding monmer energy.

The relaxed bifurcated structures, IB and IIB are 1.7 and 1.0 kcal/mole less stable than the optimal structures, IA and IIA. The crystal structures, are another 2.0 and 1.5 kcal./mole less stable than the bifurcated structures.

The small destabilization of the relaxed crystal, B, and crystal structures, C, relative to the optimal dimer are likely overcome by other interactions in the crystal such as attractions between planes and weak H-bonding between adjacent chains. For example, the amino hydrogen not involved in an H-bond in IC can form a weak interaction with a nitro-group on the adjacent chain.

The interactions between the ortho-hydrogens and the nitro groups that are manifest in several structures play an important role in defining the relative orientations of the nitroanilines in the crystal chains. In fact, there are two possible bifurcated structures for the dimer of I. Either a) one amino hydrogen can interact with two nitro oxygens, as in IB, or b) one oxygen can interact with two hydrogens, as in ID. IB, which is favored, has an additional hydrogen bond to an ortho hydrogen, while for ID, the hydrogen ortho to the nitro group is only 3.15 Å from one of the hydrogens on the amino group. What is attractive in IB becomes repulsive in ID, whose energy is 1.6 kcal./mole higher than that of IB.

The crystal structure can be rationalized by considering the influence of interactions with the neighboring chains. For I, II, and V, the bifurcated structure leaves the second amino hydrogen more available for interactions with a neighboring chain than the more energetically favored head-on dimer. The shortening of the hydrogen bond to the ortho substituent (in C vs. B-type structures) that is often apparent may also serve to better accommodate interaction with adjacent chains or more efficient packing. For II, since the crystal interacts with both amino hydrogens instead of with the ortho hydrogen, which is more energetically favored, the ortho hydrogen rather than the amino hydrogen is left free to interact with a neighboring chain.

The geometry of the nitroanilines in the dimers are substantially different from the monomers. Pertinent geometrical parameters for the monomers and for the optimal dimers are compared in Table II. The H-N-H bond angle increases by 1.6 and 4.4 degrees in the dimer over that of the monomer while the O-N-O bond angle decreases by 1.1 and 0.8 degrees for I and II respectively. This may serve to enhance the oxygen lone-pair directionality in hydrogen bonding as found by Murray-Rust and Glusker (13) and Vedani and Dunitz. (14)

Table II. Relavent Angles (degrees) in the
Monomers and Dimers

	NH2 Dihedral			H-N-H Angle		O-N-O Angle	
	Monomer	Dimer		Monomer	Dimer	Monomer	Dimer
		A	B				
I	13.9	0.2	0.3	116.9	118.5	121.7	120.6
	16.8	0.6	0.5				
II	19.3	5.3	16.6	114.4	118.6	122.0	121.2
	23.8	6.8	22.6				

Dimer A is that providing the NH2 that H-bonds (on the left in the figures). The dihedral angles refer to the plane of the aromatic ring. The valence angles refer to the NH2 and NO2 groups involved in the H-bonds. The figures are for the most stable dimer of each monomer.

Additionally, the amino groups which are pyramidal in the monomers, become substantially more planar in the dimers. For I, the amino hydrogens in the monomer are 13.9 and 16.8 degrees out of the plane, while, in IA, the hydrogen-bonding H's are only 0.2 and 0.6 degrees out of the plane. Even the amino hydrogens not involved in hydrogen bonding are 0.3 and 0.5 degrees out of the plane. For II, the H-bonding amino group's dihedrals change signifacantly more than for the non-H-bonding amino group (see Table II).

The differences in the calculated geometrical parameters of the isolated molecules and the H-bonding dimers (which are models for the gas and solid phases) serve to emphasize the potential errors that may arise upon comparison of calculated geometrical parameters for isolated molecules with crystal structural data. It is significant that the calculated optimized geometries, themselves, change when intermolecular interactions that simulate the solid phase are explicitly considered. This observation strongly suggests that application of molecular orbital methods to dimers or small aggregates may be useful for modelling the geometries of molecules in the solid phase. Molecular orbital modelling of individual molecules is properly compared with experimental observations of the molecular properties of gas phase molecules. Further work in this direction will be necessary before more definite conclusions can be reached.

Inspection of the net atomic charges in the monomers and dimers, Table III, indicates that there is increased charge alternation in the dimer. This suggests that mutual polarization might be occurring. The net charge transfer is very small (<0.01 electrons in all cases). In an infinite crystal, all units must be neutral. The small degree of charge transfer observed in the dimer supports the appropriateness of the dimer as a model for the crystal. In the cases where H-bonds to ortho hydrogens are implicated, the charge increases substantially on the ortho hydrogen upon dimer formation.

The charge polarization and the planarization of the nitroanlines upon dimer formation, suggests that increasing the H-bonded chain beyond two molecules should result in greater stabilization (per monomer-monomer interaction) since whatever energy that is required to distort the monomer to its planar, polarized state is already (at least partially) overcome for both molecules at the dimer stage. Adding an additional monomer would require only one (rather than two) additional distortions.

The crystal structure of I strongly resembles the relaxed crystal dimer structure (B). In the case of II, structure IIB would require a bend in the crystal chain that might be very difficult to accommodate.

Studies of crystals of stoichiometrical H-bonding complexes

Crystals of stoichiometric 1:1 mixtures of compounds that can complex with each other have been shown to form preferentially to pure crystals of the individual components. In some cases these crystals may have potential non-linear optical properties. An interesting example is the 1:1 mixture of p-aminobenzoic acid and 3,5-dinitrobenzoic acid. (15) A view of the crystal structure is shown in figure 3. Examination of this figure leads one to the hypothesis that the preference for the mixed crystal may be due to a) a more stable H-bonding interaction between the different benzoic acids in the hetero-dimer than in the homo-dimer; b) the ability of the mixed crystal (hetero-dimers) to H-bond between their amino and nitro groups. It is likely that both of these factors play a role in the stability of the crystal structure. Calculational modelling can aid in determining the importance of these factors.

Table III. Charge Distributions for Dimers I
and II (in units of atomic charge)

atom	Monomer	Dimer A	Dimer B	Dimer C
For para-nitroaniline, I				
C-1 A	-0.197	-0.222	-0.218	-0.215
C-1 B	-0.197	-0.209	-0.209	-0.210
C-2 A	-0.016	-0.009	-0.015	-0.013
C-2 B	-0.016	-0.005	-0.006	-0.007
C-3 A	-0.225	-0.246	-0.234	-0.236
C-3 B	-0.225	-0.236	-0.236	-0.237
C-4 A	0.139	0.175	0.172	0.173
C-4 B	0.139	0.173	0.172	0.176
C-5 A	-0.224	-0.246	-0.247	-0.248
C-5 B	-0.224	-0.237	-0.235	-0.236
C-6 A	-0.018	-0.018	-0.012	-0.014
C-6 B	-0.018	-0.018	-0.006	-0.005
For meta-nitroaniline, II				
C-1 A	-0.117	-0.14?	-0.142	-0.144
C-1 B	-0.117	-0.118	-0.116	-0.126
C-2 A	-0.099	-0.090	-0.092	-0.090
C-2 B	-0.099	-0.094	-0.094	-0.089
C-3 A	-0.159	-0.187	-0.187	-0.190
C-3 B	-0.159	-0.160	-0.159	-0.171
C-4 A	0.066	0.125	0.117	0.122
C-4 B	0.066	0.080	0.074	0.111
C-5 A	-0.144	-0.173	-0.160	-0.173
C-5 B	-0.144	-0.153	-0.150	-0.165
C-6 A	-0.086	-0.073	-0.076	-0.072
C-6 B	-0.086	-0.078	-0.080	-0.071

'A" refers to the monomer supplying the NH_2
to the H-bonding interaction (on the left in
figures 1 and 2), 'B' to the other. See figures 1 and 2 for the numbering conventions.

In order to determine whether molecular orbital methods could be used to predict and explain preferences for
cocrystalization analogous to that discussed above, we
present here AM1 calculations on the dimerization energies of variously substituted benzoic acids.
All geometrical parameters for each monomer and
dimer were individually optimized.
In this study we considered all possible dimers between p-amino, p-nitro, m,m-diamino, and m,m-dinitro ben-

zoic acids. The results (16) are indicated in table IV.
It is immediately apparent that the hetero-dimers are
generally more stable than the homo-dimers. In addition,
the m,m-dinitrobenzoic acid/p-aminobenzoic acid dimer is
the most stable of the group.

Table IV. Interaction Stabilization Energies
of Benzoic Acid Dimers (kcal /mole)

Components of Substituted Benzoic acid Dimers		Stabilization Energy
Monomer A	Monomer B	
p-nitro	p-nitro	6.1
p-amino	p-amino	6.1
3,5-dinitro	3,5-dinitro	6.1
3,5-diamino	3,5-diamino	6.3
p-amino	p-nitro	6.5
p-amino	3,5-dinitro	7.2
p-amino	3,5-diamino	6.1
3,5-diamino	p-nitro	6.6
3,5-diamino	3,5-dinitro	7.0
p-nitro	3,5-diamino	6.7

By implication, the H-bonding within the dimer seems
to be of some importance. The accuracy of this prediction was tested by mixing 3,5-dinitrobenzoic acid with p-
dimethylaminobenzoic acid to see if a 1:1 crystalline
material formed. In this case, H-bonding between the
nitro and amino groups is precluded by the methylation
of the amino groups. Apparently, a stoichiometric mixed
solid does form (as evidenced by an unmistakable change
in color to red)[14] although the structure has not yet
been determined.
In the study of hydrogen bonded dimers of various
nitroanilines (discussed, in part, above), we reported
that the charge alternation of the individual monomer
units was accentuated in each of the monomeric units of
the dimer. In contrast the local charges on the carbons
of the aromatic rings of the various benzoic acids are
virtually unchanged upon hydrogen bonding. The foregoing
is true irrespective of the substituent groups on the
benzoic acids, even when one bears nitro and the other
amino groups. One is tempted to note that since there
are six π-electrons in the cyclic H-bonding structure
formed by a nitro and an amino group (figure 1A) while
eight in that formed by two carboxylic acids (figure 4),
aromaticity might be involved. While this concept bears
further investigation, the orbitals in neither case seem
to support aromaticity.
The H-bonding energies presented here are somewhat
lower than those expected from the reported H-bonding energies of gas phase carboxylic acids. Notably, that reported for formic and acetic acid are roughly twice the

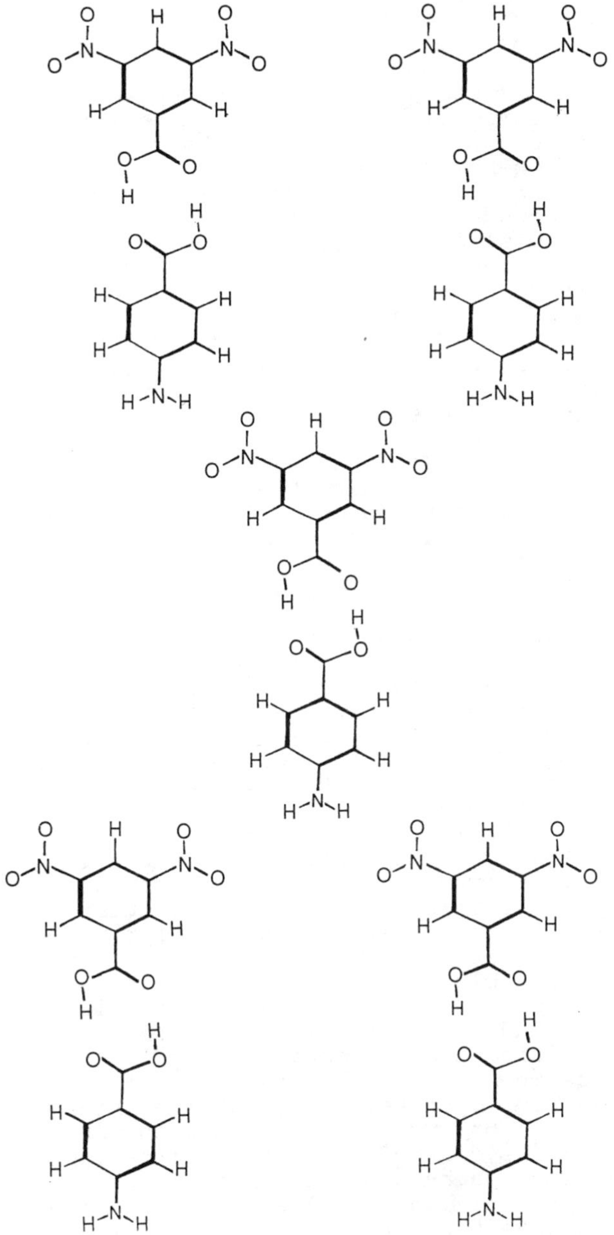

Figure 4. H-bonded aggregate of the crystal of the 1:1 complex of p-aminobenzoic acid and 3,5-dinitrobenzoic acid.

calculated values for benzoic acid. The AM1 values for
the interactions of dimers of formic and acetic acid molecules are similar to those of the benzoic acids. The experimental methods used generally only measure the
amount of dimer directly, calculating the amount of monomer by difference. It has been shown that errors can
arise from such phenomena as adsorption on surfaces. Extrapolation to zero surface area has led to lower estimates for the H-bonding interactions. Nevertheless, the
latest and presumably best (also the lowest) estimates
of the interaction energies predict an H-bonding interaction considerably higher than that calculated. This may
be due to a continued slight overestimation of H-H repul-
sion energies (as in MNDO). One should note that in the
carboxylic acid dimers, the two H's from <u>different</u> molecules approach each other in the dimer. Perhaps the re-
pulsion between these H's (as calculated by AM1)
destabilizes the dimer. In the nitroanilines and other
cases studied, the two H's that are close in the dimer
are on the same monomeric unit, therefore, since the re-
pulsion continues when the dimer is cleaved, it does not
contribute to the interaction energy.

<u>Conclusion</u>

AM1 calculations on the dimers of nitroanilines and substituted benzoic acids are of considerable value in predicting their crystal structures. In particular, the
intermolecular forces that dictate the relative orientations of the individual molecules in the crystal chains
can be understood. It is likely that this methodology
will be useful for modelling the kinds of interactions
that might occur in other crystals.

The present calculations suggest that the differences in molecular geometry between gas and solid (crystal) phases may be largely manifest in aggregates as
small as dimers. If this be the case, molecular orbital
theory may be extremely useful as a tool for understanding these differences.

We hope to be able to demonstrate the ability of the
methodology to adequately predict the intermolecular interactions in situations such as the interactions involving these two nitroaniline isomers. It is likely that
interactions between local (rather than molecular) dipoles will play an important part in these intermolecular interactions. For instance, figure 4 shows the
calculated relative charge densities in m- and p-
nitroaniline. One can hypothesize that the propensity
for adjacent rings to line up 'head to head' or 'head to
tail' might involve the interaction of the local dipoles
in the H-bonding interface of two monomers in the same
row, with the local dipole in the aromatic ring of a
monomer in the adjacent row. The charge densities give
some indication that this may be the case. Nevertheless,
detailed calculations will be necessary to verify this
effect and obtain an idea of its magnitude (thereby, its
importance).

Placing additional monomers in the same plane adjacent to the nitroaniline dimers can test the hypothesis that week H-bonds exist between adjacent rows in the same plane, and that these interactions are sufficient to distort the dimers from their most stable conformations to those observed in the crystal structures.Adding additional monomers to the head (or tail) of the dimers by increasing the H-bonding chain within a row will allow us to determine the number of molecules required to render a chain calculationally stable. That is, the point at which adding additional monomers no longer has an effect upon the others. We expect to continue our theoretical studies in these directions.

Literature Cited

1. For reviews see a) Chemla, D. S.; Zyss, J, "Nonlinar Optical Properties of Organic Molecules and Crystals," (2 volumes), **1987**, Acaemic Press; b) Williams, D. J. *Angew. Chem., Internat. ed.*, **1984**, _23_, 690.
2. Pughn D.; Morley, J. O., "Nonlinar Optical Properties of Organic Molecules and Crystals," Chemla, D. S, and Zyss, J. eds, **1987**, vol. 1, Academic Press, p 193.
3. Zyss, J.; Oudar, J. L., *Phys. Rev. A*, **1982**, _26_, 2028.
4. Oudar, J. L.; Zyss, J., *Phys Rev. A*, **1982**, _26_, 2016.
5. unpublished results of Dirk, Twieg and Wagniere cited in Nicoud, J. F.; Twieg, R. J., *"Nonlinear Optical Properties of Organic Molecules and Crystals"*, Chemla and Zyss eds, **1987**, Academic Press, 227.
6. For a detailed discussion of the crystal structures of various nitroanilines, see Panunto, T. W.; Urbanczyk-Lipkowska, Z.; Johnson, R.; Etter, M. C., *J. Am. Chem. Soc.*, **1987**, _109_, 7786.
7. Dannenberg, J. J.; Vinson, L. K., *J. Am. Chem. Soc.*, **1989**, _111_, 2777.
8. Dewar, M. J. S; Zoebisch, E. G.; Healy, E. F.; Stewart, J. J. P., *J. Am. Chem. Soc.*, **1985**, _107_, 3902.
9. Dewar, M. J. S.; Thiel, W., *J. Am. Chem. Soc.*, **1977**, _99_, 4899.
10. Dannenberg, J. J.; Vinson, L. K., *J. Phys. Chem.*, **1988**, _92_, 5635.
11. Dannenberg, J. J, *J. Phys. Chem.*, **1988**, _92_, 6869.

12. Galera, S.; Lluch, J. M.; Oliva, A.; Bertran, J., THEOCHEM, **1988**, _40_, 101.
13. Murray-Rust, P.; Glusker, J. P., *J. Am. Chem. Soc.*, **1984**, _106_, 1018.
14. Vedani, A.; Dunitz, J. D., *J. Am. Chem. Soc.*, **1985**, _197_, 7653.
15. Etter, M. C.; Frankenbach, G. M., MATERIALS, **1989**, _1_, 10.
16. Dannenberg, J. J., submitted for publication.
17. Etter, M. C.; Frankenbach, G. M., private communication.

RECEIVED August 13, 1990

Chapter 31

Strategies for Design of Solids with Polar Arrangement

R. Popovitz-Biro, L. Addadi, L. Leiserowitz, and M. Lahav

Structural Chemistry, Weizmann Institute of Science, Rehovot, 76100 Israel

Two novel methodologies for the design of solid materials with poar axes by a kinetic controlled process are described. In the first approach we demonstrate that amphiphilic molecules bearing two amide groups along the hydrocarbon chain invariably deposite to yield Z-type Langmuir-Blodgett films. Attachment of hyperpolarizable molecules to such hydrocarbon chains resulted in the formation of films displaying frequency doubling. In the second methodology, crystallographic information has been used for the design of polymeric crystallization inhibitors of a stable non-polar polymorph of PAN [N-(2-acetamido-4-nitro-phenyl)- pyrrolidene]. As predicted, addition of minute amounts of polymer 15 to a supersaturated solution of PAN results in the precipitation of the metastable polar form which displays second harmonic generation.

In the absence of a general theory of packing of molecules, the preparation of solid materials with required structures and desired physical or chemical properties is done, by and large, empirically. Solids with a polar structure in which molecules assume a head-to-tail arrangement, display pyroelectric, piezoelectric and frequency doubling properties, as a result of a constructive summation of dipoles and hyperpolarizability tensors. It often happens, however, that these arrangements are metastable and their formation is prevented by the existence of a stable, but non-polar form. Recently, our group has been developing methodologies for the preparation of thermodynamically metastable polar structures by the process of kinetic control. These include the design of amphiphilic molecules which form Langmuir-Blodgett films with a polar packing arrangement (1), and the control of crystal polymorphism with the assistance of

0097–6156/91/0455–0472$06.00/0

auxiliary molecules (2, 3). We shall illustrate these two approaches with representative examples.

Design of Amphiphilic Molecules Forming Polar (Z-Type) Langmuir-Blodgett (LB) Films

Amphiphilic molecules, composed of a hydrophilic head group and hydrophobic chain (hydrocarbon or fluorocarbon), have a tendency to aggregate at the air/water interface and, when compressed, form monomolecular Langmuir films. These films can be transferred onto solid supports, layer by layer, to form LB multilayers. This technique has recently aroused considerable interest as a method for the build-up of ordered assemblies. The most common and thermodynamically stable multilayer structures are of the Y-type, where the layers are deposited in the head-to-head, tail-to-tail fashion. X and Z-type multilayers comprising molecules of the same kind in a head-to-tail arrangement may also be formed, but the X-type films are generally unstable and have a tendency to rearrange to the more stable Y-type films. Polar Y-type films with an ABAB... arrangement have been prepared by alternate deposition of two different monolayers, using special troughs (Scheme 1) (4). Several sporadic examples of genuine Z-type depositions have been reported (5). From these examples, however, it would be difficult to make any generalizations or predictions as to which amphiphiles would tend to deposit in a polar structure. The design of new molecules with a tendency to form Z-type films, requires an understanding of the deposition behaviour on a molecular level.

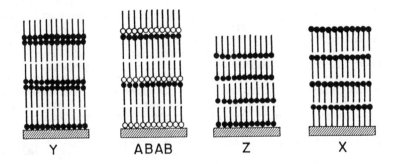

Y   ABAB   Z   X

Scheme 1

The mode of deposition of a Langmuir film from the air/water interface to a solid-support is determined by the shape of the water miniscus during transfer at the interface. This shape is determined by the wettability of the surface, i.e. the advancing and receding contact angles. Z-type multilayers are formed when deposition occurs only during withdrawal of the substrate but not during dipping. Such behaviour requires both advancing and receding contact angles to be <90°.

Wettability is a macroscopic property that has not been treated theoretically on a molecular scale (6). Experiments have demonstrated that this property is determined by the structure and polarity of groups close to the interface (7). Wettability has also been found to depend on the porosity of the film to water (8a-8c). However, in certain cases, groups with strong dipoles located as much as 10Å underneath the surface show an effect on wettability (8d). We have also shown polar groups incorporated beneath the surface may influence molecular packing as well as water penetration and retention in the film, without being part of the outermost region of the surface (9). During some stereochemical studies related to a different program on the triggering of crystal nucleation with the assistance of Langmuir and LB films, we prepared amphiphilic α-amino acids bearing both amide and ester groups within the hydrocarbon chain (10).

We noticed that while esters invariably form Y-type films, the mono and diamides have a pronounced tendency to form Z-type films. Wettability studies performed on these monolayers demonstrated that the diamides display water advancing contact angles always lower than 90° and receding contact angles lower than 45°. On the other hand, the advancing contact angle of water on surfaces coated with amphiphiles of the α-amino acids with esters in the chain is always larger than 90° (as listed in the Table). This tendency of the diamides is also preserved when polarizable groups, such as p-nitroaniline or merocyanine were attached at the ω-position of the diamide amphiphiles, thus displaying second harmonic generation. Furthermore, this tendency is not limited to the α-amino acids but is also present for diamides bearing other polar head groups such as carboxylic acids, p-nitroaniline, merocyanine, etc., some of which are listed below. Structural characterization of the various films was carried out by analytical tools including ellipsometry, infrared, second harmonic generation and X-ray analysis (1).

A direct demonstration of the formation of stable Z-type films was provided by SHG studies on LB films of 6 (measurements done by the group of G. Meredith at DuPont). Fig.1 shows the plot of the intensity of frequency double light versus the number of deposited layers. Samples were prepared at different times and with different LB troughs. The very good dependence of $I_{2\omega}$ on N^2 is a direct proof of the polar nature of the films. Their long-term stability has been demonstrated by reproducing the SHG measurements after a year from the date of sample preparation. Compared to the known value of the hyperpolarizability of p-nitroaniline, the SHG conversion efficiency is relatively low, indicating that the p-nitroaniline chromophore in the films may have a rather large average tilt from the surface normal (approximately 70°).

Multilayers of compound **8** and **9** were also studied by SHG. One would expect more efficient SHG signal from these films due to the larger molecular hyperpolarizability of the merocyanine chromophore as compared to that of p-nitroaniline. However, in these systems, we observed a sublinear dependence of $I_{2\omega}$ on the number of layers N for N<4 and a saturation of the signal for N>4. U.V. visible absorption spectra of these films indicates a substantial degree of protonation in the merocyanine moiety which apparently caused the reduction in the molecular hyperpolarizability. Some monolayers, such as **10** and **11** were not sufficiently stable at the air/water interface to allow multilayer deposition at a constant surface pressure. Improved stability was obtained in monolayer binary mixtures of these amphiphiles with **5** which do not contain an hyperpolarizable chromophore. A Z-type film containing up to twelve layers of 1:1 (molar ratio) of **10** and **5** was prepared. These films also exhibit SHG, however, with substantial in-plane inhomogeneity indicating clustering and domain formation.

Recently, with the aim of increasing the film stability, we synthesized amphiphiles bearing one amide group and a polymerizable diacetylene function. Previous studies using monolayers of diacetylenic carboxylic acids have shown that they form the usual Y-type multilayers (11). We could demonstrate that amino acids **12** and **13** bearing a diacetylene and an amide group, form stable polymeric Z-type films. Their polar structure was unambigously demonstrated by both ellipsometric and X-ray studies (9b).

Suggested Mechanism

The very low receding contact angles of the Z-type multilayers indicate that the monolayers of amide films are highly porous and that water can efficiently penetrate between the pores. Another support for the porous nature of these films come from grazing angle X-ray diffraction and X-ray reflectivity measurements (12) of palmiloyl-R-lysine **1**, compressed at the air/water interface to 20 mN/m. The diffraction data showed that this monolayer is a two-dimensional powder with a strong reflection corresponding to a d spacing of 4.5Å and a coherence length of approximately 500Å. The hydrocarbon chains in these domains are tilted at an angle of 30° to the surface normal, thus precluding close packing of the neighbouring domains. The X-ray reflectivity data indicated that the monolayer coverage at this pressure is only 90%, thus exposing bare patches of water. A compressed monolayer of the diamide **4** over water gave an X-ray diffraction peak at a d spacing of 4.5Å which suggests that these monolayers pack in a manner similar to the monoamide **1**.

This porosity which is also retained in the LB films, may explain the water penetration between the 2-D crystallites. The amide groups located at the periphery of these domains generate hydrophilic centers which may retain water molecules by binding via hydrogen bonds. In addition, the hydrophilic head groups, such as α-amino acids or carboxylic acids which (in contrast to the Y-type films), cannot form hydrogen bonded pairs, are free to bind water molecules between the layers. We suggest that when

Table: List of amphiphilic molecules, their transfer behaviour at 20 mN/m and contact angles with water

Compound	Type of multilayer	Contact angle advancing (receding)
(1) $CH_3(CH_2)_{14}CONH(CH_2)_4CH$ with CO_2^- and NH_3^+	Z	80(24)
(2) $CH_3(CH_2)_9CONH(CH_2)_{11}CO_2H$	Z	85(30)
(3) $CH_3(CH_2)_{17}OC(CH_2)_2CH$ with CO_2^- and NH_3^+	Y	95(40)
(4) $CH_3(CH_2)_9CONH(CH_2)_{11}CONH(CH_2)_4CH$ with CO_2^- and NH_3^+	Z	75(15)
(5) $CH_3(CH_2)_{10}CONH(CH_2)_{10}CONH(CH_2)_4CH$ with CO_2^- and NH_3^+	Z	82(28)
(6) O_2N-⟨benzene⟩-$NH(CH_2)_{11}CONH(CH_2)_4CH$ with CO_2^- and NH_3^+	Z	70(40)

No.	Structure		Yield
(7)	$O_2N-\langle\bigcirc\rangle-NH(CH_2)_{11}\overset{O}{\overset{\|}{C}}NH(CH_2)_{11}\overset{O}{\overset{\|}{C}}NH(CH_2)_3CH\langle^{CO_2^-}_{NH_3^+}$	Z	72(30)
(8)	$\overset{O}{\overset{\|}{\langle\bigcirc\rangle}}=N(CH_2)_{10}\overset{O}{\overset{\|}{C}}NH(CH_2)_{11}CO_2^-K^+$	Z	68(45)
(9)	$CH_3(CH_2)_4NH\overset{O}{\overset{\|}{C}}(CH_2)_{11}NH\overset{O}{\overset{\|}{C}}(CH_2)_{10}N\langle\bigcirc\rangle=O$	Z	75(30)
(10)	$O_2N-\langle\bigcirc\rangle-NH(CH_2)_{11}\overset{O}{\overset{\|}{C}}NH(CH_2)_{11}CO_2H$		
(11)	$CH_3(CH_2)_{11}NH\overset{O}{\overset{\|}{C}}(CH_2)_{11}NH-\langle\bigcirc\rangle-NO_2$		
(12)	$CH_3(CH_2)_{15}-C\equiv C-C\equiv C-(CH_2)_8\overset{O}{\overset{\|}{C}}NH(CH_2)_4CH\langle^{CO_2^-}_{NH_3^+}$	Z	89(37)
(13)	$CH_3(CH_2)_{15}-C\equiv C-C\equiv C-(CH_2)_8\overset{O}{\overset{\|}{C}}NH(CH_2)_3CH\langle^{CO_2^-}_{NH_3^+}$	Z	90(25)

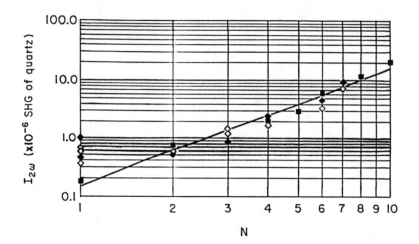

Figure 1. Intensity of second harmonic light ($I_{2\omega}$) produced by LB films of compound **6** as a function of layer number (N). Both the fundamental and second harmonic beams are p-polarized. The line (with slope 2) is the best fit according to the least square criterion.

these films are immersed in water, the voids between the domains retain water, reducing the receding contact angle close to 0°, thus permitting the deposition of the next layer on the upstroke. Independent experimental support for the role played by the occluded water on the deposition properties is provided by the fact that gentle drying of the thin film in a dessicator for 24 hours at 25°C raised the advancing contact angle close to 90° The first following layer gave Y-type deposition, followed by a Z-type deposition of the subsequent layers.

The present proposed mechanism implies that it should hold also for amphiphilic molecules bearing other polar groups along the hydrocarbon chain. Thus, incorporation of hydroxyl groups or sulfoxides as demonstrated by Taylor et al.(5d) should display a similar effect on wettability as the amides themselves. The isotherms of 12-hydroxystearic acid and 12-ketostearic acid have been reported, but not their deposition properties. In contrast to 12-ketostearic acid, which yields a Y-type film, the 12-hydroxystearic acid gave, as anticipated, four Z-type layers with good transfer ratio. The monolayers, however, were not stable enough to permit depositions of large number of layers without collapse.

In conclusion, the design of stable Z-type multilayers requires a deep understanding of the wettability properties of the surfaces on a molecular level, which obviously depends not only on the structured fraction of the films but also on their defect sites and porosity. In the absence of such knowledge, the empirical approach described here for the preparation of stable Z-type films should be valid in the foreseeable future.

Control of Crystal Polymorphism with the Assistance of "Tailor-Made" Auxiliaries

The search of polar crystals for SHG is done generally by trial and error. It happens often that the precipitation of a metastable polar polymorph is prevented by the existence of a stable non-polar form. We present here a new method for the control of polymorphism with the assistance of "tailor-made" crystal inhibitors of the stable crystalline form (2). This approach is general and applicable in particular for the induced precipitation of a polar metastable crystalline form in the presence of the thermodynamically stable polymorph, provided the two structures are not separated by too large an energy gap (13). The method is based on a general working hypothesis which assumes that nuclei of precritical size adopt structures akin to those of the mature crystals in which they evolve. We take this hypothesis one step further by assuming that supersaturated solutions of materials displaying polymorphism may contain molecular aggregates of structures resembling those of the various mature phases. Under close-to-equilibrium conditions, however, only nuclei corresponding to the thermodynamically stable phase grow into crystals (Scheme 2).

We had used the structural information of the packing arrangement of two polymorphs in order to model molecules which selectively inhibit the formation and growth of the nuclei of

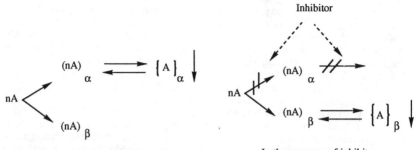

In the absence of inhibitor   In the presence of inhibitor

( ) nuclei

{ } crystals

Scheme 2

the stable polymorph. If this inhibitor does not interact with
the nuclei of the metastable crystalline form, we may anticipate
that the operation of the inhibitor will favor its precipitation.

This is illustrated in the following example:

N-(2-acetamido-4-nitrophenyl)pyrrolidene (PAN) is dimorphic.
The thermodynamically stable form is non-polar (Fig.2), (14).
The crystal space group is $P2_1$ with two independent molecules per
asymmetric unit which are related by a pseudo center of
inversion. This crystal does not display measurable frequency
doubling generation.
        Dendritic needles of the metastable-form could be
precipitated from quickly cooled supersaturated aqueous
solutions. The space group is $P2_1$ with one molecule in the
asymmetric unit (Fig.3), (14). Following simple symmetry
arguments, the α-form grows in the +b and -b directions with
identical rate, whereas the polar β-form grows at different rates
towards the +b and -b directions. Planning growth inhibitors for
the α-form, we took into consideration the fact that in the
α-form the acetamido-groups of the two independent molecules are
antiparallel and so oriented towards both the +b and -b
directions, while in the β-form the symmetry related molecules
have the acetamido group oriented exclusively towards one
direction along the b-axis.
        On the basis of this information, we predicted that
replacement of the amide group of PAN with another amide bound to
a polymer-chain will generate a most efficient crystallization
inhibitor. According to our analysis we prepared polymer 14 and

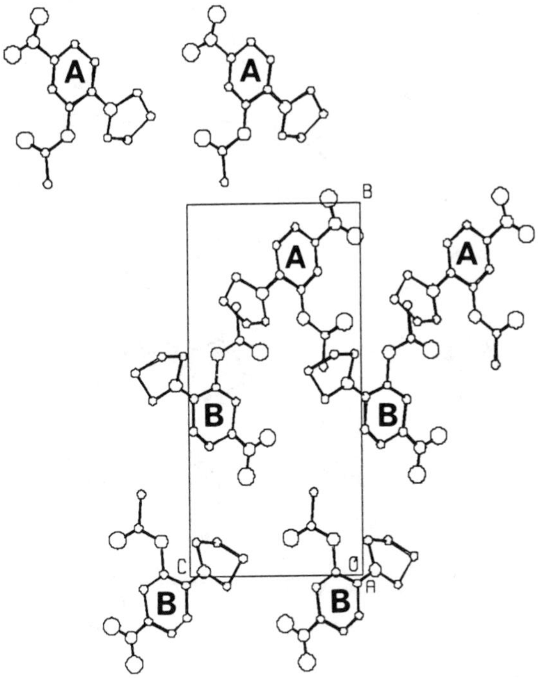

Figure 2. Packing arrangement of the stable α-form of PAN, viewed down the a axis. The two sets of molecules A and B are crystallographically independent, but are related to each other by a pseudo center of inversion. Thus the crystal is non-polar and SHG inactive.

15. These inhibitor molecules should adhere to both nuclei of the α- and of the β-form. In the α-form the additive should inhibit growth along both the +b and the -b directions, whereas in the β-form only from one of the two directions of b. We thus expected that in the presence of inhibitors of this molecular structure, nuclei of the α-form would be preferentially prevented form growing.

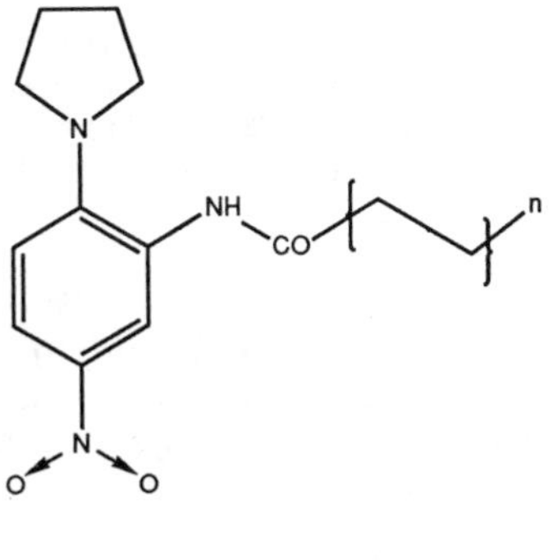

14 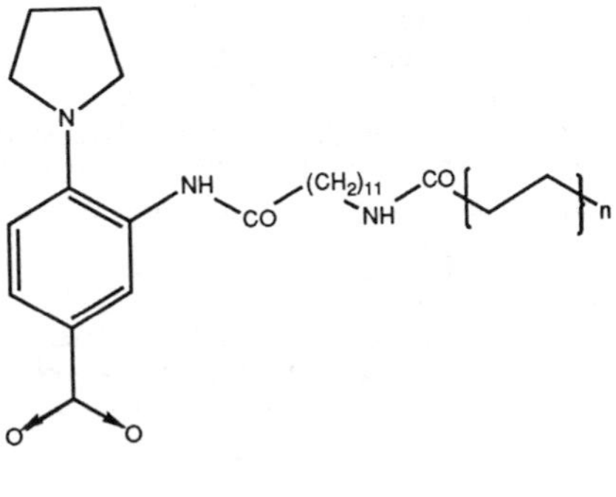   15

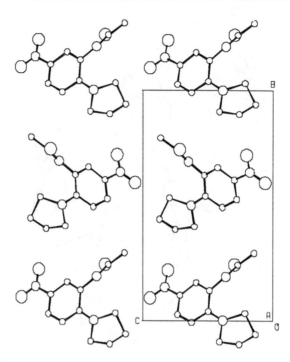

Figure 3. Packing arrangement of the metastable, polar β-form
of PAN, viewed down the a axis.

PAN precipitates from DMF at room temperature exclusively
in the α-form. Under similar conditions, additon of polymer **14**
in amounts of 1 to 2 wt% progressively induces crystallization of
the β-form and 3 wt% is sufficient to obtain an homogeneous batch
of β-needles. Even better results were obtained with polymer **15**
: while addition of 0.01 wt% yield exclusively the α-form, as low
as 0.03-0.5 wt% yield only β-needles.

The same method has been successfully applied for the
crystal growth of the metastable 𝛾-form of glycine (15), as well
as for the induced resolution of enantiomers of racemates which
generally precipitate as racemic compounds (16).

Acknowledgements
We thank Prof. J. Sagiv for introducing us to the field of
monolayers , Dr. H. Hsiung, G. Meredith and H. Vanherzeele
(DuPont), for their fruitful cooperation in the SHG studies and
Drs. K. Hill and E. Staab, Mr. D.J. Hung and Mrs. E. Shavit for
performing experiments described in this work.

References
1.   For a full report of this study see: Popovitz-Biro, R., Hill,
      K., Hung, D.J., Leiserowitz, L., Lahav, M., Sagiv, J.,
      Hsiung, H., Meredith, G., Vanherzeele, H. J. Amer. chem.
      Soc. 1990, 112, 2498.

2. Staab, E.; Addadi, L.; Leiserowitz, L.; Lahav, M. Advanced Materials 1990, 2, 40.

3. Weissbuch, I.; Lahav, M.; Leiserowitz, L.; Meredith, G.; Vanherzeele, H. Chem. Mat. 1989, 1, 114.

4. Ledoux, I.; Josse, D.; Fremaux, P.; Zyss, J.; McLean, T.; Hann, R.A.; Gordon, P.F. Allen, S. This Solid Films 1988, 160, 217.

5. (a) Daniel, M.F.; Lettington, O.C.; Small, S.M. Thin Solid Films, 1983, 99, 61.

   (b) Higashi, N.; Kunitake, T. Chem. Lett. 1986, 105.

   (c) Mumby, S.J.; Swalen, J.D.; Rabolt, J.F. Macromolecules, 1986, 19, 1054.

   (d) Taylor, D.M.; Oliveira, N.; Stirling, C.J.M.; Guo, B.Z.; Tripathi, S. Thin Solid Films, 1989, 178, 27.

6. (a) Joanny, J.F.; De Gennes, P.G. J. Chem. Phys. 1984, 81, 552.

   (b) Schwartz, L.W.; Garoff, S. Langmuir 1985, 1, 219.

7. (a) Zisman, W.A. Contact Angle, Wettability and Adhesion Advances in Chemistry, 43, Fowkes, F.M., Ed.; Amer. Chem. Soc., Washington, D.C. 1974.

   (b) Traughton, E.B.; Bain, C.D.; Whitesides, G.M.; Nuzzo, R.G.; Allara, D.; Porter, M.D. Langmuir 1988, 4, 219.

8. (a) Maoz, R.; Sagiv, J. Langmuir 1987, 3, 1034.

   (b) i b i d, 1987, 3, 1045.

   (c) Bain, C.D.; Whitesides, G.M. J. Amer. Chem. Soc. 1988, 110, 5897.

   (d) Shafrin, E.G. and Zisman, W.A., J.Phys. Chem. 1962, 66, 740.

9. (a) Popovitz-Biro, R.; Kill, K.; Landau, E.M.; Lahav, M.; Leiserowitz, L.; Sagiv, J.; Hsiung, H.; Meredith, G.R.; Vanherzeele, H. J. Amer. Chem. Soc. 1988, 110, 2672.

   (b) Popovitz-Biro, R.; Hung, D.J.; Shavit, E.; Lahav, M.; Leiserowitz, L. Thin Solid Films. 1989, 178,203.

10. (a) Landau, E.M.; Grayer-Wolf, S.; Levanon, M.; Leiserowitz, L.; Lahav, M.; Sagiv, J. J. Amer. Chem. Soc. 1989, 111, 1436.

    (b) Landau, M.; Grayer-Wolf, S.; Sagiv, J.; Deutsch, M.; Kjaer, K.; Als-Nielsen, J.; Leiserowitz, L.; Lahav, M. Pure & Appl. Chem. 1989, 61, 673.

11. (a) Leiser, G.; Tieke, B.; Wegner, G. Thin Solid Films, 1980, 68, 77.

    (b) Day, D. and Lando, J.B. Macromolecules, 1980, 13, 1478.

12. Grayer-Wolf, S.; Leiserowitz, L.; Lahav, M.; Deutsch, M.; Kjaer, K.; Als-Nielsen, J. Nature 1987, 328, 63.

13. Weissbuch, I.; Zbaida, D.; Addadi, L.; Leiserowitz, L.; Lahav, M. J. Amer. Chem. Soc. 1987, 109, 1869.

14. Hall, S.R.; Kolinsky, P.V.; Jones, R.; Allen, S.; Gordon, P.; Bothwell, B.; Bloor, D.; Norman, P.A.; Hursthouse, M.; Karalov, A.; Baldwin, J.; Goodyear, M.; Bishop, D. J. Cryst. Growth, 1986, 79, 746.

15. Weissbuch, I.; Zbaida, D.; Idelson, M. Work in Progress.

16. Albeck, S. M.Sc. thesis,1988, Feinberg Graduate School, Weizmann Institute of Science.

RECEIVED July 18, 1990

Chapter 32

Ferroelectric Liquid Crystals Designed for Electronic Nonlinear Optical Applications

David M. Walba[1], M. Blanca Ros[2], Noel A. Clark[2], Renfan Shao[2],
Kristina M. Johnson[3], Michael G. Robinson[3], J. Y. Liu[3],
and David Doroski[3]

[1]Department of Chemistry and Biochemistry, [2]Department of Physics, and
[3]Department of Electrical and Computer Engineering and Optoelectronic
Computing Systems Center, University of Colorado,
Boulder, CO 80309

The first ferroelectric liquid crystal (FLC) designed specifically for
electronic second order nonlinear optics (NLO) applications, and
possessing $|d_{eff}| > 0.074$ pm/V for frequency doubling from 1.064 μm
Nd:YAG laser light, is described. This value (a conservative lower
limit) is about 7 times larger than d_{eff} of the commercial FLC mixture
SCE9, until now the FLC material with the largest measured second
order susceptibility. Our approach to the design of FLCs for NLO, a
description of the polar order expected for the C* phase for the newly
designed materials, and data on the macroscopic spontaneous electric
polarization exhibited by these materials is given in addition to the NLO
data.

Ferroelectric liquid crystals (FLCs) are true fluids possessing thermodynamically stable
polar order. When incorporated into devices in the Clark-Lagerwall surface stabilized
ferroelectric liquid crystal (SSFLC) geometry (1), electro-optic and photo-optic light
valves with very large interaction strength (i.e. half-wave interaction lengths on the
order of 2 μm) are easily fabricated. However, since switching in the SSFLC light
valve involves large nuclear motions, the switching speeds are relatively slow (for
currently available materials rise times are $\cong 20$ μsec in response to 15 V/μm driving
field).

Even given the slow switching, SSFLC light valves are typically several orders
of magnitude faster than other liquid crystal switches. This coupled with low switching
energy, bistable memory, the strong interaction with light, and relative ease of
obtaining "single crystal" thin films on diverse substrates (glass, amorphous silicon
films, silicon integrated circuits), has led SSFLC devices to emerge as a potentially
important solution for many optoelectronic applications, ranging from spatial light
modulators for flat panel TV and optical computing to phase conjugating mirrors (2).

The advantages of SSFLC devices derive to a large extent from the spontaneous
macroscopic polarization **P** of the phase. For example the electrooptic rise time of a
prototypical SSFLC light valve is inversely proportional to the magnitude of the
polarization. In order to design new FLC materials with large P in a directed way, we

0097–6156/91/0455–0484$06.00/0

have developed a simple stereochemical model for the molecular origins of **P**, and have been testing and refining the model in a program of synthesis and evaluation of new FLC materials (3-13).

Based upon this work, we feel it should be possible to orient virtually any organic functional array along the polar axis of an FLC thin film. Naturally, this proposal suggests another application of FLC materials: Electronic second order nonlinear optics (NLO). For example, FLC films possessing large second order hyperpolarizability $\chi^{(2)}$ should provide the basis of a light switching technology where the optical interaction strength is much weaker than in SSFLC devices, but the switching speeds are much faster. Of course, for many applications, in general those where large information density is temporal rather than spatial, this tradeoff of interaction strength for speed is desirable.

In this paper, we describe initial experiments of a project aimed at the development of FLC materials with large $\chi^{(2)}$.

Background—FLCs for NLO

FLC phases in the surface stabilized geometry possess a single C_2 axis of symmetry, and therefore polar order and non-zero $\chi^{(2)}$ in the simple electronic dipolar model. Thus, it is not surprising that experiments aimed at measuring this property were first reported shortly after the Clark-Lagerwall invention. Early studies (14-15) described second harmonic generation in (S)-2-Methylbutyl 4-(4-decyloxybenzylideneamino)cinnamate, the first ferroelectric liquid crystal, also known as DOBAMBC (1).

$C_{10}H_{21}O$—

DOBAMBC (1)

A-priori, given that FLC phases possess polar order, and the DOBAMBC structure should possess a substantial molecular second order hyperpolarizability β, it seems reasonable that $\chi^{(2)}$ of a material such as DOBAMBC in the FLC phase might be large. Measured values of the second harmonic generation efficiencies of DOBAMBC and of several other FLC materials, however, indicate that in fact the d_{eff} of FLCs is very small relative to LiNbO$_3$.

For example, prior to 1990 the commercial FLC mixture ZLI 3654 was shown by Taguchi, Ouchi, Fukuda, and Takazoe to possess the largest known second order susceptibility for an FLC material (16-17). Using type I eeo phase matching, a value of $|d_{eff}| \cong 0.0025$ time $|d_{22}|$ (LiNbO$_3$) was measured for ZLI 3654 when phase matched. Recently, Johnson et al have more fully characterized the commercial FLC material SCE9, which possessed double the d_{eff} of ZLI 3654 (18). Using measured values of the refractive indices (the sample behaves uniaxially), Maker fringe measurements on wedge cells, type I eeo angular phase matching, and some assumptions regarding the ratio of d-coefficients, it was possible to obtain an excellent fit to the observed phase matching curve, and to extract the absolute values of all the non-zero components of the **d** tensor for the C_2 symmetrical sample of SCE9. The values so obtained are given in Table I. The coordinate system used for this work has y along the polar axis and z along the liquid crystal director.

Table I. Absolute values of the components of the **d** tensor for the commercial FLC
material SCE9

$d_{23} = d_{34} = 0.073(2)$ pm/V
$d_{22} = 0.027(1)$ pm/V
$d_{21} = d_{16} = 0.0026(1)$ pm/V
$d_{25} = d_{14} = d_{36} = 0.0009(1)$ pm/V

Two points of special interest concerning the data given in Table 1 are as
follows. First, the largest component is d_{23}, that reflecting the second order NLO
response of the sample along the polar axis when driven along the FLC director. The
component reflecting the response along the polar axis when driven along the polar axis
(d_{22}) is a factor of two smaller. This seems reasonable in light of the fact that the axis
of maximum linear polarizability (maximum refractive index) is along the director,
normal to the polar axis in known FLCs.

Second, given that for LiNbO$_3$ d_{15} = 5 pm/V and d_{22} = 3.2 pm/V, the values
for SCE9, with the largest second order susceptibility reported to date for an FLC, are
small.

In the context of organic second order NLO materials, where electronic
hyperpolarizabilities can be very large relative to inorganics such as LiNbO$_3$, the small
measured $\chi^{(2)}$ for FLCs seems discouraging and has led some workers in the field to
dismiss these materials as a possible solution to the general problem of obtaining thin
films with large $\chi^{(2)}$. The processibility and stability advantages of FLCs and FLC
polymers, however, would make them potentially very useful if materials with large
$\chi^{(2)}$ could be obtained. A description of our approach to solving this fascinating
problem in materials synthesis follows.

Geometry of the FLC self assembly

Figure 1 illustrates the relationship between the laboratory frame (glass bounding
plates), smectic layers, the layer normal $\hat{z}$, and the FLC director $\hat{n}$ (the long axis of the
molecules ≅ the optic axis of the sample), in a parallel-aligned SSFLC cell. Generally,
parallel alignment is favored by standard rubbed polymer alignment layers (films of
polymer about 300Å thick spun onto the glass plates and rubbed — when filled with
FLC, such surface treatment induces a "crystal" orientation where the layers are normal
to the rubbing direction). With this geometry of the FLC relative to the bounding
plates, the C$_2$ axis of the crystal is oriented normal to the plates, and the director aligns
parallel to the plates. Under these conditions the tilt plane, that plane containing both $\hat{z}$
and $\hat{n}$, is parallel to the plates with the following caveat.

Elegant studies of defects commonly occurring in SSFLC cells, and termed zig-
zag walls, have shown that the layers are in fact tilted with respect to the bounding
plates, taking on a chevron structure as indicated in Figure 1 (19-20). Thus, the polar
axis of the crystal and the tilt plane actually intersect the bounding plates at an angle. It
is important to note, however, that the chevron layer structure itself introduces no
problematic defects, but rather defects occur where two chevron domains intersect.
Since it is possible to obtain a monodomain of chevrons, essentially "single crystal"
FLC films may be relatively easily fabricated over large areas.

In addition, in the absence of the bounding plates, a spontaneous helix develops
in the director field about the layer normal along the tilt cone (indicated in the Figure)

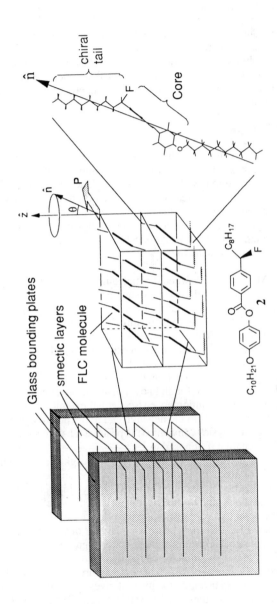

Figure 1. The geometry of a parallel-aligned SSFLC cell. Note that the spacing between the glass bounding plates ($\cong 1.5$ μm) and the smectic layer spacing ($\cong 35$ Å) are not to scale.

with a pitch of about 1 μm. If the layer thickness (spacing between the plates) is small relative to the helix pitch, then the helix is spontaneously unwound. If, however, the layer is thick relative to the helix pitch, then the sample far from the plates will exhibit the helix parallel to the surface of the plates.

Application of an electric field normal to the plates (typically the plates are coated with thin films of conducting glass such as indium-tin oxide) unwinds the helix if there is one, and also may cause the polar axis to orient normal to the plates (along the field), or even flatten the chevrons. It should be stressed that any added orientation of molecular dipoles along the field direction should be a weak secondary effect — the polar order occurring in the FLC phase is a thermodynamic property of the phase and not dependent upon applied fields.

Molecular Origins of **P** in FLCs

The existence of the layers and director tilt in the achiral smectic C liquid crystal phase are experimental facts. Given these, the <u>maximum</u> possible symmetry of the phase would be C_i, with a C_2 axis normal to the tilt plane, and a σ plane congruent with the tilt plane. In fact, there is no fundamental reason why a given C phase must possess either of these symmetry elements. But, breaking of either of the symmetry elements would afford polar symmetry, and no C phase has ever been shown to possess any property associated with polar symmetry (e.g. pyroelectricity). Therefore, we can say that all known C phases indeed possess the maximum possible symmetry consistent with the layers and tilt.

As first realized by Meyer in 1974, when the molecules making up the C phase are non-racemic, the resulting chiral C* phase can possess no reflection symmetry. Thus, the <u>maximum</u> possible symmetry of a C* phase is C_2, and the phase must possess polar order (<u>21</u>). One of the macroscopic manifestations of polar order can be a macroscopic electric dipole moment (the polarization **P**) associated with orientation of molecular dipoles along the polar axis. While the existence of polar order is not sufficient to assure an observable polarization (just as chirality does not assure optical activity), in fact many FLC materials do possess an observable **P**.

This polarization has a sign, which for all known materials is related directly to the chirality of the molecules. Thus, enantiomers possess equal magnitude but opposite sign of **P**. By convention, **P** (pointing from the negative end to the positive end of the macroscopic dipole following the physics convention) is positive when in the direction of the unit vector $\hat{z} \times \hat{n}$. **P** is negative if opposed to $\hat{z} \times \hat{n}$.

While the symmetry argument is rigorous, it gives no insights into which functional groups are actually oriented along the polar axis. For example, (S)-DOBAMBC possess an experimental polarization of -0.009 Debye/molecule. Given the large dipoles present in the molecule, it is not clear at all exactly which of those dipoles is oriented along the polar axis, or how strong the orientation is.

In order to design new FLC materials possessing large **P** or $\chi^{(2)}$, it is necessary to have some insight into the molecular origins of the functional group orientation occurring in the phase. Much of our recent research has focussed upon the development of such a model, the basics of which have been published (<u>1-13</u>). In essence, the molecular orientation occurring in the phase derives from intermolecular interactions between neighboring molecules (the "crystal" lattice). On the time average, this interaction may be considered to take the form of a "binding site" with the shape of a bent cylinder oriented in a specific way relative to the tilt plane. The C phase order results when the molecules "dock" into this binding site.

Currently, we feel that the existing data may be interpreted by assuming that the shape of this binding site is the same for known C phases or C* phases, regardless of the structure, chirality or achirality of the actual molecules comprising the phase. Thus, the molecular order occurring in the phase is the same whether the molecules are chiral or achiral. If the molecules are achiral or racemic, this order gives rise to no macroscopic polarization, while a polarization results when the molecules are non-racemic. As indicated in Figure 1, the binding site shape is such that the molecules are oriented with the tails less tilted than the core relative to the layer normal.

The macroscopic polarization of the phase is given by equations 1 and 2, where D_i is the number density of the ith conformation, $\vec{\mu}_{\perp i}$ is the component of the molecular dipole normal to the tilt plane when the ith conformation of the molecule is oriented in the rotational minimum in the binding site, ROF_i is the "rotational orientation factor", a number from zero to one reflecting the degree of rotational order for the ith conformation, and ε is a complex and unmeasured dielectric constant of the medium (local field correction).

$$P = \sum_i D_i \cdot P_i \qquad (1)$$

over all i conformations

$$P_i = \vec{\mu}_{\perp i} \cdot ROF_i \cdot 1/\varepsilon \qquad (2)$$

For this discussion, several points should be stressed here. Most importantly, there is no polar order along the director in any known liquid crystal phase, including the C* phase. Thus, functional arrays with large β along the director are not oriented along a polar axis in the FLC phase. This is our interpretation of the small $\chi^{(2)}$ of DOBAMBC and other FLC materials. There are other possible problems as well, however. For example, though DOBAMBC possesses substantial dipoles oriented normal to the director, it's observed macroscopic polarization (-0.009 D/molecule) is very small. This could be due to poor molecular orientation in the FLC phase, which in turn could represent a fundamental problem in design of FLCs for $\chi^{(2)}$.

We feel, however, that the data suggests molecular order in FLCs is in fact high. That is, when the FLC compounds are properly designed, molecular dipoles are oriented along the polar axis to a high degree. Thus, consider the 1-fluorononyl-benzoate 2 shown in Figure 1. According to our model, the only molecular dipole expected to orient anisotropically along the polar axis is the C-F dipole, with the orientation as indicated in the Figure. Thus, assuming a dipole moment of 2 D for the C-F bond, if the orientation were perfect (as in a crystal), then the component of the C-F dipole along the polar axis ($\vec{\mu}_\perp$) would be about 1.6 D, and assuming $\varepsilon \cong 3$, and of course ROF=1, then the macroscopic polarization should be about 0.5 D/molecule according to equ 2.

In fact, the experimental polarization of compound 2 = +44 nC/cm$^2 \cong$ +0.15 D/molecule assuming a density of 0.8 gm/cm^3, or fully 1/3 of that expected for the corresponding perfectly oriented solid (7). While this calculation is very crude, and depends upon estimation of a dielectric constant which cannot be measured, it still seems that good functional group orientation typically can occur in FLC phases (i.e. ROF > 0.1).

Design of FLCs for NLO

While typical high polarization FLCs may show good functional group orientation along the polar axis, these groups are normally carbon-halogen, epoxy, or ester units expected to produce only a small bulk second order hyperpolarizability. This is our interpretation of the small $\chi^{(2)}$ observed in FLCs to date. In considering design of FLCs with large $\chi^{(2)}$, the simplest approach would be to achieve good orientation of functionalized aromatic rings along the polar axis in an FLC film. In fact, recent results strongly suggest that such orientation does occur, and can be easily understood in terms of the stereochemical model.

Thus, as first demonstrated experimentally by the FLC chemistry group at Chisso (22), and later by us (11), FLCs possessing an aromatic ring with the 1-methyl-heptyloxy chiral tail, and substituted on the ring ortho to this tail, exhibit sign and magnitude of **P** consistent with good orientation of the functionalized ring along the polar axis. We felt that the o-nitro-1-methylheptyloxy aromatic system should afford appropriate geometry for orientation of the nitroalkoxy β along the polar axis, and therefore prepared the series of compounds **3 - 5**. Compounds **3** and **4** represent our first generation of FLCs designed specifically for $\chi^{(2)}$, while structure **5** serves as a control. To our knowledge, no compounds possessing an o-nitroalkoxy array similar to compounds **3** and **4** have been previously reported in the literature.

Several empirical design criteria for these materials were applied. Specifically, it is well known that the 3-ring biphenylbenzoate core system, first explored by Gray and Goodby (23), is among the best for obtaining broad temperature range smectic C phases. However, these materials also typically possess large orientational viscosity, affording slow switching speeds in SSFLC devices. A major advantage of NLO from the point of view of FLC design is that this orientational viscosity is of only secondary importance for electronic NLO since switching does not require nuclear motion.

In addition, the 3-ring biphenylbenzoates tend to show larger polarizations (better orientation) for a given chiral tail than a corresponding two ring system. Finally, in general they are relatively easily synthesized in a convergent manner.

Most importantly, of course, is the expectation that the nitroalkoxy functional array in compounds **3** and **4**, but not compound **5**, should orient along the polar axis in a geometry leading to good orientation of molecular βs for large $\chi^{(2)}$ in the FLC phase.

The rationale for this expectation was developed to interpret the sign and magnitude of **P** observed for the unsubstituted 1-methylheptyloxy FLCs prior to any of the experimental reports on o-substitution. As illustrated in Figure 2 for the nitroalkoxy system found in compound **3**, according to the model the two conformers A and B should predominate in the C* phase. In these drawings, the polar axis of the phase is (almost) parallel to the plane of the page (normal to the tilt pane), and the conformers

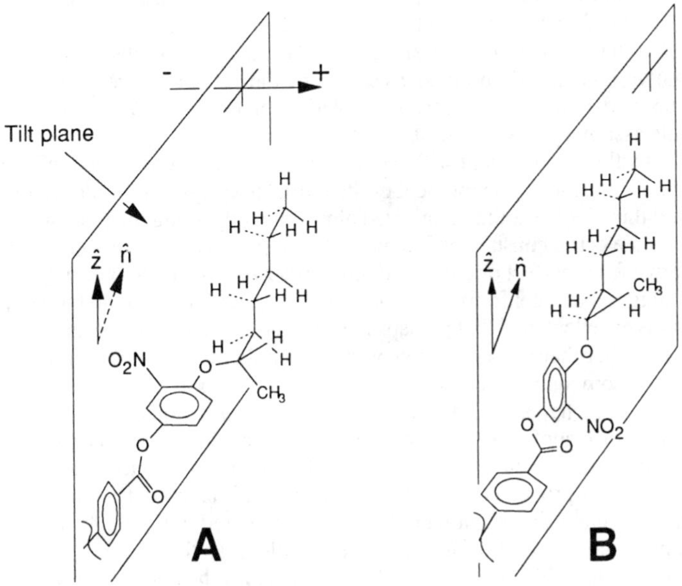

Figure 2. Preferred conformational and rotational orientation relative to the tilt plane for compound **3** in the C* phase according to the Boulder Model.

are oriented relative to the tilt plane by our expectation of how they would dock into the bent cylinder binding site. Note that there must be an equal number density of molecules in an orientation flipped by 180° about the polar axis (the C_2 axis present in all FLC phases), and also that these structures really represent large families of conformations, since many conformers differing in bonds past the stereocenter should also orient as indicated.

Experimental Evidence of Polar Orientation of the o-Nitroalkoxy System in the C* Phase

The drawings in Figure 2 illustrate that a large, negative polarization is expected for both compounds **3** and **4** for the (S) absolute configuration at the stereocenter. In agreement with the Chisso data on similar o-halo and o-cyano substituted FLCs (22), large negative **P** is indeed observed, as shown by the data in Figure 3. Specifically, at 34°C compound **3** shows a macroscopic polarization of -557 nC/cm², or about -2.1 D/molecule assuming a density of 0.8 gm/cm³.

This value is consistent with the model shown in Figure 2, though it seems quite large. Thus, the dipole moment of the nitroalkoxy system should be about 4.8 Debye, and this dipole should be oriented almost directly along the polar axis. Assuming a dielectric constant of 3, a macroscopic polarization of about 1.6 D/molecule should derive from perfect orientation of conformer A. The large observed value certainly suggests that conformer A is preferred over conformer B, that the rotational orientation is excellent, and that perhaps the appropriate dielectric constant of the medium is smaller than 3 and/or the appropriate dipole moment is larger than 4.8 D.

While compound **5** should also possess a large dipole associated with the o-nitroacyloxy grouping, as indicated in Figure 4 that dipole should not orient relative to the tilt plane, since conformers A and B in Figure 4 should be present in essentially equal number density. Due to the rapid crystallization of the monotropic C* phase of this material, it has proven possible only to obtain preliminary data. This data is, however, completely consistent with the picture presented in Figure 4. That is, an observed polarization of -79 nC/cm² (7 times smaller than that observed for compound **3**) with a tilt angle θ=39° is what would be expected for the compound with no nitro group at all (**P** for the nonyloxy homologue with no nitro substituent has been reported by the Chisso group to be -49 nC/cm² (24), but no tilt angle data is given).

We feel that these data in fact show that the o-nitroalkoxy functional array is indeed oriented along the polar axis in the FLC thin film as evidenced by the observed sign and magnitude of the macroscopic electric dipole moment of the film. This, of course, means that the molecular β associated with this functional array must also be oriented along the polar axis of the film, which should therefore possess a substantial $\chi^{(2)}$.

Preliminary Evaluation of the Nonlinear Susceptibility of Compound 3

The measurements of polarization for compounds **3** - **5** were accomplished in a thin (2 μm) parallel aligned cell with geometry as illustrated in Figure 1. In order to easily measure the second order susceptibility of, e.g. SCE9 by the SHG method, however, a different cell geometry is more preferred. That is, by appropriate surface treatments, an alignment wherein the layer normal is perpendicular to the glass bounding plates, called homeotropic alignment, may be obtained. In this geometry, the "crystal" is oriented with the polar axis parallel to the plates, and the resulting near-normal incidence of the fundamental at the phase-matching angle is convenient.

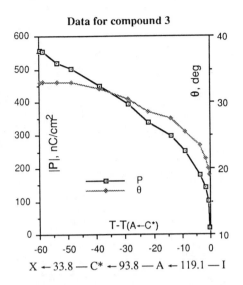

X ← 33.8 — C* ← 93.8 — A ← 119.1 — I

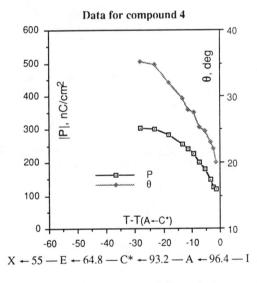

X ← 55 — E ← 64.8 — C* ← 93.2 — A ← 96.4 — I

Figure 3. Phase sequence, polarization and tilt angle data as a function of temperature for compounds **3** and **4**.

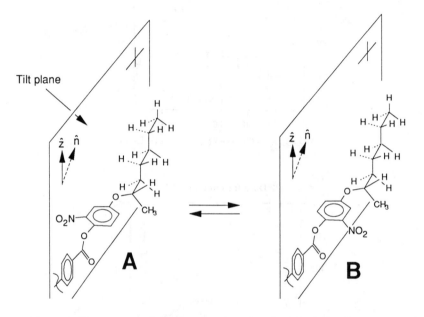

Figure 4. Important conformational and rotational orientations relative to the tilt plane for compound **5** in the C* phase according to the Boulder Model.

Unfortunately, homeotropic alignment was not possible with compound **3**, even using a magnetic field to attempt to align the sample. We believe this is due to the lack of a nematic phase in the phase sequence in combination with a fairly tight C* helix pitch. In any event, phase matched second harmonic generation using FLC material **3** has not yet been demonstrated. Encouraging preliminary results have, however, been obtained using a 10 μm thick parallel aligned cell and prisms to couple the pump light in at a steep angle relative to the bounding plates for phase-matching.

Thus, with reference to the drawing in Figure 1, Type I eeo angular phase matching was attempted using light polarized in the tilt plane (extraordinary beam), and rotating the cell about the axis normal to the director and lying in the tilt plane. At large incidence angles, the output SHG intensity began increasing dramatically, but it was never possible to hit the top of a phase-matched peak due to total internal reflection at very large incidence angles. Even so, a lower limit of 55 times the SHG efficiency of SCE9 was obtained for nitroalkoxy compound **3**, and thus d_{eff}^2 (**3**)/d_{eff}^2 (SCE9) > 55. This sets the lower limit of d_{eff} for **3** $\cong$ 0.07 pm/V — the largest observed to date for an FLC material.

It should be stated that an electric field of < 10 V/μm was applied to the cell in order to unwind the FLC helix of **3**, and the observed NLO behavior is a combination of the electric field induced SHG (EFISH) and that due to the spontaneous polar order in the phase. While other FLCs give much lower SHG efficiency with the same applied fields, and achiral smectic LC phenylbenzoates in our hands give unobservable SHG under identical conditions, we cannot completely rule out at this time the possibility that a significant amount of the response from compound **3** is due to the electrical poling. Control experiments to test for this (e.g. by SHG from compound **5** and/or racemic **3**) are in progress, as are further experiments aimed at obtaining the phase-matched SHG efficiency for **3**.

Nevertheless, at this stage we feel the best interpretation of the results obtained is that the observed response is due to the high degree of spontaneous polar orientation of the nitroalkoxy β along the polar axis in the phase. Given the expected value of β, the density of the material, and the symmetry of the system, one may expect based upon the measured value of **P** that the large coefficients of the **d** tensor for **3** (d_{23} is the largest) should be much greater than any observed to date for FLC materials. Synthesis of second generation targets is proceeding.

Problems to address in the future include: 1) Increasing the density of NLO active units in the phase; 2) Orientation of functional arrays with larger β; and 3) Developing materials with better processibility.

Finally, it should be mentioned that often <u>solids</u> are more desirable than liquids in typical applications of $\chi^{(2)}$ films. The prospects for obtaining polymer films with useful thermodynamically stable $\chi^{(2)}$ seems high given the recent demonstration that functional group orientation in FLC side chain polymers appears very similar to that observed for the low molecular weight materials (<u>10</u>). The fact that FLC polymers possess thermodynamically stable polar order in a non-crystalline solid film would appear to make this novel type of polymer glass uniquely suited for many second order NLO applications.

<u>Acknowledgments</u>

This work was supported in part by the Office of Naval Research.

Literature Cited

1. Clark, N. A.; Lagerwall, S. T. Appl. Phys. Lett. 1980, 36, 899-901.
2. Handschy, M. A.; Johnson, K. M.; Moddel, G.; Pagano-Stauffer, L. A. Ferroelectrics 1988, 85, 279-289.
3. Walba, D. M.; Slater, S. C.; Thurmes, W. N.; Clark, N. A.; Handschy, M. A.; Supon, F. J. Am. Chem. Soc. 1986, 108, 5210-5221.
4. Walba, D. M.; Vohra, R. T.; Clark, N. A.; Handschy, M. A.; Xue, J.; Parmar, D. S.; Lagerwall, S. T.; Skarp, K. J. Am. Chem. Soc. 1986, 108, 7424-7425.
5. Walba, D. M.; Clark, N. A. In Spatial Light Modulators and Applications II, Efron, U., Editor, Proc. SPIE 825, 81-87 (1988).
6. Walba, D. M.; Clark, N. A. Ferroelectrics 1988, 84, 65-72.
7. Walba, D. M.; Razavi, H. A.; Clark, N. A.; Parmar, D. S. J. Am. Chem. Soc. 1988, 110, 8686-8691.
8. Walba, D. M.; Eidman, K. F.; Haltiwanger, R. C. J. Org. Chem. 1989, 54, 4939-4943.
9. Walba, D. M.; Clark, N. A.; Razavi, H. A.; Eidman, K. F.; Haltiwanger, R. C.; Parmar, D. S. In Liquid Crystal Chemistry, Physics, and Applications, Doane, J. W. Yaniv, Z., Editor, Proc. SPIE 1080, 115-122 (1989).
10. Walba, D. M.; Keller, P.; Parmar, D. S.; Clark, N. A.; Wand, M. D. J. Am. Chem. Soc. 1989, 111, 8273-8274.
11. Walba, D. M.; Razavi, H. A.; Horiuchi, A.; Eidman, K. F.; Otterholm, B.; Haltiwanger, R. C.; Clark, N. A.; Shao, R.; Parmar, D. S.; Wand, M. D.; Vohra, R. T. Ferroelectrics, in press.
12. Walba, D. M.; Clark, N. A.; Razavi, H. A.; Parmar, D. S. In Proceedings of the 5th International Symposium on Inclusion Phenomena and Molecular Recognition, Atwood, J. L. (Ed.); Plenum Publishing Corp, in press.
13. Walba, D. M. In Advances in the Synthesis and Reactivity of Solids, Mallouk, T. E. (Ed.); JAI Press Inc., Greenwich, Connecticut, in press.
14. Vtyurin, A. N.; Ermakov, V. P.; Ostrovskii, B. I.; Shabanov, V. F. Phys. Status Solidi B 1981, 107, 397-402.
15. Shtykov, N. M.; Barnik, M. I.; Beresnev, L. A.; Blinov, L. M. Mol. Cryst. Liq. Cryst. 1985, 124, 379-390.
16. Taguchi, A.; Kajikawa, K.; Ouchi, Y.; Takezoe, H.; Fukuda, A. In Nonlinear Optics of Organic and Semiconductors, Kobayashi, T., Editor, Springer Proceedings in Physics, Vol 36, 250-253 (1989).
17. Taguchi, A.; Ouchi, Y.; Takezoe, H.; Fukuda, A. Jpn. J. Appl. Phys. 1989, 28, L 997-L 999.
18. Liu, J. Y.; Robinson, M. G.; Johnson, K. M.; Doroski, D. Optics Letters 1990, 15, 267-269.
19. Clark, N. A.; Rieker, T. P.; Maclennan, J. E. Ferroelectrics 1988, 85, 79-97.
20. Clark, N. A.; Rieker, T. P. Phys. Rev. A 1988, 37, 1053.
21. Meyer, R. B.; Liebert, L.; Strzelecki, L.; Keller, P. J. Phys., Lett. (Orsay, Fr.) 1975, 36, L-69-L71.
22. Furukawa, K.; Terashima, K.; Mitsuyoshi, I.; Saitoh, S.; Miyazawa, K.; Inukai, T. Ferroelectrics 1988, 85, 451-459.
23. Goodby, J. W.; Gray, G. W.; McDonnel, D. G. Mol. Cryst. Liq. Cryst. 1977, 34, 183-188.
24. Terashima, K.; Ichihashi, M.; Kikuchi, M.; Furukawa, K.; Inukai, T. Mol. Cryst. Liq. Cryst. 1986, 141, 237.

RECEIVED July 18, 1990

Chapter 33

Model Polymers with Distyrylbenzene Segments for Third-Order Nonlinear Optical Properties

T. E. Mates and C. K. Ober

Materials Science and Engineering, Cornell University,
Ithaca, NY 14853–1501

New main-chain poly(esters) and poly(ethers) based on four derivatives of 4,4'-(p-phenylenedi-1,2-ethenediyl) bisphenol have been prepared. The poly(esters) were made by Schotten-Baumen polymerization with aliphatic diacid chlorides; the poly(ethers) were produced by phase-transfer catalysis using dibromoalkanes. Fusibility and solubility were achieved through the flexible methylene spacers and the incorporation of methyl, methoxy, and ethoxy side groups. Most of the poly(esters) were liquid crystalline (LC) and in general had lower melting points and longer liquid-crystal ranges than the poly(ethers). The liquid-crystallinity was investigated with polarized light microscopy, differential scanning calorimetry, and X-ray diffraction. Chemical structure was analyzed by infrared and UV/visible spectroscopy, ^{1}H NMR, and elemental analysis of monomers and model compounds. Poly(esters) were drawn into fibers and were also aligned by magnetic field. Preliminary third-harmonic generation results have been obtained in a polymer thin film.

Nonlinear optical (NLO) materials, because of their broad bandwidth capabilities, could soon be used for switching signals among optical fibers. Among the most promising photonic materials which have emerged are organic molecules with extended conjugation, in which ultrafast NLO responses can be induced as virtual excitations of p-orbital electrons. Polymers are outstanding among organic solids for their mechanical integrity, and so efforts have been made at incorporat-

0097–6156/91/0455–0497$06.00/0

ing various inorganic (1) and organic (2) photonic materials into polymers, either covalently or as mixtures. Because alignment is critical to the optimization of strength and macroscopic hyperpolarizability, precursor-route polymers have been made, in which flexible polymers are shear-aligned during the elimination of side groups, leaving an aligned but intractable material. The formation of poly(phenylenevinylene), from an ionic sulfonium precursor (3) is an example of this, and while reasonably high values for conductivity (4) and $\chi^{(3)}$ (5) have been obtained for oriented specimens, the nature and prevalence of chemical and electronic irregularities in the conjugated chains (and their influence on the optical and electrical properties) are not well understood. These factors,along with polydispersity in molecular weight, give rise to a mixture of conjugation lengths, each of which makes a unique contribution to the polymers' optical and electrical performance.

Since liquid-crystalline polymers are relatively easy to align, they have been the focus of many recent organic NLO investigations. Side-chain LC polymers are especially easy to align and pole, and attempts have been made to produce side-chain materials for second-order NLO (6). Lyotropic rigid rods such as poly(p-phenylenebenzobisthiazole) (PBT), with large third-order polarizabilities (7), may be oriented by shear from solution. Magnetic (8), and electrical fields (9) may also generally be used to align LC polymers, in contrast to the non-LC "all-conjugated" polymers such as PPV, which cannot be aligned by applied fields. It is also important, however, that the well-defined conjugation lengths usually found in the mesogens of thermotropic LC polymers will give optical and electrical results which are more readily interpretable than those from "all-conjugated" polymers.

Main-chain thermotropic LC polymers such as those presented here combine the advantages of alignability and a well-characterized conjugation length with freedom from the solvent-removal problem of lyotropics. Since the conjugation in the present system can provide both liquid-crystalline and photonic behavior, it simultaneously provides us experience with a third harmonic material and a novel LC system.

The conjugated mesogen used in this study, often referred to as 1,4-distyrylbenzene, was recognized as mesogenic by Campbell and McDonald (10), who synthesized several small-molecule derivatives. Three previous references were found to the incorporation of this mesogen into a main-chain LC polymer. Memeger (11) used the Wittig reaction to produce all-hydrocarbon LC polymers, and Suzuki et al (12) produced a non-LC poly(hydrocarbon) as well as LC poly(ethers) and poly(esters) by the Heck reaction, using a palladium catalyst. Finally, Iimura et al (13) used a transesterification reaction to achieve high molecular-weight LC distyrylbenzene poly(esters), the only previous reference to the use of a standard polymerization reaction being used for the incorporation of this mesogen. Of these three references, only one route, the Heck reaction, provided nearly all-*trans* unsaturations in the as-produced polymer, yet the products of that reaction displayed serious thermal instability relative to the other two. It will be shown that the mesogenic monomers used in the present method are produced in the rodlike, all-*trans* configuration and are readily soluble, and also that the polymers are relatively thermally stable.

The preparation and characterization of the model compounds and polymers will be described, including molecular weight, physical structure, and melt behavior. Following that, methods of polymer alignment and the results achieved will be described. Finally, optical properties will be discussed, along with our interest in this system, with its short but well-defined conjugation in the main chain. Third harmonic generation measurements are to be done of polymer and model-compound solutions, as well as model-compound single crystals and aligned polymer specimens. Coupled with measurements of the real and imaginary refractive indices, this will give us a basis for understanding the optical properties of these materials.

Experimental

Sodium hydride (97%), triethyl phosphite (99%), α,α′ dibromo-p-xylene (97%), 4-hydroxy-3-methylbenzaldehyde (97%), 3-ethoxy-4-hydroxybenzaldehyde (99%), vanillin (99%), and valeryl chloride (98%), were supplied by Aldrich Chemical Co. and were used without further purification; 4-hydroxybenzaldehyde (96%, Aldrich) was resublimed prior to use. The acid chlorides were supplied by Aldrich Chemical Co. and, with the exception of sebacoyl chloride (99+%), were fractionally distilled at reduced pressure through a 6-inch Vigreux column prior to use: pimeloyl chloride 95-6°C at 1.1 torr, suberoyl 114-16°C at 2.2 torr, azelaoyl 104-6°C at 0.35 torr. Dibromoalkanes from Aldrich Chemical were fractionally distilled prior to use: 1,7-dibromoheptane 111-114°C at 1.4 torr, 1,9-dibromononane 135-137°C at 2.7 torr, 1,11-dibromoundecane 129-131°C at 0.85 torr. Tetrabutylammonium iodide (98%) was supplied by Aldrich. Reagent grade solvents were obtained from Fisher Scientific.

Bisphenol Preparation. The mesogenic bisphenols were produced in a manner similar to that of Stilz and Pommer (14). First, tetraethyl-p-xylylenediphosphonate was produced in a reaction between α,α′-dibromo-p-xylene and triethyl phosphite in xylene. The diphosphonate was then reacted with an excess of the appropriately-substituted p-hydroxybenzaldehyde. The preparative details and analytical results for the four monomers were given earlier (Mates, T.E.; Ober, C.K. J. Polym. Sci. Lett., to be published).

Ester Model Compounds and Poly(esters). The synthesis and characterization of the ester materials was described previously (Mates, T.E.; Ober, C.K. J. Polym. Sci. Lett., to be published). The model compounds were prepared by reacting the bisphenols with an excess of valeryl chloride; to make the poly(esters), the bisphenols were reacted with stoichiometric amounts of alkyl diacid chlorides as shown in Scheme 1.

Ether Model Synthesis. The ether models were prepared in two-phase reactions of the bisphenols with 1-bromobutane. In a typical reaction, 0.50g (3.7 mmol) of 1-bromobutane was dissolved in 5 mL nitrobenzene in a three-necked round-bottom flask equipped with a mechanical stirrer. To this was added 0.30g (0.96 mmol) of (II, R=H) in 30 mL of 2N aqueous NaOH (the bisphenol was not completely dissolved). Approximately 10 mg of tetrabutylammonium iodide was added and the mixture was vigorously stirred at 40-50° C for 3h. After cooling to room temperature, the mixture was poured into methanol, filtered, washed with more methanol, and recrystallized from DMF to yield 0.32g (79.2%) of platelike green crystals, mp 277°C (T_i 290°C); IR 2956m, 1605s, 1252vs, 1178s, 970m. *Anal.* Calcd. for $C_{30}H_{34}O_2$: C, 84.51; H 7.98. Found: C, 84.31; H 8.04. The others were recrystallized from ethyl acetate and were similar in appearance. The methoxy and ethoxy-substituted models were not liquid-crystalline:

3,3'-dimethyl model ether (79.2%); mp 236°C (Ti 255°C); IR 2852m, 1516vs, 1259s, 1121m, 964m; ^{1}H NMR d 0.95 (tr, 6H CH_3), 1.51(sextet, 4H CH_2), 1.76(quintet, 4H CH_2), 2.21 (s, 6H CH_3 on aromatic ring), 3.96 (tr, 4H CH_2), 6.92 (d, 2H =CH-), 7.01 (d, 2H =CH-), 7.3 (m, 6H arom), 7.43 (s, 4H arom). *Anal.* Calcd. for $C_{32}H_{38}O_2$: C, 84.58; H, 8.37. Found: C 83.93; H, 8.32.

3,3'-dimethoxy model ether (66.9%); mp 222°C; IR 2947w, 1516vs, 1259s, 1219s, 964m; ^{1}H NMR 0.96 (tr, 6H CH_3), 1.51 (sextet, 4H CH_2), 1.83 (quintet, 4H CH_2), 3.92 (s, 6H CH_3 methoxy), 4.04 (tr, 4H CH_2), 6.87 (d, 2H =CH-), 7.03 (m, 8H =CH- and arom), 7.48 (s, 4H arom). *Anal.* Calcd. for $C_{32}H_{38}O_4$: C, 79.01; H 7.82. Found: C 79.41; H 7.79.

3,3'-diethoxy model ether (74.1%); mp 220°C; IR 2935m, 1518vs, 1252vs, 1136s, 968s; ^{1}H NMR 0.96 (tr, 6H CH_3), 1.45 (tr, 6H CH_3 ethoxy), 1.53 (sextet, 4H CH_2), 1.80 (quintet, 4H CH_2), 4.00 (tr, 4H CH_2), 4.12 (quartet, 4H CH_2 ethoxy), 6.96 (m, 10H =CH- and arom), 7.43 (s, 4H arom). *Anal.* Calcd. for $C_{34}H_{42}O_4$: C, 79.37; H 8.17. Found: C 79.93; H 8.22.

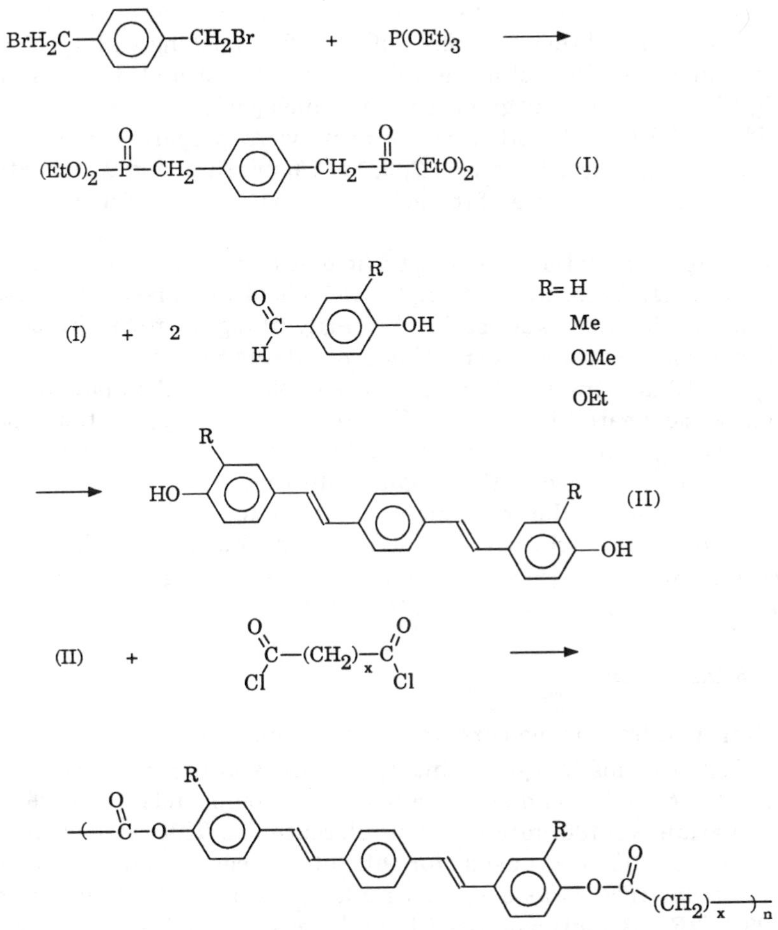
BrH2C
CH2Br
+
P(OEt)3
(EtO)2
P
CH2
O
(I)
2
C
H
R
OH
R= H
Me
OMe
OEt
HO
(II)
(CH2)x
Cl
n

Scheme 1

Poly(ether) Synthesis. The poly(ethers) were made by a two-phase reaction with dibromoalkanes. In a typical reaction, 0.75 g (2.4 mmol) of (II, R=H)) was mixed with 70 ml of 2N NaOH in a 3-necked round-bottom flask equipped with a mechanical stirrer. To this was added an equimolar amount of 1,9-dibromononane in 20 mL nitrobenzene and approximately 10 mg of tetrabutylammonium iodide and the mixture was stirred overnight at 50°C. The resulting solid mass was washed with methanol and then with 2N NaOH. After washing with 0.1N HCl, the product was Soxhlet-extracted with methanol and dried to yield 0.68g (64.7%) of a light green powder which melted to an anisotropic liquid at 290°C. The other poly(ethers) were prepared in the same manner, using spacer lengths of 7, 9, 11, 7/9 mixture, and 9/11 mixture. The yields and IR spectra of the poly(ethers) is shown in Table I.

Alignment. Orientation by magnetic field at 13.5 T was done at the MIT Bitter Magnet Laboratory. Samples were held in 3mm-wide quartz x-ray tubes. The tubes were packed by partly filling the tube with powder and then melting the powder. This process was repeated until a dense sample 5-10mm long was formed. The samples were then placed in the magnet and heated into their LC states for varying lengths of time under nitrogen by a cylindrical graphite heater in the magnet bore. The samples remained in the magnet until they reached 30°C, at which temperature molecular movement is slow enough for the sample to remain aligned indefinitely outside the field. Fibers were drawn with tweezers from the surface of the Fisher-Johns melting point apparatus. Typical draw ratios were between 10 and 20.

Characterization

Melting points of monomers and model compounds were determined on a Fisher-Johns Melting Point Apparatus, and are uncorrected. ^{1}H NMR of the methyl and ethoxy ether model compounds was performed on a Varian XL-400 spectrometer referenced to TMS at 0.00 ppm; a Varian XL-200 was used for all other compounds. Infrared spectroscopy (KBr) was done using 200 scans on an IBM Instruments IR/98 FTIR. A Leitz polarized light microscope and a Mettler FP-52 hotstage were used for the optical characterization of liquid crystal phases. Differential Scanning Calorimetry (DSC), both heating and cooling, was performed on a Perkin-Elmer DSC-2 at 20°C/min under nitrogen; polymer melting and clearing temperatures are taken as the centers of the appropriate peaks. Thermogravimetric analysis (TGA) was done on a DuPont Instruments 951 Thermogravimetric Analyzer at 10°C/min with nitrogen flow at 90 cc/min. Intrinsic viscosity was determined in nmp solution in a Cannon-Ubbelohde 100 Viscometer

Table I. Characterization of Poly(ethers)

R	x	Yield (%)	IR
H	9	64.7	2928 s, 1603 m, 1516 s, 1248 vs, 970 m
	11	80.7	2922 vs, 1516 s, 1250 vs, 1177 m, 970 m
	7/9	76.4	2934 m, 1603 m, 1514 s, 1248 vs, 962 m
	9/11	69.8	2920 m, 1605 m, 1516 s, 1250 vs, 968 m
Me	7	71.1	2922 s, 1602 s, 1248 vs, 1126 s, 960 m
	9	82.0	2924 vs, 1602 s, 1248 vs, 1128 m, 960 m
	11	83.1	2928 s, 1604 s, 1514 s, 1247 vs, 966 w
	7/9	72.6	2924 s, 1602 s, 1248 s, 1126 m, 964 m
	9/11	68.6	2930 m, 1602 s, 1247 vs, 1128 m, 964 m
MeO	9	59.3	2932 m, 1516 vs, 1254 s, 1138 m, 962 w
	11	72.9	2924 m, 1516 vs, 1254 s, 1138 vs, 959 m
	7/9	74.5	2926 s, 1514 vs, 1252 s, 1136 s, 957 m
	9/11	67.8	2932 w, 1516 vs, 1254 s, 1136 m, 959 w
EtO	7	79.4	2934 w, 1516 vs, 1254 s, 1136 m, 960 w
	9	78.1	2924 m, 1514 vs, 1252 vs, 1134 s, 960 m
	11	75.5	2926 m, 1516 vs, 1175 m, 968 w, 835 m
	7/9	62.6	2932 w, 1516 vs, 1256 s, 1136 m, 960 w

heated in a Cannon Instrument Co. constant temperature bath. The poly(esters) were measured at 35°C, the poly(ethers) at 65°. Sample solutions were filtered through a 0.5 mm Millipore filter and equilibrated 30 minutes prior to measurement. Specific viscosity was measured at 4 concentrations and extrapolated to zero concentration. Low-angle laser light scattering was performed on a Chromatix KMX-6 photometer at room temperature in nmp solution. Ultraviolet/Visible spectra were recorded on a Perkin-Elmer Lambda 4A UV/VIS spectrophotometer in chloroform. Elemental analysis was performed by the Scandinavian Microanalytical Laboratory, Herlev, Denmark. Powder x-ray patterns were recorded on a Scintag Pad V diffractometer scanning at 2 degrees/min, using a germanium detector. Transmission exposures were taken using a Statton camera using Ni-filtered Cu K-α radiation at 3.2 and 4.8 cm sample-to-film distances.

Third-Harmonic Generation. In order that the third harmonic signal not be absorbed, a laser source of wavelength greater than 1.3 μm was desired, since the UV absorption edge of these materials, discussed below, is 403 nm. To achieve this, a 1.064-mm Nd:YAG laser (3 ns pulse length at 200mJ/pulse) was equipped with a 3 MPa hydrogen cell which Raman-shifted the input signal to 1.907 mm, allowing third-harmonic to be detected at 635 nm. This input frequency also avoids absorption of the principle beam as overtones of the materials' IR resonances.

In order for the polymer films not to scatter light excessively, it was necessary to make the measurements on unoriented films of the non-liquid crystalline polymers. The films were cast from filtered nmp solutions onto 1/16 " thick quartz plates.

Results and Discussion

Structure and Properties. The weight-average molecular weight of one of the more soluble of the polymers, the ethoxy-substituted poly(ester) with a heptamethylene spacer, was determined by static low-angle laser light scattering (LALLS) to be 18,000. Its intrinsic viscosity of 0.56 is typical of the poly(esters) presented here, indicating that their average degree of polymerization is about 20. Of the poly(ethers), only those with R=MeO and EtO were sufficiently soluble for viscometry. Their intrinsic viscosities were between .300 and .500.

Linearity in the conjugated monomers was desired to enhance liquid crystallinity. Campbell and McDonald (10) noted the necessity of isomerization using iodine in obtaining the *trans, trans* form of the mesogen, when producing it from the conventional (triphenyl-phosphonium) ylide Wittig reaction. Tewari et al (15) however, found that the phosphonate used in this study yielded *trans, trans* almost exclusively. The predominant stereoisomer among the compounds presented here was identified using IR spectroscopy. The 885 cm^{-1} region can be attributed to C-H out-of-plane deformation of *cis* methine carbons in poly(1,4-phenylenevinylenes), and the 960 cm^{-1} region to the *trans* form (16).

The present compounds generally displayed a peak only in the 960 cm^{-1} region. One of the methoxy-substituted (x=5) and two of the ethoxy-substituted (x=5 and 8) poly(esters) were not LC, neither were two of the ether model compounds and several poly(ethers). As the *cis/trans* ratio was small and relatively independent of substituent, there are two likely reasons for the limitations on liquid-crystallinity:

the loss of rodlike shape inherent to the addition of substituents to the aromatic rings, and a possible loss of coplanarity. Variations in coplanarity were assessed with UV/VIS spectroscopy of the model compounds in chloroform solution. Only slight changes in the UV absorption edge (403 to 406 nm) and maximum (356 to 364 nm) were detectable as the aromatic substituent was changed from H to EtO; red shifts consistent with moderate steric interference to coplanarity (17). There was also only a small change in the fluorescence peak maximum, gradually shifting from 417 nm to 422 nm, as substituent size was increased from H to OEt. The spectra of the ethers and esters were virtually identical. Therefore, the reason for the decrease in the variety of spacer lengths at which one finds liquid crystallinity in the case of large 3,3' substituents is the loss of rodlike mesogen shape.

The ester models, as stated above, were all LC, but two of the ether model compounds, the methoxy- and ethoxy-substituted derivatives, were not, despite the fact that each of these mesogens yielded some liquid-crystallinity in the poly(ether) form. Therefore it seems that the polymers in this system tend to be "more liquid crystalline" than the related small molecules. This hypothesis is supported by the fact that Memeger(11) found liquid crystallinity in all-hydrocarbon polymers incorporating the distyrylbenzene mesogen, even in cases where the *cis/trans* ratio of the unsaturations was as large as 0.3, while Campbell and McDonald (10) noted that iodine isomerization to the all-*trans* form was essential for the observation of an LC phase in the small-molecule derivatives which they prepared.

The poly(esters) and poly(ethers) in this study began to decompose during TGA at about 320° C and their loss peaks occurred at 430-460° C, making them slightly inferior to Memeger's (11) all-hydrocarbon materials, but considerably more stable than those of Suzuki et al (12), which readily thermally crosslinked. More to the point, the thermal stability was sufficient that it was not a limiting factor during the magnetic alignment or fiber-drawing operations.

Melt Behavior. The polymer melting points are taken from DSC scans, heating at 20°C/min. Melting and clearing points of the ether model compounds are given in the Experimental section. Figure 1 shows the melting points of the poly(esters) as a function of spacer length and aromatic substituent. The "odd-even effect", the commonly-observed phenomenon of polymers with an even number of spacer units melting higher than those with an odd number, is observed in the methyl and methoxy-substituted materials, and weakly in the others. These polymers, on average, remain LC for about 60°C above their melting points before clearing to isotropic liquids. The only polymer which clearly displayed a smectic phase in addition to a nematic phase was the methyl-substituted poly(ester) with a pentamethylene spacer;

Figure 2 shows its DSC trace. Melting to a smectic LC at 210°C, it becomes nematic at 240°, and remains nematic into the vicinity of decomposition above 300°. Smectic structure has been verified by optical interference microscopy and by x-ray diffraction at the Cornell High Energy Synchrotron (CHESS) x-ray facility. A sample was heated between 25-μm Kapton sheets with the x-ray beam passing through orthogonally. A video camera was used as a "flat-film" detector. The sample was heated into the nematic range and then cooled at 10°C/min. At 220° a ring corresponding to a 22 Å spacing, the approximate molecular length, appeared, became strong by 215°C and remained to room temperature. The freezing point of this polymer at this cooling rate is 197°C. All other LC polymers were nematic; x-ray investigation into possible smectic behavior in several of these polymers is in progress.

The methyl-substituted ester model compound, shown below, also shows both smectic and nematic phases, as well as a smectic C-to-nematic transitional phase (18) from 169 to 170°C. Figure 3 shows two of the textures observed in this model.

A large length/width ratio was also important for the achievement of liquid crystallinity in the poly(ethers). All five of the R=H and all five of the R=Me materials were LC, exhibiting robust birefringence and stir opalescence. For R=MeO only three of five were LC, and for R=EtO, only the two mixed-spacer length polymers were LC.

Alignment. The x-ray powder patterns showed the expected intermolecular spacing of approximately 4.5 Å and a considerable amount of crystallinity. Fig. 4(a) and (b) show "fiber" patterns (magnetic field direction vertical) of samples of the hexamethylene-spacer, ethoxy-substituted poly(ester) in the 3mm quartz x-ray tubes. The diffraction patterns before (a) and after (b) magnetic alignment are shown. The aligned sample was exposed to the field for 10 minutes; no difference was discernable between this sample and one held in the field for an hour. The arcs visible in the oriented specimen indicate an intermolecular spacing of approximately 8 Å. Fig. 4(c) shows the alignment achieved by pulling a fiber of the same material. The key difference between magnetic and shear alignment is shown by the appearance of meridional arcs in the fiber's pattern. Shear alignment apparently has a stronger tendency to bring the mesogenic units of neighboring chains into coincidence laterally than does magnetic alignment. In general, the aligning force should be more uniform over the sample in the magnetic case, since the field simply provides a cylindrically symmetrical "director". In the case of shear, the holding arrangement can influence the structure by, for example, causing the sample to adopt a ribbon-like shape, as was the case here. The fiber's pattern shows the large spacing (inner arc) similar to that of the

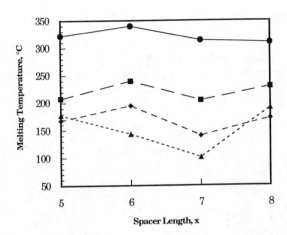

Figure 1. Melting temperature of the poly(esters) as a function of spacer length. [Substituent R: H(●), Me(■), MeO(◆), EtO(▲).]

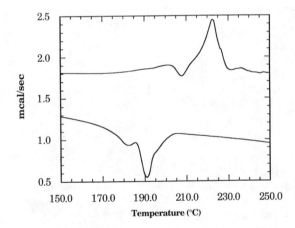

Figure 2. DSC trace of methyl-substituted poly(ester) with pentamethylene spacer, which displayed both smectic and nematic textures.

a

b

Figure 3. Liquid-crystal textures of the methyl-substituted model ester viewed through crossed polarizers. a, Smectic C-to-nematic transitional phase; and b, smectic mosaic texture at 160 °C. Original magnification, 320×.

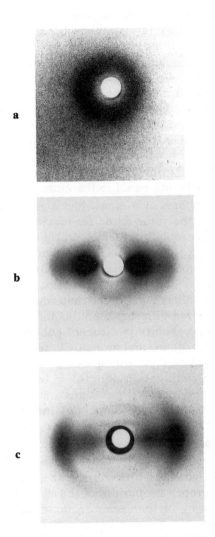

Figure 4. X-ray fiber diagrams of ethoxy-substituted, hexamethylene-spacer poly(ester). a, Before magnetic alignment; b, after magnetic alignment; and c, fiber. Sample-to-film distance, 4.7 cm.

magnetic sample, but shows a more prominent outer arc
corresponding to the 4.5 Å intermolecular spacing. The absence of this
arc from Fig. 4(b) is apparently an artifact of the quartz tube holding
arrangement. An unaligned sample with same thermal history as
that of the sample in Fig. 4(a), but simply pressed into a flat film, rather
than being melted into the 3mm tube, yielded a ring corresponding to
the outer arcs of Fig. 4(c). Optically, Fig. 5 shows the texture difference,
as viewed through crossed polarizers, between the anisotropic melt (a)
and the corresponding fiber (b) of this same polymer.

Third-Harmonic Generation. The mesogenic unit in this system
contains eleven double bonds, equal to the number found in β-carotene.
That material, in the form of a glass, has a $\chi^{(3)}$ of 1×10^{-12} esu (19),
approximately 15% of the value found for biaxially-oriented specimens
of poly(p-phenylenebenzobisthiazole) (4), a rigid, conjugated aromatic
polymer. However, the (anisotropic) molecular third-order
polarizability of the carotene suggests a value five times as large if the
molecules were aligned and the measurement was made parallel to
their length, making values for the two materials nearly equal. While
these two were measured differently and cannot strictly be compared
due to a difference in proximity to resonant enhancement, and because
of the presence of heteroatoms and aromaticity in the polymer, the point
is made that the extent of pi-electron delocalization does not increase in
a simple way with"conjugation length". In general, conjugated
molecules' minimum-energy configurations incorporate a substantial
amount of bond-length alternation, putting an upper limit to
delocalization and UV absorption edge (20). Bradley and Mori (21)
measured $\chi^{(3)}$ for poly(p-phenylenevinylene) as it was being formed
from a soluble, non-conjugated precursor. They found that the UV
edge and $\chi^{(3)}$ values changed little during the final stages of curing.
They point out that proper comparison of theory and experimental data
can only be achieved by further work on materials with well-defined
conjugation lengths.

To date, results have only been acquired for the ethoxy-
substituted poly(ester) with a pentamethylene spacer. A 2.5 μm film
produced a third-harmonic signal at 632.8 nm 29 times the intensity of
a quartz signal, corresponding to a $\chi^{(3)}$ of approximately 8×10^{-13} esu.
Degenerate four-wave mixing experiments are underway on this and
other materials, and preliminary results are comparable to the values
from third harmonic generation.

Summary

A new family of liquid-crystalline poly(esters) and poly (ethers) has

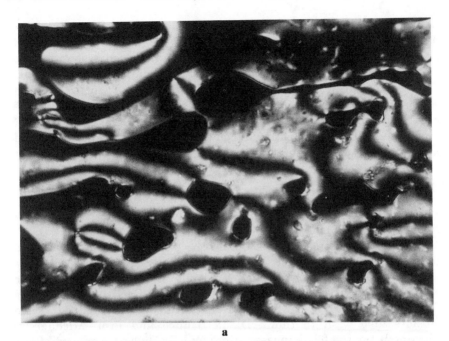

a

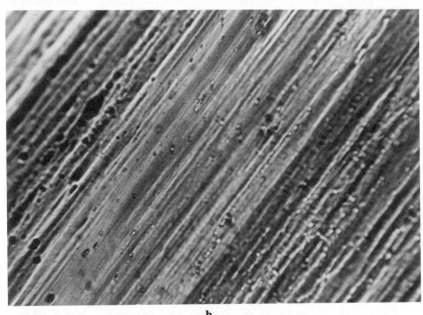

b

Figure 5. Optical textures corresponding to Figures 4a and 4c fiber diagrams. a, Unaligned at 165 °C; and b, fiber. Original magnification, 320×.

been described. Alignment by drawing and magnetic field have been demonstrated. Given their well-defined conjugation length and tractability (especially the esters), they present a good system for the study of the fundamental structure-property relationships in organic NLO materials. Preliminary third-harmonic results show a reasonable $\chi^{(3)}$ for this conjugation length of approximately 8 x 10^{-13} esu. The goal of alignment, including the growth of single-crystals of model compounds, is to maximize third harmonic generation and understand its relationship to structure.

Acknowledgments

The authors sincerely thank the Office of Naval Research (Contract # N00014-87-K-0826) for partial support of this work. We also thank Dr. T. Mourey of the Eastman Kodak Co. for the LALLS data. TM wishes to thank IBM Corp., Endicott, NY for a Resident Study fellowship. The use of the facilities of the Cornell Materials Science Center, the Bitter National Magnet Laboratory at MIT, and the Cornell High-Energy Synchrotron Source (CHESS), are gratefully acknowledged. The authors also wish to thank David Brunco and Prof. Michael O. Thompson for the THG investigation, and Dr. Robert Norwood of the Hoechst-Celanese Corp., for degenerate four-wave-mixing results.

Literature Cited

1. Wang, Y.; Suna, A.: Mahler, W. In Nonlinear Optical Properties of Polymers; Heeger, A.J.; Orenstein, J.; Ulrich, D.R., Eds.; MRS Symp. Proc., V. 109; Materials Research Society: Pittsburgh, 1988; pp 345-350.
2. Calvert, P.D.; Moyle, B.D., ibid.; pp 357-362.
3. Wessling, R.A.; Zimmerman, R.G. U.S. Patent 3 532 643, 1968.
4. Gagnon, D.R., Capistran, J.D., Karasz, F.E., Lenz, R.W., Antoun., S. Polymer 1987, 28, 567.
5. Singh, B.P., Prasad, P.N., Karasz, F.E. Polymer 1988, 29, 1940-1942.
6. Griffin, A.C.; Bhatti, A.M.; Howell, G.A.; pp 115-125; ref. 1.
7. Rao, D.N.; Swiatkiewicsz, J.; Chopra, P.; Ghoshal, S.K.; Prasad, P.N.Appl. Phys. Lett. 1986, 48, 1187-1189.
8. Moore, J.S.; Stupp, S.I. Macromolecules 1987, 20, 282-293.
9. Gilli, J.M.; Schmidt, H.W.; Pinton, J.F.; Sixou, P. Mol. Cryst. Liq. Cryst.Letters 1984, 102, 49-58.
10. Campbell, T.W.; McDonald, R.N. J. Org. Chem. 1959, 24, 1246-1251.
11. Memeger, W. Macromolecules 1989, 22(4), 1577-1588.
12. Suzuki, M.; Lim, J.C.; Saegusa, T. Macromolecules 1990, 23, 1574-1579.

13. Iimura, K.; Koide, N.; Miyabayashi, M. Jap. Patent 62 256 825, 1987.
14. Stilz, W.; Pommer, H. Ger. Pat. 1 108 219 (Cl. 12o), 1961.
15. Tewari, R.S.; Kumari, N.; Kendurkar, P.S., Indian J. Chem. 1977, 15B, 753-755.
16. Bellamy, L.J. The Infrared Spectra of Complex Molecules; Interscience Publishers, New York, 1958; pp 50-54.
17. Williams, D.H.; Fleming, I. Spectroscopic Methods in Organic Chemistry; McGraw-Hill, London, 1987; pp 27, 28.
18. Demus, D.; Richter, L. Textures of Liquid-Crystals Revised Ed., VEB Deutscher Verlag fur Grundstoffindustrie, Leipzig, 1980; p 136.
19. Hermann, J.P., Ducuing, J. J. Appl. Phys. 1974, 45, 5100-2.
20. Blythe, A.R., Electrical Properties of Polymers Cambridge Univ. Press, London, 1979; p 107.
21. Bradley, D.D.C.; Mori, Y. Jap. J. Appl. Phys. 1989, 28, 174-177.

RECEIVED August 7, 1990

COMPOSITE MATERIALS

Chapter 34

Composites

Novel Materials for Second Harmonic Generation

C. B. Aakeröy[1], N. Azoz[1], P. D. Calvert[2], M. Kadim[1], A. J. McCaffery[1], and K. R. Seddon[1]

[1]School of Chemistry and Molecular Sciences, University of Sussex, Falmer, Brighton BN1 9QJ, United Kingdom
[2]Arizona Materials Laboratory, University of Arizona, Tucson, AZ 85712

Herein is described a new class of materials for second harmonic generation (SHG), in which microcrystals of an SHG-active material (guest) are deposited within a polymer matrix (host) in an aligned fashion. The guest crystallites range from 3-nitroaniline (mNA) to a new class of hydrogen-bonded dihydrogenphosphate salts, $[AH][H_2PO_4]$ (where A = an amine). These latter materials have a range of physical properties that make them highly suited as SHG-active guest crystals. The guest crystals are aligned within the polymer matrix, using a thermal gradient technique, a method which produces transparent, non-scattering, flexible, SHG-active composites, with excellent temporal stability.

The field of nonlinear optics (NLO) is currently one of the most active in terms of research intensity and funding. However, despite many heroic efforts, both theoretical and experimental, the most common SHG–active materials that are in use today, potassium dihydrogenphosphate (KDP) and lithium niobate(V), $LiNbO_3$, were discovered when the field was in its infancy. Efforts now have to be concentrated on the development of processable, nonlinear optical materials, and this has to be achieved on a very short time scale.

Material Requirements for Second–Harmonic Generation

In order to bridge the gap between an SHG–active material and one optimized for use in an optoelectronic device, many compounds have been synthesized, characterized, modified and then ultimately rejected during the past decade (_1–3_). This is not surprising, since the ideal material must fulfil a plethora of stringent requirements (_4–7_). The most critical condition for an SHG–active material is that it must form noncentrosymmetric structures; however, thermal stability, involatility, transparency, lack of colour, mechanical strength and crystal habit are also crucial properties for materials to be incorporated into practical devices.

We present here an overview of our recent work, in which two novel approaches to new materials for SHG have been combined to yield composites exhibiting quite remarkable optical properties.

Hydrogen-Bonded Salts: Novel SHG-Active Compounds

Current Materials. Even though the nonlinear coefficients among inorganic materials are generally significantly lower than those found for organic materials, it has been possible to grow good quality single crystals of several inorganic compounds, making them available as bulk media for utilization in conversion processes in laser-operated systems. However, there has been a great increase in interest in organic materials for SHG and, as a result, many novel organic materials have been synthesized and characterized during the last fifteen years (*8–13*). There has also been much recent interest in organometallic compounds (*14,15*), but Kanis *et al*. (Boston ACS Meeting, 1990, Abstract INOR 472) have cast doubts on their practicality.

Unfortunately, various factors are hampering the efficiency of materials from these principal classes. Inorganic compounds often exhibit low $\chi^{(2)}$ values, restricted birefringence, and limited solubility. Organic molecules are, potentially, more versatile due to larger β-values, and the possibilities of specifically designing molecules for high SHG activity, *e.g.* combining large polarizability with the presence of substituents capable of charge transfer. However, they frequently suffer from volatility, low thermal stability and mechanical weakness. Organometallic compounds are usually strongly coloured.

Novel Materials. Despite the fact that many organic molecules have very high β-values, their $\chi^{(2)}$-values are often very small. The reason for this is that a high β-value is, usually, accompanied by a large molecular dipole moment. The large dipole moment encourages the molecules to form pairs, aligned in an antiparallel fashion, which usually favours centrosymmetric crystal forms, thereby ruling out the possibility of SHG-activity. If highly polarizable organic molecules, with large second-order molecular coefficients, could be prevented from forming unfavourable crystalline structures, their full potential could then be utilized. One very successful approach has been to incorporate these molecules into zeolitic frameworks (*16*). Our approach has been to incorporate anions and cations into the crystal structure which are capable of forming a strong, three-dimensional network of hydrogen bonds, in the hope that this additional lattice force would overwhelm the propensity for dipole alignment and thus increase the probability of forming noncentrosymmetric crystals.

To this end, we have designed a new range of salts $[AH][H_2PO_4]$ (*17*), combining a cation derived from an organic amine (*e.g.* A = benzylamine, 3-hydroxy-6-methylpyridine, or piperidine), with an inorganic anion, dihydrogenphosphate, which is capable of forming strong hydrogen-bonded crystal structures. The only previously known compound of this type was L-argininium dihydrogenphosphate monohydrate, $[(H_2N)_2CNH(CH_2)_3CH(NH_3)COO][H_2PO_4].H_2O$ (*18*).

In initial studies, two dozen salts of formula $[AH][H_2PO_4]$ (A = primary, secondary, or tertiary amine) were prepared and screened for SHG activity, using the powder technique (*19*). The measured SHG intensities of the organic salts of dihydrogenphosphate (*17*) are, in general, not particularly high (in the range 0.2–5, relative to α-SiO$_2$). However, this is not surprising, as the amines selected are not specifically designed to produce large nonlinear effects.

There is a high incidence (eight out of twenty-four) of SHG-active materials among this class of materials. A success rate of 33% with regard to noncentrosymmetric structures is significantly higher than the expected statistical average {oft quoted as 20%(*1,2*)}; a 50% success rate was found for a recently reported series of stilbazolium salts (*20*). Clearly, these dihydrogenphosphates must only represent a small fraction of the total number of SHG-active materials in this class of salts.   Even though the materials studied exhibit rather small nonlinear responses, they are all colourless, chemically stable, involatile and soluble, and also show a propensity for the growth of good quality crystals.

Structure and Hydrogen Bonding.   In order to provide information about the presence and extent of hydrogen bonding within these novel salts, X–ray crystallographic studies were undertaken on single crystals of five of these dihydrogenphosphate salts (*17*).   It was found that each structure was dominated by chains or sheets (*e.g.* Figure 1) of dihydrogenphosphate anions, invariably held together by short hydrogen bonds (*17*).   Not only were strong hydrogen–bonded networks between the anions detected, but the disposition of the cations was dominated by strong hydrogen bonds between the cations and the anion lattice.

Lattice Energy Calculations.   Even though the crystal structures of the dihydrogenphosphate salts contain a number of seemingly strong hydrogen–bonded interactions, no explicit information about the energetic contribution made by hydrogen bonding to the overall lattice energy of the materials can be obtained from the crystal structures alone.   In order to acquire this information, lattice energy calculations were carried out on four dihydrogenphosphate salts (Aakeröy, C.; Leslie, M.; Seddon, K.R., to be published).

The calculations were performed with the CASCADE suite of programmes, written and developed by Leslie at SERC Daresbury Laboratory (*21*), and designed specifically for the facilities of the CRAY–1 computer.   The results of these calculations are summarized in Table I.

The calculated lattice energies, U_{cal}, of the four salts show that three of them have very comparable values, whereas the lattice energy of 3–hydroxy–6–methyl–pyridinium dihydrogenphosphate is significantly lower.   This salt also has the largest unit cell volume per empirical formula unit, which is a measure of the packing efficiency throughout the structure.   The presence of the methyl group increases the bulk of the cation, and makes close–packing of the ions more difficult.

Based on a wide range of experimentally determined values for hydrogen bond energies between ions (which are significantly higher than corresponding values for hydrogen bonds between neutral molecules) (*22*), each O–H...O interaction was assigned an energy content of 35 kJ mol^{-1}, and each N–H...O interaction was assigned a value of 30 kJ mol^{-1}.   By using these values (which underestimate the probable true values by approximately 50%), combined with the appropriate number and type of hydrogen bonds in each salt, an approximate minimum estimate of the total hydrogen bond energy, E_{HB}, for each salt was obtained, Table I.

As shown in Table I, the energetic contributions made by hydrogen bonding to the total lattice energy, U_{tot}, of organic salts of dihydrogenphosphate is considerable.   The minimum contributions, δ_{HB}, lie in a range of 20–25%.

Table I   Hydrogen bond contributions (kJ mol^{-1}) to the total lattice energy of four dihydrogenphosphate salts, $[AH][H_2PO_4]$ [a]

A[b]	U_{cal}	E_{HB}	U_{tot}	δ_{HB}	V/Z[c]
Piperidine	500	130	630	21 %	0.2139
3–HOpy	545	135	680	20 %	0.1996
3–HO–6–Mepy	410	135	545	25 %	0.2159
4–HOpy	515	135	650	21 %	0.2039

[a] Energy terms are defined in the main text.   [b] 3–HOpy = 3–hydroxypyridine; 4–HOpy = 4–hydroxypyridine;   3–HO–6–Mepy = 3–hydroxy–6–methylpyridine.   [c] V = unit cell volume; Z = number of empirical formula units per unit cell; units of nm^3.

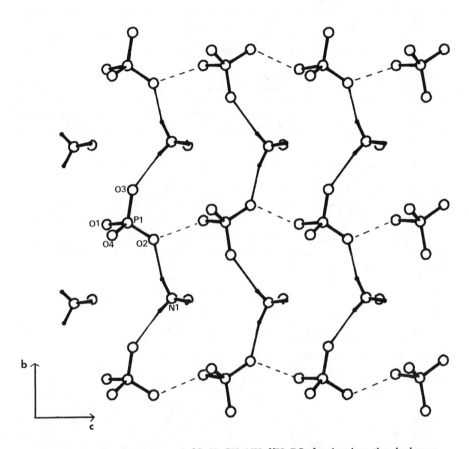

Figure 1. The structure of $[C_6H_5CH_2NH_3][H_2PO_4]$, showing the hydrogen bond network, within a plane parallel to $b–c$.

It should be emphasized that the estimated relative contributions of δ_{HB} represent the lowest possible level, as we have consistently (and quite deliberately) adopted values, at every stage of these calculations, that have minimized the magnitude of the hydrogen bond interactions; a more realistic consideration of δ_{HB} would place it significantly above 30%.

Although hydrogen bonding itself should not have a preference for symmetric or asymmetric structures, we believe the primary effect of hydrogen bonding interactions, in these salts, on the packing of a structure is to overwhelm the dipole–dipole interactions, which do have a preference (for a centrosymmetric structure). The results presented here would indicate that the hydrogen bonding, in removing a preference for centrosymmetry, will appear to favour noncentrosymmetry. Certainly, the size of its contribution to the overall lattice energy leaves beyond any reasonable doubt the fact that it must have a deterministic effect on the final structure. Indeed, the prevailing factor in the structures of these salts is the hydrogen bonding within the three–dimensional network of the anions, which itself determines the final locations of, and interactions with, the cations.

Composites from Melts: A New Class of SHG–Active Materials

Rationale. In order to prepare a processable SHG–active material, it is highly desirable to improve on the physical and chemical parameters of current materials. In nature, many composites (*e.g.* bone, teeth, and shells) exist with very high loadings of guest crystals within a host matrix (often approaching loadings of 95%) (*23*). This enables nature to combine the desirable properties of both guest and host in a new composite material. Moreover, in the natural materials, a remarkable degree of alignment of the crystals of guest material is often achieved.

Our approach was to mimic nature, and to create a new class of materials, 'tailor–designed' for a combination of optical and mechanical properties. The optical (in the cases described here, we limit the optical properties to NLO properties, and specifically SHG properties – this is not an inherent limitation on the technique, which could be employed in many other optical {and electrical} applications) properties are to be provided by the guest crystals and the physical strength and flexibility to be provided by the host polymer.

The main difficulty with such an approach lies in trying to align the SHG–active guest crystals within the polymer matrix. Unless alignment is achieved, light scattering from the microcrystals (due to disorientation, reflection and refraction) will render the composite useless. In addition, it is very important to maximize the loading degree (*i.e.* the guest/host ratio), as the total nonlinear response is proportional to the amount of SHG–active material present. Finally, the importance of matching the refractive indices of the guest and host materials cannot be overemphasized.

Early attempts at aligning molecules within a polymer matrix involved film stretching (*24,25*) or electric field poling (*26–28*), but neither method initially met with significant success. However, recent studies of SHG–active polymers (*29,30*) and low–concentration guest–host composites (*31,32*) have resulted in superior materials with greatly improved temporal stability.

Preparation of SHG–Active Composites.
We have developed a new technique, including the construction of a device, which has made it possible to grow crystals of 3–nitroaniline (*m*NA), in a matrix of poly(methyl methacrylate) (PMMA) or poly(vinylcarbazole) (PVK) in an aligned fashion, to produce transparent, SHG–active composites (*33*). This approach is based upon the Temperature

Gradient Zone Melting (TGZM) method (*34,35*), which is well known in many related applications, but does not appear to have been applied to the production of composites, particularly for electrooptic applications.

The composites were prepared from solutions of *m*NA and PMMA with varying loading degrees (between 30 wt % and 90 wt %), using toluene as a solvent. Thin films (30–40 μm) were cast on a glass slide and the solvent was allowed to evaporate. The film was then covered with a second slide and placed in the sample channel at the heated end of the thermal gradient device. For the *m*NA/PMMA composites, the heated section was maintained at 150 °C, which is above both the melting point of the *m*NA crystals (114 °C) and the glass transition point (T_g) of the polymer (105 °C), but below the decomposition point of both materials. The cooled block was kept at 20 °C, well below the the melting point of *m*NA and the T_g of PMMA.

The softened, but not completely melted, sample was then drawn slowly from the hot end across the thermal junction. By optimizing the relevant variables (*e.g.* loading degree, temperature differential, and drawing speed), the guest material crystallizes in a line within the polymer as it traverses the thermal gradient. The crystals of *m*NA adopt a needle–like habit, which is strongly aligned along the drawing axis (the direction of the thermal gradient). The degree of crystal alignment in the resulting composite is a critical function of the variables listed above, which must be individually optimized for each guest–host system.

Characterization of SHG–Active Composites.   The alignment of the guest crystals within the polymer matrix is the dominating factor in terms of eliminating light scattering from a composite. This is clearly illustrated in Figure 2, which show the angular distribution of the second harmonic (SH) intensity as a function of the alignment of the sample. Indeed, even better results were achieved when using PVK (as opposed to PMMA), as its refractive index is a better match for that of *m*NA. Well–aligned samples of *m*NA/PVK display an SHG intensity which is approximately 600 times that of a powdered sample of KDP (sample thickness = 250 μm); further improvement of the SHG–efficiency can be anticipated by using phase–matching techniques, such as birefringence or host–index modification.

These new composites have excellent chemical and optical stability. The samples prepared have shown no change in transparency, composition or SHG–efficiency (or directionality) over a period of more than two years. In contrast to electric field poled composites (whose SHG activity usually decays in a period measured in hours or days, rather than years), these materials have superior temporal stability.

Composites from Solution: A Superior Class of SHG–Active Composites

Rationale.   The composites described in the previous section represent an important new discovery. However, although extremely promising, the approach places several constraints on the guest material which will provide the nonlinear response. In addition, a prerequisite of the method is that the guest material melts cleanly at temperatures below the melting point of the polymer host. As, in a real device, the requirement for thermal stability may approach 320 °C (see Lytel and Lipscomb, elsewhere in this volume), this places an almost prohibitive restriction upon potential organic guest materials, both in terms of preparing the composites from a melt and in terms of the stability of the guest under operating and assembly conditions. For these reasons, we have developed a method for preparing composites which does not entail melting the guest material; in principle, a refractory material can be incorporated with this new methodology (*vide infra*), providing that it is soluble in a solvent in which the polymer also dissolves.

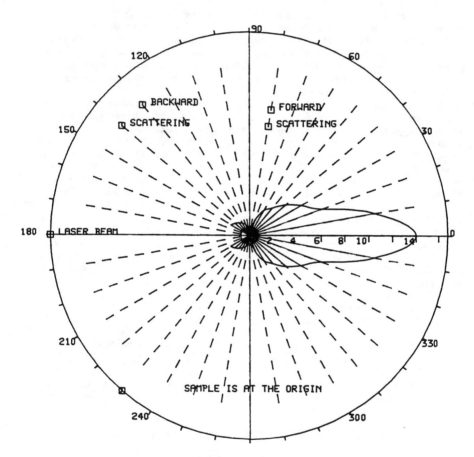

Figure 2a. Angular dependence of SHG response (relative to KDP) of an *m*NA/PMMA (50 wt %) composite; this sample contains spherulitic crystals.

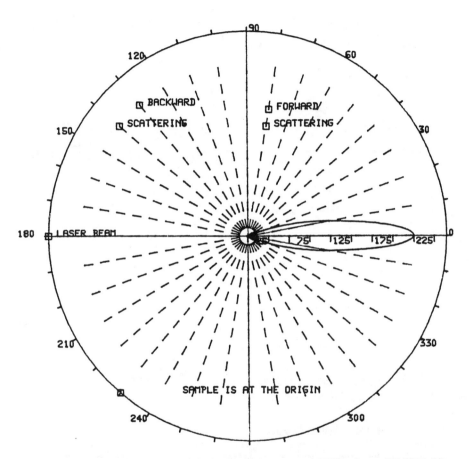

Figure 2b. Angular dependence of SHG response (relative to KDP) of an *m*NA/PVK (50 wt %) composite; this sample contains poorly aligned needle crystals.

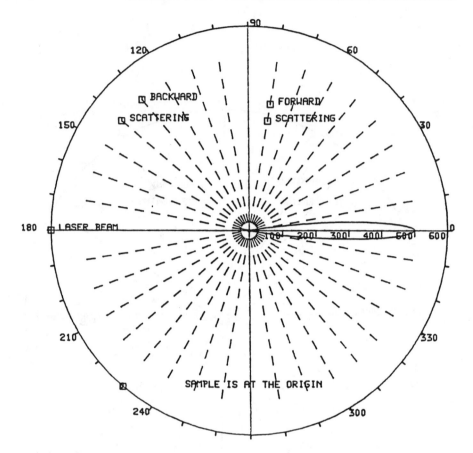

Figure 2c. Angular dependence of SHG response (relative to KDP) of an *m*NA/PVK (50 wt %) composite; this sample contains well-aligned needle crystals.

Preparation of SHG–Active Composites.   A thermal gradient apparatus was utilized to grow aligned crystals from solution (rather than from melt, as described above), by drawing the sample from the cold side (below a temperature at which the solvent evaporates at a significant rate) towards the hot end.

Using this novel solution–based method, we have been able to incorporate $[C_6H_5CH_2NH_3][H_2PO_4]$, BADP (*17*), an SHG–active material, into polymeric hosts such as poly(acrylamide) (PAA) and poly(ethylene oxide) (PEO) to produce transparent, colourless, low scattering SHG–active composites with excellent temporal stability.

Characterization of SHG–Active Composites.   The nonlinear response of a well–aligned BADP/PAA composite is fifty–times higher than that of a powdered sample of the guest material, BADP, itself.   This is the first example of a solution-grown SHG–active composite containing thermally aligned microcrystals.   In contrast to the *m*NA/PVK composites, which are yellow, these films are completely colourless.   Moreover, they are transparent, and do not seem to be damaged by the incident laser beam, even at high intensities.

The properties which make benzylammonium dihydrogenphosphate such a good guest material in the above composite are a combination of good solubility, transparency, suitable refractive index (which improves the refractive index matching between guest and host), and a propensity to form needle–like crystals.   The last factor appears to be important for obtaining an alignment of the SHG–active crystals within a polymer matrix (disc–like crystals cannot be easily aligned).

Angular distribution measurements of the SHG were carried out on BADP/PAA composites, in a manner similar to that described for mNA/PMMA. Analogous behaviour was observed, and Figure 3 illustrates the distribution obtained from a well–aligned sample.   In this case, the angular distribution of the SH is very narrow, and most of the SH flux is confined to a narrow cone in the forward direction.

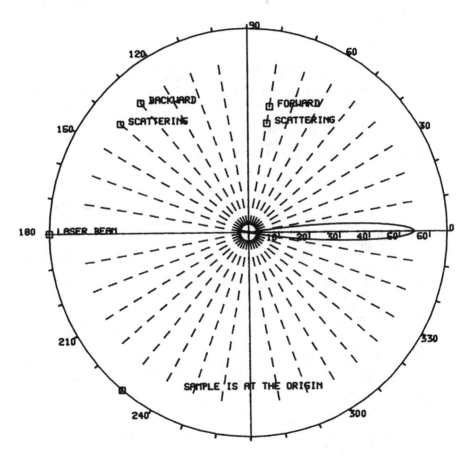

Figure 3.   Angular dependence of SHG response (relative to α–SiO$_2$) of a BADP/PAA (70 wt %) composite; this sample contains well aligned needle crystals.

Summary

The composite materials described here have a wide range of chemical and physical properties which make them prime candidates for the incorporation into optoelectronic devices. The development of two synthetic routes, melt–growth and solution–growth, have expanded the range of potential guest materials from organic molecular solids to ionic compounds. In addition, a range of different polymers can be utilized as host material. The development of novel SHG–active dihydrogenphosphate salts also means that we can achieve a fine tuning between physical properties of the guest and host.

Acknowledgments

We are indebted to BP Venture Research for funding this work, and to the Iraqi Government for two research scholarships (to N.A. and M.K.).

Literature Cited

1. *Nonlinear Optical Properties of Organic Molecules and Crystals*; Chemla, D.S.; Zyss, J., Eds.; Academic Press: Orlando, 1987; Vols 1 and 2.
2. *Nonlinear Optical Properties of Organic and Polymeric Materials*; Williams, D.J., Ed.; *ACS Symp. Ser. 233*; American Chemical Society: Washington D.C., 1983.
3. Rez, I.S. *Sov. Phys. Usp.* **1968**, *10*, 759.
4. Zyss, J. *J. Mol. Electron.* **1985**, *1*, 24.
5. Williams, D.J. *Angew. Chem. Int. Ed. Engl.* **1984**, *23*, 690.
6. Hulme, K.F. *Rep. Progr. Phys.* **1973**, *36*, 497.
7. Allen, S.; Murray, A.T. *Phys. Scr.* **1988**, *T23*, 275.
8. Nicoud, J.F. *Mol. Cryst. Liq. Cryst. Inc. Nonlin. Opt.* **1988**, *156*, 257.
9. Nicoud, J.F. In *Nonlinear Optical Properties of Organic Materials*; Khanarian, G., Ed.; SPIE: Bellingham, Washington, 1988, Vol. 971; p. 681.
10. Twieg, R.J.; Azema, A.; Jain, K.; Cheng, Y.Y. *Chem. Phys. Lett.* **1982**, *92*, 208.
11. Jain, K.; Crowley, J.I.; Hewig, G.H.; Cheng, Y.Y.; Twieg, R.J. *Opt. Laser Technol.* 1981 [Dec], 297.
12. Singer, K.D.; Sohn, J.E.; King., L.A.; Gordon, H.M.; Katz, H.E.; Dirk, C.W. *J. Opt. Soc. Am. B* **1989**, *6*, 1339.
13. Garito, A.F.; Singer, K.D. *Laser Focus* **1982**, *18*, 59.
14. Tam, W.; Wang, Y.; Calabrese, J.C.; Clement, R.A. In *Nonlinear Optical Properties of Organic Materials*; Khanarian, G., Ed.; SPIE: Bellingham, Washington, 1988, Vol. 971; p. 107.
15. Green, M.L.H.; Marder, S.R.; Thompson, M.E.; Bandy, J.A.; Bloor, D.; Kolinsky, P.V.; Jones, R.J. *Nature* **1987**, *330*, 360.
16. Cox, S.D.; Gier, T.E.; Stucky, G.D.; Bierlein, J. *J. Am. Chem. Soc.* **1988**, *110*, 2986.
17. Aakeröy, C.B.; Hitchcock, P.B.; Moyle, B.D.; Seddon, K.R. *J. Chem. Soc., Chem. Commun.* **1989**, 1856.
18. Xu, D.; Jiang, M.; Tan, Z. *Huaxue Xuebao* **1983**, *41*, 570; *Acta Chim. Sinica.*, **1983**, *2*, 230.
19. Kurtz S.K.; Perry, T.T. *J. Appl. Phys.* **1968**, *39*, 3798.
20. Marder, S.R.; Perry, J.W.; Schaefer, W.P. *Science* **1989**, *245*, 626.
21. Leslie, M. *SERC Daresbury Lab. Rept.*, DL–SCI–TM31T, **1982**.

22. Meot–Ner (Mautner), M. In *Molecular Structure and Energetics*; Liebman, J.F.; Greenberg, A., Eds.; VCH: New York, 1987, Vol.4; pp. 72–103.

23. *Biomineralization*; Mann, S.; Webb, J.; Williams, R.J.P., Eds.; VCH: Weinheim (Ger.), 1989.

24. Azoz, N.; Calvert, P.D.; Moyle, B.D. In *Organic Materials for Non–Linear Optics*; Hann, R.A; Bloor, D., Eds.; Royal Soc. Chem.: London, 1989; pp. 308–314.

25. Calvert, P.D.; Moyle, B.D. *Mat. Res. Soc. Symp.* 1988, *109*, 357.

26. Singer, K.D.; Sohn, J.E.; Lalama, S.J. *Appl. Phys. Lett.* 1986, *49*, 248.

27. Pantelis, P.; Davies, G.J. *US Patent* 4748074, 1988.

28. Pantelis, P.; Davies, G.J. *US Patent* 4746577, 1988.

29. Eich, M.; Reck, B.; Yoon, D.Y.; Willson, C.G.; Bjorklund, G.C. *J. Appl. Phys.* 1989, *66*, 3241.

30. Singer, K.D.; Kuzyk, M.G.; Holland, W.R.; Sohn, J.E.; Lalama, S.J.; Comizzoli, R.B.; Katz, H.E.; Schilling, M.L. *Appl. Phys. Lett.* 1988, *53*, 1800.

31. Miyazaki, T.; Watanabe, T.; Miyata, S. *Jpn. J. Appl. Phys.* 1988, 27, L1724.

32. Lytel, R.; Lipscomb, G.F.; Stiller, M.; Thackara, J.I.; Ticknor, A.J. In *Nonlinear Optical Properties of Organic Materials*; Khanarian, G., Ed.; SPIE: Bellingham, Washington, 1988, Vol. 971; p. 218.

33. Azoz, N.; Calvert, P.D; Kadim, M.; McCaffrey, A.J.; Seddon, K.R. *Nature*, 1990, *344*, 49.

34. Pfann, W.G. *Zone Melting*; 2nd Edit.; Wiley: New York, 1958; pp. 254–268.

35. Herington, E.F.G. *Zone Melting of Organic Compounds*; Blackwell: Oxford, 1963.

RECEIVED August 2, 1990

Chapter 35

Clathrasils

New Materials for Nonlinear Optical Applications

Hee K. Chae[1,2,3], Walter G. Klemperer[1,2,3], David A. Payne[1,2,4],
Carlos T. A. Suchicital[1,2,4], Douglas R. Wake[2], and Scott R. Wilson[3]

[1]Beckman Institute for Advanced Science and Technology, [2]Materials Research Laboratory, [3]School of Chemical Sciences, and [4]Department of Materials Science and Engineering, University of Illinois at Urbana–Champaign, Urbana, IL 61801

Hydrothermal methods were developed for the growth of large, 3 mm-sized crystals of pyridine dodecasil-3C (Py-D3C) from a pyridine-SiO_2-HF-H_2O system at 190 °C. The crystals were acentric at ambient temperature and were weak second harmonic generators. Phase transformations were observed by differential scanning calorimetry to commence at 161 and -46 °C on cooling. The crystal structure of the ambient temperature tetragonal or pseudotetragonal $17SiO_2 \cdot C_5H_5N$ phase was determined using single crystal X-ray diffraction techniques [$q_2 = 13.6620(5)$ Å, $c = 19.5669(7)$ Å, Z = 4, space group $I\bar{4}2d$-D_{2d}^{12}]. The domain structure of this phase was studied using optical microscopy, and domain configurations were manipulated by heat treatment. Scanning electron micrographs strongly suggested that the boundaries of these domains were associated with growth twin boundaries.

Clathrasils are host/guest complexes comprised of covalent guest molecules entrapped within cages formed by a silica host framework (1, 2). Like all zeolitic materials, clathrasils have enormous potential as advanced optical and electronic materials whose composite character permits synthetic manipulation of both the molecular structure of the guest species and the extended structure of the host framework (3, 4). Like other zeolites, however, clathrasils also suffer severe handicaps as advanced materials due to a reluctance to form large single crystals and a tendency to form stoichiometrically and structurally defective crystals (5 - 10).

In the course of our investigations of clathrasils as optical and electronic materials, we have discovered that certain clathrasils having the MTN framework structure (11 - 13) are acentric at ambient temperature. They thus are second harmonic generators and possibly useful nonlinear optical materials. We have succeeded in growing large, 3 mm-sized crystals of the clathrasil pyridine dodecasil-3C, $17SiO_2 \cdot C_5H_5N$, and solved the crystal structure of its ambient temperature, tetragonal phase. The availability of large single crystals has also enabled us to study and manipulate domain structures in the tetragonal phase.

0097–6156/91/0455–0528$06.00/0
© 1991 American Chemical Society

Experimental Section

<u>General Analytical Techniques</u>. Elemental analyses were carried out in the University of Illinois School of Chemical Sciences Microanalytical Laboratory. Solid state ^{13}C and ^{29}Si NMR spectra were recorded on a General Electric GN-300 WB spectrometer. ^{13}C and ^{29}Si NMR chemical shifts were referenced internally to $[(CH_3)_3Si]_4Si$, TTMS. A Du Pont 1090 thermal analyzer was used for thermal analyses. Hot and cold stage optical microscopy studies used a Leitz 1350 microscope stage. Scanning electron microscopy was carried out in a JEOL JSM-35C microscope. Crystal surfaces were coated with sputtered gold prior to examination. Powder X-ray diffraction patterns were acquired on a Rigaku D/MAX diffractometer using CuK_α radiation.

<u>Starting Materials and Reagents</u>. Fumed silica (Cab-O-Sil, grade EH-5, Cabot Corporation) and aqueous hydrofluoric acid (reagent grade, 49 wt%, Fisher) were used without further purification. Pyridine (reagent grade, Fisher) was dried over sodium hydroxide and freshly distilled prior to use. Fused quartz glass tubing (8 mm ID, 1mm thickness) was purchased from G.M. Associates Incorporated.

<u>Preparation of Pyridine Dodecasil-3C</u>. A 2.2 M aqueous HF solution was prepared by dilution of 8.84 mL of 49 wt% HF solution with 108 mL of deionized water. A 1.0 M aqueous pyridinium bifluoride solution was prepared by adding 8.7 mL of pyridine to 100 mL of the 2.2 M HF solution at 0 °C. Fumed silica (90 mg, 1.5×10^{-3} moles) and 0.90 mL (1.1×10^{-2} moles) of pyridine were added to 1.5 mL of the 1.0 M pyridinium bifluoride solution, and the resulting mixture was stirred for 2 h to obtain a pH 6, turbid solution. A 12 cm long fused quartz tube was filled to about 1/3 capacity with this solution, the solution was degassed by three freeze-pump-thaw cycles, and the tube was sealed under vacuum with the solution frozen at liquid nitrogen temperature. The reaction tube was placed in a convection oven at 190 °C for three weeks, during which a considerable amount of the quartz tube was etched away. The tube was then opened, the reaction solution was decanted, and the crystalline product was washed with deionized water and acetone. After air drying for 12 h the Py-D3C crystals were chipped away from the quartz glass wall with an awl to yield 620 mg of product (5.6×10^{-4} moles). Anal. Calcd for $17SiO_2 \cdot C_5H_5N$: C, 5.46; H, 0.46; N, 1.27; Si, 43.38. Found: C, 5.33; H, 0.62; N, 1.21; Si, 43.36. ^{13}C CPMAS NMR (TTMS): δ 119.4 (s, $meta$-C_5H_5N), δ 131.3 (s, $para$-C_5H_5N), δ 147.0 (s, $ortho$-C_5H_5N). ^{29}Si CPMAS NMR (TTMS): δ -107.2 (s,T_1), δ -114.3 (s,T_2), δ -119.5 (s, T_3), δ -119.9 (s, T_3), δ -120.4 (s, T_3).

<u>Single Crystal X-ray Diffraction Study of Pyridine Dodecasil-3C</u>. Pyridine dodecasil-3C, $17SiO_2 \cdot C_5H_5N$, crystallized in the tetragonal crystal system with a = 13.6620(5) Å, c = 19.5669(7) Å, V = 3652.2(5) Å^3 and ρ_{calc} = 2.001 g/cm^3 for Z = 4. A transparent, colorless, tabular crystal, sliced from the (1 1 0) face of a large crystal, was bound by the following faces at distances (mm) given from the crystal center: (1 1 0), 0.09; (-1 -1 0), 0.09; (0 1 1), 0.34; (1 0 -1), 0.34; (1 0 1), 0.35; (0 1 -1), 0.35; and (0 0 -1), 0.45. The large crystal volume was required to enable other (nondiffraction) experiments on the same sample. Systematic conditions suggested space group $I4_1md$ or $I\bar{4}2d$; the latter was confirmed by refinement. A total of 4551 diffraction data ($2\theta < 70°$ for $+h+k+l$: $h+k+l = 2n$) were measured at 26 °C on a Syntex P2$_1$ diffractometer using graphite monochromated Mo radiation [$\lambda(K\bar{\alpha})$ = 0.71073 Å]. These data were corrected for anomalous dispersion, absorption (maximum and minimum numerical transmission factors, 0.891 and 0.579), Lorentz and polarization effects, and then merged (R_i = 0.027) resulting in 2186 unique data. The structure was solved by direct methods (<u>14</u>); the five silicon atom positions were

deduced from an E-map. Subsequent difference Fourier syntheses revealed positions for ten oxygen atoms. One of these atoms, O_3, was disordered in two positions along the unique 2-fold axis. Anisotropic least-squares refinement (15) of the host molecule using 2132 observed [$I < 2.58\ \sigma(I)$] data converged at R = 0.052 and R_w = 0.090. Contributions from an "idealized" guest pyridine molecule (no hydrogen atoms) disordered about the $\overline{4}$ symmetry axis within the hexadecahedral cage gave final agreement factors R = 0.039 and R_w = 0.072. The final difference Fourier map (range $0.70 < e/\text{Å}^3 < 0.52$) located maximum residual electron density in the vicinity of atom O_7. The final analysis of variance between observed and calculated structure factors showed a slight dependence on sine (θ).

Second Harmonic Generation Measurements. χ^2 measurements were carried out with a flashlamp-pumped YAG laser. Clathrasil crystal grains were segregated by grain size and measured against similarly prepared ground quartz. A photomultiplier tube gated synchronously with the laser detected the second harmonic remaining after filtering of the transmitted beam.

Results and Discussion

Synthesis and Characterization. Large (3 mm) single crystals of pyridine dodecasil-3C (Py-D3C) were obtained by treating fumed silica with an aqueous solution of pyridine and hydrofluoric acid in an evacuated, sealed quartz tube for 500 h at 190 °C. The precise reaction conditions, detailed above in the experimental section, were derived from procedures originally developed by Liebau, Gerke, and Gies (16, 17). The crystals had the truncated octahedral habit shown in Figure 1. Comparison of the X-ray diffraction pattern shown in Figure 2 with published data served to identify the tetragonal MTN structural framwork (7, 10). The formulation $17SiO_2 \cdot C_5H_5N$ was established by elemental analysis (see above), and chemical homogeneity was evaluated by ^{13}C and ^{29}Si CPMAS NMR spectroscopy (see Figure 3). Analytical data indicated that, within the accuracy of measurement, all hexadecahedral cages in the MTN framework were occupied by pyridine molecules. The ^{13}C NMR spectrum revealed the presence of only pyridine (18) and no degradation products as guest molecules, and the ^{29}Si NMR spectrum displayed only resonances assignable to the tetragonal MTN framework (10,19-22).

Phase Transformations. Differential scanning calorimetry of Py-D3C (Figure 4) established the existence of three phases in the -75°C to +200°C temperature range. A transformation from the ambient temperature tetragonal phase to a higher temperature cubic phase (see 10,20) started at +164 °C, with an energy change of 1.6 J/g. The reverse transformation began at +161 °C, with a thermal hysteresis of 8 °C between peak temperatures. Low temperature differential scanning calorimetry showed a lower temperature endothermic transformation from the ambient temperature phase to a low temperature phase commencing at -46 °C, with an energy change of 4.2J/g. The reverse transformation began at -43 °C with a thermal hysteresis of 6 °C between peak temperatures.

The phase transformation behavior of Py-D3C was far simpler than that reported for tetrahydrofuran/N_2- and tetrahydrofuran/Xe-D3C in reference 10. We attribute this difference in part to impurities in the samples employed, samples that contained methanol and ethylenediamine according to ^{13}C CPMAS NMR spectroscopy. We have observed that use of $Si(OCH_3)_4$ as a silica source or ethylenediamine as a catalyst in clathrasil synthesis introduces defects that can alter phase transition temperatures by as much as 30 °C and/or introduce new phase transformations.

Figure 1. Optical micrograph of an as-grown crystal of Py-D3C exhibiting a characteristic truncated octahedral shape. Domains are apparent in transmitted cross-polarized light.

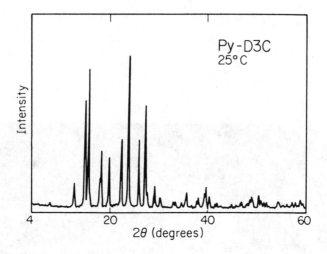

2 θ	d (Å)	I/I₀ (%)
7.88	11.21	5
12.88	6.87	17
15.16	5.84	72
15.84	5.59	91
18.14	4.89	20
18.38	4.82	38
19.96	4.44	35
22.50	3.95	47
23.70	3.75	47
23.92	3.72	100
25.96	3.43	46
27.26	3.27	69
27.62	3.23	11
28.64	3.08	9
29.26	3.05	15
30.24	2.95	9
33.00	2.71	5
33.48	2.67	5
34.70	2.58	5
35.66	2.52	12
36.84	2.44	3
37.84	2.38	5
38.12	2.36	9
39.38	2.29	10
39.62	2.27	15
40.34	2.23	9

Figure 2.   X-ray powder diffraction pattern of Py-D3C.

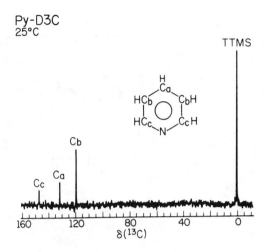

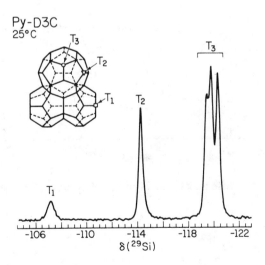

Figure 3. Solid state [13]C (top) and [29]Si (bottom) cross-polarized magic angle spinning NMR spectra of Py-D3C.

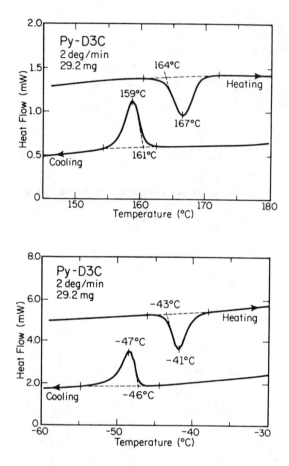

Figure 4.   Differential scanning calorimetry curves for Py-D3C.

The phase transformations of Py-D3C were also monitored by NMR spectroscopy. Distinct ^{29}Si NMR spectra were observed for all three phases as reported in references 10 and 20. ^{2}H NMR spectra of deuterated Py-D3C samples showed rapid isotropic rotation of the guest molecules in the ambient and high temperature phases but restricted rotation in the low temperature phase.

The domain structures of the three Py-D3C phases described above were observed by optical microscopy in cross-polarized light. The photomicrograph of an as-synthesized crystal shown in Figure 1 was taken at ambient temperature. Here, parallel arrays of domains were aligned along the principal axes of the pseudo-cubic crystal and intersected with other domain arrays at 60 ° angles. When thin sections (see Figure 5) were heated to just above 167 °C, these domains disappeared, but reappeared in their original configurations upon cooling below 159 °C. This "memory effect" could be eliminated, however, by quickly heating the crystal to 700 °C and cooling it to below 159 °C to obtain a new domain configuration. According to thermogravimetric analysis and ^{13}C NMR spectroscopy, this heat treatment does not involve significant loss of pyridine guest molecules. Cold stage microscopy was used to monitor the low temperature Py-D3C phase transformation. Here again, a memory effect was observed: the parallel domain arrays in the ambient temperature phase were lost upon cooling below -47 °C but returned in their original configurations upon reheating above the transformation temperature.

Crystal Structure of Tetragonal Py-D3C. The MTN framework structure (11), also known as the ZSM-39 (12), dodecasil-3C (6,13), CF-4 (7), and holdstite (23) structure, is topologically well-defined and known to contain both pentagonal dodecahedral and hexadecahedral cages. The detailed crystal structure of the ambient temperature phase remains undetermined, however, due to disorder problems. All single crystal studies to date have yielded a cubic structure in which averaged oxygen atom positions are determined. Given the large size and high purity of the Py-D3C crystals obtained in the present study, however, it was possible to excise a section from a quadrilateral crystal face (see Figure 1) whose domains were not randomly oriented to yield an averaged, cubic structure. Instead, a tetragonal structure was observed where the unique unit cell axis is oriented along the diagonal of a quadrilateral crystal face, i.e., parallel to one of the three cubic axes of the truncated cuboctahedral crystal.

The results of the single crystal X-ray diffraction study described in the Experimental Section are summarized in Table I. The structures obtained for the dodecahedral and hexadecahedral cages are shown in Figures 6a and 6b, respectively, and a cutaway spacefilling view of the pyridine environment is shown in Figure 6c. The hexadecahedral cage showed $\bar{4}$ symmetry, and the pyridine guest molecule was disordered over four symmetry-equivalent locations, only one of which is shown in Figure 6b. The ten unique oxygen atoms converged with temperature factors unusually higher than adjacent silicon atom coefficients. This has been attributed to static or dynamic disorder of the host framework (6) and, in fact, the present structural model resolved two statistically disordered sites for atom O3. In Figure 6, the coordinates of these two sites have been averaged and oxygen atom O3 is marked with an asterisk. Alternatively, these oxygen atom temperature factors may represent a twinned orthorhombic structure such that the tetragonal unit cell is the average of two orthorhombic cells.

Crystal Growth Mechanism. In order to further investigate the possibility of crystallographic twinning in Py-D3C, a crystal was isolated during the early stages of its growth and examined by scanning electron microscopy. As shown in Figure 7, the surface of an incompletely developed quadrilateral face is composed of rectangular growth steps oriented normal to the cubic axes of the crystal. Since the orientation of

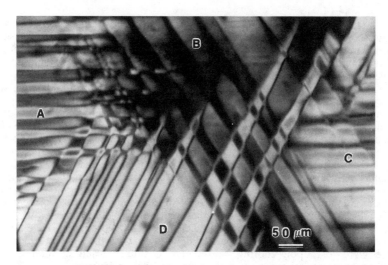

Figure 5. Optical micrograph of a thin section of a Py-D3C crystal in cross-polarized light. Regions A-D show in detail the parallel domain arrays.

Table I. Atomic Coordinates for Non-Hydrogen Atoms in Crystalline $17SiO_2 \cdot C_5H_5N$

| Atom Type | Fractional Coordinates | | | |
	x/a	y/b	z/c	U (eq)[a]
Si_{T1}	0.5	0.5	0.0	0.0161(6)
Si_{T2}	0.31931(6)	0.50859(6)	0.09160(3)	0.0137(4)
Si_{T3a}	-0.21392(6)	0.68790(6)	0.05777(4)	0.0132(4)
Si_{T3b}	-0.01353(6)	0.61527(6)	-0.00438(4)	0.0146(4)
Si_{T3c}	0.18446(6)	0.68522(6)	0.05793(4)	0.0140(4)
O_1	-0.1836(5)	0.25	0.125	0.045(3)
O_2	-0.1945(4)	0.75	0.125	0.040(3)
O_{3A}[b]	0.0	0.5	-0.0115(7)	0.027(6)
O_{3B}[c]	0.0	0.5	0.0096(8)	0.034(6)
O_4	0.0739(2)	0.6688(2)	0.0344(2)	0.035(2)
O_5	-0.1125(2)	0.6418(3)	0.0342(2)	0.039(2)
O_6	-0.4113(3)	0.4618(2)	0.0457(2)	0.036(2)
O_7	-0.0197(4)	0.6527(4)	-0.0805(1)	0.056(3)
O_8	-0.2313(2)	0.4184(2)	0.0735(2)	0.039(2)
O_9	-0.2596(3)	0.7551(3)	0.0003(2)	0.040(2)
O_{10}	-0.2879(2)	0.6022(2)	0.0757(2)	0.036(2)
N_1[d]	0.0089(19)	0.5880(30)	0.2132(13)	0.092(4)
C_2	0.0382	0.5032	0.1842	
C_3	0.0172	0.4114	0.2106	
C_4	-0.0377	0.4070	0.2706	
C_5	-0.0692	0.4931	0.3020	
C_6	-0.0441	0.5811	0.2713	

[a] 1/3 trace of the $\underline{U}$ (ij) tensor ($Å^2$). [b] site "A" occupancy 0.28(1). [c] site "B" occupancy 0.22(1). [d] isotropic group thermal parameter and "ideal" geometry imposed on pyridine.

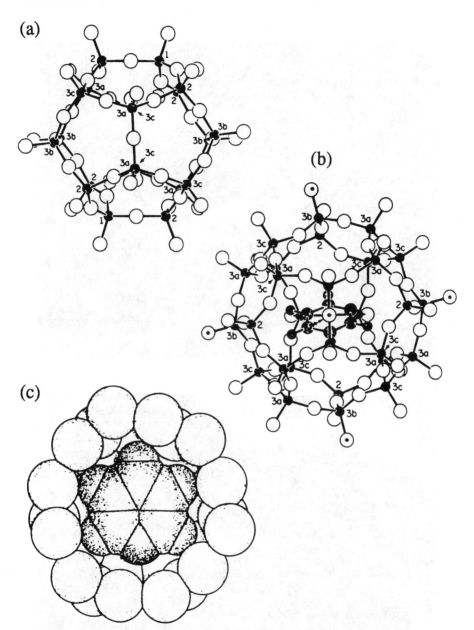

Figure 6.   Selected fragments of the tetragonal Py-D3C structure : (a) ball and stick model of the pentagonal dodecahedral cage, (b) ball and stick model of the hexadecahedral cage with a clathrated pyridine molecule, and (c) spacefilling cutaway view of the pyridine guest molecule and its oxygen environment. In (a) and (b), silicon atoms are represented by small filled circles and oxygen atoms by larger open circles, and silicon atoms are labeled by their numerical subscripts (see Table I). In (b), all the atoms in the pyridine molecule are represented by shaded spheres. The sphere radii in (c) are van der Waals radii.

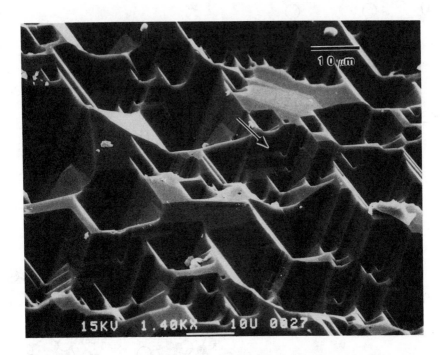

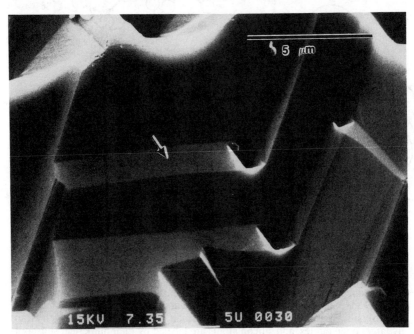

Figure 7.   SEM photomicrographs of a quadrilateral face of a Py-D3C crystal in an early stage of growth. Arrows indicate a growth step.

these steps matches that of the domains observed by optical microscopy, a simple explanation of the "memory effect" noted above is in hand if the domain boundaries coincide with growth twin boundaries. The existence of orthorhombic growth twins would also support the interpretation of crystallographic disorder in the tetragonal crystal structure as twinning of an orthorhombic structure.

Preliminary Property Measurements. Preliminary measurements of second harmonic generating activity on powder samples of Py-D3C, segregated by grain diameter, determined that χ^2 for the material is $1/85 \pm 20\%$ that of quartz. These measurements also indicated a coherence length at least as long as that of quartz.

Acknowledgments

The authors gratefully acknowledge support from the U.S. Department of Energy, Division of Materials Science, under contract DE-ACO2-76ERO1198, and central facilities of the Materials Research Laboratory of the University of Illinois, which is supported by the U.S. Department of Energy. We are also grateful to Drs. F. Liebau and H. Gies for helpful advice.

Literature Cited

1. Liebau, F. Structural Chemistry of Silicates; Springer-Verlag: Berlin, 1985; p 159, pp 240-243.
2. Gies, H. Nachr. Chem. Tech. Lab. 1985, 33, 387-391.
3. Ozin, G.A.; Kuperman, A.; Stein, A. Angew. Chem. Int. Ed. Engl. 1989, 28, 359-376.
4. Stucky, G.D.; Mac Dougall, J.E. Science, 1990, 247, 669-678.
5. Groenen, E.J.J.; Alma, N.C.M.; Bastein, A.G.T.M.; Hays, G.R.; Huis, R.; Kortbeek, A.G.T.G. J. Chem. Soc., Chem. Commun. 1983, 1360-1362.
6. Gies, H. Z. Kristallogr. 1984, 167, 73-82.
7. Long, Y.; He, H.; Zheng, P.; Wu, G.; Wang, B. J. Inclusion Phen. 1987, 5, 355-362.
8. Dewaele, N.; Vanhaele, Y.; Bodart, P.; Gabelica, Z.; Nagy, J.B. Acta Chim. Hung. 1987, 124, 93-108.
9. Dewaele, N.; Gabelica, Z.; Bodart, P.; Nagy, J.B.; Giordano, G.; Derouane, E.G. Stud. Surf. Sci. Catal. 1988, 37, 65-73.
10. Ripmeester, J.A.; Desando, M.A.; Handa, Y.P.; Tse, J.S. J. Chem. Soc., Chem. Commun. 1988, 608-610.
11. Meier, W.M.; Olson, D.H. Atlas of Zeolite Structure Types; Butterworths: London, 1987; p 104-105.
12. Schlenker, J.L.; Dwyer, F.G.; Jenkins, E.E.; Rohrbaugh, W.J.; Kokotailo, G.T.; Meier, W.M. Nature 1981, 294, 340-342.
13. Gies, H.; Liebau, F.; Gerke, H. Angew. Chem. Int. Ed. Engl. 1982, 21, 206-207.
14. Sheldrick, G.M. In Crystallographic Computing 3; Sheldrick, G.M.; Kruger, C.; Goddard, R., Eds.; Oxford University: London, 1985; SHELXS-86, pp 175-189.
15. Sheldrick, G.M. : SHELXS-76, a program for crystal structure determination, University Chemical Laboratory, Cambridge, England, 1976.
16. Gerke, H.; Gies, H.; Liebau, F. Ger. Offen. DE 3 128 988, 1983.
17. Gerke, H.; Gies, H.; Liebau, F. Ger. Offen. DE 3 201 752, 1983.
18. Pugmire, R.J.; Grant, D.M. J. Am. Chem. Soc. 1968, 90, 697-706.
19. Kokotailo, G.T.; Fyfe, C.A.; Gobbi, G.C.; Kennedy, G.J.; DeSchutter, C.T. J. Chem. Soc., Chem. Commun. 1984, 1208-1210.

20.   Strobl, H.; Fyfe, C.A.; Kokotailo, G.T.; Pasztor, C.T.; Bibby, D.M. <u>J. Am.</u> <u>Chem. Soc.</u> 1987, <u>109</u>, 4733-4734.
21.   Fyfe, C.A.; Gies, H.; Feng, Y. <u>J. Chem. Soc., Chem. Commun.</u> 1989, 1240-1242.
22.   Fyfe, C.A.; Gies, H.; Feng, Y. <u>J. Am. Chem. Soc.</u> 1989, <u>111</u>, 7702-7707.
23.   Smith, J.V.; Blackwell, C.S. <u>Nature</u> 1983, <u>303</u>, 223-225.

RECEIVED July 2, 1990

Chapter 36

Inorganic Sol–Gel Glasses as Matrices for Nonlinear Optical Materials

Jeffrey I. Zink[1], Bruce Dunn[2], R. B. Kaner[1], E. T. Knobbe[2], and J. McKiernan[1]

[1]Department of Chemistry and Biochemistry and [2]Department of Materials Science and Engineering, University of California, Los Angeles, CA 90024

Sol-gel synthesis of inorganic glasses offers a low temperature route to the microencapsulation of organic and organometallic molecules in inorganic matrices. The encapsulated molecules can be used to induce new optical properties in the material or to probe the changes at the molecular level which occur during the polymerization, aging and drying of the glass. Two different aspects of non-linear optical properties induced in the glass are discussed here. First, laser dyes including rhodamines and coumarins are encapsulated. The resulting doped gel-glasses exhibit optical gain and laser action. The non-linear response to the pulse energy of the pump laser as well as other optical characteristics of these new solid-state lasers will be discussed. Second, encapsulation of 2-ethylpolyaniline has been achieved. Degenerate four-wave mixing studies have been carried out, but the observed signal cannot be unambiguously attributed to $\chi^{(3)}$ effects.

The sol-gel process is a solution synthesis technique which provides a low temperature chemical route for the preparation of rigid, transparent matrix materials (1-8). A wide variety of organic and organometallic molecules have been incorporated, via the sol-gel technique, into SiO_2, Al_2O_3-SiO_2 and organically modified silicate (ORMOSIL) host matrices (1-8). The focus of this paper is the encapsulation of organic laser dye molecules to produce new optical materials which exhibit optical gain and laser action, and of soluble polyaniline to produce new materials having potentially large third-order susceptibilities.

In the first part of this paper, we report the results of our studies of laser action. The three types of host materials mentioned above are used to encapsulate coumarin and rhodamine laser dyes. The synthesis of the doped gels and gel-glasses is reported. The results of our studies of optical gain, laser spectral output, the output energy dependence on the pump pulse energy, and stability are discussed. The characteristics of the three types of hosts and their effects on laser action are compared. In the second part of this paper, we report the results of our studies of incorporation of 2-ethylpolyaniline in SiO_2 gels. The results of a degenerate four wave mixing study are presented and discussed.

0097–6156/91/0455–0541$06.00/0

Gel and Gel-Glass Tunable Dye Lasers

Synthesis of Laser Dye Doped SiO2 Gels. The SiO2 gels were synthesized via the
sonogel method (9). A "starter" sol was composed of a 3:1:0.03 molar ratio of water
to tetraethoxy silane (TEOS) to HCl (added as a catalyst). The initial precursors were
mixed and immediately sonicated in a ultrasonic cleaner. After several minutes of
ultrasound exposure, a drop of HCl was added to the solution followed by further
sonication. The temperature of the reaction mixture was kept below 20° C. After
about ten minutes the two-phase mixture had become a single phase solution
indicating nearly complete hydrolysis of the TEOS precursor. The resultant silanol
solution was then diluted by the addition of 5 v/o ethylene glycol and the organic dyes
were dissolved directly into the sol at room temperature with mild agitation.
Polycondensation of the doped sols was carried out at room temperature in covered
silica cuvettes with four polished sides. The samples became highly viscous within
24 hours and had become completely rigid in less than 60 hours. The gels were then
aged at room temperature for one to two weeks to promote complete condensation.
Optical measurements were performed on rigid gels which were partially dried (10%
solvent loss) and had exhibited about 10 percent volume shrinkage.

The aluminosilicate xerogels were prepared by slowly adding a solution of 10 ml
of isopropanol and 5 ml distilled water to a solution of 10 ml of the aluminosilicate
precursor diisobutoxyaluminoxytriethoxysilane (ASE) and 10 ml of isopropanol.
Rhodamine 6G (R6G) was dissolved in the isopropanol beforehand to give final
concentrations of 5 x 10^{-4} M or 1 x 10^{-3} M in the sol. The sol was placed into
polystyrene cuvettes and sealed with wax film. Gelation of the sol was complete after
~3 days. The gels were allowed to age in the sealed cuvettes for one week after
gelation. A small hole was subsequently pierced in the film to allow slow evaporation
of alcohol and water. When the gels were fully dried xerogels (~4 weeks after drying
began) the monoliths were removed from the cuvettes and then used in the
experiments *as cast*, with no polishing or surface treatment.

The organically modified silicate (ORMOSIL) gels were prepared following the
literature methods (10,11). The Coumarin 153 doped ORMOSIL was prepared by
combining 15.2 g of TMOS and 24.8 g of 3-(trimethoxysilyl)propyl methacrylate,
TMSPM, in a polyethylene beaker with 6 ml of 0.04 N aqueous HCl. The two phase
mixture was sonicated until a single phase sol was obtained. The solution was then
stirred for 30 minutes. Following this initial hydrolysis step, 10 g of
methylmethacrylate was added with stirring. After 15 minutes of continuous stirring,
2 drops of Triton X-100, a surfactant, were mixed into the sol and the mixture stirred
for another fifteen 15 minutes. The sol was tightly covered and aged at room
temperature for 1 day. After aging, C153 was dissolved into the low viscosity sol at a
starting concentration of 2.0 x 10^{-3} M.

The R6G doped ORMOSIL was prepared with epoxy functionalities so that
diols, which have good solvation properties for R6G, would be formed under the
hydrolytic conditions employed. 8.1 g of TMOS and 3.36 ml of 0.04 N aqueous HCl
were combined in a polyethylene beaker and sonicated in a chilled bath. 2.6 g of 3-
glycidoxypropyl trimethoxysilane, GPTMS, and 1 ml of 0.04 N HCl were added to
the precursor solution with stirring. 3.3 g of ethylene glycol were added to the sol, 2
drops of Triton X were stirred in, and the solution hydrolyzed for 1 day in a closed
container. The glycol aids the stabilization of the monomer form of R6G in the host
matrix. R6G chloride was dissolved into the precursor solutions at an initial
concentration of 2 x 10^{-3} M and the sol kept covered until the onset of gelation. After
gelation, small holes were introduced in the cover film to permit slow evaporation of
the alcohols. The samples were maintained at room temperature for approximately 3
days and then placed into a 70° C oven for 2 weeks. The samples were then removed

from the cuvettes and baked at 70° C for an additional 3 weeks. The resulting monoliths had faces which were nearly plane parallel and had good optical quality.

Laser Action in Doped Silicate Gels. Spectral gain measurements were performed on the C153 and R6G doped SiO_2 gels by inserting the sample into the amplifier cell of a PTI model Pl 202 nitrogen pumped tunable dye laser. The pump pulse width is about 500 psec (FWHM) at the fundamental 337 nm wavelength; the repetition rate was 2 Hz. The primary oscillator cell held an ethanol solution of the same dye, at the same concentration, as in the gel sample. A reverse biased p-i-n diode was used to measure spectral variations in the amplitudes of the probe signal S generated by the primary oscillator cell, the fluorescence emission from the amplifier cell F, the amplified spontaneous emission ASE, and the total signal gain G_t. The spectral probe signal intensity was measured without any gain medium in the amplifier cell. F was measured while pumping only the sample in the amplifier cell. The ASE was determined by pumping both of the sample cells with the grating tuned to a wavelength just outside of the oscillation region of the dye solution held in the primary oscillator. The total signal gain was then measured as a function of pump wavelength for the gel samples as well as for standard ethanol solutions. The stimulated power gain G_s due to the sample in the amplifier cell was calculated as

$$G_s = (G_t - F - ASE)/S. \qquad (1)$$

The spectral gain envelopes for C153 are shown in Figure 1. The effect of the aging from one to eleven days on the gain envelopes is shown. The gel which was aged for one day is quite similar to that of the reference solution. Prior to gelation, the two gain curves were indistinguishable. After eleven days of aging, the two spectra become distinctly different; a 19 nm red shift in the peak of the gel gain curve is observed. Significantly, the aging produced very little reduction in the peak gain of the system. Thus it is possible to obtain high values of optical gain in a well-aged gel matrix. The red shift is consistent with the chemical changes occurring during the aging process. Condensation of the hydrolyzed species produces water which enters the solvent phase. Coumarin dyes in ethanol-water solutions exhibit red shifts of about 11 nm compared to pure ethanol (12).

Free-running cavity laser oscillation was achieved in all of the coumarin doped gels. For a C153 doped gel which was aged one week, the peak output was centered about 558 nm with a FWHM oscillation band of 13 nm. The doped gel demonstrated fundamental oscillation from 545 to 572 nm in the free-running cavity as shown in Figure 2. The primary amplitude structure is probably caused by hole burning in the inhomogeneously broadened gain profile due to strong interactions with the charged silica matrix. The spectral output of the free running ethanol solution contains a great deal of structure, partially due to etalon effects at the cuvette wall interfaces. Peak emission was centered at 542 nm with a full spectral oscillation range from 529 to 554 nm.

Other laser dyes including C1, C102 and R6G exhibited optical gain and laser action in silicate gels. The results are summarized in Table I. The optical gain in R6G gels was reduced during aging because of dimerization of the dye molecule. Laser action however was readily achieved. The best performance and stability for R6G was observed using aluminosilicate gel matrices as discussed below.

Laser Action in Doped ORMOSIL Gels. The spectral gain characteristics for C153 in ethanol and in the ORMOSIL gel are shown in Figure 3. The gain envelope is red-shifted with respect to that of the ethanolic solution consistent with the red shift of the emission peak in more polar solvents. Although the peak gain value in the gel was slighly less than that in ethanol, the bandwidth is broader. The ORMOSIL gel also

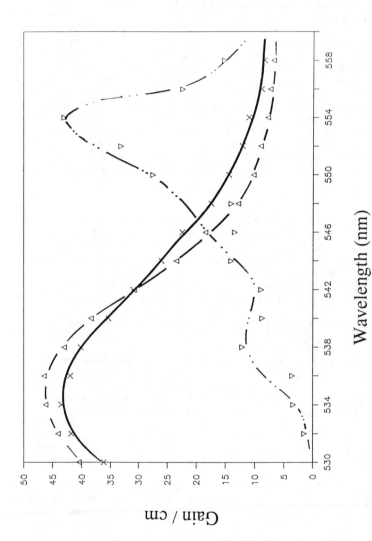

Figure 1.   Spectral gain envelope for coumarin 153 in ethanol and in aged gels. Triangles, ethanol solution; crosses, gel aged for one day; inverted triangles, gel aged for eleven days.

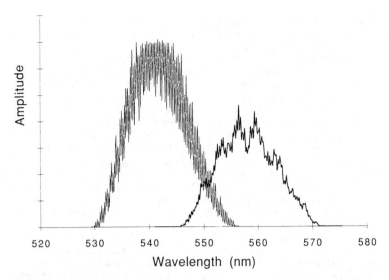

Figure 2. Spectral output of free running coumarin 153 in ethanol (left) and in aged silica gel (right.)

Table I. Laser Behavior in SiO$_2$ and ORMOSIL

Dye	Medium	Peak (nm)	FWHM (nm)	Oscillation Range (nm)
Coumarin 1	EtOH soln.	438	7	432-446
(4 x 10^{-3})	Gel, Aged	444	10	433-457
Coumarin 102	EtOH Soln	457	8	468-483
(5.9 x 10^{-3})	Gel, Aged	491	6	487-495
Coumarin 153	EtOH Soln	542	13	529-554
(3.4 x 10^{-3})	Gel, Aged	558	13	545-572
Coumarin 153	ORMOSIL	526	20	498-574
(2.0 x 10^{-3})				
Rhodamine 6G	EtOH Soln	586	12	570-615
(1.0 x 10^{-3})	Gel, Aged	589	8	570-510
Rhodamine 6G	ORMOSIL-I	571	4	559-587
(5.0 x 10^{-4})				
Rhodamine 6G	ORMOSIL-G	568	8	557-598
(5.0 x 10^{-4})				

exhibits a gain peak centered about 554 nm, a region where the gain in ethanol solution approaches zero. This secondary peak may be associated with inhomogeneous broadening caused by microscopic phase separation.

Broad band laser oscillation from Coumarin 153 doped ORMOSIL gels was easily obtained in the free-running laser cavity. The laser emission and the luminescence spectrum both peak at nearly the same wavelength as shown in Figure 4. The laser emission peak was at 526 nm with an oscillation bandwidth of approximately 20 nm FWHM. (For comparison, the reported total oscillation bandwidth in ethanol pumped at 308 nm is 75 nm.) The fluorescence spectrum has a broad peak at about 530 nm with a bandwidth (FWHM) of about 80 nm.

The laser emission peak from R6G doped ORMOSIL gels occurred at 571 nm with a bandwidth of 4 nm. The laser emisison band is narrower than the FWHM fluorescence band. The doped ORMOSIL sample exhibited a luminescence peak at 565 nm with a bandwidth of 55 nm (FWHM) In contrast to the C153 gel, the solid state rhodamine doped sample did not oscillate over the FWHM range of the fluorescence emission spectrum. The R6G samples exhibited detectable oscillation over a total range of about 38 nm (559 to 587 nm).

The laser output intensity of the C153 and R6G ORMOSIL gels was studied as a function of the number of laser pump pulses. Both materials could be pulsed for more than 3000 shots with a reduction of the emission amplitude of about a factor of four. Specifically, the C153 gel laser intensity decreased by a factor of 6 after more than 6000 pulses of 500 MW/cm^3. The plot of the intensity versus number of shots has a double exponential decay. This phenomenon is not yet completely understood, but it could be associated with microscopic phase separation in the medium. The R6G decay plot shows that the intensity undergoes a 90% reduction after 5300 laser pulses.

The dye-doped ORMOSIL materials exhibit much higher photostability than the reported polymer-doped materials. For example, optical gain was reported for several coumarin dyes in PMMA (13). It is significant to note that the best dyes reported could only be used to several hundred pulses and that most of the coumarin dyes could not even oscillate for 100 pulses. The ORMOSIL gels thus represent an improvement of at least 1-2 orders of magnitude in photostability. Optical gain was also reported for R6G doped into modified poly(methyl methacrylate) (14). These materials exhibited loss of 20% output after 180 pump pulses (1 J/cm^2) at 532 nm. The reported plots indicate that these materials would have dropped by 90% after 275-300 pulses due to rapid steady state photodegradation once the pulse count exceeded the critical pulse number. In this case the R6G doped ORMOSIL gel laser offers a useful lifetime improvement by a factor of more than 15 over the reported polymer material.

Laser Action in Doped Aluminosilicate Gel-Glasses. The spectrum of the free running laser is shown in Figure 5. The band is centered around 570nm, and the average bandwidth is 5nm, strongly narrowed when compared to the fluorescence spectrum. Although this would be considered very broad for a tunable laser, it is typical of a free running laser which, due to the lack of a selective grating or prism in the cavity, runs many different modes simultaneously. The different output wavelengths of the modes are apparent in the spectra as the variations in the peak shape; the exact mechanism which determines the relative intensities of different modes is not yet known. The bandwidth, however, stays roughly constant over time.

The intensity of the laser output decreases with the total number of pulses which have pumped the sample as shown in Figure 5. The rate of this decrease with respect to the number of pulses varies with concentration, pump power, pump rate, and from sample to sample. The plots of the laser output intensity versus the total number of pump pulses could not be fit to a single or a double exponential curve; however, all experiments performed showed a continuously decreasing non-exponential decay rate.

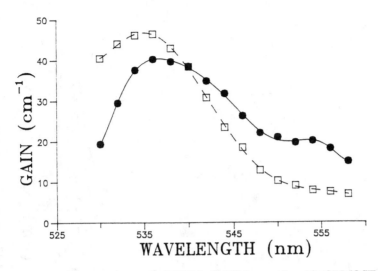

Figure 3. Spectral gain curve of C153 in EtOH (squares) and in ORMOSIL (dots) pumped at 337 nm.

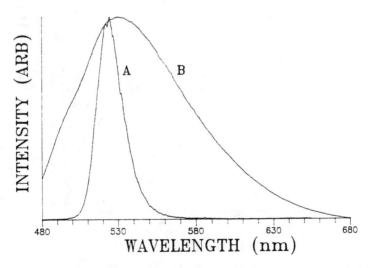

Figure 4. Free running laser oscillation (A) and fluorescence (B) of C153 in ORMOSIL. The pump wavelength for both curves is 460 nm.

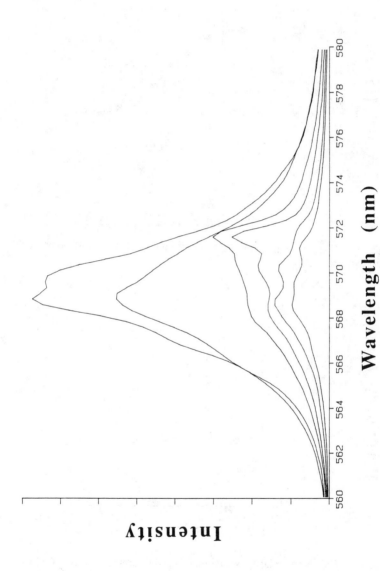

Figure 5.   Spectrum of the free-running R6G in ASE laser as a function of the number of pump laser pulses. From top to bottom the number of pulses range from 60 to 2250.

The sample pumped at 1 Hz had lost 50% of its output intensity after approximately 1500 pulses. It would have been reduced by 90% after greater than 4000 pulses due to the "levelling out" of the decay rate. In addition to the decrease in output intensity, the samples also showed visible signs of degradation (darkening) to a depth of 500um. The photodegradation was not counteracted by changing the focal point of the pump beam to a fresh area of the gel, either by translation or refocussing.

The power dependence of the laser output was determined by plotting the relative output intensity vs. the relative pump intensity. In Figure 6 this function is clearly non-linear. Separate measurements show that the absolute threshold is ~1 mJ, and the absolute output energy is 0.5 μJ/pulse when the input energy is 2mJ/pulse. Additional laser dyes which exhibited laser action in ASE are listed in Table II.

Polyaniline-Doped Gels

Many of the conducting polymers, including polyaniline (PANi), have been shown to possess large third order susceptibilities. Polyaniline has received a great deal of attention in recent years because of its stability in the presence of oxygen. In addition, polyaniline stands out among the conducting polymer family in that it exhibits unusual solubility in various organic solvents. PANi, for example can be partially dissolved into dimethyl formamide (DMF) and N-methyl pyrrolidinone (NMP). It is believed that the high molecular weight polymer has a significant degree of crystallinity, thereby limiting its solubility. It is possible to improve the overall solubility of PANi in either of two ways: alkyl substitution or swelling of the semi-crystalline matrix.

Leclerc showed that modification by suitable substituents, such as short chain alkylation, can make the polymer (mol wt ≥ 5000) completely soluble in a variety of solvents, even producing modest solubility in alcoholic solutions (16). Polymethylaniline, polyethylaniline, and polypropylaniline forms have been reported, along with significant solubility increases.

It is also possible to swell polyaniline in tetrahydrofuran (THF), thereby greatly improving the solubility of PANi. It is believed that high molecular weight PANi (MW 100,000) exists "as synthesized" in a highly convoluted morphology. Upon swelling of the polyaniline and subsequent vaporation of the THF, the semi-crystalline polymer phase is caused to relax, resulting in a much lower degree of crystallinity. The amorphous form of PANi exhibits greatly improved solubility in NMP with respect to the "as synthesized" forms. Both alkylation and THF swelling methods were used to increase the PANi concentration level within the silica gel matrix. In the studies reported here, 2-ethyl polyaniline (2-Et PANi) and THF swollen with PANi, in the partially oxidized emeraldine base (EB) form, were incorporated into rigid silica gel hosts by the sol-gel solution method.

Sample Preparation. Chemically polymerized 2-ethyl polyaniline, with reported molecular weight of 5000, was prepared by the method outlined by Leclerc et al. (15). Treatment of the insoluble product with ammonium hydroxide solution resulted in transforming the salt into the soluble EB form, which exhibited slight solubility in methanol. The soluble EB form of PANi is known to be readily protonated under acidic conditions, producing the highly insoluble ammonium salt form (16,17). In order to maintain the free amine base form in solution, it was necessary to synthesize the silica gel in the presence of a minimal amount of acid catalyst.

The initial silica sol consisted of 11 ml of tetramethoxy silane (TMOS), 5 mL of distilled water, and 1 drop of 12 N HCl. A monophasic solution was prepared by the 'sonogel' method (9). Under continuous sonication, a total of 30 ml of tetraethoxy silane (TEOS), 7.5 mL of distilled H_2O, and 11 mL of TMOS were addded in a dropwise fashion to form the precursor solution. The final silica sol was composed of a 1 : 1 : 5 mole ratio of TEOS : TMOS : H_2O. The acid catalyst was present as a

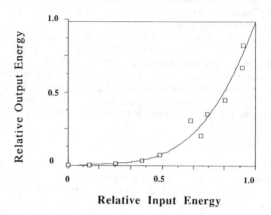

Figure 6. Relative dependence of the output power of R6G in ASE on the pump laser power.

Table II. Laser Behavior in ASE

Dye	Pump Wavelength	Peak (nm)	FWHM (nm)	Stability
Rhodamine 6G (5.0×10^{-4})	511	565	9	1800
Rhodamine 6G (5.0×10^{-4})	540	574	4	2500
Rhodamine 6G (1.0×10^{-3})	511	567	5	9000
Rhodamine 6G (5.0×10^{-4})	540	569	6	600
Fluorescein (4.0×10^{-3})	511	552	7	600
Coumarin 153 (5.0×10^{-3})	460	555	10	--

level of 1×10^{-3} : 1 molar ratio of acid to alkoxide. A minimal volume of water was used in the reaction in order to form a more suitable solvent for the 2-Et PANi.
     Doping of Et-PANi into the silica sol was accomplished by dissolving the polymer into a silica sol-compatible solvent. N-methyl pyrrolidinone (NMP) was chosen as the carrier solvent due to the good solubility of 2-Et PANi and its compatibility with the silanol solution. NMP containing approximately 2.5 weight percent 2-Et PANi was added to a silica sol-NMP mixutre. The final compostion of the solution was 7 : 1 : 2 volume ratio of silica sol : NMP : 2-Et PANi/NMP. After the addition of all components, the solution was filtered through a 0.7 μm glass fiber filter and subsequently placed into styrene containers for gelation. Two weeks after the onset of gelation the closed containers were opened, and volatile hydrolysis products were allowed to evaporate. The film samples were dried in air at room temperature for three weeks prior to optical analysis.

Optical Characterization. Degenerate four-wave mixing studies were carried out on the materials described above. (18). The pump/probe wave source was an actively Q-switched Nd:YAG laser. All measurements were performed at 1.06 μm, with Q-switched pulses of 15-20 nsec (FWHM). Nominal power output of the laser was 5 MW. Source waves were focussed to an area of approximately 8 mm^2 at the sample plane. Pump/probe wave intensities (I_1, I_2, and I_3) were approximately matched, having amplitudes of 21 MW/cm^2 each at the sample plane ("S" polarized). The Nd:YAG laser did not contain an etalon device, resulting in the generation of relatively short coherence length (20 mm) pulses. For this reason L_1, L_2, and L_3 were path length matched to within 10 mm. The counterpropagating phase conjugate wave, I_4, as detected by means of partial reflection from a beamsplitter. The signal was detected using a reverse biased p-i-n silicon photodiode, which transmitted current pulses to a Tektronix 2430 digital storage oscilloscope. Measurements were conducted on thick (1 mm) film samples, with the I_1 and I_2 waves intercepting the samples near normal incidence. The angle between I_1 and I_3 was determined to be 7 degrees. To eliminate any lensing effects due to slight warping, solid gel samples, doped and undoped, were held within a liquid gate, filled with an index matching fluid (mineral oil). The CS_2 reference liquid was measured in a 1 mm path length silica glass cuvette. Background signal amplitudes from the liquid gate and silica glsss cuvette were measured and subtracted from the observed gel and reference liquid sample response values.

Results and Discussion. The 2-ethyl polyaniline concentration in the silica gel film was determined by constructing a Beer's law calibration curve from solutions of known concentration. Assuming an average molecular weight of 5000, the 2-Et PANi concentration in the silica gel was found to be 9.6×10^{-4} M. The refractive indices of CS_2 and 2-Et PANi:SiO$_2$ were estimated to be 1.6 and 1.4 at 1.06 μm, respectively. The emeraldine base doped silica gel was found to have low losses due to scatter, and exhibited good transparency at 1.06 μm. Spectrophotometric measurements at 1.06 μm yielded absorption coefficients of 0.1 cm^{-1} (> 99% T over 1 mm pathlength) for the CS_2 reference and 4 cm^{-1} (96% T over 1 mm pathlength) for the 2-Et PANi doped silica film.
     The reflectivity of the doped gel was found to be 31% of that from the CS_2 reference. Assuming that the third-order susceptibility of carbon disulfide is 1.7×10^{-12} esu at 1.06 μm, and assuming that the entire signal arises from third order susceptibility effects, $\chi^{(3)}$ of the polyaniline doped gel was calculated to be 4.8×10^{-13} esu (6.7×10^{-21} m^2/V^2). However, a number of effects including thermal effects could contribute to the observed signal and we cannot unambiguously attribute the observed signal to $\chi^{(3)}$. The delay of the peak response of the doped gel with respect of CS_2 by about 3-4 nsec suggests that thermal effects could be important.

Acknowledgments
The support of the National Science Foundation (DMR 87-06010) is gratefully acknowledged.

Literature Cited
1. Mackenzie, J. D.; Ulrich, D. R., eds. Proc. Third Intnl. Conf. on Ultrastructure Processing Wiley, New York, 1988.
2. McKiernan, J.; Pouxviel, J.-C.; Dunn, B.; Zink, J. I. J. Phys. Chem. 1989, 93, 2129.
3. Pouxviel, J. C.; Dunn, B.; Zink, J. I. J. Phys. Chem. 1989, 93, 2134.
4. Kaufman, V.; Avnir, D. Langmuir 1986, 2, 717.
5. Avnir, D.; Levy, D.; Reisfeld, R. J. Phys. Chem. 1984, 88, 5956.
6. Avnir, D.; Kaufman, V. R.; Reisfeld. R. J. Non-Cryst. Solids, 1985, 74, 395.
7. Kaufman, V. R. Avnir, D.; Pines-Rojanski, D.; Huppert, D. J. Non-Cryst. Solids 1988, 99, 379.
8. Pouxviel, J. C.; Boilot, J. P.; Lecomte, A.; Dauger, A. J. Phys. (Paris) 1987, 48, 921.
9. Esquivas, L.; Zarzycki, J. Proc. Third Intnl. Conf. on Ultrastructure Processing of Ceramics, Glasses, and Composites, Mackenzie, J. D. and Ulrich, D. R., Eds., Wiley Interscience, 1988.
10. Capozzi, C. A.; Pye, L. D. Proc. SPIE, 1988, 970.
11. Schmidt, H.; Seiferling, B. MRS Symp. Proc. 1986, 73, 739.
12. Jones, G.; Jackson, W. R., Halpern, A. M. Chem. Phys. Lett. 1980, 72, 391.
13. Itoh, U.; Takakusa, M.; Moriya, T.; Saito, S. Jap. J. Appl. Phys., 1977, 16, 1059.
14. Gromov, D. A.; Dyumaev, K. M.; Manenkov, A. A.; Maslyukov, A. P.; Matyushin, G. A.; Nechitailo, V. S.; Prokhorov, A. M. J. Opt. Soc. Am. B, 1985, 2, 1028.
15. Leclerc, M.; Guay, J.; Ho, L. H. Macromolecules, 1989, 22, 649.
16. MacDiarmid, A. G.; Chiang, J. C.; Halpern, M.; Huang, W. S. Mol. Cryst. Liq. Cryst. 1985, 121, 173.
17. Cushman, R. J.; McManus, P. M.; Yang, S. C. J. Electroanal. Chem. 1986, 291, 335.
18. Yariv, A.; Fisher, R. A. Optical Phase Conjugation; Fisher, R. A., ed. Academic Press, New York, 1983.
19. Altman, J. C.; Elizando, P. J.; Lipscomb, G. F.; Lytel, R. Mol. Cryst. Liq. Cryst. Inc. Nonlin. Opt. 1988, 157, 515.

RECEIVED July 23, 1990

MOLECULAR AND SUPRAMOLECULAR METAL-BASED SYSTEMS

Chapter 37

Intrazeolite Semiconductor Quantum Dots and Quantum Supralattices
New Materials for Nonlinear Optical Applications

Geoffrey A. Ozin[1], Scott Kirkby[1], Michele Meszaros[1], Saim Özkar[1], Andreas Stein[1], and Galen D. Stucky[2]

[1]Lash Miller Chemical Laboratories, University of Toronto, Toronto, Ontario M5S 1A1, Canada
[2]Department of Chemistry, University of California, Santa Barbara, CA 93106

Recent developments in host-guest inclusion chemistry have paved the way to the controlled and reproducible assembly of sodalite and faujasite quantum dots and supralattices, the latter being comprised of regular arrays of monodispersed semiconductor (eg. AgX, WO_3) quantum dots confined in a dielectric material. This work has led to the synthesis of the first examples of mixed component semiconductor quantum supralattices represented by the new sodalite family of materials $(8-2n)Na,2nAg,(2-p)X,pY-SOD$. Collective electronic coupling between these encapsulated and stabilized nanostructures can be altered through judicious variations in the host structure and guest loading. When the carrier wave function is restricted to the region of the imbibed nanostructures, quantum size effects (QSE's) are observed which give rise to differences in the optical, vibrational and magnetic resonance properties of these materials with respect to those of the bulk semiconductor parent. In this regime of strong quantum confinement, one anticipates resonant and non-resonant excitonic optical nonlinearities associated with $\chi^{(3)}$ to be enhanced with respect to those of the respective quantum wire, quantum well, and bulk semiconductor materials.

In the continuing quest for new materials with superior optical nonlinearities, fast response times, high photochemical and thermal stabilities for applications in optical switching and signal processing, chemists and physicists have recently turned their attention to semiconductor ultramicrostructures exhibiting reduced charge carrier mobility in one to three dimensions. Structures exhibiting quantum size effects (QSE's) caused by carrier confinement in three dimensions, that is, zero dimensional mobility, are commonly referred to as quantum dots (QD's) (1).

0097–6156/91/0455–0554$08.00/0

At present the available literature invokes more questions than it solves, including: precisely what are the material requirements that make a QD of value in the technology of nonlinear optics (NLO)? How do the properties of an idealized QD differ from those of quantum wells (QW's, one dimensional confinement microstructures) and bulk materials? In which ways are the optical properties of QD's modified by local field effects (LFE's) arising from the embedding media? How do the optical properties of three dimensional arrays of electronically coupled QD's compare with those of isolated QD's? In addition, decisions concerning the selection of QD atomic constituents, the degree of quantum and dielectric confinement, and the extent of electronic coupling between QD's all require a detailed assessment when attempting to fabricate new semiconductor materials with attractive NLO properties.

A concern of particular importance, is the fabrication of monodispersed collections of QD's. In the experimental literature, great efforts have been expended to isolate the effects of a single particle size from those caused by even small size distributions (2). In our recent work, we have learned how to encapsulate, as clusters, the components of some well known semiconductors inside the accessible 0.66 nm and 1.3 nm void spaces of sodalite and faujasite host lattices (3, Özkar, S.; Ozin, G.A. *J. Phys. Chem.*, in press.). This kind of guest-host inclusion chemistry provides a convenient route to strongly quantum confined nanostructures with densities ranging from isolated QD's through to perfectly organized, three dimensional periodic arrays of interacting QD's. While this synthetic route gives perfect monodispersed size distributions, the choice of particle size at present is limited by the range of currently available host cavity sizes (0.66 - 1.30 nm) and channel dimensions (0.50 - 1.05 nm). This situation could however, rapidly change with the synthesis of new generation large pore zeolites.

One important question is not addressed by the current theory: whether monodispersed nanostructures this small (isolated or ordered, non-interacting or coupled), fabricated from the components of bulk semiconductors are likely to be interesting candidates as NLO materials. In order to advance current theories on linear and non-linear optical properties of semiconductor nanostructures, suitable materials must be made available.

In the present article, we will survey the known synthetic procedures to intrazeolite semiconductor QD's and quantum supralattices (QS's; this name has been chosen to refer to these structures, since they do not alter the crystallographic unit cell of the host, as the more common name of superlattice would imply). This is followed by two examples from our recent work concerning the I-VII pure and mixed halide system Ag,X-SOD (3) and the VI-VI system $n(WO_3)-M_{56}Y$ (Özkar, S.; Ozin, G.A. *J. Phys. Chem.*, in press.). Some key properties of these materials will be briefly described that relate to QSE's, LFE's, electronic and vibrational coupling between QD's. The paper concludes with a very brief survey of some of the pertinent physics behind these early observations and how they might relate to the NLO properties of these materials.

Synthesis

Intrazeolite semiconductors (IZS's) of several types, II-VI, IV-VI, I-VII, III-V, VI and VI-VI, exemplified by CdSe, PbS, AgBr, GaP, Se and WO_3, respectively, have been fabricated. These composites are normally prepared by aqueous (Ozin, G.A.; Stein, A.; Stucky, G.D.; Godber, J.P. *J. Incl. Phenom.*, in press; *4,5*) and melt (*3*) ion-exchange, metal-organic chemical vapour deposition (*6*), vapour phase impregnation (*7*) and phototopotactic (Özkar, S.; Ozin G.A. *J. Phys Chem*, in press.) methods. A major synthetic challenge with these materials concerns control of the nuclearity, population, location, distribution, dimensionality, purity, defect and doping concentration of the encapsulated semiconductor guest in the zeolite host lattice. In what follows we will focus attention on some of the interesting aspects of the use of melt ion-exchange techniques and simple binary metal carbonyl phototopotaxy for the growth and stabilization of IZS quantum nanostructures of the type exemplified by Ag,X-SOD and $n(WO_3)$-$M_{56}Y$ respectively.

Sodalite Quantum Supralattices: Preamble

Sodalite, 8M,2X-SOD (where $SOD \equiv Si_6Al_6O_{24}$, $M \equiv$ cation and $X \equiv$ anion reflect the framework, cation and anion content of the sodalite unit cell), is unique as a host material, as it consists of bcc packed cubo-octahedral cavities (called β-cages) having a free diameter of about 0.66 nm (Figure 1) (*8*). A network of SiO_4 and AlO_4 tetrahedra form densely packed cubo-octahedral cavities (non-rigid, originating from Si-O-Al angular flexibility) with eight six-ring and six four-ring openings. The negative charge on the framework is balanced by exchangeable cations at tetrahedral sites near the six-rings of the β-cage. An additional six-ring cation and an anion at the centre of each β-cage are often present as well. Thus sodalite can be viewed as the archetype QS boasting perfectly periodic arrays of all-space filling β-cages (a Federov solid) containing atomically precise, organized populations of M_4X clusters.

Class A Quantum Supralattices. As an example, silver halide exhibiting molecular behaviour has been produced inside sodalite by silver ion exchange of sodium halo-sodalites using a $AgNO_3/NaNO_3$ melt containing substoichiometric amounts of silver (*3*). Rietveld refinement of high resolution X-ray powder data for a sample containing 0.3 Ag^+ per unit cell, or an average of one AgX molecule in every eighth cage, indicates that the AgX molecules can be considered isolated. At increased Ag^+ concentrations up to complete silver exchange, the product is best described as a sodalite lattice containing expanded silver halide, that is, a three dimensional array composed of monodispersed zero dimensional silver halide dots. Rietveld refinement showed that the Ag_4X units in 8Ag,2X-SOD are perfectly ordered.

Control over the silver halide environment is possible by varying the anion and cation compositions as illustrated below and in Figure 1:

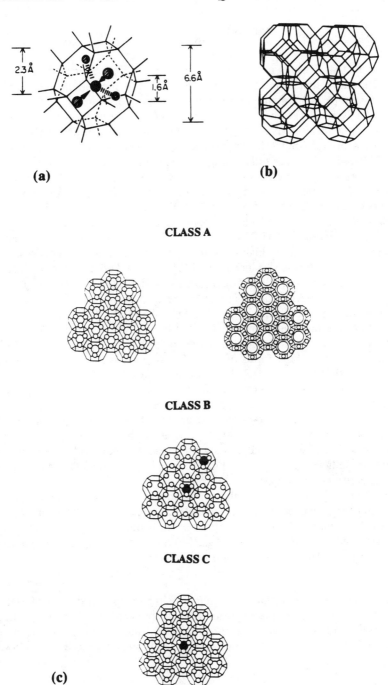

Figure 1. (a) The sodalite β-cage exhibiting the imbibed tetrahedral M_4X cluster. (b) The bcc packing arrangement of β-cages in the sodalite unit cell. (c) Quantum supralattices.

$$8\text{Na}, 2\text{X}-\text{SOD} \xrightarrow{\quad 2n\,\text{Ag}^+\text{melt}\quad} (8-2n)\text{Na}, 2n\text{Ag}, 2\text{X}-\text{SOD} + 2n\text{Na}^+ \quad (1)$$

where $n = 0\text{-}4$. Results obtained from far and mid-IR spectroscopy and powder XRD indicate that in mixed sodium-silver sodalite the cations are distributed statistically (a 3-D commensurate compositionally disordered solid-solution of $\text{Na}_{4\text{-}n}\text{Ag}_n\text{X}$, $n = 0\text{-}4$) rather than in aggregates (domains of Na_4X and Ag_4X) or in an ordered fashion ($\text{Na}_{4\text{-}n}\text{Ag}_n\text{X}$, n fixed). Both far-IR and XRD data of $(8\text{-}2n)\text{Na},2n\text{Ag},2\text{X-SOD}$ show breaks at a silver loading of $n \sim 1$ corresponding to more than one Ag^+ per sodalite cage possibly indicating a percolative threshold reflecting collective interactions between cavity contents. As the $\text{Na}_{4\text{-}n}\text{Ag}_n\text{X}$ cavity nuclearity is limited to five, no significant band shift in the electronic spectrum occurs on increasing the loadings of Ag^+ from $n = 1\text{-}4$. This is to be contrasted with most other supported and unsupported quantum size particles (e.g. glasses, micelles, vesicles, clays, LB films, and surface-capped particles respectively), where an increase in the loading of the semiconductor components results in a red-shift as the particle size increases (*9,10*) The band edges of the 8Ag,2X-SOD quantum supralattices (described in terms of the tight binding and miniband approximations (*11,12*)) all lie at higher energy than in the bulk semiconducting AgX (band-gap, $\text{VB}[\text{X}^-(\text{np}), \text{Ag}^+ (4\text{d})] \to \text{CB}[\text{Ag}^+ (5\text{s})]$ excitation). As in the bulk AgX, the band positions are affected by the type of halide. A red-shift is observed for the bigger anions in larger unit cells in the case of the isolated molecular and fcc bulk forms of AgX (*3*). In contrast, the estimated energies (computer fit using $\alpha^{\text{da}} = K(E_g\text{-}E)^{1/2}$) of the absorption edge of the silver halide quantum supralattices (Figure 2) follow the order $\text{Cl} \leq \text{Br} > \text{I}$. This is indicative of an interplay of decreasing band-gap with decreasing bandwidth down the halide series, implying that the extent of inter-β-cage electronic coupling follows the order of the observed distances between the centres of the β-cages, that is, $\text{Cl} < \text{Br} < \text{I}$ (Figure 2). The absorption edges show no temperature dependence down to 10K, therefore indicating a direct band-gap for the 8Ag,2X-SOD QS's in accordance with the estimation of E_g from α^{da}.

This investigation shows that an organized assembly ranging from isolated molecules to expanded structures stabilized inside a sodalite host matrix (Figure 1) can be readily fabricated out of a material that is normally a I-VII semiconductor. The $(8\text{-}2n)\text{Na},2n\text{Ag},2\text{X-SOD}$ sodalites might find applications as electronically tunable nonlinear optical materials (see later).

Class B Quantum Supralattices. An interesting series of sodalite materials that for the first time offer the opportunity to manipulate the degree of collective electronic and vibrational interactions between monodispersed semiconductor components in neighboring β-cavities (Figure 1) can be synthesized as follows:

Molecular Absorptions/Absorption Edges (eV)

	AgCl	AgBr	AgI	Order
Molecule	5.12,5.85	4.00,5.04	3.56,4.72	Cl > Br > I
Expanded SC	3.83	3.85	3.76	Cl ≤ Br > I
Bulk SC	3.25	2.68	2.33	Cl > Br > I
SOD Unit Cell Size (Å)	8.8708	8.9109	8.9523	Cl < Br < I

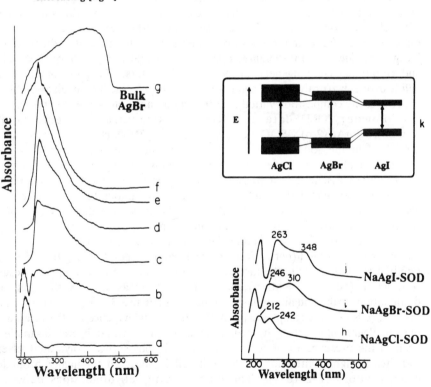

NaAgBr-SOD

Increasing [Ag$^+$]

Figure 2. UV-visible reflectance spectra of (8-2n)Na,2nAg,2Br-SOD with varying silver concentrations. Ag$^+$/uc: (a) 0; (b) 0.05; (c) 0.28; (d) 2.0; (e) 3.1; (f) 8.0; (g) bulk AgBr; (h) (8-2n)Na,2nAg,2Cl-SOD, 0.1 Ag$^+$/uc; (i) (8-2n)Na,2nAg,2Br-sodalite, 0.1 Ag$^+$/uc; (j) (8-2n)Na,2nAg,2I-SOD, 0.1 Ag$^+$/uc. Note that band-gap absorptions for 8Na,2X-SOD expanded insulators peak at 192 (Cl), 208 (Br), and 214 nm (I) (Figure 2h,i,j) and parallel the order for the bulk fcc NaX materials, 138 (Cl), 165 (Br), and 211nm (I), respectively. (k) Schematic band diagram for expanded AgX semiconductors (Reprinted from ref. 3. Copyright 1990 American Chemical Society.)

$$8 \text{ Na}, (2-p)X,pOH-SOD$$
$$\downarrow \quad -pNaOH \text{ Soxhlet}$$
$$(8-p) \text{ Na}, (2-p)X-SOD$$
$$\downarrow \quad 2nAg^+ \text{ melt}$$
$$(8-p-2n) \text{ Na}, 2nAg^+, (2-p)X-SOD \tag{2}$$

where $p = 0\text{-}2$, $2n = 0\text{-}(8\text{-}p)$. The first step involves a Soxhlet extraction of the pNaOH content of the parent sodalite which creates a sample having a predetermined mixture of $(2-p)Na_4X$ and pNa_3 units as the cavity components of the unit cell. These sodium aggregates can be progressively exchanged by silver ions to yield at full substitution the corresponding mixture of $(2-p)Ag_4X$ and pAg_3 units. Thus for a solid solution variations in p are expected to alter the mean distance between $Na_{4-n}Ag_nX$ clusters while changes in n will shift the constituents of the clusters from those of an insulator to a semiconductor type. Such controlled tunable volume filling and compositional alteration of the contents of the sodalite β-cages provides an unprecedented opportunity to adjust the extent of electronic and vibrational coupling between $Na_{4-n}Ag_nX$ clusters. Samples containing a halide in one, two or five out of eight β-cages on the average have been assembled with the cavity composition varied over the whole range of $n = 0\text{-}4$ (Figure 1).

Powder XRD data and Rietveld structural refinements of $(8-p-2n)Na,2nAg,(2-p)X\text{-}SOD$ indicate that the $(2-p)M_4X$ clusters are randomly organized in the sodalite lattice of pM_3 "spectator cationic triangles" (13). These data eliminate models based on domains (segregation) or organized (superlattice) assemblies of M_4X clusters. By example, the response of the sodalite framework to the gradual addition of Br^- anions to the unit cell of $6Na\text{-}SOD$ to eventually yield $8Na,2Br\text{-}SOD$ is shown for the fully hydrated samples in Figure 3. From inspection of the data one can immediately determine that the first Br^- to enter the Na_3 cages to form Na_4Br QD's, causes the unit cell to expand (partial loss of structural hydrogen-bonding from β-cage imbibed H_2O (13)) while the further influx of Br^- up to full loading has a cage contraction (Na_4Br formation) space filling effect as seen by the gradual diminution of the sodalite unit cell dimension (Figure 3). Upon dehydration the unit cells of these samples expand, thus providing evidence for the retention of some degree of structural hydrogen-bonding in the series of hydrated samples, even in the presence of bromide. In the analogous series of hydrated silver sodalites a slight contraction of the unit cell is observed for higher bromide concentrations (Figure 3). The difference in cell dimensions between corresponding sodium and silver sodalites is small, as these cations have nearly identical ionic radii (4-coordinate $r_{Na+} = 0.113$ nm, $r_{Ag+} = 0.114$ nm). The exceptional cell expansions after introduction of silver to bromide-free $6Na\text{-}SOD$ indicate that structural hydrogen-bonding may be ineffective in $6Ag\text{-}SOD$. The presence of Br^- is required for effective vibrational coupling between the Na^+ ions of adjacent cages. Far-IR spectra display a Br^- anion correlation couplet at 161, 68 cm^{-1}. When the population of Na_4Br clusters is

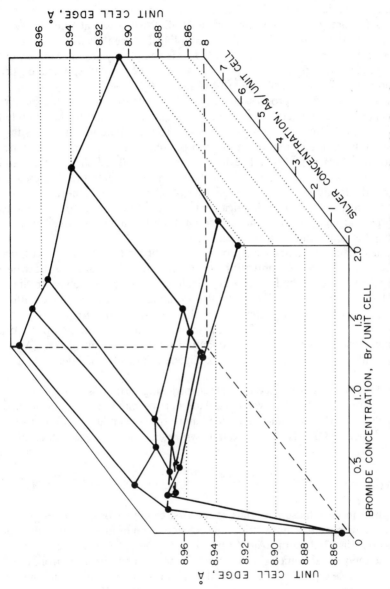

Figure 3. Unit cell size versus composition diagram for the series of hydrated sodalites of the type $(8-p-2n)Na,2nAg,(2-p)Br-SOD$, where $p = 0-2$, $2n = 0-8$.

dropped from two to zero per unit cell of sodalite, the splitting of this correlation couplet decreases as the average separation between bromides becomes greater (larger p). Its intensity subsequently goes to zero. However, the Na^+ cation correlation couplet observed at 200 and 105 cm^{-1} remains, and each band broadens (inhomogeneous distribution of Na_4Br/Na_3). Even at one Br^- every fourth cavity, the collective vibrational coupling between Na_4Br clusters has been severely suppressed. In 6Na-SOD (14) it is completely absent. Similar, but less pronounced effects can be discerned in the far-IR spectra of the corresponding fully Ag^+ exchanged (8-p)Ag,(2-p)Br-SOD.

Figure 4 shows the progression of the UV-visible spectra from the sodalite encapsulated AgBr-molecule to the isolated Ag_4Br cluster, and to the extended $(Ag_4Br)_n$ supralattice. At low silver and bromide loadings a very sharp UV-spectrum resembles that of an isolated gas phase silver bromide monomer $(Br^-(4p), Ag^+(4d) \rightarrow Ag^+(5s))$. The line broadening and red shift of the band edge observed when either the silver or the bromide concentration is increased indicates electronic coupling between $Na_{4-n}Ag_nBr$ clusters. Whereas sodium does not appear to contribute significantly to this coupling (the absorption bands are sharpest at high sodium concentrations), both bromide and silver must be mediating the communication between clusters.

These data for the first time provide compelling evidence for the existence of collective electronic and vibrational coupling interactions between $Na_{4-n}Ag_nBr$ QD's over the full Ag^+ and Br^- loading ranges, as well as, from the isolated silver bromide molecule to the expanded silver bromide quantum supralattice. Furthermore, they reveal the genesis of the $Br^-(4p),Ag^+(4d)$ mini-valence band and $Ag^+(5s)$ mini-conduction band on passing from embryonic $Na_{4-n}Ag_nBr$ clusters to a quantum supralattice built of $Na_{4-n}Ag_nBr$ QD's, and ultimately to the "parent" bulk mixed sodium-silver halides.

Class C Quantum Supralattices. Another intriguing series of sodalites, that for the first time permits the investigation of narrow band gap monodispersed semiconductor QD's embedded in a wide band-gap semiconductor QS composed of monodispersed QD's (Figure 1), can be synthesized according to the following recipe:

$$8Na,(2-p)X,pY-SOD \xrightarrow{\text{melt } 8 Ag^+} 8Ag,(2-p)X,pY-SOD + 8 Na^+ \quad (3)$$

One can view these materials as the first examples of mixed component semiconductor quantum supralattices. By consideration of Figure 1, the situation of a random distribution of pAg_4Y QD's in a $(2-p)Ag_4X$ QS can be described within the theoretical framework of the tight binding approximation, from which one can assemble a qualitative mini-band diagram of the type shown in Figure 5. Depending on the nature of the anions, the choice of the X:Y anion ratio, the degree of n- or p-doping of the QS (see later) and the extent of coupling between constituent QD's of the QS one might expect that an 8Ag,(2-p)X,pY-SOD mixed component semiconductor QS could function as

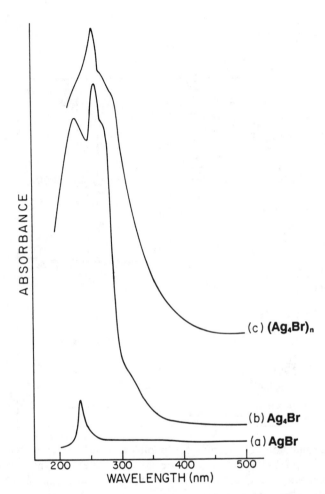

Figure 4. UV-visible spectra showing the progression from the isolated AgBr molecule to the isolated Ag_4Br cluster and to the extended $(Ag_4Br)_n$ supralattice inside hydrated sodalite $(8-2n-p)Na,2nAg,(2-p)Br-SOD$. (a) $p = 1.7$, $n = 0.05$; (b) $p = 1.7$, $n = 4$; (c) $p = 0$, $n = 4$.

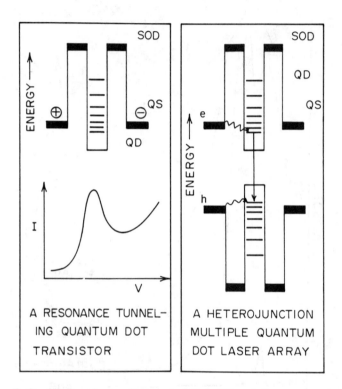

Figure 5. Qualitative electronic state (miniband) band diagrams envisaged for 8Ag,(2-p)X,pY-SOD as a new material for a chemistry approach to a resonance tunneling quantum dot transistor and a heterojunction multiple quantum dot laser array.

either a resonant tunneling QD transistor (RTQDT) or maybe even a high gain heterojunction multiple quantum dot laser (HMQDL) array. The concept behind a MQD laser involves excitation of an electron(e)-hole(h) pair in the mini-VB and mini-CB of the wide bandgap Ag_4Cl QS , e and h tunneling through the thin insulating barrier (sodalite wall) into discrete energy levels of the narrow bandgap Ag_4I QD, with e-h recombinative emission (laser action) between the lowest excited and ground states of the Ag_4I QD's (Figure 5).

Possible n-doping routes of these intrinsic QS's involve electron injection from intentionally imbibed $C_2O_4^{2-}$ or OH^- (photolytic) (Ozin, G.A.; Stein, A.; Stucky, G.D.; Godber,J.P. *J. Incl. Phenom.*, in press.), ion exchanged Cu^+ (thermal) or impregnated Na (thermal or electrochemical) (*15*). The latter technique is summarized in the following scheme:

$$8\,Na, (2-p-q)X, pY, qOH-SOD$$
$$\downarrow \;\; -q\,NaOH \;\;(\text{Soxhlet extraction of NaOH})$$
$$(8-q)\,Na, (2-p-q)X, pY-SOD$$
$$\downarrow \;\; q\,Na \;\;(\text{Reductive doping by } q\,Na^+ + q\,e^-)$$
$$8\,Na, (2-p-q)X, pY-SOD \tag{4}$$

A feasible p-doping pathway of these intrinsic QS's involves hole injection from, for example, ion exchanged Cu^{2+} (thermal).

The parent $8Na,(2-p)Cl,pI$-SOD and fully Ag^+ exchanged $8Ag,(2-p)Cl,pI$-SOD samples have been investigated over the full Cl, I concentration range $p = 0$-2. During the hydrothermal syntheses of the parent sodalites the formation of chloride-containing cages appears to be kinetically and/or thermodynamically favored, as the Cl/I ratio in the products is always greater than in the reaction mixtures. The difference is especially pronounced at iodide fractions less than 50% ($p < 1$) in the starting materials. The powder XRD, MAS-NMR, mid and far-IR results for these materials point to a 3-D commensurate, compositionally disordered solid-solution model, rather than one involving ordering, domains or complete segregation of M_4Cl and M_4I QD's ($M = Na^+, Ag^+$).

^{23}Na-MAS-NMR spectra of Class C sodalites displayed an essentially linear shift of the sodium resonance to higher field as chloride was displaced by progressively more iodide, following the mole fraction weighted mean of the shift due to each anion. This effect is related to increased shielding of the sodium cations by the less electronegative iodide ions. An additional contribution to the shielding of sodium ions may arise from the flexing of the framework as the unit cell size increases. As the Si-O-Al angle α becomes larger, the charge density from the sodalite cage lattice six-ring oxygens to Na^+ (3s) decreases as that to Si^{IV} and Al^{III} increases.

When the M_4Cl QD's exist at defect concentration levels in a QS of M_4I QD's and vice-versa, the optical reflectance and mid,far-IR spectra appear to be dominated by the spectral diagnostics of the QS. On the other hand, when the concentration of M_4Cl in M_4I (or M_4I in M_4Cl) exceeds about 10-20%, new spectral signatures can be discerned which shift continuously over the $p = 0$-2 range and appear to be associated with different populations and distributions of M_4Cl and M_4I QD's contained in the sodalite lattice. By illustration, the far-IR cation and anion translatory modes provide insight into the effect of changing the Cl:I ratio on the intra and intercavity vibrational coupling characteristics of the M_4Cl/M_4I QD's. One finds that the FWHH of the high frequency Na^+ cation mode around 200 cm^{-1} increases when more than one type of halide ion is present. This most likely originates from an order-disorder-order transition on passing from from 8Na,2Cl-SOD to 8Na,(2-p)Cl,pI-SOD to 8Na,2I-SOD, respectively. The enhanced line width observed for the mixed halo-sodalites is expected for a compositionally disordered solid-solution (glassy) model and likely corresponds to an inhomogeneous broadening effect caused by a distribution of microenvironments around each Na_4Cl and Na_4I intracavity oscillator. A smooth decrease in the Cl$^-$ anion correlation splitting with increasing p, is manifested by a gradual shift of the higher frequency Cl$^-$ anion translatory mode toward its low frequency coupled Cl$^-$ anion partner. The effect is less dramatic in case of the I$^-$ anion high frequency mode. By contrast to the halide modes, the Na^+ translatory mode correlation splitting increases with p. A simple way to understand this phenomenon involves consideration of the increasing distance between the centres of Na_4Cl and Na_4I QD's (Rietveld refinement gives interanion distances: Na,Cl-SOD, 0.768nm; Na,I-SOD, 0.780nm (*16,17*); Ag,Cl-SOD; 0.768nm, Ag,I-SOD, 0.776nm (Stein, A., unpublished results)) but decreasing distance between adjacent intercavity cations (intercage cation distances: Na,Cl-SOD, 0.4797nm; NaI-SOD, 0.4682nm (*16,17*); Ag,Cl-SOD, 0.492(2)nm; Ag,I-SOD 0.482(2)nm (Stein, A., unpublished results)) as p goes from 0-2. Thus the housing of Na_4I QD's in a Na_4Cl QS decreases the coupling between Na_4Cl units (loose cage) but at the same time enhances that between Na_4I units (tight cage) compared to that of the Na_4I QS. This pattern continues as the population of Na_4I in Na_4Cl grows, and maintains the same trend even after the mixture inverts to one of Na_4Cl QD's in a QS of Na_4I. Meanwhile, as the sodalite unit cell expands the response of the Na^+ cations is to move increasingly into the plane of the six-ring thereby shortening $Na^+...Na^+$ intercavity distances with a concomitant increase in the Na^+ correlation coupling. Similar effects also appear to exist for the measureable far-IR chloride modes of the 8Ag,(2-p)Cl,pI-SOD analogues.

Preliminary optical reflectance results obtained for low loading of Ag^+ (2n = 0.1) in (8-n)Na,nAg,(2-p)Cl,pI-SOD (corresponding to the "isolated AgX" molecule regime described earlier for the Class A QS) reveal new UV-spectral features which appear best ascribed to Na_3AgCl coexisting with Na_4I units, and Na_3AgI coexisting with Na_4Cl units in the same sodalite unit cell. The corresponding data for the more interesting fully exchanged 8Ag,(2-p)Cl,pI-SOD series is shown in Figure 6. The gradual conversion from Ag_4I QD's in a Ag_4Cl QS to Ag_4Cl QD's in a Ag_4I QS is apparent from inspection of these optical reflectance data. The UV-region of these samples

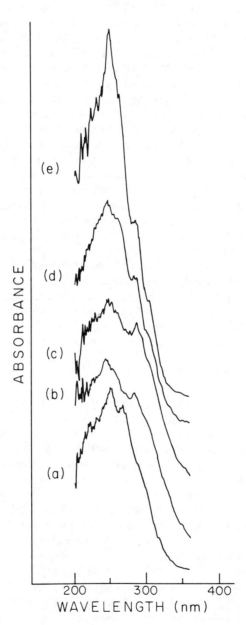

Figure 6. Optical reflectance data for the fully exchanged silver sodalite series: 8Ag,(2-p)Cl,pI-SOD. (a) p = 2; (b) p = 1.54; (c) p = 0.90; (d) p = 0.02; (e) p = 0.

(above the $O(2p) \rightarrow Si(IV), Al(III)$ sodalite framework band-gap excitations around 190-220nm) contains excitations of the $Cl^-(3p), Ag^+(4d) \rightarrow Ag^+(5s)$ and $I^-(5p), Ag^+(4d) \rightarrow Ag^+(5s)$ type, with a cluster molecular orbital discription of states for the Ag_4X QD's and a mini-VB, mini-CB picture for the Ag_4X QS's. Serious overlap complications clearly exist between the interband excitations of the QS's and interstate excitations of the QD's in this mixed halide $8Ag,(2-p)Cl,pI$-SOD system (Figure 6). Increasing the population of Ag_4I QD's in an Ag_4Cl QS appears to show optical reflectance spectra that are essentially a convolution of the main bands of the individual QD/QS components. It will require site selective luminescence emission/excitation spectroscopy of these samples, probably at low temperature and with time resolution, to precisely pinpoint optical spectral signatures associated with component Ag_4Cl/Ag_4I QD's and QS's. In this way it should prove possible to begin to evaluate the molecular orbital and mini-band architecture of these mixed semiconductor QD's and QS's, to probe collective electronic coupling between QD's, as well as to contemplate experiments designed to probe their potential for resonance tunnelling, laser emission and excitonic optical nonlinearities.

From the above data it is becoming clear that collective electronic and vibrational coupling effects exist between M_4X and M_4Y QD's in the $p = 0.05$-1.95 loading range. However, more sensitive spectroscopic and/or electrical transport techniques will be required in order to explore the interesting "defect" regime above and below these limits. These kinds of investigations are currently underway in our laboratory.

Intrazeolite Metal Carbonyl Phototopotaxy

One of the key issues in the microfabrication of semiconductor quantum nanostructures is the maintenance of sharp boundaries or interfaces among organized assemblies of quantum objects with precisely defined size and dimensionality (called quantum wells, wires or dots). By using a photon source to dissociate one or more of the precursors, deposition and epitactic growth of 2-D quantum confined layered superlattices can often be accomplished at lower substrate temperatures than by thermal epitaxy, thereby preventing atomic diffusion and deleterious interfacial effects at boundaries. In this section, attention is briefly focused on the use of simple intrazeolite binary metal carbonyls for the photo-oxidative nucleation, growth and stabilization of monodispersed semiconductor quantum nanostructures. The rationale for selecting this particular group of precursor molecules relates to their volatility, molecular dimensions, ease of purification, availability, facile and quantitative photo-conversion to the respective metal oxide materials and gaseous CO_2, with minimal carbon contamination. The volume filling of the zeolite by the metal carbonyl precursor can be precisely controlled. Convenient chemical and physical methods have been developed that can show that the precursor is internally confined and homogeneously dispersed in the zeolite crystals. The photo-oxidative synthesis is mild, clean and quantitative and produces no Brønsted protons. It is clear that phototopotactic routes to intrazeolite QD's and supralattices of QD's offer all of the advantages of photoepitactic methods to quantum wells and superlattices built of these wells.

Intrazeolite Tungsten(VI) Oxide Quantum Dots to a Zero Dimensional Quantum Supralattice. In this section we will briefly outline our recent studies of the intrazeolite topotactic photo-oxidative reaction of α-cage encapsulated hexacarbonyltungsten(0) in $Na_{56}Y$ and $H_{56}Y$, denoted $n[W(CO)_6]$-$Na_{56}Y(H_{56}Y)$ with O_2 to yield α-cage located tungsten(VI) oxide denoted $n(WO_3)$-$Na_{56}Y(H_{56}Y)$. (Recall that zeolite Y has 1.3nm cages called α- or supercages of overall tetrahedral symmetry interconnected by 0.8nm windows and 0.5nm β-cages interconnected by 0.3 nm windows, see Figure 7, (*18*).) Interestingly, the α-cages are arranged in the form of a diamond lattice while the β-cages are packed in a fcc array. The experimental details are succinctly summarized in Scheme 1. This approach permits the mild, clean and controlled intrazeolite nucleation, growth and organization of zero dimensional $(WO_3)_{2,4}$ QD's into a zero dimensional QS built of $(WO_3)_{2,4}$ QD's. New solid-state materials of this genre might find applications in quantum electronics and nonlinear optics.

A sequential saturation loading/photo-oxidation scheme of the type laid out in Figure 8 accounts for the observations recorded for the conversion of $n[W(CO)_6]$-$Na_{56}Y(H_{56}Y)$ to $n(WO_3)$-$Na_{56}Y(H_{56}Y)$. The discovery of essentially complete consumption of α-cage Brønsted acid sites (hydrogen-bonded) in a sample that analyses closely for $32(WO_3)$-$H_{56}Y$, nicely demonstrates that the original α-cage encapsulated WO_3 product, is equivalently dispersed throughout the supercages of the zeolite Y crystals. The intrazeolite aggregation process that leads to the imbibed WO_3 species, yields what appears to be a single kind of WO_3 cluster product, which in view of the elemental analysis figures obtained must be formulated as $n(WO_3)$-$Na_{56}Y(H_{56}Y)$.

Intimate details of the aggregation picture for the conversion of $n[W(CO)_6]$-$Na_{56}Y$ to $n(WO_3)$-$Na_{56}Y$ have been obtained by performing the aforementioned sequential saturation impregnation/photo-oxidation experiments in $H_8Na_{48}Y$, $H_{16}Na_{40}Y$ and $H_{56}Y$ containing one, two and four α-cage protons respectively. Here the protons act as *in situ* hydrogen-bonding probes capable of counting the number of WO_3 units contained in each α-cage. A preliminary aggregation picture that emerges from these studies is illustrated in Figure 8. Here one notes that the intrazeolite WO_3 aggregate that forms in the range up to half loading is best ascribed to the dimer $(WO_3)_2$-$Na_{56}Y(H_{56}Y)$ which at $n = 16$ can be considered to exist as a QS of dimers, $8(WO_3)_2$-$Na_{56}Y(H_{56}Y)$. At $n > 16$ it appears that the WO_3 dimers are gradually converted to WO_3 tetramers which at $n = 32$ can be considered to exist as a QS of tetramers, $8(WO_3)_4$-$Na_{56}Y(H_{56})$. With the available data it is not possible to differentiate whether the $(WO_3)_4$ tetramer is actually best formulated as either a dimer-of-dimers or an authentic tetramer. The mutual spatial requirements (evaluated by CHEM-X space filling models (e.g. Figure 7), spectroscopic properties and EXAFS (Ozin,G.A.; Ozkar, S.; Bein, T.; Moller, K., *J. Am. Chem.*

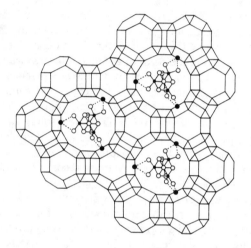

Figure 7. Chem-X model for a $(WO_3)_4$ cubane-cluster encapsulated in an α-cage of zeolite Y, projected down a 3-fold axis of the host structure, displaying tungsten-dioxo anchoring to extraframework site II Na^+ cation six-ring binding sites.

INTRAZEOLITE METAL CARBONYL
PHOTOTOPOTAXY

$$n[W(CO)_6] - Na_{56}Y(H_{56}Y) + \frac{9}{2}n\,O_2 \xrightarrow{h\nu} n\,(WO_3) - Na_{56}Y(H_{56}Y) + 6n\,CO_2$$

- **NO ROOM-TEMPERATURE DARK REACTION**

- **EFFICIENT $\lambda > 240$ PHOTOINDUCED REACTION**

- **ELEMENTAL ANALYSIS:**

  W to Evolved CO_2 Ratio is 1:6

  W to Non-Framework O is 1:3

  W to Na to Al Bulk Ratio Agrees with above Stoichiometry

- **NO DETECTABLE REACTION INTERMEDIATES OR INTERMEDIATE OXIDATION STATES**

- **XPS DATA SHOW:**

  $W^0 \longrightarrow W^{VI}$

  No Carbon Contamination (at the 1000 ppm Level)

  W:Na:Al Surface Atomic Ratio Agrees with Bulk Ratio

  No Deposition or Migration to the Surface

- **POWDER XRD: ZEOLITE LATTICE INTACT, NO BULK WO_3**

- **^{29} Si MAS-NMR: ZEOLITE LATTICE INTACT, SLIGHT FRAMEWORK PERTURBATION**

- **SEQUENTIAL SATURATION IMPREGNATION / PHOTOOXIDATION GIVES: $n = 16 \rightarrow 24 \rightarrow 28 \rightarrow 30 \rightarrow \cdots \rightarrow 32$ (Maximum 32 / Unit Cell in $Na_{56}Y$ or $H_{56}Y$)**

SCHEME 1

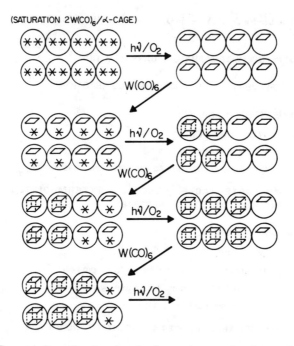

Figure 8. Proposed mechanism for the formation, nucleation and growth of $n(WO_3)$-$Na_{56}Y(H_{56}Y)$.

Soc., in press) and Rietveld data (Ozin,G.A.; Ozkar, S.; Stucky, G.D.; Bein, T.; Moller, K., unpublished results) of the proposed WO_3 cluster guest and the α-cage of the host, with its four tetrahedrally disposed α-cage extraframework sodium cations for $Na_{56}Y$ or Brønsted protons for $H_{56}Y$, are consistent with the proposed cluster inclusion models.

Evidence for both quantum confinement in $n(WO_3)$-$Na_{56}Y$ and dielectric confinement in $n(WO_3)$-$Na_{56}Y$ compared with $n(WO_3)$-$H_{56}Y$ can be found in both the optical and X-ray photoelectron spectra of these materials as one traverses the loading range $n = 0$-32, and alters the supercage cation respectively (Figure 9 and Table I).

Table Ia. Quantum Confinement in $n(WO_3)$-$Na_{56}Y$

Size Regime	Band Gap Energy (ev)	$W(4d_{5/2})$ Binding Energy (eV)	$W(4f_{7/2})$ Binding Energy (eV)
Bulk WO_3	2.7	246.8	35.9
Quantum Supralattice $(n = 32)$	3.3	247.8	36.9
Percolated Clusters $(8 \leq n < 32)$	3.3	247.8	36.9
Isolated Clusters $(n < 8)$	3.5	247.8	36.9

Table Ib. Zeolite Induced Matrix Effects (for $n < 8$) in $n(WO_3)$-$Na_{56}Y$ Compared With $n(WO_3)$-$H_{56}Y$

Zeolite Type	Band Gap Energy (ev)	$W(4d_{5/2})$ Binding Energy (eV)	$W(4f_{7/2})$ Binding Energy (eV)
$Na_{56}Y$	3.5	247.8	36.9
$H_{56}Y$	3.1	248.8	37.9

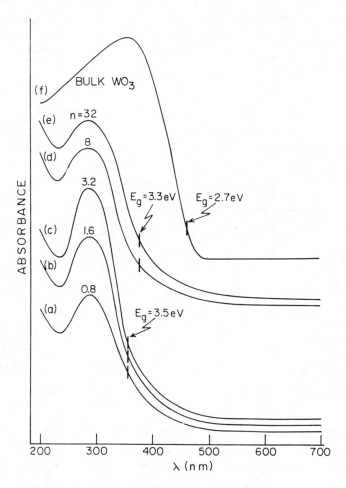

Figure 9. UV-visible reflectance spectra of n(WO$_3$)-Na$_{56}$Y (a) n = 0.8, (b) n = 1.6, (c) n = 3.2, (d) n = 8.0, (e) n = 32.0 compared to that of (f) bulk WO$_3$.

The data favour a nucleation and growth model for the intrazeolite phototopotactic oxidation of $n[W(CO)_6]$-$Na_{56}Y(H_{56}Y)$ by O_2 which involves the initial creation of zeolite encapsulated and isolated $(WO_3)_2$ clusters namely, $(n/2)(WO_3)_2$-$Na_{56}Y(H_{56}Y)$, $(n < 8)$. As the size of the proposed $(WO_3)_2$ dimer is expected to be considerably smaller than that of the bulk monoclinic m-WO_3 exciton, electron and hole length scales, this loading regime is best described as having produced strongly quantum confined tungsten(VI) oxide having an estimated $E_g = 3.5(3.1)$ eV (Figure 9). At $n \geq 8$ where one begins to observe red-shifting of the absorption edge to $E_g = 3.3(2.9)$ eV with minimal alteration of the intensity of the band-gap absorption, it appears that some kind of coupling between $(WO_3)_2$ dimers has ensued (Figure 9). In the range $8 \leq n < 32$, connectivity between $(WO_3)_{2,4}$ clusters evolves through the embryonic growth stages towards the genesis of an expanded tungsten(VI) oxide quantum supralattice. For full volume filling at $n = 32$, one has effectively created perfectly organized assemblies of α-cage encapsulated $(WO_3)_4$ tetramers running throughout the entire supercage void-volume of the zeolite Y host lattice (Figure 7), that is $8(WO_3)_4$-$Na_{56}Y(H_{56}Y)$. The latter can be considered to be either an expanded tungsten (VI) oxide quantum supralattice or a zero dimensional quantum supralattice composed of tungsten (VI) oxide clusters.

In the context of dielectric confinement, it is worth noting the approximately 0.4 eV red shift observed on the optical threshold energy, together with the roughly 1.0 eV blue shift observed on the W(4f) and W(4d) XPS core level ionization potentials on passing from $(WO_3)_{2,4}$-$Na_{56}Y$ to $(WO_3)_{2,4}$-$H_{56}Y$. These "zeolite induced matrix shifts" are consistent with a scheme in which the α-cage local environment of the $(WO_3)_{2,4}$ cluster is changed from that of sodium cations in $Na_{56}Y$ to that of Brønsted protons in $H_{56}Y$. In $H_{56}Y$ one envisages that the combined effect of hydrogen-bonding and protonation of the α-cage encapsulated $(WO_3)_{2,4}$ cluster by $\alpha-$ and β-cage Brønsted acid sites respectively, will be to deplete the $(WO_3)_{2,4}$ cluster of electron-density relative to its counterpart in $Na_{56}Y$. This will result in an overall weakening of skeletal tungsten-oxygen bonds, causing a red-shift in those cluster electronic excitations of the LMCT type, as observed in the optical spectra on passing from $(WO_3)_{2,4}$-$Na_{56}Y$ to $(WO_3)_{2,4}$-$H_{56}Y$ (Table Ib). In addition, the electronic manifold of the entire cluster will stabilize, inducing a blue-shift in the W core-level ionization potentials, as observed in the XPS on passing from $(WO_3)_{2,4}$-$Na_{56}Y$ to $(WO_3)_{2,4}$-$H_{56}Y$ (Table Ib).

A final point of considerable interest focuses attention on the 1.0(2.0) eV blue-shift in the W(4d) and the W(4f) core level ionization potentials observed on passing from $(WO_3)_{2,4}$-$Na_{56}Y(H_{56}Y)$ to that of bulk m-WO_3 respectively (Table Ib). This effect almost certainly can be traced to the extremely small size of the intrazeolite $(WO_3)_{2,4}$ clusters. It relates to the expected size dependence of initial (Koopman) and final (relaxation) state effects on the core-level binding energies of small clusters when in the quantum size regime. Thus on passing from bulk m-WO_3 to either a $(WO_3)_{2,4}$ cluster or a quantum supralattice composed of $(WO_3)_{2,4}$ clusters one anticipates higher ionization potentials as a consequence of (a) lower valence electron charge and lower coordination number (initial-state effect) and (b) lower valence electron screening (final-state effect), (Table Ib).

Basic Concepts of the NLO Properties of Semiconductor QD's and QS's

The interest in semiconductor QD's as NLO materials has resulted from the recent theoretical predictions of strong optical nonlinearities for materials having three dimensional quantum confinement (QC) of electrons (e) and holes (h) (1,19,20). QC whether in one, two or three dimensions increases the stability of the exciton compared to the bulk semiconductor and as a result, the exciton resonances remain well resolved at room temperature. The physics framework in which the optical nonlinearities of QD's are couched involves the third order term of the electrical susceptibility (called $\chi^{(3)}$) for semiconductor nanocrystallites (these particles will be referred to as nanocrystallites because of the perfect uniformity in size and shape that distinguishes them from other clusters where these characteriestics may vary, but these crystallites are definitely of molecular size and character and a cluster description is the most appropriate) exhibiting QC in all three dimensions. (Second order nonlinearites are not considered here since they are generally small in the systems under consideration.)

Optically excited semiconductor nanostructures show effects of QC if at least one spatial dimension of the material becomes comparable to, or smaller than the characteristic length scale (the classical Bohr radius) of an e-h pair. Different regimes of QC have been defined which depend on the semiconductor nanocrystallite size R relative to that of the Bohr radius of the exciton a_B, the electron a_e or the hole a_h:

$$a_B = \frac{\hbar^2 \varepsilon_2}{\mu \, e^2} \qquad a_e = \frac{\hbar^2 \varepsilon_2}{m_e \, e^2} \qquad a_h = \frac{\hbar^2 \varepsilon_2}{m_h \, e^2} \qquad (5)$$

Three regimes of QC have been distinguished (21): $R > a_B$, weak confinement; $a_e > R > a_h$, moderate confinement; and $a_e, a_h > R$, strong confinement. The linear optical properties of QD's are well understood. With increasing 3-D QC, the continuous interband absorption of the bulk semiconductor gradually transforms into the discrete lines of the QD (1). Using the effective mass approximation (EMA, which views the delocalization of charge carriers by assuming that they are free quasi-particles with the appropriate effective mass (11)) for the QD, one computes a blue shift of the e-h pair ground state energy proportional to R^{-2}, having different prefactors for the distinct QC regimes (19). The calculations also predict a concentration of the bulk semiconductor oscillator strength into single spectral lines of the QD (proportional to the inverse of the QD volume (1) up to the limit of the sum of the oscillator strengths of the constituent atoms (22)). Coupled with their small volumes, this means that QD's are likely to display enhanced linear and nonlinear optical responses to "interband absorption". It turns out that for strong QC, saturation of the lowest transition is a straightforward case of a two-level system dominated by phase-space filling and Pauli exclusion effects (*1*). Coulombic screening and fermion exchange effects prove to be minimal in the strong QC regime. Interband saturation for QD's can induce large changes in absorption and refractive index. Moreover, by embedding the QD in matrices of different dielectric constant (dielectric confinement) one is able to engineer LFE's of high enough strength to produce optical bistability with an intrinsic feedback mechanism (*1*).

An important consideration when estimating the enhancement factor of NLO absorption of QD's compared to quantum well and bulk semiconductor analogues, concerns the broadening mechanisms of the linear absorption properties. Here one is dealing mainly with intrinsic phonon-coupling effects (which spreads the oscillator strength thus reducing the NLO effects) and inhomogeneities in size and shape of the QD (*1*). This is the analogue of the inhomogenious broadening of excitonic absorption in quantum wells due to random fluctuations in constituent layer thickness. Improved fabrication techniques can help control the broadening arising from variations in crystallite size and shape for QD's. This is the distinct advantage of the sodalite and zeolite systems. The rigid size and shape discrimination provided by the host eleminates the complex engineering that will be required to improve upon other semiconductor QD systems.

Clearly it is inadequate to simply describe semiconductor QD's as nanostructures composed of those constituents that make up semiconductors in the bulk. One must specify material dependant size regimes in which the theoretical models are valid for predicting and interpreting the NLO properties of QD's. By computing $\chi^{(3)}$ in different QC regimes one can access the magnitude of the expected optical nonlinearities for a variety of semiconducting materials. As intrazeolite semiconductor QD's most likely fall in the strong confinement regime we will refer mainly to this case.

To obtain estimates of the expected nonlinearities in the regime of strong QC one evaluates from a normalized $\chi^{(3)}$ ($\tilde{\chi}^{(3)}$) the changes in absorption $\Delta\alpha$ and refractive index Δn accompanying excitonic optical absorption in a simple two level system (*19*):

$$\frac{\Delta\alpha}{I} = K_1 \left(\frac{a_B}{R}\right)^3 \operatorname{Im}\tilde{\chi}^{(3)} \qquad \frac{\Delta n}{I} = K_2 \left(\frac{a_B}{R}\right)^3 \operatorname{Re}\tilde{\chi}^{(3)} \qquad (6)$$

where I is the intensity "inside" the QD and K_1 and K_2 are prefactors scaling the absorption and dispersion changes. Immediately one sees that size quantized semiconductor materials with large bulk exciton lengths a_B (narrow band-gap) are attractive candidates for the observation of large NLO effects. (Equation (6) is valid only under the EMA and thus fails for very small isolated clusters.) Some representative materials data from the literature are tabulated below (*19*):

Table II. Representative Data for $\widetilde{\chi}^{(3)}$ of Semiconductor
Materials

	$E_g(eV)$	$E_R(meV)$	a_B (nm)	m_e (nm)	m_h (nm)	$K_1(cm^{-1})$	K_2
GaAs	1.519	4.2	12.4	0.00665	0.052	181	1.18×10^{-3}
CdS	2.583	27	3.01	0.0235	0.135	6	2.30×10^{-5}
CuCl	3.354	152	0.75	0.0500	0.200	0.223	6.88×10^{-6}

These data clearly show that for the above examples, isolated GaAs QD's have the largest NLO effects for a given dot size and that $\Delta\alpha$ and Δn scale with a_B^3/R^3. One finds from the calculations that although $\widetilde{\chi}^{(3)}$ is larger for medium than strong QC, it is the a_B^3/R^3 term that enhances the optical nonlinearities for the strong QC regime (19).

It is concerning this last part that an intrinsic limitation is spotted for the region of strong QC. For the aforementioned semiconductor QD's, strong QC demands dimensions of R(GaAs)~1.4 nm, R(CdS)~0.45 nm, and R(CuCl)~0.15 nm (19). Although one can usefully employ zeolite encapsulation techniques to fabricate and stabilize such truly ultramicroscopic structures, one cannot expect the theory of bulk semiconductors with lattice periodicity, bands and effective electron and hole masses to apply to this novel class of solids. By contrast, the materials requirements for the regime of medium QC are found to be far less stringent than the above, and one can expect the effective mass approximation to be valid in this size regime (19).

Clearly achieving a narrow cluster size distribution is of paramount importance for minimizing inhomogenious broadening of interband excitonic optical absorption, but an equally important consideration is the limitation imposed on the theoretical models of $\widetilde{\chi}^{(3)}$ when the QD's become so minute that their internal properties change compared to the bulk semiconductor parent.

There exists a specific range of sizes for which the effects of QC on the extended electronic states may be large yet the structure remains crystalline in all directions. Here one assumes the EMA is valid for electrons and holes which requires several lattice constants for the spatial extent of the QD. This structure may not be satisfied for many strongly QC semiconducting materials.

While it is true that high population densities of these ultra-small QD's can be assembled in zeolite cavities, the QD may lose the properties of bulk crystallinity with respect to the parent bulk semiconductor. Furthermore the extent of lattice periodicity within a QD is likely to be too small for the EMA to be valid. Indeed a pivotal, and still unanswered question, concerns precisely the number of unit cells required for application of the EMA (it has been shown suitable for crystallites containing as few as 100 atoms (20)). These QD's are therefore not expected to behave like bulk crystallite QD's. However, the presence of uniform arrays of these minute single-size QD's in a host zeolite lattice reintroduces a different kind of periodicity, envisioned in terms of a zero dimensional supralattice of QD's (that is, a 3-D array of QD's). Using the tight binding approximation one can create a miniband picture of such a periodic array of QD's and hence an insight into the collective electronic behaviour of variously coupled QD's (11). Here however, one needs to examine $\chi^{(3)}$ for the "expanded semiconductor" supralattice and how e-h states and effective exciton, electron and hole length scales are modified compared to the bulk semiconductor. A tunneling supralattice composed of QD's may be analogous to the bulk system but with localization effects.

At this early stage of development of the field, one can speculate that it may prove useful to construct an array of zero dimensional QD's into a supralattice to actually further enhance the optical nonlinearities. Here one builds a supralattice with, for instance, narrow minibands. The miniband gaps are likely to be smaller than for the isolated QD's. Presumably one can engineer large non-parabolicities into these narrow bands (11). A new kind of lattice periodicity is created and so the EMA may become applicable to the e-h carriers tunneling or hopping between coupled QD's. The supralattice excitonic length will presumably be less than a_B. Possibly one can gain the benefit of the strong QC of the individual QD's with a beneficial perturbation arising from the collective electronic behaviour of a high population density of coupled QD's.

The primary advantages of IZS systems are the ordering, the size monodispersity, and the high particle number density of the clusters that can be readily attained in the zeolite lattice. Since most of the theory for the NLO effects in the strong QC regime use a two level model, it would seem reasonable to include IZS clusters with these projections if species of sufficient oscillator strength and appropriate electronic energy spacing can be fabricated.

A final point worth mentioning is the effect of local fields on the optical nonlinearities of strongly QC nanostructures. These arise from embedding QD's in a medium of different dielectric constant (1). One requires to know how the field intensity inside the particle varies at saturation in excitonic absorption. This has been approached theoretically by defining a local field factor f such that $E_{in} = f\, E_{out}$ (1). The factor f depends on the shape of the QD and the dielectric constant of the QD $\varepsilon = \varepsilon_1 + i\varepsilon_2$ relative to that of the surrounding medium. Here $f = [1 + 1/3(\varepsilon-1)]^{-1}$ for a spherical QD which yields the local field intensity factor F, such that, $I_{in} = FI_{out}$ and $F = |f|^2 = 9/[(\varepsilon_1 + 2)^2 + \varepsilon_2^2]$. This is used in the expression for the dielectric constant of the QD including the effect of saturation band filling, remembering the relationship of optical constants (α, n) and the complex dielectric constant $(n^2 - k^2 = \varepsilon_1;\ 2nk = \varepsilon_2;\ \alpha = 2\omega k/c)$ (1):

$$\varepsilon = \varepsilon_\infty + \frac{\beta\,(\delta + i)}{1 + \delta^2 + F(I_{out}/I_{sat})} \tag{7}$$

where I_{sat} is the saturation intensity given by $I_{sat} = \hbar/\beta\,\tau V$ with β a line shape factor, τ the recombination lifetime and V the volume of the QD.

Only in the region of saturation absorption does one observe steep variations in F. On one side of the exciton resonance the field is very strongly concentrated inside the particle and on the other side the field hardly penetrates into the particle (this line shape is called the Fano interference profile (*1*)). Clearly these dramatic alterations in local fields at an excitonic absorption will enhance any sensitivity of the dielectric constant to the applied field. In the above equation one realizes an intrinsic feedback mechanism that can lead to regions of optical bistability and therefore possible applications in optical switching. Because of saturation band filling effects the dielectric constant inside the QD depends on I_{out}. Hence changes in I_{out} will modify F, which will in turn change I_{in}, ε, and so on. This implies a strong modulation of the absorption for a narrow range of intensities and can lead to the characteristic S-shaped ε_2 versus I/I_{sat} functions typical of an optically bistable switching transition (*1*). These ideas apply to a single QD embedded in a solid matrix. Collective effects are expected to appear in a zero dimensional superlattice and have been described using conventional electromagnetic theory (*1*).

Summary

All of the existing theories in the literature of the NLO properties of semiconductor QD's concern atomically "perfect" usually isolated QD's. Now that the goal of perfect monodispersity and organization has been realized for intrazeolite QD's and QS's, pivotal questions have been raised regarding the all-optical nonlinearities of such strongly quantized particles and their (interacting) periodic arrays in dielectric supports. The possibility of unveiling new and desirable optical properties for intrazeolite semiconductor QD's and QS's will hopefully encourage further research in this exciting new dimension of solid-state chemistry.

Acknowledgments

We wish to acknowledge the Natural Sciences and Engineering Research Council of Canada's Operating and Strategic Grants Programmes, the Office of Naval Research (GDS) as well as Alcan Canada for generous financial support of this work. (SO) expresses his gratitude to the Middle East Technical University for granting him an extended leave of absence to conduct his research at the University of Toronto. (AS,SK) would like to thank N.S.E.R.C. for graduate scholarships. Acknowledgements also go to Drs. Andrew Holmes, Richard Prokopowicz, Peter Macdonald, Hellmut Eckert, Jim MacDougall, Bob Ramik, David Creber, Bob Lazier, Peter Lea, Battista Calvieri, Stuart McIntire and Bill Mercer for their expertise and advice in carrying out some of the analytical measurements referred to in this paper. We also thank all of our coworkers at Toronto for many stimulating and enlightening discussions during the course of this work.

Literature Cited

1. Schmitt-Rink, S; Miller, D.A.B.; Chemla, D.S. *Phys. Rev B* **1987**, *35*, 8113.
2. Alivisatos, A.P.; Harris, A.L.; Levinos, N.J.; Steigerwald, M.L.; Brus, L.E. *J. Chem. Phys.* **1988**, *89*, 4001; (and references cited therein).
3. Stein, A.; Ozin, G.A.; Stucky, G.D. *J. Am. Chem. Soc.* **1990**, *112*, 904.
4. Wang, Y.; Herron, N. *J. Phys. Chem.* **1987**, *91*, 257.
5. Herron, N.; Wang, Y.; Eddy, M.M.; Stucky, G.D.; Cox, D.E.; Moller, K.; Bein, T. *J. Am. Chem. Soc.* **1989**, *111*, 530.
6. MacDougall, J.E.; Eckert, H.; Stucky, G.D.; Herron, N.; Wang, Y.; Moller, K.; Bein, T. *J. Am. Chem. Soc.* **1989**, *111*, 8006.
7. Bogomolov, V.N.; Kholodevich, S.V.; Romanov, S.G.; Agroskin, L.S. *Solid State Commun.* **1983**, *47*, 181.
8. Barrer, R.M. *Hydrothermal Chemistry of Zeolites*; Academic Press: London, U.K., 1982.
9. Steigerwald, M.L.; Brus, L.E. *Ann. Rev. Mater. Sci.* **1989**, *19*, 471; (and references cited therein).
10. Henglein, A. *Topics in Current Chemistry* **1988**, *143*, 113.
11. Jaros, M. *Physics and Applications of Semiconductor Microstructures*; Oxford University Press: Oxford, U.K., 1989.
12. Hoffman, R. *Solids and Surfaces: A Chemist's View of Bonding in Extended Structures*; VCH Publishers Inc.: New York, U.S.A., 1988.
13. Felsche, J.; Luger, S. *Ber. Bunsenges. Phys. Chem.* **1986**, *90*, 731.
14. Godber, J.; Ozin, G.A. *J. Phys. Chem.* **1988**, *92*, 4980.
15. Smeulders, J.B.A.F.; Hefni, M.A.; Klaasen, A.A.K.; de Boer, E. *Zeolites* **1987**, *7*, 347.
16. Hassan, I.; Grundy, H.D. *Acta. Cryst.* **1984**, *B40*, 6.
17. Dempsey, M.J.; Taylor, D. *Phys. Chem. Miner.* **1980**, *6*, 107.
18. Breck, D.W. *Zeolite Molecular Sieves*; Krieger Publishing Co.: Malabar, U.S.A. 1984; via reference (4).
19. Banyai, L.; Hu, Y.Z.; Lindberg, M.; Koch, S.W. *Phys. Rev. B* **1988**, *38*, 8142.
20. Takagahara, T. *Phys. Rev. B* **1989**, *39*, 10206.
21. Efros, Al.L.; Efros, A.L. *Soviet Physics Semiconductor* **1982**, *16*, 772.
22. Green, B.I.; Orenstein, J.; Schmitt-Rink, S. *Science* **1990**, *247*, 679.

RECEIVED August 2, 1990

Chapter 38

Small Semiconductor Particles

Preparation and Characterization

Norman Herron

Central Research and Development Department, E. I. du Pont
de Nemours and Company, Wilmington, DE 19880–0328

The construction of discreet particles of
semiconductors, either inside porous hosts or as
free-standing, surface-capped clusters, has
generated a new class of materials where
confinement effects on the semiconductor optical
properties are pronounced. In porous zeolite
hosts, in addition to the size-quantization effects,
novel intercluster phenomena become manifest
as the individual semiconductor clusters reach a
volume density above the percolation limit and
begin to interact three-dimensionally. This
interaction is modulated by the zeolite
framework topology and hence leads to an
ordered array of clusters in what we have
termed cluster crystals. Novel absorption,
emission and excitation behaviors of these
materials, dominated by defect sites, result.
Detailed characterization of the semiconductor
species responsible (by x-ray powder diffraction
and EXAFS) reveal a cubane like $(CdS)_4$ unit as
the basic building block of the structure. The
random porosity but good optical properties of
sol gel glasses allow the generation of optical
materials containing related quantum-dot
semiconductor clusters (prepared by
organometallic means) where now, nonlinear
optical (χ^3) properties have been estimated.
Finally, non-resonant non-linearity has been
observed in free-standing surface-capped

0097–6156/91/0455–0582$06.00/0
© 1991 American Chemical Society

semiconductor clusters of CdS, CdSe and CdTe whose surfaces have been terminated and passivated using thiophenolate ligands. These latter materials are highly soluble in polar organics and therefore processable into thin films or polymer composites. The control of cluster size available via synthesis conditions in these materials makes possible a unique and systematic study of optical properties as a function of cluster size and thus of quantum confinement.

Small metal and semiconductor clusters, having hybrid molecular and bulk properties, represent a new class of materials and are under intensive investigation1. The basic problem facing researchers in this area is the control of surface reactions of such particles so as to arrest their growth at the small cluster stage. Many approaches have been explored for the preparation of these small clusters including the use of micelles2, colloids3, polymers4 and glasses5 to control the aggregation problem. In all cases, however, the cluster sizes and crystallinities are poorly defined and one would like to find an approach to this class of materials which produces a mono-dispersion of cluster sizes in a well defined and characterizable array. These criteria would seem well met by an inclusion type approach using a porous host lattice as the template within which the clusters could be constructed and confined. Alternately, one may use synthetic chemistry to control the cluster surface such that it is terminated by capping groups which both passivate the cluster electronically and prevent its further aggregation and growth. This paper describes our efforts in both directions and includes: 1) the synthesis and characterization of CdS clusters in zeolites Y, X and A; 2) the preparation of a variety of semiconductors in the "poor-man's zeolite" - porous glass and 3) use of thiophenol capping chemistry to generate free-standing passivated soluble clusters of CdS and CdSe. The resulting effects of size-confinement on the semiconductor optical and nonlinear optical properties will be described.

Why small semiconducting particles are interesting
It is important to understand why there is the current interest in very small particles of semiconductors1 and what limitations this interest places on the nature of the materials.

The concept of an all-optical or opto-electronic computer technology has attracted attention because of its potential for extreme speeds and parallel processing capabilities in such areas as image recognition. Such a technology requires several basic optical materials for the construction of devices which mimic their electronic counterparts. One such fundamental computing element is the optical transistor or bistable device which acts as a light switch or valve/amplifier. Basic requirements placed on materials for such a device are that they have a very rapid switching speed (ideally picosecond) and extreme photostability in order to perform trillions of switching operations/sec for years at a time. One realization of such a material could involve the use of third-order nonlinear optical properties, χ^3, to effect a transient refractive index change. Illumination of such a material with intense laser light will cause a change of its refractive index leading to a switch from an opaque to transmissive state in an interferometer-type bistable device. While semiconductor materials themselves will perform this kind of switching at their band-edge wavelengths, the speed of the effect is slow - usually as a consequence of a long free-carrier lifetime. This speed can be increased by providing more sites for efficient removal of these free-carriers - in other words more defect sites. One can view surface sites on a semiconductor particle as such defect sites and one simple way to increase their concentration is to go to very small particles. The commercial color filters of Schott and Corning based on CdS/Se nanoparticulates in a silica matrix have verified the utility of this kind of material for nonlinear- optical devices 5. We would like to explore a wide range of other semiconductors and matrices for these purposes and the zeolite host provides an almost ideal starting point.

CdS in zeolite Y6

The zeolite Y occurs naturally as the mineral faujasite and consists of a porous network of aluminate and silicate tetrahedra linked through bridging oxygen atoms (Figure 1). The structure consists of truncated octahedra called sodalite units arranged in a diamond net and linked through double six-rings7. This gives rise to two types of cavity within the structure - the sodalite cavity of ~6.6Å free diameter with access through ~2.3Å windows and the supercage of ~13Å diameter with access through ~7.5Å windows. Whenever an aluminum atom occurs in the framework this introduces one negative charge onto the zeolite skeleton which is compensated by loosely attached cations which give rise to the well known ion-exchange properties of zeolites.

Cadmium ion-exchange of the zeolite is carried out by slurrying 10g of zeolite LZY-52 (sodium zeolite Y from Linde) in 1L of distilled water and the pH is adjusted to 5 with nitric acid. A calculated amount of cadmium nitrate designed to give a specific exchange level is stirred into the slurry and the mixture is stirred at room temperature overnight. Collection of the exchanged zeolite by filtration and washing with distilled water is followed by drying and calcination. The powder is heated to 400°C at 3°/min in flowing dry oxygen (100mL/min) then cooled in vacuo to 100°C. The zeolite is then exposed to flowing hydrogen sulfide (40mL/min) at 100°C for 30min. Finally the still white zeolite is evacuated at 100°C for 30mins then sealed and transferred to an inert atmosphere dry box for handling and storage. The zeolite turns pale yellow/cream during the final evacuation step. All zeolites prepared in this manner are moisture sensitive becoming deep yellow (zeolite Y or X) or pale yellow (zeolite A) on prolonged exposure to atmospheric humidity.

Chemical analysis confirms Cd and S are present in from 0 to 25wt% depending on exchange conditions. XPS shows that there is no detectable Cd on the exterior surface of the zeolite crystallites. IR spectra show no SH groups but there are the expected OH groups attached to the zeolite framework6. The exact nature of the CdS cluster units is revealed by a combined application of optical spectroscopies and x-ray techniques.

Structure of CdS in Y and its Optical Consequences.
Detailed analysis of the powder x-ray diffraction data on a series of CdS loaded zeolite Y samples reveals the fundamental CdS cluster present consists of interlocking tetrahedra of Cd and S atoms (although some of the S atoms are occasionally replaced by O atoms6) forming a distorted cube (Cd-S = 2.47Å) (Figure 2). This structure, which is heavily dictated by the zeolite symmetry, is confirmed by EXAFS data at the Cd edge which reveals the local symmetry and coordination environment of the Cd6. To our initial surprise, these Cd_4S_4 cubes were not located in the supercages of the Y structure but were instead sited within the smaller sodalite cages. In retrospect this location is entirely reasonable since these cages are the preferred sites for the original Cd ions upon exchange8

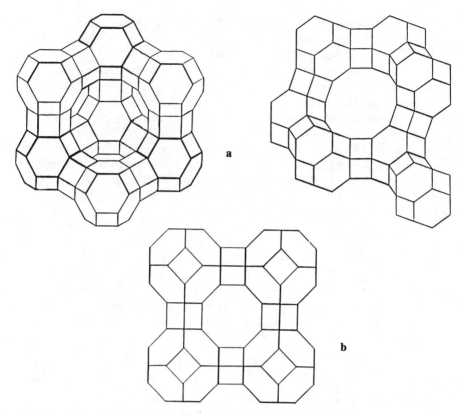

Figure 1.   Representative zeolite structures where the open framework is represented by sticks joining the Si or Al atoms. Oxygen bridge atoms lie roughly at the mid-point of these atoms and are omitted for clarity. a) zeolite Y   b) zeolite A

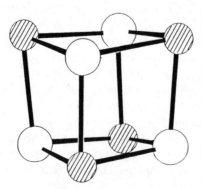

Figure 2.   Structure of the $(CdS)_4$ unit located within the zeolite sodalite units. (hatched circles = Cd; open circles = S).

and all that needs to occur upon exposure to the H_2S is that the cube zips together. The Cd ions of the ideal cluster are in octahedral coordination to 3 sulfur atoms of the cube and 3 oxygen atoms of the zeolite framework six-ring window (Figure 3). The sodalite cage seems to have been made for this CdS cluster!. (for structural details of this material, the reader is referred to ref. 6)

The evolution of the optical spectra as a function of CdS loading density is particularly revealing. At loading densities of <5wt% the sample has an optical absorption spectrum with a shoulder at 280nm. This represents a shift of the band edge by ~230nm and correlates well9 with the extremely small clusters Cd_4S_4 found in the x-ray analysis. However, as the loading rises above 5wt% a new absorption feature at 350nm begins to appear and grows as the loading increases up to the maximum yet attained of ~25wt%. While still quantum confined CdS, the structural nature of the material responsible for this new absorption feature was unknown. The band edge observed is similar to that of colloidal CdS particles with 25-30Å diameters9 yet there are no void spaces of this kind of dimension within zeolite Y. The real clue as to the nature of this species comes from the threshold loading density of 4±1wt% - this corresponds to the concentration at which, statistically, CdS cubes must now populate adjacent sodalite units. Simple calculations based on the bulk density of CdS and the sodalite pore volume of Y reveal that the 4wt% loading density corresponds to filling of ~14% of the available sodalite volume. Percolation theory10 predicts the percolation threshold in such systems to be at ~15volume% ie. above 5wt% of CdS the individual clusters must begin to interact with one another in adjacent sodalite cages in a percolative fashion. In the limit, when all sodalite units are occupied (~28wt% CdS) the structure would be such as that represented in Figure 3 and a cluster crystal of CdS dictated by the zeolite Y topology results.

This unique arrangement of individual clusters into an interconnected network gives rise to the luminescence and excitation spectra shown in Figure 4. As anticipated, the luminescence is not band-gap recombination in nature but rather is dominated by Cd related defects.

Since the novel optical behavior at higher loading densities of CdS is a consequence of the interaction between

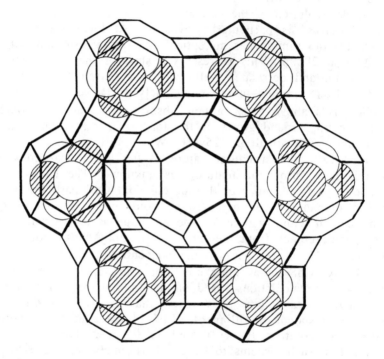

Figure 3.   The hyperlattice arrangement of CdS clusters in adjacent sodalite units of the zeolite Y. (hatched = Cd; open = S).

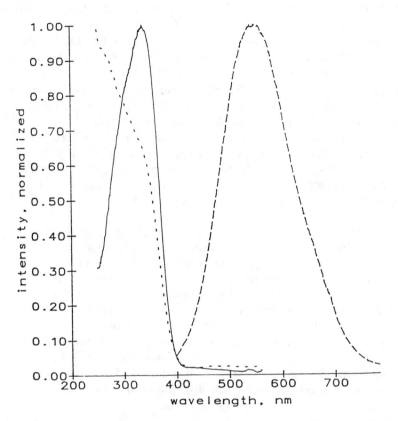

Figure 4.   Absorption (dotted), excitation (solid) and emission (dashed) spectra of 6.5wt% CdS in zeolite Y. The absorption spectrum was taken at room temp and the others at 77K.

individual clusters as modulated by the zeolite structure it would be interesting to study a zeolite with a different connectivity between the same kinds of CdS/sodalite clusters. In this case the spectral behavior should be very similar to zeolite Y at low loading densities but differ significantly as the interconnected CdS cluster crystal develops. Zeolite A is just such a zeolite system7. In this case, the sodalite cages are connected via double 4-rings (Figure 1) in a cubic arrangement and the distance between Cd_4S_4 cubes is projected at ~9Å ie. the cluster-cluster interaction must be much weaker in A than in Y. The optical behavior follows these predictions precisely. While at low loading the absorption spectra of A and Y look very similar (isolated clusters) as the loading increases there is no shift in the absorption edge of the A zeolite material indicating a much weaker interaction between clusters than in Y and so no significant development of an electronically cogent hyperlattice.

Zeolites provide a novel host for the generation of semiconductor hyperlattices within their pore volume. The control of the connectivity between the clusters of semiconductor is unparalleled in any other host medium and so has allowed a detailed study of the optical consequences of such connectivity6. However, from the practical standpoint, such materials have some severe drawbacks - most notably the lack of single crystals of sufficient size to produce viable optical devices such as optical transistors or spatial light modulators. We have therefore moved on to look at more practical/processable quantum-dot materials such as semiconductor-doped porous glasses.

Porous glass
Porous glass is available from a variety of synthetic approaches and we have explored sol-gel derived silica11 . This material consist of macroscopic (centimeter dimensions) monoliths of optically clear (when treated correctly) silica glass, into the pores of which semiconductors may be introduced using chemistry very similar to that which was used for the zeolites described above12. The pores of such glasses range in size from the zeolite range (~10Å) to as large as several hundred Å controlled largely by the pH of hydrolysis during synthesis13. Unlike zeolites this pore structure is not part of a crystalline matrix and therefore has a distribution of sizes and connectivities (Figure 5). One is

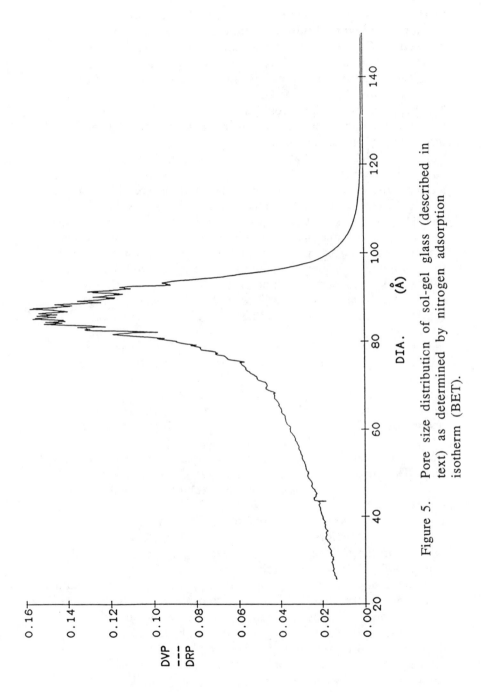

Figure 5.   Pore size distribution of sol-gel glass (described in text) as determined by nitrogen adsorption isotherm (BET).

therefore forced to compromise regularity of pore size and topology in favor of ultimate optical and device related properties. One synthetic advantage to the sol-gel porous glasses is that metal ions may be introduced to the original fluid/sol stage of the synthesis and remain available for conversion to the semiconductor after the rigid glass matrix has been formed. This provides for excellent uniformity for the semiconductor dispersion throughout the glass over macroscopic pieces of material.

A typical preparation11,14 of a glass/semiconductor composite begins with the glass synthesis as follows. 22.31ccs tetraethylortho- silicate and 23.48ccs ethanol are mixed in a plastic beaker and stirred vigorously. A second solution consisting of 7ccs water, 0.25ccs hydrofluoric acid (40%) and optionally a soluble metal ion nitrate salt (eg cadmium, zinc, lead etc nitrates) is added to the first and the mixture stirred for 5minutes at room temperature. The resultant clear fluid solution is then cast into polypropylene vials and tightly capped (vials are ~2cm diameter and are typically filled to a depth of ~5mm). After standing thus at room temperature for 24 hours the sol has set to a rigid wet gel . Drying of the gel is performed by loosening (but not removing) the caps and placing the gels in a forced draft furnace set at 50°C. Over a period of 2 days the gels dry and shrink to ~ 30% of their original volume. The resultant disks are finally calcined to remove all residual solvent, unreacted organics and to decompose any nitrate salts added. A typical temperature profile uses a controlled ramp to 600°C over 12 hours in flowing dry air followed by a 6 hour soak at 600°C also in dry air . The recovered disks are typically ~1cm diameter and 2-3mm thick and are usually colorless (unless a colored metal ion was included) except for a slight bluish haze produced by light scattering from the pores of the structure.

Semiconductors have been produced in this type of glass by one of three different ,but related, routes: 1) If a metal ion was introduced into the original sol of the glass synthesis it may be converted to the appropriate semiconductor by simple annealing of the glass disk in an atmosphere of reactive gas. For example CdS may be prepared this way by taking a Cd containing glass and heating in 500torr hydrogen sulfide gas. The temperature of synthesis and length of time at that temperature dictates to some extent how large the semiconductor crystallites will grow ie how marked is the quantum size effect. Of course the largest size of cluster possible in a given glass is ultimately controlled by the glass pore size itself. X-ray powder diffraction of the product gives a good estimate of the cluster size produced. Materials produced by this method can have extremely interesting optical behaviors - for example size-quantized dilute magnetic semiconductors such as Mn doped ZnS can be produced in this manner (see Figure 6).   2) A pure silica disk may have a semiconductor generated in the pores by an impregnation approach. For example CdS may be produced by first impregnating a disk with a methanolic solution of cadmium nitrate, drying to 100°C in flowing air then exposure to 500torr hydrogen sulfide exactly as in 1 above. While this approach can lead to very high loading densities of the semiconductor in the glass pores, it suffers from severe problems of homogeneity (or rather lack of it) - it is very difficult to maintain a uniform dispersion of the semiconductor throughout a large monolith using an impregnation approach. 3) The final method is one we14 and others12 have developed to introduce III-V semiconductors such as GaAs. It is a variation of method 2 above and makes use of MOCVD-type chemistry which is widespread in the semiconductor industry for production of nanostructures (quantum wells and epitaxial materials) of III-V material. A pure silica disk is first exposed to a the vapor of a volatile group-III trialkyl, eg. trimethyl gallium, on a vacuum line followed by annealing in an atmosphere of phosphine or arsine gas.   Again x-ray diffraction of the resultant material reveals the presence of small semiconductor crystallites (Figure 7).

Once the semiconductor/glass composite has been produced by these means it is desirable to remove all of the residual porosity and produce a dense material. This has several benefits 1) removing porosity removes the residual light scattering from the pores themselves and vastly improves the optical clarity11. 2) mechanical properties are likewise vastly improved and the previously very fragile porous glass becomes sufficiently robust to survive cutting and polishing operations as would normal glass. 3) contaminants from the atmosphere are no longer absorbed into the pores and the semiconductor clusters are encased in an inert matrix where they are much more stable with respect to oxidation and aggregation. The approach we have taken is to impregnate the porous glass/ semiconductor composite in

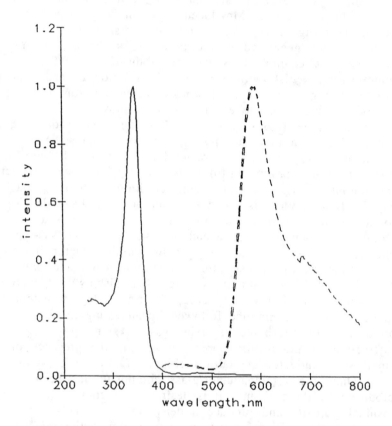

Figure 6.   Excitation (solid line) and emission (dotted line) spectra of $Zn_{0.93}Mn_{0.07}S$ prepared in porous sol-gel glass by method 1 in text.

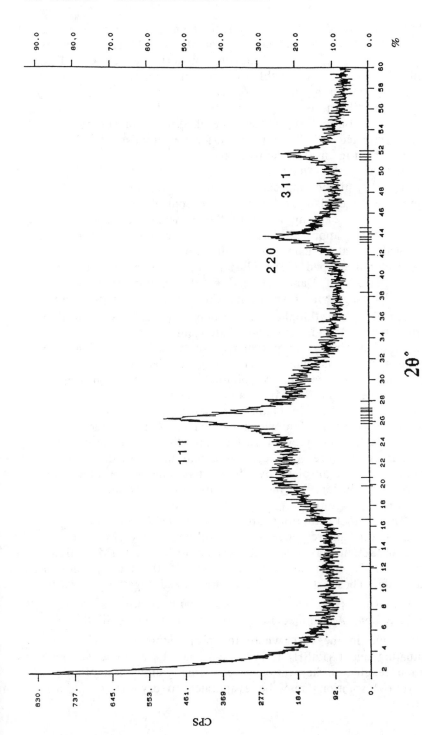

Figure 7. Powder x-ray diffractogram of ~80Å crystallites of InP in porous sol-gel glass.

methylmethacrylate monomer and to then polymerize the MMA using VAZO-64 radical initiator at 60°C11. This produces a dense polymer/glass/semiconductor material which can be machined into a form suitable for optical devices.

While research into the optical behaviors of these materials is still in its early stages we have begun to look at the nonlinear optical properties via absorption bleaching. Results are listed in Table I and offer the promise of materials having useful non-linearities.

Surface capped clusters

As a final synthetic approach to soluble, quantum-confined clusters we have also explored cluster control via covalent attachment of a capping agent to the growing cluster surface15. This approach has previously been explored by Steigerwald et al.16 using a micellar approach to benzeneselenol capped CdSe clusters and even earlier in the pioneering work of Dance et al.17 on the rational synthesis of a molecular fragment of sphalerite CdS where a $Cd_{10}S_4$ core was capped by 16 thiophenolate groups. We have viewed the reaction between cadmium and chalcogenide ions as an example of an inorganic polymerization with an initiation step forming a small nucleus which then grows larger in a propagation step before precipitating as a termination step. With this viewpoint it is interesting to consider what will happen if a propagation chain terminator - some chemical such as thiophenolate ion which can attach to the surface of a growing cluster and prevent further growth - is added to the polymerizing mixture. One can anticipate that growth of the clusters will be controlled by the relative kinetics of attack on the growing clusters by the propagation reagents and the termination reagent.

Three stock solutions are prepared inside an inert (nitrogen) atmosphere glove box as follows: A) 0.1M cadmium acetate in 80%methanol and 20% acetonitrile; B) 0.1M sodium sulfide in 50% water and 50% methanol; C) 0.2M thiophenol in acetonitrile. These stock solutions are mixed together in ratios such that $[Cd^{2+}]x2 = [S^{2-}]x2 + [PhSH]$ with vigorous stirring in the glove box. An immediate yellow color develops in the solution and in the case where the propagating agent (S^{2-}) to terminating agent (PhSH) ratio is less than 1 a stable yellow solution results, higher ratios give a fine yellow precipitate. The yellow solutions may be evaporated to dryness giving a bright yellow solid which has excellent solubility in DMF, DMSO, acetonitrile, acetone or THF. Chemical analyses of these powders15 reveal that they consist entirely of Cd, S and PhSH. Simply by varying the S/SPh ratio of this preparation it is possible to produce a continuous range of materials having steadily red- shifting band-edges in the visible spectrum as the S/SPh ratio increases (Figure 8) and whose particle sizes can be reasonably estimated using the broadening of the x-ray powder diffraction lines15. Using this synthetic approach we have found that soluble particles with a sphalerite core size up to 40Å can be produced in large quantities (10's of grams) and so represent one of the few examples of a bottleable 'reagent' consisting of a controlled-size, quantum-confined semiconductor.

By exercising extreme care in controlling the mixing conditions (low concentrations and vigorous mixing) and component ratios during these types of syntheses it is possible to generate solutions of CdS, CdSe and CdTe clusters all capped by thiophenolate ion and which exhibit very sharp absorption features at the band-edge in their electronic spectra (Figure 9). Such sharp, probably excitonic, features are of great interest with respect to inducing large optical non-linearities in these materials and probably indicate that this synthetic approach is capable of producing clusters with a very narrow size distribution. The fact that such clusters can be kept in solutions of organic solvents has also meant that high quality thin films may be spin-coated onto quartz substrates from solutions containing a soluble polymer such as PMMA. This highly desirable processability makes further investigations of this class of material very attractive.

We have taken advantage of the high solubility of these clusters and the variability of the cluster size by simple synthetic manipulation to investigate their non-resonant non-linearity as a function of cluster size18. The experiment involves frequency tripling a 1.91μm incident beam to 637nm through a long path length (5cm) of solution. Results are shown in table II and indicate that the intrinsic nonresonant χ^3 of these CdS clusters is very large and that the nonlinearity increases with the cluster size. Measurements of this type are still in progress to determine where the maximum in cluster size vs. nonlinearity occurs.

Table I.   Photobleaching of semiconductors in sol-gel glass. All measurements are done with a laser power of between 1 and 3 MW/cm^2

Sample	Wavelength(nm)	abs change (%)
GaP	540	-10.7
PbS	625	-6.8
In_2S_3	625	-6.3
In_2Se_3	540	+5.0
	570	+2.6
	625	-5.7
CdS	450	-22.0
CdS in Corning glass	500	-29.0

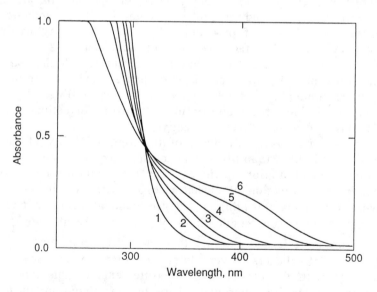

Figure 8.   Absorption spectra of CdS/SPh clusters in acetonitrile solution with [Cd] = 4 x 10^{-3}M in 1mm path-length cells. (1) $(NMe_4)_4(Cd_{10}S_4SPh_{16})$; (2) idealized S/SPh ratio used in preparation, (see text) 0.33; (3) 0.5; (4) 0.75; (5) 1.17; (6) 2.0.

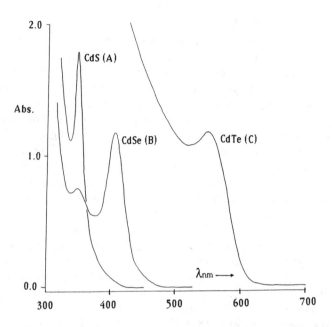

Figure 9.   Solution spectra of A) CdS/SPh; B) CdSe/SPh; C) CdTe/SPh   all in DMF solvent in 1 mm path length cell at [Cd] = 1.25 x 10^{-4} M.

Table II.   Third-order nonresonant optical nonlinearity of thiophenolate-capped CdS clusters. Units of χ^3 and γ are in esu. The χ^3 of pure acetonitrile and DMF are 2.8 x 10^{-14} and 5.8 x 10^{-14} esu, respectively

Sample	χ^3 comp	Mole frac.	χ^3 CdS	γ x 10^{36}
~30Å CdS in acetonitrile	3.3 x 10^{-14}	1.84 x 10^{-5}	~3.2 x 10^{-10}	~10^4
~15Å CdS in DMF	9.1 x 10^{-14}	1.23 x 10^{-3}	~2.3 x 10^{-11}	~730
Cd$_{10}$S$_4$(SPh)$_{16}$ in DMF	8.3 x 10^{-14}	5.34 x 10^{-3}	4.7 x 10^{-12}	200

Conclusions and Directions

Several approaches to quantum-confined semiconductor clusters have been explored and have emphasized the importance of careful control of the cluster surface chemistry. Control based on physical confinement in zeolites or porous glass gives composite materials revealing fundamental details of the cluster/cluster interactions and proffer the hope of generating practical materials for optical switching devices. Clusters having covalently attached surface caps can be prepared as soluble size-graded reagents and allow detailed characterization of both resonant and nonresonant nonlinearities while maintaining a flexibility with respect to device construction via incorporation in a support matrix such as a polymer or other porous support. Further research in this area will continue to explore 1) preparation of a range of materials with steadily increasing cluster sizes so that NLO properties can be assessed as a function of this parameter and 2) methods of synthesis or separation which allow the production of extremely narrow size distributions of these variously sized clusters. In this way, the intrinsic properties of a specific cluster dimension rather than an average over a range of dimensions can be assessed. From the perspective of a synthetic chemist the challenges of this area present a rich vista for the future.

Acknowledgments:

I would like to acknowledge the excellent collaboration with Dr. Y.Wang, whose contributions to the photophysics presented in the above manuscript were immense. The fine technical assistance of J. B. Jensen and S. H.Harvey is gratefully acknowledged. Collaborations with Dr. J. E. MacDougall on synthesis and photocatalysis, Drs. M. M. Eddy and G. D. Stucky on x-ray powder analysis, Drs. K. Moller and T. Bein on EXAFS, Dr H. Eckert on MAS nmr and Dr. L. Abrams on pore size distribution by BET were key contributions to this work.

References.

1)   Brus, L. E. J. Phys. Chem 1986, 90, 2555 and references
       therein
       Brus, L. E. J. Chem. Phys. 1984, 80, 4403.

2)   Weller, H; Schmidt, H. M.; Koch, U.; Fojtik, A.; Baral, S.;
       Henglein,   A; Kunath, W.; Weiss, K.; Dieman, E. Chem.
       Phys. Lett., 1986, 124, 557.

3)   Tricot, Y-M.; Fendler, J. H. J. Phys. Chem. 1986, 90, 3369.

4)   Wang, Y.; Mahler, W. Opt. Comm., 1987, 61, 233.

5)   Borelli, N. F.; Hall, D. W.; Holland, H. J.; Smith, D. W. J. Appl. Phys. 1987, 61, 5399.

6)   Wang, Y.; Herron, N. J. Phys. Chem., 1988, 92, 4988. Herron, N.; Wang, Y.; Eddy, M.; Stucky, G. D.; Cox, D. E.; Bein, T.; Moller, K. J. Am. Chem. Soc., 1989, 111, 530.

7)   Breck, D. W. Zeolite Molecular Sieves; Wiley: New York, 1974.

8)   Calligaris, M.; Mardin, G.; Randaccio, L.; Zangrando, E. Zeolites, 1986, 6, 439.

9)   Rossetti, R.; Hull, R.; Gibson, J. M.; Brus, L. E. J. Chem. Phys. 1985, 83, 1406.

10) Kirkpatrick, S., Rev. Mod. Phys., 1973, 45, 574 and references therein.

11) Pope, E. J. A.; Asami, M.; Mackenzie, J. D. J. Mater. Res. 1989, 4, 1018.

12) Luong, J. C.; Borelli, N. F. Mat. Res. Soc. Symp. Proc. 1989, 144, 695.

13) Shafer, M. W.; Awschalom, D. D.; Warnock, J. J. Appl. Phys. 1987, 61, 5438.

14) Herron, N.; Wang, Y. U.S. Patent Application #07/315,630 filed 2/24/89.

15) Herron, N.; Wang, Y.; Eckert, H. J. Am. Chem. Soc. 1990, 112, 1322.

16) Steigerwald, M. L.; Alivasatos, A. P.; Gibson, J. M.; Harris, T. D.; Kortan, R.; Muller, A. J.; Thayer, A. M.; Duncan, T. M.; Douglass, D. C.; Brus, L. E. J. Am. Chem. Soc. 1988, 110, 3046.

17) Dance, I. G.; Choy, A.; Scudder, M. L. J. Am. Chem. Soc. 1984, 106, 6285.

18) Cheng, L-T.; Herron, N.; Wang, Y. J. Appl. Phys. 1989, 66, 3417.

RECEIVED August 24, 1990

Chapter 39

Synthetic Approaches to Polymeric Nonlinear Optical Materials Based on Ferrocene Systems

Michael E. Wright and Steven A. Svejda

Department of Chemistry and Biochemistry, Utah State University,
Logan, UT 84322–0300

Synthetic strategy has been developed that permits the
efficient preparation of organometallic polymers which
have structural characteristics necessary for
nonlinear optical behavior. Complex $\{\eta^5\text{-}C_5H_4CH_2OH\}$-
$\{\eta^5\text{-}C_5H_4CH=C(CN)CO_2Et\}Fe$ (4) was prepared and
homopolymerized through a thermally induced
transesterification-polycondensation reaction.
Compound 4 was also attached to chloromethylated
polystyrene through the hydroxyl moiety to afford a
pendant-type ferrocenyl NLO polymer.

The importance and relevance of nonlinear optical (NLO) materials
has been discussed in detail in previous sections of this symposium
series and will not be duplicated herein. NLO materials can be
divided into several main types such as organic, polymeric,
organometallic, semiconductor, and so on. The topic of this paper
is polymeric-organometallic NLO materials.
    Certain crystalline ferrocene complexes were very recently
shown to have second harmonic generation (SHG) activity 200 times
that of urea (1-3). Ferrocene serves as the electron-donating
portion in the NLO complex. The attractive features of ferrocene
based NLO materials are: a) air- and thermal-stability; b) photochemical stability; c) synthetic versatility. An undesirable
feature associated with nearly all organometallic NLO materials is
their often strong metal to ligand charge transfer (MLCT) absorption
band observed in the visible light region.
    Polymeric NLO materials are easily fabricated into useful NLO
devices and thus have certain advantages over other NLO materials.
The incorporation of NLO organometallic species into a polymeric
material has yet to be reported. This paper illustrates effective
synthetic strategies to incorporate ferrocenyl based NLO groups.
The strategy permits both pendant-type polymeric materials and main-
chain homopolymers to be synthesized.

0097–6156/91/0455–0602$06.00/0
© 1991 American Chemical Society

Ferrocene Based Main-Chain Homopolymer Synthesis

We recently published synthetic methodology to selectively functionalize the cyclopentadienyl rings of ferrocene (**4**) based on the Seyferth transmetalation reaction (**5**). Starting from complex 1 (**4**) monomer 4 was prepared in 65% overall yield. Ferrocene 4 is isolated as long purple needles (mp 81-2 °C) after crystallization from chloroform/pentane [λ_{max} in CH_2Cl_2, 515 nm ($\epsilon = 2.24 \times 10^3$) and 322 nm ($\epsilon = 1.37 \times 10^4$)].

Complex 4 was melt-casted (sodium chloride plates) with concomitant homopolymerization at 200 °C for 36 h under a nitrogen atmosphere. Polymer 5 was insoluble in common organic solvents. Polymer 5 exhibited a T_m of 360 °C (sealed capillary tube) with no apparent signs of decomposition.

Pendant-Type Ferrocene Based NLO Polymers

Complex **2** also proved valuable in attaching the ferrocene unit to chloromethylated polystyrene. Treatment of chloromethylated styrene (18% of the phenyl rings modified) with **2** under basic conditions afforded polymer **6**. Polymer **7** was then prepared and found to have UV-Vis spectroscopic data analogous to monomer **4** and polymer **5**.

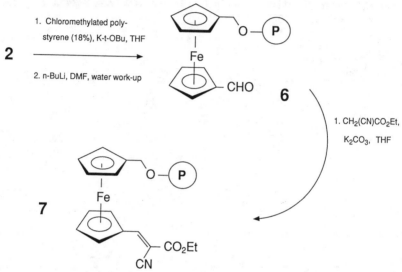

This work provides synthetic routes into polymeric-organometallic NLO materials based on some new and exciting ferrocene chemistry. Electric poling and SHG studies on these new materials will allow us to contrast the relative effectiveness of the two approaches (i.e. pendant versus main-chain).

Acknowledgment

MEW is grateful to the donors of the Petroleum Research Fund, administrated by the American Chemical Society, for their generous support of this work.

References

1.   Green, M. L. H.; Marder, S. R.; Thompson, M. E. Nature 1987, 330, 360.
2.   Bandy, J. A.; Bunting, H. E.; Green, M. L. H.; Marder, S. R.; Thompson, M. E.; Bloor, D.; Kolinsky, P. V.; Jones, R. J. Organic Materials for Non-linear Optics; Hann, R. A., Bloor, D., Eds.; Spec. Publ. No. 69, The Royal Society of Chemistry: London, 1989, p. 219.
3.   Marder, S. R.; Tiemann, B. G.; Perry, J. W. In New Materials for Nonlinear Optics, Marder, S. R.; Stucky, G.; Sohn, J., Eds.; ACS Symposium series, this book.
4.   Wright, M. E. Organometallics 1990, 9, 853.
5.   Seyferth, D.; Weiner, M. A. J. Am. Chem. Soc. 1961, 83, 3583.

RECEIVED August 2, 1990

Chapter 40

Transition Metal Acetylides for Nonlinear Optical Properties

Todd B. Marder[1], Gerry Lesley[1], Zheng Yuan[1], Helen B. Fyfe[1],
Pauline Chow[1], Graham Stringer[1], Ian R. Jobe[1], Nicholas J. Taylor[1],
Ian D. Williams[2], and Stewart K. Kurtz[2]

[1]Department of Chemistry, University of Waterloo, Waterloo,
Ontario N2L 3G1, Canada
[2]Materials Research Laboratory, Pennsylvania State University,
University Park, PA 16802

Transition metal acetylide complexes represent a class of linear conjugated molecules which can exhibit both second- and third-order optical nonlinearities. We have prepared several classes of such complexes including both symmetrically (X = Y) and unsymmetrically (X ≠ Y) substituted bis(acetylide) complexes $trans$-[Pt(PMe$_2$Ph)$_2$(C≡C-X)(C≡C-Y)]. Complexes where X is a strong π-donor and Y is a strong π-acceptor have the fundamental electronic and structural relationships required for good $\chi^{(2)}$ materials. Binuclear and polymeric rhodium compounds containing bridging M-C≡C-(C$_6$H$_4$-4-)$_n$-C≡C-M units (n = 0, 1, 2) have also been prepared. Second Harmonic Generation was observed in powder samples of all unsymmetric platinum bis(acetylides), with the largest powder efficiencies being on the order of urea. Single-crystal X-ray diffraction studies on several symmetric and unsymmetric compounds demonstrate both intra- and intermolecular parallel alignment of all acetylide units. The complex $trans$-[Pt(PMe$_2$Ph)$_2$(C≡C-C$_6$H$_4$-4-OMe)(C≡C-C$_6$H$_4$-4-NO$_2$)], for example, crystallizes with point group symmetry 1 (space group P1, Z = 1), which gives rise to parallel alignment of all molecular dipoles.

The second- and third-order nonlinear optical properties of organic molecules and polymers are the subject of much current interest (1-5). The molecular properties which are critical include a delocalized π-system. For second-order effects, an accessible charge-transfer transition which gives rise to a large change in dipole moment, and a non-centrosymmetric molecular structure are required. In addition, for a bulk material, such as a single-crystal, to exhibit optimal second-order effects, a non-centrosymmetric arrangement of molecules is required in which there is a large degree of alignment of the molecular dipoles.

Acetylenes R-C≡C-R' and higher oligomers, such as diynes R-(C≡C)$_2$-R', triynes R-(C≡C)$_3$-R', and tetraynes R-(C≡C)$_4$-R', have the requisite conjugation required for both $\chi^{(2)}$ and $\chi^{(3)}$ effects. With suitable substitution patterns, R = π-

0097–6156/91/0455–0605$06.00/0

donor and R' = π-acceptor, low-lying charge transfer transitions and molecular asymmetry can be incorporated to enhance molecular hyperpolarizability, β. The remaining requirement for a crystalline material, i.e. bulk dipole alignment in a crystal lattice, is somewhat more difficult to control.

Recently, several groups have demonstrated ($\underline{6\text{-}13}$) large $\chi^{(2)}$ values for donor-acceptor substituted diphenyl acetylenes 4-D-C_6H_4-C≡C-C_6H_4-4-A and their higher oligomers 4-D-C_6H_4-(C≡C)$_n$-C_6H_4-4-A. In one case ($\underline{13}$), phase-matched second harmonic generation (SHG) was confirmed for the compound 4-MeO-C_6H_4-C≡C-C_6H_4-4-NO_2. Organotransition metal complexes, including square-planar compounds of the general form $trans$-[M(PR$_3$)$_2$(X)(C_6H_4-4-A)] (M = Pd, Pt), have also been shown ($\underline{3,14\text{-}24}$) to exhibit large second-order optical nonlinearities. SHG values as high as ca. 220x urea have now been reported for certain ferrocene derivatives ($\underline{3,22\text{-}24}$). In addition, a series of symmetrically substituted square-planar palladium and platinum acetylides, and their polyyne polymers, have been reported ($\underline{25\text{-}28}$) to exhibit significant $\chi^{(3)}$ values, and their incorporation into electro-optic devices has now been achieved ($\underline{29}$).

Transition metal acetylides combine the properties of acetylenes with those of the transition metals, offering flexibility in the tuning of structural and electronic properties of both the organic and inorganic constituents. Optimization of the molecular and bulk crystalline properties is envisaged to lead to a new class of useful nonlinear optical materials.

The metal acetylides are inherently linear molecules due to the ca. 180° bond angles in an L$_n$M-(C≡C)$_n$-R unit. Linearity can be extended by use of $trans$-bis(acetylides), R-(C≡C)$_n$-ML$_n$-(C≡C)$_n$-R', in which the (C≡C)-M-(C≡C) angle is also 180°. The transition metal acetylide σ-bonds are extremely strong, giving rise to the thermodynamic stability in complexes with ≤ 8 d-electrons. Of importance in terms of optimizing optical nonlinearities is the degree of interaction between the metal d-block orbitals and the acetylene p-π and p-π* systems. Molecular orbital calculations ($\underline{31,33}$) on several L$_n$M-C≡C-H complexes indicated weak π-donor and π-acceptor properties for the -C≡C-H group. However, calculations on diynes and on acetylide groups with strong π-acceptor substituents, -C≡C-A (Marder, T.B.; unpublished results), suggest that such perturbations can significantly alter the π and π* orbital energies and atomic coefficients resulting in excellent π-acceptor properties. Recent electrochemical experiments ($\underline{34}$), on a series of (chelate)Rh-C≡C-A complexes, have confirmed this hypothesis. In contrast, substitution with strong π-donors should raise the energies of the filled acetylide π-orbitals allowing for the design of good π-donor ligands.

The combination of a good π-donor acetylide and a good π-acceptor acetylide in $trans$-positions on a metal center would provide the push-pull relationship resulting in low-lying intramolecular charge transfer transitions, and lack of a molecular center of symmetry. Thus, the molecular hyperpolarizability, β, should be large, and if the crystal packing can be controlled, large $\chi^{(2)}$ values should be achieved.

During the course of our studies ($\underline{30\text{-}32}$) of the synthesis and structures of rhodium acetylide and hydrido-acetylide complexes, we developed ($\underline{32}$) a step-wise route to $trans$-bis(acetylides) of the general form mer-$trans$-[Rh(PMe$_3$)$_3$(H)(C≡CR)(C≡CR')]. Unfortunately, scrambling processes have thus far precluded the preparation of the unsymmetrically substituted complexes (R ≠ R') in the absence of their symmetric (R = R') counterparts. However, single crystal X-ray diffraction studies (32) on three of the symmetric hydrido-bis(acetylides) demonstrated an interesting feature of the crystal packing of such complexes. We observed that, in each case, all acetylide moieties are aligned in a parallel fashion throughout the crystal lattice. The possibility that such a packing arrangement might be a general feature of bis(acetylides) led us to prepare a series of symmetrically substituted *trans*-bis(acetylide) complexes of platinum. In contrast to the rhodium systems, which are octahedral d^6-compounds, the platinum complexes are square-planar d^8-complexes. We have developed a successful synthetic route to the unsymmetrically substituted platinum bis(acetylides) which avoids the formation of significant quantities of the symmetric counterparts.

Finally, we developed routes to both linear bimetallic and polymeric acetylide complexes of Rh, Pd, and Pt, which will allow tuning of the electronic nature of the linker groups.

This paper reports the synthesis, representative single-crystal X-ray structural data, optical spectra, and preliminary measurements of powder SHG efficiencies of the new compounds.

Preparation of Terminal Alkynes and Diynes

The terminal alkynes (35,36) 4-R-C$_6$H$_4$-C≡CH (R = NO$_2$, 1a; CN, 1b; MeO, 1c; MeS, 1d; H$_2$N, 1e; Me$_2$N, 1f), 4-ethynylpyridine 2 (37), and ethynylferrocene (FcC≡CH) 3 (38), were prepared by modifications of literature routes. Compound 1d, which was not previously reported, was prepared by coupling of 4-MeS-C$_6$H$_4$Br with Me$_3$SiC≡CH, by analogy with the other *para* substituted phenyl acetylenes. Terminal diynes, R-C≡C-C≡CH (R = Ph, 4a; Fc, 4b), were prepared by coupling of the appropriate alkyne, R-C≡CH with *cis*-ClHC=CHCl, followed by treatment with base (39). Diacetylene (5) was prepared by treatment of ClCH$_2$C≡CCH$_2$Cl with base (40), and the compounds 4-HC≡C-(C$_6$H$_4$)$_n$-C≡CH (n = 1, 6a; n = 2, 6b) were prepared *via* coupling (41) of either Me$_3$SiC≡CH or HC≡C-C(OH)Me$_2$ with the 1,4-C$_6$H$_4$Br$_2$ or 4,4'-BrC$_6$H$_4$C$_6$H$_4$Br, followed by deprotection. We also prepared and structurally characterized 9,10-bis(trimethylsilylethynyl) anthracene; however, clean deprotection to 9,10-bis(ethynylanthracene) has not yet been achieved.

Preparation of Symmetric and Unsymmetric *trans*-Bis(acetylide) Complexes of Platinum

The symmetric bis(acetylides), *trans*-[Pt(PMe$_2$Ph)$_2$(C≡C-R)$_2$] (R = MeO, 7a; NO$_2$, 7b; 4-py, 7c), can be prepared conveniently, and in high yield, by room temperature (*ca.* 1-2hr) reaction of two equivalents of R-C≡CH with [Pt(PMe$_2$Ph)$_2$Cl$_2$] in Et$_2$NH solution in the presence of CuI, a route initially reported by Hagihara *et al.* (42). The bis(acetylides), *trans*-[Pt(PMe$_2$Ph)$_2$(C≡C-C≡C-R)$_2$] (R = Ph, 8a; Fc, 8b), were also prepared by this route. Preparation of the unsymmetric complexes was more difficult in that the two acetylides must be attached sequentially, and scrambling must be avoided. Refluxing a solution of [Pt(PMe$_2$Ph)$_2$Cl$_2$] and 1a or 1b in CHCl$_3$ in the presence of Et$_2$NH for 3 days yielded the mono-acetylide species *trans*-[Pt(PMe$_2$Ph)$_2$(Cl)C≡C-C$_6$H$_4$-4-A)] (9a, A = NO$_2$, 88%; 9b, A = CN, 98%). This general procedure, reported by Furlani *et al.* (43), avoids undesired formation of significant quantities of the symmetric bis(acetylide) complexes. We find that with careful control of reaction time and reagent stoichiometry, the mono-acetylides (**9a,b**) react with the donor-alkynes **1c-f, 2**, and **3** in CHCl$_3$ in the presence of small amounts of Et$_2$NH and CuI, yielding the novel unsymmetrically substituted complexes *trans*-[Pt(PMe$_2$Ph)$_2$(C≡C-C$_6$H$_4$-4-A)(C≡C-D)] (**10,11**). Using short reaction times (≤ 1 hr, room temperature) and relatively small amounts of Et$_2$NH (*ca.* 100 equiv.) and CuI (*ca.* 0.1 equiv.), acetylide scrambling is avoided.

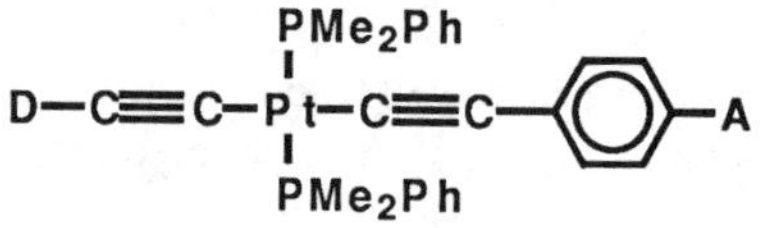

D

$$D—C≡C-\underset{\underset{PMe_2Ph}{|}}{\overset{\overset{PMe_2Ph}{|}}{Pt}}-C≡C—\hspace{-2pt}\bigcirc\hspace{-2pt}—A$$

a	4-MeO-C$_6$H$_4$	
b	4-MeS-C$_6$H$_4$	
c	4-H$_2$N-C$_6$H$_4$	
d	4-Me$_2$N-C$_6$H$_4$	
e	4-C$_5$H$_4$N	
f	ferrocenyl	
g	C$_6$H$_5$	

10   or   **11**

(A = NO$_2$)   (A = CN)

Preparation of Dinuclear and Polymeric Rhodium Complexes Linked by Acetylides

Our previously reported synthetic strategies (30-32), based on oxidative addition of terminal alkyne C-H bonds to Rh$^{(I)}$ centers, could be extended to include bifunctional alkynes HC≡C-(C$_6$H$_4$-4-)$_n$-C≡CH (n = 0-2). Thus, reaction of 2 equiv. of [(PMe$_3$)$_4$Rh]Cl with **5** or **6a,b** in THF suspension gave the dinuclear hydrido-acetylides *cis-cis*-[{Rh(PMe$_3$)$_4$(H)}$_2${μ-C≡C-(C$_6$H$_4$-4-)$_n$-C≡C-}]$^{2+}$[Cl$_2$]$^{2-}$ (n = 0, **12a**; n = 1, **12b**; n = 2, **12c**) in high yields as white precipitates.

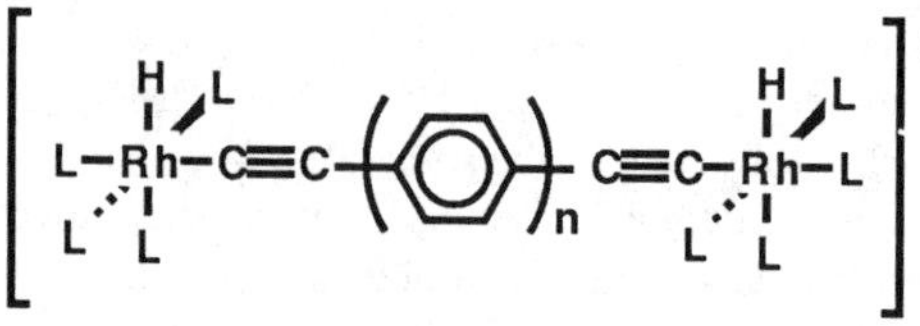

Reaction of [(PMe$_3$)$_4$RhMe] with two equiv. of **6a,b** yields the bis(acetylide)-hydride complexes *mer-trans*-[Rh(PMe$_3$)$_3$(H){-C≡C-(C$_6$H$_4$-4-)$_n$-C≡CH}$_2$] (n = 1, **13a**; n = 2, **13b**) in which one terminal hydrogen still remains on each end for subsequent reactions.

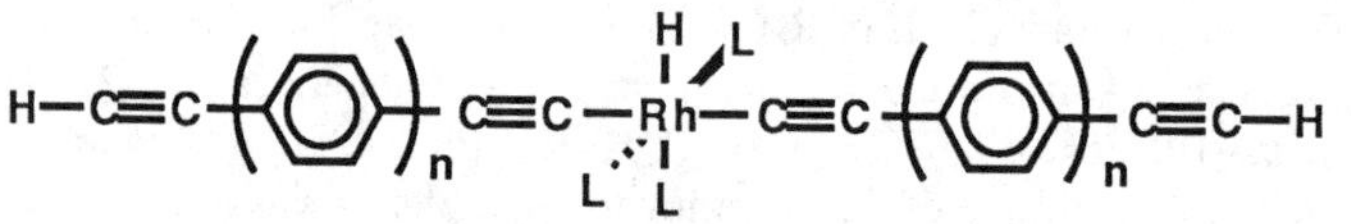

A similar reaction conducted with 1:1 stoichiometry yields the rigid-rod polymeric species *mer-trans*-[Rh(PMe$_3$)$_3$(H){C≡C-(C$_6$H$_4$-4-)$_n$C≡C}-]$_x$ (n = 1, **14a**; n = 2, **14b**) directly as insoluble powders.

14a,b

These new complexes and polymers are related to the square-planar palladium and platinum polyynes which have recently been shown (25-28) to exhibit interesting $\chi^{(3)}$ behavior. We have not yet measured the $\chi^{(3)}$ properties of our rhodium complexes, or the molecular weights of the polymers. Current synthetic work is directed towards understanding the influence of the linker groups on electronic communication between the metal centers, and on designing new linkers with low-lying π^* levels to improve conjugation.

X-Ray Crystal Structure Determinations

Single crystal X-ray studies on representative examples of the platinum bis(acetylides) were carried out to gain information on both the molecular geometry and nature of the packing of the acetylide moieties in the crystal. Selected crystal data for the various compounds are given in Table I. Full details of the structural analyses will be reported elsewhere.

Table I. Crystal Data Collection and Refinement Parameters

Compound	7a	7b	10a	10d	8a
Crystal System	triclinic	triclinic	triclinic	monoclinic	monoclinic
a (Å)	8.063(1)	8.762(1)	5.985(1)	20.961(5)	10.076(2)
b (Å)	9.782(2)	9.579(1)	9.065(2)	5.693(1)	8.530(1)
c (Å)	11.240(1)	11.110(2)	14.105(3)	28.551(8)	18.260(3)
α (°)	92.72(1)	113.05(1)	84.27(2)	-	-
β (°)	102.97(1)	95.73(1)	88.12(2)	110.29(2)	94.70(2)
γ (°)	112.89(1)	112.55(1)	78.35(2)	-	-
Volume (Å³)	768.8(2)	757.4(2)	745.7(3)	3196(1)	1564.1(5)
Space Group	P$\bar{1}$	P$\bar{1}$	P1	Cc or C/2c	P2₁/c
Z	1	1	1	4	2
T (K)	295	295	170	170	295
2θ Range (°)	3.5 - 50.0	3.5 - 60.0	3.5 - 70.0	3.5 - 60.0	3.5 - 50.0
Observed Data (I ≥ 3σ(I))	2778	4202	12,978	2899	1866
R	0.0235	0.0269	0.0286	0.0353	0.0232
R_w	0.0245	0.0295	0.0292	0.0373	0.0237

One important feature which has emerged is the proclivity of the acetylide units to align in a parallel manner. This has now been found to hold for the symmetric platinum bis(acetylides), [Pt(PMe$_2$Ph)$_2$(C≡CC$_6$H$_4$X)$_2$] (**7a,b**), which crystallize in the centrosymmetric triclinic space group P$\bar{1}$, with one molecule per unit cell and the platinum atom residing on an inversion center. Although the two compounds are not isostructural, they have in common the fact that all the acetylides in the crystal are in perfect parallel alignment. This arises as a consequence of: a) the linear nature of each individual acetylide; b) the *trans-* configuration at each metal (compatible with their $\bar{1}$ site symmetry; and, c) the fact that the only other symmetry elements in the triclinic system are pure translations.

The unsymmetrically substituted bis(acetylide) (**10a**) was also found to crystallize in the triclinic system, with Z = 1. Unlike its symmetric analogues, the lack of molecular symmetry precludes the platinum atom from sitting at a true inversion. Collection of all diffraction data including Friedel pairs, hkl and hkl, allowed successful refinement in space group P1, despite the severe pseudosymmetry. The lack of significant residual electron density in the region of the acetylide end groups OMe and NO$_2$, and the observation (*vide infra*) of a second harmonic signal for the material, supported this choice of space group and indicated little or no end group disorder in the crystal. The packing diagram of **10a** is shown in Figure 1 as representative of these triclinic materials.

Another unsymmetric analogue, **10d** (D = C$_6$H$_4$NMe$_2$, A = NO$_2$), was found to crystallize in the monoclinic system. The systematic absences were consistent with the space groups C2/c or Cc. Again, the centric choice would require molecular end-to-end disorder in the crystal. Structure refinement in the non centric group Cc has so far proved unstable, possibly due to the similarity between N(CH$_3$)$_2$ and NO$_2$ substituents. A degree of disorder cannot be ruled out in spite of an observed SHG signal. Once again, the packing of all acetylides moieties in the crystal is found to be parallel (Figure 2).

Extension of the linear chain system is possible by use of diacetylide (C$_4$R) rather than acetylide (C$_2$R) ligands. We have now carried out a single crystal structure determination of the symmetric bis(acetylide) complex *trans*-[Pt(PMe$_2$Ph)$_2$(C≡C-C≡C-Ph)$_2$] (**8a**). The structure is the first of any transition metal diacetylide, and the geometry is shown in Figure 3. Interestingly, the molecular packing for this compound does not result in the parallel alignment of all the acetylide moieties; a herringbone pattern is found (Figure 4).

Second Harmonic Generation Measurements

All samples were tested on a second harmonic analyzer based on a modified version of that designed by Dougherty and Kurtz (44). The instrument comprises a Nd-glass laser rod operating at 1.06 μm. Fundamental and second harmonic signals are compared for time correlation to eliminate the possibility that the second harmonic signals are spurious. The system was calibrated using a quartz standard, *ca.* 10 mg graded 62 μm quartz powder immersed in two drops of an index matching liquid (Cargille Inc., Cedar Grove, NJ) with n = 1.544. Approximate SHG powder efficiencies were obtained using the Kurtz powder technique (45). The compounds were ground or roughly crushed, depending on the crystallinity of the sample, to produce powders with particle sizes ranging approximately from 50-100 μm. The indices of refraction were estimated (Becke line method) to be close to 1.50 for **10a**. The same index matching fluid, n = 1.544, was used throughout.

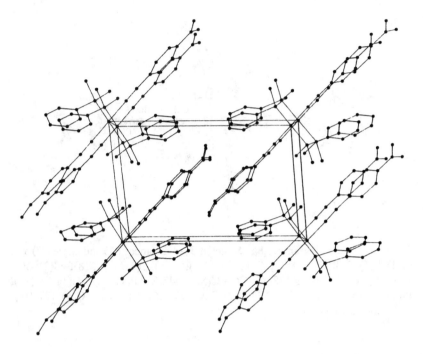

Figure 1. Packing diagram for **10a** viewed down the a axis.

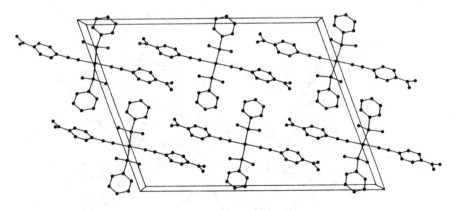

Figure 2. Packing diagram for **10d** viewed down the b axis.

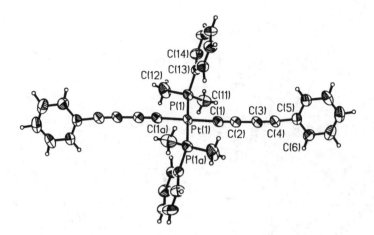

Figure 3.   Ortep diagram for **8a**. Selected bond distances (Å) and angles (°) are: Pt(1)-P(1) 2.308(2), Pt(1)-C(1) 2.009(5), C(1)-C(2) 1.175(8), C(2)-C(3) 1.406(8), C(3)-C(4) 1.188(8), C(4)-C(5) 1.442(8), P(1)-Pt(1)-C(1) 88.7(2), Pt(1)-C(1)-C(2) 177.8(5), C(1)-C(2)-C(3) 177.1(6), C(2)-C(3)-C(4) 177.1(6), C(3)-C(4)-C(5) 174.9(6).

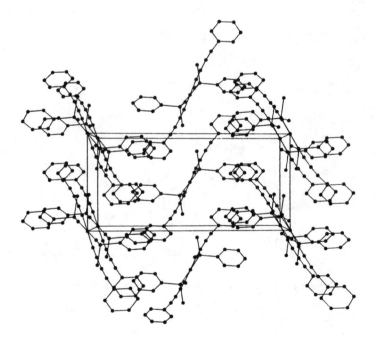

Figure 4.   Packing diagram for **8a** viewed down the a axis.

In the case of the asymmetrically substituted platinum acetylides, SHG was detected in all samples (Table II). The symmetrically substituted bis(pyridylacetylide) complex **7c** was found to be centrosymmetric. Signals were quantified by peak height analysis of several laser pulses.

Table II. Second Harmonic Generation and UV-VIS Absorption Data
in Platinum Acetylides

Compound	SHG [a]	λ max (nm) [b]	ε $(l \cdot cm^{-1} \cdot mol^{-1})$
9a	0.005	368	17,250
9b	0.01	326	19,500
11b	0.33	342	30,300
10d	0.40	378	19,200
11g	0.50	338	30,200
11a	0.50	350	26,250
10a	1.00	386	38,800

[a] SHG (1.064 μm → 0.532 μm) relative to optimized, index-matched, 62 μm quartz. Average of 3 measurements. For comparison, an unoptimized urea sample gave a signal of 1.5 x quartz under these conditions.
[b] In CH_3CN. Lowest energy absorption band.

The unsymmetric bis(acetylides) (**10,11**) are significantly more effective than the mono-acetylides **9a,b**. The SHG efficiencies obtained by this method are probably not of sufficient accuracy to warrant detailed comparisons among samples of **10** and **11**. It is interesting, however, that **11g**, in which D = Ph, has a comparable powder SHG efficiency to **11a,b** in which the donor groups contain MeO or MeS substituents in the *para* position. Preliminary X-ray powder diffraction studies indicate that **11g** is not isostructural with **11a,b**. Thus, an alternative packing arrangement may be responsible for the magnitude of the SHG signal for **11g**.

One important conclusion is that the entire class of compounds seems to crystallize in non centrosymmetric space groups. The completely parallel alignments observed for **10a,d** are not optimum for SHG; however, this is an advantageous arrangement for second-order electro-optic effects (46). With the exception of **10f**, which has a significant absorption tail into the region around 532 nm, the compounds are essentially transparent in this region. The lowest energy electronic transitions (Table II) are in the range of *ca.* 325-390 nm. Complete experimental and spectroscopic details for all new compounds will be reported elsewhere.

Acknowledgments

We thank the Natural Sciences and Engineering Research Council of Canada, the Ontario Centre for Materials Research, and the University of Waterloo for support, Johnson Matthey Ltd. for a loan of precious metal salts, the DuPont Company for a donation of materials and supplies, and Drs. S.R. Marder and A. Stiegman (JPL-Caltech), W. Tam and L.T. Cheng (DuPont) for helpful discussions and preprints of several publications.

Literature Cited

1. Williams, D.J., Angew. Chem. Int. Ed. Engl. 1984, 23, 690.
2. Nonlinear Optical Properties of Organic and Polymeric Materials; Williams, D.J., Ed.; ACS Symposium Series No. 233; American Chemical Society: Washington, D.C., 1983.
3. Nonlinear Optical Properties of Organic Molecules and Crystals, Vol. 1, 2; Chemla, D.S., Zyss, J., Eds.; Academic Press: Orlando, FL, 1987.
4. Nonlinear Optical Properties of Organic Materials; Proc. SPIE No. 971; The International Society for Optical Engineering: Washington, DC, 1988.
5. Organic Materials for Non-linear Optics; Hann, R.A., Bloor, D., Eds.; Spec. Publ. No. 69, The Royal Society of Chemistry: London, England, 1989.
6. Tabei, H.; Kurihara, T.; Kaino, T. Appl. Phys. Lett. 1987, 50, 1855.
7. Fouquey, C.; Lehn, J.-M.; Malthête, J. J. Chem. Soc., Chem. Commun. 1987, 1424.
8. Perry, J.W.; Stiegman, A.E.; Marder, S.R., Coulter, D.R. In ref. 5; p. 189.
9. Perry, J.W.; Stiegman, A.E.; Marder, S.R.; Coulter, D.R.; Beratan, D.N.; Brinza, D.E.; Klavetter, F.L.; Grubbs, R.H. In ref. 4; p. 17.
10. Tam, W.; Wang, Y; Calabrese, J.C.; Clement, R.A. In ref. 4; p. 107.
11. Stiegman, A.E.; Miskowski, V.M.; Perry, J.W.; Coulter, D.R. J. Am. Chem. Soc. 1987, 109, 5884.
12. Graham, E.M.; Miskowski, V.M.; Perry, J.W.; Coulter, D.R.; Stiegman, A.E.; Schaefer, W.P.; Marsh, R.E. J. Am. Chem. Soc. 1989, 111, 8771.
13. Kurihara, T.; Tabei, H.; Kaino, T. J. Chem. Soc., Chem. Commun. 1987, 959.
14. Bandy, J.A.; Bunting, H.E., Garcia, M.H.; Green, M.L.H.; Marder, S.R.; Thompson, M.E.; Bloor, D.; Kolinsky, P.V.; Jones, R.J. In ref. 5; p. 225.
15. Eaton, D.F.; Anderson, A.G.; Tam, W.; Wang, Y. J. Am. Chem. Soc. 1987, 109, 1886.
16. Tam, W.; Calabrese, J.C. Chem. Phys Lett. 1988, 144, 79.
17. Calabrese, J.C.; Tam, W. Chem. Phys Lett. 1987, 133, 244.
18. Frazier, C.C.; Harvey, M.A.; Cockerham, M.P.; Hand, H.M.; Chauchard, E.A.; Lee, C.H. J. Phys. Chem. 1986, 90, 5703.
19. Coe, B.J.; Jones, C.J.; McCleverty, J.A.; Bloor, D.; Kolinsky, P.V.; Jones, R.J. J. Chem. Soc., Chem. Commun. 1989, 1485.
20. Anderson, A.G.; Calabrese, J.C.; Tam, W.; Williams, I.D. Chem. Phys. Lett. 1987, 134, 392.
21. Tam, W.; Eaton, D.F., Calabrese, J.C.; Williams, I.D.; Wang, Y.; Anderson, A.G. Chem. Mater. 1989, 1, 128.
22. Green, M.L.H.; Marder, S.R.; Thompson, M.E.; Bandy, J.A.; Bloor, D.; Kolinsky, P.V.; Jones, R.J. Nature 1987, 330, 360.
23. Bandy, J.A.; Bunting, H.E.; Green, M.L.H.; Marder, S.R.; Thompson, M.E.; Bloor, D.; Kolinsky, P.V.; Jones, R.J. In ref. 5; p. 219.
24. Marder S.R.; Perry, J.W. J. Am. Chem. Soc., submitted.
25. Frazier, C.C.; Guha, S.; Chen, W.P.; Cockerham, M.P.; Porter, P.L.; Chauchard, E.A.; Lee, C.H. Polymer 1987, 28, 553.
26. Frazier, C.C.; Chauchard, E.A.; Cockerham, M.P.; Porter, P.L. Mat. Res. Soc. Symp. Proc. 1988, 109, 323.

27. Frazier, C.C.; Guha, S.; Porter, P.L.; Cockerham, P.M.; Chauchard, E.A. In ref. 4; p. 186.

28. Guha, S.; Frazier, C.C.; Kang, K.; Finberg, S.E. Optics Lett. 1989, 14, 952.

29. Frazier, C.C.; Guha, S.; Chen, W. P.C.T. Int. Appl. WO 89 01,182, Feb. 1989; U.S. Appl. 81,785, Aug. 1987; Chem. Abstr. 1989, 111, 105446p.

30. Marder, T.B.; Zargarian, D.; Calabrese, J.C.; Herskowitz, T.; Milstein, D. J. Chem. Soc., Chem. Commun. 1987, 1484.

31. Zargarian, D.; Chow, P.; Taylor, N.J.; Marder, T.B. J. Chem. Soc., Chem. Commun. 1989, 540.

32. Chow, P.; Zargarian, D.; Taylor, N.J.; Marder, T.B. J. Chem. Soc., Chem. Commun. 1989, 1545.

33. Kostic, N.M.; Fenske, R.F. Organometallics 1982, 1, 974.

34. Bianchini, C.; Laschi, F.; Ottaviani, M.F.; Peruzzini, M.; Zanello, P.; Zanobini, F. Organometallics 1989, 8, 893.

35. Takahashi, S.; Kuroyama, Y.; Sonogashira, K.; Hagihara, N. Synthesis 1980, 627.

36. Austin, W.B.; Bilow, N.; Kelligan, W.J.; Lau, K.S.Y. J. Org. Chem. 1981, 46, 2280.

37. Ciana, L.D.; Haim, A. J. Heterocyclic Chem. 1984, 21, 607.

38. Rosenblum, M.; Brawn, N.; Papermeier, J.; Applebaum, M. J. Organomet. Chem. 1966, 6, 173.

39. Kende, A.S.; Smith, C.A. J. Org. Chem. 1988, 53, 2655.

40. Bradsma, L.; Verkruijsse, H.D. Synthesis of Acetylenes, Allenes and Cumulenes, A Laboratory Manual; Elsevier: Amsterdam, 1981; p. 146.

41. See for example: Neenan, T.X.; Whitesides, G.M. J. Org. Chem. 1989, 53, 2489.

42. Sonogashira, K.; Fujikura, Y.; Yataki, T.; Tokoshima, N.; Takahashi, S.; Hagihara, N. J. Organomet. Chem. 1978, 145, 101.

43. Furlani, A.; Licoccia, S.; Russo, M.V.; Chiesi-Villa, A.; Gustani, C. J. Chem. Soc., Dalton Trans. 1984, 2197.

44. Dougherty, J.P.; Kurtz, S.K. J. Appl. Cryst. 1976, 9, 145.

45. Kurtz, S.K.; Perry, T.T. J. Appl. Phys. 1968, 39, 3798.

46. Zyss, J.; Oudar, J.L. Phys. Rev. A 1982, 26, 2028.

RECEIVED July 18, 1990

Chapter 41

Third-Order Near-Resonance Nonlinearities in Dithiolenes and Rare Earth Metallocenes

C. S. Winter[1], S. N. Oliver[1], J. D. Rush[1], R. J. Manning[1], C. Hill[2], and A. Underhill[2]

[1]British Telecom Research Laboratories, Martlesham Heath, Ipswich IP5 7RE, England
[2]Department of Chemistry, Bangor University, Bangor, Wales

The nonlinear refractive index, n_2, linear absorption, α, and two photon absorption coefficient, β, of metal dithiolenes and rare earth metallocenes have been measured at 1064 nm using 100 ps and 10 ns pulses. These measurements have shown resonance enhanced molecular nonlinearities of 10^3-10^5 times those of carbon disulphide can be obtained. Figures of merit based on the nonlinear refractive index and the linear and two photon absorption coefficients are within the limits required for devices.

The increasing use of optical fibre in the telecommunications network will, ultimately, require all-optical signal processing to exploit the full bandwidth available. This has led to a search for materials with fast, large third order optical nonlinearities. Most of the current materials either respond in the nanosecond regime or the nonlinearity is too small (1-3). Organic materials are attractive because of their ultra-fast, broadband responses and low absorption. However the main problem in the materials studied to date, e.g. polydiacetylenes (4) and aromatic main chain polymers (5), has been the small nonlinear coefficients.

A recent analysis (1) of certain classes of waveguide devices suggests that polydiacetylenes are marginally acceptable, based on the ratio of the nonlinearity to linear absorption; however recent attempts to fabricate waveguide structures from polydiacetylene have shown that in a suitable device configuration the nonlinearity is dominated by two photon absorption rather than nonlinear refraction (6). Measurements of both two photon absorption and linear absorption are critical to analysing the suitability of a material for device applications (3,7). Two photon absorption is particularly important since the phenomenon displays the same dependence on device length, waveguide diameter and input power in a waveguide device as nonlinear refraction. Thus trade-offs between the device parameters are not

0097–6156/91/0455–0616$06.00/0

available, unlike the trade-offs that are possible with linear absorption (7). The ratio $2\lambda\beta/n_2$ defines which process will dominate, where β is the two photon absorption coefficient and n_2 the nonlinear refractive index. If this ratio is greater than unity, standard waveguide devices such as couplers and Mach-Zender interferometers will not function suitably (7). Unlike the limits imposed by the nonlinearity and linear loss - which may be traded off with device dimensions and possibly by the use of the recent fibre-based laser amplifiers - this limit is a fundamental materials limitation causing both loss and pulse shape modification.

There are two possible approaches towards making feasible devices. One approach is to develop device structures that need smaller coefficients to operate, *e.g* lossy Fabry-Perot cavities (8), the other approach is to trade-off some of the speed and low loss in current organics for larger responses, for instance by tuning into resonances. In this paper we will explore the latter route and show how this can be achieved in some materials whilst maintaining, at acceptable levels, the critical figures of merit relating the nonlinear refraction to linear loss and two photon absorption to nonlinear refraction.

One of the major problems with utilising single photon near-resonance phenomena is the telecommunications requirement that a material responds at 1300-1550 nm. It is difficult to obtain small organic molecules or conjugated polymers that possess absorption bands or band edges at these low energies. In order to examine the potential for resonance enhancement we have thus begun to examine organometallic systems possessing both reasonable polarisability and tunable near IR absorption bands.

Two classes of material will be described here - the metal dithiolenes and rare earth metallocenes. In the metal dithiolenes a strong, low energy pi-pi* transistion occurs in the near IR (9,10). This can be tuned from about 700 nm to 1400 nm by altering the metal ion, substituents or charge state of the dithiolene. The dithiolenes are particularly attractive because of their optical stability which has been exploited in their use as laser Q-switch materials. In the rare earth complexes the near IR band is provided by *f-f* transistions of the rare earth ion rather than the cyclopentadienyl ring structure; various nonlinear optical phenomena have been observed in glasses incorporating similar ions. Previous studies have shown that dicyclopentadienyl complexes such as ferrocene have off-resonant nonlinearities similar to nitrobenzene or carbon disulphide (11-13)

For both classes of materials initial trial compounds were selected for their potential at 1064 nm, these included Yb *tris*-cyclopentadienyl and dithiolenes with absorption bands from 700-950 nm. However the results observed here may be extended to useful wavelengths by appropriate selection of derivatives.

<u>EXPERIMENTAL</u>

Tris-cyclopentadienyl ytterbium, $Yb(cp)_3$, was prepared from sodium cyclopentadienylide and ytterbium chloride in tetrahydrofuran (THF) under nitrogen (14). The crude dark green product was then twice sublimed at 150 C at 5×10^{-5} mBar and thereafter handled under inert or vacuum conditions. Solutions of $Yb(cp)_3$ were made up in THF which had been doubly distilled under nitrogen from lithium aluminium hydride, stored over 4A molecular sieves and distilled into 2mm cuvettes containing the Yb(cp)$_3$ under nitrogen. The solutions were dark green and showed no deterioration in the sealed cuvettes over several months.

bis [1,2-diphenyl-1,2-ethenedithiolato(2-)-S,S'] Nickel ("phenyl dithiolene") and *bis* [1,2-dimethyl-1,2-ethenedithiolato(2-)-S,S'] Nickel ("methyl dithiolene") were synthesized by the method of Schrauzer and Mayweg (15). Solutions of approximately10^{18} molecs/cc were made up with dichloromethane. Linear absorption spectra were measured in a Perkin-Elmer λ9 spectrophotometer.

Nonlinear optical studies were carried out using a combination of nonlinear absorption, self-focusing and degenerate four wave mixing measurements. The measurements were made using a Quantel Nano-Pico system that permits operation at either 10 ns or 100 ps pulse lengths at 1064 nm. The 10 ns pulses were TEM$_{00}$ mode and temporally smoothed to a near gaussian (11).

Two photon absorption and measurements of the critical power for self-focusing were carried out using 10 ns pulses. In the self-focusing (or optical power limiter) experiments a laser beam is tightly focused into a solution of the material and the onset of self-focusing is observed from the pinning of the throughput of a second lens focused onto a pin-hole (16,17). The critical power at which self-focusing occurs is related to n$_2$ (see below). Self-focusing can also be observed by the onset of streamers - seen as bright flashes in the solution when viewed with an IR viewer. Two photon effects can be distinguished either by varying the lens focal length or measuring the total fluence. Since self-focusing is a *power* dependent phenomenon and two photon absorption an *intensity* dependent phenomenon, this experimental arrangement can distinguish the two contributions - the focal spot size varying with the lens chosen. A more accurate value for the two photon absorption coefficient can be obtained by choosing a lens such that the depth of focus of the lens is greater than the sample width and measuring the nonlinear absorption of the collimated beam directly (18). The experiments described below were all carried out using 1-5mm cuvettes and 50-500mm singlet 'best-form' lenses. The laser beam profile was measured directly using a silicon detector array and found to be TEM$_{00}$, with a diameter of 5mm. The focused or collimated beam sizes were measured by observing the spot focused onto a microscope graticule using a camera and microscope objective. This gives the spot size with an accuracy of 10-15%.

A retroreflection degenerate four wave mixing (DFWM) arrangement was used to measure the picosecond characteristics (19,20). In this technique the pump beam passes through the sample and is then reflected back to form the second pump beam. A probe beam is split off from the first pump beam. The ratio of this probe to the phase conjugate signal, which propagates back along the path of the probe beam, gives the nonlinearity. Temporal overlap is achieved by passing the first pump and the probe beams through time delays and confirmed by autocorrelation techniques using a second harmonic generating (SHG) crystal. The signal-noise ratio was improved by focusing the part of the probe onto an SHG crystal and detecting the second harmonic generated. All signals were logged into a EG&G boxcar. To allow for the effects of any non-Gaussian profile of the beam, the sample signal was compared with that generated by an identical length of carbon disulphide.

RESULTS AND DISCUSSION

Figure 1 shows the schematic chemical structure of the two dithiolenes reported here. The near IR absorption peaks and linear loss at 1064 nm are listed in Table I below. The sharp, relatively weak, absorption band in $Yb(cp)_3$ causes a very low absorption loss even when the irradiating wavelength is less than 30nm from the absorption maxima. The broader absorption bands of the dithiolene cause larger linear absorption loss further from the absorption maxima. BDN, a similar Ni dithiolene, shows saturable absorption at 1064 nm (20), however this has not been observed in these materials.

Two photon absorption was observed in two materials - $Yb(cp)_3$ and phenyl dithiolene. Figure 2 shows the nonlinear transmission through the dithiolene in a collimated beam. The solid curve is the best fit two photon absorption (18) obtained from -

$$T = T_L \frac{2}{\pi^{1/2}Q} \int_0^\infty \ln[1 + Qe^{-x^2}]dx \qquad (1)$$

where T_L is the linear transmission and Q is a dimensionless parameter defined as -

$$Q = \beta(1-R)I_{Inc}(1-e^{-\alpha l})/\alpha \qquad (2)$$

By tightly focusing into the solution, but omitting the refocusing optics and pinhole, a similar two photon absorption curve is obtained. However at sufficiently high powers the data deviates from two photon absorption due to self-focusing. This can be confirmed by using an IR viewer to observe the 'streamers' in the material. The output power at this point is used to assign an approximate value, allowing for absorption (21), to the nanosecond n_2 given in Table I below.

No two photon absorption was observed in methyl diothiolene. The optical power limiter (self-focusing) arrangement showed power limiting at low incident powers. Figure 3 shows the power in-power out plot for the methyl derivative. The clear break in the plot shows the onset of self-focusing; at this point 'streamers' are observable in the solution when its viewed through an IR viewer. The absorption in the visible region of the spectrum prevents the streamers being visible to the eye as occurs in nitrobenzene or CS_2 (16). Due to the extreme sensitivity of self-focusing to minute fluctuations in laser power and beam profile this region is often the noisiest experimentally. The critical power, P_c, for self-focusing is related to n_2 by -

$$P_c = 3.72 \ c\lambda^2/32\pi^2 n_{2,E} \qquad (3)$$

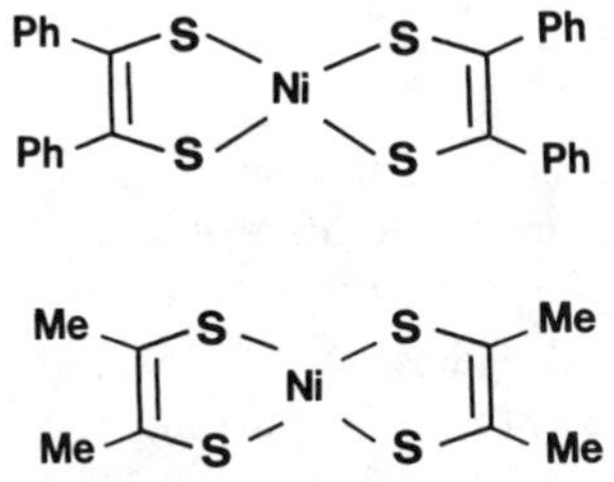

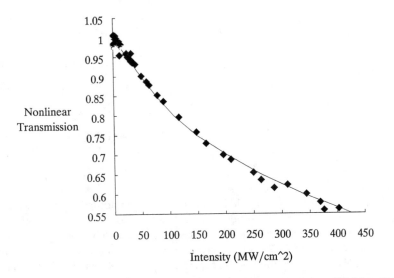

Figure 1. Schematic diagrams of *bis* [1,2-diphenyl-1,2-ethenedithiolato(2-)-S,S'] Nickel and *bis* [1,2-dimethyl-1,2-ethenedithiolato(2-)-S,S'] Nickel.

Figure 2. Nonlinear Transmission through *bis* [1,2-diphenyl-1,2-ethenedithiolato(2-)-S,S'] Nickel.

The n_2 here is in esu, and defined from

$$n = n_0 + n_{2,E} <E^2> \qquad (4)$$

whereas the normal device n_2 is in cm^2/kW and defined by

$$n = n_0 + n_{2,I}I. \qquad (5)$$

Interconversion is via the relationship (22) -

$$n_{2,E} \ (esu) = 0.2387 \ n_0 \ n_{2,I} \ (cm^2/kW). \qquad (6)$$

The origin of the n_2 measured using the 10 ns pulses could be electronic or molecular rotation. These can be distinguished by measuring the ratio of the critical power for self-focusing for linear and circular polarised light. The observed ratio of 2.1 is consistent with a molecular rotation (11-13,16) and relates to the anisotropic polarisability of the molecule. The rotational relaxation time, calculated from the Debye formula (11), is about 0.5-2 ns, consistent with these results.

The similarity of the results for the $Yb(cp)_3$ may indicate the molecular origin is electronic, similar to the results observed in simple metallocenes (11-13). However interpretation of the nanosecond data requires picosecond confirmation.

Table I: Absorption and nanosecond n_2 measurements

Material	conc. (molecs/cc)	λ_{max} (nm)	α (cm^{-1})	β (cm/GW)	P_c (kW)	$n_{2,E}$ (esu)
$Yb(cp)_3$	$3.0x10^{19}$	1030	1.1	4.1	9.1^1	$4.3x10^{-11}$
$Yb(cp)_3$					8.7^c	$4.5x10^{-11}$
Phenyl-Di.	$6.9x10^{17}$	865	0.875	2.92	9.0	$8.9x10^{-11}$
Methyl-Di.	$7.3x10^{17}$	770	0.045	<0.01	1.5^1	$2.7x10^{-10}$
Methyl-Di.					3.2^c	$1.2x10^{-10}$

The superscripts, l and c, in Table I identify linear and circular polarised light respectively. Since the nanosecond measurements were dominated by molecular rotation in the dithiolene complexes, degenerate four wave mixing measurement were carried out using 100 ps pulses to measure the electronic contribution directly.

Figure 4 shows the result of one such measurement on the phenyl dithiolene. The two lines on the plot are for carbon disulphide and the dithiolene. The relative slopes

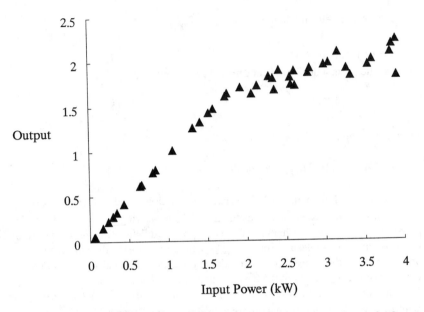

Figure 3. Optical power limiting due to self-focusing in *bis* [1,2-dimethyl-1,2-ethenedithiolato(2-)-S,S'] Nickel (linear polarised light).

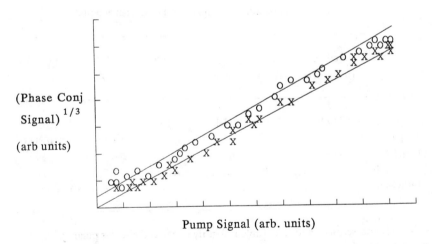

Figure 4. Degenerate four wave mixing in Carbon disulphide (x) and *bis* [1,2-diphenyl-1,2-ethenedithiolato(2-)-S,S'] Nickel (o).

of the lines are important; however, on this plot, the intersect is a function of the DC noise in the boxcar. Because of the detection system (see above) used the slopes of the two plots are related to the $\chi^{(3)}$'s of the two materials by the relationship -

$$\chi_{org}^{(3)} = \chi_{CS_2}^{(3)} \left[\frac{S_{org}}{S_{CS_2}} \right]^{3/2} \tag{7}$$

where S is the gradient taken fron figure 4. The relationship between nonlinear refraction and $\chi^{(3)}$ is $n_{2,E} = 4\pi\chi^{(3)}/n_0$ (esu).

Using the values of Witte et al (23) or Williams et al (17) for the $n_{2,E}$ of CS_2 of 1.3×10^{-11} esu, gives the $n_{2,E}$ values shown in Table II.

Table II: Picosecond n_2 and Figures of Merit

Material	$n_{2,E}$ (esu)	W	$2\lambda\beta/n_{2,I}$
Methyl Dithiolene.	0.3×10^{-11}	2.2	<0.2
Phenyl Dithiolene	1.2×10^{-11}	0.4	18.0
Yb(cp)$_3$	-	1.2	3.0

The ratio of the nonlinearity to the various absorption coefficients have been calculated using the conversion factor in equation 6 above. W is the figure of merit defined in reference 3, and given by -

$$W = I_{sat} n_{2,I} / \alpha \lambda \tag{8}$$

where I_{sat} is the intensity required to maximise the change in refractive index, in these materials it corresponds to the damage threshold of 500-1000 MW/cm². For Yb(cp)$_3$ the values are based on the nanosecond results and thus need further verification. A W>1 is required for device demonstrations, and preferably >2.5 (3). However for certain Fabry-Perot devices a value as low as 0.25-0.5 may be acceptable (8).

For nonlinear refraction to dominate over two photon absorption requires the appropriate ratio in Table II to be less than one; this ratio is important since there is no trade-off possible between β and n_2 in either intensity or device length (7).

To directly compare these results with those obtained on polymer systems it is necessary to extrapolate from the solution concentrations given in Table I to a solid state value of about 2×10^{21} molecs/cc. If this extrapolation is valid, the ratios in Table II will be unaltered but the value of n_2 will increase approximately 3000x. This will give a value of $n_{2,1}$ of $2.5 \times 10^{-8} cm^2/kW$ for the methyl dithiolene. The other compounds, although initially interesting due to the size of the coefficient or the linear loss, would be dominated by two photon effects if fabricated into waveguides. The difference in two photon absorption may hold important implications for the design of materials and the degree of tuning into an absorption band that can be exploited. Preliminary studies show that the two photon effects are strongly enhanced in phenyl substituents. This may be due either to the longer wavelength absorption bands in these materials compared with methyl derivatised compounds or due to the phenyl rings stabilising a two photon state. We are currently examining different derivatives with and without phenyl rings to explore the effects of shifting the absorption bands closer to 1064 nm.

A simple two level model was proposed for BDN (20) to explain the origin of the nonlinearity obtained by tuning onto the resonance in that material. A similar model may explain the results observed here, although in the model proposed in reference 20 a high quantum yield for fluorescence is important, whereas these materials do not fluoresce (9).

The Yb(cp)$_3$ complex is both unstable and a strong two photon absorber. Similar absorption has been observed in polydiacetylene (6) and lead glasses (7) at 1064 nm. This may limit usefulness of these materials unless there is a difference in the dispersion of n_2 and β out to 1300-1550 nm. However other rare earth derivatives are being examined to see if lower values of β can be obtained whilst enhancing n_2.

CONCLUSIONS

The measured solution value of n_2 for methyl dithiolene of 0.3×10^{-11} esu at 7×10^{17} molecs/cc, and the ratios of two photon absorption to nonlinear refraction ($2\lambda\beta/n_2 = 0.1$) and the Stegemann figure of merit (W = 2.2) show that it is possible to use near-resonance phenomena to obtain substantial values for the nonlinear refractive index whilst maintaining absorption coefficients below the levels required for devices. Dithiolene derivatives with absorption bands at 1200-1400 nm could be exploited to shift the resonance-enhancement to telecommunications wavelengths. The optical stability of dithiolenes overcomes the risk of degradation due to long term irradiation on the edge of an absorption band.

These materials may also be included in guest-host or side-chain polymer systems, similar to those exploited in electro-optic polymer studies. This would improve processability for waveguide devices. The coefficients quoted above show that such a doped polymer could function at reasonable power levels and waveguide dimensions with an active region 1-2 mm long.

Further studies are now in progress to analyse the effects of changing substituents, metal ions and the charge state of the dithiolene, and to study polymeric dithiolene systems.

LITERATURE CITED

1. Stegeman, G.I.; Zanoi, R.; Seaton, C.T., MRS Proc. 1988, 109, 53.
2. Gibbs, H.M, Optical Bistability: Controlling Light with Light; Academic Press, Orlando, 1985; p305.
3. Chang, T.Y., Opt. Eng. 1981, 20, 220.
4. Carter, G.M.; Chen, Y.J.; Tripathy, S.K., Opt. Eng. 1985, 24, 609.
5. Rao, D.N.; Swiatkiewicz, J.; Chopra, P.; Ghoshal, S.K.; Prasad, P.N., Appl. Phys. Lett. 1986, 48, 1187.
6. Townsend, P.D.; Jackel, J.L.; Baker, G.L.; Shelburne, J.A.; Etemad, S., Appl. Phys. Lett 1989, 55, 1829.
7. Mizrahi, V.; DeLong, K.W.; Stegeman, G.I., Saifi, M.A., Andrejco M.J., Opt. Lett. 1989, 14, 1140.
8. Garmire, E, IEEE J. Quant. Elect. 1989, 25, 289.
9. McCleverty, J.A.; Prog. Inorg. Chem. 1968, 10, 49.
10. Mueller-Westerhoff, U.T.; Vance, B., in Comprehensive Co-ordination Chemistry; Wilkinson, G., Ed.; Pergamon Press, Oxford, 1970; Vol 2, 595.
11. Winter, C.S.; Oliver, S.N.; Rush, J.D., Opt. Commun. 1988, 69, 45.
12. Winter, C.S.; Oliver, S.N.; Rush, J.D., NATO ASI series E 1988, 162, 247.
13. Winter, C.S.; Oliver, S.N.; Rush, J.D., RSC Spec. Pub. 1989, 69, 232.
14. Wilkinson, G.; Birmingham, J.M., J. Amer. Chem. Soc. 1956, 78, 42.
15. Schauzer, G.N.; Mayweg, V.P., J. Am. Chem. Soc. 1965, 87, 1483.
16. Soileau, M.J.; Williams, W.E.;Van Stryland, E.W.,IEEE J. Quant. Elect. 1983, QE-19, 731.
17. Williams, W.E.; Soileau, M.J.; Van Stryland, E.W., Opt. Commun.1984, 50, 256.
18. Smith, W.L., Handbook of Laser Science and Technology; Weber, M.J, Ed.; CRC Press, Boca Raton, 1986; Vol III, Part 1, 229.
19. Pepper, D.M.; Yariv, A., in Optical Phase Conjugation; Fisher, R.A., Ed.; Academic Press, London, 1983, p23.
20. Maloney, C.; Blau, W., J. Opt. Soc. Am. B 1987, 4, 1035.
21. Mohebi, M.; Soileau, M.J.; Van Stryland, E.W., Opt. Lett. 1988, 13, 758.
22. Smith, W.L., Handbook of Laser Science and Technology; Weber, M.J, Ed.; CRC Press, Boca Raton, 1986; Vol III, Part 1, 259.
23. Witte, K.J.; Galanti, M; Volk, R., Opt. Commun. 1980, 34, 278.

RECEIVED July 10, 1990

Chapter 42

Nonlinear Optical Properties of Substituted Phthalocyanines

James S. Shirk, J. R. Lindle, F. J. Bartoli, Zakya H. Kafafi, and Arthur W. Snow

Naval Research Laboratory, Washington, DC 20375

The third order optical susceptibility was measured for a series of transition metal tetrakis(cumylphenoxy)phthalocyanines at 1.064 μm. Metal substitution caused a dramatic variation in the third order susceptibility. The largest $<\gamma_{xxxx}>$'s were found in the Co, Ni, and Pt complexes. Metal substitution introduces low lying electronic states which can enhance the susceptibility in these phthalocyanines. A strategy for enhancing the figure of merit, $\chi^{(3)}/\alpha$, of centrosymmetric nonlinear optical materials is suggested.

In a recent communication we reported that the third order nonlinear optical susceptibility of Pt, Pb, and H_2 tetrakis(cumylphenoxy)phthalocyanines was large and varied substantially with the metal substituent. (1) The structure of these compounds is shown in Fig. 1. The susceptibility was measured by degenerate four-wave mixing at 1.064 μm, a wavelength far from the main absorption bands of phthalocyanines near 650 nm. The nonlinear susceptibility of the Pt phthalocyanine was about a factor of 9 larger than that of the Pb phthalocyanine and a factor of 45 larger than the metal free compound.

This paper is a more extensive survey of the influence of the metal on the hyperpolarizability of a series of the transition metal tetrakis(cumylphenoxy)-phthalocyanines ($MPcCP_4$). The compounds chosen were those most closely related to $PtPcCP_4$, the compound which showed the largest hyperpolarizibility in the previous study. Specifically, phthalocyanines substituted with the last four members of the first row transition metal series (Co, Ni, Cu, and Zn) and also with the Ni, Pd, Pt triad were prepared and studied. The near IR spectra of these tetrakis(cumylphenoxy)-phthalocyanines are briefly discussed. Speculation on how metal substitution can influence the third order susceptibility of a near centrosymmetric structure, like that of the phthalocyanines, is presented.

Experimental

The third order optical susceptibility was measured by degenerate four-wave mixing (DFWM). A single pulse at 1.064 μm with a full width at half maximum of 35 ps was selected from the output of a passively mode locked Nd/YAG laser and split into three

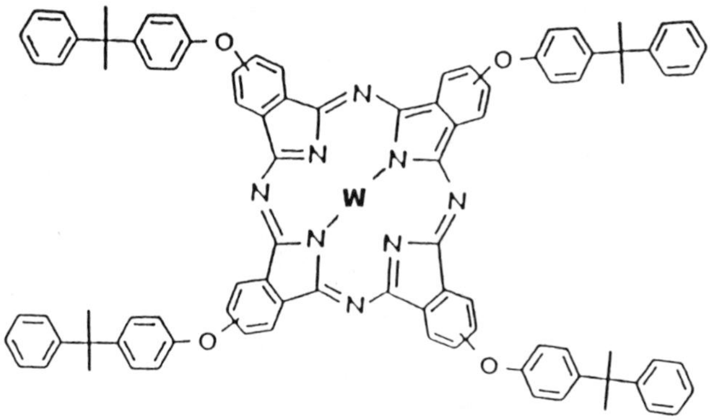

Figure 1. The structure of the metal tetrakis(cumylphenoxy)phthalocyanine (MPcCP$_4$). This is one resonance form; in the metal complexes, the phthalocyanine moiety has D$_{4h}$ symmetry.

beams. The beams were overlapped in the sample using a counter-propagating pump geometry. Time delays could be introduced into either the probe or the backward pump beam. The beams were weakly focussed onto the sample contained in a 0.2 mm thick glass or quartz cell. The laser intensities at the sample were ca 0.2 to 20 GW/cm^2 in each of the pump beams and 0.05 to 5 GW/cm^2 in the probe beam. The phase-conjugate reflection was detected with a Si photodiode. The temporal dependence of this signal was measured by delaying the arrival time of the back pump beam. All beams were polarized parallel to each other.

The preparation of the tetrakis(cumylphenoxy)phthalocyanines has been described.(2) The cumylphenoxy derivative was chosen because of its solubility in common organic solvents. The four-wave mixing experiments were performed on CHCl$_3$ solutions of the phthalocyanines with concentrations in the range of 5 x 10^{-3} M to 0.1 M. The concentration was chosen so that the solution $\chi^{(3)}$ was dominated by the phthalocyanine and the sample transmission was > 0.8. Most measurements were performed on solutions with concentrations near 10^{-2} M ($\sim$1% by weight). The measured transmission at 1.064 μm of the samples used for the four-wave mixing experiments ranged from .8 (for 10^{-2} M NiPcCP$_4$) to >0.99 for similar concentrations of PdPcCP$_4$, H$_2$PcCP$_4$ and ZnPcCP$_4$. Absorption spectra were recorded on a Cary/Varian Model 2300 spectrophotometer. Path lengths of 5 mm or 1 cm were used when necessary to obtain accurate absorbance measurements for weak bands.

Results

Spectroscopy A spectrum of PtPcCP$_4$, which is typical of these phthalocyanines is shown in Fig. 2. The most intense band is the Q band which occurs between 640 nm and 680 nm for the different metal phthalocyanines. It is the lowest allowed π - π* transition of the phthalocyanine ring. In dilute solution, the Q band of the monomer typically had a molar extinction coefficient of 2 x 10^5 l/mole-cm in agreement with previous reports. (2) Additional bands, which have been assigned to phthalocyanine aggregates (2)(3), were observed on the short wavelength side of the Q band in the relatively concentrated solutions used for the nonlinear optical studies.

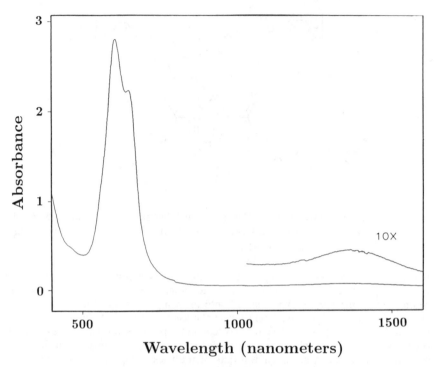

Wavelength (nanometers)

Figure 2. The spectrum of Pt tetrakis(cumylphenoxy)phthalocyanine (PtPcCP$_4$), 2.0 x 10^{-3} M in CHCl$_3$ solution. The inset shows the near IR region on an expanded absorbance scale. (Reproduced with permission from reference 1)

Some of the metal tetrakis(cumylphenoxy)phthalocyanines were found to have weak absorptions in the region 1.1 - 1.5 μm. The band for PtPcCP$_4$ can be seen on an expanded scale in Fig. 2. The λ_{max} and the molar extinction coefficient, ε, for the near IR band in each of the metal phthalocyanines studied here are given in Table 1. The near IR bands were very much weaker than the Q band. The strongest absorptions, in the Pt, Ni, and Co complexes, were more than 2 orders of magnitude weaker than the Q band. The Cu and Pb phthalocyanine absorptions were about four orders of magnitude weaker than the Q band. The Zn, Pd, and H$_2$ (metal-free) complexes showed no distinct bands in this region. The nonlinear optics experiments were performed at 1.064 μm, far from resonance with the Q band and except for CoPcCP$_4$, above the near IR band.

Nonlinear Optical Measurements The magnitude of the phase-conjugate reflection for each of the phthalocyanines in CHCl$_3$ solution was measured as a function of laser intensity. In each case the signal was much larger than that observed for pure solvent. For the Co, Ni, Cu, and Pt phthalocyanines the intensity of the phase-conjugate reflection was found to depend upon the laser intensity to the 3.0 ± 0.3 power with no evidence of saturation up to approximately 15 - 20 GW/cm^2 in each pump beam. For Pd, Zn, and H$_2$PcCP$_4$, the slope was significantly greater than a pure cubic dependence. Such deviations from a cubic dependence might occur because at the highest powers used in these experiments, the two photon transition probability can become significant. If two photon absorption produces a population grating in the sample, the signal arising from diffraction by such a population grating will depend upon the laser intensity to the fifth power.(4) In this paper, we are primarily interested in the third order response, so the signal for Pd, Zn and H_2PcCP_4 was fit to a curve of the form $a_3I^3 + a_5I^5$, and the cubic contribution was used to obtain $\chi^{(3)}$. The phase-conjugate signal as a function of laser intensity for $PdPcCP_4$ is shown in Fig. 3. The best fit to this data shows that the major part of the observed signal was due to the cubic term. In H_2PcCP_4 the fifth power term was more significant.

$\chi^{(3)}$ for each solution was obtained by comparison with a CS_2 reference using(5):

$$\chi^{(3)} = \chi^{(3)}_{ref} \sqrt{\frac{S}{S_{ref}}} \left(\frac{n}{n_{ref}}\right)^2 \left(\frac{l_{ref}}{l}\right) \left(\frac{\alpha l}{e^{-\left(\frac{\alpha l}{2}\right)} \left(1-e^{-\alpha l}\right)}\right) \tag{1}$$

where $S = a_3$, the coefficient of the cubic term of a least squares fit of the phase-conjugate signal vs the laser intensity, l is the sample path length, n the refractive index, and α is the absorption coefficient at 1.064 μm. The subscript " *ref* " refers to CS_2, for which a value of $\chi^{(3)}_{xxxx} = 4 \times 10^{-13}$ esu was used.(6)(7)

Table 1
Optical Properties of the metallo-phthalocyanines
at 1.064 μm

	$<\gamma_{xxxx}>$	$\chi^{(3)}_{xxxx}$ [a]	λ_{max} [b]	ϵ_{max}	$\sigma_{1.064}$	$\chi^{(3)}_{xxxx}/\alpha$ [c]
	(esu)	(esu)	(μm)	($M^{-1}cm^{-1}$)	(cm^2)	(esu-cm)
$CoPcCP_4$	5×10^{-32}	8×10^{-11}	1.15	400	1.5×10^{-18}	1×10^{-13}
			1.03	shoulder		
$NiPcCP_4$	4×10^{-32}	6×10^{-11}	1.20	1500	1.6×10^{-18}	1×10^{-13}
$CuPcCP_4$	3×10^{-32}	4×10^{-11}	1.10	~40	1.7×10^{-19}	5×10^{-13}
$ZnPcCP_4$	5×10^{-33}	7×10^{-12}			$<2 \times 10^{-20}$	$>7 \times 10^{-13}$
$PdPcCP_4$	1×10^{-32}	2×10^{-11}			2×10^{-19}	3×10^{-13}
$PtPcCP_4$	1×10^{-31}	2×10^{-10}	1.38	1600	2.4×10^{-18}	2×10^{-13}
$PbPcCP_4$	1×10^{-32}	2×10^{-11}	1.23	10	4×10^{-20}	2×10^{-12}
H_2PcCP_4	2×10^{-33}	4×10^{-12}			$<2 \times 10^{-20}$	$>4 \times 10^{-13}$

a) normalized to the concentration of the pure material
b) for the near IR band
c) measured in approximately 10^{-2} M solution

The molecular hyperpolarizibilities, γ, given in Table 1 were derived from the measured $\chi^{(3)}$ of the solutions using the expression:

$$<\gamma_{xxxx}> = \chi^{(3)}_{xxxx} / L^4 N \tag{2}$$

where N is the number density of the phthalocyanine and L is the local field factor. L is assumed to be a constant, in this case due to the solvent and given by:

$$L = (n^2 + 2)/3 \tag{3}$$

where n = 1.44 the refractive index of the solvent. Another procedure that has been used to compare the nonlinearities in new materials (8)(9) is to extrapolate to the $\chi^{(3)}$ of the pure solid material. Table 1 includes such an extrapolated $\chi^{(3)}_{xxxx}$ that was calculated using:

$$\chi^{(3)}_{xxxx} = (C_0/C) \chi^{(3)}_{xxxx \text{ solution}} \tag{4}$$

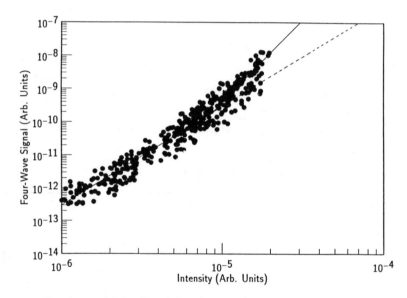

Figure 3. The observed intensity of the phase-conjugate reflection as a function of laser intensity for PdPcCP$_4$ (6 x 10^{-3} M in CHCl$_3$). The line is a fit to a curve of the form Signal = a$_3$I^3 + a$_5$I^5. The dashed line is the cubic term only. The deviations from a cubic dependence can be seen at high intensity.

where C is the concentration of the solution and C_0 the concentration of the phthalo-cyanine in the pure solid (0.73 moles/liter). In such an extrapolation for $\chi^{(3)}$, L is implicitly assumed to be constant and due to other phthalocyanine molecules rather than solvent. The experimental ratio $\chi^{(3)}_{xxxx}/N = L^4 < \gamma_{xxxx} >$ was found to be approximately constant for the Pt, Pb and H_2 phthalocyanines over the experimentally accessible range of concentrations up to about 1-10% by weight.(1) This data is consistent with the assumption that L is constant over the concentration range used in the current experiments but it does not establish which assumption about L is more realistic. Table 1 also includes the linear absorption cross section, σ, for the phthalocyanine at the laser wavelength, at the concentrations used in the four-wave mixing experiments and a figure of merit, $\chi^{(3)}/\alpha$, for the solutions measured. This figure of merit, the nonlinear response that can be achieved for a given absorption loss, is useful for comparing nonlinear materials in regions of absorption.

The temporal response of the phase-conjugate signal was measured by delaying the arrival time of the back pump beam. The response showed evidence for two components. The largest, accounting for 80 - 90% of the signal amplitude had a 35 psec FWHM, indicating a pulse width limited response time. The remaining signal had a decay time >500 psec, the maximum delay currently available in our apparatus.

Discussion

The experimental results reported in Table 1 give striking evidence that metal substitution can significantly enhance the third order susceptibility of these phthalocyanines. The $\chi^{(3)}_{pure}$ for the metal tetrakis(cumylphenoxy)phthalocyanines range from 7 x 10^{-12} esu to 2 x 10^{-10} esu. $\chi^{(3)}$'s of 5 x 10^{-11} esu and 2.5 x 10^{-11} esu at 1.06 μm were previously reported for the fluoro-aluminum and chloro-gallium phthalocyanines respectively from the third harmonic generation efficiencies.(10) The $\chi^{(3)}$'s for the phthalocyanines measured by THG are in the same range as those measured by four-wave mixing and reported in Table 1.

The data in Table 1 reveal systematic variations in the measured third order susceptibilities of these phthalocyanines with the metal. There is a monotonic variation of γ in the series Co, Ni, Cu, Zn. The nonlinear susceptibility decreases as the d orbitals of the metal become filled. There is also a qualitative correlation between a large hyperpolarizability and the presence of a weak, near IR transition. However, the variation of the figure of merit, $\chi^{(3)}/\alpha$, shows that the correlation between γ and the absorption coefficient is not linear. No clear trend is seen in the triad Ni, Pd, Pt; although PtPcCP$_4$ does have a larger hyperpolarizability as might be expected for a larger, more polarizable metal ion.

These phthalocyanines are known to form aggregates in solution, and metal substitution influences the tendency of these phthalocyanines to aggregate (11). Aggregation does not seem to account for the variations in the hyperpolarizibility reported here. Pt, Pd, and Ni tetrakis(cumylphenoxy)phthalocyanines are the most aggregated, forming, on the average, trimers and tetramers in the solutions studied here. The hyperpolarizibility of PdPcCP$_4$ is much smaller than that of the Pt or Ni complexes. On the other hand CoPcCP$_4$, which forms principally dimers, has a hyperpolarizibility comparable to NiPcCP$_4$ and much larger than the Pd complex. The other phthalo-cyanines form dimers except for Pb which is monomeric. For any one phthalocyanine the $\chi^{(3)}$ can be measured over about a factor of 50 in concentration. In a previous study, no significant variations in γ_{xxxx} with the degree of aggregation were found over the concentration range that could be studied.(1)

It seems more likely that the variation in γ are due to differences in the electronic structure of the metal phthalocyanines. Substitution of a metal ion for the hydrogens in a phthalocyanine introduces new low lying electronic states into the phthalocyanine electronic manifold. The electronic structure and spectroscopy of the metal phthalocyanines have been discussed extensively. For NiPc, as an example, the calculations of Schaffer, Gouterman, and Davidson (12) show several new ligand-to-metal charge transfer (LMCT) states which give allowed transitions in the spectrum. There should also be metal-to-ligand charge transfer (MLCT) states with a lower energy than the upper state of the Q band which involve transitions of the type d --> $e_g(\pi^*)$. Since these states are of g symmetry, transitions from the ground state will be forbidden. There are also states in the metal phthalocyanines due to d-d transitions on the metal ion itself. Some of these new states, especially those giving allowed transitions, have been identified in phthalocyanine spectra.(13)

The near IR spectra of the tetrakis(cumylphenoxy)phthalocyanines have not been reported before. The absorption in the Cu complex and one of the absorptions in the Co complex lie close to bands which have been tentatively assigned to trip-multiplet transitions in other phthalocyanines.(14)   However, the other absorption bands shown in Table 1 have not been previously reported for phthalocyanines with no peripheral substitution. The small absorption cross sections of these bands in the cumylphenoxy phthalocyanines suggest that they are forbidden transitions. Possible assignments for these bands include: a symmetry forbidden electronic transition (like the MLCT transitions in NiPc discussed above) becoming vibronically allowed, d-d transitions on the metal ion, or trip-multiplet transitions. Spectroscopic studies are in progress to provide a more definitive assignment of these absorptions.

The states that are introduced into the electronic manifold of the phthalocyanine by metal substitution will also affect the nonlinear optical properties. Such new states can alter the electronic part of the hyperpolarizability, and, in those cases where the metal introduces some absorption at 1.064 μm, can give an optically pumped grating that can contribute to the $\chi^{(3)}$ measured by four-wave mixing. However a population grating between the ground state and an excited state does not easily account for the $\chi^{(3)}$'s reported here. First, most of the observed signal has a pulse width limited decay time of <35 psec, much shorter than the lifetimes that have been reported for states in the near IR for an number of phthalocyanines. For example, a phosphorescence lifetime of 7 $micro$seconds with a quantum efficiency of 10^{-2} has been reported near 9.4 μm for Pt phthalocyanine with no peripheral substitution.(15)   Further, the $\chi^{(3)}$ due to an excited state grating can be related to the saturation intensity, I_s, of the optically pumped transition.(16) No saturation was observed at intensities as high as 30-40 GW/cm^2 for these cumylphenoxy phthalocyanines. For most of these phthalocyanines, an I_s > 40 GW/cm^2 implies a population grating $\chi^{(3)}$ substantially smaller than the observed $\chi^{(3)}$. For PtPcCP$_4$, for example, the implied population grating $\chi^{(3)}$ < 3 x 10^{-11} esu, compared to the measured value of 2 x 10^{-10} esu.in table 1. We conclude that a simple population grating is probably not the predominant mechanism for the 35 psec component of the observed $\chi^{(3)}$.

Another potential source of the enhancement in the hyperpolarizibility of phthalocyanines is the contributions of the new electronic states introduced by metal substitution. The usual theoretical expression for the hyperpolarizibility involves a sum of the contributions of each electronic state.(17) Introducing new electronic states expands the number of terms which may contribute to the hyperpolarizibility. A rigorous calculation of the contribution of the new states to the $\chi^{(3)}$ of these large molecules, or its variation with metal, would be difficult. It is interesting, however, that opening up the d shell occupancy in the transition metal phthalocyanines (as in the sequence: Zn, Cu, Ni, Co) increases the number of electronic states that may contribute to the hyperpolarizibility and it also enhances the observed $\chi^{(3)}$.

The hyperpolarizibility of PdPcCP$_4$ was anomalously low compared to the Ni and Pt complexes. This low hyperpolarizibility may be related to the differences in the low lying electronic states. There is, for example, no near IR band in PdPcCP$_4$ corresponding to those in the Ni and Pt complexes. The details of and the reason for these differences in electronic structure among the d^8 cumylphenoxy phthalocyanines will require further clarification.

The contributions of optically forbidden electronic states to the $\chi^{(3)}$ of centrosymmetric structures are of particular interest.(18) Each of the terms in a sum-over-states calculation of $\chi^{(3)}$ involves the product of transition moments between a sequence of four states. There are symmetry selection rules that govern which states which can contribute to the individual terms. In a centrosymmetric molecule the symmetry of the contributing states must be in a sequence g --> u --> g --> u --> g.(19) This means that all the non-zero terms in the summation which determines the hyperpolarizibility must include an excited electronic state of g symmetry (or the ground state) as an intermediate state. The tetrakis(cumylphenoxy)phthalocyanines are approximately centrosymmetric and many of the new electronic states in a metal phthalocyanine will be of g symmetry. Such states may well contribute to the dependence of the hyperpolarizibility on metal substitution.

If, in centrosymmetric molecules, states to which a transition is forbidden in the normal absorption spectrum can make important contributions to $\chi^{(3)}$, this suggests a strategy for enhancing the figure of merit, $\chi^{(3)}/\alpha$, of such a nonlinear material. Chemically introducing low lying states with gerade symmetry and thus small or zero absorption cross sections has the potential to enhance the $\chi^{(3)}$ but not increase the absorption probability, α.

Conclusions

The third order optical hyperpolarizibilities of the tetrakis(cumylphenoxy)-phthalocyanines substituted with four sequential members of the first transition series, Co, Ni, Cu, and Zn, and the Ni, Pd, Pt triad were measured at 1.064 µm, a wavelength far from the main absorption band. The third order susceptibilities were remarkably large and varied dramatically with the metal. A monotonic variation of γ with the metal was observed in the series Co, Ni, Cu, Zn where substitution with metals with a more open d shell gave larger nonlinearities. There was a qualitative correlation between a large hyperpolarizibility and the presence of a weak, near IR transition but the correlation between γ and the absorption coefficient was not linear. These absorptions are a manifestation of the new low lying electronic states introduced into the phthalocyanine electronic manifold by metal substitution. A population grating from optical pumping of the weak absorption at 1.064µm in some of the tetrakis(cumylphenoxy)-phthalocyanines does not easily account for the magnitude and time response of the observed $\chi^{(3)}$. Contributions of these new electronic states to the electronic hyperpolarizibility may account for a variation in $\chi^{(3)}$ with the metal. A quantitative estimate of the magnitude of their contributions is difficult and requires a more complete knowledge of the electronic states than is currently available.

This work suggests a way to influence the figure of merit, $\chi^{(3)}_{xxxx}/\alpha$, of centrosymmetric or near centrosymmetric nonlinear materials like the phthalocyanines.

Symmetry considerations show that chemical substitution to introduce low lying, one photon forbidden states into the molecule has the potential to enhance the $\chi^{(3)}$ without increasing the absorption probability, α.

Acknowledgments

This work was supported by the Office of Naval Research, the Office of Naval Technology and the Strategic Defense Initiative Organization, Innovative Science and Technology Program.

Literature Cited

1. J.S. Shirk, J.R. Lindle, F.J. Bartoli, C.A. Hoffman, Z.K. Kafafi and A.W. Snow, Appl. Phys. Lett. 55, 1287 (1989)

2. A.W. Snow and N.L. Jarvis, J. Am. Chem. Soc. 106, 4706, (1984)

3. M.J. Stillman and T.Nyokong, in "Phthalocyanines" ed. C.C. Leznoff and A.B.P. Lever, VCH, New York, p. 133-290 (1989)

4. G.C. Bjorkland, D.M. Burland, and D.C. Alvarez, J. Chem. Phys. 73, 4321 (1980)

5. R.G. Caro and M.C. Gower, IEEE J. Quant. Electr. QE-18, 1376 (1982)

6. M.J. Moran, C.S. She, R.L. Carman, IEEE J. Quant. Electr. QE-11, 259 (1975)

7. R.W. Hellwarth, Prog. Quant. Electr., 5, 1 (1977)

8. S.A.Jenekhe, S.K.Lo, S.R. Flom, Appl. Phys. Lett. 54, 2524 (1989)

9. D. Ricard, P. Roussignol, C. Flytzanis, Opt. lett. 10, 511 (1985);

10. Z.Z. Ho, C.Y. Ju, and W.M. Heatherington, J. Appl. Phys. 62, 716 (1987)

11. A.W. Snow and N.L. Jarvis, J. Am. Chem. Soc. 106, 4706 (1984)

12. A.M. Schaffer, M. Gouterman, and E.R. Davidson; Theor. Chim. Acta 30, 9 (1973)

13. for a recent review see: M.J. Stillman and T.Nyokong, in "Phthalocyanines" ed. CC Leznoff and A.B.P. Lever, VCH, New York, p. 133-290 (1989)

14. A.B.P. Lever, S.R. Pickens, P.C. Minor, S.Licoccia, B.S. Ramaswamy, and K. Magnell; J. Am. Chem. Soc. 103, 6800 (1981)

15. P.S. Vincett, E.M. Voigt, K.E. Reickhoff, J. Chem. Phys. 55, 4131, (1971)

16. M.A. Kramer, W.R. Tompkin, and R.W. Boyd; Phys. Rev. A, 34, 2026 (1986)

17. J.F. Ward, Rev. Mod. Phys 37, 1 (1965)

18. M.G. Kuzyk and C.W. Dirk, Phys. Rev. A, 41, 5098 (1990)

19. J.W. Wu, J.R.Heflin, R.A. Norwood, K.Y. Wong, O. Zamani-Khamiri, A.F. Garito, P. Kalyanaraman and J. Sounik; J. Opt. Soc. B, 6, 707 (1989)

RECEIVED September 4, 1990

SIGMA AND PI DELOCALIZED THIRD-ORDER NONLINEAR OPTICAL MATERIALS

Chapter 43

Nonlinear Optical Properties of Substituted Polysilanes and Polygermanes

R. D. Miller, F. M. Schellenberg[1], J.-C. Baumert, H. Looser, P. Shukla,
W. Torruellas, G. C. Bjorklund, S. Kano[2], and Y. Takahashi[2]

Almaden Research Center, IBM Research Division,
San Jose, CA 95120–6099

Polysilane high polymers, which contain only silicon in the polymer
backbone show interesting electronic properties which may be
ascribed to extensive sigma electron delocalization. Catenated silicon
linkages constitute therefore a highly polarizable yet thermally and
oxidatively stable alternative to pi electron conjugation. We have
measured $\chi^{(3)}$ values for third harmonic generation in a variety of
polysilane derivatives and one polygermane and find values in the
10^{-11}-10^{-12} esu range. Although the measured values depend on
polymer orientation and to some extent on film thickness, the
nonresonant numbers are relatively insensitive to backbone
conformation even though significant changes in the linear absorption
spectra are observed. The substituted silane polymers are also
characterized by strong two-photon absorption which often leads by
anisotropic photodestruction to a strong induced birefringence. The
spectral response of the two-photon induced birefringence in PDN6S
is identical to that determined by two-photon fluorescence excitation.
The $I_m \chi^{(3)}$ ($-\omega; \omega, -\omega, \omega$) associated with this resonant transition
is large ($\sim 6 \times 10^{-10}$ esu) and is associated with a significant value for
the nonlinear refractive index in the region around 570 nm.

Organic polymers are of interest for nonlinear optical studies because of their ease
of processing, intrinsically large nonlinearities, rapid response times and their high
threshold damage levels. The most commonly studied materials are conjugated
carbon backbone polymers such as polyenes and polydiacetylenes (1,2). Although
the nonlinearities of these materials are very large, they are often oxidatively and/or
thermally labile, sometimes difficult to process and always have strong absorptions
in the visible spectral region. While polyaryl and polyheteroaryl derivatives are
thermally and oxidatively stable and are amenable to polymer alignment techniques
because of their backbone rigidity, these materials often suffer from limited
processibility, nonideal spectral properties and reduced nonlinearities relative to the
polyacetylenes and polydiacetylenes.

[1] Current address: Ginzton Labs, Stanford University, Stanford, CA 94305
[2] Current address: IBM Tokyo Research Lab, 5-19 Sanban-cho, Chiyoda-Ku, Tokyo, 102 Japan

0097–6156/91/0455–0636$07.25/0
© 1991 American Chemical Society

The polymers described above are all characterized by extensive pi electron delocalization, a feature which is responsible for the remarkable spectral and electronic properties. On the other hand, polysilanes 1a and polygermanes 1b which contain only silicon or germanium in the polymer backbone also show very unusual electronic properties which have been attributed to extensive *sigma* electronic delocalization in the polymer backbone. In recent times, the synthesis of soluble silicon and germanium homo and copolymers has stimulated renewed scientific interest in these materials which has resulted in a number of potential applications (3). The polysilanes have been examined as (i) thermal precursors to silicon carbide, (ii) broad spectrum photoinitiators for vinyl polymerizations, (iii) a new class of amorphous polymers for photoconduction and charge transport, and (iv) new radiation sensitive materials for microlithography. Most recently, the polysilanes and polygermanes have been demonstrated to have interesting nonlinear optical (NLO) properties (4-8) and we will concentrate on these in this paper.

$$\{R^1R^2Si\}_n \qquad\qquad \{R^1R^2Ge\}_n$$
$$\underline{1a} \qquad\qquad\qquad \underline{1b}$$

Optical Response of Dielectric Materials

We must begin with the general description of electromagnetic propagation through a dielectric medium (9). This is well described classically by Maxwell's equations. The driving term for the propagation of the oscillating electromagnetic field is the bulk material polarization $\vec{P}$. This can be expanded as a Taylor series, relating the vector components of the polarizability to the various applied fields through susceptibility tensors $[\chi^{(n)}]$:

$$P_j(\omega_4) = \left\{ [\chi_{ij}^{(1)}]E_j(\omega_1) \right.$$
$$+ [\chi_{ijk}^{(2)}]E_j(\omega_1)E_k(\omega_2) \qquad\qquad (1)$$
$$\left. + [\chi_{ijkl}^{(3)}]E_j(\omega_1)E_k(\omega_2)E_l(\omega_3) + \cdots \right\}.$$

In this notation, the subscripts i, j, k, l etc. represent spatial (xyz) orientations, so the various $[\chi^{(n)}]$ susceptibility tensors are matrices relating the kth, jth, etc., components applied field(s) to the ith component of the resulting output field. All the elements of the $[\chi^{(n)}]$ tensors are therefore essentially transfer functions between the applied electric field(s) and the generated field. For materials with inversion symmetry, all terms of the $[\chi^{(even)}]$ tensor are zero, leaving only $[\chi^{(1)}]$, $[\chi^{(3)}]$, etc., with nonzero elements.

The second rank linear susceptibility $[\chi^{(1)}]$ is more familiar as the linear dielectric response of a material. The nonlinear term $[\chi_{ijkl}^{(3)}]$ presents a more complicated situation. This is a fourth rank tensor, with three independent frequency arguments for each of the applied fields. Although the values of the elements of $[\chi^{(3)}]$ are usually smaller by many orders of magnitude than those of $[\chi^{(1)}]$, application of strong DC fields or intense laser pulses can provide E fields high enough to generate a non-negligible polarizability in the material. Many different nonlinear effects can be so produced, depending on the frequency and orientation of the various applied fields.

One of the more commonly measured components of $\chi^{(3)}$ is third harmonic generation. For this interaction, a single laser beam of the necessary intensity is focused into the nonlinear material, and can be viewed as three simultaneously superimposed input fields. These combine through the nonlinear tensor to produce higher energy photons at the third harmonic, which are separated from the fundamental photons by filters and counted. The strength of the third harmonic therefore depends on the magnitude of the nonlinear coefficient.

Other commonly measured $\chi^{(3)}$ interactions are those occurring with degenerate frequencies. Such interactions include phenomena such as four-wave-mixing, self-focusing, two-photon absorption, and CARS. If a single laser beam is focused into a material, the three applied fields interacting through the nonlinear tensor are coincident, and the interpretation is very similar to that for the dielectric tensor $[\chi^{(1)}]$; the imaginary part of the tensor elements $\operatorname{Im} \chi^{(3)}$ corresponds to an intensity dependent absorption, while the real part $\operatorname{Re} \chi^{(3)}$ corresponds to an intensity dependent index of refraction. In both cases, the electromagnetic field couples two energy levels separated in energy by $2\hbar\omega$. One therefore generally writes these as corrections to the linear absorption and refractive index:

$$\alpha_{tot} = \alpha_0 + \beta I ; \quad n = n_0 + n_2 I \tag{2}$$

where $\alpha_0 \propto \operatorname{Im}(1 + 4\pi\chi^{(1)})^{1/2}$ and $n_0 \propto \operatorname{Re}(1 + 4\pi\chi^{(1)})^{1/2}$ correspond to the linear absorption and refractive index, $\beta \propto \operatorname{Im} \chi^{(3)}$, and $n_2 \propto \operatorname{Re} \chi^{(3)}$ are the nonlinear absorption and refractive indices, and I is the intensity $\propto E^2$ of the incident optical field. Since nonlinear effects are usually negligible at intensities below several MW/cm^2, typical units for the nonlinear coefficients are cm/MW for β and cm^2/MW for n_2. In general however, CGS units are used to describe nonlinear coefficients and values for nonlinear optics, so $[\chi^{(3)}]$ itself is often reported in esu (electrostatic units). Although $\chi^{(3)}$ interactions are usually smaller by orders of magnitude when compared to $\chi^{(1)}$ effects, the nonlinear effects are highly frequency dependent and can be significantly larger "on resonance." Application of DC fields or intense laser pulses can then provide E fields high enough to generate a non-negligible polarization in materials. For a more complete description of nonlinear effects, see the excellent texts by Shen ([10]) or Levenson ([11]).

The nonlinear absorption between two states separated by energy $2\hbar\omega$ is better known as two-photon absorption. Two-photon transitions are highly allowed only if the initial and final states of the transition are the same parity. Two-photon spectroscopy is therefore inherently complimentary to spectra observed with $\chi^{(1)}$ interactions, which couple only levels of opposite parity. These two-photon resonances can also significantly increase the nonlinear response of the material in other $\chi^{(3)}$ interactions such as four wave mixing or nonlinear waveguide switching.

Polysilanes and Polygermanes

Recent studies of polymeric Group IV catenates (in particular, polysilanes and polygermanes, 1a,b have demonstrated that there is significant sigma electronic delocalization along the polymer backbone which is responsible for many of the curious electronic properties of this class of materials [3].

The chromophore is the polymer backbone itself and the longest wavelength electronic transition is best described as $\sigma\sigma^*$. This transition is strongly polarized along the polymer backbone and is very intense ($\varepsilon/SiSi \sim 5000 - 25,000$).

The transition energies depend on the nature of the substituent ([12]), the polymer molecular weight ([12]) and surprisingly even the conformation of the backbone itself ([13]). The latter effect was unanticipated for a sigma bonded system but has been rationalized computationally using both ab initio ([14]) and semiempirical techniques (15). The effect of backbone conformation on the absorption spectra is dramatically illustrated for poly(di-n-hexylsilane)(PDN6S) and poly(di-n-hexylgermane)(PDN6G) in Figure 1. For these materials, side chain crystallization enforces a trans planar backbone conformation below the side chain melting transition (42°C and 12°C for PDN6S (14) and PDN6Ge (17) respectively). Above the transition, the side chains melt and the polymer backbones become conformationally disordered, and only the short wavelength transition is observed.

Early MNDO calculations suggested that the molecular polarizability of the ground state of a linear polysilane chain should exceed that of a fully conjugated polyene of comparable length (17). More recent *ab initio* studies have generally supported these initial conclusions and have suggested that the polarizability of a polysilane catenate along the chain axis should be larger than that of a conjugated polyene for chain lengths containing up to 75 pairs of silicon atoms or double bonds (18). Beyond this, it is predicted that the polarizability of the polyene will exceed that of the polysilane. In most respects, the electronic properties of silicon and germanium catenates actually more closely resemble those of conjugated polyenes than those of saturated carbon backbone polymers (3).

The polysilanes and polygermanes are (i) soluble in common organic solvents from which they produce high optical quality films, (ii) oxidatively and thermally stable and (iii) extremely strongly absorbing in the UV spectral region but transparent in the visible. Furthermore, they are easily oriented by standard polymer techniques (e.g., stretching, flow extension, etc.) (15,19,20) and are imageable to high resolution by UV light and ionizing radiation (21,22). Irradiation causes chain scission to occur reducing the molecular weight of the polymer chain which results in a rapid spectral bleaching of the long wavelength absorption band. The model of substituted silane polymers which is emerging is one of isolated and partially decoupled chromophores comprised of segments of the polymer backbone (15). Calculations suggest that these segments may consist of trans or nearly trans segments of varying lengths which are partially electronically decoupled from one another by conformational kinks or twists. Furthermore, these calculations suggest that the spectral characteristics will vary with segment length and that excitation energy can be localized in these segments even when they are quite short. The excitation energy finds its way to the longest segments either by direct absorption or by rapid energy transfer. Subsequent chain scission of the longest segments results in a reduction in length and a continual blue shifting of the absorption maximum with absorbed dose. This unusual collection of chemical and electronic properties suggests that group IV catenates should exhibit a variety of interesting nonlinear optical (NLO) properties.

All of the polysilanes used in this study were prepared by the modified Wurtz-type coupling of substituted dichlorosilanes by sodium dispersion in an inert aromatic solvent as previously described (23,24). Poly(di-n-hexylgermane) was similarly prepared from di-n-hexyldichlorogermane (16). All of the polymer samples were of high molecular weight ($\overline{M}_W \sim 1 \times 10^{-5} - 3 \times 10^6$ Daltons) and were coated on 1 mm thick quartz substrates by solution spinning techniques. The film thickness of the samples varied from 0.05-1.2 μm.

Third Harmonic Generation $\chi^{(3)}(- 3\omega; \omega, \omega, \omega)$

We have studied the third harmonic generation for a variety of polysilanes and a single polygermane (5). A high power Q-switched laser is focused onto a thin film of the polymeric material on a quartz substrate which was mounted in a vacuum chamber on a temperature controlled stage which could be aligned relative to the laser beam and polarization using a computer controlled 3-axis goniometer.

A fundamental wavelength at 1.064 μm was provided by a Nd:YAG laser operated at a repetition rate of 10 Hz and a pulse duration of 15 ns. Pulses at 1.907 μm were generated by Raman shifting the 1.064 μm pulses in a high pressure

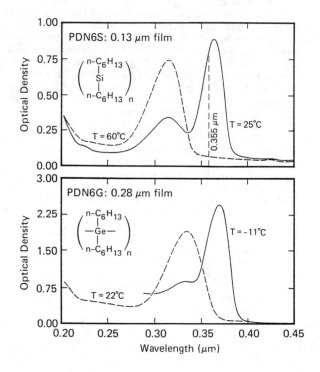

Figure 1. Variable temperature absorption spectra of PDN6S (top) and PDN6G (bottom).

(140 kPa) hydrogen cell. With typical peak power densities of $8MW/cm^2$ at the sample, third harmonic signals at 355 or 636 nm could easily be detected with good signal to noise ratios.

The third order susceptibilities of a number of polysilanes and a polygermane for frequency tripling $[\chi^{(3)}(-3\omega; \omega, \omega, \omega)]$ were measured by Maker fringe techniques. This technique allows the measurement of the $\chi^{(3)}$ value of the polymer relative to the known value for quartz ($\chi_q^3 = 3.1 \times 10^{-14}$ esu at 1.064 μm) by monitoring the fluctuating intensity of the third harmonic signal generated as a function of the angle of incidence of the polarized laser light on the sample ([25]). The fringes are created by the rotation of the sample along an axis which is parallel to the fundamental beam polarization. Since many of the samples have strong absorptions at 3ω, care was taken to include accurate measurements of both the polymer refractive index and absorption coefficient at 3ω in the analysis. Failure to do so would have resulted in a value for $\chi^{(3)}$ which was 100% too large ([5]).

$\chi^{(3)}$ values for a number of amorphous polysilanes as measured by the Maker fringe technique are reported in Table I. These values are quite large for a sigma-bonded system as predicted by the highly polarizable nature of the polymer backbone and the intense $\sigma\sigma^*$ electronic transition in the ultraviolet. In some cases, the numbers are as large as those reported for a number of pi conjugated polymers ([26-29]). The measured $\chi^{(3)}$ value for poly(methylphenylsilane)(PMPhS) at 1.064 μm is significantly larger than at 1.097 μm due to the absorption resonance of the third harmonic. A similar conclusion has recently been reported by Morichère et al. based on Kerr effect studies on PMPhS ([30]). It is assumed that the $\chi^{(3)}$ values measured for the other aryl substituted polysilanes in Table I are similarly larger than expected for the nonresonant values based on the similarity of their absorption spectra to PMPhS, although these materials were not studied at 1.907 μm. We have also observed that the $\chi^{(3)}$ value measured for PMPhS was somewhat larger in thin films. This variation with film thickness was also observed for other polysilanes (vide infra, Table II, entries 1a-c) and *may* result from some orientation produced in the thin films during spin coating but this effect is not well understood at this time. Anisotropy in thin films of certain polysilanes produced by spin coating has also been detected by IR-dichroism studies ([31]).

Since many of the aromatic polymers studied [e.g., poly(n-hexylphenylsilane)] are also quite rigid in solution and optical microscopy studies on concentrated solutions often show signs of long range order, a partially oriented sample of this material was prepared by shear flow extension. Third harmonic measurements at 1.064 μm on partially oriented films prepared in this manner were quite anisotropic ($\chi_{\parallel}^{(3)} = 9.2 \times 10^{-12}$ esu vs $\chi_{\perp}^{(3)} = 1.6 \times 10^{-13}$ esu) confirming that polymer orientation can significantly enhance the observed $\chi^{(3)}$ values ([32,33]). Similar orientational anisotropy for third harmonic generation is also observed in polysilane multilayers prepared by Langmuir-Blodgett techniques ([34]).

Since the electronic absorption spectra of polysilanes and polygermanes also depend strongly on conformation ([3]), these materials provide a unique opportunity to study the effect of backbone conformation on the NLO properties, particularly since varying alkyl substituents can dramatically influence the backbone structure through intermolecular interactions (e.g., side chain crystallization) while causing little substituent perturbation of the electronic structure. For this reason, $\chi^{(3)}$ measurements were performed on a number of polymers of established structure and the results are shown in Tables II and III.

In the cases of PDN6S and PDN6G (Table II), the polymer backbones are known to be predominantly trans planar below their respective transition temperatures ([13,16]). Similarly, the diaryl derivative poly(bis-p-butylphenylsilane)

Table I. $\chi^{(3)}(-3\omega; \omega, \omega, \omega)$ values for a number of amorphous polysilane derivatives at room temperature

Entry	λ_ω (nm)	Temp. °C	ℓ_p^a (nm)	λ_{max} (nm)	$\chi^{(3)} \times 10^{12}$ (esu)
Poly(methylphenylsilane)(PMPhS)					
1a	1064	23	120	339	7.2
1b	1907	23	120	339	4.2
1c	1907	23	1200	339	1.9
Poly(ethylphenylsilane)(PEPhS)					
2	1064	23	120	338	5.3
Poly(n-hexylphenylsilane)(PN6PhS)					
3	1064	23	120	346	6.2
Poly(t-butylphenylmethylsilane)(PT4PhMS)					
4	1064	23	120	332	4.9
Poly(n-hexyl-n-pentylsilane)(PN6N5S)					
5	1064	23	120	318	2.3
Poly(4-methylpentylsilane)(P4MPS)					
6	1064	23	120	319	1.8

aPolymer film thickness. The film was spin cast onto a 1 mm thick quartz substrate.

Table II $\chi^{(3)}(-3\omega; \omega, \omega, \omega)$ values for a PDN6S and PDN6G
as a function of fundamental frequency and temperature

Entry	λ_ω (nm)	Transition Temp. °C	Backbone Conformation	ℓ_p^a (nm)	Temp. °C	λ_{max} (nm)	$\chi^{(3)} \times 10^{12}$ (esu)
Poly(di-n-hexylsilane)(PDN6S)							
1a	1064	42	Trans Planar	50	23	372	11.0
1b	1064	42	Trans Planar	120	21	372	5.5
1c	1064	42	Trans Planar	240	23	372	4.6
1d	1064	42	Amorphous	120	50	318	2.0
1e	1907	42	Trans Planar	240	23	372	1.3
1f	1907	42	Amorphous	240	53	318	0.9
Poly(di-n-hexylgermane)(PDN6G)							
2a	1064	12	Trans Planar	295	-11	372	6.5
2b	1064	12	Amorphous	295	23	338	3.3
2c	1907	12	Trans Planar	295	-11	372	1.4
2d	1907	12	Amorphous	295	23	338	1.1

[a]Polymer film thickness. The film was spin cast onto a 1 mm thick quartz substrate.

Table III. $\chi^{(3)}(-3\omega; \omega, \omega, \omega)$ values for a number of polysilane derivatives with regular backbone structures as a function of fundamental frequency and temperature

Entry	λ_ω (nm)	Transition Temp. °C	Backbone Conformation	ℓ_p^a (nm)	Temp. °C	λ_{max} (nm)	$\chi^{(3)} \times 10^{12}$ (esu)
Poly(bis-p-n-butylphenylsilane)(PBN4PhS)							
1a	1064	—	Trans Planar	215	25	398	5.2
1b	1907	—	Trans Planar	800	25	398	0.6
1c	1907	—	Trans Planar	800	100	398	0.9
Poly(di-n-butylsilane)(PDN4S)							
2a	1064	83	7/3 Helix	220	28	314	2.2
2b	1064	83	Amorphous	220	98	314	1.8
2c	1097	83	7/3 Helix	220	28	314	0.45
2d	1097	83	Amorphous	220	98	314	0.48
Poly(di-n-tetradecylsilane)(PDN14S)							
3a	1064	55	TGTG′	235	25	344	4.5
3b	1064	55	Amorphous	235	65	322	0.9
3c	1097	55	TGTG′	2330	25	344	0.23
3d	1097	55	Amorphous	2330	65	322	0.19
Poly(di-n-hexylsilane)(PDN6S)							
4a	1064	42	Trans Planar	240	23	372	5.0
4b	1064	42	Amorphous	240	50	318	2.1
4c	1907	42	Trans Planar	240	23	372	1.3
4d	1907	42	Amorphous	240	50	318	0.9

aPolymer film thickness.
The film was spin cast onto a 1 mm thick quartz substrate.

(PD4PhS) is suspected from its spectral properties (35) and from light scattering studies (36) to contain long trans segments, even in solution. The polymer backbone of the dialkyl derivatives poly(di-n-butylsilane) (PDN4S) and poly (di-n-tetradecylsilane) (PDN14S) are helical (7/3 helix) and TGTG' respectively in the solid state (3). The latter material is also strongly thermochromic with a transition temperature of 55°C. PDN4S is not thermochromic, but the absorption band (λ_{max} 315 nm) broadens significantly above 80°C.

It is interesting to note that the $\chi^{(3)}$ values measured for PDN6S and PDN6G are very similar both below their respective transition temperatures where the backbone is predominantly trans planar and above where the polymer is extended, but the backbone is conformationally mobile and disordered. For these examples at least, there seems to be little difference between the silicon and the germanium backbone polymer. It is probably not just a coincidence that the λ_{max} of the long wavelength transition for the trans planar form of both PDN6S and PDN6G is also practically identical (16). The convergence observed for the long wavelength absorption of high molecular weight catenates of silicon and germanium has also been predicted theoretically (37). The same level of theory suggests, however, that significant differences should be observed in the absorption characteristics for tin and lead catenates relative to silicon and germanium implying that the NLO properties of the former materials might be interesting.

For PDN6S, the phase change at 42°C can be detected by monitoring the change in the intensity of the third harmonic signal as a function of temperature both at 1.064 and 1.907 µm as shown in Figure 2 (38). This behavior is reversible and a characteristic hysteresis loop is results. The hysteresis observed upon cooling is due to the tendency of the polymer to supercool prior to side chain crystallization which initiates the backbone conformational change.

Table III shows $\chi^{(3)}$ values for other structurally regular substituted silane high polymers measured both at 1.064 and 1.907 µm. Examination of this data suggests relatively little difference between the polysilanes with nonplanar, yet regular structures and trans planar PDN6S which is included in the table for comparison. This result is a little surprising given that changes in backbone conformation can cause spectral absorption shifts of more than 60 nm.

In summary, it appears that polysilanes and polygermanes have large third order nonlinear susceptibilities, particularly for materials containing only a saturated sigma bonded polymer backbone. This is consistent with their large polarizabilities and the extensive sigma delocalization along the backbone. In some instances, the values measured for $\chi^{(3)}$ are comparable to those of a variety of pi-conjugated polymers (see Table IV for comparison). The picosecond response time recently measured for PMPhS by Kerr gate techniques is consistent with a purely electronic effect (7). The nonlinearities observed for these silicon and germanium catenates are dominated by the polymer backbone itself which comprises a relatively extended chromophore. For comparable substituents, there seems to be relatively little difference between the third order susceptibilities of the polysilanes and polygermanes. At 1.064 µm, polysilanes and polygermanes with trans planar backbones appear to have significantly larger $\chi^{(3)}$ values than their helical or atactic counterparts. Unfortunately, meaningful interpretation of these differences is complicated by resonance effects as revealed by the high temperature measurements and studies conducted at 1.907 µm. From the data, however, it seems safe to conclude that non-resonant values of $\chi^{(3)}(-3\omega; \omega, \omega, \omega)$ for the polysilanes are *not* particularly sensitive to backbone conformation, at least for the limited range of samples surveyed.

As expected, the third order susceptibilities vary significantly with polymer orientation. It seems unlikely however that this feature alone will ever increase the values by more than an order of magnitude and further significant improvements will probably require more highly polarizable substituents, the introduction of

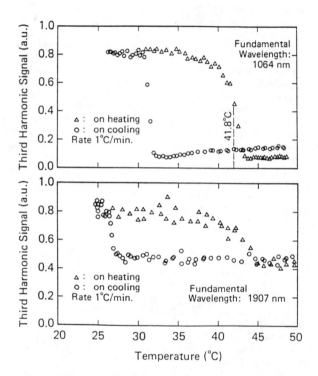

Figure 2. Hystereses curves for third harmonic generation in PDN6S vs temperature at two different fundamental wavelengths. At 1.9 μm there is a change of intensity of about a factor of two whereas at 1.06 μm there is a change of intensity of about six due to resonance effects.

multidimensional skeletal bonding, or the synthesis of block copolymers where the respective blocks comprise donor-acceptor units. These approaches are currently under investigation.

Degenerate Nonlinear Interactions: $\chi^{(3)}(-\omega; \omega, -\omega, \omega)$

Since initial experiments by third harmonic generation indicated $\chi^{(3)}$ is large in these polymers, we have subsequently studied some of the other nonlinear properties. For these studies, we have concentrated on poly(di-n-hexylsilane) (PDN6S). In an initial experiment designed to measure the nonlinear index of the material by a conventional four-wave mixing arrangement, it was found that exposure at intensities high enough to excite the nonlinear interaction induced permanent, anisotropic refractive index changes in the polymer films, with the axes of anisotropy corresponding to the polarization of the exposing laser. Spectroscopic examination of the exposed area also showed a shift in the UV absorption maximum to higher energy, indicative of chain scission, and a quadratic dependence of the effect on exposing power, consistent with absorption by a two-photon process (41).

This two-photon induced birefringence is pronounced, and the change in refractive index ($\Delta n \approx 0.03$, 632.8 nm) produced is large enough to form birefringent lithographic patterns for both passive and active waveguide devices (42,43). This effect has also proven to be a useful tool for the investigation of the nonlinear spectrum of the polymer (8). For these experiments a polarized, pulsed laser is focused onto the polymer film, which induces chain scission through two-photon absorption. A low power He-Ne laser polarized at $+45°$ relative to the polarization of the pulsed laser is focused to the center of the pulsed laser spot. The He-Ne light transmitted through the film then passes through an analyzer, oriented at $-45°$. When the polymer is isotropic, the light emerges from the film linearly polarized and is rejected by the analyzer. However, if the film becomes birefringent through photoexposure, the light emerges elliptically polarized. The component transmitted through the analyzer can then be related to the film birefringence through a straightforward calculation (8).

Typical data from such an experiment for PDN6S are seen in Figure 3. For multiphoton exposure, the birefringence initially rises until a saturation value is reached, then begins to diminish. This corresponds to the initial scission of the polymer chain segments aligned with the pulsed laser exposure and subsequent scission of polymer chain segments of other orientations.

This behavior is not unique to poly(di-n-hexylsilane). Detectable birefringence was produced in several polymers, but not in those compounds with grossly unsymmetrical sidechains, such as poly(methylphenylsilane) (PMPhS) (8).

We have modeled this growth in birefringence with a simple mathematical model, which assumes the scission probability for a polymer chain segment is proportional to the component of the pulsed laser polarization parallel to a particular backbone chain segment (44). The agreement has proven to be qualitatively quite good, with some deviation observed at long exposure times. The model may, therefore, require some small correction, such as a change in film thickness with exposure, or inclusion of the possibility of multiple scissions occurring in a single chain segment. However, the calculations generally support a model of localized scission for PDN6S.

Because two-photon induced chain scission produces such a pronounced effect, it is surprising that birefringence has not been reported from chain scission induced by single photon UV transitions. In fact, upon closer examination, exposure to polarized UV light can indeed produce a small degree of birefringence in thin films of poly(di-n-hexylsilane), as shown by the data marked (▲) in Figure 3. However, the saturation value with UV exposure is much lower than that observed with two-photon exposure, and the birefringence subsequently decreases much more

Table IV. $\chi^{(3)}$ values for a variety of

	Polymer	Fundamental
$\left[-\equiv-\overset{R}{\underset{R}{C}}=\right]_n$ $R = CH_2OTS$	Single Crystal	1.89 μm
$\left[-\equiv-\overset{R}{\underset{R}{C}}-\right]_n$ $R = (CH_2)_4OCONHC_4H_9$	Cast film	585 nm 605 nm
$\left[-\equiv-\overset{R}{\underset{R'}{C}}=\right]_n$ $R = CH_3(CH_2)_{15}$ $R' = -(CH_2)_8CO_2M$	Polymerized LB film	1.64 μm
(benzothiazole–phenylene polymer)	Film (biaxial)	585 nm 604 nm
$\left[\langle\bigcirc\rangle-CH=CH-\right]_n$	Film (amorphous)	1.85 μm
$[-\overset{Ph}{\underset{}{C}}=CH+$	Film (amorphous)	1.06 μm
$[(C_6H_{13})_2Si\rightleftharpoons]_n$	Film (unoriented) 23°C	1.06 μm
$[(C_6H_{13})_2Ge\rightleftharpoons]_n$	Film (unoriented) -11°C	1.06 μm
$(PMPhS)_n$	Film (amorphous) 23°C	1.06 μm

π-conjugated and Σ-conjugated polymers

$\lambda^{(3)}$ (esu)	Technique	Reference
5×10^{-10}	4-wave mixing	40
4×10^{-10} (red) 2.5×10^{-11} (yellow)	4-wave mixing	41
1.3×10^{-12} (red) 1.3×10^{-12} (blue)	Maker fringes	27
9×10^{-12}	4 wave mixing	28
7.8×10^{-12}	Maker fringes	29
7×10^{-12}	Maker fringes	30
4.6×10^{-12}	Maker fringes	5
6.5×10^{-12}	Maker fringes	5
7.2×10^{-12} 1.5×10^{-12}	Maker fringes	5 4

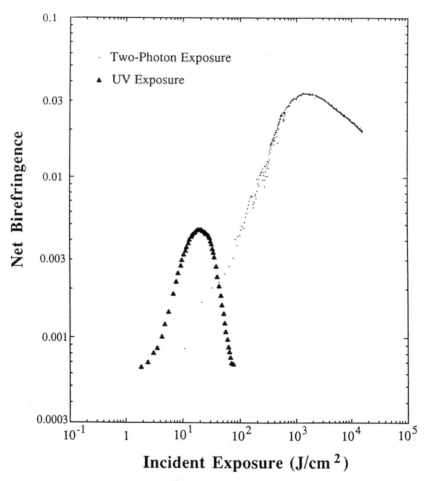

Figure 3. Birefringence (Δn) versus incident exposure in (J/cm^2) for polarized UV(▲) and polarized pulsed laser (●) exposure. The continuous 325 nm power density was 0.2 W/cm^2, while the 575 nm pulsed laser intensity (8 ns pulses at 10 Hz) was 1.6 GW/cm^2. (Reproduced with permission from Ref. 8. Copyright 1990 Elsevier Science Publishers.)

rapidly with continued exposure. Earlier work in solution has suggested that energy transfer processes between chain segments can occur in these polymers, resulting in rapid depolarization of fluorescence (15,45). It is therefore possible that the difference between the single-photon and two-photon birefringence growth behavior is the presence of fast energy transfer to (and subsequent chain scission of) chain segments of random orientations in the single photon case, while little energy transfer is necessary (as predicted by the simple projection model) to adequately describe the behavior with two-photon exposure.

Measurement of such birefringence growth curves for many pulsed laser wavelengths can be made to determine a spectrum of the birefringence effect. At low values of birefringence, the scission of nonaligned chain segments excited by energy transfer is negligible, and Δn is proportional to the number of aligned polymer backbone segments undergoing scission p_{sc} per unit area in time t. This in turn is related to multi-photon absorption by

$$\Delta n \propto p_{sc} \propto q\, p_{ab} \propto [\, 1 - e^{-\beta(\lambda)Iz}\,]It \approx q\beta(\lambda)IzIt\,, \qquad (3)$$

where q is the scission quantum yield (generally low, ≈ 0.01 in the solid state), p_{abs} is the number of photons absorbed per unit area, I is the laser intensity in W/cm^2, β is the two-photon absorption coefficient in cm/W, t is the exposure time in s, and z is the film thickness in cm. Assuming the scission yields are similar for both UV and two-photon excitation, this result can be calibrated by comparing the number of chain segments broken against the scission produced by a known UV exposure dose. The two-photon spectrum of a solid film of poly(di-n-hexylsilane), taken over the range 543-720 nm, can be seen in Figure 4. The main feature is a broad band, about 420 meV in width. This is comparable in width to the broad band single photon absorption. There is also a sharp spike at 579 nm (corresponding to a two-photon energy of 4.28 eV), with maximum value of $\beta = 1.2$ cm/MW (corresponding to Im $\chi^{(3)} \sim 6 \times 10^{-10}$ esu). The spike is quite reproducible, and, although the data are too sparse to determine the exact lineshape, a least squares Gaussian fit gives a linewidth of 33 meV; more than 10 times smaller than the overall linewidth of the broad two-photon absorption. This feature is unlike any typically observed in high molecular weight polysilanes in the solid state, although some sharp absorption features have been observed in samples of low molecular weight poly(di-n-hexylsilane) (46).

With single-photon exposure, excitations may decay either through a variety of processes including chain scission and fluorescence (47). We would therefore expect to observe fluorescence from two-photon excitation as well. To observe the fluorescence, we used a Spectra Physics mode locked dye laser system, operating with Rhodamine 560 dye. This was focused onto the polymer film, and the emitted light collected into a spectrometer with a Princeton Instruments Optical Multichannel Analyzer (OMA) attachment.

The spectrum of the emission was found to be identical to that from UV excitation: a 10 nm broad band centered at the low energy edge of the optical absorption (8,48). Measurement of the integrated intensity of this emission as a function of two-photon wavelength is shown in Figure 5. For the measurements at 77K, we observe the same broad band and sharp spike that were observed in the birefringence experiment at room temperature, even though little scission was observed at low temperature. At room temperature, where significant polymer scission is observed, the fluorescence is significantly reduced, as shown by the curve marked ($\blacktriangle$) in the figure. This suggests that scission and fluorescence are two competitive decay processes for the two-photon process, just as is found for single photon excitation (48-49).

These measurements have concentrated on the determination of β, which is proportional to Im $\chi^{(3)}$ (50). Associated with this transition should be a significant

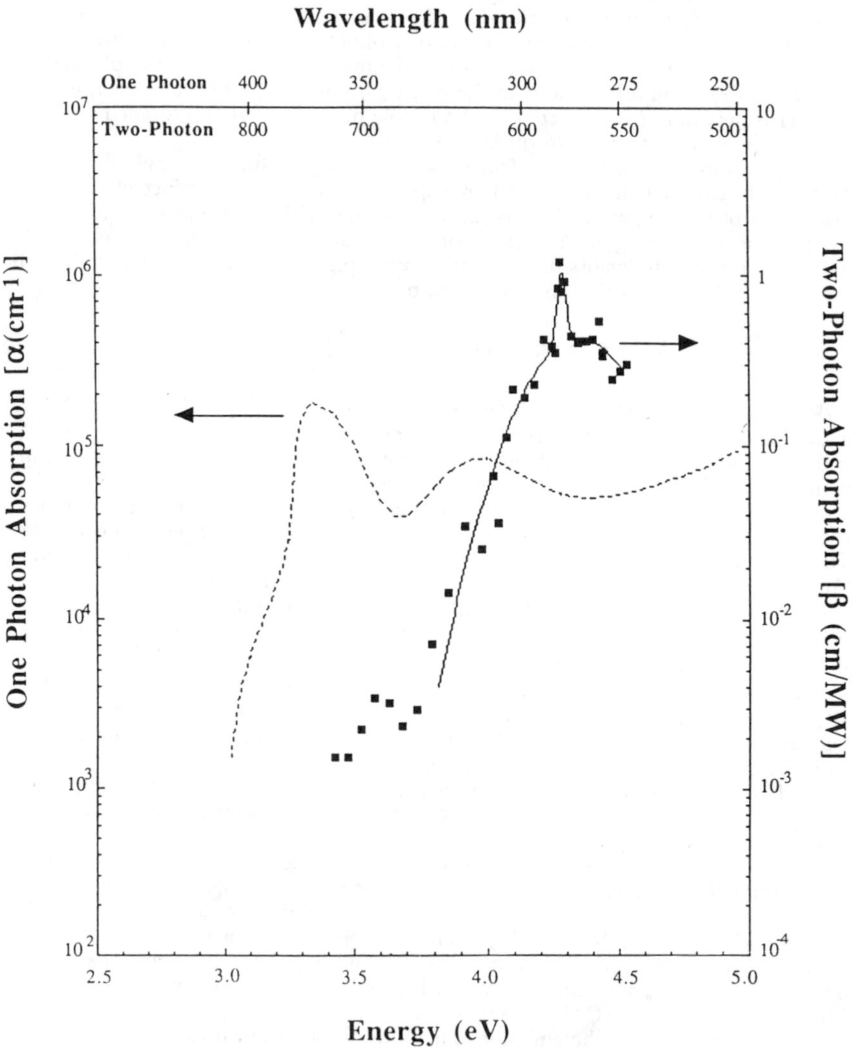

Figure 4. Spectrum of the two-photon absorption coefficient β in cm/MW for PDN6S, calculated from birefringenece growth curves as a function of two-photon energy (■) compared with single-photon absorption spectrum (– – –). Solid line is a least-squares fit to β using the sum of two Gaussians, one broad (~412 meV), the other narrow (~33 meV). (Reproduced with permission from Ref. 8. Copyright 1990 Elsevier Science Publishers.)

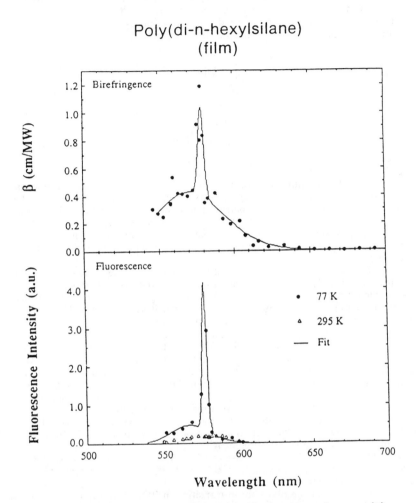

Figure 5. (upper) Plot of β calculated from birefringence growth curves (●) on a linear scale as a function of exposing wavelength. (lower) Linear plot of fluorescence vs exposing wavelength for both 77K (●) and room temperature (△). The solid line is a least-squares fit of the sum of two Gaussians to the low temperature data, adding one broad feature (~235 meV) and one narrow (~25 meV). (Reproduced with permission from Ref. 8. Copyright 1990 Elsevier Science Publishers.)

increase in $\mathrm{Re}\,\chi^{(3)}$ as well. Because these two components of $\chi^{(3)}$ are related through the standard Kramers-Kronig relationship (51), it should be possible to estimate the value of the nonlinear index of refraction n_2 for this polymer due to the two-photon resonance. This is given in Equation 4.

$$\mathrm{Re}\chi^{(3)}(\omega) = \frac{2}{\pi}\left[\lim_{\mathrm{E}\to 0}\int_0^{\omega-\mathrm{E}} f(\omega,\omega')d\omega' + \int_{\omega+\mathrm{E}}^{\infty} f(\omega,\omega')d\omega'\right] \qquad (4)$$

where

$$f(\omega,\omega') = \frac{\omega'^2\,\mathrm{Im}\chi^{(3)}(\omega')}{\omega'^2 - \omega^2} \qquad (5)$$

Using the computer program Mathematica, the calculation of Equation 4 is straightforward and produces the nonlinear dispersion shown in Figure 6. The curve shows the characteristic dispersion lineshapes due to the broad band and the sharp spike, and furthermore indicates that there is a spectral region with low two-photon absorption where the nonlinear index is significantly enhanced. Table V compares the n_2 value calculated with those of other common organic and inorganic materials.

The experimental value reported for PDN6S in Table V is very preliminary and is based on only three experimental points derived from prism-coupled waveguide reflectivity measurements. At this point, spurious effects due to heating, photoinduced birefringence etc. can not be conclusively ruled out. This preliminary data is presented only to provide a comparison with the calculated value.

This spectral region (640-700 nm) corresponds to a wavelength region where commercial diode lasers are available. Although these lasers are generally low power devices, when focused into a waveguide with a few microns cross sectional area, the intensity can be high enough to drive several proposed nonlinear switching devices. The possibility of fast intensity modulation of these laser sources, combined with the ease of fabricating and patterning polysilane waveguides, suggests that the polysilanes may have applications as materials for the demonstration of several nonlinear integrated optical devices that currently exist only on paper due to the lack of an appropriate nonlinear material. In this regard, Stegeman et al. (58) have proposed a figure of merit for a nonlinear directional coupler ($T = 2\beta\lambda/n_2 < 1$). In the case of PDN6S, using measured values for β and calculated n_2's, T values ranging from 0.1-1.0 for the wavelength range 580-650 nm have been calculated.

The nature of the two-photon transition in PDN6S is a question of some interest. Although the highest occupied molecular orbital of dialkyl polysilanes has been conclusively determined to involve the delocalized σ-bonded conjugated backbone (15,19,20), assignment of the excited states are still uncertain. Mintmire (14) and others (59) have extensively modeled these polymers as one-dimensional infinite chain semiconductors. In this model, the lowest energy $\sigma\sigma^*$ optical absorption at $\approx$ 3-4 eV corresponds to the direct band gap of these materials. These band structure calculations also predict the existence of bands approximately 1 eV above the $\sigma\sigma^*$ transition, coincident with the energy difference we observe in two-photon absorption. Mintmire has also shown that the calculated energy shifts observed for various regular backbone conformations also correlate well with those actually measured for a number of polysilanes of known polymer structure, lending some credibility to the model and the methods used.

On the other hand, the polysilanes are not strictly infinite one-dimensional chains. Several experimental studies (15,45,47) in solution have suggested that the polysilanes are trans or nearly trans planar only over 20-35 silicon atoms, restricted

Two-Photon Energy (eV)

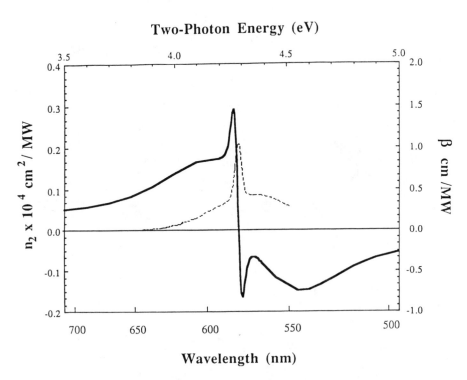

Figure 6. Calculated values for $n_2(-)$; generated from a Kramers-Kronig transformation of the fit to the two-photon absorption measurement $(- -)$.

Table V. Comparison of n_2 values for PDN6S with a number of other monomers and polymers. (a) See text for explanation

	$\lambda(\mu m)$	$n_2 \times 10^{12}$ (cm^2/W)	Reference
Silica	1.064	0.0006	53
$LiNbO_3$	1.064	2.5	53
GaAs	10.0	100	53
Sodium (vapor)	0.59	50	53
2-Methyl-4-nitroaniline	1.32	1.7	53
Lucite	1.064	0.0008	54
$R = -C_{16}H_{33}$, $R' = -(CH_2)_8-CO_2H$ (polymerized monolayer)	1.064	~1.0	57
$R = -CH_2OTs$ (single crystal)	2.62	26	59
$R = -(CH_2)_4OCONHCH_2CO_2C_4H_9$ (film - red form)	0.604	42	56
(film)	0.604	1.2	55
(film	1.064	5.0	58
PDN6S (calculated)	0.65	16	....
PDN6S (experimental)[a]	0.59	64	....

at each end by conformal kinks that partially electronically isolate the chain segment. Several recent papers have discussed the similarity of the excitations in polysilanes to π-conjugated carbon systems such as the polydiacetylenes (60), in which the lower energy transitions appear to be excitonic in nature. Kepler et al. (61) have suggested that the first direct UV transition in PMPhS may be due to a tightly bound exciton, in which an electron-hole pair created by the photon moves in concert down the polymer chain. Hochstrasser et al. (48) have proposed that the two-photon excitation might produce a more loosely bound exciton (charge transfer), in which the electron is transferred one or two silicon atoms from the hole. Recent electroabsorption measurements also lend some credibility to this excitonic model (62). Although the electron and hole in this charge transfer exciton case would still travel as a correlated pair, it is less likely that this type of exciton would remain bound during transfer through conformal kinks at the end of a chain segment. If correct, this may provide a possible explanation for the very different birefringence results for the two types of excitation, in which one appears to exhibit energy transfer while the other does not. Further studies are, however, necessary to determine if this excitonic model is, in fact, a satisfactory explanation of the electronic structure of the polysilanes.

Conclusions

In summary, the polysilanes and polygermanes constitute a new class of organic polymers with interesting nonlinear properties. They are soluble in common organic solvents and form high optical quality films. In addition, they are imageable with UV light to high resolution; a feature which could prove useful in the generation of patterned waveguides and nonlinear optical devices. These materials show large nonlinearities as measured by third harmonic generation consistent with electron delocalization in the sigma backbone. The nonlinearity resides primarily in the backbone which serves as an extended chromophore. The nonlinear response is very fast which is consistent with an electronic effect. Although the measured $\chi^{(3)}(-3\omega; \omega, \omega, \omega)$ depends on polymer orientation and to some extent on film thickness, the nonresonant values are relatively insensitive to the backbone conformation even though significant changes in the linear absorption spectra are observed. Furthermore, for simple alkyl substitution, there seems to be little difference between polysilanes and polygermanes for third harmonic generation.

The polysilanes are characterized by strong two-photon absorptions which lead to the production of a strong induced birefringence ($\Delta n \sim 0.03$; 632.8 nm) caused by anisotropic photodestruction. The spectral response of the two-photon induced birefringence identical to that determined by two-photon fluorescence excitation. In each case a strong resonance occurs around 570 nm (~ 4.28 eV two-photon energy). The Im $\chi^{(3)}(-\omega; \omega, -\omega, \omega)$ associated with this resonant transition is quite large $\sim (6 \times 10^{-10}$ esu) and the Kramers-Kronig transformation of the spectral response of $\beta(cm/MW)$ yields the anomalous dispersion curve for the nonlinear refractive index (n_2). Consideration of the magnitude of both $\beta(cm/W)$ and n_2 slightly off resonance suggests that PDN6S might be useful for optical switching using visible light if the two-photon photodecomposition could be suppressed by the incorporation of stabilizing additives.

Literature Cited

1. *Nonlinear Optical Properties of Organic and Polymeric Materials*; Williams, D. J., Ed.; ACS Symposium Series No. 233, American Chemical Society: Washington, D.C., 1983.
2. *Nonlinear Optical Properties of Organic Materials and Crystals*; Chemla, D. S., Zyss, J. eds., Vols I and II; Academic Press Inc.: New York, 1987.

3.   For a recent review of substituted silane high polymers, see:
     Miller, R. D.; Michl, J. Chem. Rev. 1989, 89, 1359.
4.   Kazjar, F.; Messier, J.; Rosilio, C. J. Appl. Phys. 1986, 60, 3040.
5.   Baumert, J.-C.; Bjorklund, G. C.; Jundt, D. H.; Jurich, M. C.; Looser, H.;
     Miller, R. D.; Rabolt, J.; Sooriyakumaran, R.; Swalen, J. D.; Twieg, R. J.
     Appl. Phys. Lett. 1988, 53, 1147.
6.   McGraw, D. J.; Siegman, A. E.; Wallraff, G. M.; Miller, R. D. Appl.
     Phys. Lett., 1989, 84, 1713.
7.   Yang, L.; Wang, Q. Z.; Ho, P. P.; Dorsenville, R.; Alfano, R. R.;
     Zou, W. K.; Yang, N. L. Appl. Phys. Lett. 1988, 53, 1245.
8.   Schellenberg, F. M.; Byer, R. L.; Miller, R. D. Chem. Phys. Lett. 1990,
     166, 331.
9.   For a general detailed description see Optical Waves in Layered Media;
     Yeh, P. Ed. J. Wiley and Sons Inc., New York, 1988.
10. Shen, Y. The Principles of Nonlinear Optics; J. Wiley and Sons, Inc.;
     New York, 1984.
11. Levenson, M.; Kano, S. Introduction to Nonlinear Spectroscopy,;
     Academic Press Inc.; New York, 1988.
12. Trefonas, P.; West, R.; Miller, R. D.; Hofer, D. J. Polym. Sci., Polym.
     Lett. Ed. 1983, 21, 823.
13. Miller, R. D.; Rabolt, J. F.; Sooriyakumaran, R.; Fleming, W.;
     Fickes, G. N.; Farmer, B. L.; Kuzmany, H. In Inorganic and
     Organometallic Polymers; Zeldin, M.; Wynne, K. J., Allcock, H. R., Eds.;
     ACS Symposium Series No. 360, American Chemical Society:
     Washington, D.C., 1987, Chap. 4 and references cited therein.
14. Mintmire, J. W. Phys. Rev. 1989, B39, 350.
15. Michl, J.; Downing, J. W.; Karatsu, T.; Klingensmith, K. A.;
     Wallraff, G. M.; Miller, R. D. In Inorganic and Organometallic
     Polymers; Zeldin, M., Wynne, K. J., Allcock, H. R., Eds.; ACS
     Symposium Series No. 360, American Chemical Society: Washington,
     D.C., 1987, Chap. 5.
16. Miller, R. D.; Sooriyakumaran, R. J. Polym. Sci., Polym. Chem. Ed. 1987,
     25, 111.
17. Bigelow, R. W.; McGrane, K. M. J. Polym. Sci., Polym. Phys. Ed. 1986,
     24, 1233.
18. Dehalle, J.; Champagne, B.; Dory, M.; Fripiat, J. G.; André, J. M. Bull.
     Soc. Chem. Belg. 1989, 98, 811.
19. Harrah, L. A.; Zeigler, J. M. Macromolecules 1987, 20, 601.
20. McCrary, V. R.; Sette, F.; Chen, C. T.; Lovinger, A. J.; Robin, M. B.;
     Stöhr, J.; Zeigler, J. M. J. Chem. Phys. 1988, 88, 5925.
21. Miller, R. D.; Hofer, D.; Rabolt, J.; Sooriyakumaran, R.; Willson, C. G.;
     Fickes, G. N.; Guillet, J. E.; Moore, J. In Polymers for High Technology:
     Electronics and Photonics; Bowden, M. J., Turner, S. R., Eds.; ACS
     Symposium Series No. 346, American Chemical Society: Washington,
     D.C., 1986, Chapter 15.
22. Miller, R. D.; Wallraff, G. M.; Clecak, N.; Sooriyakumaran, R.; Michl, J.;
     Karatsu, T.; McKinley, A. J.; Klingensmith, K. A.; Downing, J. In
     Polymers in Microlithography: Materials and Processes; Reichmanis, E.,
     MacDonald, S. A., Iwayanagi, T., Eds.; ACS Symposium Series No. 412,
     American Chemical Society: Washington, D.C., 1989, p. 115.
23. Trefonas III, P.; Djurovich, P. I. Zhang, X.-H.; West R.; Miller, R. D.;
     Hofer, D. J. Polym. Sci. Polym. Lett. Ed. 1983, 21, 819.
24. Miller, R. D.; Hofer, D.; McKean, D. R.; Willson, C. G.; West, R.;
     Trefonas III, P. T. In Materials for Microlithography;
     Thompson, L. F., Willson, C. G., Fréchet, J. M. J., Eds.; ACS
     Symposium Series No. 266, American Chemical Society: Washington,
     D.C., 1984, p. 293.

25. See Kajzar, F.; Messier, J. In Nonlinear Optical Properties of Organic Molecules and Crystals, Chemla, D. S., Zyss, J., Eds.; Academic Press, Inc.: New York, 1987, Vol. 2, Chap. III-2 for an excellent description of the experimental details of the Maker fringe technique.
26. Kajzar, F.; Messier, J.; Zyss, J. J. Phys. France 1983, C3-709, 44.
27. Rao, D. N.; Swiatkiewicz, J.; Chopra, P.; Ghoshal, S. K.; Prasad, P. N. Appl. Phys. Lett. 1986, 48, 1187.
28. Kaino, T.; Kubodera, K.; Tomaru, S.; Kurihara, T.; Saito, S.; Tsutsui, T.; Tokito, S. Electron. Lett. 1987, 23, 1095.
29. Neher, D.; Wolf, A.; Bubeck, C.; Wegner, G. Chem. Phys. Lett. 1989, 163, 116.
30. Morichère, D.; Dentan, V.; Kajzar, K.; Robin, P.; Lévy, Y.; Dumont, M. Optical Commun. 1989, 74, 69.
31. Miller, R. D.; Sooriyakumaran, R.; Farmer, B. L. Bull. Am. Phys. Soc. 1987, 32, 886.
32. Singh, B. P.; Prasad, P. N.; Karusz, F. E. Polymer 1988, 29, 1940.
33. Tomaru, S.; Kubaodera, K.; Zembutsu, S.; Takeda, K.; Hasegawa, M. Electronic Lett. 1987, 23, 595.
34. Wegner, G.; Neher, D.; Embs, F.; Willson, C. G.; Miller, R. D. (unpublished results).
35. Miller, R. D.; Sooriyakumaran, R. J. Polym. Sci., Polym. Lett. Ed. 1987, 25, 321.
36. Cotts, P. H.; Miller, R. D.; Sooriyakumaran, R. In Silicon-Based Polymer Science; Zeigler, J. M., Fearon, F. W. G., Eds.; Advances in Chemistry Series No. 224, American Chemical Society: Washington, D.C., 1990, p. 397.
37. Pitt, C. G. In Homoatomic Rings, Chains and Macromolecules of Main Group Elements; Rheingold, A. L.; Ed.; Elsevier: Amsterdam, 1977, pp. 203-234.
38. See Reference 40 for another example of a polymer phase change detected by NLO techniques.
39. Sauteret, C.; Hermann, J. P.; Frey, R.; Pradère, F.; Ducuing, J.; Baughman, R. H.; Chance, R. R. Phys. Rev. Lett. 1976, 36, 956.
40. Rao, D. N.; Chopra, P.; Ghosal, S. K.; Swiatkiewicz, J.; Prasad, P. N. J. Chem. Phys. 1986, 84, 7049.
41. Schellenberg, F. M.; Byer, R. L.; Miller, R. D.; Sooriyakumaran, R. XVI International Conference on Quantum Electronics Technical Digest, Japan Soc. Appl. Phys., Tokyo, 1988, 702.
42. Schellenberg, F. M.; Byer, R. L.; Zavislan, J.; Miller, R. D. In Nonlinear Optics of Organics and Semiconductors; Kobayashi, T., Ed.; Springer-Verlag: Berlin, 1989, p. 192.
43. Schellenberg, F. M.; Byer, R. L.; Miller, R. D. Optics Lett. 1990, 15, 242.
44. Schellenberg, F. M.; Schiller, S. (unpublished results).
45. Kim, Y. R.; Lee, M.; Thorne, J. R. G.; Hochstrasser, R. M. Chem. Phys. Lett. 1988, 145, 75.
46. Schellenberg, F. M.; Byer, R. L.; Miller, R. D.; Takahashi, Y.; Kano, S. LEOS NLO'90 Technical Digest, 1990, 34.
47. Michl, J.; Downing, J. W.; Karatsu, T.; McKinley, A. J.; Poggi, G.; Wallraff, G. M.; Sooriyakumaran, R.; Miller, R. D. Pure Appl. Chem. 1988, 60, 959.
48. Two-photon fluorescence excition spectra for PDN6S have also been measured by Hochstrasser et al.; Thorne, J. R. G.; Ohsako, Y.; Zeigler, J. M.; Hochstrasser, R. M. Chem. Phys. Lett. 1989, 162, 455.
49. Schellenberg, F. M.; Byer, R. L.; Miller, R. D.; Kano, S. Mol. Cryst. Liq. Cryst. 1990, 183, 197.
50. See Ref. 11, Ch. 12.
51. See Ref. 9, pp. 44-50.

52.   Glass, A. M. Science 1984, 226, 657.
53.   Moran, M. J.; She, C.-Y.; Carman, R. L. J. Quantum Electron. 1975, QE-11(6), 259.
54.   Ho, P. P.; Dorsinville, R.; Yang, N. L.; Odian, G.; Eichmann, G.; Jimbo, T.; Wang, Q. Z.; Tang, C. C.; Chen, D. D.; Zou, W. K.; Li, Y.; Alfano, R. R. Proc. SPIE 1986, 682, 36.
55.   Prasad, P. N. Proc. SPIE 1986, 682, 120.
56.   Carter, G. M.; Chen, T.; Tripathy, S. K. Appl. Phys. Lett. 1983, 43, 891.
57.   Fukaya, T.; Heinämäki, A.; Stubb, H. J. Mol. Electronics 1989, 5, 187.
58.   K. W. Delong, K. B. Rochford, and G. I. Stegeman, Appl. Phys. Lett. 1989, 55, 1823.
59.   Takeda, K.; Shiraishi, K. Phys. Rev. B 1989, 39, 11028.
60.   See Ref. 2, Vol. II, Ch. III - 1.4.
61.   Kepler, R. G.; Zeigler, J. M.; Harrah, L. A.; Kurz, S. R. Phys. Rev. B 1987, 35, 2818.
62.   Tachibana, H.; Kawabata, Y.; Koshihara, S.; Tokura, Y. Solid State Comm. 1990 (in press).

RECEIVED September 3, 1990

Chapter 44

Design of New Nonlinear Optic–Active Polymers

Use of Delocalized Polaronic or Bipolaronic Charge States

Charles W. Spangler and Kathleen O. Havelka

Department of Chemistry, Northern Illinois University, DeKalb, IL 60115

Polaronic and bipolaronic charge states are well
known in electroactive polymers, and can be observed
in model oligomers with overall delocalization
lengths as small as 16 atoms. It has been suggested
that localized charge states may be involved in
oligomers and polymers having enhanced $\chi^{(3)}$ properties. In this paper we would like to suggest how
such charge states may be generated and stabilized
in formal copolymers so that both the delocalization
lengths and the optical absorption characteristics
are controllable and predictable. For the model
oligomer systems studied to date, there appears to be
differences in the bipolarons formed by either
protonic or oxidative doping mechanisms. Finally,
we will describe the first generation of copolymers
currently under investigation and the evaluations of
their intrinsic $\chi^{(3)}/\alpha$.

Over the past decade the chemistry and physics of delocalized pi-
electron polymers have attracted the attention of workers in a
number of different fields, and inspired a remarkable degree of
interdisciplinary collaboration in studying such seemingly
disparate phenomena as the insulator — conductor transition in
conducting polymers and the design of new organic materials having
enhanced nonlinear optical properties. While it was originally
assumed that long pi-conjugation sequences were necessary to
observe such phenomena, recent work has indicated that this may
not be the case. In addition, it is now recognized that although
the remarkable changes in conductivity in electroactive polymers
upon either chemical or electrochemical doping are indeed
fascinating, the unique nonlinear optical properties of these
materials may prove to be a more important intrinsic property (1-
3).

0097–6156/91/0455–0661$06.00/0

$\chi^{(3)}$ ACTIVITY IN CONJUGATED PI-ELECTON SYSTEMS

Large third order susceptibilities have recently been observed for
trans-polyacetylene, heteroaromatic polymers, and poly[p-phenylene
vinylene] (4-11). Such nonlinear phenomena in electroactive
polymers due to intense laser irradiation has been linked to the
photogeneration of charged solitons on time scales of the order
10^{-13} s, and values of $\chi^{(3)}$ ($3\omega = \omega + \omega + \omega$) = 4 x 10^{-10} esu for
PA and 4.4 x 10^{-10} esu for PPV have recently been measured (Dury,
M. Proc. First Int. Symp. on Nonlinear Optical Polymers for
Soldier Survivability, in Press). Electroactivity in these
polymers has also been explained on the basis of mobile solitons,
polarons and bipolarons, and these charge states have been studied
spectroscopically in doped polymer thin films or in well-
characterized oligomers (11-15). With the recent surge of
interest in the concept of multifunctional materials, it would be
indeed intriguing if the same type of charge carriers that give
rise to electroactivity, would also lead to enhanced NLO
properties. Flytzanis and coworkers (16-17) originally suggested
that third order susceptibilities might be related to the sixth
power of the electron delocalization length. Dalton and coworkers
(18-19) have discussed the consequences of such enhanced electron
delocalization in ladder polymers, pointing out that either
oxidation or reduction of conjugated pi-electron polymers will
have significant effects upon the optical properties and possibly
on the nonlinear optical properties as well.

Several workers have also attempted to model the consequences
of increasing conjugation (or delocalization) on third-order NLO
properties. Beratan, et al. (20) have shown that the third
order hyperpolarizability γ increases rapidly for trans-polyenes
as conjugation increases to 10-15 repeat units, and then more
slowly up to 40 repeat units. This suggests that very long
conjugation sequences may not be required for high NLO activity,
and that oligomeric segments could be mixed with nonactive
segments to maximize both NLO activity and desirable physical
properties.

Hurst and coworkers (21) have also calculated second
hyperpolarizability tensors via ab initio coupled-perturbed
Hartree-Fock Theory for a series of polyenes up to $C_{22}H_{24}$. They
found that the γ_{xxxx} was proportional to chain length, with a
power dependence of 4.0, but that this dependence tapered off as
N increased. More recently, Garito and coworkers (22) calculated
a power law dependence of γ_{xxxx} on chain length on the order of
4.6 ± 0.2. They also suggest that large values of $\chi^{(3)}$ should be
attainable with conjugation sequence of intermediate length (100
Å). Prasad concurs with the conclusion that γ/N levels off
with increasing N (23). In addition, Prasad has measured γ for a
series of polythiophene oligomers by degenerate four wave mixing
(DFWM) in solution and found a power law dependence for γ of 4.
$\chi^{(3)}$ measurements for poly(3-dodecyl-thiophene) prepared by
either chemical or electrochemical means were approximately the
same, even though the molecular weights and number of repeat units
were substantially different. Prasad concludes that effective
conjugation for NLO purposes does not extend much beyond 10 repeat

units, and that similar measurements poly(p-phenylene) shows a
leveling off $\chi^{(3)}$ at the terphenyl level (N = 3) (23). The above
theoretical predictions and experimental results thus seem to
indicate that efforts should be directed in the design of new NLO
polymers to include relatively short, or oligomeric, subunits in
copolymer type structures in which the non-NLO-active portion is
designed to improve physical properties.

DESIGN OF CONJUGATED PI-ELECTRON POLYMERS

Typical pi-electron polymers such as polyacetylene, polythiophene
and PPV often contain a mixture of various conjugation lengths,
and this gives rise to broad absorption bands.   If one
contemplates the use of doped polymers for NLO applications, then
one must contend with the fact that upon chemical or
electrochemical doping a decrease in intensity of the original π-
π^* transition, a shift of λ_{max} to higher energy, and the
appearance of the new polaronic and bipolaronic absorptions in the
gap are typical observations. This often results in optical non-
transparancy over an extremely broad spectral region with
consequent effects on the observed $\chi^{(3)}/\alpha$. However, this may not
be an unavoidable characteristic of conjugated pi-electron
systems. Dalton and coworkers (18,19) have shown the extent of
pi-electron delocalization in polyacetylene and heteroaromatic
ladder polymers by advanced magnetic resonance techniques such as
ENDOR and ESE. They display electron self-delocalization, and the
polaronic domains are limited to only 20-30 atoms. More recently,
Kamiya and Tanaka (24) have estimated that individual polaron
domains in iodine-doped polyacetylene contain only 15-20 atoms,
and that the size of these polaronic domains are independent of
dopant identity.   Thus, <u>long conjugation lengths are not
necessary for high $\chi^{(3)}$ activity or for high conductivity</u>.   In
addition it might be possible to increase delocalization, and thus
$\chi^{(3)}$, by preferentially stabilizing bipolaronic states which may
not be subject to the same degree of self-delocalization as
polaronic states. Polaronic and bipolaronic charge states can
then be induced by chemical or electrochemical redox techniques in
copolymer structures, and electron delocalization stabilized by
mesomerically interacting functionalities. Thus, for example,
electron-donating substituents could stabilize (+)(+) bipolarons,
while electron-withdrawing substitutents could stabilize (-)(-)
bipolarons. This type of copolymer can be schematically
envisioned as follows:

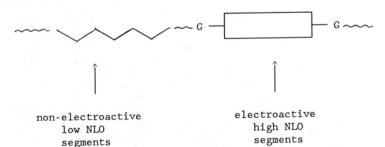

non-electroactive   electroactive
low NLO   high NLO
segments   segments

G = Mesomerically interactive functional group
capable of stabilizing P or BP charge states

MODELING ELECTROACTIVE SEGMENTS: FORMATION AND STABILIZATION OF
POLARONIC AND BIPOLARONIC CHARGE STATES

During the past three years we have been studying the chemical
($SbCl_5$) oxidation of well-characterized oligomers of
polyacetylene, poly[p-phenylene vinylene] (PPV) and poly[2,5-
thienylene vinylene] (PTV) in order to determine how polaron and
bipolaron states can be preferentially formed and stabilized.

$$R \longrightarrow (\!-\!\! CH\!=\!CH)\!\!- R \qquad\qquad n = 7,8,9,10; \ R = Me, \ Ph$$

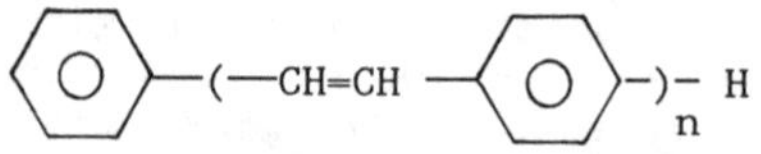

$$n = 2,3,4,5$$

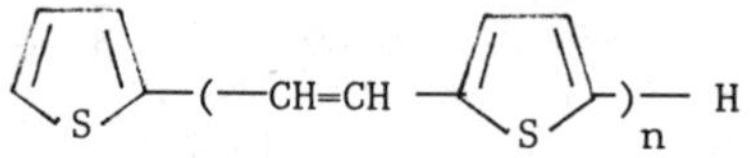

$$n = 2,3,4,5$$

In all cases, the oxidations proceed _via_ two consecutive one-
electron transfers forming polaron and bipolarons states
consecutively, with the bipolaron being the more stable state in
all cases. A typical bipolaron formed from a bis-(p-methoxy
phenyl) polyene is shown below (n = 4). In fact, both as the
conjugation

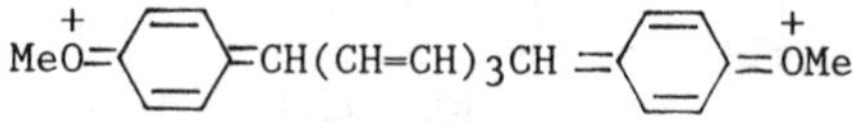

delocalizaton over 18 atoms

length and the electron-donating strength of the mesomerically
interactive functional group increases, the stability of the
bipolaron increases. For example, when n = 6 and EDG = Me_2N, the
bipolaron is stable for several days in solution (10^{-5} M in
CH_2Cl_2) in contact with moist air as evidenced by the lack of
decay in the optical signal (13). We have investigated a variety
of substiutents and have previously reported on the spectroscopic
properties of polaron (P) and bipolaron (BP) states (13,25).
These results are summarized in Table I.

    In all cases, formation of a bipolaron state coincides with a
complete bleaching of the original π-π^* polyene transition, which
in effect moves the optical absorption "window" several hundred nm
to the red. More recently we have discovered that polaronic state
formation (EDG = MeO) can be preferentially controlled if the
quantity of oxidizing agent is carefully monitored (25).
However we have not yet been able to accomplish differentiation
for EDG = Me_2N. Thus by controlling the size of the oligomeric
segment we predetermine the delocalization length of the polaronic
or bipolaronic domain as well as the absorption characteristics of
the polymer in either pristine or oxidized form.

Table I. Stabilized Polaron and Bipolaron Absorption Spectra

$$p\text{-EDG-}C_6H_4(CH=CH)_nC_6H_4\text{-}p'\text{-EDG}$$

EDG	n	λ_{max}^a P (nm)	λ_{max}^a BP (nm)
H	5	[717][b]	612, 564
F	5	[720][b]	615, 567
Cl	5	[727][b]	622, 567
Br	5	1127, 1033, 740	640, 587
MeO	5	1200, 1073, 797	755, 692, 627
Me$_2$N	5	c	773, 713, 646
H	6	[770][b]	685, 615
F	6	[1123, 733][b]	687, 627
Cl	6	1240, 1120, 787	687, 630
Br	6	1273, 1113, 780	693, 640
MeO	6	1300, 1175, 853	818, 741, 680
Me$_2$N	6	c	833, 748, 700

[a]CH_2Cl_2 solvent; [b]unstable transient absorption which decays to BP in less than 10 minutes; [c]no polaron spectrum observed, but weak transient ESR spectrum observed.

It must be emphasized that the identification of the polaron and bipolaron charge states in solution is essentially based upon the optical spectra. These oligomers have limited solubility (ca. 10^{-5}-10^{-6} M in CH_2Cl_2), and nmr spectra of either the parent neutral molecule (N) or of the oxidized species have not been observed. The use of guest-host systems, as opposed to the synthesis of formal copolymers, also suffers from this lack of solubility, particularly if the guest-host systems need to be spin-coated to obtain optical quality films. While we are continuously pursuing additional characterization of these charge states, it is proving to be an extremely difficult task.

OXIDATIVE VERSUS PROTONIC DOPING

Han and Elsenbaumer (26) have recently reported on the protonic doping of alkoxy substituted PPV. This technique involves the in situ doping of the polymer as formed in acid solution, such as CF_3COOH. We have recently compared this nonoxidative approach to the formation of bipolaronic states for a series of methoxy-substituted α,ω-diphenylpolyenes. The pure polyenes, synthesized via Wittig methodology developed in our laboratory (27), were dissolved in neat CF_3COOH to yield 10^{-5} M solutions, and the optical changes monitored as a function of time. Protonic doping is significantly slower than oxidative doping, with up to four hours required for stabilized bipolaron formation versus less than one minute for oxidative doping with excess $SbCl_5$ (13). Han and Elsenbaumer (26) suggested a doping mechanism that involved initial proton attack at the positions ortho to the vinylene repeat units:

Han and Elsenbaumer also note that the BP initially formed can react with neutral polymer to form two distinctly different polarons _via_ interchain electron transfer. After twenty-four hours, our optical spectra are unchanged, and have measurable ESR activity. However, in contrast to alkoxy-PPV polymer, we do not observe a typical polaronic absorption spectrum, but rather one almost identical to the bipolaron obtained from $SbCl_5$ doping of 10^{-5} M solutions in CH_2Cl_2. A possible interpretation is one which allows for P and BP states coexisting in dynamic equilibrium, with the bipolaron dominating the optical absorption. The absorption characteristics of the protonically doped polyenes are shown in Table II compared to the same samples doped with $SbCl_5$.

Table II. Stabilized Bipolarons from Protonic and $SbCl_5$ Doping

Substituents	cmpd	n	H⁺ Doping		SbCl₅ Doping		
			λ_{max} BP (nm) [a]		λ_{max} BP [b]		
4,4'-(OMe)$_2$	1	5	664, 727		627, 692, 755		
4-4'-(OMe)$_2$	2	6	713, 776		680, 741, 818		
2,2',5,5'-(OMe)$_4$	3	5	653		c		
2,2',5,5'-(OMe)$_4$	4	6	713		755		
2,2'4,4'5,5'-(OMe)$_6$	5	5	691, 726		762		
2,2'4,4',5,5'-(OMe)$_6$	6	6	707, 767		797		

[a] CF_3COOH solvent, absorption maxima after 24h.   [b] CH_2Cl_2, absorption after 2h.   [c] Broad absorption

In order to determine the P and BP contributions to the observed
optical spectra 10^{-4} M solutions of the polyenes shown in Table II
were prepared in CF_3COOH and allowed to equlibrate at room
temperature for twenty-four hours. All of the samples displayed
ESR activity, and the spin concentrations are shown in Table III.

Table III.   ESR Spectra of Protonically Doped Polyenes

Compound	Polyene conc. (M)	Spin conc. (M)	g Value
1	9.9×10^{-5}	4.8×10^{-8}	2.00056
2	1.0×10^{-4}	2.0×10^{-7}	2.00196
4	1.0×10^{-4}	8.3×10^{-7}	2.00321
5	1.0×10^{-4}	4.5×10^{-7}	2.00295
6	1.1×10^{-4}	1.4×10^{-6}	2.00360

One can conclude from this data that although polaron states
are present, they are present to the extent of 1% or less. Thus
it is not surprising that the dominant optical absorption is
bipolaronic. Since Han and Elsenbaumer (26) do not report a
quantitative spin concentration for their protonically doped PPV,
it is not clear at the present time whether their bipolaron ⟶
polaron conversion is quantitative. With our model compounds, it
would appear that BP and P states are in equilibrium, with the BP
overwhelmingly dominant.

An additional inconsistency in comparing protonic versus
SbCl$_5$ doping is the fact that the BP absorption maxima for the
protonically doped species is blue-shifted from the SbCl$_5$-doping
maxima. One possible explanation is that the position of proton
attack, as proposed by Han and Elsenbaumer (26) is different in
our system. The PPV oligomers would be resistant to protonation
at the 4-position due to the disruption of the chain conjugation
sequence, and position of attack is dominanted by the bipolaronic
stability. However, in our model compounds, several positions are
available for protonation. For compound 6, we can envision attack
at five positions:

$$\text{RO} - \overset{b \quad a}{\underset{d \quad \quad e}{\langle O \rangle}} - (- C{=}C)_6 \sim$$

Attack at position (a) leads to a bipolaron with a delocalization
path of 22 atoms, but no mesomeric stabilization by OR. Attack at
positions (b) or (d) lead to cross-conjugated bipolarons which
would not be expected to have the same absorption as a non-cross
conjugated system. Protonation at position (e) leads to the
longest delocalization path (24 atoms) but attack at (c) is less
hindered and yields a mesomerically stabilized path of 20 atoms.
Since SbCl$_5$ oxidation yields a 22 atom delocalization path, this
difference (22 vs. 20 atoms) may account for the observed blue
shift. We are currently investigating protonic doping in CF_3COOD
in an attempt to distinguish between these possibilities.

COPOLYMER DESIGN AND SYNTHESES

In principle almost any combination of electroactive and spacer
groups in formal copolymer structures is possible. We are
currently in the initial stages of our copolymer synthesis
program which incorporates our previous experience in modeling P
and BP states. Our first efforts along these lines has been quite
successful and were recently reported at the Pacific Basin
Societies Meeting in Hawaii (Spangler, C. W.; Polis, D. W.; Hall,
T. J.; Dalton, L. R. Int. Chem. Congress Pac. Basin Socs.,
Polymers in Photonics Symp., Honolulu, HA, 1989). We have
incorporated the following electroactive segments in a polyamide
repeat structure via interfacial polymerization.

$x = 3,4$

In order to facilitate optical quality film formation, the
electroactive-diacid monomer was mixed with varying percentages of
a saturated acid, with the electroactive acid incorporation varied
from 5-50%. $\chi^{(3)}$ was measured by DFWM on thin films (ca. 1 micron)
coated Pyrex at 532 nm (near the band edge) by Professor Robert
Hellwarth's group at the University of Southern California. The
following results are typical of this approach, and we will report
a more comprehensive study of these copolymers in the future.

10% incorporation, $\chi^{(3)}/\alpha = 1.4 \times 10^{-13}$ esu-cm

10% incorporation, $\chi^{(3)}/\alpha = 0.62 \times 10^{-13}$ esu-cm

We are now concentrating our synthetic efforts on repeat units capable of oxidative and protonic doping in order to determine $\Delta\chi^{(3)}$ upon going from a neutral to bipolaronic (or polaronic) state.

These copolymers have electron-withdrawing substituents linking the electroactive segments to the spacer group. Thus, these copolymers must be reduced electrochemically to the (-)(-) BP to determine $\Delta\chi^{(3)}$ upon going from a N to a BP state. We are currently synthesizing copolymers with identical electroactive segments, but with the amide group reversed so that electro-chemical or chemical oxidation would produce the (+)(+) BP. Comparison of these two approaches will then give us an idea as to which BP state would be more productive in enhancing $\chi^{(3)}$ and in shifting the optical absorption window. These experiments are currently in progress, and we have recently been encouraged by the report of Cao, et al. that third order nonlinearity in doped ladder polymers are enhanced by bipolaronic states, thus lending support to our approach to the design of high $\chi^{(3)}$ polymers (Cao, X. F.; Jiang, J. P.; Hellwarth, R. W.; Yu, L. P.; Dalton, L. R. Proc. SPIE, 1990 <u>1337</u> (in press)).

CONCLUSIONS

Bipolaronic charge states, and in some cases polaronic charge states, can be supported in stable form in solution for small oligomers of electroactive polymers. These oligomeric segments can also be incorporated as part of a copolymer sequence. Large shifts in the optical absorption spectra are also produced by either oxidative or protonic doping of the oligomers in solution. What remains to be seen is whether these large optical shifts are accompanied by significant increase or decrease in the polymer $\chi^{(3)}$ for the same oligomeric segments in P or BP states. We hope to be able to supply definitive experimental evidence for $\Delta\chi^{(3)}$ in the near future.

There is no doubt that polaronic and bipolaronic charge states can be supported in stable form in small oligomers in solution, or incorporated as part of a copolymer sequence. Enhanced $\chi^{(3)}$ properties can derive from either N, P or BP states as a function of increased delocalization length, and we anticipate several families of copolymers based on the modeling studies discussed in this paper to become available in the near future to test these proposals.

ACKNOWLEDGMENTS

I would like to thank Professor Larry R. Dalton for many helpful discussions on the problems of NLO polymer design, and Dr. David W. Polis for collaborating on copolymer design and synthesis. I would also like to thank Tom J. Hall and Pei-Kang Liu for their synthetic assistance for the model compounds, and Paul Bryson and Linda S. Sapochak for carrying out the ESR measurements. In addition I would like to thank Professor Robert Hellwarth and his group for preliminary $\chi^{(3)}$ evaluation of the first generation copolymers. Financial support as a Visiting Scholar at the

University of Southern California was provided in part by Air
Force Office of Scientific Research contracts F49620-87-C-0010
and F49620-88-C-0071 and is gratefully acknowledged by Charles W.
Spangler. Current support by Air Force Office of Scientific
Research Grant AROSR-90-0060 is also acknowledged.

LITERATURE CITED

1. Williams, D. J. Nonlinear Optical Properties of Organic
    Polymeric Materials; American Chemical Society: Symposium
    Series #233; New York, 1983.
2. Chemla, D. S.; Zyss, J., Eds. Nonlinear Optical Properties
    of Organic Molecules and Crystals; Academic Press: New York,
    1987; Vol. 2 .
3. Prasad, P. N.; Ulrich, D. R., Eds. Nonlinear Optical and
    Electroactive Polymers; Plenum: New York, 1988.
4. Prasad, P. N. Proc. SPIE, 1986, 682, 120.
5. Garito, A.; Wang, K.; Cai, Y.; Man, H.; Zamani-Khamiri, O.
    Proc. SPIE, 1986, 682, 2.
6. Etemad, S.; Baker, G.; Jaye, D. Proc. SPIE, 1986, 682, 44.
7. Heeger, A. J.; Moses, D.; Sinclair, N. Synthetic Metals,
    1987, 17, 343.
8. Sinclair, M.; Moses, D.; Heeger, A. J.; Vilhemsson, K.; Valk,
    B.; Salour, M. Solid State Commun., 1987, 61, 221.
9. Kajzar, F.; Messier, J.; Sentein, C.; Elsenbaumer, R. L.;
    Miller, G. G. Proc. SPIE, 1989, 1147, 36.
10. Yu, L.; Vac, R.; Dalton, L. R.; Hellworth, R. W. Proc. SPIE,
    1989, 1147, 142.
11. Bradley, D. C. J. Phys. D., Appl. Phys., 1987, 20, 1389.
12. Fichou, D.; Garnier, F.; Charra, F.; Kajzar, F.; Messier, J.
    In Organic Materials for Nonlinear Optics; Hann, R. A.;
    Bloor, D., Eds.; The Royal Society of Chemistry Spec. Pub.
    #69: London, 1989, pp. 176-182.
13. Spangler, C. W.; Sapochak, L. S.; Gates, B. D.   In Organic
    Materials for Nonlinear Optics; Hann, R. A. and Bloor, D.,
    Eds.; The Royal Society of Chemistry Spec. Pub. #69: London,
    1989, pp. 57-62.
14. Patil, A. O.; Heeger, A. J.; Wudl, F. Chem. Rev. 1988, 88,
    183.
15. Spangler, C. W.; Hall, T. J.; Sapochak, L. S.; Liu, P-K
    Polymer 1989, 30, 1166.
16. Agrawal. G. P.; Flytzanis, C. Chem. Phys. Lett. 1976, 44,
    366.
17. Agrawal, G. P.; Cojan, C.; Flytzanis, C. Phys. Rev. B. 1978,
    17, 776.
18. Dalton, L. R.; Thomson, J.; Nalwa, H. S. Polymer 1987, 28,
    543.
19 Dalton, L. R. in Ref. 3, pp. 243-273.
20. Beratan, D. N; Onuchic, J. N.; Perry, J. W. J. Phys. Chem.,
    1987, 91, 2696.

21. Hurst, G. J. B.; Duplis, M.; Clementi, E. J. Chem. Phys., 1988, 89, 385.
22. Garito, A. F.; Heflin, J. R.; Wong, K. Y.; Zamani-Khamiri, O. In Organic Materials for Non-Linear Optics; Hann, R. A.; Bloor, D., Eds.; Royal Society of Chemistry Spec. Pub. #69; London, 1989, pp 16-17.
23. Prasad, P. N. In Organic Materials forNon-Linear Optics; Hann, R. A.; Bloor, D., Eds.; Royal Society of Chemistry Spec. Pub. #69; London, 1989, pp. 264-274.
24. Kamiya, K.; Tanaka, J. Synthetic Metals, 1988, 88, 183.
25. Spangler, C. W.; Havelka, K. O. Polymer Preprints 1990, 31(1), 396.
26. Hann, C. C.; Elsenbaumer, R. L. Synthetic Metals, 1989, 30, 123
27. Spangler, C. W.; McCoy, R. K.; Dembek, A.; Sapochak, L. S.; Gates, B. D. J. Chem. Soc Perkin 1 1989, 151.

RECEIVED July 10, 1990

Chapter 45

New Polymeric Materials with Cubic Optical Nonlinearities Derived from Ring-Opening Metathesis Polymerization

R. H. Grubbs[1], C. B. Gorman[1], E. J. Ginsburg[1], Joseph W. Perry[2], and Seth R. Marder[2]

[1]The Arnold and Mabel Beckman Laboratory of Chemical Synthesis, Division of Chemistry and Chemical Engineering, California Institute of Technology, Pasadena, CA 91125
[2]Jet Propulsion Laboratory, California Institute of Technology, Pasadena, CA 91109

Partially substituted derivatives of polyacetylene are synthesized via the ring-opening metathesis polymerization (ROMP) of cyclooctatetraene (COT) and its derivatives. Certain poly-COT derivatives afford soluble, highly conjugated polyacetylenes. These materials exhibit large third-order optical nonlinearities and low scattering losses.

Organic materials are currently under intense investigation with respect to their potential for nonlinear optical applications (1-5). While the overall prospects for organic materials and their potential merits for nonlinear optical applications have been discussed, the detailed material property requirements for specific device applications are only beginning to be enumerated (6-8). Recent experimental (9-11) and theoretical (12-14) studies indicate that extended electron delocalization leads to large cubic susceptibilities. Materials research efforts are now faced with the challenge to develop materials that have high nonlinear activity and also satisfy stringent requirements, such as low optical absorption and scattering loss, ease of fabrication, and high mechanical, thermal and environmental stability. Polyacetylene, the simplest fully-conjugated organic polymer, displays large third-order optical nonlinearities and high iodine-doped conductivities. Unfortunately, since polyacetylene is an insoluble, unprocessable material with a morphology which is largely fixed during its synthesis, it is difficult to fully exploit all the properties of this potentially useful material. We have shown that ring-opening metathesis polymerization (ROMP) of cyclooctatetraene (COT) produces poly-cyclooctatetraene, a new form of polyacetylene (Figure 1) (15). In this paper, we discuss the ROMP of substituted COTs to form partially substituted polyacetylenes with large third-order optical nonlinearities and greatly improved materials properties relative to polyacetylene.

Synthesis of Polymers

Polymerizations of substituted COTs are readily accomplished on gram scales in a nitrogen drybox. In a typical polymerization, the tungsten catalyst (16) (2 mg, 2.5 μmol) is dissolved in a solution containing 20 μL of tetrahydrofuran and the monomer (yellow liquid, 100 mg, 0.6 mmol). The yellow solution polymerizes over the course

0097–6156/91/0455–0672$06.00/0

of 1-2 minutes, during which time it may be cast onto a variety of substrates. Typically, it is transferred by pipette onto a glass slide where it spreads out to form a film which is 20-200 μm thick depending on the viscosity of the reaction mixture at the time of the transfer. Polymerization in dilute solution is avoided since a decrease in the monomer concentration at the catalyst center encourages "back-biting" reactions to produce benzene and/or substituted benzenes (Figure 2). This chain transfer reaction does not terminate the polymerization, but it does reduce the molecular weight. The films which are formed can be iodine doped to a conductive state with typical conductivities of 0.1 - 50 Ω^{-1}cm^{-1} (Table I).

Nascent poly-COT has a high cis content. Differential scanning calorimetry reveals an irreversible exotherm at 150 °C (15) corresponding to cis/trans isomerization in polyacetylene. Three of the four cis bonds in the monomer are expected to retain their geometric configuration during polymerization to give a polymer with at least 75% cis configuration. However, although cis/trans isomerization is slow at room temperature (17), the polymerization is exothermic and may induce some isomerization.

Properties of Partially Substituted Polyacetylenes

Polymerization of substituted COT derivatives results in partially substituted polymers that, in several cases, are soluble and still highly conjugated (18-19). Substitution of polyacetylene via the polymerization of substituted acetylenes results in materials with low effective conjugation lengths as evidenced by their high-energy visible absorption spectra and comparatively low iodine-doped conductivities (20-23). This low conjugation length is presumably due to twisting around the single bonds in the backbone resulting from steric repulsions between the side groups (Figure 3a) (24). Chien has prepared copolymers of acetylene and methyl-acetylene. However, extension of this method to other copolymerizations requires mixing a gas (acetylene) and a liquid (R-acetylene), and this two-phase system is not expected to be well-behaved (25). In contrast, polymers of substituted COT derivatives have on the average a substituent on every eighth carbon. Thus, in these systems, the predominant steric interaction is between the substituent and a hydrogen on the β-carbons of the backbone (Figure 3b).

n-Alkyl derivatives of COT polymerize to give red materials that are soluble. Upon standing, the polymer solutions turn blue and gel or precipitate if not diluted. (Table I) (19). The color change is proposed to be due to cis-trans isomerization of the polymer in solution (18). A thin film of isomerized poly-*n*-octylCOT has a broad absorption centered around 650 nm which is comparable to that observed for a thin film of polyacetylene (26). Moreover, in contrast to poly-COT (polyacetylene), which shows large optical scattering due to its crystallinity, the alkylCOT polymers are amorphous and show low scattering losses. Only amorphous halos are observed in the wide angle X-ray profile of these polymers. In general, poly-*n*-alkylCOTs are soluble in the cis form, amorphous, and highly conjugated as determined by electronic and Raman spectroscopy (Table I). However, these polymers are only barely soluble in the trans form.

Placing a secondary or tertiary group adjacent to the polymer chain reduces the effective conjugation length somewhat but affords solubility in both the cis and trans forms of the polymer. Poly-*t*-butylCOT is freely soluble but yellow-orange in color, indicating a low effective conjugation length. Freely soluble poly-trimethylsilylCOT and poly-*sec*-butylCOT are red in the cis form and purple in the trans form indicative of high conjugation. These polymers can be contrasted with poly-neopentylCOT where the *t*-butyl group is spaced one methylene unit away from the polymer chain. The solubility and effective conjugation length of this polymer resemble that of the *n*-alkyl substituted polymers. Alkoxy substituted polymers such as poly-*t*-butoxyCOT are also not completely soluble in the trans form.

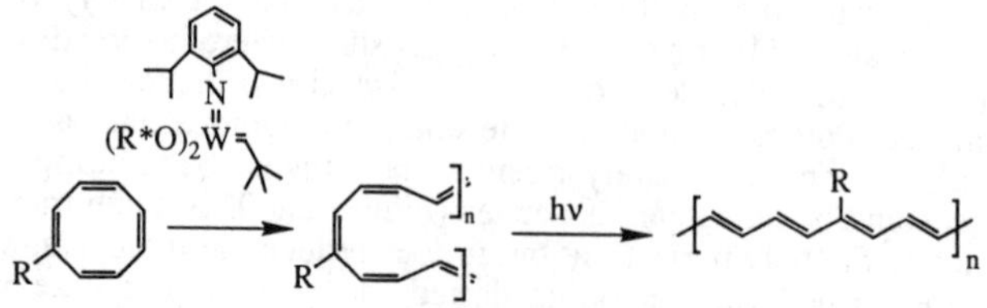

$R*O = (CF_3)_2CH_3CO$, R = see text

Figure 1. Polymerization of cyclooctatetraenes.

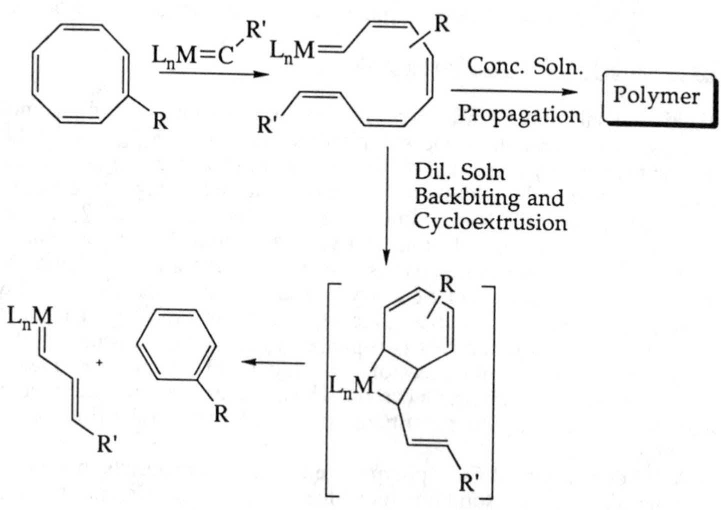

L_nM = (see Experimental),
R denotes any monosubstituted COT, R' = polymer tail or t-Bu

Figure 2. Cycloextrusion in dilute solution polymerization of cyclooctatetraenes.

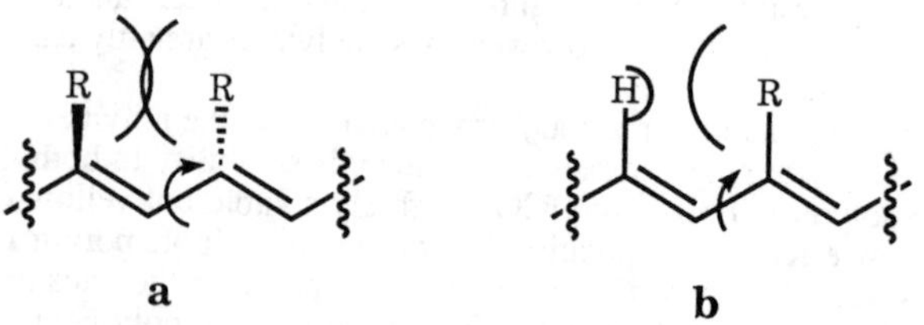

Figure 3. Chain twisting in a, substituted polyacetylene and b, substituted polyCOT.

Table I. Data for RCOT Polymers

R	Abs. Max. after synthesis[a] ,[b]	Abs. Max. after isom[a,c]	Raman v_1 (Ag C-C str) cm^{-1} [d]	Raman v_2 (Ag C=C str) cm^{-1} [d]	σ (S/cm)[e]	y[f]
Methyl	522 [g]	----	1126-1132	1516	15-44	----
n-Butyl	462	614 (6-12 hrs)	1132	1514	0.25-0.7	0.10-0.13
n-Octyl	480	632 (6-12 hrs)	1114-1128	1485	15-50	0.11-0.19
n-Octadecyl	538	630 (2-3 hrs)	---	---	0.60-3.65	0.13-0.16
Phenyl	522	620 (6-12 hrs)	---[h]	---	0.3-0.6	0.19-0.28
t-Butyl	302	432 (2-3 wks)	1147	1539-1547	< 10^{-8}	≈ 0.03
s-Butyl	418	556 (2-3 wks)	1125-1128	1512	0.03[i]	0.31
TMS	380	512 (2-3 wks)	1132	1532	0.2[i]	0.12
Neopentyl	412, 628	634 (6-12 hrs)	1131	1509	0.2-1.5	0.10-0.18
Neopentyl[j]					15-21	0.12-0.17

[a]Spectra taken in tetrahydrofuran. Wavelength in nm. [b]Spectra obtained ca. 15 minutes after polymer synthesis. [c]Time after synthesis is shown in parentheses. [d]Excitation wavelength of 488.0 nm. [e]After iodine-doping. [f]Based on the molecular formula (C[H/R]I$_y$)$_x$. [g]Very small amount of material that leached out of the film. [h]Stretches obscured by peaks due to phenyl stretching. [i]Before isomerization/recasting, conductivity is < 10^{-4} S/cm. [j]Film produced by isomerization of poly-neopentylCOT in solution followed by recasting.

Side groups of the optimal size are thought to induce a twist in the polymer chain which reduces conjugation length slightly but permits enough conformational mobility to induce solubility. A trimer of the polymer chain was modelled using a molecular mechanics calculation with the MM2 force field available in Batchmin (W. C. Still, Columbia University). Minimization was via the OS variable metric method using derivative convergence. For model compounds in which two angles Θ_1 and Θ_2 (Θ_1 and Θ_2 are the supplements of the dihedral angles in degrees. See Figure 4.) are both substantial, the analogous polymers are soluble (Table II). Note that in the model of poly-t-butoxyCOT, Θ_1 is large, but Θ_2 is not, and, experimentally, the polymer is not soluble. Qualitatively similar results are seen using an MM2 calculation which includes a π contribution (PC-Model 1.0, Serena Software using an MM2/MMX force field with a π calculation.)

Table II. Computed Twist Angles for RCOT Polymers

R	n-Bu	s-Bu	t-Bu	MeO	t-BuO	TMS	Np
$\Theta_1°$	5.03	21.95	52.74	16.48	27.77	28.35	15.72
$\Theta_2°$	6.50	12.41	14.40	0.88	2.67	13.26	3.10

Isomerization from cis to trans does not occur with equal ease in all polymers. Qualitatively, the more conjugated poly-n-octylCOT isomerizes more quickly in solution than the less conjugated poly-TMSCOT. Differential scanning calorimetry was performed on all of the polymers (Table III).

Table III. Isomerization Temperatures for Films of Poly-RCOT

R	Me	n-Bu	n-Oct	n-C$_{18}$	C$_6$H$_5$	t-Bu	s-Bu	TMS	Np
T^a	103	107	102	102	114	164	122	150	110

^{a} Temperature ($^\circ$C) of irreversible cis-trans isomerization exotherm.

All films display an irreversible exotherm between 100-165 °C that does not correspond to any weight loss as shown by thermal gravimetric analysis. Most of the films isomerize below 150 °C, which is the cis/trans isomerization temperature reported for polyacetylene (27) and observed for poly-COT (i.e. R = H). Since any side group on the polymer renders it amorphous (*vide supra*), we conclude that crystalline polyacetylene is harder to isomerize than amorphous polyacetylene. This conclusion is supported by the thermal analysis of another amorphous polyacetylene, that produced by the precursor route of Feast and Edwards (28). This form of polyacetylene is reported to have an isomerization temperature of 117 °C (29). The soluble polymers where conjugation length is reduced, however, have higher isomerization temperatures than the less-twisted derivatives. This behavior is understandable given the notion that a longer conjugation length polyene sequence should be easier to isomerize than a shorter conjugation length sequence. The photochemically induced cis/trans isomerization of the soluble polymers has been monitored by UV/Vis spectroscopy (Figure 5). Moreover, ^{1}H NMR (18) shows a smooth conversion from the predominantly cis isomer to the trans isomer.

Several attempts have been made to correlate the effective conjugation length of a polyene with its absorbance maximum. A solution of poly-n-octylCOT has an absorption maximum at 620 nm in THF. The absorption maximum shifts to slightly lower energy absorption in the solid state. A thin film of poly-n-octylCOT, like polyacetylene (26) has a broad absorption centered around 650 nm. Based on the extrapolation of polyene absorption data obtained from a variety of workers (30-36) to the band gap of polyacetylene (26), an absorption maximum of 600 nm implies an effective conjugation length of at least 25 double bonds. Similar results have been obtained by R. Chance, Exxon Corp. Poly-TMSCOT has a higher energy absorption maximum (~530 nm) than poly-n-octylCOT, indicating a conjugation length of greater than 15 double bonds.

Nonlinear Optical Properties of COD/COT Copolymers and Poly-RCOTs

When COT is copolymerized with 1,5-cyclooctadiene (COD), a monomer of similar reactivity, a random copolymer results (15). Incorporation of COD into the polymer interrupts conjugation, allowing the distribution of conjugation lengths to be varied. This control has been used to study the dependence of the third-order nonlinear optical properties on conjugation length (*vide infra*). The optical spectra of COD/COT copolymer solutions (*vide supra*), an example of which is shown in Figure 6, indicate that the copolymers contain segments with 5, 9 and 13 double bonds (37). Nonlinear optical properties of the polymer mixtures were studied as a function of the composition. The third-order susceptibilities of the copolymer solutions were determined using wedged cell third harmonic generation (THG) techniques (38-39). The 1907 nm Raman shifted (H$_2$ gas) output from a Q-switched Nd:YAG laser was

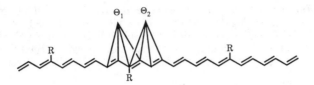

Figure 4. The model compound used in MM2/MM2Π computations.

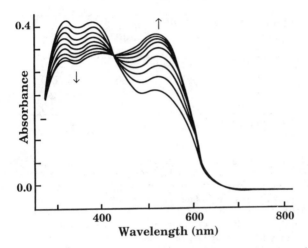

Figure 5. UV-Visible spectra of poly(TMSCOT) in carbon tetrachloride (10^{-6} M) obtained between eight periods of photolysis (10 sec each).

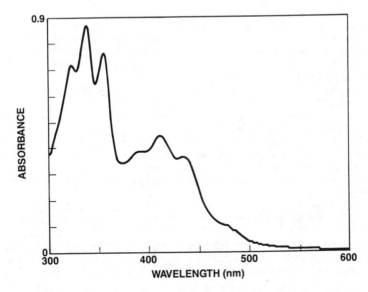

Figure 6. UV-Visible spectrum of copolymer derived from 50% cyclooctatetraene, 50% 1,5-cyclooctadiene monomer mixture.

used as the fundamental for THG measurements. Wedge THG interference fringes were observed by translating the cell normal to the laser beam. Table IV summarizes the results of the THG studies. The third-order susceptibility of the polymer solution, $|\chi^{(3)}_p|$, and the hyperpolarizability per monomer unit, γ_p, values listed in Table IV are based on the total mole fraction of monomer incorporated into the polymer.

Table IV. Summary of Composition and Third-Order Optical Nonlinearities at 1907 nm of COT/COD Copolymers

Mol. fract. COT in polymer	$\chi^{(3)}_p\ 10^{-14}$ esu[a],	$\gamma_p\ 10^{-36}$ esu[a]	$\gamma'_p\ 10^{-36}$ esu[a]
0.08	33	20	210
0.15	60	36	220
0.27	130	81	280
0.32	160	100	300

[a]Nominal uncertainty ± 25 %

The $\chi^{(3)}$ and γ_p values of the copolymers increase substantially with increasing fraction of COT. This increase reflects both the increasing concentration of conjugated units and the increasing conjugation length with higher fraction of COT. It is expected that the nonlinearity of the units of COD in the polymer is negligible compared to that of the nonlinearity of the units of COT; assuming so, one can calculate the hyperpolarizability per unit of COT in the polymer, γ'_p, as listed in Table IV. The fact that γ'_p increases with increasing fraction of COT in the polymer shows that the presence of the increased conjugation lengths (segments of 9 and 13 double bonds) results in enhanced nonlinearity. From the solution results, we have estimated the $\chi^{(3)}$ of a copolymer film with 32% COT to be ~2 x 10^{-12} esu. By comparison, a solution measurement on β- carotene (11 double bonds) gave a value of $\chi^{(3)}$ = 9 x 10^{-11} esu. Measurements on neat polyacetylene have given a value of 1.3 x 10^{-9} esu (enhanced by three-photon resonance) at 1907 nm (9-11). Transparent uniform films of these soluble polymers with low scattering losses can be prepared by spin coating. Thus while the $\chi^{(3)}$ of the copolymer is modest, this work suggested that the ROMP methodology could be used to produce materials with substantial nonlinearities and is flexible enough to allow tailoring of materials properties.

Accordingly, we studied the nonlinear optical properties of some partially substituted polyacetylenes prepared by ROMP. The linear and nonlinear optical properties of films of poly-*n*-butylCOT were examined. These films were typically prepared by polymerizing the neat monomer and casting the polymerizing mixture either between glass slides, resulting in films of about 20 μm thickness, or between the fused silica windows of a 100 μm pathlength demountable optical cuvette. Films cast between substrates were easily handled in air and were very stable for long periods of time (months). In addition, such assemblies were convenient for examination of the optical properties.

THG measurements on poly-*n*-butylCOT films, referenced to a bare fused silica plate, were made using 1064 nm pulses. These measurements showed that the $|\chi^{(3)}|$ values of films of poly-*n*-butylCOT, ~1x10^{-10} esu, were comparable to that for unoriented polyacetylene at the same wavelength (39). However, comparison of the linear transmission spectra of these materials in the near infrared shows that the partially substituted polyacetylene has greatly improved optical quality. (See Figure 7.)

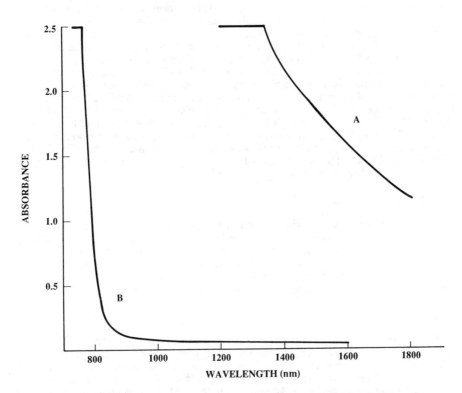

Figure 7. UV-Visible-Near IR spectra of films of poly(COT) (A) and poly(n–butylCOT) (B).

Absorption spectra of polyCOT films show high optical density (1-3 for 20 μm thick films) even below the true absorption edge (40) in the near IR. The apparent absorption decreases with increasing wavelength but extends out beyond 2000 nm. This apparent absorption is actually due to scattering as shown by laser light scattering observations. We estimate the loss coefficient of poly-COT films to be > 500 cm^{-1} at 1500 nm. The origin of this scattering is certainly due to internal optical inhomogeneities in the polymer associated with the semi-crystalline, fibrillar morphology. In contrast, films of poly-n-butylCOT show very clean transmission in the near IR. Films 100 μm thick show a sharp absorption edge at ~900 nm and little absorption beyond 1000 nm. For poly-n-butylCOT films, we estimate the loss coefficient to be < 0.2 cm^{-1} at 1500 nm. The greatly reduced scattering loss indicates that partial substitution of polyacetylene with n-butyl groups has resulted in a more homogeneous morphology, approaching that of an amorphous polymer.

We have also examined films of poly-TMSCOT. As discussed above, this polymer is completely soluble and can be converted to a fully trans conformation in solution. Films of the trans form of the polymer are then easily produced from solution by casting or spin-coating. THG measurements at 1064 nm on films of poly-TMSCOT give $|\chi^{(3)}| = 2 \pm 1 \times 10^{-11}$ esu. This value is somewhat lower than that of poly-n-butylCOT or polyacetylene, consistent with the reduced effective conjugation length inferred from the energy of the absorption maximum, as discussed earlier. The films of poly-TMSCOT prepared from solution are of good optical quality and show scattering losses at least as low as the poly-n-butylCOT films.

Conclusions

Ring-opening metathesis polymerization of substituted cyclooctatetraene derivatives yields partially substituted polyacetylenes, many of which are soluble and highly conjugated. Highly conjugated polymers obtained exhibit high optical nonlinearities and low scattering losses. Given the ability to fabricate these polymers into uniform, high quality films with optical nonlinearities comparable to that of polyacetylene, these polymers may be of interest for nonlinear waveguiding experiments.

Acknowledgments

The research described in this paper was performed, in part, by the Jet Propulsion Laboratory, California Institute of Technology as part of its Center for Space Microelectronics Technology which is supported by the Strategic Defense Initiative Organization, Innovative Science and Technology Office through an agreement with the National Aeronautics and Space Administration (NASA). RHG acknowledges financial support from the Office of Naval Research. SRM thanks the National Research Council and NASA for a Resident Research Associateship at JPL. EJG thanks IBM for a research fellowship. CBG thanks the JPL for a research fellowship. The authors thank Dr. L. Khundkar, B. G. Tiemann and K. J. Perry for technical assistance.

Literature Cited

1.   Nonlinear Optical Properties of Organic and Polymeric Materials; Williams, D. J., Ed.; ACS Symposium Series No. 233; American Chemical Society: Washington, DC, 1983.

2.   Molecular and Polymeric Optoelectronic Materials: Fundamentals and Applications; Khanarian, G., Ed.; Proc. SPIE Int. Soc. Opt. Eng., No. 682, 1987.

3. Nonlinear Optical Properties of Polymers; Heeger, A. J.; Orenstein, J.; Ulrich, D. R.; Eds. Materials Research Society Symposium Proceedings, Vol. 109, Materials Research Society: Pittsburgh, PA, 1988.
4. Nonlinear Optical Properties of Organic Molecules and Crystals; Chemla, D. S.; Zyss, J.; Eds.; Academic: Orlando, FL 1987, Vols. 1 and 2.
5. Nonlinear Optical and Electroactive Polymers; Prasad, P. N.; Ulrich, D. R.; Eds.; Plenum: New York, NY 1988,
6. Stegeman, G. I.; Zanoni, R.; Seaton, C. T. in reference 3, p. 53.
7. DeMartino, R.; Haas, D.; Khanarian, G.; Leslie, T.; Man, H. T.; Riggs, J.; Sansone, M.; Stamatoff, J.; Teng, C.; Yoon, H. in reference 3, p. 65
8. Thackara, J. I.; Lipscomb, G. F.; Lytel, R. S.; Ticknor, A. J. in reference 3, p. 19.
9. Sauteret, C.; Hermann, J. P.; Frey, R.; Pradere, F.; Ducuing, J.; Baughman, R. H.; Chance, R. R. Phys. Rev. Lett. 1976, 36, 956-9.
10. Carter, G. M.; Chen, Y. J.; Tripathy, S. K. Appl. Phys. Lett. 1983, 43, 891-3.
11. Kajzar, F.; Etemad, S.; Baker, G. L.; Messier, J. Synth. Met. 1987, 17, 563-7.
12. Agrawal, G. P.; Cojan, C.; Flytzanis, C. Phys. Rev. B 1978, 17, 776-89.
13. Beratan, D. N.; Onuchic, J. N.; Perry, J. W. J. Phys. Chem. 1987, 91, 2696-8.
14. Garito, A. F.; Heflin, J. R.; Wong, K. Y.; Zamani, K. O. in reference 3, p. 91.
15. Klavetter, F. L.; Grubbs, R. H. J. Am. Chem. Soc. 1988, 110, 7807-13.
16. Schaverien, C. J.; Dewan, J. C.; Schrock, R. R. J. Am. Chem. Soc. 1986, 108, 2771-3.
17. Chien, J. C. W.; Karasz, F. E.; Wnek, G. E. Nature (London) 1980, 285, 390-2.
18. Ginsburg, E. J.; Gorman, C. B.; Marder, S. R.; Grubbs, R. H. J. Am. Chem. Soc. 1989, 111, 7621-2.
19. Gorman, C. B.; Ginsburg, E. J.; Marder, S. R.; Grubbs, R. H. Angew. Chem. Adv. Mater. 1989, 101, 1603.
20. Zeigler, J. M. U. S. Pat. Appl. 760 433 AO, 21 November 1986; Chem. Abstr. 1986, 20, No. 157042.
21. Zeigler, J. M. Polym. Prepr. 1984, 25, 223-4.
22. Okano, Y.; Masuda, T.; Higashimura, T. J. Polym. Sci., Polym. Chem. Ed. 1984, 22, 1603-10.
23. Masuda, T.; Higashimura, T. In Adv. Polym. Sci.; Okamura, S., Ed.; Springer- Verlag: Berlin, 1986; Vol. 81, pp 121-165.
24. Leclerc, M.; Prud'homme, R. E. J. Polym. Sci., Polym. Phys. Ed. 1985, 23, 2021-30.
25. Chien, J. C. W.; Wnek, G. E.; Karasz, F. E.; Hirsch, J. A. Macromolecules 1981, 14, 479-85.
26. Patil, A. O.; Heeger, A. J.; Wudl, F. Chem. Rev. 1988, 88, 183-200.
27. Ito, T.; Shirakawa, H.; Ikeda, S. J. Polym. Sci., Polym. Chem. Ed. 1975, 13, 1943-50.
28. Edwards, J. H.; Feast, W. J.; Bott, D. C. Polymer 1984, 25, 395-8.
29. Bott, D. C. Polym. Prepr. 1984, 25, 219-20.
30. Bohlmann, M. Chem. Ber. 1952, 85, 386-389.
31. Bohlmann, M. Chem. Ber. 1953, 86, 63-69.
32. Bohlmann, M.; Kieslich Chem. Ber. 1954, 87, 1363-1372.
33. Nayler, P.; Whiting, M. C. J. Chem. Soc. Chem. Comm. 1955, 3037-3046.
34. Sondheimer, F.; Ben-Efrian, D.; Wolovsky, R. J. Am. Chem. Soc 1961, 83, 1675-1681.
35. Karrer, P.; Eugster, C. H. Helv. Chim. Acta 1951, 34, 1805-1814.
36. Winston, A.; Wichacheewa, P. Macromolecules 1973, 6, 200-5.

37. Marder, S. R.; Perry, J. W.; Klavetter, F. L.; Grubbs, R. H. Chem. Mater. 1989, 1, 171-3.
38. Meredith, G. R.; Buchalter, B.; Hanzlik, C. J. Chem. Phys. 1983, 78, 1543-51.
39. Kajzar, F.; Messier, J. J. Opt. Soc. Am. B: Opt. Phys. 1987, 4, 1040-6.
40. Weinberger, B. R.; Roxlo, C. B.; Etemad, S.; Baker, G. L.; Orenstein, J. Phys. Rev. Lett. 1984, 53, 86-9.

RECEIVED August 2, 1990

Chapter 46

Polymers and an Unusual Molecular Crystal with Nonlinear Optical Properties

F. Wudl, P.-M. Allemand, G. Srdanov, Z. Ni, and D. McBranch

Departments of Chemistry and Physics, University of California, Santa Barbara, CA 93106

In the recent past, conjugated polymers were found to have very fast, subpicosecond nonlinear optic response and $\chi^{(3)}$ on the order of 10^{-9} esu. We have been working on the synthesis of processible conjugated polymers in relation to their electrical conductivity properties. Once processibility was established, we were able to prepare thin films which were suitable for optical measurements. The syntheses of these conjugated polymers and of the monomers will be described. In a different project involving organic ferromagnetism, we found two compounds whose solid state structure was non-centrosymmetric. In one case the molecule is polar with two dipolar moieties (nitronylnitroxide and nitro) pointing in the same crystallographic direction over the whole lattice. In another case the molecule is symmetrical, yet the lattice is polar. The properties of these molecular solids are described including second harmonic generation (SHG).

This presentation is divided into two parts, one dealing with third harmonic generation (THG) non linear optical (NLO) materials and the other dealing with second harmonic generation (SHG) materials.

<u>A New Conjugated Polymer for THG Applications</u>

Conjugated polymers are very fast response NLO materials. The most studied are the poly(diacetylenes) and polyacetylene. Poly(para-phenylenevinylene) (PPV) is a conjugated backbone polymer which can be processed through a water soluble precursor polymer by use of the Wessling-Zimmerman method(1-3). Once the conjugated backbone is obtained, the yellow polymer has excellent mechanical properties but is intractable. In the recent past, soluble conjugated PPV's have been obtained by the introduction of long chain alkoxy groups(4,5). In our hands even the dioctyloxy substituted, high molecular weight, PPV was soluble only in hot chlorobenzene. In order to be able to fabricate optically smooth films by spin casting, we required a genuinely, ambient temperature soluble PPV in solvents such as cyclopentanone. We reasoned that if the two alkoxy groups were of disparate size and if one of the alkoxy groups had a branch, the solubility of the polymer would be

0097–6156/91/0455–0683$06.00/0

enhanced considerably because the ensuing asymmetry built onto the macromolecule's stiff backbone would prevent it from packing in an ordered fashion. Noting that hydroquinone monomethyl ether is commercially available and that the 2-ethylhexyl moiety has been used extensively in the past as part of plasticizer ingredients in commercial polymer blends, we decided that the title polymer should be obtained relatively easily and should have the desired properties.

Results and Discussion

The target polymer was produced by the "traditional" precursor approach(1) as depicted in Scheme I, below.

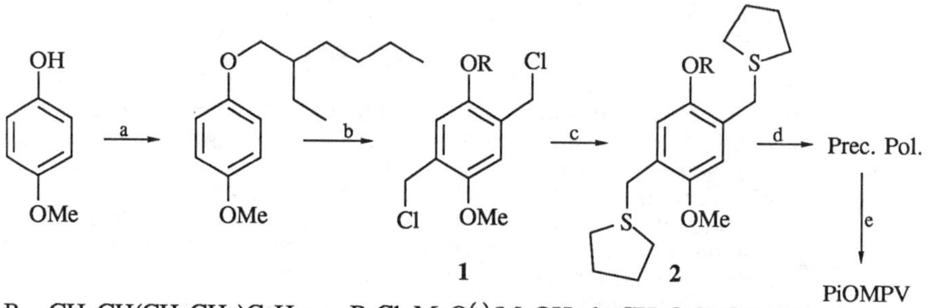

$R = CH_2CH(CH_2CH_3)C_4H_9$; a, R-Cl, $MeO^{(-)}$/MeOH; b, $CH_2O \cdot H_3O^+Cl^-$/dioxane; c, THT/MeOH; d, NaOH/MeOH; e, Δ/1,2,4-trichlorobenzene.

Scheme I

The polymer obtained by this procedure is a red powder, insoluble in methanol and ethanol but soluble in THF, benzene, chlorobenzene, cyclopentanone and other nonpolar organic solvents. The polymer had a molecular weight of ~300,000 with a polydispersity of ~4 as determined by GPC relative to polystyrene standard. All spectroscopic properties are in accord with the proposed structure. Some of the properties are solvent dependent. For example, the polymer is thixsotropic in benzene.

Very smooth films can be cast from THF. Free standing films have the appearance of "red cellophane".

Preparation of the Precursor Polymer

A solution of 200 mg (0.39 mmol) of the monomer salt (2)(6)in 1.2 mL dry methanol was cooled to 0° C for 10 min and a cold degassed solution of 28 mg (1.7 equivalents) of sodium hydroxide in 0.7 mL methanol was added slowly. After 10 min the reaction mixture became yellow and viscous. The above mixture was maintained at 0° C for another 2-3 h and then the solution was neutralized. A very thick, gum-like material was transferred into a Spectrapore membrane (MW cutoff 12,000-14,000) and dialysed in degassed methanol containing 1 % of water for 3 days. After drying in vacuo, 70 mg (47 %) of "plastic" yellow material was obtained. UV (CHCl$_3$) 365. IR (film) 740, 805, 870, 1045, 1075, 1100, 1125, 1210, 1270, 1420, 1470, 1510, 2930, 2970, 3020. Soluble in C_6H_5Cl, $C_6H_3Cl_3$, CH_2Cl_2, $CHCl_3$, Et_2O, THF. Insoluble in MeOH.

Preparation of Poly[3-Methoxy-6-(2-Ethyl-Hexyloxy)phenylene vinylene

A solution of 385 mg (1 mmol) of the precursor polymer prepared above in 120 mL 1,2,4-trichlorobenzene was allowed to reflux under N_2 for 48 h. After cooling to R.T., 300-400 mL of cold MeOH was added, the mixture was centrifuged and 230 mg (92 %) of the solid was obtained. UV (CHCl$_3$) 500 nm. IR (film) 695, 850, 960, 1035, 1200, 1250, 1350, 1410, 1460, 1500, 2840, 2900, 2940, 3040 cm^{-1}. El. Anal. Calculated for $C_{17}H_{24}O_2$: C, 78.46; H, 9.25. Found: C, 78.34; H 9.26. NMR shows no resonance due to THT. Maximum conductivity for non-stretched, I_2 doped films: 60 S/cm.

Molecular Crystals With SHG Properties

Two molecular crystals were prepared with the intent to prepare organic ferromagnets. These are the nitronyl nitroxide 1 and 1,3,5-tris(tricyanovinyl)benzene (2).

Recently, the discovery of short range ferromagnetic interactions (SRFM) (Weiss temperature, $\theta \sim 1K$ from $1/\chi$ vs T, magnetization saturation curves corresponding to S = 2, rather than S = 1/2) in crystals of 1(7) was reported(7,8).

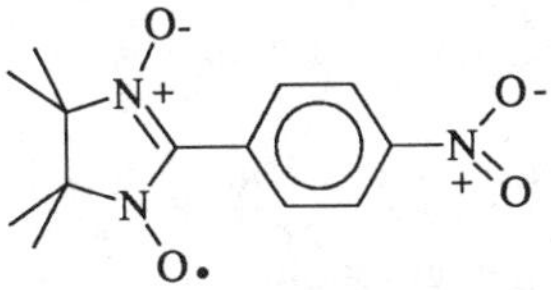

1

We, in the process of repeating Awaga's discovery, found that this molecule crystallizes in three different polymorphs. One of these is a polar structure(7) (orthorhombic, F2dd) and shows an SHG efficiency equivalent to quartz. The efficiency would probably be considerably higher if the fundamental (1.060μm) were of a different wavelength, since the solid absorbs the second harmonic.

A much more interesting discovery is that of SHG by crystals of 2. The latter is devoid of a large dipole moment but crystallizes in a polar space group (P2$_1$2$_1$2$_1$).

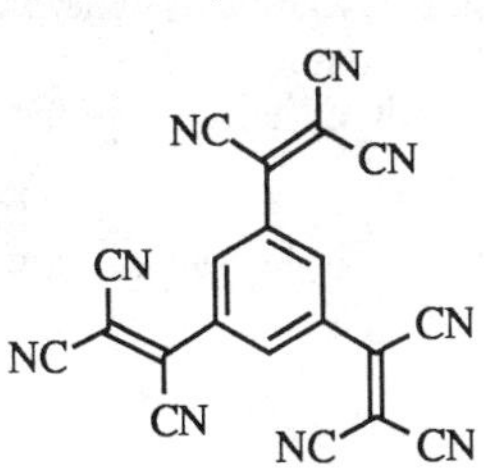

2

This material is a white solid which exhibits a powder SHG efficiency of ~100x quartz. This observation is rather unusual because most molecules designed for the improvement of SHG properties require an extended dipolar structure and consequently have an approximately D_{2h} symmetry, yet 2 has a distorted C_{3v} symmetry. Elsewhere in this symposium, J.-M. Lehn reported that crystals of 1,3,5-triamino-2,5,6-trinitro benzene (TATNB), another molecule of similar symmetry to 2, exhibit a large SHG efficiency. However TATNB can, in principle, distort to a molecule with a large dipole moment (as shown below), whereas **2** cannot. Therefore **2** is indeed unique in its NLO behavior.

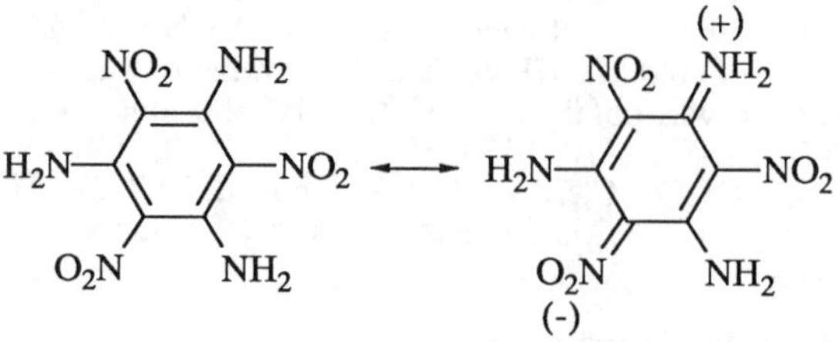

Conclusion

We have shown that a new stable, processible polymer can be produced by pushing dissymmetric substitution on the PPV skeleton to an extreme. We have also described our discovery of a molecular crystal which does not absorb in the visible (λmax 300 nm) and has no obvious large molecular dipole moment, yet shows a substantial signal for the second harmonic of Nd-YAG laser light.

Acknowledgments

We thank the Air Force Office for Scientific Research (AF49620-88-C-0138), the Office for Naval Research (N00014-83-K-0450) and the National Science Foundation (Grant DMR 88-20933) for support of this research.

Literature Cited
1.   Wessling, R.A.; Zimmerman, R.G. U. S. Patent, 3401152 (1968); 3404132 (1968); 3532643 (1970); 3705677 (1972).
2.   Wessling, R.A. J. Polym. Chem.; Polym. Symp., 1985, 72, 55.
3.   Lahti, P.M. Modarelli; D.A. Denton III; F.R. Lenz, R.W.; Karasz, F.E. J. Am. Chem. Soc. 1988, 110, 7259 and references within to the U. of Massachusetts work.
4.   Askari, S. H.; Rughooputh, S. D.; Wudl, F. Proceedings of the ACS Division of Polymeric Materials : Science and Engineering, 1988, 59, 1068.
5.   Han, C. C.; Jen, K. Y.; Elsenbaumer, R. L. Synth. Met., 1989, 30, 123.
6 .   The salt was prepared by standard literature procedures (see references 1-3, above).
7.   Awaga, K.; Maruyama, Y. Chem. Phys. Lett. 1989, 158, 556.
8 .   Awaga, K.; Maruyama, Y. J. Chem. Phys. 1989, 91, 2743.

RECEIVED July 18, 1990

Chapter 47

Quadratic Electrooptic Effect in Small Molecules

C. W. Dirk[1,3] and M. G. Kuzyk[2,4]

[1]AT&T Bell Laboratories, 600 Mountain Avenue,
Murray Hill, NJ 07974
[2]AT&T Bell Laboratories, P.O. Box 900, Princeton, NJ 08540

An attempt is made to fit quadratic electrooptic (QEO) results to a two-level model for the microscopic third order susceptibility, γ. The results are to some extent inconclusive and suggest that a two-photon state may have to be included. Also reported here are some further major improvements in molecular second order nonlinearities of particular importance to poled-polymer electrooptic applications (EO). Thus, it is found that appropriate replacement of benzene moieties with that of thiazole in certain azo dyes results in a factor of three increase in $\mu \cdot \beta$, the molecular dipole (μ_0) projected molecular second order nonlinear optical susceptibility, β.

There is great interest in preparing materials which could facilitate the development of electrooptic devices. Such devices could permit broad band optical signal encoding so that telephone, data, television, and even higher frequency transmissions could simultaneously be sent down a single optical fiber. The nonlinear optical process which makes this possible is the linear electrooptic effect (EO). It is based on the first field nonlinearities ($\vec{E}^2$) of the molecular dipole moment, $\vec{p}$,

$$\vec{p} = \vec{\mu}_0 + \tilde{\alpha}\vec{E} + \tilde{\beta}\vec{E}^2 + \tilde{\gamma}\vec{E}^3 + \cdots \quad , \tag{1}$$

and the macroscopic polarization, $\vec{P}$,

---

[3]Current address: Department of Chemistry, University of Texas, El Paso, TX 79968
[4]Current address: Department of Physics, Washington State University, Pullman, WA 99164–2814

$$\vec{P} = \vec{M}_0 + \tilde{\chi}^{(1)}\vec{E} + \tilde{\chi}^{(2)}\vec{E}^2 + \tilde{\chi}^{(3)}\vec{E}^3 + \cdots \quad , \tag{2}$$

as governed by the second order tensors, $\tilde{\beta}$ and $\tilde{\chi}^{(2)}$, respectively. The tensor, $\tilde{\beta}$, is responsible for the magnitude of the microscopic (molecular) effect, while the bulk macroscopic effect is dictated by $\tilde{\chi}^{(2)}$. The even order tensors are exactly zero when an inversion operation can be applied, so that second order nonlinear optical materials must be noncentrosymmetric. Odd order tensors (i.e. $\tilde{\alpha}$, $\tilde{\gamma}$) are unaffected by inversion symmetry. For typical laser, or electrical modulation fields there are at least several orders of magnitude difference between the largest second order and third order polarizations, $\vec{P}^{(2)}$ ($\tilde{\beta}\vec{E}^2$ or $\tilde{\chi}^{(2)}\vec{E}^2$) and $\vec{P}^{(3)}$ ($\tilde{\gamma}\vec{E}^3$ or $\tilde{\chi}^{(3)}\vec{E}^3$), respectively. Additionally, optimization of the second order process is well understood, while the structure/property relationship for third order nonlinearities has remained more mysterious. Consequently, despite the annoying restriction to noncentrosymmetry, the present incipient electrooptic modulation technology relies on the linear electrooptic effect (mediated by $\tilde{\beta}$) rather than the quadratic electrooptic effect (mediated by $\tilde{\gamma}$). This chapter explores further significant optimization of EO, then focuses on the problem of QEO. The main goal of the QEO work is to provide a model for γ in terms of simple physical-organic parameters such as λ_{max}, integrated absorption (oscillator strength), solvatochromatic behavior, etc.. This would then provide a tool that organic chemists could readily apply to optimize γ, or to at least broaden the class of materials that have relatively large γ.

Optimizing $\chi^{(2)}$ Nonlinearities for Electrooptics

Understanding second order nonlinearities in terms of simple well known physical-organic parameters requires starting from the standard perturbation theory expressions and then deriving the more limited expressions which can be related to simple physical observables. It is best to approach perturbation theory from a phenomenological direction, since this can ultimately provide a more intuitive understanding of the physics. We start with the second harmonic generation process.

Second harmonic generation (SHG) involves the mixing of two photons at frequency ω, and producing one photon at frequency 2ω. This is frequently referred to as a three-wave mixing process. Third order nonlinearities are four-wave mixing processes.

Nonlinear optics is a scattering process. As each photon "arrives" or "leaves", it induces a virtual dipole allowed transition ($\int \psi_\iota e\vec{r}\psi_\kappa d\tau$, frequently abbreviated as $\vec{\mu}_{\iota\kappa}$) between states (ψ_ι, ψ_κ). For SHG, the first photon at ω stimulates a transition between the ground state g (or "zero", 0) and some excited state m, the next photon at ω stimulates a transition between state m and state n. The departing photon, 2ω, stimulates a transition from n back to the ground state, g (Figure 1). Thus, this single microscopic event involves the tensor product of

$$F_{EO}(\omega) = \frac{6\omega_{01}^2 - 2\omega^2}{(\omega_{01}^2 - \omega^2)^2}.$$ (7)

Optimization of the two-level model involves either increasing the change in dipole moment ($\Delta\mu_{01} = \mu_e - \mu_g$) between the ground state, g (designated "0"), and first excited state e (designated "1"), increasing the transition moment (μ_{01}) between those states, or operating closer to the molecular electronic resonance, ω_{01}, with either the fundamental, ω, or second harmonic, 2ω (in the case of SHG). The preferable course is to increase the moments terms, $\Delta\mu_{01}$ and μ_{01}. Increasing the nonlinearity by resonance is easier, and can lead to substantial enhancements, though this is usually accompanied by linear absorption or damping of the second harmonic. Note from Equation (7), that in the case of EO, one has much more latitude in the use of resonance to enhance β.

Past increases in β have been accomplished by two main avenues: Increasing the length of the conjugation path between the donor and acceptor, or by increasing the electron donating and accepting abilities of the donor and acceptor.

Increasing the molecular length increases the vector, $\vec{r}$, of the dipole operator, guaranteeing an increase in the excited state dipole moment, μ_e. The ground state dipole moment μ_g, also increases, though since the ground state is far less charge separated than the excited state, the increase is less for μ_g, so that there is still a significant increase in $\Delta\mu_{01}$ (Figure 2). One other consequence of increasing the molecular length is an increase in the transition moment, μ_{01}, supplying yet another boost to β. At some point $\Delta\mu_{01}$ and μ_{01} saturate, and an increase in molecular length does not result in useful increases in β.

It is considered important to not further increase the molecular length in order to improve β. In addition to saturation of the electronic moments with increasing length, molecules of the size of the commonly used stilbenes and azobenzene dyes(7) seem to be optimal in terms of solubility properties. Further increases in molecular size would probably induce aggregation in poled polymer systems. Katz(8) has shown that β can be further increased by improving the electron accepting ability of the acceptor moiety, and there-by presumably increasing μ_e. Thus, replacement of nitro with dicyanovinyl greatly improves β without significantly changing the molecular length. He has demonstrated an excellent correlation with Hammett σ constants in explaining this enhancement. Undoubtedly, Hammett constants will provide at least a qualitative guide for further improvements in β (9).

In the absence of significantly better donors and acceptors, and keeping in mind the restrictions on molecular size, we have decided to investigate the effect of changing aromaticity in the conjugating group separating the donor and acceptor. Note in Figure 3, the nitroaniline ground state must localize to a cyclohexatriene structure in order to reach the more charge-separated quinoid excited state. The delocalization energy between benzene and cyclohexatriene is quite large, 36Kcal/mole. It might be postulated that replacing a benzene ring with

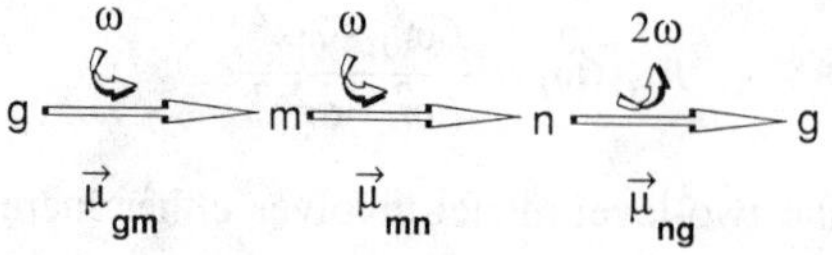

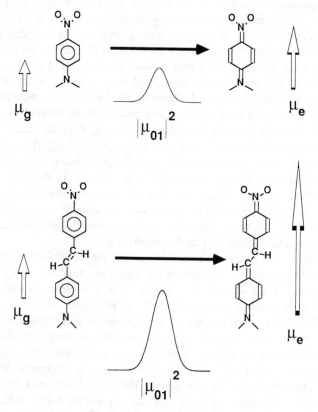

Figure 1. The optical scattering leading to a single microscopic nonlinear optical event.

Figure 2. Lengthening the molecule increases both $\Delta\mu_{01}$ and the integrated absorption ($\propto |\mu_{01}|^2$).

three dipole transitions, $\vec{\mu}_{gm}\vec{\mu}_{mn}\vec{\mu}_{ng}$. Since the transition frequencies (ω_{gm}, ω_{gn}) of the states m and n can be arbitrarily different, one must generally weight this term with a product of terms resonant with either ω or 2ω,

$$\frac{\vec{\mu}_{gm}\vec{\mu}_{mn}\vec{\mu}_{ng}}{(\omega_{gm}-2\omega)(\omega_{gn}-\omega)}, \quad \text{or,} \quad \frac{\vec{\mu}_{gm}\vec{\mu}_{mn}\vec{\mu}_{ng}}{(\omega_{gm}-\omega)(\omega_{gn}-2\omega)}. \tag{3}$$

The states ψ_m and ψ_n could be any state in the molecule, so the full molecular second order SHG polarization, $\vec{P}_{SHG}^{(2)}$, must be represented as the sum of all possible microscopic three-wave scattering events(1):

$$\tilde{\beta}_{SHG} = \frac{1}{2} \sum_{m,n=0} \left[\frac{\vec{\mu}_{gm}\vec{\mu}_{mn}\vec{\mu}_{ng}}{(\omega_{gm}-2\omega)(\omega_{gn}-\omega)} + \frac{\vec{\mu}_{gm}\vec{\mu}_{mn}\vec{\mu}_{ng}}{(\omega_{gm}+\omega)(\omega_{gn}+2\omega)} + \frac{\vec{\mu}_{gm}\vec{\mu}_{mn}\vec{\mu}_{ng}}{(\omega_{gm}+\omega)(\omega_{gn}-\omega)} \right]. \tag{4}$$

This expression, referred to as a sum-over-states (SOS), can be used to calculate molecular $\tilde{\beta}$ tensors, presuming one has first calculated the transition moments and energies using, for example, a molecular orbital program. Since much of the susceptibility arises from π-electrons, it is frequently sufficient to only include a single $p-\pi$ orbital per atom capable of donating a π-electron. Calculations of this type have been shown to be relatively accurate (2,3).

It has been known experimentally(4) that much of the second order susceptibility generally arises from the lowest singlet excited state. For any particular molecule, the recently introduced Missing States Analysis (MSA)(5,6). can show, via calculation, to what extent β is dominated by the first excited state. For instance, the β of p-nitroaniline has been shown by MSA to be heavily dominated by the first excited state, at least with a PPP (Pariser-Pople-Parr) Hamiltonian and standard basis. The result of these findings is that one can often approximate Equation (4) by including only one excited state in the sum, there-by arriving at a two-level model:

$$\beta_{TL} = |\mu_{01}|^2 |\Delta\mu_{01}| F_{SHG}(\omega), \tag{5}$$

where the SHG dispersion factor $F_{SHG}(\omega)$ is given by(7)

$$F_{SHG}(\omega) = \frac{3\omega_{01}^2}{2(\omega_{01}^2-\omega^2)(\omega_{01}^2-4\omega^2)}. \tag{6}$$

For the linear electrooptic effect (EO), the two-level model only differs in dispersion, with the dispersion factor, $F_{EO}(\omega)$, given by a heterocycle could result in an improvement in β by allowing easier access to the charge separated excited state. In Table 1 is a comparison of our best azobenzene EO dye with an analogous one incorporating a thiazole moiety (C. W. Dirk, H. E. Katz, M. L. Schilling, L. A. King, Submitted to *J. Am. Chem. Soc.*). The increase in μ·β (the appropriate quantity to compare when considering applications using EO dyes in poled-polymers) is substantial. We have found much of this increase to be due to dispersion. However, examination of Figure 4 shows that the thiazole dye has a more narrow transition, so that absorption and damping are relatively constant. Thus, use of this heterocycle has resulted in a useful increase in β. The stability and solubility properties of this dye are not significantly different from the earlier benzene analog, so it should presently be among the best available for EO applications involving poled-polymers.

Optimizing $\chi^{(3)}$ Nonlinearities for Quadratic Electrooptics

In general, the optimization of organic molecules for third order nonlinear optical applications has enjoyed much less success than for second order optical nonlinearities. The major reason for this has been the questionable validity of the two-level model for γ, and the difficult assessment of the contribution of two-photon states for the more acceptable three-level model.

Using a syllogistic approach analogous to the earlier construction of the $\vec{P}_{SHG}^{(2)}$ perturbation summation (Equation 4), we can "derive" the general third order perturbation theory expression(1),

$$\tilde{\gamma} = 4K'I_{-\sigma,1,2,3} \times \left[\sum_{l,m,n>0} \frac{\vec{\mu}_{gl}\vec{\mu}'_{lm}\vec{\mu}_{mn}\vec{\mu}_{ng}}{(\omega_{gl}-\omega_\sigma)(\omega_{gm}-\omega_1-\omega_2)(\omega_{gn}-\omega_1)} \right.$$

$$\left. - \sum_{m,n>0} \frac{\vec{\mu}_{gl}\vec{\mu}_{lg}\vec{\mu}_{gn}\vec{\mu}_{ng}}{(\omega_{gl}-\omega_\sigma)(\omega_{gn}-\omega_1)(\omega_{gn}+\omega_2)} \right] , \qquad (8)$$

where K' is a constant that depends on the optical process (i.e THG, DFWM, QEO, etc.), $\omega_\sigma=\omega_1+\omega_2+\omega_3$, $I_{-\sigma,1,2,3}$ is the average of the 48 terms obtained by permuting $-\omega_\sigma$, ω_1, ω_2, ω_3, and, $\vec{\mu}_{\iota\kappa}=\vec{\mu}_{\iota\kappa}-\vec{\mu}_{gg}$. Note that there are one-photon, $(\omega_{gn}-\omega_1)$, two-photon, $(\omega_{gm}-\omega_1-\omega_2)$, and three-photon, $(\omega_{gl}-\omega_\sigma)$, resonant denominators, respectively. One- and three-photon states are accessible from the ground state by use of the dipole operator, $e\vec{r}$, and can be the same state (i.e. $l=n$). Pure two-photon states are not similarly accessible from the ground state, so that $m \neq l,n$. Thus, the minimal *formal* approximation that we can make to Equation (8) is to sum over two excited states, yielding a three-level model(10),

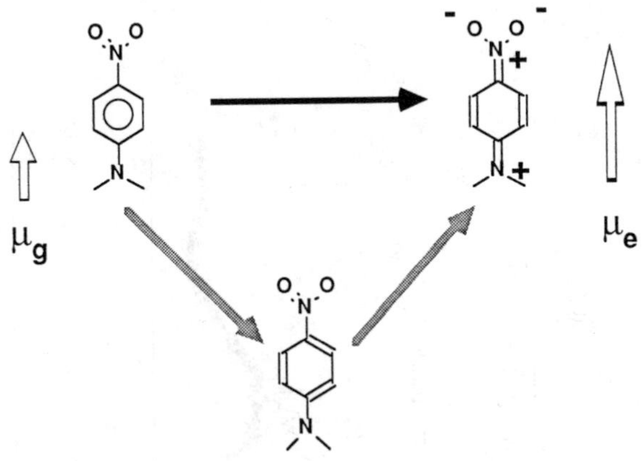

Figure 3. Bond localization necessary to reach quinoid excited state from aromatic ground state.

Table 1. Second-Order Nonlinearities of Thiazole Dyes

$$\mu \cdot \beta$$
$$(10^{-30} \ D \ cm^5 \ esu^{-1})$$

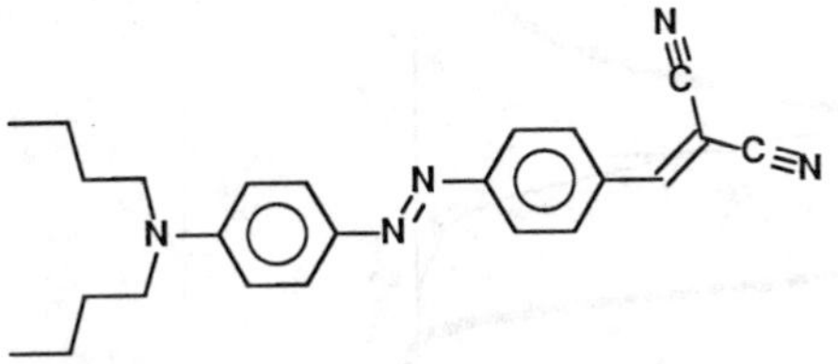

1880

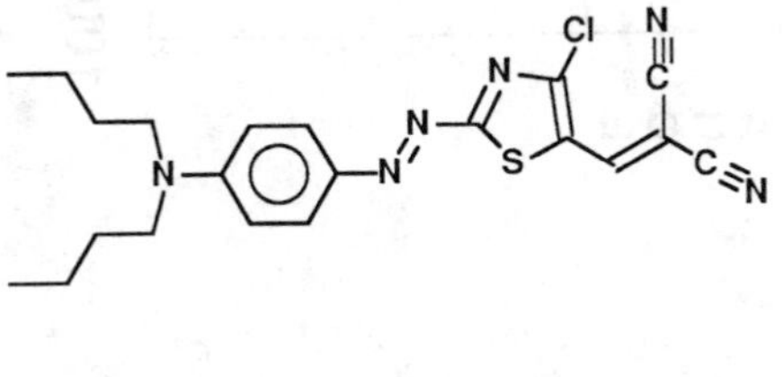

5320

Note that these results are local field corrected

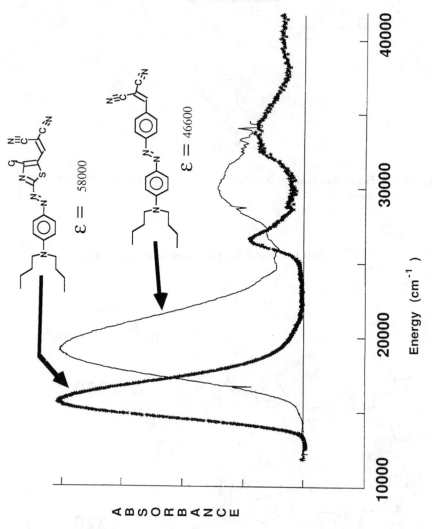

Figure 4. Electronic spectra of analogous benzene amd thiazole azo dyes.

$$\gamma = K' \left[-\mu_{01}^4 D_{11} + \mu_{01}^2 (\Delta\mu_{01})^2 D_{111} - \mu_{02}^2 D_{22} + \mu_{02}^2 (\Delta\mu_{02}) D_{222} - 2\mu_{01}^2 \mu_{02}^2 D_{12} \right.$$

$$\left. + \mu_{12}^2 \mu_{01}^2 D_{121} + \mu_{12}^2 \mu_{02}^2 D_{212} + 2\mu_{01}\mu_{02}\mu_{12}(\Delta\mu_{01}) D_{211} + 2\mu_{01}\mu_{02}\mu_{12}(\Delta\mu_{02}) D_{122} \right] ,(9)$$

where D_{lmn} and D_{mn} are dispersion terms and refer to the permuted sums of the triple and double sums, respectively, of Equation (8). This expression may be too complex to be of much direct experimental use, but a further simplification is possible by presuming the second excited state to be pure two-photon in nature, implying $\mu_{02} \equiv 0$. This yields:

$$\gamma = K' \left[-\mu_{01}^4 D_{11} + \mu_{01}^2 (\Delta\mu_{01})^2 D_{111} + \mu_{12}^2 \mu_{01}^2 D_{121} \right], \qquad \text{or} \qquad (10)$$

$$\gamma = \gamma_c + \gamma_n + \gamma_{TP} , \qquad (11)$$

where,

$$\gamma_c = -K'\mu_{01}^4 D_{11},$$

$$\gamma_n = K'\mu_{01}^2 (\Delta\mu_{01})^2 D_{111},$$

$$\gamma_{TP} = K'\mu_{12}^2 \mu_{01}^2 D_{121}.$$

This approximation may only be true for centrosymmetric molecules where by analogy to the particle-in-a-box, the selection rules are now more strict. As one increases the asymmetry, selection rules begin to break down and it is possible that μ_{02} will become more allowed. For the same reason, as $|\mu_{02}|$ increases, we might expect $|\mu_{12}|$ to decrease, so that in the extreme of large $|\Delta\mu|$ s (i.e. molecules with large second order nonlinearities, β), we might anticipate a small γ_{TP}. Thus, either in the extreme of centrosymmetry with a small transition moment between excited states (μ_{12}) and/or if D_{121} is small relative to D_{11} and D_{111}, or in the case of large β it would be possible to *mathematically* impose a two-level model,

$$\gamma = \gamma_c + \gamma_n , \qquad (12)$$

on the observed data. Under what circumstances might this be applicable?

Since two-photon states are difficult to characterize, it would be difficult to generally ascertain the magnitude of the excited state transition moment, μ_{12}. This remains an unknown in the system, thus we must depend on the characteristics of the dispersion terms $(D_{11}, D_{111}, D_{121})$ to define a limiting situation leading to Equation (12). The two-photon resonance of D_{121} consists of $1/(\omega_{02}-\omega_1-\omega_2)$. To minimize its contribution, we must maximize the magnitude of the denominator. This can be done by setting $\omega_1=\omega_2=0$. If $\omega_3=\omega\neq0$, this restriction defines the process $\gamma(-\omega;0,0,\omega)$. This is the QEO process.

For the QEO process, in order to maximize the contributions of the D_{11} and D_{111} terms relative to that of the D_{121} term, it is necessary to operate close (on the low energy side) to the first excited frequency, ω_{01}, with the QEO probe, ω. As long as the two-photon state frequency, ω_{02} is not too close to ω_{01}, and/or $|\mu_{01}| \geq |\mu_{12}|$, the term based on D_{121} could possibly be ignored for the QEO process. The general validity of Equation (12) remains in doubt, because of the uncertainty over the magnitude of μ_{12}, and the unknown frequency difference, $|\omega_{02}-\omega_{01}|$. However, these conditions are likely the closest one can get toward imposing a two-level model for γ.

If we check the preceding argument against the CNDO and PPP results of Heflin et. al(11) and Soos et. al(12), respectively, for the *trans*-octatetraene molecule, we can judge how good this approximation is for at least one centrosymmetric system. From their transition moments, and transition energies, it is possible to calculate γ_c and γ_{TP} for a hypothetical QEO measurement as proposed above. For this hypothetical measurement, we have set the probe $2000cm^{-1}$ below the lowest excited state. In an actual measurement it is possible to measure γ_{QEO} closer to resonance, and even right on resonance, though damping corrections necessary to correct D_{11} (for γ_c) and D_{121} (for γ_{TP}) become uncertain so that experimental data cannot be easily related to anything. Actual measurements are generally done far enough off resonance ($1000cm^{-1}$ to $3000cm^{-1}$, depending on peak width) that effective damping corrections can be made. The values $(\times10^{-34}cm^7esu^{-2})$, as calculated for the hypothetical measurement on *trans*-octatetraene are given below in Table 2:

Table 2. Calculated Third-Order Susceptibilities

	Heflin et. al	Soos et. al
$Re[\gamma_c^*]$	−4.9	−6.6
$Re[\gamma_n^*]$	≡0	≡0
$Re[\gamma_{TP}^*]$	+10.7	+15.2

For both calculations (based on the MO and VB calculations of Garito or Soos, respectively), the two-photon state of *trans*-octatetraene (ψ_6) is significantly removed ($\nu_1-\nu_6\approx22000cm^{-1}$) from the lowest excited state, ψ_1, so that D_{121} (actually D_{161}; states ψ_2-ψ_5 appear irrelevant) should be poorly competitive with D_{11}. Despite this, $|\gamma_{TP}|$ is *still* larger than $|\gamma_c|$. Thus, in spite of a best case scenario for a two-level model, we should see a significant two-photon contribution for a centrosymmetric molecule. This leaves unresolved the case of a molecule possessing a large β, which is experimentally dealt with below.

QEO Measurements

Third order NLO measurements are frequently plagued by artifacts. Among the most difficult contributions to eliminate are the slow orientational contributions which mask the smaller fast electronic component. For the QEO process, two major orientational contributions must be dealt with, often referred to as "$\mu\beta$" and "$\alpha\alpha$", respectively. The "$\mu\beta$" component arises from the electric field coupling to the molecular dipole moment, orienting the molecule, permitting a field induced electrooptic effect. The "$\alpha\alpha$" effect, on the other hand, involves a field induced dipole moment, $(\tilde{\alpha}\vec{E})$ via the polarizability tensor, $\tilde{\alpha}$, then coupling of this *induced* dipole with the electric field reorienting the birefringent molecule, thereby changing the refractive index. The magnitude of the "$\mu\beta$" contribution, $\gamma_{\mu\beta}$, to the molecular QEO susceptibility, γ_{QEO}, depends on the magnitude of the dipole moment $\vec{\mu}_0$ and the dipole projection of $\tilde{\beta}$, while the magnitude of the "$\alpha\alpha$" contribution, $\gamma_{\alpha\alpha}$, depends on the magnitude and anisotropy of the molecular polarizability, $\tilde{\alpha}$ (13,14). These two contributions are of opposite sign, and for molecules with large β, normally the same order of magnitude. The resulting mutual cancellation of $\gamma_{\mu\beta}$ and $\gamma_{\alpha\alpha}$ leads to a total orientational contribution, γ_{OR}^{tot}, which is comparable to, or smaller than the electronic QEO susceptibilities measured here (*vide infra*).

In order to limit orientational contributions, we have created a new procedure to measure γ_{QEO}. This involves dissolving the molecule of interest in a solution containing poly(methylmethacrylate) (PMMA), and spinning a thin (2–3μM) film onto an ITO glass electrode, placing two of these films face to face and heating briefly under compression to effect an optical contact between the film surfaces (Figure 5a). One then places the sample into a Mach-Zehnder interferometer, oscillate an electric field (at 4000Hz) across the ITO electrodes, monotonically "delay" the signal in the other arm of the interferometer (Figure 5b), and lock-in on the fringes being created at 8000Hz. The fringe magnitude provides the real part of the quadratic electrooptic coefficient, $|\,\mathrm{Re}[s]\,|$, while the imaginary part, $|\,\mathrm{Im}[s]\,|$ is measured from the offset. Details are provided elsewhere(13,15).

The high viscosity of the PMMA damps out orientational contributions so that the γ_{QEO} that is measured is thought to be $\approx$60–90% electronic. This has been ascertained by measuring the electric field induced second harmonic generation (EFISH) below the T_g of the polymer. From this can be obtained the microscopic elastic constant, which can in turn be used to estimate the magnitude of the two orientational contributions to γ_{QEO}. Details are provided elsewhere(13,16).

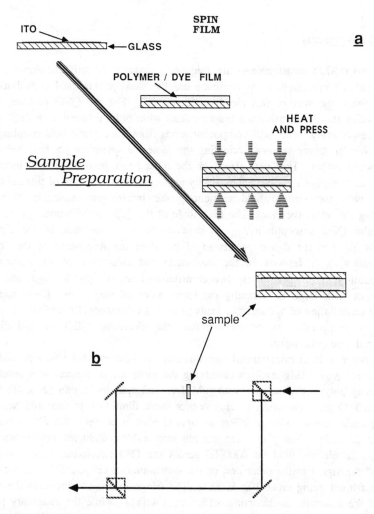

Figure 5. (a)QEO sample preparation. (b) QEO interferometric measurement.

QEO Results

The QEO susceptibility results for several molecules (structures shown in Figure 6) are summarized in Table 3 (17).

Table 3. QEO Susceptibility Results

molecule	$\mid Re[\gamma_{QEO}] \mid$ ‡	$Re[\gamma_c]$ §	$Re[\gamma_n]$ §	$Re[\gamma_c]+Re[\gamma_n]$
1	29	-6.8	+45	+38
2	7.1	-1.2	+9.3	+7.1
3	7.3	-0.5		
4	1.8	-0.2		
5	2.8	-0.06		
6	6.2	-0.4		
7	18	-3.8		
8	4.3	-0.6		
9	68	-11	≡0	-11

Note that all susceptibilities are $\times 10^{-34} cm^7 esu^{-2}$.
‡ ±25%; § γ_c ±20%; γ_n ±30-35%

Note that the squarylium dye, **9**, has quite a large susceptibility. At 1% in PMMA, the bulk susceptibility is $13 \times 10^{-14} esu$ (18). At a hypothetical "100%", this susceptibility is comparable to the THG and DFWM susceptibilities of some polydiacetylenes, though this comparison must be viewed with some circumspection considering the difficulty in accounting for the differences in dispersion between different processes.

Optimizing QEO Materials

Is it possible to enhance γ_{QEO}?
Answering this question involves an examination of Equation (12). The largest QEO nonlinearities are available when the QEO probe frequency, ω, is close to resonance. This is the restriction that can potentially allow the two-level model to be applicable. Note that the two-level model depends on the competition between γ_c and γ_n, which have opposite sign. It is important to note that under centrosymmetry, γ_c should be the only contribution to γ_{QEO}, while the γ_{QEO} for noncentrosymmetric molecules consists of $\mid \gamma_c + \gamma_n \mid$. A first order correction to this model will require the additional term γ_{TP}. It is important to judge what effect this term may have on this speculation and the potential interpretation. In general, it should be to lower (However, note as in the case of *trans*-octatetraene,

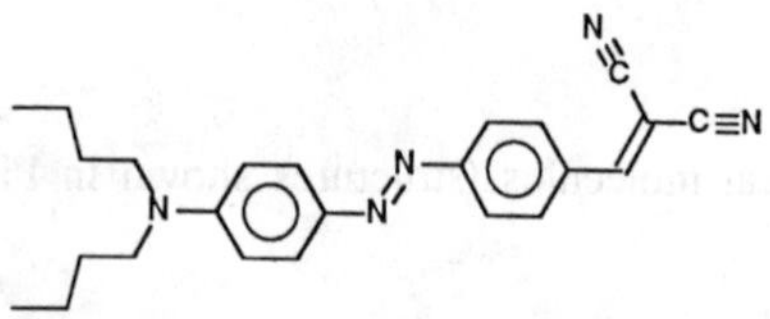

Figure 6. Structures of molecules reported in Table 3.

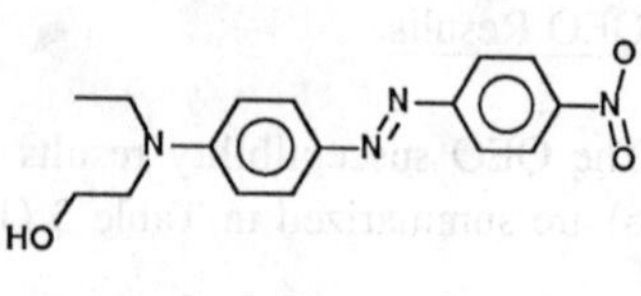

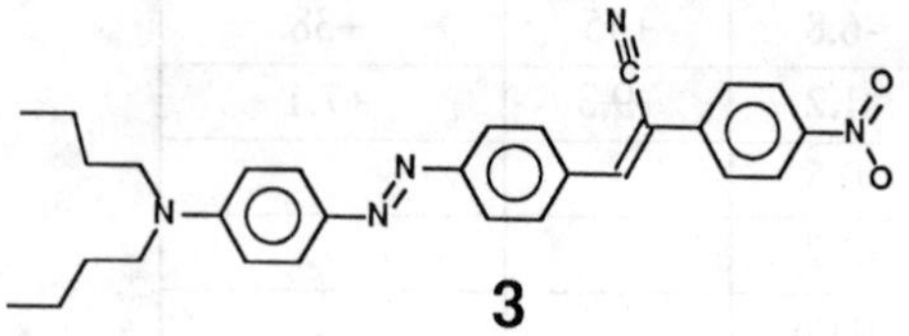

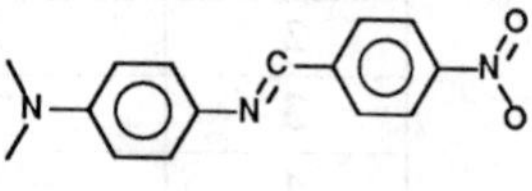

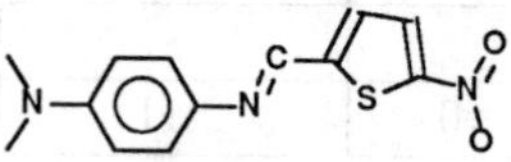

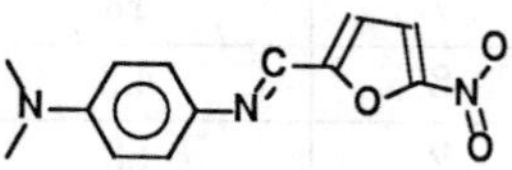

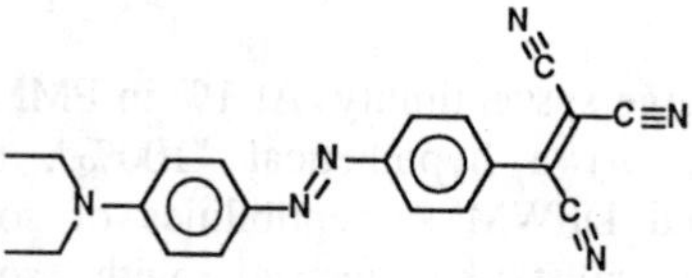

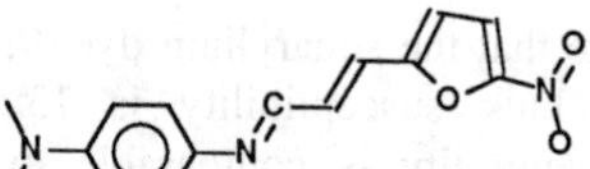

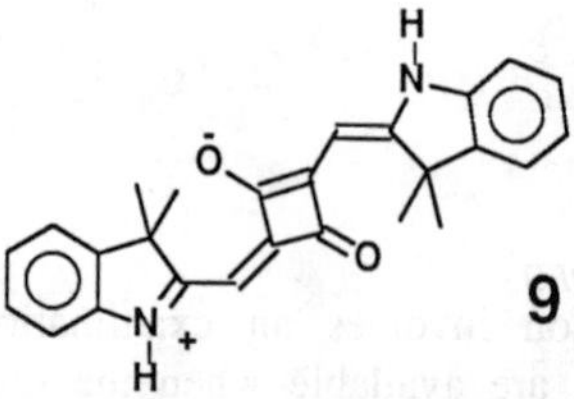

the magnitude of γ_{TP} is sufficiently large that despite the opposite sign to γ_c, $|\gamma_c + \gamma_{TP}| > |\gamma_c|$) the expected $|\gamma_{QEO}|$ for centrosymmetric molecules, but increase the susceptibility magnitude expected in noncentrosymmetric molecules.

For our measurements, the quantity, D_{111}/D_{11} approximately equals two. As one operates the QEO probe, ω, closer to resonance (i.e. $|\omega_{01} - \omega| \leq 500cm^{-1}$), this ratio can increase greatly. If the breadth of the molecular excitation, ω_{01}, is sufficiently narrow ($\leq 700cm^{-1}$ at half-height), then considerable enhancements ($>20\times$) are possible in γ_n when close to resonance, which could lead to much larger γ_{QEO}s. It would then appear that one has two options in increasing γ_{QEO} as mediated through Equation (12): increase the integrated absorption ($\propto \mu_{01}^2 \propto \gamma_c$) of centrosymmetric molecules, or prepare noncentrosymmetric molecules with large β ($\propto \Delta\mu_{01} \propto \gamma_n$) and especially narrow electronic absorptions.

Transition moments, μ_{01}, to the first excited state can be calculated from the integrated absorption of the linear electronic spectrum. This can be used to calculate $-K'\mu_{01}^4 D_{11}$, the first term (defined here as γ_c) of the two-level model. The second term ($K'\mu_{01}^2 \Delta\mu_{01}^2 D_{111}$) from Equation (12) (defined here γ_n) involves $\Delta\mu_{01}$, which can be determined directly from solvatochromism(19), or from a two-level analysis of a molecular EFISH measurement of β.

Shown in Table 3 are the Re[γ_c] results(17) along with some preliminary values for Re[γ_n] as determined from EFISH. It can be seen that γ_{QEO} is not well accounted for by γ_c. For the two results for which we have γ_n data, $|\gamma_c + \gamma_n|$ does reasonably well account for γ_{QEO}. However, note that for molecule 9, $\gamma_n \equiv 0$ in its centrosymmetric conformation, so that γ_{QEO} is not well accounted for by $|\gamma_c + \gamma_n|$. There are several possible explanations for this: (1) There may be significant noncentrosymmetric conformations that exist for 9 in solution leading to $\gamma_n \neq 0$. (2) If molecule 9 is indeed dominated by the centrosymmetric conformation, $\gamma_{\mu\beta}$ should be small, and may not fully cancel $\gamma_{\alpha\alpha}$. Thus, there could be a much larger total orientational contribution, γ_{OR}^{tot}, than is anticipated. (3) Finally, the two-photon contribution, γ_{TP}, is unknown. As pointed out earlier, for centrosymmetric systems, γ_{TP} may be the most significant contributing term to the measured γ_{QEO}.

If γ_{TP} represents a significant contribution to γ_{QEO} for molecule 9, then it is curious that it appears to be less important for the two dipolar dyes, 1 and 2. In keeping with the earlier discussion, this could possibly reflect the effect of breaking symmetry so that for dyes with large β, the second excited state is not purely two-photon in nature with $|\mu_{12}| < |\mu_{01}|$.

Conclusions

We report the largest known useful microscopic NLO susceptibilities for the linear and quadratic electrooptic effects. The QEO susceptibility of 9 might be large enough to explore simple primitive QEO modulation experiments, though perhaps is not nearly large enough to be of commercial importance.

With the permutation-symmetry-corrected results(17) it appears that we cannot successfully fit our QEO data to a two-level model, at least for centrosymmetric structures. We have only one data point (9) to make this judgement, however, and have not eliminated the problem of an uncompensated "$\alpha\alpha$" orientational contribution. In the case of molecules with large β, our two results for 1 and 2 indicate a reasonable fit to Equation (12). These results are consistent with the observations of Garito et. al. where they show γ_n to be a dominant contribution in donor-acceptor substituted polyenes(20). If Equation (11) is a more accurate overall representation to cover the extremes between centrosymmetry and large $\tilde{\beta}$, it would appear that despite the large value of $|\text{Re}[\gamma_{QEO}]|$ for 9, an attractive route to especially large QEO susceptibilities might be noncentrosymmetric molecules with large β and narrow electronic transitions. Any significant γ_{TP} will only end up supplementing the potentially very large susceptibility offered by γ_n when very close to resonance. While there has been little success (or effort?) at narrowing electronic transitions of molecules with large $\tilde{\beta}$, the thiazole dye reported earlier in this chapter is a distinct improvement in this regard over others previously reported. Preliminary QEO measurements indicate this dye to possess a γ_{QEO} larger than that of molecule 1.

It would appear that optimizing molecules for a large EO effect (via γ_n) will also optimize them for a large QEO effect. However, compared to EO, obtaining the largest QEO susceptibilities may require operating even closer to resonance with the probe frequency.

While it appears that a good deal of the QEO susceptibility may be accounted for by three terms, γ_c, γ_n, and γ_{TP}, we do not intend to ignore the potential importance of other terms in Equation (9). The strategy will be to see where a reasonable cut-off of terms of Equation (9) can still lead to an adequate explanation of the structure/property trend. We are presently working on solvatochromatic and EFISH determinations of $\Delta\mu_{01}$, with the aim to more fully characterize molecules with regard to γ_n, and have plans to experimentally determine μ_{12} in order to calculate γ_{TP}. There are also efforts underway to more accurately determine orientational contributions.

Literature Cited

1. Orr, B. J.; Ward, J. F. Mol. Phys. 1971, 20, 513-526.

2. Dirk, C. W.; Twieg, R. J.; Wagniere, G. J. Am. Chem. Soc. 1986, 108, 5387-

3. Li, D.; Ratner, M. A.; Marks, T. J. J. Am. Chem. Soc. 1988, 110, 1707-

4. Oudar, J. L. J. Chem. Phys., 1977, 67, 446-457.

5. Dirk, C. W.; Kuzyk, M. G. Physical Review A, 1989, 39, 1219-1226.

6. Dirk, C. W.; Kuzyk, M. G. SPIE Proceedings 1988, 971, 11-16.

7. Singer, K. D.; Sohn, J. E.; King, L. A.; Gordon, H. M.; Katz, H. E.; Dirk, C. W. J. Opt. Soc. B, 1989, 6, 1339-1350.

8. Katz, H. E.; Singer, K. D.; Sohn, J. E.; Dirk, C. W.; King, L. A.; Gordon, H. M. J. Am. Chem. Soc., 1987, 109, 6561-

9. For some further discussion on the applicability of linear free energy relationships to second order nonlinear optics, see: Ulman, A. J. Phys. Chem. **1988**, 92, 2385-2390.

10. Kuzyk, M. G.; Dirk; C. W. Physical Review A, **1990**, 41, 5098-5109.

11. Heflin, J. R.; Wong, K. Y.; Zamani-Khamiri, O.; Garito, A. F. Phys. Rev. B **1988**, 38, 1573-.

12. Soos, Z. G.; Ramasesha, S. J. Chem. Phys. **1989**, 90, 1067-.

13. Kuzyk, M. G.; Dirk, C. W.; Sohn, J. E. J. Opt. Soc. B, **1990**, 7, 842-.

14. Kuzyk, M. G.; Moore, R. C.; Sohn, J. E.; King, L. A.; Dirk, C. W. SPIE Proceedings, **1989**, 1147, 198-209.

15. Kuzyk, M. G.; Dirk, C. W. Appl. Phys. Lett. **1989**, 54, 1628-1630.

16. Kuzyk, M. G.; Moore, R. C.; King, L. A. J. Opt. Soc. B, **1990**, 7, 64-72.

17. Dirk, C. W.; Kuzyk, M. G. SPIE Proceedings **1989**, 1147, 18-25. Note that the Re[γ_c] in this reference are incorrect (too high by a factor of six) due to an error in properly applying the permutation symmetry operation.

18. Dirk, C. W.; Kuzyk, M. G. Chem. Materials, **1990**, 2, 4-6.

19. Paley, m. S.; Harris, J. M.; Looser, H.; Baumert, J. C.; Bjorklund, G. C. Jundt, D.; Twieg R. J. J. Org. Chem. **1989**, 54, 3774-3778.

20. Garito, A. F.; Heflin, J. R.; Wong, K. Y.; Zamani-Khamiri, O. SPIE Proceedings **1988**, 971, 2-10.

RECEIVED July 10, 1990

Chapter 48

Third-Order Nonlinear Optical Properties of Organic Materials

Toshikuni Kaino[1], Takashi Kurihara[1], Ken-ichi Kubodera[2], and Hirohisa Kanbara[2]

[1]NTT Opto-electronics Laboratories, Tokai, Naka-gun, Ibaraki, 319–11 Japan
[2]NTT Opto-electronics Laboratories, Morinosato Atsugi-shi, Kanagawa, 243–01 Japan

Third-order nonlinear optical (NLO) organic materials with good processabilities are discussed. THG measurements are demonstrated on poly(arylene vinylene)s, dye attached polymers, cyanine dye doped polymers and charge transfer organic crystals. These materials possess a third order optical nonlinearity, $\chi^{(3)}$, of the order of 10^{-10} to 10^{-11} esu. Resonant effects of poly(arylene vinylene)s and intramolecular charge transfer effects derived from substituted donors and acceptors of dye-attached polymers are revealed to contribute to the increment of the third-order optical nonlinearity. It is emphasized that a high $\chi^{(3)}$ value is possible even in short π-electron conjugated systems by modifying the molecular structure.

The development of highly efficient third-order nonlinear optical (NLO) materials with processability is required for optical signal processing systems. Recently, organic NLO materials have been energetically studied to achieve higher efficiency (1,2). Some organic materials have been reported to possess large optical nonlinearities and fast response times(3,4). Organic materials have other merits compared with semiconductor NLO materials. They have a capability in the design of their chemical structures and absorption wavelengths to achieve highly efficient functions for NLO device use. Therefore, they can be applied to many optical devices such as thin film waveguides and fiber waveguides. In particular, thin films which exhibit third-order optical nonlinearities have many useful applications in integrated optics such as optical switching and optical data processing. Most of these materials reported so far have π-electron conjugated systems which are the main origin of their optical nonlinearities. Various kinds of π-conjugated organic materials have been investigated as NLO materials, showing efficient intensity-dependent refractive indices with very fast

0097–6156/91/0455–0704$06.00/0

response times and low absorption losses. Third-order NLO susceptibility, $\chi^{(3)}$, has been evaluated by third-harmonic generation (THG) measurements or by degenerate four wave mixing (DFWM) measurements.

Among them the most studied to date are delocalized π-conjugated polymers such as polydiacetylenes (PDA) and polyacetylene, which have been reported to show larger $\chi^{(3)}$s of the order of 10^{-10} to 10^{-9} esu (5,6). The nonlinear electronic polarization of π-conjugated polymers originates from mesomeric effects which depend on the size of the π-conjugated systems. Therefore, conducting polymers have been of special interest because of their delocalized π-conjugated systems along a chain direction. Unfortunately, these π-electron conjugated polymers are usually in crystalline states and they are difficult to process because they are rarely fusible and are insoluble in almost all solvents. For example, PTS-polydiacetylene (PTS-PDA), one of the representative π-electron conjugated polymers, shows a large $\chi^{(3)}$ of around 10^{-9} esu, however, it has problems in processability and stability. Therefore, amorphous or low crystallinity π-conjugated materials with good processability are expected to be an excellent material for use in NLO devices. Thus, $\chi^{(3)}$ has been recently investigated for various amorphous and low crystallinity polymers such as poly(phenylene benzobisthiazole) (PBT), poly(arylene vinylene)s (PAVs), dye attached polymers and polysilanes. Low crystallinity conducting polymers such as PAVs and nBCMU-PDA, have easy processing for making high-quality thin films, and high $\chi^{(3)}$ values ($\sim10^{-10}$ esu)(7,8).

In the first part of this paper, PAVs which are one group of the promising processable $\chi^{(3)}$ materials, will be discussed. The wavelength dependence of THG $\chi^{(3)}$ reveals the enhancement of $\chi^{(3)}$ due to the three-photon resonant effect. In the second part, dye attached polymeric materials with large $\chi^{(3)}$ characteristics, which also possess a good processability are discussed. In the third part, dye doped polymeric NLO materials are also presented. In the final part, NLO crystals which possess the highest $\chi^{(3)}$ values among low molecular weight compounds are discussed and optical Kerr shutter experiments using solution of the crystal will be discussed.

Poly(Arylene Vinylene)s

In the view of the practical use, polymeric materials are expected to overcome the disadvantages of organic molecular crystals in mechanical properties and processability. Most polymers are usually in a symmetric structure as a whole, so they should be considered as third-order rather than second-order NLO materials, except for poled polymer systems.

PAVs, where electrically conductive films with good transparency are easily obtained from their precursor polymers, are one group of π-conjugated polymers and regarded as an alternating copolymer of acetylene and arylene. They have variations in chemical structure such as poly(p-phenylene vinylene), (PPV), poly(2,5-thienylene vinylene), (PTV), and poly(2,5-dimethoxy-p-phenylene vinylene), (MOPPV)(9-11). They have good processability in that conventional solvent casting or spin coating techniques are applicable for fabricating thin films of these polymers. PAVs are recently of special interest as NLO materials because of their processability achieved without chemical modifications of the conjugated polymers(12-14).

The PAV thin films used in our experiments were prepared through water or organic solvent soluble precursor polymers, mainly prepared through new precursor polymers which were soluble in organic solvents(15). The new precursor polymers possess a methoxy group in place of a sulphonium salt group. The new precursor polymers are chemically stable in air below 100°C. Thin precursor polymer films were obtained by spinning from organic solutions. Uniform, dense, and tough PAV films were obtained from the precursor polymer films by heat treatment.

For example, PTV film was fabricated as follows. Polymerization of a sulphonium monomer, 2,5-thienylene bis(methylene-dimethyl-sulphonium chloride) was carried out in a methanol-water mixture at -20°C by adding a methanol solution of tetramethyl-ammonium hydroxide. The reaction was quenched by an addition of hydrochloric acid. A yellow precipitate (precursor polymer) appeared as the solution was warmed to room temperature. The precipitated precursor polymer was completely soluble in dich-loromethane. A precursor polymer thin film was obtained by spin-coating of the dichloromethane solution of the precursor polymer onto a fused silica glass substrate under inert atmosphere to prevent oxidation with air. The film was heated at 200~250°C in a vacuum of 10^{-2} Torr for 5 hours, to give a tough, flexible PTV film. The resulting PTV thin film was chemically stable in air.

The thickness of the PAV films were from 0.5 to 0.02 μm in this experiment. The optical absorption spectrum was measured with a spectro-photometer and maxima were revealed at 0.52, 0.50 and 0.42 μm for PTV, MOPPV and PPV films, respectively. X-ray diffraction patterns indicate that the PAV films have low crystallinity.

$\chi^{(3)}$ values were evaluated from a THG measurement as follows.(16) THG intensities in these thin films were measured at a fundamental wavelength between 1.475 μm and 2.10 μm, which was obtained using a difference-frequency generation of a Q-switched Nd:YAG laser beam and a tunable dye laser beam. The pulse duration and the pulse repetition rate were 5.5 nsec and 10 Hz, respectively. The peak power density was 400 MW/cm^2.

The sample was mounted on a goniometer and rotated about an axis perpendicular to the laser beam. The laser beam was linearly polarized in a direction parallel to the rotational axis. The generated third harmonic was passed through a fundamental wave cutting filter and was detected by a photomultiplier tube. Third harmonic intensities were measured as a function of the incident angle for PAV films and a silica glass standard whose $\chi^{(3)}$ was reported to be 2.8 x 10^{-14} esu(17). The details of the experiments were reported elsewhere(18).

The thickness of the film prepared is much less than the coherence length, so the output intensity is a monotonic function of the incident angle.

$\chi^{(3)}$ was calculated using the following equation(18),

$$\chi^{(3)}=2/\pi \cdot \{(I_{3\omega}-I_{3\omega,s}/2)/I_{3\omega,s}\}^{1/2} \cdot \chi^{(3)}_s \cdot l_{c,s}/l$$

where l is the sample thickness, $\chi^{(3)}_s$ is the $\chi^{(3)}$ of the standard fused silica, and $l_{c,s}$ is its coherent length. $I_{3\omega}$ is a center value of the envelopes for maximum amd minimum of the superimposed intensity pattern in the polymers. $I_{3\omega,s}$ is a peak intensity value for the fused silica.

The $\chi^{(3)}$ values of these PAV films at 1.85 μm wavelength were found to be 7.8 x 10^{-12} esu, 3.2 x 10^{-11} esu, and 5.85 x 10^{-11} for PTV, MOPPV and PPV films, respectively[7,19,20]. These $\chi^{(3)}$ values are almost the same as that of processable nBCMU-PDA thin film[19].

The fundamental wavelength dependence of $\chi^{(3)}$ has been measured on a PPV film. As a consequence, the resonant effects on $\chi^{(3)}$ were clearly revealed. Figure 1 shows the measured $\chi^{(3)}$ of the PPV film as a function of the incident light wavelength together with the PPV absorption spectrum. The $\chi^{(3)}$ value compensated for the effects of absorption is also shown. As the wavelength decreases, $\chi^{(3)}$ increases. This may be caused by enhancement effects (three photon resonance) due to the overlap of the harmonic wavelength with a polymer absorption band. As is shown in this figure, strong exitonic absorption exists as a shoulder on the transition absorption peak, near the wavelength region of third harmonic resonance. A maximum $\chi^{(3)}$ of 1.4 x 10^{-10} esu was obtained at a 1.475 μm wavelength. The resonant $\chi^{(3)}$ of the PPV film is about one order stronger than that of nonresonant values which was reported earlier[7]. Almost the same result was obtained for a PPV thin film fabricated through a tetramethylene sulfonium chloride precursor method[14].

The $\chi^{(3)}$ values were almost the same for PPV, PTV and MOPPV in the absorption wavelength region. Dispersion measurements are required in this energy range in order to clarify the nature of the resonance and to determine the maximum resonantly enhanced $\chi^{(3)}$. The evaluated susceptibilities of the PAV films are almost the same order of magnitude as that of PDA films. This result indicates that PAV thin films may be effective for NLO device application, due to the ease in processing films as well as possible good optical quality.

Dye Attached Polymers

Expecting that polymer systems containing NLO materials in high concentrations would show large third order NLO effects, azo dyes or stilbene dyes attached to polymers were synthesized and their optical nonlinearities were evaluated by a THG measurement. Large $\chi^{(3)}$ amorphous polymers can be obtained by attaching a nonlinearity generating unit to the transparent polymer chain [21,22]. This requires the introduction of high concentrations of the nonlinear optical units into the polymer chain. Two types of amorphous polymers were synthesized with π-electron conjugated systems in their main chain or side chain.

<u>Side Chain Polymers</u>. Dye-substituted polymers were synthesized with azo or stilbene dyes incorporated into a transparent polymer matrix at the side chain of methacrylate or acrylate polymer with high dye contents. The azo and the stilbene dyes have a π-electron conjugated system with acceptor and donor substituent groups at both ends of their molecules.

The attached dyes are different in their conjugation length as follows.

One example is an azo benzene system which consists of three aromatic rings bonded through nitrogen-nitrogen double bonds (bisazo dye), and another example is a two ring azo benzene system (monoazo dye). These polymers are referred to as 3R for bisazo dye attached polymer and 2R for monoazo dye attached one. The dyes all have a nitro group as an electron acceptor and an amino group as an electron donor. The 3R was synthesized by a diazo coupling reaction. The azo dye attached polymers were copolymerized with methylmethacrylate by radical polymerization process.

These copolymers were soluble in conventional solvents and the films of these copolymers were fabricated by spinning the solution onto a glass substrate. The thickness of the film ranged from 0.2 to 0.3 μm. The $\chi^{(3)}$ of the 3R copolymer film as a function of fundamental wavelengths is shown in Fig.2 along with the molecular structure. The dye content was 7.0 mol%. As the fundamental wavelength decreases, the $\chi^{(3)}$ increases. This is caused by the enhancement due to near resonant effect for the $\chi^{(3)}$. Because the $\chi^{(3)}$ of polymethylmethacrylate (PMMA) is negligibly small (4.0 x 10^{-14} esu at 1.9 μm), only the dye moiety of the polymer contributes to the $\chi^{(3)}$. The $\chi^{(3)}$ of the 3R copolymer is more than three times higher than that of the 2R copolymer over the fundamental wavelength between 1.5 to 2.0 μm(23). In a two level model, the third-order molecular hyperpolarizability decreases as the charge transfer contribution increases in these donor-acceptor systems(24). So, the donor-acceptor contribution of the 3R polymer to the $\chi^{(3)}$ is thought to be comparable to that of 2R polymer. This indicates that the $\chi^{(3)}$ of π-conjugated dye-attached polymeric materials is raised through increasing the delocalization length of the dye as is common for one dimensional π-conjugated systems. Its absorptional maximum is at 500 nm, about 30 nm longer wavelength than the 2R polymer. This is explained by the change in the transition energy due to π-conjugation elongation. The $\chi^{(3)}$ dependence on the azo dye content in the copolymer revealed that the $\chi^{(3)}$ increases linearly with the dye content up to 17 mol%. This suggests that no strong interaction between attached dye molecules occurs. The azo dye attached polymers have larger $\chi^{(3)}$ values than the stilbene dye attached polymers(22).

<u>Main Chain Polymers</u>. Another system of dye attached polymers is a polyamic acid and a polyimide with a π-electron conjugation in the main chain. The π-electron conjugated system is not an intramolecular charge transfer system, unlike the azo and stilbene dye mentioned above. The polyamic acids (PAAs) were obtained through the reaction of a carboxylic acid anhydride and a diamine. A π-electron conjugated system exists in the diamine compound. These polyamic acids were soluble in conventional solvents and

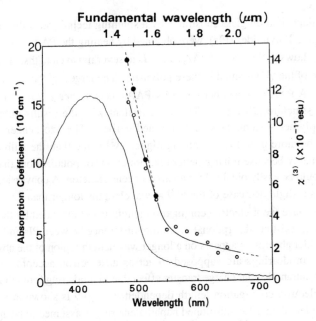

Figure 1. $\chi^{(3)}$ and absorption spectra of PPV thin film.

○ ;measured value, ● ;compensated for 3ω wave absorption

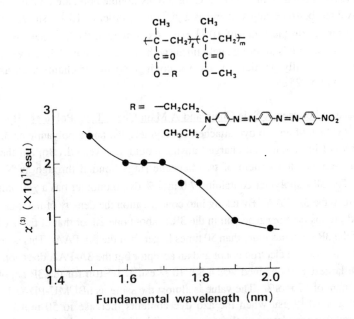

Figure 2. $\chi^{(3)}$ wavelength dependence for bisazo dye attached copolymer.

they were deposited on glass substrates by a spinning technique. Film thickness was 0.2 to 0.3 μm. Polyimides (PIs) were obtained by heating the PAAs.

Table 1 shows $\chi^{(3)}$ values of PAAs and PIs at several wavelengths. The wavelength dependence of the $\chi^{(3)}$ is small in these polymers. The largest $\chi^{(3)}$ is 1.2×10^{-11} esu for 4AS-PAA. A PAA with azo benzene (2A-PAA) has a larger $\chi^{(3)}$ than a stilbene with equivalent structure (2S-PAA). This is consistent with the results obtained for dye attached polymer systems as was mentioned earlier. The azo benzene structure is thought to be more useful for obtaining higher $\chi^{(3)}$ value than the stilbene structure. This is probably because it has a better conformation for polarization, that is, a planer structure could be easily obtained for an azo benzene skeleton. A conversion from PAAs to PIs gives a slight decrease of the $\chi^{(3)}$ at wavelengths longer than 2.05 μm. This is probably because the π-electron conjugation length is not extended by the conversion. Actually there exist no absorptional maximum difference between PIs and PAAs. Only a little shoulder absorption appears on a longer wavelength region of the absorption peak for PAAs. Imide rings are supposed to act as an electron acceptor but the result indicates no intramolecular charge transfer effect due to imide ring formation.

The π-electron conjugation length dependence of $\chi^{(3)}$ is shown in Fig.3. The π-conjugation length in a 1,4-substituted naphthalene ring is assumed to be approximately equivalent to a benzene ring, and a C=C double bond is equivalent to an N=N double bond. The $\chi^{(3)}$ value was calculated by considering a number density of the π-conjugated systems. $\chi^{(3)}$ values of π-electron conjugated systems containing two benzene rings bonded through a C=C or a N=N double bond are on the order of 10^{-13} esu. A three-benzene ring system has a $\chi^{(3)}$ on the order of 10^{-12} esu. A four-benzene ring system is on the order of 10^{-11} esu. As is shown, the $\chi^{(3)}$ dependence on the conjugation length is approximately linear with an exponent of 4.5. This value is close to the theoretically predicted value for conjugated linear chains such as polyenes (exponent of 5)(25).

Comparison between A Side Chain and A Main Chain Type Polymer. By comparing the $\chi^{(3)}$ values of an azo dye-attached copolymer, 3R, and a polyamic acid, 3A-PAA, the effect of intramolecular charge transfer should be revealed. Both of the polymers have a π-conjugated system of two benzene rings bonded through an N=N double bond. The 3R copolymer containing 7.0 mol % dye monomer has a $\chi^{(3)}$ value 8 times larger than the 3A-PAA. By taking into consideration the density of the π-electron conjugated groups (azo benzene unit in the 3R is about one 7th of that of the 3A-PAA), the $\chi^{(3)}$ of the 3R becomes more than 50 times larger than the 3A-PAA. This is because that the 3R possesses an electron donor and an acceptor but the 3A-PAA does not.

The largest $\chi^{(3)}$ obtained was 4.8×10^{-11} esu at 1.5 μm for the 3R copolymer at a dye content of 17 mol%. The value is almost the same as nBCMU-PDA. It will reach around 2×10^{-10} esu when the azo dye contents increase to 50 mol%. It is also suggested that even larger nonlinearity is possible if the molecular hyperpolarizability of

Table 1. $\chi^{(3)}$ of polyamic acid and polyimide

polymer	$\chi^{(3)}$ (10^{-12} esu)				
	1.85μm	1.9μm	2.0μm	2.05μm	2.15μm
2A-PAA	2.1	0.73			
3A-PAA		2.7	3.1	2.8	2.8
4A-PAA		12			
3A-PI	2.5	2.8	2.5	2.0	
2S-PAA		0.54			1.8

2A-PAA, : R = —⟨benzene⟩—N=N—⟨benzene⟩—

2S-PAA, : R = —⟨benzene⟩—CH=CH—⟨benzene⟩—

3A-PAA, 3A-PI : R = —⟨benzene⟩—N=N—⟨benzene⟩—N=N—⟨benzene⟩—

4AS-PAA : R = —⟨naphthalene⟩—N=N—⟨benzene(SO₃H)⟩—CH=CH—⟨benzene(SO₃H)⟩—N=N—⟨naphthalene⟩—

Polyamic Acid (PAA)

Polyimide (PI)

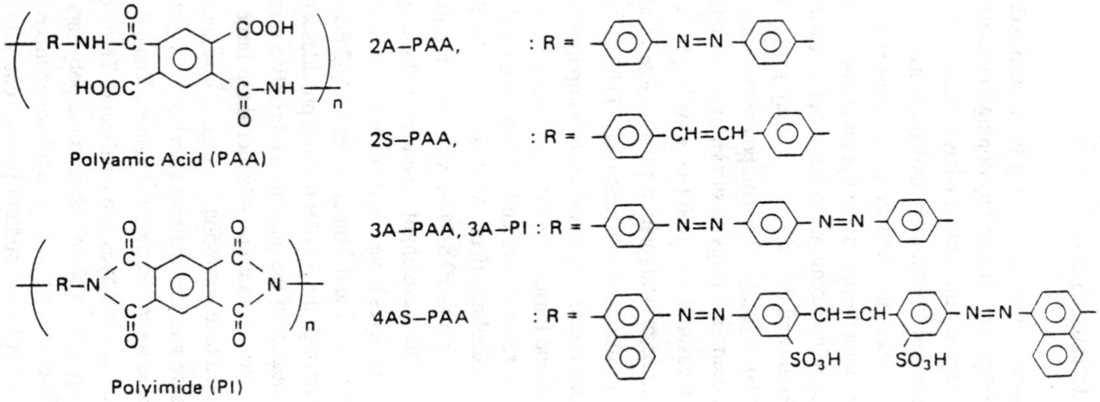

the attached dye is increased. So, a larger $\chi^{(3)}$ is expected if the π-conjugation is spread over four-benzene rings connected through azo groups (triazo dye).

<u>Dye Doped Polymers</u>

Polymeric materials having large third-order NLO properties and good processability had been also obtained by doping low molecular weight compounds into polymers. In this regard, the authors have measured THG intensities for various π-electron conjugated compounds and find that certain cyanine dyes possess powder THG intensities larger than the crystal powder of PTS-PDA. The $\chi^{(3)}$ of polymer thin films doped with these cyanine dyes was also investigated.

Cyanine compounds have been reported as possible third-order optical NLO materials and the effects of π-conjugation length upon the NLO properties have been calculated(25,26). The compounds investigated in this paper are symmetrical cyanines with quinoline rings as shown in Fig.4, referred as Q-1 and Q-2. The THG intensity was measured at a 1.06 μm wavelength using a conventional powder technique for powdered crystals with a particle size of about 100 μm. A para-nitroaniline crystal powder was used as a reference. THG $\chi^{(3)}$ measurements were also performed for the cyanine doped polymer films at pump light wavelengths from 1.85 to 2.1 μm.

Second harmonic generation was not observed from these dye powders, suggesting that the result described here is a pure third-order optical process, without any second-order cascading effect. At all wavelengths, these cyanine dyes have larger powder THG intensities than PTS-PDA which was reported to have the highest pure electronic $\chi^{(3)}$ value. The wavelength dependence of the cyanine dye powder THG intensity revealed that these dyes have a maximum intensity at around 2.0 μm, where the THG is 9 times larger for Q-1 and 4 times larger for Q-2 than PTS-PDA.

Although the powder method provides a convenient preliminary check of the optical nonlinearity of the materials, it is only a semi-quantitative evaluation method. So, it is necessary to ascertain whether these large powder THG intensities of cyanine dyes obtained here correspond to high $\chi^{(3)}$ value. As cyanine dyes usually have a low vapor pressure and low thermal stability, it is difficult to obtain vapor deposited thin films required for measuring $\chi^{(3)}$ values. Thus, the authors investigated a high concentration doping of these dyes into a polymer. These cyanine dyes are ionic compounds, so they are expected to be soluble in ionic polymers such as a sodium polystyrene sulphonate. A solvent system, acetic acid-water mixed solution, allowed us to dissolve a large amount of both a sodium polystyrene sulphonate and a cyanine dye. Consequently, polymers with highly doped cyanine dyes could be obtained. The film, 0.2 to 0.3 μm in thickness, was obtained from the solution by a spinning technique. The films have a good transparency. Figure 4 also shows a relationship between the $\chi^{(3)}$ and a concentration of the dyes in the polymer. The $\chi^{(3)}$ increases with the concentration which suggests that no aggregation of the dyes occurs in the polymer at up to 50 %

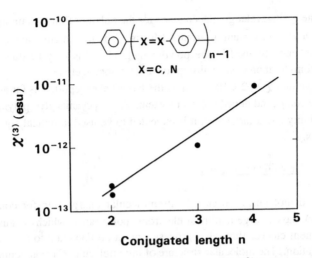

Figure 3. π-electron conjugation length dependence of $\chi^{(3)}$ for azobenzene and stilbene dyes.

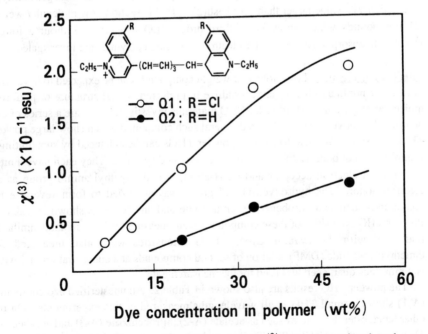

Figure 4. Dye concentration dependence of $\chi^{(3)}$ for cyanine dye doped polymers.

doping. The $\chi^{(3)}$ wavelength dependence for the polymer film containing 50 wt% Q-1 dye shows a $\chi^{(3)}$ maximum at around 1.9 μm. The maximum wavelength is a little shorter than that obtained in the powder method. This may be due to a trade-off between a third harmonic wave absorption and a resonant effect.

The maximum $\chi^{(3)}$, 2×10^{-11} esu, is the largest ever reported as far as we know for any organic compound doped polymer system. As this cyanine dye-polymer system has a processability and a large $\chi^{(3)}$, it is expected to be used in optical signal processing experiments.

<u>Low Molecular Weight Compounds</u>

π-conjugated molecules, so called intramolecular charge transfer compounds (CT compounds), have a large nonlinear electronic polarization which is caused by a large dipole moment change through a process from a ground state to an excited state by optical radiation. The molecular structure of intramolecular CT compounds distorts the distribution of π-electrons, resulting in the enhancement of optical nonlinearity. They are expected to possess an advantages such as high solubility in organic solvents and matrix polymers and the ability of easy sample preparation, as well as their possibility to give a high $\chi^{(3)}$ value larger than 10^{-12} esu(27). In this section, low molecular weight CT compounds which show large third-order NLO properties without a longer wavelength absorption as found in π-electron conjugated polymers are investigated.

<u>Effect of Aggregation</u>. Low molecular weight compounds can be expected to have large third-order nonlinearities if they crystallize in preferred crystal structure or preferred molecular alignments. Third-order NLO materials investigated here are a series of azo benzene derivatives. We synthesized several such compound which show large powder THG intensities. We also found that powder THGs can be enhanced by substituting a group that is not bonded directly to the π-conjugated systems. They each have a nitro group as an electron acceptor and an N-ethyl, N-hydroxyethyl amino group as an electron donor. The hydroxyethyl end group was esterified to form several ester compounds such as a nitrobenzoate, an acrylate and an acetate as shown in Table 2. Powder THG intensities of these compounds were measured using a method similar to that used with the cyanine dyes. THG intensities were also measured for dimethylformamide (DMF) solution of several compounds at a concentration of 2 wt%. A 1 mm thick quartz cell was used for the measurements.

The powder THG results are also shown in Table 2. An unesterified azo compound (A-1) shows a small THG at all wavelengths measured in this experiments. On the other hand, ester derivatives such as acetate (A-2), nitrobenzoate (A-3) and acrylate (A-4) show a very large THG. The THG intensity is more than 10 times larger than that of the unesterified compound at 2.05 μm wavelength. The enhancement of THG in powders by esterification is observed not only in azo benzene derivatives but also in stilbene and benzylidene aniline derivative(28). These results have not been expected due to effects on an electronic state because esterification occurs at a hydroxy group which is far from the π-electron conjugated systems. Indeed, the esterification gives little change in all derivatives for solution THG. For example, A-2, whose powder THG intensity is 11 times that of A-1, shows a slightly weaker solution THG than A-1. Thus, enhancement in THG intensity is only observed in powdered crystal, not in solution. This suggests that the change in crystal structures or molecular alignments due to the esterification produce the enhancement of powder THG intensity.

The absorption spectra of the compounds in powders and solution are shown in Fig. 5. In solution, the absorption maximum was the same for esterified and unesterified derivatives. On the other hand, large changes are observed in powders. Shoulders appear on a longer wavelength region of the absorption peak for esterified derivatives. The molar absorptivity measured in powders also seems to be increased by esterification. The difference in absorption spectra between powders and solution is consistent with the result of the THG measurements. The change in absorption spectra means that the electronic states are different in crystals of the esterified and unesterified compounds. This is thought to be due to the change in crystal structures or molecular alignments. The changes in electronic states can be provided by the variations in substituent or crystallization conditions. It was reported that absorption spectra, dependent on alkyl chain length, reflect an aggregation of azo benzenes(28). The absorption in the longer wavelength region is produced by the head-to-tail orientation, in other words by a so called J aggregation, of the azo benzene derivatives. We speculate that the esterification in this study similarly results in the head-to-tail orientation of the azo benzenes. The esterification of NLO dyes possesses another merits, i.e., it is an effective method to solve these dyes in a high concentration into organic solvents or polymer systems.

A New Molecular Crystal: DEANST. Another example of a low molecular weight compound with a large third-order NLO property is a styrene derivative. By selecting an appropriate π-conjugation length and chemical structures of donor and acceptor, several new organic molecular crystals such as stilbene and azo benzene derivatives which exhibit $\chi^{(3)}$ larger than 10^{-12} esu can be developed. The $\chi^{(3)}$ of 4-(N,N-diethylamino)-4'-nitrostilbene (DEANS), an intramolecular CT compound, has been reported in a solution, in a polymer matrix and in a crystalline state(27,29,30).

Here, a new CT compound, 4-(N,N-diethylamino)-4'-nitrostyrene (DEANST) will be presented which shows an efficient optical Kerr shutter operation. We have synthesized this new intramolecular CT compound and succeeded in obtaining a large single crystal with good optical quality. The thin crystal of the DEANST was used to measure its $\chi^{(3)}$ and the orientation of the molecule was evaluated using polarized UV-vis spectroscopy. The maximum $\chi^{(3)}$ value in the direction of the molecular long axis of the DEANST thin crystal was detected. The DEANST crystal was obtained from chloroform using a slow evaporation of the organic solvent. The thickness of the DEANST crystal can be controlled by the concentration of the DEANST-chloroform solution.

A powder THG measurement revealed that the DEANST crystal has a third-order

Table 2   THG intensities for Azo compounds

$$O_2N-\langle\bigcirc\rangle-N=N-\langle\bigcirc\rangle-N\begin{smallmatrix}CH_2CH_3\\CH_2CH_2R\end{smallmatrix}$$

sample	R	THG intensity in powder form (vs p-NA)			THG in solution (arb. unit)
		1.06μm	1.9μm	2.05μm	1.9μm
A-1	OH	2	2	4.8	1
A-2	OCOCH₃	11	22		0.86
A-3	OCO-⟨◯⟩-NO₂	27	31	57	
A-4	OCOCH=CH₂	17		60	0.88

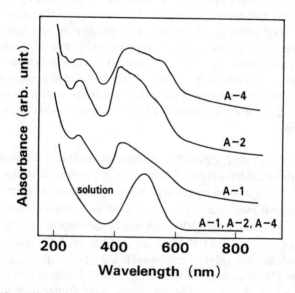

Figure 5. Absorption spectra for substituted azobenzene compounds.
A-1; unsubstituted, A-2; acetate substituted, A-4; Acrylate substituted

optical nonlinearity about 720 times lager than para-nitro aniline and 20 times larger than DEANS at a 1.06 μm fundamental wavelength. Using a THG maker fringe method for a 150 μm thick single crystal of DEANST at 1.06 μm fundamental wavelength, an anisotropy of a $\chi^{(3)}$ for the crystal was revealed. The DEANST crystal was mounted on a goniometer so that the molecular long-axis of the crystal was horizontal and rotated around the Y axis. The direction of the molecular long axis was determined by measuring a polarized electronic absorption spectrum for the DEANST thin crystal through the angular dependence of the absorption edges. The laser beam was horizontally polarized. In this setup, the extent of the interaction between the DEANST molecules and the laser beam is a function of the incident angle for the crystal. Details of the experiment will be published elsewhere(31).

The measured third-harmonic output pattern shows a correlation between molecular orientation and the NLO susceptibility of the crystal. Using a THG intensity at incident angle of about 20 degree, the $\chi^{(3)}$ value of the DEANST crystal for the molecular long axis direction was estimated to be 7.4 x 10^{-12} esu which was about 7 times larger than that for the molecular short axis direction. The value will be about one order larger by compensating for the effect of absorption at the third-harmonic wavelength (354 nm).

This DEANST crystal has advantages such as high solubility in organic solvents and matrix polymers. Figure 6 shows the wavelength dependence of a PMMA film doped with a 33.5 wt % DEANST crystal. A $\chi^{(3)}$ of around 2 x 10^{-12} esu was obtained at a 1.06 μm wavelength. Because the doped film is not only processable but also has a good transparency, this DEANST-PMMA system may be useful for fabricating devices with waveguide structures.

Kerr Shutter Experiment. The NLO properties of DEANST was also investigated through an optical Kerr shutter experiment. The optical Kerr shutter has the advantages of simple construction and fast response time. DEANST dissolves easily in many solvents. A saturated sample in a DMF (40 wt%) or in a nitrobenzene (30 wt%) was investigated in the Kerr shutter experiment. Powders of DEANST were dissolved in these solvents. The sample solution was poured into a silica glass cell of 1 mm thickness.

The sample cell was positioned between a polarizer and an analyzer consisting of a cross nicol configuration. A gate beam with a wavelength of 0.70 μm generated by a Q-switched Nd:YAG laser-pumped dye laser impinged on the sample cell with an incident angle of about 5 degrees. The duration of the gate beam was 6 nsec and its repetition rate was 10 Hz. A cw laser diode of a wavelength of 0.83 μm was used as a probe beam. The polarization of the probe beam was set 45 degrees with respect to the gate beam. The measured diameters of the gate beam and the probe beam at the sample were about 250 μm. A transmitted output signal pulse was detected by a photomultiplier with a 2 nsec response time. No signal was observed when the gate beam was blocked. Cross talk between the gate beam and the signal beam was not observed. Details of the experiments were reported elsewhere (32).

The transmission T value of the Kerr shutter operation was measured and corresponding phase difference angle, Φ, was calculated by the method reported earlier(32) as a function of power density of the gate beam, I_{in} for the Kerr shutter operation of a 40 wt% DEANST-DMF solution and a CS_2 liquid. The values of T increase in proportion to the square of I_{in}. When I_{in} was 100 MW/cm^2, the maximum value of T was 1.5 % and Φ was 20 degrees. The DEANST solution gave a transmission value of four times that of CS_2. The Kerr $\chi^{(3)}$ value was two times that of the CS_2. The absolute Kerr $\chi^{(3)}$ value at a 40 wt% concentration was calculated to be 3.0×10^{-12} esu, compared with the reported Kerr $\chi^{(3)}$ value of 1.55×10^{-12} esu for CS_2(33). The signal intensity and the Kerr $\chi^{(3)}$ dependences on DEANST concentration in DMF are shown in Fig. 7. The signal intensity, including the signal of DMF itself is approximately proportional to the square of DEANST concentration. The Kerr $\chi^{(3)}$ value of the DEANST itself is proportional to its concentration. A linear increase in the Kerr $\chi^{(3)}$ value with increasing a molecular density was confirmed up to a 40 wt% saturation concentration. This result shows that the large solubility is effective in increasing the Kerr $\chi^{(3)}$ value. It has already been reported that in the case of CS_2, the Kerr $\chi^{(3)}$ value is nine times the $\chi^{(3)}$ value determined by THG measurement (THG $\chi^{(3)}$). This is due to the enhancement of Kerr $\chi^{(3)}$ induced by the molecular orientation. The result of this experiment indicates that the DEANST solution has a similar orientational effect as CS_2, and that the Kerr $\chi^{(3)}$ values are several times the THG $\chi^{(3)}$ values.

It has been reported that the THG $\chi^{(3)}$ values increase as the dielectric constant of solvent increases due to the enhancement of nonlinear polarization(34). In the present study, the similar enhancement effect for the Kerr $\chi^{(3)}$ value was observed. DEANST was dissolved in several kinds of solvent with different dielectric constants, to evaluate the Kerr $\chi^{(3)}$ value of DEANST in solution. The solvents were categorized into two groups, aromatic and non-aromatic solvents. The aromatic solvents are thought to enhance the charge-transfer efficiency of DEANST, interacting with the conjugated system of DEANST. The solution concentration was the same value of 1.7 mol/l for each solvent. The value is the saturation point of benzene, which had the lowest solubility among the solvents used in this experiment. The Kerr $\chi^{(3)}$ values of DEANST solution increases with the dielectric constant of the solvent. This result shows that the Kerr $\chi^{(3)}$ enhancement is caused by the enlarged nonlinear polarization of the molecules due to the large charge transfer in a higher dielectric field as well as the superimposition of solvent $\chi^{(3)}$. Considering these, DEANST was saturated in a nitrobenzene (30 wt%). This combination was the most effective material used in this experiment. This sample gave a Kerr $\chi^{(3)}$ value of 3.6×10^{-12} esu, which is more than two times that of CS_2 and the largest value of any organic material.

Conclusion

In conclusion, processable organic materials with large third-order NLO properties are

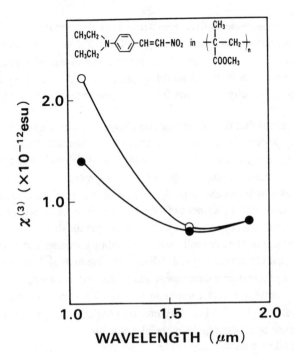

Figure 6. $\chi^{(3)}$ *wavelength dependence of a DEANST doped PMMA.*
● *;measured value,* ○ *;compensated for* 3ω *wave absorption*

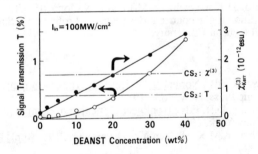

Figure 7. Kerr signal and $\chi^{(3)}$ *dependence on DEANST concentration.*

presented. THG measurements have been demonstrated on the PAV thin films prepared through a new precursor route. A resonant $\chi^{(3)}$ of the PPV thin film was evaluated to be 1.4×10^{-10} esu at 1.475 wavelength. In an absorption wavelength region, the $\chi^{(3)}$ was almost the same value as PTV and MOPPV thin films. PAVs are easy to fabricate as tough, good optical quality thin films from their precursor polymers by conventional spinning techniques.

It is also revealed that the intramolecular charge transfer derived from substituted donor and acceptor contributes to the increase in the third-order optical nonlinearity of dye-attached polymer systems. $\chi^{(3)}$ values larger than 10^{-10} esu can be achieved even in relatively short π-electron conjugated dye-attached polymers.

Cyanine dyes which possess large third-order optical nonlinearities are found and they are incorporated into a polymer to fabricate a thin film for device applications. The $\chi^{(3)}$ is 2×10^{-11} esu for a polymer film with 50 wt% cyanine dye.

The enhancement in THG intensity was observed by the esterification of azo benzene derivatives. A new CT organic crystal ,DEANST, with high $\chi^{(3)}$ was investigated and the high efficiency Kerr shutter experiment was performed using a solution of the crystal which shows a solvent effect for obtaining high Kerr $\chi^{(3)}$ values. It should be emphasized that a high THG intensity can be possible even in shorter π-electron conjugated systems by modifying the molecular structure.

The organic NLO materials reported here have good processabilities, so they have an advantage for developing new nonlinear optical organic devices.

Acknowledgments

The authors would like to acknowledge the collaboration of S.Matsumoto, M.Amano, S.Tomaru, Y.Shuto, H.Kobayashi, Y.Mori and N.Ooba of NTT Opto-electronics Laboratories in addition to the contributions made by Professors S.Saito and T.Tsutsui, and Dr.S.Tokito all Kyushu University.

Literature Cited

1. Chemla,D.S.; Zyss J. Eds., Nonlinear Optical Properties of Organic Molecules and Crystals; Academic Press, New York, 1987
2. Prasad, P.N.; .Ulrich, D.R. Eds., Nonlinear Optical and Electroactive Polymers; Plenum Press, New York, 1988
3. Carter,G.M.; Thakur, M.K.; J.Chen,Y.; Hryniewicz,J.V. Appl. Phys. Lett. 1985, 47, 457
4. Hattori, T.; Kobayashi,T. Chem. Phys. Lett. 1987, 133, 230
5. Sautertt, C.S.; Hermann, J.P.; Frey, R.; Pradere, F.; Ducuing, J.; Baughman, R.H.; Chance, R. Phys. Rev. Lett., 1976,. 36, 956
6. Heeger,A.J.; Moses,D.; Sinclair, M. Synthetic Metals, 1986, 15, 95
7. Kaino,T.; Kubodera,K; Tomaru, S.; Kurihara,T.; Saito, S.; Tsutsui, T.; Tokito, S. Electron.Lett.,1987, 23, 1095
8. Baker, G.L.; .Etemad, S; Kajzar,F. Proc. SPIE, 1987, 824, 102
9. Wnek, G.E.; Chien, J.C.W.; Karasz, F.E.; Lillya, C.P. Polymer, 1979, 20, 1441
10.Murase, I.; Ohnishi, T.; Noguchi, T.; Hirooka,M. Polymer Commun., 1985, 26, 362
11. Momii, S,; Tokito, S.; Tsutsui, T.; Saito, S. Chem. Lett., 1988, p1201

12. Bradley, D.D.C.; Mori, Y. Jpn. J. Appl. Phys. 1989, 28, 174
13. Bubeck, C.; Kaltbeitzel, A.; Lenz, R.W.; Neher, D.; Stengersmith, J.D.; Wegner, G. In Nonlinear Optical Effects in Organic Polymers, Messier, J.; Kajzer, F.; Prasad, P.; Ulrich, D. ,Eds.; NATO ASI Series, E. vol.162 1989, p143
14. Singh, B.; Prasad, P.N.; Karasz, F.E. Polymer 1988, 29, 1940
15. Yamada, S.; S.Tokito, S.; Tsutsui, T.; Saito, S. J. Chem. Soc., Chem. Commun. 1987, p148
16. Tomaru, S.; Kubodera, K.; Zembutu, S.; Takeda, S. Electron. Lett., 1987, ,595
17. Meredith, G.R.; Buchalter, B.; Hanzlik, C. J. Chem. Phys., 1983, 78, 1533
18. Kubodera, K.; Kaino, T. In Nonlinear Optics of Organics and Semiconductors, Kobayashi, T. ,Ed.; Springer-Verlag: Berlin, 1989, p163
19. Kaino, T.; Kubodera, K.; Kobayashi, H.; Kurihara, T.; Saito, S.;Tsutsui, T.; Tokito, S.; Murata, S. Appl. Phys. Lett., 1988, 53, 2002
20. Kaino, T.; Kobayashi, H.; Kubodera, K.; Kurihara, T.; Saito, S.; Tsutsui , T,; Tokito, S. Appl. Phys. Lett., 1989, 54, 1619
21. Matsumoto, S.; Kurihara, T.; Kubodera, K.; Kaino, T. Appl. Phys. Lett., 1987, 51, 1
22. Matsumoto, S.; Kurihara, T.; K.Kubodera, K.; Kaino, T. Mol. Cryst. Liq. Cryst., 1990, 182A, 115
23. Amano, M.; Kaino, T. Chem. Phys. Lett., 1990, 170, 515
24, Barzoukas, M.; Fremaux, P.; Josse, D.; Kajzer, F.; Zyss, J.; In Nonlinear optical Properties of Polymers, Heeger, A.J.; Orenstein, J.; Ulrich, D.R. Eds.; Material Research Society: Pittsburgh, 1988, p171
25. Rustagi, K.C.; Ducuing, K.C. Opt. Commun., 1974, 10, 258
26. Malony, C.; Blau, W. J. Opt. Soc. Am., 1987, B4, 1035
27. Kurihara, T.; Kobayashi, H.; Kubodera, K.; Kaino, T. Chem. Phys. Lett., 1990, 2.3, 171
28. Matsumoto, S.; Kubodera, K.; Kurihara, T.; Kaino, T. In Nonlinear Optics of Organics and Semiconductors, Kobayashi, T. ,Ed.; Springer-Verlag: Berlin, 1989, p236
29. Uchiki, H.; Kobayashi, T. J. Appl. Phys., 1988, 64, 2625
30. Kobayashi, T.; Uchiki, H.; Minoshima, K. In Nonlinear Optics of Organics and Semiconductors, Kobayashi, T. ,Ed.; Springer-Verlag: Berlin, 1989, p140
31. Kurihara, T.; Kobayashi, H.; Kubodera, K.; Kaino, T. Optics Commun.. submitted for publication.
32. Kambara, H,; Kobayashi, H.; Kubodera, K. IEEE Photonics Tech. Lett., 1989, 1, 149
33. Paillette, M. Ann. Phys. (Paris) vol. 4, 1969, p671
34. Kurihara, T.; Matsumoto, S.; Kaino, T.; Kubodera, K. 57th. Nat. Mtg Chem. Soc. Jpn. 1988 Abstr. No.3XIIC14, (in Japanese)

RECEIVED October 2, 1990

INDEXES

Author Index

Affiliation Index

Subject Index

Production: Victoria L. Contie and Paula M. Befard
Indexing: Deborah H. Steiner
Acquisition: Cheryl Shanks

Books printed and bound by Maple Press, York, PA
Dust jackets printed by Sheridan Press, Hanover, PA

Paper meets minimum requirements of American National Standard
for Information Sciences—Permanence of Paper for Printed Library
Materials, ANSI Z39.48–1984   ∞

Other ACS Books